RSMeans

Square Foot Costs

23rd Annual Edition

2002

Senior Editor
Barbara Balboni

Contributing Editors
Robert A. Bastoni
Howard M. Chandler
John H. Chiang, PE
Jennifer L. Curran
Stephen E. Donnelly
J. Robert Lang
Robert C. McNichols
Robert W. Mewis, CCC
Melville J. Mossman, PE
John J. Moylan
Jeannene D. Murphy
Peter T. Nightingale
Stephen C. Plotner
Michael J. Regan
Marshall J. Stetson
Phillip R. Waier, PE

Manager, Engineering Operations
John H. Ferguson, PE

Senior Vice President
Charles Spahr

Vice President and General Manager
Roger J. Grant

Vice President, Sales and Marketing
John M. Shea

Production Manager
Michael Kokernak

Production Coordinator
Marion E. Schofield

Technical Support
Thomas J. Dion
Michael H. Donelan
Jonathan Forgit
Mary Lou Geary
Gary L. Hoitt
Paula Reale-Camelio
Robin Richardson
Kathryn S. Rodriguez
Sheryl A. Rose
Elizabeth Testa

Book & Cover Design
Norman R. Forgit

First Printing

Foreword

R.S. Means Co., Inc. is a subsidiary of CMD (Construction Market Data), a leading provider of construction information, products, and services in North America and globally. CMD's project information products include more than 100 regional editions, national construction data, sales leads, and over 70 local plan rooms in major business centers. CMD PlansDirect provides surveys, plans and specifications. The First Source suite of products consists of *First Source for Products*, SPEC-DATA™, MANU-SPEC™, CADBlocks, Manufacturer Catalogs and First Source Exchange (www.firstsourceexchange.com) for the selection of nationally available building products. CMD also publishes ProFile, a database of more then 20,000 U.S. architectural firms. R.S. Means provides construction cost data, training, and consulting services in print, CD-ROM and online. CMD is headquartered in Atlanta and has 1,400 employees worldwide. CMD is owned by Cahners Business Information (www.cahners.com), a leading provider of critical information and marketing solutions to business professionals in the media, manufacturing, electronics, construction and retail industries. Its market-leading properties include more than 135 business-to-business publications, over 125 Webzines and Web portals, as well as online services, custom publishing, directories, research and direct-marketing lists. Cahners is a member of the Reed Elsevier plc group (NYSE: RUK and ENL)—a world-leading publisher and information provider operating in the science and medical, legal, education and business-to-business industry sectors.

Our Mission

Since 1942, R.S. Means Company, Inc. has been actively engaged in construction cost publishing and consulting throughout North America.

Today, over 50 years after the company began, our primary objective remains the same: to provide you, the construction and facilities professional, with the most current and comprehensive construction cost data possible.

Whether you are a contractor, an owner, an architect, an engineer, a facilities manager, or anyone else who needs a fast and reliable construction cost estimate, you'll find this publication to be a highly useful and necessary tool.

Today, with the constant flow of new construction methods and materials, it's difficult to find the time to look at and evaluate all the different construction cost possibilities. In addition, because labor and material costs keep changing, last year's cost information is not a reliable basis for today's estimate or budget.

That's why so many construction professionals turn to R.S. Means. We keep track of the costs for you, along with a wide range of other key information, from city cost indexes . . . to productivity rates . . . to crew composition . . . to contractor's overhead and profit rates.

R.S. Means performs these functions by collecting data from all facets of the industry, and organizing it in a format that is instantly accessible to you. From the preliminary budget to the detailed unit price estimate, you'll find the data in this book useful for all phases of construction cost determination.

The Staff, the Organization, and Our Services

When you purchase one of R.S. Means' publications, you are in effect hiring the services of a full-time staff of construction and engineering professionals.

Our thoroughly experienced and highly qualified staff works daily at collecting, analyzing, and disseminating comprehensive cost information for your needs. These staff members have years of practical construction experience and engineering training prior to joining the firm. As a result, you can count on them not only for the cost figures, but also for additional background reference information that will help you create a realistic estimate.

The Means organization is always prepared to help you solve construction problems through its five major divisions: Construction and Cost Data Publishing, Electronic Products and Services, Consulting Services, Insurance Services, and Educational Services.

Besides a full array of construction cost estimating books, Means also publishes a number of other reference works for the construction industry. Subjects include construction estimating and project and business management; special topics such as HVAC, roofing, plumbing, and hazardous waste remediation; and a library of facility management references.

In addition, you can access all of our construction cost data through your computer with Means CostWorks 2002 CD-ROM, an electronic tool that offers over 50,000 lines of Means detailed construction cost data, along with assembly and whole building cost data. You can also access Means cost information from our Web site at www.rsmeans.com

What's more, you can increase your knowledge and improve your construction estimating and management performance with a Means Construction Seminar or In-House Training Program. These two-day seminar programs offer unparalleled opportunities for everyone in your organization to get updated on a wide variety of construction-related issues.

Means also is a worldwide provider of construction cost management and analysis services for commercial and government owners and of claims and valuation services for insurers.

In short, R.S. Means can provide you with the tools and expertise for constructing accurate and dependable construction estimates and budgets in a variety of ways.

Robert Snow Means Established a Tradition of Quality That Continues Today

Robert Snow Means spent years building his company, making certain he always delivered a quality product.

Today, at R.S. Means, we do more than talk about the quality of our data and the usefulness of our books. We stand behind all of our data, from historical cost indexes... to construction materials and techniques... to current costs.

If you have any questions about our products or services, please call us toll-free at 1-800-334-3509. Our customer service representatives will be happy to assist you or visit our Web site at www.rsmeans.com

Table of Contents

RESIDENTIAL
- INTRODUCTION
- WORKSHEET
- ECONOMY
- AVERAGE
- CUSTOM
- LUXURY
- ADJUSTMENTS

COMM./INDUST./INSTIT.
- INTRODUCTION
- EXAMPLES
- BUILDING TYPES
- DEPRECIATION

ASSEMBLIES
- SUBSTRUCTURE   A
- SHELL   B
- INTERIORS   C
- SERVICES   D
- EQUIPMENT & FURNISHINGS   E
- SPECIAL CONSTRUCTION & DEMO.   F
- BUILDING SITEWORK   G

REFERENCE INFORMATION
- GENERAL CONDITIONS
- LOCATION FACTORS
- COST INDEXES
- GLOSSARY

How the Book Is Built: An Overview

A Powerful Construction Tool

You have in your hands one of the most powerful construction tools available today. A successful project is built on the foundation of an accurate and dependable estimate. This book will enable you to construct just such an estimate.

For the casual user the book is designed to be:

- quickly and easily understood so you can get right to your estimate
- filled with valuable information so you can understand the necessary factors that go into the cost estimate

For the regular user, the book is designed to be:

- a handy desk reference that can be quickly referred to for key costs
- a comprehensive, fully reliable source of current construction costs for typical building structures when only minimum dimensions, specifications and technical data are available. It is specifically useful in the conceptual or planning stage for preparing preliminary estimates which can be supported and developed into complete engineering estimates for construction.
- a source book for rapid estimating of replacement costs and preparing capital expenditure budgets

To meet all of these requirements we have organized the book into the following clearly defined sections.

How To Use the Book: The Details

This section contains an in-depth explanation of how the book is arranged . . . and how you can use it to determine a reliable construction cost estimate. It includes information about how we develop our cost figures and how to completely prepare your estimate.

Residential Section

Model buildings for four classes of construction—economy, average, custom and luxury—are developed and shown with complete costs per square foot.

Commercial, Industrial, Institutional Section

This section contains complete costs for 70 typical model buildings expressed as costs per square foot.

Assemblies Section

The cost data in this section has been organized in an "Assemblies" format. These assemblies are the functional elements of a building and are arranged according to the 7 divisions of the UNIFORMAT II classification system. The costs in this section include standard mark-ups for the overhead and profit of the installing contractor. The assemblies costs are those used to calculate the total building costs in the Commercial, Industrial, Institutional section.

Reference Section

This section includes information on General Conditions, Location Factors, Historical Cost Indexes, a Glossary, and a listing of Abbreviations used in this book.

Location Factors: Costs vary depending upon regional economy. You can adjust the "national average" costs in this book to 930 locations throughout the U.S. and Canada by using the data in this section.

Historical Cost Indexes: These indexes provide you with data to adjust construction costs over time. If you know costs for a project completed in the past, you can use these indexes to calculate a rough estimate of what it would cost to construct the same project today.

Glossary: A listing of common terms related to construction and their definitions is included.

Abbreviations: A listing of the abbreviations used throughout this book, along with the terms they represent, is included.

The Scope of This Book

This book is designed to be as comprehensive and as easy to use as possible. To that end we have made certain assumptions and limited its scope in two key ways:

1. We have established material prices based on a "national average."
2. We have computed labor costs based on a 30-city "national average" of union wage rates.

All costs and adjustments in this manual were developed from the *Means Building Construction Cost Data, Means Assemblies Cost Data, Means Residential Cost Data,* and *Means Light Commercial Cost Data.* Costs include labor, material, overhead, and profit, and general conditions and architect's fees where specified. No allowance is made for sales tax, financing, premiums for material and labor, or unusual business and weather conditions.

What's Behind the Numbers? The Development of Cost Data

The staff at R.S. Means continuously monitors developments in the construction industry in order to ensure reliable, thorough and up-to-date cost information.

While *overall* construction costs may vary relative to general economic conditions, price fluctuations within the industry are dependent upon many factors. Individual price variations may, in fact, be opposite to overall economic trends. Therefore, costs are continually monitored and complete updates are published yearly. Also, new items are frequently added in response to changes in materials and methods.

Costs—$ (U.S.)

All costs represent U.S. national averages and are given in U.S. dollars. The Means Location Factors can be used to adjust costs to a particular location. The Location Factors for Canada can be used to adjust U.S. national averages to local costs in Canadian dollars.

Factors Affecting Costs

Due to the almost limitless variation of building designs and combinations of construction methods and materials, costs in this book must be used with discretion. The editors have made every effort to develop costs for "typical" building types. By definition, a square foot estimate, due to the short time and limited amount of detail required, involves a relatively lower degree of accuracy than a detailed unit cost estimate, for example. The accuracy of an estimate will increase as the project is defined in more detailed components. The user should examine closely and compare the specific requirements of the project being estimated with the components included in the models in this book. Any differences should be accounted for in the estimate.

In many building projects, there may be factors that increase or decrease the cost beyond the range shown in this manual. Some of these factors are:

* Substitution of building materials or systems for those used in the model.
* Custom designed finishes, fixtures and/or equipment.
* Special structural qualities (allowance for earthquake, future expansion, high winds, long spans, unusual shape).
* Special foundations and substructures to compensate for unusual soil conditions.
* Inclusion of typical "owner supplied" equipment in the building cost (special testing and monitoring equipment for hospitals, kitchen equipment for hotels, offices that have cafeteria accommodations for employees, etc.).
* An abnormal shortage of labor or materials.

* Isolated building site or rough terrain that would affect the transportation of personnel, material or equipment.
* Unusual climatic conditions during the construction process.

We strongly urge that, for maximum accuracy, these factors be considered each time a structure is being evaluated.

If users require greater accuracy than this manual can provide, the editors recommend that the *Means Building Construction Cost Data* or the *Means Assemblies Cost Data* be consulted.

Final Checklist

Estimating can be a straightforward process provided you remember the basics. Here's a checklist of some of the items you should remember to do before completing your estimate.

Did you remember to . . .

* include the Location Factor for your city
* mark up the entire estimate sufficiently for your purposes
* include all components of your project in the final estimate
* double check your figures to be sure of your accuracy
* call R.S. Means if you have any questions about your estimate or the data you've found in our publications

Remember, R.S. Means stands behind its publications. If you have any questions about your estimate . . . about the costs you've used from our books . . . or even about the technical aspects of the job that may affect your estimate, feel free to call the R.S. Means editors at 1-800-334-3509.

Residential Section

Table of Contents

Introduction to Residential Section

The residential section of this manual contains costs per square foot for four classes of construction in seven building types. Costs are listed for various exterior wall materials which are typical of the class and building type. There are cost tables for Wings and Ells with modification tables to adjust the base cost of each class of building. Non-standard items can easily be added to the standard structures.

Cost estimating for a residence is a three-step process:
  (1) Identification
  (2) Listing dimensions
  (3) Calculations

Guidelines and a sample cost estimating procedure are shown on the following pages.

Identification

To properly identify a residential building, the class of construction, type, and exterior wall material must be determined. Page 6 has drawings and guidelines for determining the class of construction. There are also detailed specifications and additional drawings at the beginning of each set of tables to further aid in proper building class and type identification.

Sketches for eight types of residential buildings and their configurations are shown on pages 8 and 9. Definitions of living area are next to each sketch. Sketches and definitions of garage types are on page 10.

Living Area

Base cost tables are prepared as costs per square foot of living area. The living area of a residence is that area which is suitable and normally designed for full time living. It does not include basement recreation rooms or finished attics, although these areas are often considered full time living areas by the owners.

Living area is calculated from the exterior dimensions without the need to adjust for exterior wall thickness. When calculating the living area of a 1-1/2 story, two story, three story or tri-level residence, overhangs and other differences in size and shape between floors must be considered.

Only the floor area with a ceiling height of six feet or more in a 1-1/2 story residence is considered living area. In bi-levels and tri-levels, the areas that are below grade are considered living area, even when these areas may not be completely finished.

Base Tables and Modifications

Base cost tables show the base cost per square foot without a basement, with one full bath and one full kitchen for economy and average homes and an additional half bath for custom and luxury models. Adjustments for finished and unfinished basements are part of the base cost tables. Adjustments for multi family residences, additional bathrooms, townhouses, alternate roofs, and air conditioning and heating systems are listed in modification tables below the base cost tables.

Costs for other modifications, including garages, breezeways and site improvements, are in the modification tables on pages 58 to 61.

Listing of Dimensions

To use this section of the manual only the dimensions used to calculate the horizontal area of the building and additions and modifications are needed. The dimensions, normally the length and width, can come from drawings or field measurements. For ease in calculation, consider measuring in tenths of feet, i.e., 9 ft. 6 in. = 9.5 ft., 9 ft. 4 in. = 9.3 ft.

In all cases, make a sketch of the building. Any protrusions or other variations in shape should be noted on the sketch with dimensions.

Calculations

The calculations portion of the estimate is a two-step activity:
  (1) The selection of appropriate costs from the tables
  (2) Computations

Selection of Appropriate Costs

To select the appropriate cost from the base tables, the following information is needed:
  (1) Class of construction
      (a) Economy
      (b) Average
      (c) Custom
      (d) Luxury
  (2) Type of residence
      (a) One story
      (b) 1-1/2 story
      (c) 2 story
      (d) 3 story
      (e) Bi-level
      (f) Tri-level
  (3) Occupancy
      (a) One family
      (b) Two family
      (c) Three family
  (4) Building configuration
      (a) Detached
      (b) Town/Row house
      (c) Semi-detached
  (5) Exterior wall construction
      (a) Wood frame
      (b) Brick veneer
      (c) Solid masonry
  (6) Living areas

Modifications are classified by class, type and size.

Computations

The computation process should take the following sequence:
  (1) Multiply the base cost by the area
  (2) Add or subtract the modifications
  (3) Apply the location modifier

When selecting costs, interpolate or use the cost that most nearly matches the structure under study. This applies to size, exterior wall construction and class.

How to Use the Residential Square Foot Cost Pages

The following is a detailed explanation of a sample entry in the Residential Square Foot Cost Section. Each bold number below corresponds to the item being described on the facing page with the appropriate component of the sample entry following in parenthesis.

Prices listed are costs that include overhead and profit of the installing contractor. Total model costs include an additional mark-up for General Contractors' overhead and profit, and fees specific to class of construction.

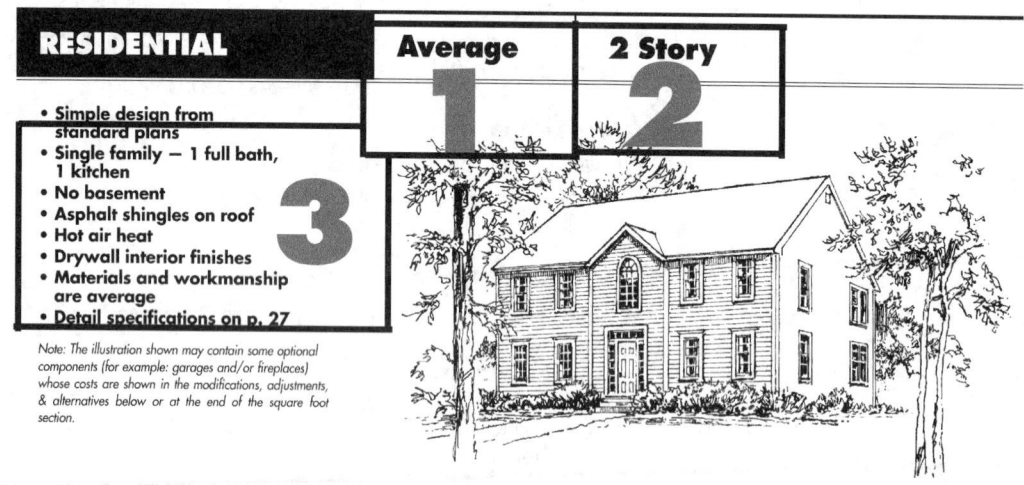

RESIDENTIAL   **Average**   **2 Story**

1   **2**

- Simple design from standard plans
- Single family — 1 full bath, 1 kitchen
- No basement
- Asphalt shingles on roof
- Hot air heat
- Drywall interior finishes
- Materials and workmanship are average
- Detail specifications on p. 27

3

Note: The illustration shown may contain some optional components (for example: garages and/or fireplaces) whose costs are shown in the modifications, adjustments, & alternatives below or at the end of the square foot section.

Base cost per square foot of living area

5

Exterior Wall	1000	1200	1400	1600	1800	Living Area 2000	2200	2600	3000	3400	3800
4							**6**				
Wood Siding - Wood Frame	87.65	79.45	76.00	73.55	70.90	68.10		62.80	59.20	57.70	56.25
Brick Veneer - Wood Frame	95.55	86.75	82.80	80.05	77.10	74.05		67.90	63.90	62.20	60.50
Stucco on Wood Frame	88.05	79.80	76.30	73.85	71.25	68.40		63.05	59.45	57.90	56.45
Solid Masonry	107.20	97.55	92.90	89.70	86.15	82.75	80.25	75.40	70.85	68.75	66.75
Finished Basement, Add	15.20	14.60	14.15	13.80	13.45	13.25		12.50	12.20	12.00	11.80
Unfinished Basement, Add	5.70	5.35	5.05	4.85	4.60	4.50	**8**	4.05	3.85	3.75	3.60

7

Modifications

Add to the total cost

Upgrade Kitchen Cabinets	$	+ 2206
Solid Surface Countertops		+ 896
Full Bath - including plumbing, wall and floor finishes		+ 3861
Half Bath - including plumbing, wall and floor finishes		+ 2405
One Car Attached Garage		+ 8424
One Car Detached Garage		+ 11,040
Fireplace & Chimney		+ 4160

9

Adjustments

For multi family - add to total cost

Additional Kitchen	$	+ 4061
Additional Bath		+ 3861
Additional Entry & Exit		+ 1128
Separate Heating		+ 1211
Separate Electric		+ 1410

For Townhouse/Rowhouse - Multiply cost per square foot by

Inner Unit	.90
End Unit	.95

10

Alternatives

Add to or deduct from the cost per square foot of living area

Cedar Shake Roof	+ 1.06
Clay Tile Roof	2.27
Slate Roof	+ 3.89
Upgrade Walls to Skim Coat Plaster	+ .39
Upgrade Ceilings to Textured Finish	+ .39
Air Conditioning, in Heating Ductwork	+ 1.56
In Separate Ductwork	+ 4.00
Heating Systems, Hot Water	+ 1.34
Heat Pump	+ 2.05
Electric Heat	− .50
Not Heated	− 2.46

11

Additional upgrades or components

Kitchen Cabinets & Countertops	Page 58
Bathroom Vanities	59
Fireplaces & Chimneys	59
Windows, Skylights & Dormers	59
Appliances	60
Breezeways & Porches	60
Finished Attic	60
Garages	61
Site Improvements	61
Wings & Ells	37

12

1 Class of Construction (Average)

The class of construction depends upon the design and specifications of the plan. The four classes are economy, average, custom and luxury.

2 Type of Residence (2 Story)

The building type describes the number of stories or levels in the model. The seven building types are 1 story, 1-1/2 story, 2 story, 2-1/2 story, 3 story, bi-level and tri-level.

3 Specification Highlights (Hot Air Heat)

These specifications include information concerning the components of the model, including the number of baths, roofing types, HVAC systems, and materials and workmanship. If the components listed are not appropriate, modifications can be made by consulting the information shown lower on the page or the Assemblies Section.

4 Exterior Wall System (Wood Siding - Wood Frame)

This section includes the types of exterior wall systems and the structural frame used. The exterior wall systems shown are typical of the class of construction and the building type shown.

5 Living Areas (2000 SF)

The living area is that area of the residence which is suitable and normally designed for full time living. It does not include basement recreation rooms or finished attics. Living area is calculated from the exterior dimensions without the need to adjust for exterior wall thickness. When calculating the living area of a 1-1/2 story, 2 story, 3 story or tri-level residence, overhangs and other differences in size and shape between floors must be considered. Only the floor area with a ceiling height of six feet or more in a 1-1/2 story residence is considered living area. In bi-levels and tri-levels, the areas that are below grade are considered living area, even when these areas may not be completely finished. A range of various living areas for the residential model are shown to aid in selection of values from the matrix.

6 Base Costs per Square Foot of Living Area ($68.10)

Base cost tables show the cost per square foot of living area without a basement, with one full bath and one full kitchen for economy and average homes and an additional half bath for custom and luxury models. When selecting costs, interpolate or use the cost that most nearly matches the

residence under consideration for size, exterior wall system and class of construction. Prices listed are costs that include overhead and profit of the installing contractor, a general contractor markup and an allowance for plans that vary by class of construction. For additional information on contractor overhead and architectural fees, see the Reference Section.

7 Basement Types (Finished)

The two types of basements are finished or unfinished. The specifications and components for both are shown on the Building Classes page in the Introduction to this section.

8 Additional Costs for Basements ($13.25)

These values indicate the additional cost per square foot of living area for either a finished or an unfinished basement.

9 Modifications and Adjustments (Solid Surface Countertops $896)

Modifications and Adjustments are costs added to or subtracted from the total cost of the residence. The total cost of the residence is equal to the cost per square foot of living area times the living area. Typical modifications and adjustments include kitchens, baths, garages and fireplaces.

10 Multiplier for Townhouse/ Rowhouse (Inner Unit .90)

The multipliers shown adjust the base costs per square foot of living area for the common wall condition encountered in townhouses or rowhouses.

11 Alternatives (Skim Coat Plaster $.39)

Alternatives are costs added to or subtracted from the base cost per square foot of living area. Typical alternatives include variations in kitchens, baths, roofing, air conditioning and heating systems.

12 Additional Upgrades or Components (Wings & Ells page 37)

Costs for additional upgrades or components, including wings or ells, breezeways, porches, finished attics and site improvements, are shown in other locations in the Residential Square Foot Cost Section.

Building Classes

Economy Class

An economy class residence is usually built from stock plans. The materials and workmanship are sufficient to satisfy building codes. Low construction cost is more important than distinctive features. The overall shape of the foundation and structure is seldom other than square or rectangular.

An unfinished basement includes 7' high 8" thick foundation wall composed of either concrete block or cast-in-place concrete.

Included in the finished basement cost are inexpensive paneling or drywall as the interior finish on the foundation walls, a low cost sponge backed carpeting adhered to the concrete floor, a drywall ceiling, and overhead lighting.

Custom Class

A custom class residence is usually built from plans and specifications with enough features to give the building a distinction of design. Materials and workmanship are generally above average with obvious attention given to construction details. Construction normally exceeds building code requirements.

An unfinished basement includes a 7'-6" high 10" thick cast-in-place concrete foundation wall or a 7'-6" high 12" thick concrete block foundation wall.

A finished basement includes painted drywall on insulated 2" x 4" wood furring as the interior finish to the concrete walls, a suspended ceiling, carpeting adhered to the concrete floor, overhead lighting and heating.

Average Class

An average class residence is a simple design and built from standard plans. Materials and workmanship are average, but often exceed minimum building codes. There are frequently special features that give the residence some distinctive characteristics.

An unfinished basement includes 7'-6" high 8" thick foundation wall composed of either cast-in-place concrete or concrete block.

Included in the finished basement are plywood paneling or drywall on furring that is fastened to the foundation walls, sponge backed carpeting adhered to the concrete floor, a suspended ceiling, overhead lighting and heating.

Luxury Class

A luxury class residence is built from an architect's plan for a specific owner. It is unique both in design and workmanship. There are many special features, and construction usually exceeds all building codes. It is obvious that primary attention is placed on the owner's comfort and pleasure. Construction is supervised by an architect.

An unfinished basement includes 8' high 12" thick foundation wall that is composed of cast-in-place concrete or concrete block.

A finished basement includes painted drywall on 2" x 4" wood furring as the interior finish, suspended ceiling, tackless carpet on wood subfloor with sleepers, overhead lighting and heating.

Configurations

Detached House

This category of residence is a free-standing separate building with or without an attached garage. It has four complete walls.

Semi-Detached House

This category of residence has two living units side-by-side. The common wall is a fireproof wall. Semi-detached residences can be treated as a row house with two end units. Semi-detached residences can be any of the building types.

Town/Row House

This category of residence has a number of attached units made up of inner units and end units. The units are joined by common walls. The inner units have only two exterior walls. The common walls are fireproof. The end units have three walls and a common wall. Town houses/row houses can be any of the building types.

Building Types

One Story

This is an example of a one-story dwelling. The living area of this type of residence is confined to the ground floor. The headroom in the attic is usually too low for use as a living area.

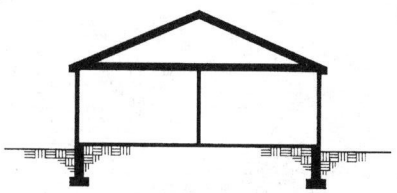

One-and-a-half Story

The living area in the upper level of this type of residence is 50% to 90% of the ground floor. This is made possible by a combination of this design's high-peaked roof and/or dormers. Only the upper level area with a ceiling height of 6 feet or more is considered living area. The living area of this residence is the sum of the ground floor area plus the area on the second level with a ceiling height of 6 feet or more.

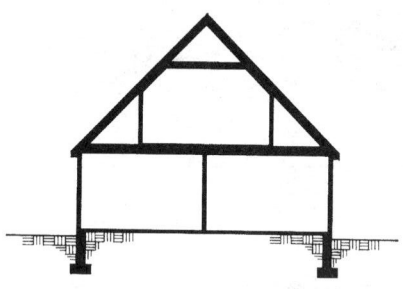

One Story with Finished Attic

The main living area in this type of residence is the ground floor. The upper level or attic area has sufficient headroom for comfortable use as a living area. This is made possible by a high peaked roof. The living area in the attic is less than 50% of the ground floor. The living area of this type of residence is the ground floor area only. The finished attic is considered an adjustment.

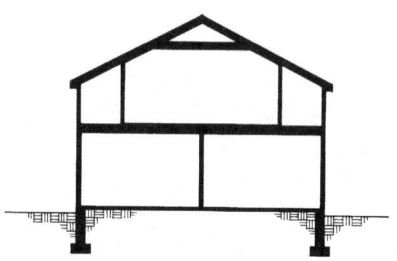

Two Story

This type of residence has a second floor or upper level area which is equal or nearly equal to the ground floor area. The upper level of this type of residence can range from 90% to 110% of the ground floor area, depending on setbacks or overhangs. The living area is the sum of the ground floor area and the upper level floor area.

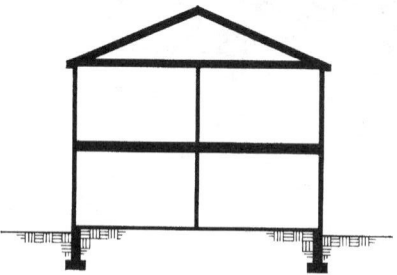

Two-and-one-half Story

This type of residence has two levels of equal or nearly equal area and a third level which has a living area that is 50% to 90% of the ground floor. This is made possible by a high peaked roof, extended wall heights and/or dormers. Only the upper level area with a ceiling height of 6 feet or more is considered living area. The living area of this residence is the sum of the ground floor area, the second floor area and the area on the third level with a ceiling height of 6 feet or more.

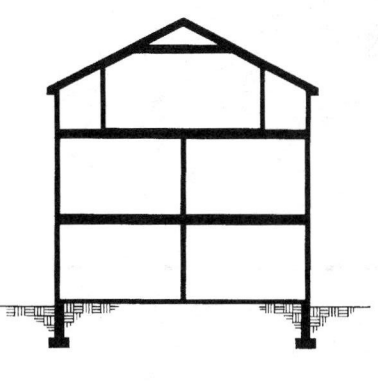

Bi-level

This type of residence has two living areas, one above the other. One area is about 4 feet below grade and the second is about 4 feet above grade. Both areas are equal in size. The lower level in this type of residence is originally designed and built to serve as a living area and not as a basement. Both levels have full ceiling heights. The living area is the sum of the lower level area and the upper level area.

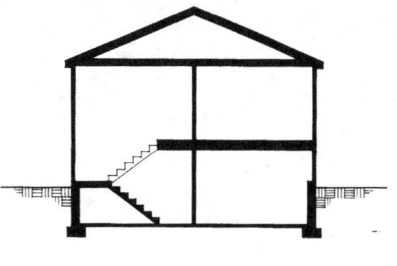

Three Story

This type of residence has three levels which are equal or nearly equal. As in the 2 story residence, the second and third floor areas may vary slightly depending on setbacks or overhangs. The living area is the sum of the ground floor area and the two upper level floor areas.

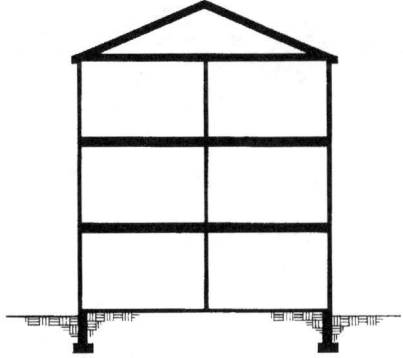

Tri-level

This type of residence has three levels of living area, one at grade level, one about 4 feet below grade and one about 4 feet above grade. All levels are originally designed to serve as living areas. All levels have full ceiling heights. The living area is a sum of the areas of each of the three levels.

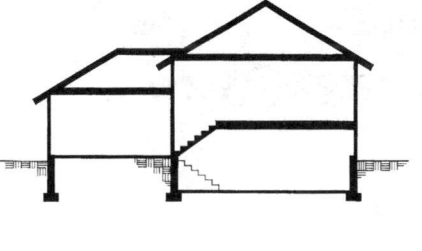

Garage Types

Attached Garage

Shares a common wall with the dwelling. Access is typically through a door between dwelling and garage.

Basement Garage

Constructed under the roof of the dwelling but below the living area.

Built-In Garage

Constructed under the second floor living space and above basement level of dwelling. Reduces gross square feet of living area.

Detached Garage

Constructed apart from the main dwelling. Shares no common area or wall with the dwelling.

Building Components

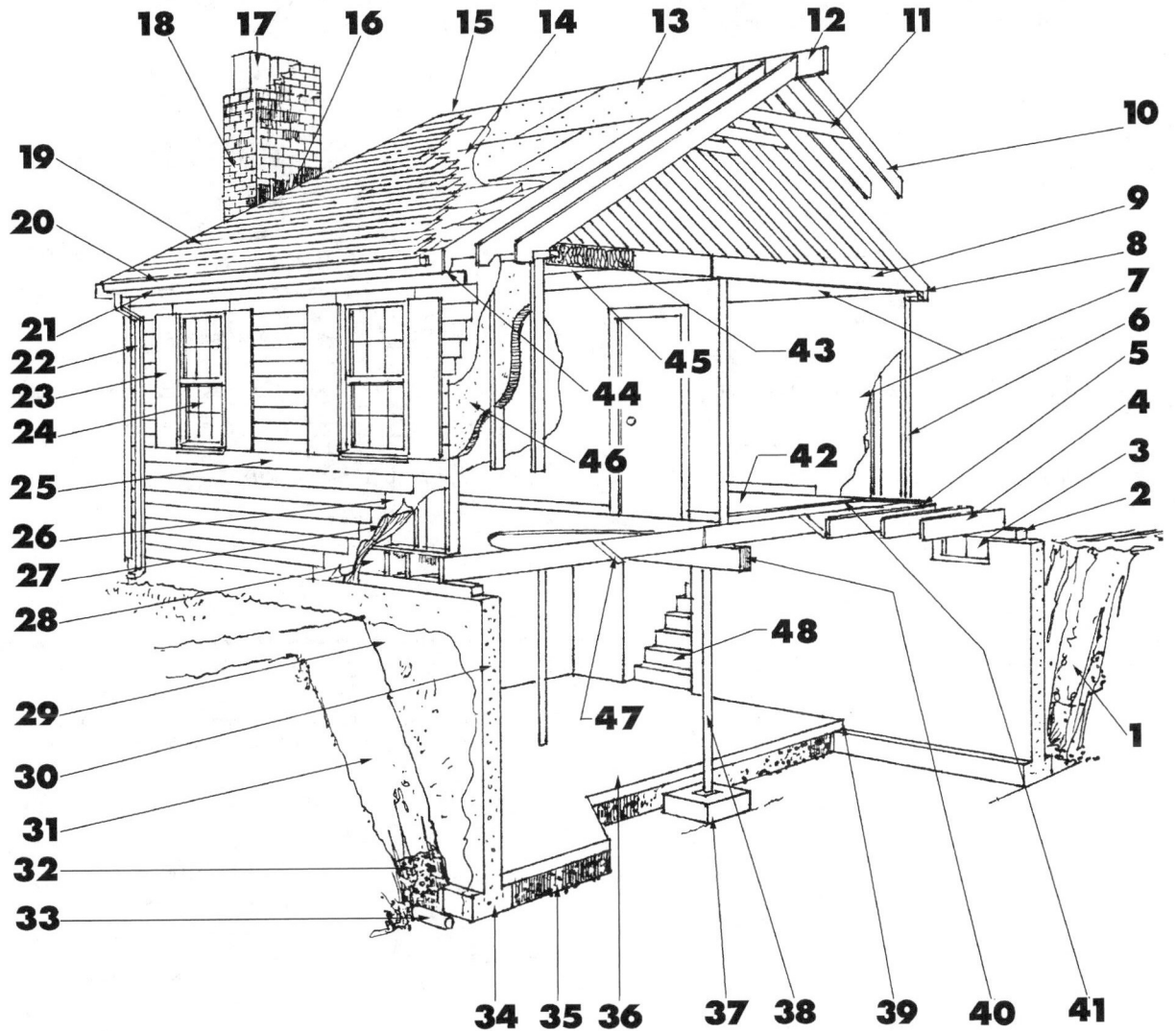

1. Excavation
2. Sill Plate
3. Basement Window
4. Floor Joist
5. Shoe Plate
6. Studs
7. Drywall
8. Plate
9. Ceiling Joists
10. Rafters

11. Collar Ties
12. Ridge Rafter
13. Roof Sheathing
14. Roof Felt
15. Roof Shingles
16. Flashing
17. Flue Lining
18. Chimney
19. Roof Shingles
20. Gutter

21. Fascia
22. Downspout
23. Shutter
24. Window
25. Wall Shingles
26. Building Paper
27. Wall Sheathing
28. Fire Stop
29. Dampproofing
30. Foundation Wall

31. Backfill
32. Drainage Stone
33. Drainage Tile
34. Wall Footing
35. Gravel
36. Concrete Slab
37. Column Footing
38. Pipe Column
39. Expansion Joint
40. Girder

41. Sub-floor
42. Finish Floor
43. Attic Insulation
44. Soffit
45. Ceiling Strapping
46. Wall Insulation
47. Cross Bridging
48. Bulkhead Stairs

Exterior Wall Construction

Typical Frame Construction

Typical wood frame construction consists of wood studs with insulation between them. A typical exterior surface is made up of sheathing, building paper and exterior siding consisting of wood, vinyl, aluminum or stucco over the wood sheathing.

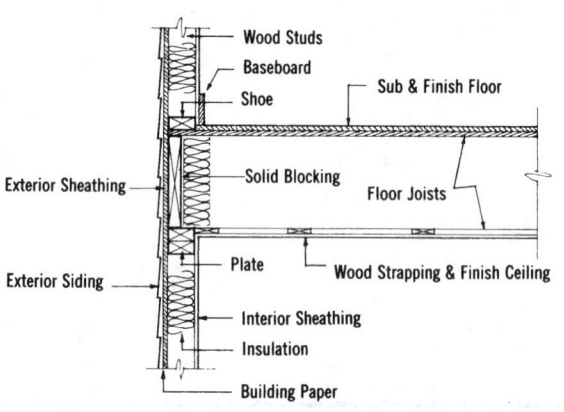

Brick Veneer

Typical brick veneer construction consists of wood studs with insulation between them. A typical exterior surface is sheathing, building paper and an exterior of brick tied to the sheathing, with metal strips.

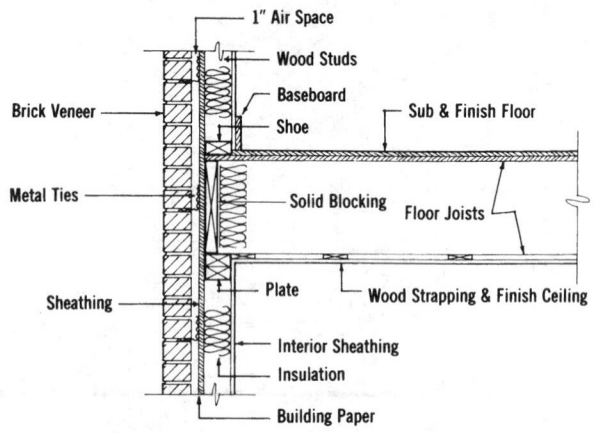

Stone

Typical solid masonry construction consists of a stone or block wall covered on the exterior with brick, stone or other masonry.

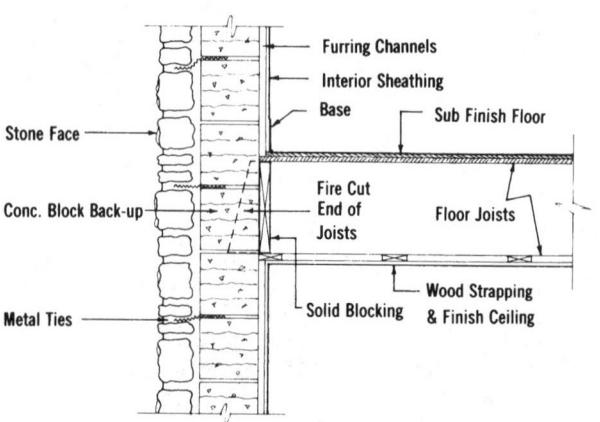

Residential Cost Estimate Worksheets

Worksheet Instructions

The residential cost estimate worksheet can be used as an outline for developing a residential construction or replacement cost. It is also useful for insurance appraisals. The design of the worksheet helps eliminate errors and omissions. To use the worksheet, follow the example below.

1. Fill out the owner's name, residence address, the estimator or appraiser's name, some type of project identifying number or code, and the date.
2. Determine from the plans, specifications, owner's description, photographs or any other means possible the class of construction. The models in this book use economy, average, custom and luxury as classes. Fill in the appropriate box.
3. Fill in the appropriate box for the residence type, configuration, occupancy, and exterior wall. If you require clarification, the pages preceding this worksheet describe each of these.
4. Next, the living area of the residence must be established. The heated or air conditioned space of the residence, not including the basement, should be measured. It is easiest to break the structure up into separate components as shown in the example. The main house (A), a one-and-one-half story wing (B), and a one story wing (C). The breezeway (D), garage (E) and open covered porch (F) will be treated differently. Data entry blocks for the living area are included on the worksheet for your use. Keep each level of each component separate, and fill out the blocks as shown.
5. By using the information on the worksheet, find the model, wing or ell in the following square foot cost pages that best matches the class, type, exterior finish and size of the residence being estimated. Use the *modifications, adjustments, and alternatives* to determine the adjusted cost per square foot of living area for each component.

6. For each component, multiply the cost per square foot by the living area square footage. If the residence is a townhouse/rowhouse, a multiplier should be applied based upon the configuration.
7. The second page of the residential cost estimate worksheet has space for the additional components of a house. The cost for additional bathrooms, finished attic space, breezeways, porches, fireplaces, appliance or cabinet upgrades, and garages should be added on this page. The information for each of these components is found with the model being used, or in the *modifications, adjustments, and alternatives* pages.
8. Add the total from page one of the estimate worksheet and the items listed on page two. The sum is the adjusted total building cost.
9. Depending on the use of the final estimated cost, one of the remaining two boxes should be filled out. Any additional items or exclusions should be added or subtracted at this time. The data contained in this book is a national average. Construction costs are different throughout the country. To allow for this difference, a location factor based upon the first three digits of the residence's zip code must be applied. The location factor is a multiplier that increases or decreases the adjusted total building cost. Find the appropriate location factor and calculate the local cost. If depreciation is a concern, a dollar figure should be subtracted at this point.
10. No residence will match a model exactly. Many differences will be found. At this level of estimating, a variation of plus or minus 10% should be expected.

Adjustments Instructions

No residence matches a model exactly in shape, material, or specifications.

The common differences are:

1. Two or more exterior wall systems
   a. Partial basement
   b. Partly finished basement

2. Specifications or features that are between two classes

3. Crawl space instead of a basement

EXAMPLES

Below are quick examples. See pages 15-17 for complete examples of cost adjustments for these differences:

1. Residence "A" is an average one-story structure with 1,600 S.F. of living area and no basement. Three walls are wood siding on wood frame, and the fourth wall is brick veneer on wood frame. The brick veneer wall is 35% of the exterior wall area.

   Use page 28 to calculate the Base Cost per S.F. of Living Area.
   Wood Siding for 1,600 S.F.   = $70.85 per S.F.
   Brick Veneer for 1,600 S.F.   = $82.30 per S.F.
   .65 ($70.85) + .35 ($82.30) = $74.86 per S.F. of Living Area.

2. a. Residence "B" is the same as Residence "A"; However, it has an unfinished basement under 50% of the building. To adjust the $74.86 per S.F. of living area for this partial basement, use page 28.
   $74.86 + .50 ($7.35) = $78.54 per S.F. of Living Area.

b. Residence "C" is the same as Residence "A"; However, it has a full basement under the entire building. 640 S.F. or 40% of the basement area is finished.

   Using Page 28:
   $74.86 + .40 ($21.15) + .60 ($7.35) = $87.73 per S.F. of Living Area.

3. When specifications or features of a building are between classes, estimate the percent deviation, and use two tables to calculate the cost per S.F.

   A two-story residence with wood siding and 1,800 S.F. of living area has features 30% better than Average, but 70% less than Custom.

   From pages 30 and 42:

   Custom 1,800 S.F. Base Cost   = $89.05 per S.F.
   Average 1,800 S.F. Base Cost   = $70.90 per S.F.
   DIFFERENCE   = $18.15 per S.F.

   Cost is $70.90 + .30 ($18.15) = $76.35 per S.F. of Living Area.

4. To add the cost of a crawl space, use the cost of an unfinished basement as a maximum. For specific costs of components to be added or deducted, such as vapor barrier, underdrain, and floor, see the "Assemblies" section, pages 233 to 436.

Model Residence Example

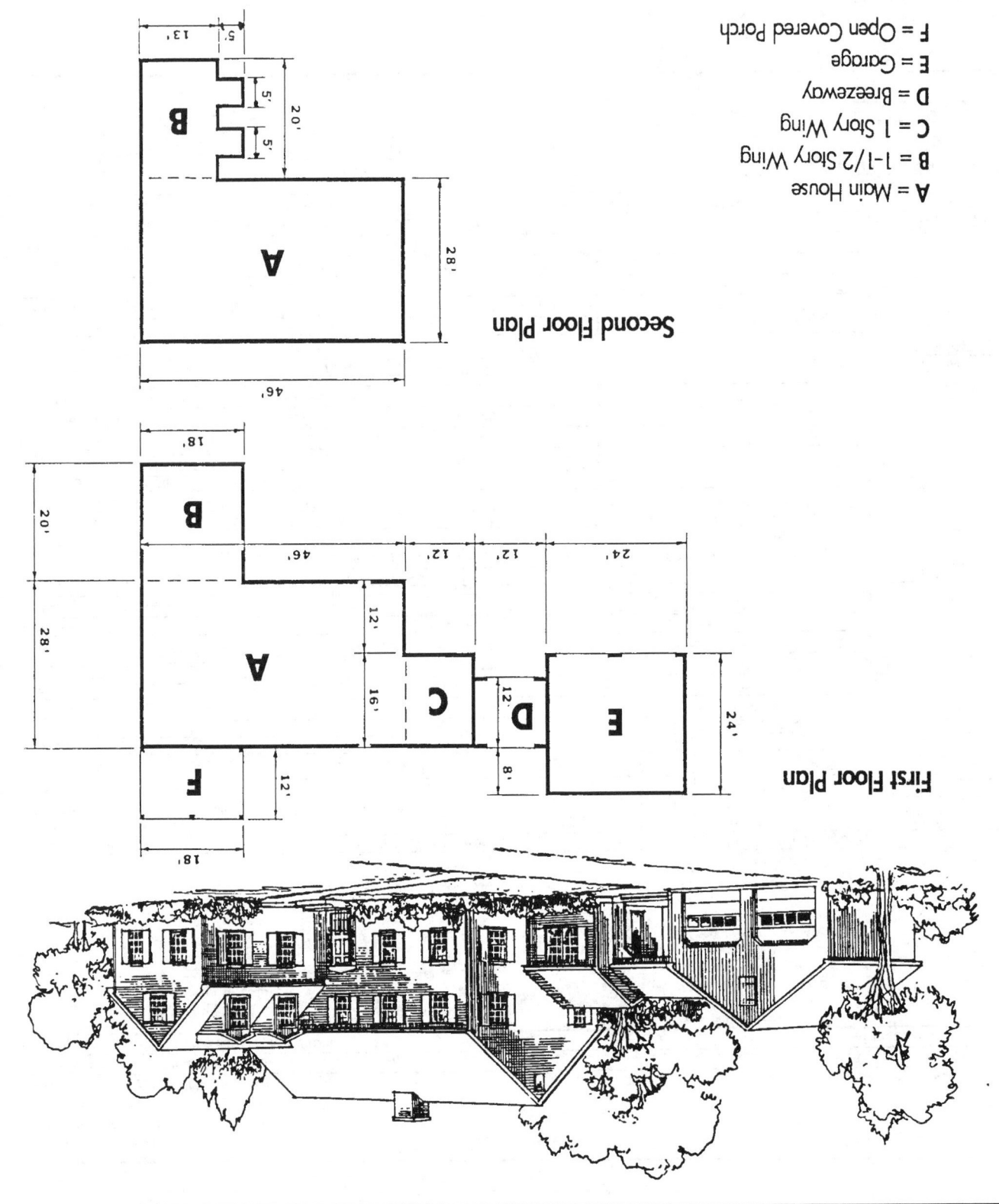

Second Floor Plan

First Floor Plan

A = Main House
B = 1-1/2 Story Wing
C = 1 Story Wing
D = Breezeway
E = Garage
F = Open Covered Porch

RESIDENTIAL
COST ESTIMATE

OWNERS NAME:	**Albert Westenberg**		APPRAISER:	**Nicole Wojtowicz**
RESIDENCE ADDRESS:	**300 Sygiel Road**		PROJECT:	**# 55**
CITY, STATE, ZIP CODE:	**Three Rivers, MA 01080**		DATE:	**Jan. 2, 2002**

CLASS OF CONSTRUCTION	RESIDENCE TYPE	CONFIGURATION	EXTERIOR WALL SYSTEM
☐ ECONOMY	☐ 1 STORY	☑ DETACHED	☑ WOOD SIDING - WOOD FRAME
☑ AVERAGE	☐ 11/2 STORY	☐ TOWN/ROW HOUSE	☐ BRICK VENEER - WOOD FRAME
☐ CUSTOM	☑ 2 STORY	☐ SEMI-DETACHED	☐ STUCCO ON WOOD FRAME
☐ LUXURY	☐ 2 1/2 STORY		☐ PAINTED CONCRETE BLOCK
	☐ 3 STORY	OCCUPANCY	☐ SOLID MASONRY (AVERAGE & CUSTOM)
	☐ BI-LEVEL	☑ ONE FAMILY	☐ STONE VENEER - WOOD FRAME
	☐ TRI-LEVEL	☐ TWO FAMILY	☐ SOLID BRICK (LUXURY)
		☐ THREE FAMILY	☐ SOLID STONE (LUXURY)
		☐ OTHER	

* LIVING AREA (Main Building)			* LIVING AREA (Wing or Ell)	(B)		* LIVING AREA (Wing or Ell)	(C)
First Level	1288 S.F.		First Level	360 S.F.		First Level	192 S.F.
Second level	1288 S.F.		Second level	310 S.F.		Second level	S.F.
Third Level	S.F.		Third Level	S.F.		Third Level	S.F.
Total	2576 S.F.		Total	670 S.F.		Total	192 S.F.

* Basement Area is not part of living area.

MAIN BUILDING			COSTS PER S.F. LIVING AREA	
Cost per Square Foot of Living Area, from Page	30		$	62.80
Basement Addition: ___ % Finished, 100 % Unfinished			+	4.05
Roof Cover Adjustment: **Cedar Shake** Type, Page 30 (Add or Deduct)			( )	1.06
Central Air Conditioning: ☐ Separate Ducts ☑ Heating Ducts, Page 30			+	1.56
Heating System Adjustment: ___ Type, Page ___ (Add or Deduct)			( )	
Main Building: Adjusted Cost per S.F. of Living Area			$	69.47

MAIN BUILDING TOTAL COST	$ 69.47 /S.F.	x	2,576 S.F.	x	_____	=	$ 178,955
	Cost per S.F. Living Area		Living Area		Town/Row House Multiplier (Use 1 for Detached)		TOTAL COST

WING OR ELL (B) 1 - 1/2 STORY			COSTS PER S.F. LIVING AREA	
Cost per Square Foot of Living Area, from Page	37		$	59.05
Basement Addition: 100 % Finished, ___ % Unfinished			+	18.65
Roof Cover Adjustment: ___ Type, Page ___ (Add or Deduct)			( )	
Central Air Conditioning: ☐ Separate Ducts ☑ Heating Ducts, Page 30			+	1.56
Heating System Adjustment: ___ Type, Page ___ (Add or Deduct)			( )	
Wing or Ell (B): Adjusted Cost per S.F. of Living Area			$	79.26

WING OR ELL (B) TOTAL COST	$ 79.26 /S.F.	x	670 S.F.	=	$ 53,104
	Cost per S.F. Living Area		Living Area		TOTAL COST

WING OR ELL (C) 1 STORY			COSTS PER S.F. LIVING AREA	
Cost per Square Foot of Living Area, from Page	37 (WOOD SIDING)		$	88.10
Basement Addition: ___ % Finished, ___ % Unfinished			+	—
Roof Cover Adjustment: ___ Type, Page ___ (Add or Deduct)			( )	—
Central Air Conditioning: ☐ Separate Ducts ☐ Heating Ducts, Page ___			+	—
Heating System Adjustment: ___ Type, Page ___ (Add or Deduct)			( )	—
Wing or Ell (C) Adjusted Cost per S.F. of Living Area			$	88.10

WING OR ELL (C) TOTAL COST	$ 88.1 /S.F.	x	192 S.F.	=	$ 16,915
	Cost per S.F. Living Area		Living Area		TOTAL COST

	TOTAL THIS PAGE	248,974

Page 1 of 2

RESIDENTIAL
COST ESTIMATE

		QUANTITY	UNIT COST		
Total Page 1				$	248,974
Additional Bathrooms: ___2___ Full, ___1___ Half 2 @ 3861 1 @ 2405				+	10,127
Finished Attic: __N/A__ Ft. x _____ Ft.			S.F.		
Breezeway: ☑ Open ☐ closed ___12___ Ft. x ___12___ Ft.	144	S.F.	16.20	+	2,333
Covered Porch: ☑ Open ☐ Enclosed ___18___ Ft. x ___12___ Ft.	216	S.F.	24.10	+	5,206
Fireplace: ☑ Interior Chimney ☐ Exterior Chimney					
☑ No. of Flues (2) ☑ Additional Fireplaces 1 - 2nd Story				+	7,020
Appliances:				+	—
Kitchen Cabinets Adjustments: (±)					—
☑ Garage ☐ Carport: ___2___ Car(s) Description **Wood, Attached** (±)					14,340
Miscellaneous:				+	

ADJUSTED TOTAL BUILDING COST $ 288,000

REPLACEMENT COST		
ADJUSTED TOTAL BUILDING COST	$	288,000
Site Improvements		
(A) Paving & Sidewalks	$	
(B) Landscaping	$	
(C) Fences	$	
(D) Swimming Pools	$	
(E) Miscellaneous	$	
TOTAL	$	288,000
Location Factor	x	1.05
Location Replacement Cost	$	302,400
Depreciation	-$	30,240
LOCAL DEPRECIATED COST	$	272,160

INSURANCE COST		
ADJUSTED TOTAL BUILDING COST	$	
Insurance Exclusions		
(A) Footings, sitework, Underground Piping	-$	
(B) Architects Fees	-$	
Total Building Cost Less Exclusion	$	
Location Factor	x	
LOCAL INSURABLE REPLACEMENT COST	$	

SKETCH AND ADDITIONAL CALCULATIONS

Page 2 of 2

17

ECONOMY

1 Story

1-1/2 Story

2 Story

Bi-Level

Tri-Level

Components

1 Site Work

Site preparation for slab or excavation for lower level; 4' deep trench excavation for foundation wall.

2 Foundations

Continuous reinforced concrete footing, 8" deep x 18" wide; dampproofed and insulated 8" thick reinforced concrete block foundation wall, 4' deep; 4" concrete slab on 4" crushed stone base and polyethylene vapor barrier, trowel finish.

3 Framing

Exterior walls—2 x 4 wood studs, 16" O.C.; 1/2" insulation board sheathing; wood truss roof 24" O.C. or 2 x 6 rafters 16" O.C. with 1/2" plywood sheathing, 4 in 12 or 8 in 12 roof pitch.

4 Exterior Walls

Beveled wood siding and #15 felt building paper on insulated wood frame walls; sliding sash or double hung wood windows; 2 flush solid core wood exterior doors.
Alternates:
• Brick veneer on wood frame, has 4" veneer of common brick.
• Stucco on wood frame has 1" thick colored stucco finish.
• Painted concrete block has 8" concrete block sealed and painted on exterior and furring on the interior for the drywall.

5 Roofing

20 year asphalt shingles; #15 felt building paper; aluminum gutters, downspouts, drip edge and flashings.

6 Interiors

Walls and ceilings—1/2" taped and finished drywall, primed and painted with 2 coats; painted baseboard and trim; rubber backed carpeting 80%, asphalt tile 20%; hollow core wood interior doors.

7 Specialties

Economy grade kitchen cabinets—6 L.F. wall and base with plastic laminate counter top and kitchen sink; 30 gallon electric water heater.

8 Mechanical

1 lavatory, white, wall hung; 1 water closet, white; 1 bathtub, enameled steel, white; gas fired warm air heat.

9 Electrical

100 Amp. service; romex wiring; incandescent lighting fixtures, switches, receptacles.

10 Overhead and Profit

General Contractor overhead and profit.

Adjustments

Unfinished Basement:
7' high 8" concrete block or cast-in-place concrete walls.

Finished Basement:
Includes inexpensive paneling or drywall on foundation walls. Inexpensive sponge backed carpeting on concrete floor, drywall ceiling, and lighting.

- **Mass produced from stock plans**
- **Single family — 1 full bath, 1 kitchen**
- **No basement**
- **Asphalt shingles on roof**
- **Hot air heat**
- **Drywall interior finishes**
- **Materials and workmanship are sufficient to meet codes**
- **Detail specifications on page 19**

Note: The illustration shown may contain some optional components (for example: garages and/or fireplaces) whose costs are shown in the modifications, adjustments, & alternatives below or at the end of the square foot section. Planners, Inc.

©Home Planners, Inc.

Base cost per square foot of living area

Exterior Wall	Living Area										
	600	800	1000	1200	1400	1600	1800	2000	2400	2800	3200
Wood Siding - Wood Frame	77.65	70.55	65.15	60.70	56.75	54.30	53.15	51.50	48.05	45.55	43.95
Brick Veneer - Wood Frame	84.30	76.45	70.50	65.55	61.20	58.40	57.15	55.25	51.50	48.75	46.90
Stucco on Wood Frame	75.80	68.90	63.65	59.35	55.55	53.15	52.05	50.45	47.10	44.70	43.15
Painted Concrete Block	78.95	71.70	66.15	61.65	57.65	55.05	53.95	52.20	48.75	46.15	44.50
Finished Basement, Add	20.85	19.65	18.75	18.00	17.35	16.95	16.70	16.40	15.90	15.55	15.25
Unfinished Basement, Add	9.45	8.45	7.75	7.15	6.65	6.35	6.15	5.90	5.50	5.25	5.00

Modifications

Add to the total cost

Upgrade Kitchen Cabinets	$ + 287
Solid Surface Countertops	+ 665
Full Bath - including plumbing, wall and floor finishes	+ 3076
Half Bath - including plumbing, wall and floor finishes	+ 1916
One Car Attached Garage	+ 8010
One Car Detached Garage	+ 10,343
Fireplace & Chimney	+ 3655

Adjustments

For multi family - add to total cost

Additional Kitchen	$ + 2167
Additional Bath	+ 3076
Additional Entry & Exit	+ 1128
Separate Heating	+ 1211
Separate Electric	+ 748

For Townhouse/Rowhouse -
Multiply cost per square foot by

Inner Unit	.95
End Unit	.97

Alternatives

Add to or deduct from the cost per square foot of living area

Composition Roll Roofing	− .56
Cedar Shake Roof	+ 2.40
Upgrade Walls and Ceilings to Skim Coat Plaster	+ .53
Upgrade Ceilings to Textured Finish	+ .39
Air Conditioning, in Heating Ductwork	+ 2.55
In Separate Ductwork	+ 4.92
Heating Systems, Hot Water	+ 1.40
Heat Pump	+ 1.70
Electric Heat	− 1.20
Not Heated	− 3.16

Additional upgrades or components

Kitchen Cabinets & Countertops	Page 58
Bathroom Vanities	59
Fireplaces & Chimneys	59
Windows, Skylights & Dormers	59
Appliances	60
Breezeways & Porches	60
Finished Attic	60
Garages	61
Site Improvements	61
Wings & Ells	25

- Mass produced from stock plans
- Single family — 1 full bath, 1 kitchen
- No basement
- Asphalt shingles on roof
- Hot air heat
- Drywall interior finishes
- Materials and workmanship are sufficient to meet codes
- Detail specifications on page 19

Note: The illustration shown may contain some optional components (for example: garages and/or fireplaces) whose costs are shown in the modifications, adjustments, & alternatives below or at the end of the square foot section.

ECONOMY

Base cost per square foot of living area

Exterior Wall	Living Area										
	600	800	1000	1200	1400	1600	1800	2000	2400	2800	3200
Wood Siding - Wood Frame	87.25	73.00	65.20	61.70	59.25	55.30	53.45	51.40	47.25	45.70	44.10
Brick Veneer - Wood Frame	96.55	79.75	71.45	67.55	64.85	60.35	58.25	56.00	51.25	49.50	47.60
Stucco on Wood Frame	84.70	71.10	63.50	60.15	57.75	53.90	52.15	50.15	46.10	44.65	43.10
Painted Concrete Block	89.05	74.30	66.45	62.85	60.35	56.30	54.40	52.30	48.05	46.40	44.75
Finished Basement, Add	16.00	13.60	13.00	12.55	12.20	11.75	11.45	11.20	10.70	10.50	10.25
Unfinished Basement, Add	8.20	6.35	5.85	5.50	5.20	4.85	4.65	4.45	4.05	3.90	3.70

Modifications

Add to the total cost

Upgrade Kitchen Cabinets	$ + 287
Solid Surface Countertops	+ 665
Full Bath - including plumbing, wall and floor finishes	+ 3076
Half Bath - including plumbing, wall and floor finishes	+ 1916
One Car Attached Garage	+ 8010
One Car Detached Garage	+ 10,343
Fireplace & Chimney	+ 3655

Adjustments

For multi family - add to total cost

Additional Kitchen	$ + 2167
Additional Bath	+ 3076
Additional Entry & Exit	+ 1128
Separate Heating	+ 1211
Separate Electric	+ 748

For Townhouse/Rowhouse -
Multiply cost per square foot by

Inner Unit	.95
End Unit	.97

Alternatives

Add to or deduct from the cost per square foot of living area

Composition Roll Roofing	– .39
Cedar Shake Roof	+ 1.71
Upgrade Walls and Ceilings to Skim Coat Plaster	+ .53
Upgrade Ceilings to Textured Finish	+ .39
Air Conditioning, in Heating Ductwork	+ 1.91
In Separate Ductwork	+ 4.28
Heating Systems, Hot Water	+ 1.34
Heat Pump	+ 1.88
Electric Heat	– .95
Not Heated	– 2.91

Additional upgrades or components

Kitchen Cabinets & Countertops	Page 58
Bathroom Vanities	59
Fireplaces & Chimneys	59
Windows, Skylights & Dormers	59
Appliances	60
Breezeways & Porches	60
Finished Attic	60
Garages	61
Site Improvements	61
Wings & Ells	25

- **Mass produced from stock plans**
- **Single family — 1 full bath, 1 kitchen**
- **No basement**
- **Asphalt shingles on roof**
- **Hot air heat**
- **Drywall interior finishes**
- **Materials and workmanship are sufficient to meet codes**
- **Detail specifications on page 19**

Note: The illustration shown may contain some optional components (for example: garages and/or fireplaces) whose costs are shown in the modifications, adjustments, & alternatives below or at the end of the square foot section.

Base cost per square foot of living area

Exterior Wall	Living Area										
	1000	1200	1400	1600	1800	2000	2200	2600	3000	3400	3800
Wood Siding - Wood Frame	69.95	63.20	60.25	58.20	56.20	53.70	52.20	49.30	46.25	44.95	43.85
Brick Veneer - Wood Frame	77.15	69.85	66.50	64.15	61.80	59.10	57.35	53.90	50.55	49.05	47.70
Stucco on Wood Frame	67.95	61.35	58.55	56.60	54.70	52.20	50.80	48.00	45.05	43.85	42.75
Painted Concrete Block	71.35	64.50	61.50	59.35	57.30	54.75	53.20	50.20	47.10	45.75	44.60
Finished Basement, Add	10.90	10.45	10.10	9.85	9.55	9.40	9.20	8.85	8.60	8.45	8.30
Unfinished Basement, Add	5.10	4.70	4.40	4.20	4.00	3.90	3.70	3.45	3.25	3.10	3.00

Modifications

Add to the total cost

Upgrade Kitchen Cabinets	$ + 287
Solid Surface Countertops	+ 665
Full Bath - including plumbing, wall and floor finishes	+ 3076
Half Bath - including plumbing, wall and floor finishes	+ 1916
One Car Attached Garage	+ 8010
One Car Detached Garage	+ 10,343
Fireplace & Chimney	+ 4035

Adjustments

For multi family - add to total cost

Additional Kitchen	$ + 2167
Additional Bath	+ 3076
Additional Entry & Exit	+ 1128
Separate Heating	+ 1211
Separate Electric	+ 748

For Townhouse/Rowhouse -
Multiply cost per square foot by

Inner Unit	.93
End Unit	.96

Alternatives

Add to or deduct from the cost per square foot of living area

Composition Roll Roofing	– .28
Cedar Shake Roof	+ 1.20
Upgrade Walls and Ceilings to Skim Coat Plaster	+ .54
Upgrade Ceilings to Textured Finish	+ .39
Air Conditioning, in Heating Ductwork	+ 1.53
In Separate Ductwork	+ 3.89
Heating Systems, Hot Water	+ 1.30
Heat Pump	+ 1.99
Electric Heat	– .79
Not Heated	– 2.75

Additional upgrades or components

Kitchen Cabinets & Countertops	Page 58
Bathroom Vanities	59
Fireplaces & Chimneys	59
Windows, Skylights & Dormers	59
Appliances	60
Breezeways & Porches	60
Finished Attic	60
Garages	61
Site Improvements	61
Wings & Ells	25

- **Mass produced from stock plans**
- **Single family — 1 full bath, 1 kitchen**
- **No basement**
- **Asphalt shingles on roof**
- **Hot air heat**
- **Drywall interior finishes**
- **Materials and workmanship are sufficient to meet codes**
- **Detail specifications on page 19**

Note: The illustration shown may contain some optional components (for example: garages and/or fireplaces) whose costs are shown in the modifications, adjustments, & alternatives below or at the end of the square foot section.

ECONOMY

Base cost per square foot of living area

Exterior Wall	Living Area										
	1000	1200	1400	1600	1800	2000	2200	2600	3000	3400	3800
Wood Siding - Wood Frame	65.25	58.85	56.20	54.35	52.55	50.20	48.90	46.25	43.40	42.30	41.30
Brick Veneer - Wood Frame	70.65	63.85	60.90	58.80	56.75	54.25	52.70	49.75	46.65	45.35	44.20
Stucco on Wood Frame	63.75	57.50	54.95	53.10	51.40	49.10	47.85	45.30	42.55	41.45	40.55
Painted Concrete Block	66.30	59.85	57.10	55.20	53.40	50.95	49.60	46.95	44.05	42.90	41.90
Finished Basement, Add	10.90	10.45	10.10	9.85	9.55	9.40	9.20	8.85	8.60	8.45	8.30
Unfinished Basement, Add	5.10	4.70	4.40	4.20	4.00	3.90	3.70	3.45	3.25	3.10	3.00

Modifications

Add to the total cost

Upgrade Kitchen Cabinets	$ + 287
Solid Surface Countertops	+ 665
Full Bath - including plumbing, wall and floor finishes	+ 3076
Half Bath - including plumbing, wall and floor finishes	+ 1916
One Car Attached Garage	+ 8010
One Car Detached Garage	+ 10,343
Fireplace & Chimney	+ 3655

Adjustments

For multi family - add to total cost

Additional Kitchen	$ + 2167
Additional Bath	+ 3076
Additional Entry & Exit	+ 1128
Separate Heating	+ 1211
Separate Electric	+ 748

For Townhouse/Rowhouse -
Multiply cost per square foot by

Inner Unit	.94
End Unit	.97

Alternatives

Add to or deduct from the cost per square foot of living area

Composition Roll Roofing	– .28
Cedar Shake Roof	+ 1.20
Upgrade Walls and Ceilings to Skim Coat Plaster	+ .51
Upgrade Ceilings to Textured Finish	+ .39
Air Conditioning, in Heating Ductwork	+ 1.53
In Separate Ductwork	+ 3.89
Heating Systems, Hot Water	+ 1.30
Heat Pump	+ 1.99
Electric Heat	– .79
Not Heated	– 2.75

Additional upgrades or components

Kitchen Cabinets & Countertops	Page 58
Bathroom Vanities	59
Fireplaces & Chimneys	59
Windows, Skylights & Dormers	59
Appliances	60
Breezeways & Porches	60
Finished Attic	60
Garages	61
Site Improvements	61
Wings & Ells	25

ECONOMY

- Mass produced from stock plans
- Single family — 1 full bath, 1 kitchen
- No basement
- Asphalt shingles on roof
- Hot air heat
- Drywall interior finishes
- Materials and workmanship are sufficient to meet codes
- Detail specifications on page 19

Note: The illustration shown may contain some optional components (for example: garages and/or fireplaces) whose costs are shown in the modifications, adjustments, & alternatives below or at the end of the square foot section.

Base cost per square foot of living area

Exterior Wall	Living Area										
	1200	1500	1800	2000	2200	2400	2800	3200	3600	4000	4400
Wood Siding - Wood Frame	60.25	55.50	51.95	50.50	48.40	46.65	45.35	43.60	41.40	40.70	39.00
Brick Veneer - Wood Frame	65.20	59.95	56.00	54.35	52.10	50.10	48.65	46.65	44.25	43.45	41.60
Stucco on Wood Frame	58.90	54.25	50.80	49.40	47.35	45.70	44.40	42.70	40.60	39.90	38.30
Solid Masonry	61.25	56.35	52.70	51.25	49.10	47.30	45.95	44.15	41.95	41.25	39.50
Finished Basement, Add*	13.05	12.50	11.95	11.75	11.50	11.30	11.10	10.85	10.60	10.50	10.30
Unfinished Basement, Add*	5.60	5.15	4.75	4.60	4.40	4.20	4.05	3.85	3.65	3.60	3.45

*Basement under middle level only.

Modifications

Add to the total cost

Upgrade Kitchen Cabinets	$ + 287
Solid Surface Countertops	+ 665
Full Bath - including plumbing, wall and floor finishes	+ 3076
Half Bath - including plumbing, wall and floor finishes	+ 1916
One Car Attached Garage	+ 8010
One Car Detached Garage	+ 10,343
Fireplace & Chimney	+ 3655

Adjustments

For multi family - add to total cost

Additional Kitchen	$ + 2167
Additional Bath	+ 3076
Additional Entry & Exit	+ 1128
Separate Heating	+ 1211
Separate Electric	+ 748

For Townhouse/Rowhouse -
Multiply cost per square foot by

Inner Unit	.93
End Unit	.96

Alternatives

Add to or deduct from the cost per square foot of living area

Composition Roll Roofing	– .39
Cedar Shake Roof	+ 1.71
Upgrade Walls and Ceilings to Skim Coat Plaster	+ .46
Upgrade Ceilings to Textured Finish	+ .39
Air Conditioning, in Heating Ductwork	+ 1.27
In Separate Ductwork	+ 3.63
Heating Systems, Hot Water	+ 1.24
Heat Pump	+ 2.07
Electric Heat	– .70
Not Heated	– 2.66

Additional upgrades or components

Kitchen Cabinets & Countertops	Page 58
Bathroom Vanities	59
Fireplaces & Chimneys	59
Windows, Skylights & Dormers	59
Appliances	60
Breezeways & Porches	60
Finished Attic	60
Garages	61
Site Improvements	61
Wings & Ells	25

ECONOMY

1 Story — Base cost per square foot of living area

Exterior Wall	Living Area							
	50	100	200	300	400	500	600	700
Wood Siding - Wood Frame	107.55	82.15	71.05	59.70	56.10	53.90	52.50	53.00
Brick Veneer - Wood Frame	124.35	94.15	81.05	66.40	62.10	59.50	57.80	58.15
Stucco on Wood Frame	102.90	78.85	68.30	57.85	54.45	52.40	51.00	51.60
Painted Concrete Block	110.80	84.50	73.00	61.05	57.25	55.00	53.55	54.00
Finished Basement, Add	31.40	25.70	23.30	19.40	18.60	18.10	17.80	17.55
Unfinished Basement, Add	17.35	13.00	11.15	8.15	7.55	7.15	6.95	6.75

1-1/2 Story — Base cost per square foot of living area

Exterior Wall	Living Area							
	100	200	300	400	500	600	700	800
Wood Siding - Wood Frame	84.05	67.45	57.30	51.65	48.60	47.15	45.25	44.75
Brick Veneer - Wood Frame	99.05	79.40	67.30	59.45	55.80	53.95	51.70	51.15
Stucco on Wood Frame	79.90	64.10	54.55	49.50	46.60	45.30	43.50	43.00
Painted Concrete Block	86.95	69.75	59.25	53.15	50.00	48.45	46.55	46.00
Finished Basement, Add	21.00	18.65	17.10	15.35	14.85	14.55	14.25	14.20
Unfinished Basement, Add	10.75	8.95	7.75	6.40	6.05	5.80	5.55	5.55

2 Story — Base cost per square foot of living area

Exterior Wall	Living Area							
	100	200	400	600	800	1000	1200	1400
Wood Siding - Wood Frame	85.25	63.40	53.75	44.85	41.70	39.85	38.60	39.25
Brick Veneer - Wood Frame	102.05	75.40	63.75	51.55	47.70	45.45	43.90	44.40
Stucco on Wood Frame	80.65	60.10	51.00	43.05	40.05	38.30	37.10	37.85
Painted Concrete Block	88.50	65.75	55.70	46.20	42.85	40.95	39.65	40.25
Finished Basement, Add	15.70	12.90	11.70	9.70	9.30	9.05	8.90	8.80
Unfinished Basement, Add	8.70	6.50	5.60	4.05	3.75	3.60	3.45	3.40

Base costs do not include bathroom or kitchen facilities. Use Modifications/Adjustments/Alternatives on pages 58-61 where appropriate.

AVERAGE

1 Story

1-1/2 Story

2 Story

2-1/2 Story

Bi-Level

Tri-Level

Components

1 Site Work	Site preparation for slab or excavation for lower level; 4' deep trench excavation for foundation wall.
2 Foundations	Continuous reinforced concrete footing, 8" deep x 18" wide; dampproofed and insulated 8" thick reinforced concrete foundation wall, 4' deep; 4" concrete slab on 4" crushed stone base and polyethylene vapor barrier, trowel finish.
3 Framing	Exterior walls—2 x 4 wood studs, 16" O.C.; 1/2" plywood sheathing; 2 x 6 rafters 16" O.C. with 1/2" plywood sheathing, 4 in 12 or 8 in 12 roof pitch, 2 x 6 ceiling joists; 1/2" plywood subfloor on 1 x 2 wood sleepers 16" O.C.; 2 x 8 floor joists with 5/8" plywood subfloor on models with more than one level.
4 Exterior Walls	Beveled wood siding and #15 felt building paper on insulated wood frame walls; double hung windows; 3 flush solid core wood exterior doors with storms. **Alternates:** • Brick veneer on wood frame, has 4" veneer of common brick. • Stucco on wood frame has 1" thick colored stucco finish. • Solid masonry has a 6" concrete block load bearing wall with insulation and a brick or stone exterior. Also solid brick or stone walls.
5 Roofing	25 year asphalt shingles; #15 felt building paper; aluminum gutters, downspouts, drip edge and flashings.
6 Interiors	Walls and ceilings—1/2" taped and finished drywall, primed and painted with 2 coats; painted baseboard and trim, finished hardwood floor 40%, carpet with 1/2" underlayment 40%, vinyl tile with 1/2" underlayment 15%, ceramic tile with 1/2" underlayment 5%; hollow core and louvered interior doors.
7 Specialties	Average grade kitchen cabinets—14 L.F. wall and base with plastic laminate counter top and kitchen sink; 40 gallon electric water heater.
8 Mechanical	1 lavatory, white, wall hung; 1 water closet, white; 1 bathtub, enameled steel, white; gas fired warm air heat.
9 Electrical	100 Amp. service; romex wiring; incandescent lighting fixtures, switches, receptacles.
10 Overhead and Profit	General Contractor overhead and profit.

Adjustments

Unfinished Basement:
7'-6" high cast-in-place concrete walls 8" thick or 8" concrete block.

Finished Basement:
Includes plywood paneling or drywall on furring on foundation walls, sponge backed carpeting on concrete floor, suspended ceiling, lighting and heating.

AVERAGE

- Simple design from standard plans
- Single family — 1 full bath, 1 kitchen
- No basement
- Asphalt shingles on roof
- Hot air heat
- Drywall interior finishes
- Materials and workmanship are average
- Detail specifications on p. 27

Note: The illustration shown may contain some optional components (for example: garages and/or fireplaces) whose costs are shown in the modifications, adjustments, & alternatives below or at the end of the square foot section.

©Home Planners, Inc.

Base cost per square foot of living area

Exterior Wall	Living Area										
	600	800	1000	1200	1400	1600	1800	2000	2400	2800	3200
Wood Siding - Wood Frame	99.05	90.40	83.80	78.35	73.85	70.85	69.30	67.30	63.30	60.35	58.35
Brick Veneer - Wood Frame	113.30	103.85	96.60	90.60	85.60	82.30	80.60	78.35	73.95	70.75	68.50
Stucco on Wood Frame	106.30	97.60	90.95	85.50	81.00	77.95	76.40	74.40	70.35	67.45	65.40
Solid Masonry	123.15	112.65	104.60	97.80	92.15	88.40	86.50	83.90	79.05	75.45	72.85
Finished Basement, Add	26.25	24.70	23.55	22.55	21.70	21.15	20.85	20.40	19.75	19.30	18.85
Unfinished Basement, Add	10.45	9.45	8.75	8.15	7.65	7.35	7.15	6.90	6.55	6.25	6.00

Modifications

Add to the total cost

Upgrade Kitchen Cabinets	$ + 2206
Solid Surface Countertops	+ 896
Full Bath - including plumbing, wall and floor finishes	+ 3861
Half Bath - including plumbing, wall and floor finishes	+ 2405
One Car Attached Garage	+ 8424
One Car Detached Garage	+ 11,040
Fireplace & Chimney	+ 3730

Adjustments

For multi family - add to total cost

Additional Kitchen	$ + 4061
Additional Bath	+ 3861
Additional Entry & Exit	+ 1128
Separate Heating	+ 1211
Separate Electric	+ 1410

For Townhouse/Rowhouse -
Multiply cost per square foot by

Inner Unit	.92
End Unit	.96

Alternatives

Add to or deduct from the cost per square foot of living area

Cedar Shake Roof	+ 2.12
Clay Tile Roof	+ 4.53
Slate Roof	+ 7.77
Upgrade Walls to Skim Coat Plaster	+ .33
Upgrade Ceilings to Textured Finish	+ .39
Air Conditioning, in Heating Ductwork	+ 2.62
In Separate Ductwork	+ 5.05
Heating Systems, Hot Water	+ 1.45
Heat Pump	+ 1.75
Electric Heat	- .67
Not Heated	- 2.63

Additional upgrades or components

Kitchen Cabinets & Countertops	Page 58
Bathroom Vanities	59
Fireplaces & Chimneys	59
Windows, Skylights & Dormers	59
Appliances	60
Breezeways & Porches	60
Finished Attic	60
Garages	61
Site Improvements	61
Wings & Ells	37

- Simple design from standard plans
- Single family — 1 full bath, 1 kitchen
- No basement
- Asphalt shingles on roof
- Hot air heat
- Drywall interior finishes
- Materials and workmanship are average
- Detail specifications on p. 27

Note: The illustration shown may contain some optional components (for example: garages and/or fireplaces) whose costs are shown in the modifications, adjustments, & alternatives below or at the end of the square foot section.

©By Designer

AVERAGE

Base cost per square foot of living area

Exterior Wall	Living Area										
	600	800	1000	1200	1400	1600	1800	2000	2400	2800	3200
Wood Siding - Wood Frame	109.40	92.55	83.10	78.95	75.95	71.25	69.05	66.55	61.70	59.85	57.80
Brick Veneer - Wood Frame	119.60	99.95	89.95	85.35	82.10	76.80	74.30	71.55	66.10	64.05	61.65
Stucco on Wood Frame	109.90	92.90	83.45	79.25	76.25	71.50	69.30	66.80	61.90	60.05	57.95
Solid Masonry	133.40	109.95	99.20	94.00	90.35	84.30	81.40	78.35	72.05	69.65	66.90
Finished Basement, Add	21.40	18.25	17.45	16.85	16.45	15.80	15.45	15.15	14.50	14.20	13.85
Unfinished Basement, Add	8.90	7.05	6.55	6.20	5.90	5.55	5.35	5.15	4.75	4.60	4.40

Modifications

Add to the total cost

Upgrade Kitchen Cabinets	$ + 2206
Solid Surface Countertops	+ 896
Full Bath - including plumbing, wall and floor finishes	+ 3861
Half Bath - including plumbing, wall and floor finishes	+ 2405
One Car Attached Garage	+ 8424
One Car Detached Garage	+ 11,040
Fireplace & Chimney	+ 3730

Adjustments

For multi family - add to total cost

Additional Kitchen	$ + 4061
Additional Bath	+ 3861
Additional Entry & Exit	+ 1128
Separate Heating	+ 1211
Separate Electric	+ 1410

For Townhouse/Rowhouse -
Multiply cost per square foot by

Inner Unit	.92
End Unit	.96

Alternatives

Add to or deduct from the cost per square foot of living area

Cedar Shake Roof	+ 1.50
Clay Tile Roof	+ 3.27
Slate Roof	+ 5.61
Upgrade Walls to Skim Coat Plaster	+ .38
Upgrade Ceilings to Textured Finish	+ .39
Air Conditioning, in Heating Ductwork	+ 1.96
In Separate Ductwork	+ 4.39
Heating Systems, Hot Water	+ 1.37
Heat Pump	+ 1.94
Electric Heat	- .58
Not Heated	- 2.54

Additional upgrades or components

Kitchen Cabinets & Countertops	Page 58
Bathroom Vanities	59
Fireplaces & Chimneys	59
Windows, Skylights & Dormers	59
Appliances	60
Breezeways & Porches	60
Finished Attic	60
Garages	61
Site Improvements	61
Wings & Ells	37

AVERAGE

- **Simple design from standard plans**
- **Single family — 1 full bath, 1 kitchen**
- **No basement**
- **Asphalt shingles on roof**
- **Hot air heat**
- **Drywall interior finishes**
- **Materials and workmanship are average**
- **Detail specifications on p. 27**

Note: The illustration shown may contain some optional components (for example: garages and/or fireplaces) whose costs are shown in the modifications, adjustments, & alternatives below or at the end of the square foot section.

Base cost per square foot of living area

Exterior Wall	Living Area										
	1000	1200	1400	1600	1800	2000	2200	2600	3000	3400	3800
Wood Siding - Wood Frame	87.65	79.45	76.00	73.55	70.90	68.10	66.35	62.80	59.20	57.70	56.25
Brick Veneer - Wood Frame	95.55	86.75	82.80	80.05	77.10	74.05	71.95	67.90	63.90	62.20	60.50
Stucco on Wood Frame	88.05	79.80	76.30	73.85	71.25	68.40	66.60	63.05	59.45	57.90	56.45
Solid Masonry	107.20	97.55	92.90	89.70	86.15	82.75	80.25	75.40	70.85	68.75	66.75
Finished Basement, Add	15.20	14.60	14.15	13.80	13.45	13.25	13.00	12.50	12.20	12.00	11.80
Unfinished Basement, Add	5.70	5.35	5.05	4.85	4.60	4.50	4.35	4.05	3.85	3.75	3.60

Modifications

Add to the total cost

Upgrade Kitchen Cabinets	$ + 2206
Solid Surface Countertops	+ 896
Full Bath - including plumbing, wall and floor finishes	+ 3861
Half Bath - including plumbing, wall and floor finishes	+ 2405
One Car Attached Garage	+ 8424
One Car Detached Garage	+ 11,040
Fireplace & Chimney	+ 4160

Adjustments

For multi family - add to total cost

Additional Kitchen	$ + 4061
Additional Bath	+ 3861
Additional Entry & Exit	+ 1128
Separate Heating	+ 1211
Separate Electric	+ 1410

For Townhouse/Rowhouse -
Multiply cost per square foot by

Inner Unit	.90
End Unit	.95

Alternatives

Add to or deduct from the cost per square foot of living area

Cedar Shake Roof	+ 1.06
Clay Tile Roof	+ 2.27
Slate Roof	+ 3.89
Upgrade Walls to Skim Coat Plaster	+ .39
Upgrade Ceilings to Textured Finish	+ .39
Air Conditioning, in Heating Ductwork	+ 1.56
In Separate Ductwork	+ 4.00
Heating Systems, Hot Water	+ 1.34
Heat Pump	+ 2.05
Electric Heat	– .50
Not Heated	– 2.46

Additional upgrades or components

Kitchen Cabinets & Countertops	Page 58
Bathroom Vanities	59
Fireplaces & Chimneys	59
Windows, Skylights & Dormers	59
Appliances	60
Breezeways & Porches	60
Finished Attic	60
Garages	61
Site Improvements	61
Wings & Ells	37

- **Simple design from standard plans**
- **Single family — 1 full bath, 1 kitchen**
- **No basement**
- **Asphalt shingles on roof**
- **Hot air heat**
- **Drywall interior finishes**
- **Materials and workmanship are average**
- **Detail specifications on p. 27**

Note: The illustration shown may contain some optional components (for example: garages and/or fireplaces) whose costs are shown in the modifications, adjustments, & alternatives below or at the end of the square foot section.

AVERAGE

Base cost per square foot of living area

Exterior Wall	Living Area										
	1200	1400	1600	1800	2000	2400	2800	3200	3600	4000	4400
Wood Siding - Wood Frame	86.60	81.80	74.60	73.50	71.00	67.05	63.95	60.70	59.10	56.10	55.20
Brick Veneer - Wood Frame	95.05	89.35	81.60	80.50	77.60	73.00	69.60	65.85	63.95	60.60	59.60
Stucco on Wood Frame	87.05	82.15	74.90	73.80	71.30	67.30	64.20	60.90	59.35	56.30	55.40
Solid Masonry	106.40	99.65	91.00	89.95	86.50	81.00	77.30	72.75	70.55	66.75	65.50
Finished Basement, Add	12.90	12.30	11.85	11.80	11.50	11.00	10.80	10.45	10.25	10.05	9.90
Unfinished Basement, Add	4.75	4.40	4.10	4.05	3.90	3.60	3.50	3.25	3.15	3.00	2.95

Modifications

Add to the total cost

Upgrade Kitchen Cabinets	$ + 2206
Solid Surface Countertops	+ 896
Full Bath - including plumbing, wall and floor finishes	+ 3861
Half Bath - including plumbing, wall and floor finishes	+ 2405
One Car Attached Garage	+ 8424
One Car Detached Garage	+ 11,040
Fireplace & Chimney	+ 4730

Adjustments

For multi family - add to total cost

Additional Kitchen	$ + 4061
Additional Bath	+ 3861
Additional Entry & Exit	+ 1128
Separate Heating	+ 1211
Separate Electric	+ 1410

For Townhouse/Rowhouse -
Multiply cost per square foot by

Inner Unit	.90
End Unit	.95

Alternatives

Add to or deduct from the cost per square foot of living area

Cedar Shake Roof	+ .90
Clay Tile Roof	+ 1.96
Slate Roof	+ 3.37
Upgrade Walls to Skim Coat Plaster	+ .36
Upgrade Ceilings to Textured Finish	+ .39
Air Conditioning, in Heating Ductwork	+ 1.43
In Separate Ductwork	+ 3.86
Heating Systems, Hot Water	+ 1.21
Heat Pump	+ 2.10
Electric Heat	– .91
Not Heated	– 2.87

Additional upgrades or components

Kitchen Cabinets & Countertops	Page 58
Bathroom Vanities	59
Fireplaces & Chimneys	59
Windows, Skylights & Dormers	59
Appliances	60
Breezeways & Porches	60
Finished Attic	60
Garages	61
Site Improvements	61
Wings & Ells	37

AVERAGE

- Simple design from standard plans
- Single family — 1 full bath, 1 kitchen
- No basement
- Asphalt shingles on roof
- Hot air heat
- Drywall interior finishes
- Materials and workmanship are average
- Detail specifications on p. 27

Note: The illustration shown may contain some optional components (for example: garages and/or fireplaces) whose costs are shown in the modifications, adjustments, & alternatives below or at the end of the square foot section.

Base cost per square foot of living area

Exterior Wall	Living Area										
	1500	1800	2100	2500	3000	3500	4000	4500	5000	5500	6000
Wood Siding - Wood Frame	80.75	73.35	70.35	68.00	63.25	61.30	58.45	55.25	54.30	53.15	52.00
Brick Veneer - Wood Frame	88.65	80.65	77.20	74.50	69.20	66.90	63.55	59.95	58.85	57.55	56.10
Stucco on Wood Frame	81.10	73.65	70.65	68.30	63.50	61.55	58.70	55.45	54.50	53.35	52.20
Solid Masonry	99.30	90.50	86.40	83.30	77.15	74.45	70.45	66.30	65.00	63.45	61.65
Finished Basement, Add	11.10	10.70	10.40	10.15	9.80	9.60	9.30	9.10	9.00	8.85	8.75
Unfinished Basement, Add	3.85	3.65	3.45	3.30	3.05	2.95	2.80	2.65	2.60	2.50	2.45

Modifications

Add to the total cost

Upgrade Kitchen Cabinets	$ + 2206
Solid Surface Countertops	+ 896
Full Bath - including plumbing, wall and floor finishes	+ 3861
Half Bath - including plumbing, wall and floor finishes	+ 2405
One Car Attached Garage	+ 8424
One Car Detached Garage	+ 11,040
Fireplace & Chimney	+ 4730

Adjustments

For multi family - add to total cost

Additional Kitchen	$ + 4061
Additional Bath	+ 3861
Additional Entry & Exit	+ 1128
Separate Heating	+ 1211
Separate Electric	+ 1410

For Townhouse/Rowhouse -
Multiply cost per square foot by

Inner Unit	.88
End Unit	.94

Alternatives

Add to or deduct from the cost per square foot of living area

Cedar Shake Roof	+ .71
Clay Tile Roof	+ 1.51
Slate Roof	+ 2.59
Upgrade Walls to Skim Coat Plaster	+ .35
Upgrade Ceilings to Textured Finish	+ .39
Air Conditioning, in Heating Ductwork	+ 1.43
In Separate Ductwork	+ 3.86
Heating Systems, Hot Water	+ 1.21
Heat Pump	+ 2.10
Electric Heat	– .71
Not Heated	– 2.67

Additional upgrades or components

Kitchen Cabinets & Countertops	Page 58
Bathroom Vanities	59
Fireplaces & Chimneys	59
Windows, Skylights & Dormers	59
Appliances	60
Breezeways & Porches	60
Finished Attic	60
Garages	61
Site Improvements	61
Wings & Ells	37

- **Simple design from standard plans**
- **Single family — 1 full bath, 1 kitchen**
- **No basement**
- **Asphalt shingles on roof**
- **Hot air heat**
- **Drywall interior finishes**
- **Materials and workmanship are average**
- **Detail specifications on p. 27**

Note: The illustration shown may contain some optional components (for example: garages and/or fireplaces) whose costs are shown in the modifications, adjustments, & alternatives below or at the end of the square foot section.

AVERAGE

Base cost per square foot of living area

Exterior Wall	Living Area										
	1000	1200	1400	1600	1800	2000	2200	2600	3000	3400	3800
Wood Siding - Wood Frame	82.05	74.25	71.10	68.85	66.55	63.90	62.35	59.15	55.85	54.50	53.25
Brick Veneer - Wood Frame	87.95	79.75	76.25	73.75	71.15	68.35	66.55	63.00	59.35	57.85	56.40
Stucco on Wood Frame	82.30	74.50	71.35	69.10	66.75	64.10	62.55	59.35	56.00	54.65	53.40
Solid Masonry	95.95	87.10	83.15	80.35	77.40	74.35	72.20	68.15	64.15	62.40	60.70
Finished Basement, Add	15.20	14.60	14.15	13.80	13.45	13.25	13.00	12.50	12.20	12.00	11.80
Unfinished Basement, Add	5.70	5.35	5.05	4.85	4.60	4.50	4.35	4.05	3.85	3.75	3.60

Modifications

Add to the total cost

Upgrade Kitchen Cabinets	$ + 2206
Solid Surface Countertops	+ 896
Full Bath - including plumbing, wall and floor finishes	+ 3861
Half Bath - including plumbing, wall and floor finishes	+ 2405
One Car Attached Garage	+ 8424
One Car Detached Garage	+ 11,040
Fireplace & Chimney	+ 3730

Adjustments

For multi family - add to total cost

Additional Kitchen	$ + 4061
Additional Bath	+ 3861
Additional Entry & Exit	+ 1128
Separate Heating	+ 1211
Separate Electric	+ 1410

For Townhouse/Rowhouse -
Multiply cost per square foot by

Inner Unit	.91
End Unit	.96

Alternatives

Add to or deduct from the cost per square foot of living area

Cedar Shake Roof	+ 2.12
Clay Tile Roof	+ 2.27
Slate Roof	+ 3.89
Upgrade Walls to Skim Coat Plaster	+ .36
Upgrade Ceilings to Textured Finish	+ .39
Air Conditioning, in Heating Ductwork	+ 1.57
In Separate Ductwork	+ 4.00
Heating Systems, Hot Water	+ 1.34
Heat Pump	+ 2.05
Electric Heat	– .50
Not Heated	– 2.46

Additional upgrades or components

Kitchen Cabinets & Countertops	Page 58
Bathroom Vanities	59
Fireplaces & Chimneys	59
Windows, Skylights & Dormers	59
Appliances	60
Breezeways & Porches	60
Finished Attic	60
Garages	61
Site Improvements	61
Wings & Ells	37

- **Simple design from standard plans**
- **Single family — 1 full bath, 1 kitchen**
- **No basement**
- **Asphalt shingles on roof**
- **Hot air heat**
- **Drywall interior finishes**
- **Materials and workmanship are average**
- **Detail specifications on p. 27**

Note: The illustration shown may contain some optional components (for example: garages and/or fireplaces) whose costs are shown in the modifications, adjustments, & alternatives below or at the end of the square foot section.

©Design Basics, Inc.

Base cost per square foot of living area

Exterior Wall	Living Area										
	1200	1500	1800	2100	2400	2700	3000	3400	3800	4200	4600
Wood Siding - Wood Frame	78.05	72.30	68.05	64.35	61.90	60.50	59.00	57.55	55.10	53.15	52.10
Brick Veneer - Wood Frame	83.45	77.25	72.50	68.40	65.65	64.15	62.45	60.90	58.20	56.05	54.90
Stucco on Wood Frame	78.30	72.55	68.25	64.50	62.05	60.70	59.15	57.70	55.25	53.30	52.25
Solid Masonry	90.80	83.90	78.50	73.85	70.80	69.10	67.10	65.35	62.35	59.95	58.70
Finished Basement, Add*	17.40	16.65	15.95	15.40	15.05	14.85	14.55	14.35	14.05	13.80	13.70
Unfinished Basement, Add*	6.35	5.90	5.50	5.15	4.95	4.80	4.65	4.55	4.35	4.20	4.15

*Basement under middle level only.

Modifications

Add to the total cost

Upgrade Kitchen Cabinets	$ + 2206
Solid Surface Countertops	+ 896
Full Bath - including plumbing, wall and floor finishes	+ 3861
Half Bath - including plumbing, wall and floor finishes	+ 2405
One Car Attached Garage	+ 8424
One Car Detached Garage	+ 11,040
Fireplace & Chimney	+ 4160

Adjustments

For multi family - add to total cost

Additional Kitchen	$ + 4061
Additional Bath	+ 3861
Additional Entry & Exit	+ 1128
Separate Heating	+ 1211
Separate Electric	+ 1410

For Townhouse/Rowhouse -
Multiply cost per square foot by

Inner Unit	.90
End Unit	.95

Alternatives

Add to or deduct from the cost per square foot of living area

Cedar Shake Roof	+ 1.50
Clay Tile Roof	+ 3.27
Slate Roof	+ 5.61
Upgrade Walls to Skim Coat Plaster	+ .31
Upgrade Ceilings to Textured Finish	+ .39
Air Conditioning, in Heating Ductwork	+ 1.31
In Separate Ductwork	+ 3.74
Heating Systems, Hot Water	+ 1.29
Heat Pump	+ 2.13
Electric Heat	– .41
Not Heated	– 2.37

Additional upgrades or components

Kitchen Cabinets & Countertops	Page 58
Bathroom Vanities	59
Fireplaces & Chimneys	59
Windows, Skylights & Dormers	59
Appliances	60
Breezeways & Porches	60
Finished Attic	60
Garages	61
Site Improvements	61
Wings & Ells	37

- Post and beam frame
- Log exterior walls
- Simple design from standard plans
- Single family — 1 full bath, 1 kitchen
- No basement
- Asphalt shingles on roof
- Hot air heat
- Drywall interior finishes
- Materials and workmanship are average
- Detail specification on page 27

Note: The illustration shown may contain some optional components (for example: garages and/or fireplaces) whose costs are shown in the modifications, adjustments, & alternatives below or at the end of the square foot section.

SOLID WALL

Base cost per square foot of living area

Exterior Wall	Living Area										
	600	800	1000	1200	1400	1600	1800	2000	2400	2800	3200
6" Log - Solid Wall	97.00	87.70	80.40	74.70	70.10	66.10	64.20	62.80	58.50	55.20	53.40
8" Log - Solid Wall	96.30	87.00	79.80	74.10	69.50	65.60	63.80	62.40	58.20	54.60	52.90
Finished Basement, Add	22.30	20.90	20.00	19.10	18.50	18.00	17.70	17.30	16.80	16.40	16.10
Unfinished Basement, Add	9.50	8.70	8.10	7.50	7.10	6.70	6.60	6.40	6.00	5.70	5.50

Modifications

Add to the total cost

Upgrade Kitchen Cabinets	$ + 2206
Solid Surface Countertops	+ 896
Full Bath - including plumbing, wall and floor finishes	+ 3861
Half Bath - including plumbing, wall and floor finishes	+ 2405
One Car Attached Garage	+ 8424
One Car Detached Garage	+ 11,040
Fireplace & Chimney	+ 3730

Adjustments

For multi family - add to total cost

Additional Kitchen	$ + 4061
Additional Bath	+ 3861
Additional Entry & Exit	+ 1128
Separate Heating	+ 1211
Separate Electric	+ 1410

*For Townhouse/Rowhouse -
Multiply cost per square foot by*

Inner Unit	.92
End Unit	.96

Alternatives

Add to or deduct from the cost per square foot of living area

Cedar Shake Roof	+ 2.12
Air Conditioning, in Heating Ductwork	+ 2.62
In Separate Ductwork	+ 5.05
Heating Systems, Hot Water	+ 1.45
Heat Pump	+ 1.75
Electric Heat	– .67
Not Heated	– 2.63

Additional upgrades or components

Kitchen Cabinets & Countertops	Page 58
Bathroom Vanities	59
Fireplaces & Chimneys	59
Windows, Skylights & Dormers	59
Appliances	60
Breezeways & Porches	60
Finished Attic	60
Garages	61
Site Improvements	61
Wings & Ells	37

- **Post and beam frame**
- **Log exterior walls**
- **Simple design from standard plans**
- **Single family — 1 full bath, 1 kitchen**
- **No basement**
- **Asphalt shingles on roof**
- **Hot air heat**
- **Drywall interior finishes**
- **Materials and workmanship are average**
- **Detail specification on page 27**

Note: The illustration shown may contain some optional components (for example: garages and/or fireplaces) whose costs are shown in the modifications, adjustments, & alternatives below or at the end of the square foot section.

Base cost per square foot of living area

Exterior Wall	Living Area										
	1000	1200	1400	1600	1800	2000	2200	2600	3000	3400	3800
6" Log-Solid	97.20	80.10	76.40	73.70	70.70	67.90	65.90	62.00	57.50	56.00	54.50
8" Log-Solid	93.30	83.70	79.60	77.00	73.90	70.70	68.80	64.60	60.20	58.60	56.80
Finished Basement, Add	12.90	12.50	12.00	11.80	11.50	11.30	11.10	10.70	10.40	10.30	10.10
Unfinished Basement, Add	5.20	4.90	4.60	4.40	4.30	4.00	4.00	3.70	3.50	3.40	3.30

Modifications

Add to the total cost

Upgrade Kitchen Cabinets	$ + 2206
Solid Surface Countertops	+ 896
Full Bath - including plumbing, wall and floor finishes	+ 3861
Half Bath - including plumbing, wall and floor finishes	+ 2405
One Car Attached Garage	+ 8424
One Car Detached Garage	+ 11,040
Fireplace & Chimney	+ 4160

Adjustments

For multi family - add to total cost

Additional Kitchen	$ + 4061
Additional Bath	+ 3861
Additional Entry & Exit	+ 1128
Separate Heating	+ 1211
Separate Electric	+ 1410

For Townhouse/Rowhouse -
Multiply cost per square foot by

Inner Unit	.92
End Unit	.96

Alternatives

Add to or deduct from the cost per square foot of living area

Cedar Shake Roof	+ 1.50
Air Conditioning, in Heating Ductwork	+ 1.56
In Separate Ductwork	+ 4.00
Heating Systems, Hot Water	+ 1.34
Heat Pump	+ 2.05
Electric Heat	– .50
Not Heated	– 2.46

Additional upgrades or components

Kitchen Cabinets & Countertops	Page 58
Bathroom Vanities	59
Fireplaces & Chimneys	59
Windows, Skylights & Dormers	59
Appliances	60
Breezeways & Porches	60
Finished Attic	60
Garages	61
Site Improvements	61
Wings & Ells	37

1 Story   Base cost per square foot of living area

Exterior Wall	Living Area							
	50	100	200	300	400	500	600	700
Wood Siding - Wood Frame	132.10	101.55	88.10	75.25	70.85	68.20	66.45	67.40
Brick Veneer - Wood Frame	143.60	107.80	92.15	75.65	70.55	67.45	65.40	66.15
Stucco on Wood Frame	132.80	102.00	88.45	75.40	71.00	68.35	66.55	67.50
Solid Masonry	175.40	132.45	113.85	92.40	86.30	82.65	81.30	81.65
Finished Basement, Add	41.35	33.45	30.20	24.70	23.60	22.95	22.50	22.20
Unfinished Basement, Add	18.35	14.00	12.20	9.15	8.55	8.20	7.95	7.75

1-1/2 Story   Base cost per square foot of living area

Exterior Wall	Living Area							
	100	200	300	400	500	600	700	800
Wood Siding - Wood Frame	107.85	86.30	73.60	66.95	63.15	61.40	59.05	58.30
Brick Veneer - Wood Frame	146.55	110.60	91.95	81.10	75.50	72.60	69.25	68.05
Stucco on Wood Frame	130.85	98.05	81.50	72.95	67.95	65.45	62.55	61.35
Solid Masonry	168.75	128.35	106.75	92.65	86.15	82.65	78.80	77.45
Finished Basement, Add	28.00	24.70	22.55	20.10	19.45	19.00	18.65	18.55
Unfinished Basement, Add	11.60	9.80	8.60	7.25	6.90	6.65	6.40	6.40

2 Story   Base cost per square foot of living area

Exterior Wall	Living Area							
	100	200	400	600	800	1000	1200	1400
Wood Siding - Wood Frame	106.50	79.65	67.70	57.40	53.50	51.15	49.60	50.70
Brick Veneer - Wood Frame	148.90	104.80	84.65	68.70	63.05	59.70	57.45	58.05
Stucco on Wood Frame	131.35	92.25	74.20	61.70	56.80	53.85	51.85	52.70
Solid Masonry	173.75	122.55	99.45	78.55	71.95	68.00	65.35	65.70
Finished Basement, Add	22.15	18.20	16.55	13.80	13.30	12.95	12.75	12.60
Unfinished Basement, Add	9.30	7.10	6.20	4.70	4.40	4.20	4.05	4.00

Base costs do not include bathroom or kitchen facilities. Use Modifications/Adjustments/Alternatives on pages 58-61 where appropriate.

1 Story

1-1/2 Story

2 Story

2-1/2 Story

Bi-Level

Tri-Level

CUSTOM

Components

1 Site Work	Site preparation for slab or excavation for lower level; 4' deep trench excavation for foundation wall.
2 Foundations	Continuous reinforced concrete footing, 8" deep x 18" wide; dampproofed and insulated 8" thick reinforced concrete foundation wall, 4' deep; 4" concrete slab on 4" crushed stone base and polyethylene vapor barrier, trowel finish.
3 Framing	Exterior walls—2 x 6 wood studs, 16" O.C.; 1/2" plywood sheathing; 2 x 8 rafters 16" O.C. with 1/2" plywood sheathing, 4 in 12, 6 in 12 or 8 in 12 roof pitch, 2 x 6 or 2 x 8 ceiling joists; 1/2" plywood subfloor on 1 x 3 wood sleepers 16" O.C.; 2 x 10 floor joists with 5/8" plywood subfloor on models with more than one level.
4 Exterior Walls	Horizontal beveled wood siding and #15 felt building paper on insulated wood frame walls; double hung windows; 3 flush solid core wood exterior doors with storms. **Alternates:** • Brick veneer on wood frame, has 4" veneer high quality face brick or select common brick. • Stone veneer on wood frame has exterior veneer of field stone or 2" thick limestone. • Solid masonry has an 8" concrete block wall with insulation and a brick or stone facing. It may be a solid brick or stone structure.
5 Roofing	30 year asphalt shingles; #15 felt building paper; aluminum gutters, downspouts and drip edge; copper flashings.
6 Interiors	Walls and ceilings—5/8" drywall, skim coat plaster, primed and painted with 2 coats; hardwood baseboard and trim; sanded and finished, hardwood floor 70%, ceramic tile with 1/2" underlayment 20%, vinyl tile with 1/2" underlayment 10%; wood panel interior doors, primed and painted with 2 coats.
7 Specialties	Custom grade kitchen cabinets—20 L.F. wall and base with plastic laminate counter top and kitchen sink; 4 L.F. bathroom vanity; 75 gallon electric water heater; medicine cabinet.
8 Mechanical	Gas fired warm air heat/air conditioning; one full bath including bathtub, corner shower, built in lavatory and water closet; one 1/2 bath including built in lavatory and water closet.
9 Electrical	100 Amp. service; romex wiring; incandescent lighting fixtures, switches, receptacles.
10 Overhead and Profit	General Contractor overhead and profit.

Adjustments

Unfinished Basement:
7'-6" high cast-in-place concrete walls 10" thick or 12" concrete block.

Finished Basement:
Includes painted drywall on 2 x 4 wood furring with insulation, suspended ceiling, carpeting on concrete floor, heating and lighting.

- **A distinct residence from designer's plans**
- **Single family — 1 full bath, 1 half bath, 1 kitchen**
- **No basement**
- **Asphalt shingles on roof**
- **Forced hot air heat/air conditioning**
- **Drywall interior finishes**
- **Materials and workmanship are above average**
- **Detail specifications on page 39**

Note: The illustration shown may contain some optional components (for example: garages and/or fireplaces) whose costs are shown in the modifications, adjustments, & alternatives below or at the end of the square foot section.

©Design Basics, Inc.

Base cost per square foot of living area

Exterior Wall	Living Area										
	800	1000	1200	1400	1600	1800	2000	2400	2800	3200	3600
Wood Siding - Wood Frame	119.90	109.40	100.90	94.05	89.35	86.90	83.85	78.00	73.80	70.80	67.65
Brick Veneer - Wood Frame	132.65	121.65	112.65	105.40	100.40	97.85	94.55	88.40	83.95	80.75	77.40
Stone Veneer - Wood Frame	137.95	126.45	117.00	109.35	104.15	101.40	97.90	91.45	86.80	83.40	79.85
Solid Masonry	138.45	126.90	117.40	109.70	104.45	101.75	98.15	91.75	87.05	83.60	80.05
Finished Basement, Add	37.45	35.65	34.00	32.75	31.90	31.45	30.70	29.75	29.05	28.40	27.85
Unfinished Basement, Add	14.30	13.45	12.70	12.05	11.65	11.40	11.10	10.60	10.25	9.95	9.65

Modifications

Add to the total cost

Upgrade Kitchen Cabinets	$ + 526
Solid Surface Countertops	+ 1280
Full Bath - including plumbing, wall and floor finishes	+ 4575
Half Bath - including plumbing, wall and floor finishes	+ 2850
Two Car Attached Garage	+ 17,118
Two Car Detached Garage	+ 19,464
Fireplace & Chimney	+ 3925

Adjustments

For multi family - add to total cost

Additional Kitchen	$ + 8491
Additional Full Bath & Half Bath	+ 7425
Additional Entry & Exit	+ 1128
Separate Heating & Air Conditioning	+ 4133
Separate Electric	+ 1410

For Townhouse/Rowhouse -
Multiply cost per square foot by

Inner Unit	.90
End Unit	.95

Alternatives

Add to or deduct from the cost per square foot of living area

Cedar Shake Roof	+ 1.78
Clay Tile Roof	+ 4.19
Slate Roof	+ 7.43
Upgrade Ceilings to Textured Finish	+ .39
Air Conditioning, in Heating Ductwork	Base System
Heating Systems, Hot Water	+ 1.48
Heat Pump	+ 1.77
Electric Heat	– 1.98
Not Heated	– 3.30

Additional upgrades or components

Kitchen Cabinets & Countertops	Page 58
Bathroom Vanities	59
Fireplaces & Chimneys	59
Windows, Skylights & Dormers	59
Appliances	60
Breezeways & Porches	60
Finished Attic	60
Garages	61
Site Improvements	61
Wings & Ells	47

- A distinct residence from designer's plans
- Single family — 1 full bath, 1 half bath, 1 kitchen
- No basement
- Asphalt shingles on roof
- Forced hot air heat/air conditioning
- Drywall interior finishes
- Materials and workmanship are above average
- Detail specifications on page 39

Note: The illustration shown may contain some optional components (for example: garages and/or fireplaces) whose costs are shown in the modifications, adjustments, & alternatives below or at the end of the square foot section.

●Donald A. Gardner Architects, Inc.

Base cost per square foot of living area

	Living Area										
Exterior Wall	1000	1200	1400	1600	1800	2000	2400	2800	3200	3600	4000
Wood Siding - Wood Frame	108.25	101.45	96.60	90.30	86.90	83.35	76.65	73.75	70.95	68.75	65.65
Brick Veneer - Wood Frame	114.10	107.00	101.85	95.10	91.45	87.65	80.45	77.35	74.25	71.95	68.65
Stone Veneer - Wood Frame	119.70	112.20	106.80	99.60	95.70	91.70	84.05	80.75	77.40	75.00	71.45
Solid Masonry	120.15	112.65	107.25	100.00	96.10	92.10	84.35	81.05	77.70	75.25	71.70
Finished Basement, Add	24.90	23.95	23.25	22.30	21.75	21.25	20.25	19.75	19.20	18.95	18.50
Unfinished Basement, Add	9.65	9.20	8.90	8.40	8.15	7.90	7.40	7.20	6.90	6.80	6.60

Modifications

Add to the total cost

Upgrade Kitchen Cabinets	$ + 526
Solid Surface Countertops	+ 1280
Full Bath - including plumbing, wall and floor finishes	+ 4575
Half Bath - including plumbing, wall and floor finishes	+ 2850
Two Car Attached Garage	+ 17,118
Two Car Detached Garage	+ 19,464
Fireplace & Chimney	+ 3925

Adjustments

For multi family - add to total cost

Additional Kitchen	$ + 8491
Additional Full Bath & Half Bath	+ 7425
Additional Entry & Exit	+ 1128
Separate Heating & Air Conditioning	+ 4133
Separate Electric	+ 1410

For Townhouse/Rowhouse -
Multiply cost per square foot by

Inner Unit	.90
End Unit	.95

Alternatives

Add to or deduct from the cost per square foot of living area

Cedar Shake Roof	+ 1.25
Clay Tile Roof	+ 3.02
Slate Roof	+ 5.37
Upgrade Ceilings to Textured Finish	+ .39
Air Conditioning, in Heating Ductwork	Base System
Heating Systems, Hot Water	+ 1.40
Heat Pump	+ 1.87
Electric Heat	– 1.72
Not Heated	– 3.04

Additional upgrades or components

Kitchen Cabinets & Countertops	Page 58
Bathroom Vanities	59
Fireplaces & Chimneys	59
Windows, Skylights & Dormers	59
Appliances	60
Breezeways & Porches	60
Finished Attic	60
Garages	61
Site Improvements	61
Wings & Ells	47

CUSTOM

- **A distinct residence from designer's plans**
- **Single family — 1 full bath, 1 half bath, 1 kitchen**
- **No basement**
- **Asphalt shingles on roof**
- **Forced hot air heat/air conditioning**
- **Drywall interior finishes**
- **Materials and workmanship are above average**
- **Detail specifications on page 39**

Note: The illustration shown may contain some optional components (for example: garages and/or fireplaces) whose costs are shown in the modifications, adjustments, & alternatives below or at the end of the square foot section.

Base cost per square foot of living area

Exterior Wall	Living Area										
	1200	1400	1600	1800	2000	2400	2800	3200	3600	4000	4400
Wood Siding - Wood Frame	102.10	96.60	92.65	89.05	85.10	79.40	74.60	71.35	69.45	67.55	65.75
Brick Veneer - Wood Frame	108.40	102.45	98.25	94.30	90.20	84.00	78.75	75.25	73.25	71.10	69.20
Stone Veneer - Wood Frame	114.35	108.00	103.55	99.30	95.00	88.35	82.75	79.00	76.80	74.45	72.45
Solid Masonry	114.85	108.50	104.05	99.75	95.45	88.70	83.05	79.30	77.15	74.75	72.75
Finished Basement, Add	20.00	19.25	18.75	18.20	17.85	17.05	16.40	16.00	15.75	15.40	15.20
Unfinished Basement, Add	7.75	7.40	7.15	6.90	6.75	6.35	6.05	5.85	5.70	5.55	5.45

Modifications

Add to the total cost

Upgrade Kitchen Cabinets	$ + 526
Solid Surface Countertops	+ 1280
Full Bath - including plumbing, wall and floor finishes	+ 4575
Half Bath - including plumbing, wall and floor finishes	+ 2850
Two Car Attached Garage	+ 17,118
Two Car Detached Garage	+ 19,464
Fireplace & Chimney	+ 4430

Adjustments

For multi family - add to total cost

Additional Kitchen	$ + 8491
Additional Full Bath & Half Bath	+ 7425
Additional Entry & Exit	+ 1128
Separate Heating & Air Conditioning	+ 4133
Separate Electric	+ 1410

For Townhouse/Rowhouse -
Multiply cost per square foot by

Inner Unit	.87
End Unit	.93

Alternatives

Add to or deduct from the cost per square foot of living area

Cedar Shake Roof	+ .89
Clay Tile Roof	+ 2.10
Slate Roof	+ 3.72
Upgrade Ceilings to Textured Finish	+ .39
Air Conditioning, in Heating Ductwork	Base System
Heating Systems, Hot Water	+ 1.37
Heat Pump	+ 2.08
Electric Heat	– 1.72
Not Heated	– 2.88

Additional upgrades or components

Kitchen Cabinets & Countertops	Page 58
Bathroom Vanities	59
Fireplaces & Chimneys	59
Windows, Skylights & Dormers	59
Appliances	60
Breezeways & Porches	60
Finished Attic	60
Garages	61
Site Improvements	61
Wings & Ells	47

- A distinct residence from designer's plans
- Single family — 1 full bath, 1 half bath, 1 kitchen
- No basement
- Asphalt shingles on roof
- Forced hot air heat/air conditioning
- Drywall interior finishes
- Materials and workmanship are above average
- Detail specifications on page 39

Note: The illustration shown may contain some optional components (for example: garages and/or fireplaces) whose costs are shown in the modifications, adjustments, & alternatives below or at the end of the square foot section.

CUSTOM

Base cost per square foot of living area

Exterior Wall	Living Area										
	1500	1800	2100	2400	2800	3200	3600	4000	4500	5000	5500
Wood Siding - Wood Frame	101.35	91.35	85.85	82.45	78.20	73.85	71.60	67.75	65.90	64.15	62.25
Brick Veneer - Wood Frame	107.85	97.40	91.25	87.55	83.05	78.25	75.80	71.65	69.60	67.60	65.60
Stone Veneer - Wood Frame	114.05	103.10	96.35	92.40	87.70	82.45	79.75	75.35	73.05	70.95	68.75
Solid Masonry	114.60	103.60	96.75	92.80	88.10	82.80	80.10	75.65	73.40	71.25	69.05
Finished Basement, Add	15.90	15.20	14.40	14.00	13.65	13.10	12.80	12.45	12.20	11.95	11.75
Unfinished Basement, Add	6.25	5.90	5.50	5.30	5.15	4.85	4.75	4.55	4.40	4.35	4.20

Modifications

Add to the total cost

Upgrade Kitchen Cabinets	$ + 526
Solid Surface Countertops	+ 1280
Full Bath - including plumbing, wall and floor finishes	+ 4575
Half Bath - including plumbing, wall and floor finishes	+ 2850
Two Car Attached Garage	+ 17,118
Two Car Detached Garage	+ 19,464
Fireplace & Chimney	+ 5005

Adjustments

For multi family - add to total cost

Additional Kitchen	$ + 8491
Additional Full Bath & Half Bath	+ 7425
Additional Entry & Exit	+ 1128
Separate Heating & Air Conditioning	+ 4133
Separate Electric	+ 1410

For Townhouse/Rowhouse -
Multiply cost per square foot by

Inner Unit	.87
End Unit	.94

Alternatives

Add to or deduct from the cost per square foot of living area

Cedar Shake Roof	+ .75
Clay Tile Roof	+ 1.81
Slate Roof	+ 3.22
Upgrade Ceilings to Textured Finish	+ .39
Air Conditioning, in Heating Ductwork	Base System
Heating Systems, Hot Water	+ 1.24
Heat Pump	+ 2.14
Electric Heat	− 3.05
Not Heated	− 2.88

Additional upgrades or components

Kitchen Cabinets & Countertops	Page 58
Bathroom Vanities	59
Fireplaces & Chimneys	59
Windows, Skylights & Dormers	59
Appliances	60
Breezeways & Porches	60
Finished Attic	60
Garages	61
Site Improvements	61
Wings & Ells	47

- A distinct residence from designer's plans
- Single family — 1 full bath, 1 half bath, 1 kitchen
- No basement
- Asphalt shingles on roof
- Forced hot air heat/air conditioning
- Drywall interior finishes
- Materials and workmanship are above average
- Detail specifications on page 39

Note: The illustration shown may contain some optional components (for example: garages and/or fireplaces) whose costs are shown in the modifications, adjustments, & alternatives below or at the end of the square foot section.

CUSTOM

Base cost per square foot of living area

Exterior Wall	Living Area										
	1500	1800	2100	2500	3000	3500	4000	4500	5000	5500	6000
Wood Siding - Wood Frame	100.90	91.05	86.65	83.05	77.00	74.15	70.35	66.35	65.05	63.50	62.05
Brick Veneer - Wood Frame	107.65	97.30	92.50	88.65	82.10	79.00	74.75	70.40	68.95	67.30	65.60
Stone Veneer - Wood Frame	114.10	103.25	98.05	93.95	86.90	83.55	78.90	74.25	72.65	70.85	68.95
Solid Masonry	114.65	103.80	98.60	94.40	87.35	83.95	79.25	74.60	72.95	71.20	69.25
Finished Basement, Add	14.00	13.35	12.85	12.45	11.90	11.55	11.15	10.80	10.65	10.45	10.25
Unfinished Basement, Add	5.45	5.15	4.95	4.75	4.50	4.35	4.10	3.95	3.90	3.80	3.70

Modifications

Add to the total cost

Upgrade Kitchen Cabinets	$ + 526
Solid Surface Countertops	+ 1280
Full Bath - including plumbing, wall and floor finishes	+ 4575
Half Bath - including plumbing, wall and floor finishes	+ 2850
Two Car Attached Garage	+ 17,118
Two Car Detached Garage	+ 19,464
Fireplace & Chimney	+ 5005

Adjustments

For multi family - add to total cost

Additional Kitchen	$ + 8491
Additional Full Bath & Half Bath	+ 7425
Additional Entry & Exit	+ 1128
Separate Heating & Air Conditioning	+ 4133
Separate Electric	+ 1410

For Townhouse/Rowhouse -
Multiply cost per square foot by

Inner Unit	.85
End Unit	.93

Alternatives

Add to or deduct from the cost per square foot of living area

Cedar Shake Roof	+ .59
Clay Tile Roof	+ 1.40
Slate Roof	+ 2.48
Upgrade Ceilings to Textured Finish	+ .39
Air Conditioning, in Heating Ductwork	Base System
Heating Systems, Hot Water	+ 1.24
Heat Pump	+ 2.14
Electric Heat	– 3.05
Not Heated	– 2.78

Additional upgrades or components

Kitchen Cabinets & Countertops	Page 58
Bathroom Vanities	59
Fireplaces & Chimneys	59
Windows, Skylights & Dormers	59
Appliances	60
Breezeways & Porches	60
Finished Attic	60
Garages	61
Site Improvements	61
Wings & Ells	47

Important: See the Reference Section for Location Factors (to adjust for your city) and Estimating Forms

- **A distinct residence from designer's plans**
- **Single family — 1 full bath, 1 half bath, 1 kitchen**
- **No basement**
- **Asphalt shingles on roof**
- **Forced hot air heat/air conditioning**
- **Drywall interior finishes**
- **Materials and workmanship are above average**
- **Detail specifications on page 39**

Note: The illustration shown may contain some optional components (for example: garages and/or fireplaces) whose costs are shown in the modifications, adjustments, & alternatives below or at the end of the square foot section.

CUSTOM

Base cost per square foot of living area

Exterior Wall	Living Area										
	1200	1400	1600	1800	2000	2400	2800	3200	3600	4000	4400
Wood Siding - Wood Frame	96.35	91.15	87.45	84.20	80.40	75.15	70.75	67.80	66.05	64.30	62.60
Brick Veneer - Wood Frame	101.05	95.50	91.65	88.15	84.20	78.60	73.90	70.75	68.90	66.95	65.15
Stone Veneer - Wood Frame	105.50	99.70	95.65	91.90	87.85	81.85	76.85	73.50	71.55	69.50	67.60
Solid Masonry	105.90	100.05	96.00	92.20	88.15	82.10	77.10	73.75	71.80	69.70	67.80
Finished Basement, Add	20.00	19.25	18.75	18.20	17.85	17.05	16.40	16.00	15.75	15.40	15.20
Unfinished Basement, Add	7.75	7.40	7.15	6.90	6.75	6.35	6.05	5.85	5.70	5.55	5.45

Modifications

Add to the total cost

Upgrade Kitchen Cabinets	$ + 526
Solid Surface Countertops	+ 1280
Full Bath - including plumbing, wall and floor finishes	+ 4575
Half Bath - including plumbing, wall and floor finishes	+ 2850
Two Car Attached Garage	+ 17,118
Two Car Detached Garage	+ 19,464
Fireplace & Chimney	+ 3925

Adjustments

For multi family - add to total cost

Additional Kitchen	$ + 8491
Additional Full Bath & Half Bath	+ 7425
Additional Entry & Exit	+ 1128
Separate Heating & Air Conditioning	+ 4133
Separate Electric	+ 1410

For Townhouse/Rowhouse -
Multiply cost per square foot by

Inner Unit	.89
End Unit	.95

Alternatives

Add to or deduct from the cost per square foot of living area

Cedar Shake Roof	+ .89
Clay Tile Roof	+ 2.10
Slate Roof	+ 3.72
Upgrade Ceilings to Textured Finish	+ .39
Air Conditioning, in Heating Ductwork	Base System
Heating Systems, Hot Water	+ 1.37
Heat Pump	+ 2.08
Electric Heat	- 1.73
Not Heated	- 2.78

Additional upgrades or components

Kitchen Cabinets & Countertops	Page 58
Bathroom Vanities	59
Fireplaces & Chimneys	59
Windows, Skylights & Dormers	59
Appliances	60
Breezeways & Porches	60
Finished Attic	60
Garages	61
Site Improvements	61
Wings & Ells	47

- **A distinct residence from designer's plans**
- **Single family — 1 full bath, 1 half bath, 1 kitchen**
- **No basement**
- **Asphalt shingles on roof**
- **Forced hot air heat/air conditioning**
- **Drywall interior finishes**
- **Materials and workmanship are above average**
- **Detail specifications on page 39**

Note: The illustration shown may contain some optional components (for example: garages and/or fireplaces) whose costs are shown in the modifications, adjustments, & alternatives below or at the end of the square foot section.

©Design Basics, Inc.

Base cost per square foot of living area

Exterior Wall	Living Area										
	1200	1500	1800	2100	2400	2800	3200	3600	4000	4500	5000
Wood Siding - Wood Frame	99.30	90.85	84.45	79.10	75.45	72.95	69.80	66.40	65.10	61.80	60.05
Brick Veneer - Wood Frame	104.00	95.10	88.25	82.60	78.70	76.05	72.70	69.10	67.75	64.20	62.35
Stone Veneer - Wood Frame	108.40	99.10	91.90	85.90	81.80	79.05	75.45	71.65	70.20	66.50	64.50
Solid Masonry	108.80	99.45	92.15	86.20	82.10	79.30	75.70	71.85	70.40	66.70	64.70
Finished Basement, Add*	24.90	23.70	22.65	21.80	21.25	20.85	20.30	19.85	19.60	19.15	18.85
Unfinished Basement, Add*	9.55	8.95	8.45	8.05	7.75	7.60	7.30	7.10	7.00	6.75	6.60

*Basement under middle level only.

Modifications

Add to the total cost

Upgrade Kitchen Cabinets	$ + 526
Solid Surface Countertops	+ 1280
Full Bath - including plumbing, wall and floor finishes	+ 4575
Half Bath - including plumbing, wall and floor finishes	+ 2850
Two Car Attached Garage	+ 17,118
Two Car Detached Garage	+ 19,464
Fireplace & Chimney	+ 4430

Adjustments

For multi family - add to total cost

Additional Kitchen	$ + 8491
Additional Full Bath & Half Bath	+ 7425
Additional Entry & Exit	+ 1128
Separate Heating & Air Conditioning	+ 4133
Separate Electric	+ 1410

For Townhouse/Rowhouse -
Multiply cost per square foot by

Inner Unit	.87
End Unit	.94

Alternatives

Add to or deduct from the cost per square foot of living area

Cedar Shake Roof	+ 1.25
Clay Tile Roof	+ 3.02
Slate Roof	+ 5.37
Upgrade Ceilings to Textured Finish	+ .39
Air Conditioning, in Heating Ductwork	Base System
Heating Systems, Hot Water	+ 1.32
Heat Pump	+ 2.16
Electric Heat	– 1.54
Not Heated	– 2.78

Additional upgrades or components

Kitchen Cabinets & Countertops	Page 58
Bathroom Vanities	59
Fireplaces & Chimneys	59
Windows, Skylights & Dormers	59
Appliances	60
Breezeways & Porches	60
Finished Attic	60
Garages	61
Site Improvements	61
Wings & Ells	47

Important: See the Reference Section for Location Factors (to adjust for your city) and Estimating Forms

CUSTOM

1 Story   Base cost per square foot of living area

Exterior Wall	Living Area							
	50	100	200	300	400	500	600	700
Wood Siding - Wood Frame	158.55	121.75	105.70	89.85	84.60	81.40	79.30	80.40
Brick Veneer - Wood Frame	174.35	133.10	115.15	96.15	90.20	86.70	84.30	85.20
Stone Veneer - Wood Frame	189.35	143.80	124.05	102.10	95.60	91.70	89.05	89.80
Solid Masonry	190.65	144.75	124.85	102.60	96.00	92.15	89.50	90.20
Finished Basement, Add	60.35	49.05	44.35	36.55	34.95	34.00	33.40	32.90
Unfinished Basement, Add	42.00	28.40	22.15	16.80	15.35	14.45	13.90	13.45

1-1/2 Story   Base cost per square foot of living area

Exterior Wall	Living Area							
	100	200	300	400	500	600	700	800
Wood Siding - Wood Frame	128.15	103.75	89.15	81.30	76.85	74.85	72.10	71.20
Brick Veneer - Wood Frame	142.30	115.05	98.55	88.65	83.60	81.25	78.15	77.20
Stone Veneer - Wood Frame	155.65	125.75	107.50	95.60	90.05	87.30	83.90	82.90
Solid Masonry	156.85	126.70	108.30	96.20	90.60	87.85	84.40	83.40
Finished Basement, Add	40.15	35.50	32.35	28.90	27.95	27.35	26.70	26.65
Unfinished Basement, Add	24.75	18.45	15.55	13.15	12.25	11.70	11.20	11.00

2 Story   Base cost per square foot of living area

Exterior Wall	Living Area							
	100	200	400	600	800	1000	1200	1400
Wood Siding - Wood Frame	127.70	96.20	82.30	70.05	65.55	62.75	60.95	62.25
Brick Veneer - Wood Frame	143.55	107.50	91.70	76.35	71.15	68.05	66.00	67.10
Stone Veneer - Wood Frame	158.55	118.20	100.65	82.30	76.55	73.05	70.75	71.70
Solid Masonry	159.85	119.15	101.45	82.80	76.95	73.50	71.15	72.05
Finished Basement, Add	30.20	24.55	22.20	18.25	17.50	17.05	16.75	16.50
Unfinished Basement, Add	21.00	14.20	11.10	8.40	7.70	7.25	6.95	6.75

Base costs do not include bathroom or kitchen facilities. Use Modifications/Adjustments/Alternatives on pages 58-61 where appropriate.

1 Story

1-1/2 Story

2 Story

2-1/2 Story

Bi-Level

Tri-Level

LUXURY

	Components
1 Site Work	Site preparation for slab or excavation for lower level; 4' deep trench excavation for foundation wall.
2 Foundations	Continuous reinforced concrete footing, 8" deep x 18" wide; dampproofed and insulated 12" thick reinforced concrete foundation wall, 4' deep; 4" concrete slab on 4" crushed stone base and polyethylene vapor barrier, trowel finish.
3 Framing	Exterior walls—2 x 6 wood studs, 16" O.C.; 1/2" plywood sheathing; 2 x 8 rafters 16" O.C. with 1/2" plywood sheathing, 4 in 12, 6 in 12 or 8 in 12 roof pitch, 2 x 6 or 2 x 8 ceiling joists; 1/2" plywood subfloor on 1 x 3 wood sleepers 16" O.C.; 2 x 10 floor joists with 5/8" plywood subfloor on models with more than one level.
4 Exterior Walls	Face brick veneer and #15 felt building paper on insulated wood frame walls; double hung windows; 3 flush solid core wood exterior doors with storms. **Alternates:** • Wood siding on wood frame, has top quality cedar or redwood siding or hand split cedar shingles or shakes. • Solid brick may have solid brick exterior wall or brick on concrete block. • Solid stone concrete block with selected fieldstone or limestone exterior.
5 Roofing	Red cedar shingles; #15 felt building paper; aluminum gutters, downspouts and drip edge; copper flashings.
6 Interiors	Walls and ceilings—5/8" drywall, skim coat plaster, primed and painted with 2 coats; hardwood baseboard and trim; sanded and finished, hardwood floor 70%, ceramic tile with 1/2" underlayment 20%, vinyl tile with 1/2" underlayment 10%; wood panel interior doors, primed and painted with 2 coats.
7 Specialties	Luxury grade kitchen cabinets—25 L.F. wall and base with plastic laminate counter top and kitchen sink; 6 L.F. bathroom vanity; 75 gallon electric water heater; medicine cabinet.
8 Mechanical	Gas fired warm air heat/air conditioning; one full bath including bathtub, corner shower, built in lavatory and water closet; one 1/2 bath including built in lavatory and water closet.
9 Electrical	100 Amp. service; romex wiring; incandescent lighting fixtures, switches, receptacles.
10 Overhead and Profit	General Contractor overhead and profit.

Adjustments

Unfinished Basement:
8" high cast-in-place concrete walls 12" thick or 12" concrete block.

Finished Basement:
Includes painted drywall on 2 x 4 wood furring with insulation, suspended ceiling, carpeting on subfloor with sleepers, heating and lighting.

LUXURY

- **Unique residence built from an architect's plan**
- **Single family — 1 full bath, 1 half bath, 1 kitchen**
- **No basement**
- **Cedar shakes on roof**
- **Forced hot air heat/air conditioning**
- **Double drywall interior**
- **Many special features**
- **Extraordinary materials and workmanship**
- **Detail specifications on page 49**

Note: The illustration shown may contain some optional components (for example: garages and/or fireplaces) whose costs are shown in the modifications, adjustments, & alternatives below or at the end of the square foot section.

eHome Planners, Inc.

Base cost per square foot of living area

Exterior Wall	Living Area										
	1000	1200	1400	1600	1800	2000	2400	2800	3200	3600	4000
Wood Siding - Wood Frame	138.45	128.45	120.35	114.90	111.85	108.30	101.50	96.65	93.20	89.60	86.55
Brick Veneer - Wood Frame	144.30	133.70	125.20	119.40	116.20	112.40	105.25	100.10	96.40	92.55	89.35
Solid Brick	154.20	142.65	133.30	127.00	123.55	119.30	111.55	105.90	101.75	97.60	94.05
Solid Stone	155.15	143.50	134.10	127.70	124.25	119.95	112.15	106.45	102.30	98.05	94.50
Finished Basement, Add	42.40	40.40	38.80	37.75	37.20	36.35	35.15	34.20	33.45	32.75	32.15
Unfinished Basement, Add	15.20	14.35	13.65	13.20	12.95	12.55	12.05	11.60	11.25	10.95	10.70

Modifications

Add to the total cost

Upgrade Kitchen Cabinets	$	+ 695
Solid Surface Countertops		+ 1284
Full Bath - including plumbing, wall and floor finishes		+ 5307
Half Bath - including plumbing, wall and floor finishes		+ 3306
Two Car Attached Garage		+ 19,788
Two Car Detached Garage		+ 22,357
Fireplace & Chimney		+ 5545

Adjustments

For multi family - add to total cost

Additional Kitchen	$	+ 10,887
Additional Full Bath & Half Bath		+ 8613
Additional Entry & Exit		+ 1718
Separate Heating & Air Conditioning		+ 4133
Separate Electric		+ 1410

For Townhouse/Rowhouse -
Multiply cost per square foot by

Inner Unit	.90
End Unit	.95

Alternatives

Add to or deduct from the cost per square foot of living area

Heavyweight Asphalt Shingles	– 1.78
Clay Tile Roof	+ 2.41
Slate Roof	+ 5.65
Upgrade Ceilings to Textured Finish	+ .39
Air Conditioning, in Heating Ductwork	Base System
Heating Systems, Hot Water	+ 1.59
Heat Pump	+ 1.92
Electric Heat	– 1.72
Not Heated	– 3.57

Additional upgrades or components

Kitchen Cabinets & Countertops	Page 58
Bathroom Vanities	59
Fireplaces & Chimneys	59
Windows, Skylights & Dormers	59
Appliances	60
Breezeways & Porches	60
Finished Attic	60
Garages	61
Site Improvements	61
Wings & Ells	57

- **Unique residence built from an architect's plan**
- **Single family — 1 full bath, 1 half bath, 1 kitchen**
- **No basement**
- **Cedar shakes on roof**
- **Forced hot air heat/air conditioning**
- **Double drywall interior**
- **Many special features**
- **Extraordinary materials and workmanship**
- **Detail specifications on page 49**

Note: The illustration shown may contain some optional components (for example: garages and/or fireplaces) whose costs are shown in the modifications, adjustments, & alternatives below or at the end of the square foot section.

©Larry E. Belk Designs

LUXURY

Base cost per square foot of living area

Exterior Wall	Living Area										
	1000	1200	1400	1600	1800	2000	2400	2800	3200	3600	4000
Wood Siding - Wood Frame	126.60	118.45	112.50	105.25	101.20	97.00	89.25	85.80	82.50	79.95	76.40
Brick Veneer - Wood Frame	133.40	124.80	118.60	110.75	106.40	101.95	93.60	89.95	86.35	83.65	79.80
Solid Brick	144.80	135.50	128.80	120.00	115.20	110.30	100.95	96.90	92.80	89.85	85.60
Solid Stone	145.95	136.50	129.75	120.90	116.00	111.10	101.70	97.60	93.40	90.45	86.15
Finished Basement, Add	29.75	28.60	27.75	26.55	25.85	25.25	23.95	23.40	22.75	22.35	21.90
Unfinished Basement, Add	10.90	10.40	10.05	9.50	9.20	8.95	8.40	8.15	7.85	7.70	7.45

Modifications

Add to the total cost

Upgrade Kitchen Cabinets	$ + 695
Solid Surface Countertops	+ 1284
Full Bath - including plumbing, wall and floor finishes	+ 5307
Half Bath - including plumbing, wall and floor finishes	+ 3306
Two Car Attached Garage	+ 19,788
Two Car Detached Garage	+ 22,357
Fireplace & Chimney	+ 5545

Adjustments

For multi family - add to total cost

Additional Kitchen	$ + 10,887
Additional Full Bath & Half Bath	+ 8613
Additional Entry & Exit	+ 1718
Separate Heating & Air Conditioning	+ 4133
Separate Electric	+ 1410

For Townhouse/Rowhouse -
Multiply cost per square foot by

Inner Unit	.90
End Unit	.95

Alternatives

Add to or deduct from the cost per square foot of living area

Heavyweight Asphalt Shingles	– 1.25
Clay Tile Roof	+ 1.77
Slate Roof	+ 4.14
Upgrade Ceilings to Textured Finish	+ .39
Air Conditioning, in Heating Ductwork	Base System
Heating Systems, Hot Water	+ 1.52
Heat Pump	+ 2.13
Electric Heat	– 1.72
Not Heated	– 3.29

Additional upgrades or components

Kitchen Cabinets & Countertops	Page 58
Bathroom Vanities	59
Fireplaces & Chimneys	59
Windows, Skylights & Dormers	59
Appliances	60
Breezeways & Porches	60
Finished Attic	60
Garages	61
Site Improvements	61
Wings & Ells	57

- **Unique residence built from an architect's plan**
- **Single family — 1 full bath, 1 half bath, 1 kitchen**
- **No basement**
- **Cedar shakes on roof**
- **Forced hot air heat/air conditioning**
- **Double drywall interior**
- **Many special features**
- **Extraordinary materials and workmanship**
- **Detail specifications on page 49**

Note: The illustration shown may contain some optional components (for example: garages and/or fireplaces) whose costs are shown in the modifications, adjustments, & alternatives below or at the end of the square foot section.

Base cost per square foot of living area

Exterior Wall	Living Area										
	1200	1400	1600	1800	2000	2400	2800	3200	3600	4000	4400
Wood Siding - Wood Frame	118.15	111.65	107.00	102.65	98.20	91.50	85.95	82.25	80.05	77.80	75.75
Brick Veneer - Wood Frame	125.40	118.40	113.45	108.75	104.05	96.80	90.75	86.75	84.40	81.90	79.70
Solid Brick	137.60	129.80	124.35	119.00	113.95	105.75	98.90	94.35	91.70	88.80	86.35
Solid Stone	138.80	130.90	125.40	120.00	114.85	106.55	99.65	95.05	92.45	89.40	87.00
Finished Basement, Add	23.90	22.95	22.35	21.70	21.25	20.25	19.45	18.95	18.60	18.25	17.95
Unfinished Basement, Add	8.75	8.35	8.10	7.80	7.60	7.20	6.85	6.60	6.50	6.30	6.20

Modifications

Add to the total cost

Upgrade Kitchen Cabinets	$ + 695
Solid Surface Countertops	+ 1284
Full Bath - including plumbing, wall and floor finishes	+ 5307
Half Bath - including plumbing, wall and floor finishes	+ 3306
Two Car Attached Garage	+ 19,788
Two Car Detached Garage	+ 22,357
Fireplace & Chimney	+ 6075

Adjustments

For multi family - add to total cost

Additional Kitchen	$ + 10,887
Additional Full Bath & Half Bath	+ 8613
Additional Entry & Exit	+ 1718
Separate Heating & Air Conditioning	+ 4133
Separate Electric	+ 1410

For Townhouse/Rowhouse -
Multiply cost per square foot by

Inner Unit	.86
End Unit	.93

Alternatives

Add to or deduct from the cost per square foot of living area

Heavyweight Asphalt Shingles	– .89
Clay Tile Roof	+ 1.21
Slate Roof	+ 2.83
Upgrade Ceilings to Textured Finish	+ .39
Air Conditioning, in Heating Ductwork	Base System
Heating Systems, Hot Water	+ 1.48
Heat Pump	+ 2.25
Electric Heat	– 1.54
Not Heated	– 3.11

Additional upgrades or components

- **Unique residence built from an architect's plan**
- **Single family — 1 full bath, 1 half bath, 1 kitchen**
- **No basement**
- **Cedar shakes on roof**
- **Forced hot air heat/air conditioning**
- **Double drywall interior**
- **Many special features**
- **Extraordinary materials and workmanship**
- **Detail specifications on page 49**

Note: The illustration shown may contain some optional components (for example: garages and/or fireplaces) whose costs are shown in the modifications, adjustments, & alternatives below or at the end of the square foot section.

©Larry W. Garnett & Associates, Inc

Base cost per square foot of living area

Exterior Wall	Living Area										
	1500	1800	2100	2500	3000	3500	4000	4500	5000	5500	6000
Wood Siding - Wood Frame	116.10	104.65	98.30	93.60	87.00	82.20	77.40	75.25	73.20	71.05	69.00
Brick Veneer - Wood Frame	123.65	111.60	104.50	99.45	92.35	87.00	81.90	79.45	77.25	74.90	72.70
Solid Brick	136.35	123.25	114.95	109.30	101.35	95.20	89.45	86.65	84.05	81.40	78.90
Solid Stone	137.55	124.40	115.95	110.25	102.15	95.95	90.15	87.30	84.75	82.05	79.50
Finished Basement, Add	19.05	18.20	17.15	16.60	15.85	15.20	14.75	14.40	14.15	13.85	13.65
Unfinished Basement, Add	7.05	6.65	6.20	6.00	5.65	5.35	5.15	5.00	4.90	4.80	4.70

Modifications

Add to the total cost

Upgrade Kitchen Cabinets	$ + 695
Solid Surface Countertops	+ 1284
Full Bath - including plumbing, wall and floor finishes	+ 5307
Half Bath - including plumbing, wall and floor finishes	+ 3306
Two Car Attached Garage	+ 19,788
Two Car Detached Garage	+ 22,357
Fireplace & Chimney	+ 6645

Adjustments

For multi family - add to total cost

Additional Kitchen	$ + 10,887
Additional Full Bath & Half Bath	+ 8613
Additional Entry & Exit	+ 1718
Separate Heating & Air Conditioning	+ 4133
Separate Electric	+ 1410

For Townhouse/Rowhouse -
Multiply cost per square foot by

Inner Unit	.86
End Unit	.93

Alternatives

Add to or deduct from the cost per square foot of living area

Heavyweight Asphalt Shingles	– .75
Clay Tile Roof	+ 1.06
Slate Roof	+ 2.47
Upgrade Ceilings to Textured Finish	+ .39
Air Conditioning, in Heating Ductwork	Base System
Heating Systems, Hot Water	+ 1.33
Heat Pump	+ 2.32
Electric Heat	– 3.05
Not Heated	– 3.11

Additional upgrades or components

Kitchen Cabinets & Countertops	Page 58
Bathroom Vanities	59
Fireplaces & Chimneys	59
Windows, Skylights & Dormers	59
Appliances	60
Breezeways & Porches	60
Finished Attic	60
Garages	61
Site Improvements	61
Wings & Ells	57

- Unique residence built from an architect's plan
- Single family — 1 full bath, 1 half bath, 1 kitchen
- No basement
- Cedar shakes on roof
- Forced hot air heat/air conditioning
- Double drywall interior
- Many special features
- Extraordinary materials and workmanship
- Detail specifications on page 49

Note: The illustration shown may contain some optional components (for example: garages and/or fireplaces) whose costs are shown in the modifications, adjustments, & alternatives below or at the end of the square foot section.

LUXURY

Base cost per square foot of living area

Exterior Wall	Living Area										
	1500	1800	2100	2500	3000	3500	4000	4500	5000	5500	6000
Wood Siding - Wood Frame	115.35	103.95	98.90	94.60	87.65	84.30	79.95	75.45	73.90	72.15	70.50
Brick Veneer - Wood Frame	123.15	111.20	105.65	101.05	93.55	89.90	85.05	80.10	78.40	76.50	74.60
Solid Brick	136.30	123.40	117.05	111.95	103.40	99.20	93.55	88.00	86.00	83.85	81.50
Solid Stone	137.60	124.60	118.15	113.00	104.35	100.10	94.35	88.75	86.75	84.55	82.10
Finished Basement, Add	16.75	15.95	15.30	14.85	14.15	13.75	13.20	12.80	12.60	12.40	12.15
Unfinished Basement, Add	6.15	5.85	5.55	5.40	5.05	4.90	4.65	4.45	4.40	4.30	4.20

Modifications

Add to the total cost

Upgrade Kitchen Cabinets	$ + 695
Solid Surface Countertops	+ 1284
Full Bath - including plumbing, wall and floor finishes	+ 5307
Half Bath - including plumbing, wall and floor finishes	+ 3306
Two Car Attached Garage	+ 19,788
Two Car Detached Garage	+ 22,357
Fireplace & Chimney	+ 6645

Adjustments

For multi family - add to total cost

Additional Kitchen	$ + 10,887
Additional Full Bath & Half Bath	+ 8613
Additional Entry & Exit	+ 1718
Separate Heating & Air Conditioning	+ 4133
Separate Electric	+ 1410

For Townhouse/Rowhouse -
Multiply cost per square foot by

Inner Unit	.84
End Unit	.92

Alternatives

Add to or deduct from the cost per square foot of living area

Heavyweight Asphalt Shingles	– .59
Clay Tile Roof	+ .80
Slate Roof	+ 1.88
Upgrade Ceilings to Textured Finish	+ .39
Air Conditioning, in Heating Ductwork	Base System
Heating Systems, Hot Water	+ 1.33
Heat Pump	+ 2.32
Electric Heat	– 3.05
Not Heated	– 3.01

Additional upgrades or components

Kitchen Cabinets & Countertops	Page 58
Bathroom Vanities	59
Fireplaces & Chimneys	59
Windows, Skylights & Dormers	59
Appliances	60
Breezeways & Porches	60
Finished Attic	60
Garages	61
Site Improvements	61
Wings & Ells	57

- **Unique residence built from an architect's plan**
- **Single family — 1 full bath, 1 half bath, 1 kitchen**
- **No basement**
- **Cedar shakes on roof**
- **Forced hot air heat/air conditioning**
- **Double drywall interior**
- **Many special features**
- **Extraordinary materials and workmanship**
- **Detail specifications on page 49**

Note: The illustration shown may contain some optional components (for example: garages and/or fireplaces) whose costs are shown in the modifications, adjustments, & alternatives below or at the end of the square foot section.

LUXURY

Base cost per square foot of living area

Exterior Wall	Living Area										
	1200	1400	1600	1800	2000	2400	2800	3200	3600	4000	4400
Wood Siding - Wood Frame	111.80	105.65	101.25	97.35	92.95	86.85	81.80	78.30	76.25	74.20	72.30
Brick Veneer - Wood Frame	117.20	110.70	106.10	101.90	97.40	90.80	85.40	81.70	79.50	77.25	75.25
Solid Brick	126.40	119.25	114.25	109.55	104.80	97.50	91.45	87.40	85.00	82.45	80.25
Solid Stone	127.25	120.05	115.05	110.30	105.50	98.15	92.05	87.90	85.55	82.95	80.70
Finished Basement, Add	23.90	22.95	22.35	21.70	21.25	20.25	19.45	18.95	18.60	18.25	17.95
Unfinished Basement, Add	8.75	8.35	8.10	7.80	7.60	7.20	6.85	6.60	6.50	6.30	6.20

Modifications

Add to the total cost

Upgrade Kitchen Cabinets	$ + 695
Solid Surface Countertops	+ 1284
Full Bath - including plumbing, wall and floor finishes	+ 5307
Half Bath - including plumbing, wall and floor finishes	+ 3306
Two Car Attached Garage	+ 19,788
Two Car Detached Garage	+ 22,357
Fireplace & Chimney	+ 5545

Adjustments

For multi family - add to total cost

Additional Kitchen	$ + 10,887
Additional Full Bath & Half Bath	+ 8613
Additional Entry & Exit	+ 1718
Separate Heating & Air Conditioning	+ 4133
Separate Electric	+ 1410

For Townhouse/Rowhouse -
Multiply cost per square foot by

Inner Unit	.89
End Unit	.94

Alternatives

Add to or deduct from the cost per square foot of living area

Heavyweight Asphalt Shingles	− .89
Clay Tile Roof	+ 1.21
Slate Roof	+ 5.65
Upgrade Ceilings to Textured Finish	+ .39
Air Conditioning, in Heating Ductwork	Base System
Heating Systems, Hot Water	+ 1.48
Heat Pump	+ 2.25
Electric Heat	− 1.54
Not Heated	− 3.11

Additional upgrades or components

Kitchen Cabinets & Countertops	Page 58
Bathroom Vanities	59
Fireplaces & Chimneys	59
Windows, Skylights & Dormers	59
Appliances	60
Breezeways & Porches	60
Finished Attic	60
Garages	61
Site Improvements	61
Wings & Ells	57

- **Unique residence built from an architect's plan**
- **Single family — 1 full bath, 1 half bath, 1 kitchen**
- **No basement**
- **Cedar shakes on roof**
- **Forced hot air heat/air conditioning**
- **Double drywall interior**
- **Many special features**
- **Extraordinary materials and workmanship**
- **Detail specifications on page 49**

Note: The illustration shown may contain some optional components (for example: garages and/or fireplaces) whose costs are shown in the modifications, adjustments, & alternatives below or at the end of the square foot section.

©Home Planners, Inc

LUXURY

Base cost per square foot of living area

Exterior Wall	Living Area										
	1500	1800	2100	2400	2800	3200	3600	4000	4500	5000	5500
Wood Siding - Wood Frame	105.90	98.40	92.25	87.95	84.95	81.35	77.40	75.85	72.05	70.10	67.80
Brick Veneer - Wood Frame	110.80	102.80	96.25	91.70	88.55	84.65	80.50	78.85	74.85	72.70	70.30
Solid Brick	119.05	110.20	103.00	98.00	94.60	90.30	85.75	83.95	79.55	77.15	74.50
Solid Stone	119.85	110.90	103.65	98.65	95.20	90.85	86.25	84.45	80.00	77.55	74.85
Finished Basement, Add*	28.25	26.95	25.85	25.20	24.70	24.05	23.40	23.15	22.55	22.15	21.80
Unfinished Basement, Add*	10.15	9.55	9.10	8.80	8.60	8.30	8.05	7.90	7.65	7.45	7.30

*Basement under middle level only.

Modifications

Add to the total cost

Upgrade Kitchen Cabinets	$ + 695
Solid Surface Countertops	+ 1284
Full Bath - including plumbing, wall and floor finishes	+ 5307
Half Bath - including plumbing, wall and floor finishes	+ 3306
Two Car Attached Garage	+ 19,788
Two Car Detached Garage	+ 22,357
Fireplace & Chimney	+ 6075

Adjustments

For multi family - add to total cost

Additional Kitchen	$ + 10,887
Additional Full Bath & Half Bath	+ 8613
Additional Entry & Exit	+ 1718
Separate Heating & Air Conditioning	+ 4133
Separate Electric	+ 1410

For Townhouse/Rowhouse -
Multiply cost per square foot by

Inner Unit	.86
End Unit	.93

Alternatives

Add to or deduct from the cost per square foot of living area

Heavyweight Asphalt Shingles	– 1.25
Clay Tile Roof	+ 1.77
Slate Roof	+ 4.14
Upgrade Ceilings to Textured Finish	+ .39
Air Conditioning, in Heating Ductwork	Base System
Heating Systems, Hot Water	+ 1.43
Heat Pump	+ 2.34
Electric Heat	– 1.37
Not Heated	– 3.01

Additional upgrades or components

Kitchen Cabinets & Countertops	Page 58
Bathroom Vanities	59
Fireplaces & Chimneys	59
Windows, Skylights & Dormers	59
Appliances	60
Breezeways & Porches	60
Finished Attic	60
Garages	61
Site Improvements	61
Wings & Ells	57

LUXURY

1 Story — Base cost per square foot of living area

Exterior Wall	Living Area							
	50	100	200	300	400	500	600	700
Wood Siding - Wood Frame	178.10	137.30	119.50	101.95	96.10	92.55	90.20	91.45
Brick Veneer - Wood Frame	196.40	150.35	130.35	109.20	102.60	98.60	96.00	97.00
Solid Brick	227.15	172.35	148.65	121.40	113.65	108.90	105.75	106.45
Solid Stone	230.10	174.45	150.40	122.55	114.60	109.85	106.70	107.35
Finished Basement, Add	74.25	59.90	53.90	43.90	41.90	40.70	39.90	39.35
Unfinished Basement, Add	30.25	23.60	20.85	16.20	15.25	14.70	14.35	14.10

1-1/2 Story — Base cost per square foot of living area

Exterior Wall	Living Area							
	100	200	300	400	500	600	700	800
Wood Siding - Wood Frame	142.20	115.05	98.90	90.20	85.25	83.05	80.05	79.00
Brick Veneer - Wood Frame	158.50	128.10	109.75	98.70	93.10	90.45	87.00	85.95
Solid Brick	185.95	150.10	128.00	112.95	106.25	102.90	98.80	97.60
Solid Stone	188.55	152.20	129.80	114.30	107.55	104.10	99.90	98.75
Finished Basement, Add	49.10	43.10	39.10	34.70	33.50	32.70	31.95	31.85
Unfinished Basement, Add	19.45	16.65	14.80	12.75	12.20	11.85	11.50	11.45

2 Story — Base cost per square foot of living area

Exterior Wall	Living Area							
	100	200	400	600	800	1000	1200	1400
Wood Siding - Wood Frame	140.15	105.20	89.80	76.30	71.30	68.20	66.20	67.65
Brick Veneer - Wood Frame	158.40	118.25	100.65	83.55	77.80	74.30	72.00	73.25
Solid Brick	189.20	140.20	118.95	95.75	88.85	84.60	81.75	82.65
Solid Stone	192.10	142.30	120.75	96.95	89.80	85.55	82.70	83.55
Finished Basement, Add	37.20	30.00	27.00	22.00	21.00	20.40	20.00	19.70
Unfinished Basement, Add	15.15	11.80	10.45	8.10	7.65	7.35	7.20	7.05

Base costs do not include bathroom or kitchen facilities. Use Modifications/Adjustments/Alternatives on pages 58-61 where appropriate.

Kitchen cabinets - Base units, hardwood (Cost per Unit)

	Economy	Average	Custom	Luxury
24" deep, 35" high,				
One top drawer,				
One door below				
12" wide	$122	$162	$215	$285
15" wide	161	215	285	375
18" wide	186	248	330	435
21" wide	178	237	315	415
24" wide	204	272	360	475
Four drawers				
12" wide	274	365	485	640
15" wide	236	315	420	550
18" wide	240	320	425	560
24" wide	259	345	460	605
Two top drawers,				
Two doors below				
27" wide	229	305	405	535
30" wide	244	325	430	570
33" wide	244	325	430	570
36" wide	255	340	450	595
42" wide	274	365	485	640
48" wide	289	385	510	675
Range or sink base				
(Cost per unit)				
Two doors below				
30" wide	200	266	355	465
33" wide	212	283	375	495
36" wide	222	296	395	520
42" wide	236	315	420	550
48" wide	248	330	440	580
Corner Base Cabinet				
(Cost per unit)				
36" wide	202	269	360	470
Lazy Susan (Cost per unit)				
With revolving door	311	415	550	725

Kitchen cabinets - Wall cabinets, hardwood (Cost per Unit)

	Economy	Average	Custom	Luxury
12" deep, 2 doors				
12" high				
30" wide	$130	$173	$230	$ 305
36" wide	149	198	265	345
15" high				
30" wide	138	184	245	320
33" wide	152	202	270	355
36" wide	157	209	280	365
24" high				
30" wide	164	219	290	385
36" wide	180	240	320	420
42" wide	197	263	350	460
30" high, 1 door				
12" wide	119	158	210	275
15" wide	132	176	235	310
18" wide	144	192	255	335
24" wide	158	210	280	370
30" high, 2 doors				
27" wide	194	258	345	450
30" wide	188	251	335	440
36" wide	214	285	380	500
42" wide	229	305	405	535
48" wide	255	340	450	595
Corner wall, 30" high				
24" wide	138	184	245	320
30" wide	162	216	285	380
36" wide	175	233	310	410
Broom closet				
84" high, 24" deep				
18" wide	345	460	610	805
Oven Cabinet				
84" high, 24" deep				
27" wide	499	665	885	1165

Kitchen countertops (Cost per L.F.)

	Economy	Average	Custom	Luxury
Solid Surface				
24" wide, no backsplash	$83	$111	$150	$195
with backsplash	91	121	160	210
Stock plastic laminate, 24" wide				
with backsplash	14	18	25	30
Custom plastic laminate, no splash				
7/8" thick, alum. molding	20	27	35	45
1-1/4" thick, no splash	24	32	45	55
Marble				
1/2" - 3/4" thick w/splash	37	50	65	85
Maple, laminated				
1-1/2" thick w/splash	36	49	65	85
Stainless steel				
(per S.F.)	70	93	125	165
Cutting blocks, recessed				
16" x 20" x 1" (each)	61	81	110	140

Vanity bases *(Cost per Unit)*

	Economy	Average	Custom	Luxury
2 door, 30" high, 21" deep				
24" wide	155	206	275	360
30" wide	180	240	320	420
36" wide	236	315	420	550
48" wide	281	375	500	655

Solid surface vanity tops *(Cost Each)*

	Economy	Average	Custom	Luxury
Center bowl				
22" x 25"	$281	$303	$328	$354
22" x 31"	325	351	379	409
22" x 37"	370	400	432	466
22" x 49"	460	497	537	579

Fireplaces & chimneys *(Cost per Unit)*

	1-1/2 Story	2 Story	3 Story
Economy (prefab metal)			
Exterior chimney & 1 fireplace	$3655	$4035	$4420
Interior chimney & 1 fireplace	3495	3890	4075
Average (masonry)			
Exterior chimney & 1 fireplace	3730	4160	4730
Interior chimney & 1 fireplace	3495	3920	4270
For more than 1 flue, add	275	455	765
For more than 1 fireplace, add	2645	2645	2645
Custom (masonry)			
Exterior chimney & 1 fireplace	3925	4430	5005
Interior chimney & 1 fireplace	3685	4165	4490
For more than 1 flue, add	310	530	725
For more than 1 fireplace, add	2820	2820	2820
Luxury (masonry)			
Exterior chimney & 1 fireplace	5545	6075	6645
Interior chimney & 1 fireplace	5295	5790	6120
For more than 1 flue, add	455	765	1070
For more than 1 fireplace, add	4365	4365	4365

Windows and skylights *(Cost Each)*

	Economy	Average	Custom	Luxury
Fixed Picture Windows				
3'-6" x 4'-0"	$ 390	$ 421	$ 455	$ 491
4'-0" x 6'-0"	617	667	720	778
5'-0" x 6'-0"	793	856	925	999
6'-0" x 6'-0"	806	870	940	1015
Bay/Bow Windows				
8'-0" x 5'-0"	1157	1250	1350	1458
10'-0" x 5'-0"	1629	1759	1900	2052
10'-0" x 6'-0"	1715	1852	2000	2160
12'-0" x 6'-0"	2036	2199	2375	2565
Palladian Windows				
3'-2" x 6'-4"		1435	1550	1674
4'-0" x 6'-0"		1481	1600	1728
5'-5" x 6'-10"		1852	2000	2160
8'-0" x 6'-0"		2269	2450	2646
Skylights				
46" x 21-1/2"	342	369	513	554
46" x 28"	451	487	637	688
57" x 44"	559	604	789	852

Dormers *(Cost/S.F. of plan area)*

	Economy	Average	Custom	Luxury
Framing and Roofing Only				
Gable dormer, 2" x 6" roof frame	$19	$21	$24	$39
2" x 8" roof frame	20	23	25	41
Shed dormer, 2" x 6" roof frame	12	14	15	25
2" x 8" roof frame	13	15	17	26
2" x 10" roof frame	14	16	18	27

ADJUSTMENTS

Appliances *(Cost per Unit)*

	Economy	Average	Custom	Luxury
Range				
30" free standing, 1 oven	$ 320	$ 873	$1149	$ 1425
2 oven	1475	1563	1606	1650
30" built-in, 1 oven	400	925	1188	1450
2 oven	1200	1600	1800	2000
21" free standing				
1 oven	355	493	561	630
Counter Top Ranges				
4 burner standard	262	426	508	590
As above with griddle	485	588	639	690
Microwave Oven	174	385	490	595
Combination Range,				
Refrigerator, Sink				
30" wide	995	1498	1749	2000
60" wide	2925	3364	3584	3803
72" wide	3325	3824	4074	4323
Comb. Range, Refrig., Sink,				
Microwave Oven & Ice Maker	5147	5919	6305	66910
Compactor				
4 to 1 compaction	445	483	501	520
Deep Freeze				
15 to 23 C.F.	500	590	635	680
30 C.F.	930	1053	1114	1175
Dehumidifier, portable, auto.				
15 pint	223	257	273	290
30 pint	247	284	303	321
Washing Machine, automatic	470	748	886	1025
Water Heater				
Electric, glass lined				
30 gal.	325	400	438	475
80 gal.	680	865	958	1050
Water Heater, Gas, glass lined				
30 gal.	455	560	613	665
50 gal.	685	855	940	1025
Water Softener, automatic				
30 grains/gal.	585	673	717	761
100 grains/gal.	760	874	931	988
Dishwasher, built-in				
2 cycles	375	440	473	505
4 or more cycles	415	468	494	520
Dryer, automatic	440	655	763	870
Garage Door Opener	273	324	350	375
Garbage Disposal	82	161	201	240
Heater, Electric, built-in				
1250 watt ceiling type	155	192	211	229
1250 watt wall type	187	218	233	248
Wall type w/blower				
1500 watt	213	232	241	250
3000 watt	380	420	440	460
Hood For Range, 2 speed				
30" wide	112	364	489	615
42" wide	262	606	778	950
Humidifier, portable				
7 gal. per day	111	128	136	144
15 gal. per day	180	207	221	234
Ice Maker, automatic				
13 lb. per day	650	748	796	845
51 lb. per day	1050	1208	1286	1365
Refrigerator, no frost				
10-12 C.F.	535	683	756	830
14-16 C.F.	550	615	648	680
18-20 C.F.	605	758	834	910
21-29 C.F.	805	2003	2601	3200
Sump Pump, 1/3 H.P.	202	281	321	360

Breezeway *(Cost per S.F.)*

Class	Type	Area (S.F.)			
		50	100	150	200
Economy	Open	$ 16.60	$14.15	$11.85	$11.60
	Enclosed	79.65	61.55	51.10	44.75
Average	Open	21.10	18.55	16.20	14.75
	Enclosed	89.55	66.55	54.55	48.00
Custom	Open	30.40	26.70	23.25	21.30
	Enclosed	124.80	92.80	75.90	66.70
Luxury	Open	31.55	27.70	25.00	24.20
	Enclosed	127.25	94.35	76.30	68.35

Porches *(Cost per S.F.)*

Class	Type	Area (S.F.)				
		25	50	100	200	300
Economy	Open	$ 47.60	$ 31.90	$24.85	$21.10	$18.00
	Enclosed	95.15	66.30	50.15	39.10	33.50
Average	Open	59.25	37.75	28.95	24.10	24.10
	Enclosed	117.05	79.30	60.00	46.50	39.35
Custom	Open	77.55	51.45	38.80	33.85	30.35
	Enclosed	154.20	105.35	80.20	62.35	53.80
Luxury	Open	83.65	54.70	40.40	36.30	32.30
	Enclosed	163.75	115.00	85.40	66.25	57.10

Finished attic *(Cost per S.F.)*

Class	Area (S.F.)				
	400	500	600	800	1000
Economy	$12.75	$12.30	$11.75	$11.55	$11.20
Average	20.10	19.55	19.15	18.85	18.35
Custom	24.75	24.15	23.70	23.25	22.80
Luxury	31.35	30.70	29.95	29.25	28.75

Alarm system *(Cost per System)*

Class	Burglar Alarm	Smoke Detector
Economy	$ 350	$ 52
Average	405	64
Custom	684	133
Luxury	1025	164

Sauna, prefabricated
(Cost per unit, including heater and controls—7' high)

Size	Cost
6' x 4'	$4600
6' x 5'	4075
6' x 6'	5375
6' x 9'	5275
8' x 10'	6825
8' x 12'	8275
10' x 12'	9025

Garages *

(Costs include exterior wall systems comparable with the quality of the residence. Included in the cost is an allowance for one personnel door, manual overhead door(s) and electrical fixture.)

Class	Type									
	Detached			Attached			Built-in		Basement	
	One Car	Two Car	Three Car	One Car	Two Car	Three Car	One Car	Two Car	One Car	Two Car
Economy										
Wood	$10,343	$15,719	$21,095	$ 8010	$13,807	$19,183	$-1374	$-2747	$1018	$1280
Masonry	14,108	20,388	26,667	10,343	17,031	23,311	-1913	-3827		
Average										
Wood	11,040	16,548	22,056	8424	14,340	19,848	-1537	-3073	1242	1674
Masonry	14,200	20,503	26,806	10,401	17,112	23,415	-1925	-3357		
Custom										
Wood	12,841	19,464	26,088	9957	17,118	23,741	-3731	-4172	1771	2732
Masonry	15,531	22,783	30,036	11,616	19,393	26,645	-4078	-4875		
Luxury										
Wood	14,496	22,357	30,218	11,402	19,788	27,649	-3808	-4336	2399	3675
Masonry	18,754	27,671	36,587	14,059	23,500	32,417	-4333	-5386		

*See the Introduction to this section for definitions of garage types.

Swimming pools (Cost per S.F.)

Residential (includes equipment)	
In-ground	$ 18 - 45
Deck equipment	1.30
Paint pool, preparation & 3 coats (epoxy)	2.78
Rubber base paint	2.56
Pool Cover	.66
Swimming Pool Heaters (Cost per unit)	
(not including wiring, external piping, base or pad)	
Gas	
155 MBH	$ 2600
190 MBH	3500
500 MBH	8500
Electric	
15 KW 7200 gallon pool	2050
24 KW 9600 gallon pool	2700
54 KW 24,000 gallon pool	4025

Wood and coal stoves

Wood Only	
Free Standing (minimum)	$ 1250
Fireplace Insert (minimum)	1260
Coal Only	
Free Standing	$ 1424
Fireplace Insert	1559
Wood and Coal	
Free Standing	$ 2928
Fireplace Insert	3000

Sidewalks (Cost per S.F.)

Concrete, 3000 psi with wire mesh	4" thick	$ 2.52
	5" thick	3.07
	6" thick	3.45
Precast concrete patio blocks (natural)	2" thick	7.50
Precast concrete patio blocks (colors)	2" thick	8.00
Flagstone, bluestone	1" thick	10.87
Flagstone, bluestone	1-1/2" thick	14.35
Slate (natural, irregular)	3/4" thick	10.46
Slate (random rectangular)	1/2" thick	15.57
Seeding		
Fine grading & seeding includes lime, fertilizer & seed	per S.Y.	1.71
Lawn Sprinkler System	per S.F.	.68

Fencing (Cost per L.F.)

Chain Link, 4' high, galvanized	$ 10.80
Gate, 4' high (each)	116.00
Cedar Picket, 3' high, 2 rail	9.25
Gate (each)	117.00
3 Rail, 4' high	10.25
Gate (each)	126.00
Cedar Stockade, 3 Rail, 6' high	10.50
Gate (each)	125.00
Board & Battens, 2 sides 6' high, pine	15.35
6' high, cedar	22.00
No. 1 Cedar, basketweave, 6' high	12.15
Gate, 6' high (each)	145.00

Carport (Cost per S.F.)

Economy	$ 5.99
Average	9.31
Custom	14.03
Luxury	16.06

ADJUSTMENTS

Insurance Exclusions

Insurance exclusions are a matter of policy coverage and are not standard or universal. When making an appraisal for insurance purposes, it is recommended that some time be taken in studying the insurance policy to determine the specific items to be excluded. Most homeowners insurance policies have, as part of their provisions, statements that read "the policy permits the insured to exclude from value, in determining whether the amount of the insurance equals or exceeds 80% of its replacement cost, such items as building excavation, basement footings, underground piping and wiring, piers and other supports which are below the undersurface of the lowest floor or where there is no basement below the ground." Loss to any of these items, however, is covered.

Costs for Excavation, Spread and Strip Footings and Underground Piping

This chart shows excluded items expressed as a percentage of total building cost.

Class	Number of Stories			
	1	1-1/2	2	3
Luxury	3.4%	2.7%	2.4%	1.9%
Custom	3.5%	2.7%	2.4%	1.9%
Average	5.0%	4.3%	3.8%	3.3%
Economy	5.0%	4.4%	4.1%	3.5%

Architect/Designer Fees

Architect/designer fees as presented in the following chart are typical ranges for 4 classes of residences. Factors affecting these ranges include economic conditions, size and scope of project, site selection, and standardization. Where superior quality and detail is required or where closer supervision is required, fees may run higher than listed. Lower quality or simplicity may dictate lower fees.

The editors have included architect/designers fees from the "mean" column in calculating costs.

Listed below are average costs expressed as a range and mean of total building cost for 4 classes of residences.

Class			Range	Mean
Luxury	—	Architecturally designed and supervised	5%–15%	10%
Custom	—	Architecturally modified designer plans	$2000–$3500	$2500
Average	—	Designer plans	$800–$1500	$1100
Economy	—	Stock plans	$100–$500	$300

ADJUSTMENTS

Depreciation as generally defined is "the loss of value due to any cause." Specifically, depreciation can be broken down into three categories: **Physical Depreciation, Functional Obsolescence,** and **Economic Obsolescence.**

Physical Depreciation is the loss of value from the wearing out of the building's components. Such causes may be decay, dry rot, cracks, or structural defects.

Functional Obsolescence is the loss of value from features which render the residence less useful or desirable. Such features may be higher than normal ceilings, design and/or style changes, or outdated mechanical systems.

Economic Obsolescence is the loss of value from external features which render the residence less useful or desirable. These features may be changes in neighborhood socio-economic grouping, zoning changes, legislation, etc.

Depreciation as it applies to the residential portion of this manual deals with the observed physical condition of the residence being appraised. It is in essence *"the cost to cure."*

The ultimate method to arrive at *"the cost to cure"* is to analyze each component of the residence.

For example:

    Component, Roof Covering — Cost = $2400
    Average Life = 15 years
    Actual Life = 5 years
    Depreciation = 5 ÷ 15 = .33 or 33%

    $2400 × 33% = $792 = Amount of Depreciation

The following table, however, can be used as a guide in estimating the % depreciation using the actual age and general condition of the residence.

Depreciation Table — Residential

Age in Years	Good	Average	Poor
2	2%	3%	10%
5	4	6	20
10	7	10	25
15	10	15	30
20	15	20	35
25	18	25	40
30	24	30	45
35	28	35	50
40	32	40	55
45	36	45	60
50	40	50	65

ADJUSTMENTS

Commercial/Industrial/Institutional Section

Table of Contents

Introduction to the Commercial/ Industrial/Institutional Section

General

The Commercial/Industrial/Institutional section of this manual contains base building costs per square foot of floor area for 70 model buildings. Each model has a table of square foot costs for combinations of exterior wall and framing systems. This table is supplemented by a list of common additives and their unit costs. A breakdown of the component costs used to develop the base cost for the model is on the opposite page. The total cost derived from the base cost and additives must be modified using the appropriate location factor from the Reference Section.

This section may be used directly to estimate the construction cost of most types of buildings knowing only the floor area, exterior wall construction and framing systems. To adjust the base cost for components which are different than the model, use the tables from the Assemblies Section.

Building Identification & Model Selection

The building models in this section represent structures by use. Occupancy, however, does not necessarily identify the building, i.e., a restaurant could be a converted warehouse. In all instances, the building should be described and identified by its own physical characteristics. The model selection should also be guided by comparing specifications with the model. In the case of converted use, data from one model may be used to supplement data from another.

Adjustments

The base cost tables represent the base cost per square foot of floor area for buildings without a basement and without unusual special features. Basement costs and other common additives are listed below the base cost table. Cost adjustments can also be made to the model by using the tables from the Assemblies Section. This table is for example only.

Dimensions

All base cost tables are developed so that measurement can be readily made during the inspection process. Areas are calculated from exterior dimensions and story heights are measured from the top surface of one floor to the top surface of the floor above. Roof areas are measured by horizontal area covered and costs related to inclines are converted with appropriate factors. The precision of measurement is a matter of the user's choice and discretion. For ease in calculation, consideration should be given to measuring in tenths of a foot, i.e., 9 ft. 6 in. = 9.5 ft., 9 ft. 4 in. = 9.3 ft.

Floor Area

The expression "Floor Area" as used in this section includes the sum of floor plate at grade level and above. This dimension is measured from the outside face of the foundation wall. Basement costs are calculated separately. The user must exercise his own judgment, where the lowest level floor is slightly below grade, whether to consider it at grade level or make the basement adjustment.

How to Use the Commercial/Industrial/Institutional Section

The following is a detailed explanation of a sample entry in the Commercial/Industrial/Institutional Square Foot Cost Section. Each bold number below corresponds to the item being described on the following page with the appropriate component or cost of the sample entry following in parenthesis.

Prices listed are costs that include overhead and profit of the installing contractor and additional mark-ups for General Conditions and Architects' Fees.

COMMERCIAL/INDUSTRIAL/INSTITUTIONAL

1 M.010 **2** Apartment, 1-3 Story

Costs per square foot of floor area

Exterior Wall	S.F. Area	8000	12000	15000	19000	22500	25000	4	32000	36000
	L.F. Perimeter	213	280	330	350	400	433		480	520
Face Brick with Concrete Block Back-up	Wood Joists	133.90	120.05	114.50	107.05	104.30	102.75	99.05	98.00	96.50
	Steel Joists	132.65	119.55	114.30	107.70	105.05	103.65		99.40	98.05
Stucco on Concrete Block	Wood Joists	119.10	106.65	101.65	95.70	93.25	91.85		88.10	86.85
	Steel Joists	125.70	113.15	108.15	102.10	99.65	98.2		94.40	93.15
Wood Siding	Wood Frame	117.60	105.35	100.40	94.65	92.25	90.90	88.20	87.20	86.00
Brick Veneer	Wood Frame	124.80	111.60	106.30	99.60	96.95	95.50	20	91.20	89.80
Perimeter Adj., Add or Deduct	Per 100 L.F.	14.20	9.50	7.60	5.95	5.00	4.50	3.90	3.55	3.15
Story Hgt. Adj., Add or Deduct	Per 1 Ft.	2.70	2.35	2.20	1.85	1.80	1.75	1.50	1.50	1.45

For Basement, add $21.65 per square foot of basement area

The above costs were calculated using the basic specifications shown on the facing page. These costs be adjusted where necessary for design alternatives and owner's requirements. Reported completed project costs, for this type of structure, from $36.15 to $134.55 per S.F.

Common additives

Description	Unit	$ Cost
Appliances		
Cooking range, 30" free standing		
1 oven	Each	340 - 1475
2 oven	Each	1475 - 1650
30" built-in		
1 oven	Each	420 - 1550
2 oven	Each	1225 - 2100
Counter top cook tops, 4 burner	Each	281 - 625
Microwave oven	Each	203 - 650
Combination range, refrig. & sink, 30" wide	Each	1225 - 2450
72" wide	Each	3725
Combination range, refrigerator, sink, microwave oven & icemaker	Each	5500
Compactor, residential, 4-1 compaction	Each	465 - 550
Dishwasher, built-in, 2 cycles	Each	495 - 740
4 cycles	Each	530 - 1000
Garbage disposer, sink type	Each	128 - 287
Hood for range, 2 speed, vented, 30" wide	Each	199 - 760
42" wide	Each	350 - 1100
Refrigerator, no frost 10-12 C.F.	Each	555 - 860
18-20 C.F.	Each	630 - 950

Description	Unit	$ Cost
Closed Circuit Surveillance, One station		
Camera and monitor	Each	1375
For additional camera stations, add	Each	750
Elevators, Hydraulic passenger, 2 stops		
2000# capacity	Each	42,725
2500# capacity	Each	43,425
3500# capacity	Each	47,225
Additional stop, add	Each	3800
Emergency Lighting, 25 watt, battery operated		
Lead battery	Each	227
Nickel cadmium	Each	660
Laundry Equipment		
Dryer, gas, 16 lb. capacity	Each	725
30 lb. capacity	Each	2800
Washer, 4 cycle	Each	810
Commercial	Each	1225
Smoke Detectors		
Ceiling type	Each	151
Duct type	Each	405

1 ## Model Number (M.010)

"M" distinguishes this section of the book and stands for model. The number designation is a sequential number.

2 ## Type of Building (Apartment, 1-3 Story)

There are 43 different types of commercial/industrial/institutional buildings highlighted in this section.

3 ## Exterior Wall Construction and Building Framing Options (Face Brick with Concrete Block Back-up and Open Web Steel Bar Joists)

Three or more commonly used exterior walls, and in most cases, two typical building framing systems are presented for each type of building. The model selected should be based on the actual characteristics of the building being estimated.

4 ## Total Square Foot of Floor Area and Base Perimeter Used to Compute Base Costs (22,500 Square Feet and 400 Linear Feet)

Square foot of floor area is the total gross area of all floors at grade, and above, and does not include a basement. The perimeter in linear feet used for the base cost is for a generally rectangular economical building shape.

5 ## Cost per Square Foot of Floor Area ($105.05)

The highlighted cost is for a building of the selected exterior wall and framing system and floor area. Costs for buildings with floor areas other than those calculated may be interpolated between the costs shown.

6 ## Building Perimeter and Story Height Adjustments

Square foot costs for a building with a perimeter or floor to floor story height significantly different from the model used to calculate the base cost may be adjusted, add or deduct, to reflect the actual building geometry.

7 ## Cost per Square Foot of Floor Area for the Perimeter and/or Height Adjustment ($5.00 for Perimeter Difference and $1.80 for Story Height Difference)

Add (or deduct) $5.00 to the base square foot cost for each 100 feet of perimeter difference between the model and the actual building. Add (or deduct) $1.80 to the base square foot cost for each 1 foot of story height difference between the model and the actual building.

8 ## Optional Cost per Square Foot of Basement Floor Area ($21.65)

The cost of an unfinished basement for the building being estimated is $21.65 times the gross floor area of the basement.

9 ## Range of Cost per Square Foot of Floor Area for Similar Buildings ($36.15 to $134.55)

Many different buildings of the same type have been built using similar materials and systems. Means historical cost data of actual construction projects indicates a range of $36.15 to $134.55 for this type of building.

10 ## Common Additives

Common components and/or systems used in this type of building are listed. These costs should be added to the total building cost. Additional selections may be found in the Assemblies Section.

The following is a detailed explanation of a specification and costs for a model building in the Commercial/Industrial/Institutional Square Foot Cost Section. Each bold number below corresponds to the item being described on the following page with the appropriate component of the sample entry following in parenthesis.

Prices listed are costs that include overhead and profit of the installing contractor.

INTRODUCTION

Model costs calculated for a 3 story building with 10' story height and 22,500 square feet of floor area **1**   **2** **Apartment, 1-3 Story**

				Unit	Unit Cost	Cost Per S.F.	% Of Sub-Total
A. SUBSTRUCTURE							
1010	Standard Foundations	Poured concrete; strip and spread footings		S.F. Ground	4.20	1.40	
1030	Slab on Grade	4" reinforced concrete with vapor barrier and granular base		S.F. Slab	3.42	1.14	
2010	Basement Excavation	Site preparation for slab and trench for foundation wall and footing		S.F. Ground	1.09	.36	5.0%
2020	Basement Walls	4' foundation wall		L.F. Wall	49	1.02	
B. SHELL							
B10 Superstructure							
1010	Floor Construction	Open web steel joists, slab form, concrete, interior steel columns		S.F. Floor	12.19	8.13	12.8%
1020	Roof Construction	Open web steel joists with rib metal deck, interior steel columns		S.F. Roof	5.49	1.83	
B20 Exterior Enclosure							
2010	Exterior Walls	Face brick with concrete block backup	88% of wall	S.F. Wall	15.79	7.41	
2020	Exterior Windows	Aluminum horizontal sliding	12% of wall	Each	314	1.34	11.5%
2030	Exterior Doors	Aluminum and glass		Each	1142	.20	
B30 Roofing							
3010	Roof Coverings	Built-up tar and gravel with flashing; perlite/EPS composite insulation		S.F. Roof	4.11	1.37	1.8%
3020	Roof Openings	N/A		—	—	—	
C. INTERIORS **3**					**6**		
1010	Partitions	Gypsum board and sound deadening board on metal studs	10 S.F. of Floor/L.F. Partition	S.F. Partition	4.50	4.00	
1020	Interior Doors	15% solid core wood, 85% hollow core wood	80 S.F. Floor/Door	Each	394	4.93	
1030	Fittings	Kitchen Cabinets		S.F. Floor	1.87		
2010	Stair Construction	Concrete filled metal pan	**4**	Flight			26.7%
3010	Wall Finishes	70% paint, 25% vinyl wall covering, 5% ceramic	**5**	S.F. Surface		**7** **8**	
3020	Floor Finishes	60% carpet, 30% vinyl composition tile, 10%		S.F. Floor			
3030	Ceiling Finishes	Painted gypsum board on resilient channels		S.F. Ceiling	2.61	2.61	**9**
D. SERVICES							
D10 Conveying							
1010	Elevators & Lifts	One hydraulic passenger elevator		Each	66,150	2.94	4.1%
1020	Escalators & Moving Walks	N/A		—	—	—	
D20 Plumbing							
2010	Plumbing Fixtures	Kitchen, bathroom and service fixtures, supply and drainage	1 Fixture/200 S.F. Floor	Each	1452	7.26	
2020	Domestic Water Distribution	Gas fired water heater		S.F. Floor	2.10	2.10	12.0%
2040	Rain Water Drainage	Roof drains		S.F. Roof	.72	.24	
D30 HVAC							
3010	Energy Supply	Oil fired hot water, baseboard radiation		S.F. Floor	4.91	4.91	
3020	Heat Generating Systems	N/A		—	—	—	
3030	Cooling Generating Systems	Chilled water, air cooled condenser system		S.F. Floor	6.57	6.57	14.8%
3050	Terminal & Package Units	N/A		—	—	—	
3090	Other HVAC Sys. & Equipment	N/A		—	—	—	
D40 Fire Protection							
4010	Sprinklers	Wet pipe sprinkler system		S.F. Floor	1.81	1.81	2.3%
4020	Standpipes	Standpipe		S.F. Floor	—	—	
D50 Electrical							
5010	Electrical Service/Distribution	600 ampere service, panel board and feeders		S.F. Floor	1.59	1.59	
5020	Lighting & Branch Wiring	Incandescent fixtures, receptacles, switches, A.C. and misc. power		S.F. Floor	4.74	4.74	9.0%
5030	Communications & Security	Alarm systems and emergency lighting		S.F. Floor	.51	.51	
5090	Other Electrical Systems	Emergency Generator, 11.5KW		S.F. Floor	.15	.15	
E. EQUIPMENT & FURNISHINGS							
1010	Commercial Equipment	N/A		—	—	—	
1020	Institutional Equipment	N/A		—	—	—	
1030	Vehicular Equipment	N/A		—	—	—	0.0%
1090	Other Equipment	N/A		—	—	—	
F. SPECIAL CONSTRUCTION & DEMOLITION							
1020	Integrated Construction	N/A		—	—	—	
1040	Special Facilities	N/A		—	—	—	0.0%
G. BUILDING SITEWORK **N/A**						**10**	
					Sub-Total	77.83	100%
	CONTRACTOR FEES (General Requirements: 10%, Overhead: 5%, Profit: 10%) **11**	**12**				25%	19.46	
	ARCHITECT FEES					8%	7.76	
					Total Building Cost	105.05	

1
**Building Description
(Model costs are calculated for a three-story apartment building with 10' story height and 22,500 square feet of floor area)**

The model highlighted is described in terms of building type, number of stories, typical story height and square footage.

2
**Type of Building
(Apartment, 1-3 Story)**

3
**Division C Interiors
(C1020 Interior Doors)**

System costs are presented in divisions according to the 7-division UNIFORMAT II classifications. Each of the component systems are listed.

4
**Specification Highlights
(15% solid core wood;
85% hollow core wood)**

All systems in each subdivision are described with the material and proportions used.

5
**Quality Criteria
(80 S.F. Floor/Door)**

The criteria used in determining quantities for the calculations are shown.

6
Unit (Each)

The unit of measure shown in this column is the unit of measure of the particular system shown that corresponds to the unit cost.

7
Unit Cost ($394)

The cost per unit of measure of each system subdivision.

8
Cost per Square Foot ($4.93)

The cost per square foot for each system is the unit cost of the system times the total number of units divided by the total square feet of building area.

9
% of Sub-Total (26.7%)

The percent of sub-total is the total cost per square foot of all systems in the division divided by the sub-total cost per square foot of the building.

10
Sub-Total ($77.83)

The sub-total is the total of all the system costs per square foot.

11
**Project Fees
(Contractor Fees) (25%)
(Architect Fees) (8%)**

Contractor Fees to cover the general requirements, overhead and profit of the General Contractor are added as a percentage of the sub-total. Architect Fees, also as a percentage of the sub-total, are also added. These values vary with the building type.

12
Total Building Cost ($105.05)

The total building cost per square foot of building area is the sum of the square foot costs of all the systems plus the General Contractor's general requirements, overhead and profit, and the Architect fees. The total building cost is the amount which appears shaded in the Cost per Square Foot of Floor Area table shown previously.

Examples

Example 1

This example illustrates the use of the base cost tables. The base cost is adjusted for different exterior wall systems, different story height and a partial basement.

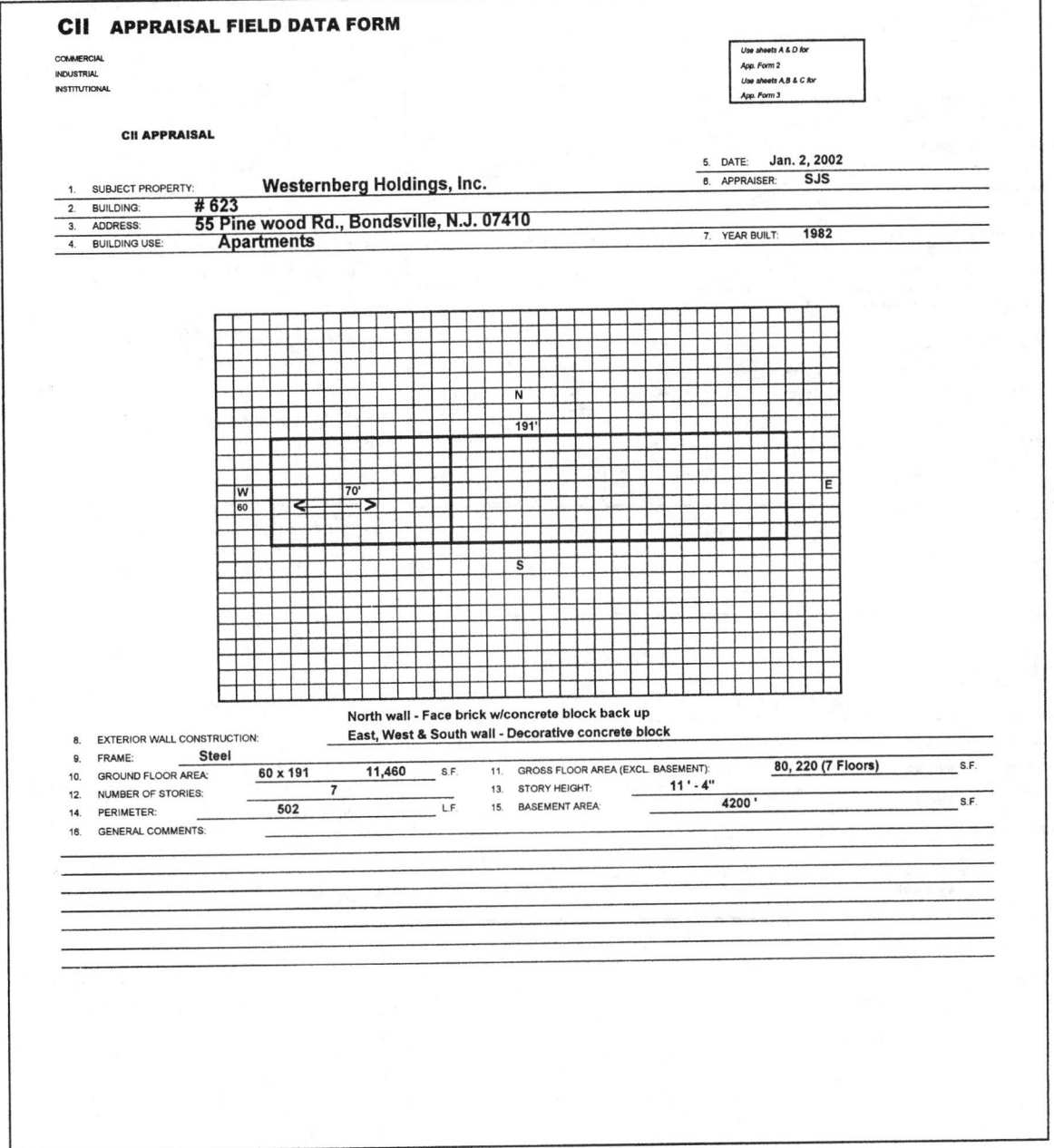

Example 1 (continued)

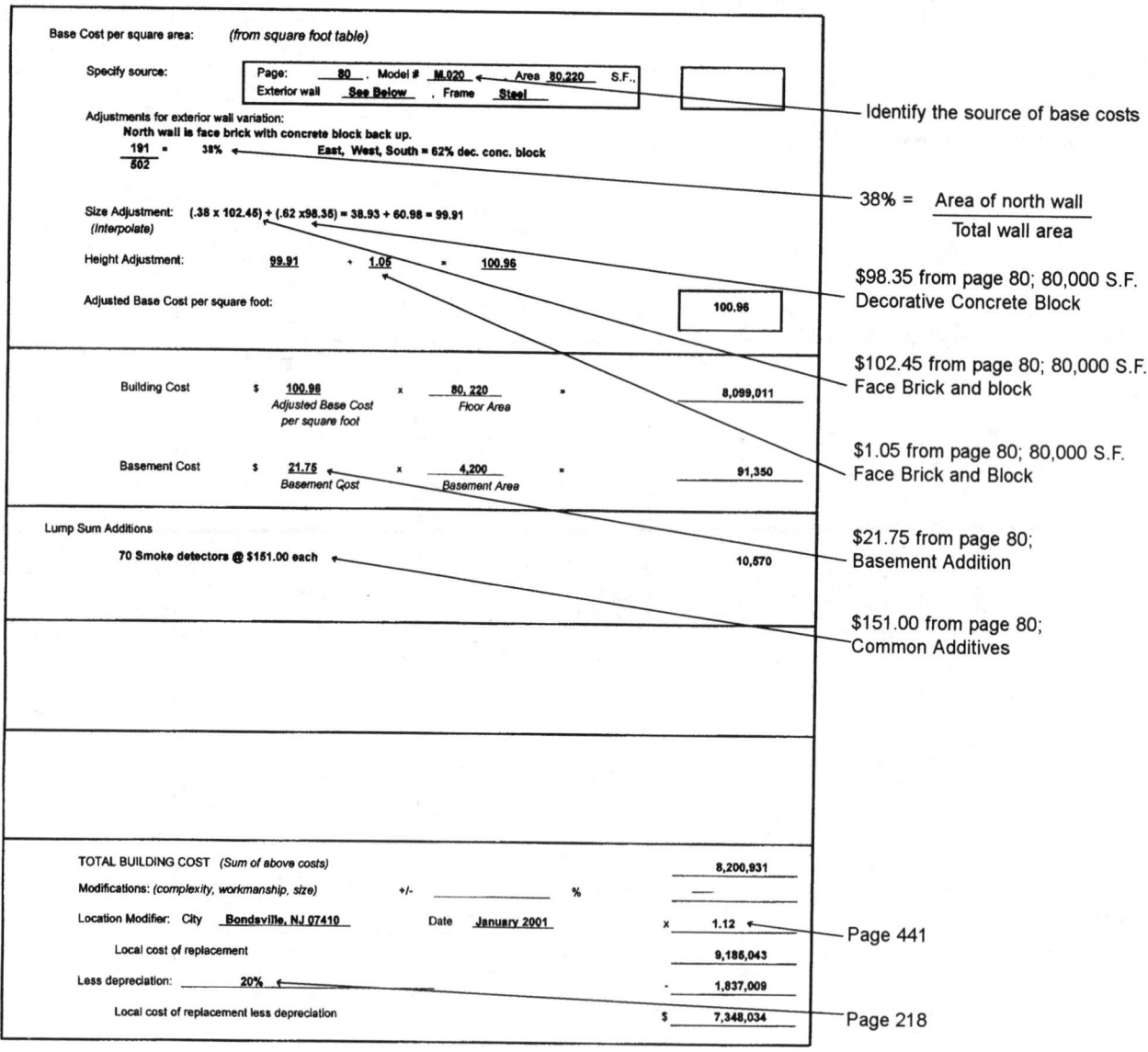

Base Cost per square area: *(from square foot table)*

Specify source:
Page: **80**, Model # **M.020** ← Area **80,220** S.F.,
Exterior wall **See Below**, Frame **Steel**

— Identify the source of base costs

Adjustments for exterior wall variation:
North wall is face brick with concrete block back up.
$\frac{191}{502}$ = **38%** ← East, West, South = 62% dec. conc. block

— 38% = $\frac{\text{Area of north wall}}{\text{Total wall area}}$

Size Adjustment: (.38 x 102.45) + (.62 x98.35) = 38.93 + 60.98 = 99.91
(Interpolate)

Height Adjustment: **99.91** + **1.05** = **100.96**

Adjusted Base Cost per square foot: **100.96**

$98.35 from page 80; 80,000 S.F.
Decorative Concrete Block

$102.45 from page 80; 80,000 S.F.
Face Brick and block

Building Cost $ **100.96** x **80, 220** = **8,099,011**
               Adjusted Base Cost   Floor Area
               per square foot

$1.05 from page 80; 80,000 S.F.
Face Brick and Block

Basement Cost $ **21.75** x **4,200** = **91,350**
                  Basement Cost   Basement Area

$21.75 from page 80;
Basement Addition

Lump Sum Additions

70 Smoke detectors @ $151.00 each ← **10,570**

$151.00 from page 80;
Common Additives

TOTAL BUILDING COST *(Sum of above costs)* **8,200,931**

Modifications: *(complexity, workmanship, size)* +/- _____ % ——

Location Modifier: City **Bondsville, NJ 07410** Date **January 2001** x **1.12** ←
— Page 441

Local cost of replacement **9,185,043**

Less depreciation: ___**20%**___ ← - **1,837,009**

Local cost of replacement less depreciation $ **7,348,034**
— Page 218

Example 2

This example shows how to modify a model building.

Model #M.020, 4–7 Story Apartment, page 80, matches the example quite closely. The model specifies a 6-story building with a 10′ story height.

The following adjustments must be made:
- add stone ashlar wall
- adjust model to a 5-story building
- change partitions
- change floor finish

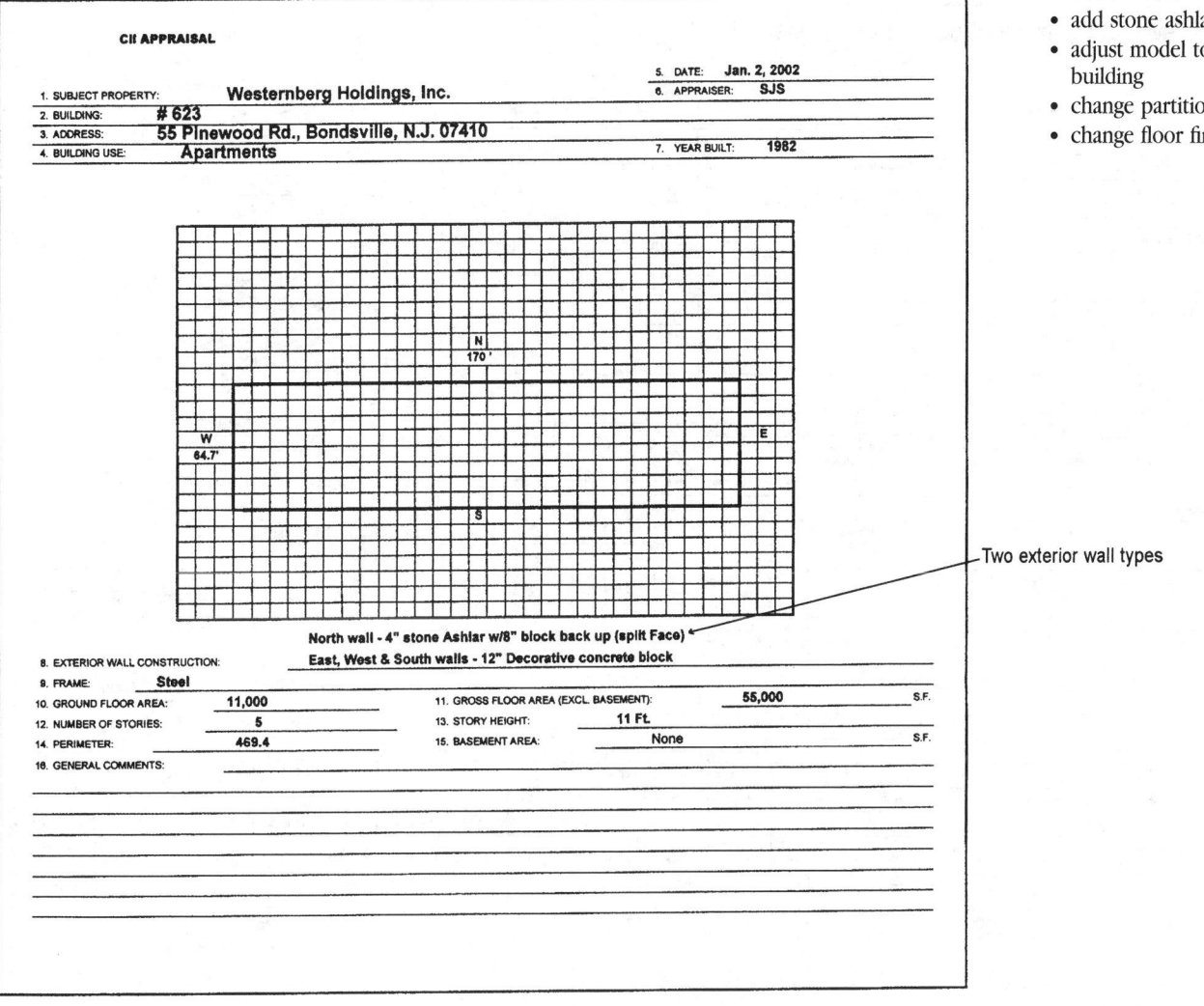

CII APPRAISAL

1. SUBJECT PROPERTY:	Westernberg Holdings, Inc.	5. DATE:	Jan. 2, 2002
2. BUILDING:	# 623	6. APPRAISER:	SJS
3. ADDRESS:	55 Pinewood Rd., Bondsville, N.J. 07410		
4. BUILDING USE:	Apartments	7. YEAR BUILT:	1982

N 170′

W 64.7′

E

S

Two exterior wall types

North wall - 4" stone Ashlar w/8" block back up (split Face)
East, West & South walls - 12" Decorative concrete block

8. EXTERIOR WALL CONSTRUCTION:

9. FRAME: **Steel**

10. GROUND FLOOR AREA:	11,000	11. GROSS FLOOR AREA (EXCL. BASEMENT):	55,000	S.F.
12. NUMBER OF STORIES:	5	13. STORY HEIGHT:	11 Ft.	
14. PERIMETER:	469.4	15. BASEMENT AREA:	None	S.F.

16. GENERAL COMMENTS:

EXAMPLES

Example 2 (continued)

Square foot area (excluding basement) from item 11 ___55,000___ S.F.

Perimeter from item 14 ___469.4___ Lin. Ft.

Item 17-Model square foot costs (from model sub-total) $79.78

NO.	DESCRIPTION			UNIT	UNIT COST	NEW SF COST	MODEL SF COST	+/- CHANGE
A SUBSTRUCTURE								
A 1010	Standard Foundations	Bay size: Same as model		S.F. Gnd.	4.80	0.96	.80	+ 0.16
A 1030	Slab on Grade	Material:	Thickness: Same as model	S.F. Slab	3.42	0.68	.57	+ 0.09
A 2010	Basement Excavation Depth:	Area:11,000 S.F.		S.F. Gnd.	1.05	0.21	.17	+ 0.04
A 2020	Basement Walls			L.F. Walls	52.00	0.45	.44	0.01
B SHELL								
B10 Superstructure								
B 1010	Floor Construction	Elevated floors: Same as model		S.F. Floor	14.76	11.81	12.30	- 0.49
				S.F. Floor				
B 1020	Roof Construction			S.F. Roof	5.16	1.03	0.86	+ 0.17
B20 Exterior Closure								
B 2010	Exterior walls Material: North wall- 4" stone ashlar on 8" CMU	Thickness: 36% of wall		S.F. Walls	17.38	7.02	6.79	+ 0.23
	Material: East, West, & South walls- 12" split rib block	Thickness: 64% of wall		S.F. Walls				
B 2020	Exterior Windows	Type:	% of wall	S.F. Wind.				
		Type:	14% of wall	Each	314	1.38	1.47	- 0.09
B 2030	Exterior Doors	Type: Same as model	Number: 4	Each	2715	0.20	.18	+ 0.02
		Type:	Number:	Each				
B30 Roofing								
B 3010	Roof Coverings	Material: Same as model		S.F. Roof	4.02	0.80	0.67	+ 0.13
B 3020	Roof Openings	N/A		S.F. Opng.				
C INTERIORS								
C 1010	Partitions: High	Material: Gyp board on metal studs Density: 8 S.F. Floor/L.F. Part.		S.F. Part.	4.70	5.29	4.70	+ 0.59
		Material:	Density:					
C 1020	Interior Doors	Type:	Number:	Each	472	5.91	5.91	---
C 1030	Fittings			Each	2.24	2.24	2.24	---
C 2010	Stair Construction	13 Flights		Flight	4100	0.97	1.09	- 0.12
C 3010	Wall Finishes	Material: Same as model % of Wall	9'-0" High	S.F. Walls	2.29	2.68	2.29	+ 0.29
		Material:						
C 3020	Floor Finishes	Material: Hardwood - 60% of floor		S.F. Floor	4.42	6.75	4.42	+ 2.33
		Material: Carpet - 30% of floor						
		Material: Tile -10% of floor						
C 3030	Ceiling Finishes	Material: Same as model		S.F. Ceil.	2.83	2.83	2.83	---

From page 81, Model #M.020 Sub-total

$$\frac{\$1.05/\text{S.F. ground area}}{5 \text{ stories}}$$

$$\frac{\$1.05/\text{S.F. ground area}}{6 \text{ stories}}$$

$$\frac{\text{Area of Elevated Floors}}{\$14.76/\text{S.F.} \times 4 \times 11,000 \text{ S.F.}}$$
$$\frac{55,000}{\text{Building Floor Area}}$$

$$\frac{\$5.16/\text{S.F. roof}}{5 \text{ stories}}$$

Windows are 14% of wall area, 15 S.F./window

$$\frac{314}{15} \times .14 \times 469.4 \times 11 \times 5 - 55,000 = 1.38$$

From Table C1010 124 page 325 and C1010 128 page 327

$$\frac{\$4.70/\text{S.F. partition} \times 9 \text{ ft. high}}{8 \text{ S.F. floor/L.F. partition}}$$

From Table C3020-410 page 340 & 341

Oak Strip, Fin. (max. price)		@$7.76/S.F.
Carpet (26 oz.)		@$3.66 S.F.
Ceramic Tile		@$7.67/S.F.

.60 x 7.76 = 4.66
.36 x 3.66 = 1.32
.10 x 7.67 = .77
6.75

North wall is 36% of wall area, e.g., $\frac{170 \times 11}{469.4 \times 11}$

From Table B2010 123 2350 page 289

Stone ashlar @ $24.20/S.F. wall

East, west and south walls are 64% of wall area

From Table B2010 113 1530 page 283

Split ribbed block @ $13.55/S.F. wall

New wall cost
.36 x $24.20/S.F. + .64 x $13.55/S.F. = $17.38/S.F. wall

Convert Cost per S.F. wall to cost per S.F. of floor
Windows are 14% of wall area

$$\$7.02 \text{ S.F. floor} = \frac{\$17.38/\text{S.F. wall} \times .86 \times 469.4 \text{ L.F.} \times 11 \text{ ft. high} \times 5 \text{ stories}}{55,000 \text{ S.F. of floor}}$$

Example 2 (Continued)

	NO.	SYSTEM/COMPONENT	DESCRIPTION	UNIT	UNIT COST	NEW S.F. COST	MODEL S.F. COST	+/- CHANGE
D SERVICES								
	D10 Conveying							
D	1010	Elevators & Lifts 2	Type: *Same as model* Capacity: Stops:	Each	134,400	4.87	4.48	+ 0.40
D	1020	Escalators & Moving Walks	Type:	Each				
	D20 Plumbing							
D	2010	Plumbing	*Kitchen, bath, + service fixtures* *1 Fixture/200 S.F. Floor*	Each	1425	8.64	6.63	+ 0.60
D	2020	Domestic Water Distribution		S.F. Floor	2.03	2.03	2.03	---
D	2040	Rain Water Drainage		S.F. Roof	0.84	0.84	0.84	---
	D30 HVAC							
D	3010	Energy Supply		Each	4.26	4.26	4.26	
D	3020	Heat Generating Systems	Type: *Same as model*	S.F. Floor				
D	3030	Cooling Generating Systems	Type: *Same as model*	S.F. Floor	6.12	6.12	6.12	---
D	3090	Other HVAC Sys. & Equipment		Each				
	D40 Fire Protection							
D	4010	Sprinklers	*Same as model*	S.F. Floor	1.51	1.51	1.51	---
D	4020	Standpipes	*Same as model*	S.F. Floor	0.29	0.29	0.29	---
	D50 Electrical							
D	5010	Electrical Service/Distribution	*Same as model*	S.F. Floor	1.51	1.51	1.51	---
D	5020	Lighting & Branch Wiring		S.F. Floor	4.72	4.72	4.72	---
D	5030	Communications & Security		S.F. Floor	0.22	0.22	0.22	---
D	5090	Other Electrical Systems		S.F. Floor	0.14	0.14	0.14	---
E EQUIPMENT & FURNISHINGS								
E	1010	Commercial Equipment		Each				
E	1020	Institutional Equipment		Each				
E	1030	Vehicular Equipment		Each				
E	1090	Other Equipment		Each				
F SPECIAL CONSTRUCTION								
F	1020	Integrated Construction	N/A	S.F.				
F	1040	Special Facilities	N/A	S.F.				
G BUILDING SITEWORK								

ITEM				Total Change	+3.89
17	Model 020 total	$ 79.78			
18	Adjusted S.F. cost	$ 80.99	item 17 +/- changes		
19	Building area - from item 11	55,000	S.F. x adjusted S.F. cost	$	4,627,700
20	Basement area - from item 15		S.F. x S.F. cost $	$	
21	Base building sub-total - item 20 + item 19			$	4,627,700
22	Miscellaneous addition (quality, etc.)			$	
23	Sub-total - item 22 + 21			$	4,627,700
24	General conditions -25 % of item 23			$	1,156,925
25	Sub-total - item 24 + item 23			$	5,784,625
26	Architects fees	5.5	% of item 25	$	318,154
27	Sub-total - item 26 + item 27			$	6,102,779
28	Location modifier	zip code 07410		x	1.12
29	Local replacement cost - item 28 x item 27			$	6,835,113
30	Depreciation	20	% of item 29	$	1,367,023
31	Depreciated local replacement cost - item 29 less item 30			$	5,468,095
32	Exclusions			$	
33	Net depreciated replacement cost - item 31 less item 32			$	5,468,095

From page 441

Use table on page 218

Costs per square foot of floor area

Exterior Wall	S.F. Area	8000	12000	15000	19000	22500	25000	29000	32000	36000
	L.F. Perimeter	213	280	330	350	400	433	442	480	520
Face Brick with Concrete Block Back-up	Wood Joists	133.90	120.05	114.50	107.05	104.30	102.75	99.05	98.00	96.50
	Steel Joists	132.65	119.55	114.30	107.70	105.05	103.65	100.40	99.40	98.05
Stucco on Concrete Block	Wood Joists	119.10	106.65	101.65	95.70	93.25	91.85	89.05	88.10	86.85
	Steel Joists	125.70	113.15	108.15	102.10	99.65	98.25	95.35	94.40	93.15
Wood Siding	Wood Frame	117.60	105.35	100.40	94.65	92.25	90.90	88.20	87.20	86.00
Brick Veneer	Wood Frame	124.80	111.60	106.30	99.60	96.95	95.50	92.20	91.20	89.80
Perimeter Adj., Add or Deduct	Per 100 L.F.	14.20	9.50	7.60	5.95	5.00	4.50	3.90	3.55	3.15
Story Hgt. Adj., Add or Deduct	Per 1 Ft.	2.70	2.35	2.20	1.85	1.80	1.75	1.50	1.50	1.45
For Basement, add $21.65 per square foot of basement area										

The above costs were calculated using the basic specifications shown on the facing page. These costs should be adjusted where necessary for design alternatives and owner's requirements. Reported completed project costs, for this type of structure, range from $36.15 to $134.55 per S.F.

Common additives

Description	Unit	$ Cost
Appliances		
Cooking range, 30" free standing		
1 oven	Each	340 - 1475
2 oven	Each	1475 - 1650
30" built-in		
1 oven	Each	420 - 1550
2 oven	Each	1225 - 2100
Counter top cook tops, 4 burner	Each	281 - 625
Microwave oven	Each	203 - 650
Combination range, refrig. & sink, 30" wide	Each	1225 - 2450
72" wide	Each	3725
Combination range, refrigerator, sink,		
microwave oven & icemaker	Each	5500
Compactor, residential, 4-1 compaction	Each	465 - 550
Dishwasher, built-in, 2 cycles	Each	495 - 740
4 cycles	Each	530 - 1000
Garbage disposer, sink type	Each	128 - 287
Hood for range, 2 speed, vented, 30" wide	Each	199 - 760
42" wide	Each	350 - 1100
Refrigerator, no frost 10-12 C.F.	Each	555 - 860
18-20 C.F.	Each	630 - 950

Description	Unit	$ Cost
Closed Circuit Surveillance, One station		
Camera and monitor	Each	1375
For additional camera stations, add	Each	750
Elevators, Hydraulic passenger, 2 stops		
2000# capacity	Each	43,925
2500# capacity	Each	44,625
3500# capacity	Each	48,525
Additional stop, add	Each	3800
Emergency Lighting, 25 watt, battery operated		
Lead battery	Each	227
Nickel cadmium	Each	660
Laundry Equipment		
Dryer, gas, 16 lb. capacity	Each	725
30 lb. capacity	Each	2800
Washer, 4 cycle	Each	810
Commercial	Each	1225
Smoke Detectors		
Ceiling type	Each	151
Duct type	Each	405

Important: See the Reference Section for Location Factors

Model costs calculated for a 3 story building with 10' story height and 22,500 square feet of floor area

BUILDING TYPES

				Unit	Unit Cost	Cost Per S.F.	% Of Sub-Total
A.	**SUBSTRUCTURE**						
	1010	Standard Foundations	Poured concrete; strip and spread footings	S.F. Ground	4.20	1.40	
	1030	Slab on Grade	4" reinforced concrete with vapor barrier and granular base	S.F. Slab	3.42	1.14	
	2010	Basement Excavation	Site preparation for slab and trench for foundation wall and footing	S.F. Ground	1.09	.36	5.0%
	2020	Basement Walls	4' foundation wall	L.F. Wall	49	1.02	
B.	**SHELL**						
	B10	**Superstructure**					
	1010	Floor Construction	Open web steel joists, slab form, concrete, interior steel columns	S.F. Floor	12.19	8.13	12.8%
	1020	Roof Construction	Open web steel joists with rib metal deck, interior steel columns	S.F. Roof	5.49	1.83	
	B20	**Exterior Enclosure**					
	2010	Exterior Walls	Face brick with concrete block backup 88% of wall	S.F. Wall	15.79	7.41	
	2020	Exterior Windows	Aluminum horizontal sliding 12% of wall	Each	314	1.34	11.5%
	2030	Exterior Doors	Aluminum and glass	Each	1142	.20	
	B30	**Roofing**					
	3010	Roof Coverings	Built-up tar and gravel with flashing; perlite/EPS composite insulation	S.F. Roof	4.11	1.37	1.8%
	3020	Roof Openings	N/A	—	—	—	
C.	**INTERIORS**						
	1010	Partitions	Gypsum board and sound deadening board on metal studs 10 S.F. of Floor/L.F. Partition	S.F. Partition	4.50	4.00	
	1020	Interior Doors	15% solid core wood, 85% hollow core wood 80 S.F. Floor/Door	Each	394	4.93	
	1030	Fittings	Kitchen Cabinets	S.F. Floor	1.87	1.87	
	2010	Stair Construction	Concrete filled metal pan	Flight	5075	1.13	26.7%
	3010	Wall Finishes	70% paint, 25% vinyl wall covering, 5% ceramic tile	S.F. Surface	2.08	1.85	
	3020	Floor Finishes	60% carpet, 30% vinyl composition tile, 10% ceramic tile	S.F. Floor	4.42	4.42	
	3030	Ceiling Finishes	Painted gypsum board on resilient channels	S.F. Ceiling	2.61	2.61	
D.	**SERVICES**						
	D10	**Conveying**					
	1010	Elevators & Lifts	One hydraulic passenger elevator	Each	66,150	2.94	3.8%
	1020	Escalators & Moving Walks	N/A	—	—	—	
	D20	**Plumbing**					
	2010	Plumbing Fixtures	Kitchen, bathroom and service fixtures, supply and drainage 1 Fixture/200 S.F. Floor	Each	1452	7.26	
	2020	Domestic Water Distribution	Gas fired water heater	S.F. Floor	2.10	2.10	12.3%
	2040	Rain Water Drainage	Roof drains	S.F. Roof	.72	.24	
	D30	**HVAC**					
	3010	Energy Supply	Oil fired hot water, baseboard radiation	S.F. Floor	4.91	4.91	
	3020	Heat Generating Systems	N/A	—	—	—	
	3030	Cooling Generating Systems	Chilled water, air cooled condenser system	S.F. Floor	6.57	6.57	14.8%
	3050	Terminal & Package Units	N/A	—	—	—	
	3090	Other HVAC Sys. & Equipment	N/A	—	—	—	
	D40	**Fire Protection**					
	4010	Sprinklers	Wet pipe sprinkler system	S.F. Floor	1.81	1.81	2.3%
	4020	Standpipes	Standpipe	S.F. Floor	—	—	
	D50	**Electrical**					
	5010	Electrical Service/Distribution	600 ampere service, panel board and feeders	S.F. Floor	1.59	1.59	
	5020	Lighting & Branch Wiring	Incandescent fixtures, receptacles, switches, A.C. and misc. power	S.F. Floor	4.74	4.74	9.0%
	5030	Communications & Security	Alarm systems and emergency lighting	S.F. Floor	.51	.51	
	5090	Other Electrical Systems	Emergency Generator, 11.5KW	S.F. Floor	.15	.15	
E.	**EQUIPMENT & FURNISHINGS**						
	1010	Commercial Equipment	N/A	—	—	—	
	1020	Institutional Equipment	N/A	—	—	—	
	1030	Vehicular Equipment	N/A	—	—	—	0.0%
	1090	Other Equipment	N/A	—	—	—	
F.	**SPECIAL CONSTRUCTION & DEMOLITION**						
	1020	Integrated Construction	N/A	—	—	—	
	1040	Special Facilities	N/A	—	—	—	0.0%
G.	**BUILDING SITEWORK**	**N/A**					

			Sub-Total	77.83	100%
CONTRACTOR FEES (General Requirements: 10%, Overhead: 5%, Profit: 10%)			25%	19.46	
ARCHITECT FEES			8%	7.76	
		Total Building Cost		**105.05**	

BUILDING TYPES

Costs per square foot of floor area

Exterior Wall	S.F. Area	40000	45000	50000	55000	60000	70000	80000	90000	100000
	L.F. Perimeter	366	400	433	466	500	566	505	550	594
Face Brick with Concrete Block Back-up	Steel Frame	109.50	108.55	107.80	107.20	106.70	105.90	102.45	101.85	101.40
	R/Conc. Frame	116.15	115.10	114.30	113.60	113.10	112.15	108.15	107.50	106.95
Decorative Concrete Block	Steel Frame	104.35	103.50	102.85	102.30	101.85	101.15	98.35	97.85	97.40
	R/Conc. Frame	107.15	106.30	105.60	105.05	104.60	103.90	101.00	100.50	100.10
Precast Concrete Panels	Steel Frame	106.70	105.80	105.10	104.45	104.00	103.20	99.95	99.40	98.95
	R/Conc. Frame	109.30	108.40	107.65	107.05	106.55	105.75	102.50	101.95	101.45
Perimeter Adj., Add or Deduct	Per 100 L.F.	4.40	3.90	3.50	3.15	2.90	2.45	2.20	1.95	1.80
Story Hgt. Adj., Add or Deduct	Per 1 Ft.	1.50	1.45	1.40	1.35	1.35	1.30	1.05	1.00	1.00

For Basement, add $21.75 per square foot of basement area

The above costs were calculated using the basic specifications shown on the facing page. These costs should be adjusted where necessary for design alternatives and owner's requirements. Reported completed project costs, for this type of structure, range from $43.35 to $125.70 per S.F.

Common additives

Description	Unit	$ Cost		Description	Unit	$ Cost
Appliances				Closed Circuit Surveillance, One station		
Cooking range, 30" free standing				Camera and monitor	Each	1375
1 oven	Each	340 - 1475		For additional camera stations, add	Each	750
2 oven	Each	1475 - 1650		Elevators, Electric passenger, 5 stops		
30" built-in				2000# capacity	Each	101,400
1 oven	Each	420 - 1550		3500# capacity	Each	107,400
2 oven	Each	1225 - 2100		5000# capacity	Each	111,900
Counter top cook tops, 4 burner	Each	281 - 625		Additional stop, add	Each	5675
Microwave oven	Each	203 - 650		Emergency Lighting, 25 watt, battery operated		
Combination range, refrig. & sink, 30" wide	Each	1225 - 2450		Lead battery	Each	227
72" wide	Each	3725		Nickel cadmium	Each	660
Combination range, refrigerator, sink,				Laundry Equipment		
microwave oven & icemaker	Each	5500		Dryer, gas, 16 lb. capacity	Each	725
Compactor, residential, 4-1 compaction	Each	465 - 550		30 lb. capacity	Each	2800
Dishwasher, built-in, 2 cycles	Each	495 - 740		Washer, 4 cycle	Each	810
4 cycles	Each	530 - 1000		Commercial	Each	1225
Garbage disposer, sink type	Each	128 - 287		Smoke Detectors		
Hood for range, 2 speed, vented, 30" wide	Each	199 - 760		Ceiling type	Each	151
42" wide	Each	350 - 1100		Duct type	Each	405
Refrigerator, no frost 10-12 C.F.	Each	555 - 860				
18-20 C.F.	Each	630 - 950				

Important: See the Reference Section for Location Factors

Model costs calculated for a 6 story building with 10'-4" story height and 60,000 square feet of floor area

Apartment, 4-7 Story

				Unit	Unit Cost	Cost Per S.F.	% Of Sub-Total
A. SUBSTRUCTURE							
1010	Standard Foundations	Poured concrete; strip and spread footings and 4' foundation wall		S.F. Ground	4.80	.80	
1030	Slab on Grade	4" reinforced concrete with vapor barrier and granular base		S.F. Slab	3.42	.57	
2010	Basement Excavation	Site preparation for slab and trench for foundation wall and footing		S.F. Ground	1.05	.17	2.5%
2020	Basement Walls	4' foundation wall		L.F. Wall	52	.44	
B. SHELL							
	B10 Superstructure						
1010	Floor Construction	Open web steel joists, slab form, concrete, steel columns		S.F. Floor	14.76	12.30	
1020	Roof Construction	Open web steel joists with rib metal deck, steel columns		S.F. Roof	5.16	.86	16.5%
	B20 Exterior Enclosure						
2010	Exterior Walls	Face brick with concrete block backup	86% of wall	S.F. Wall	15.28	6.79	
2020	Exterior Windows	Aluminum horizontal sliding	14% of wall	Each	314	1.47	10.6%
2030	Exterior Doors	Aluminum and glass		Each	2715	.18	
	B30 Roofing						
3010	Roof Coverings	Built-up tar and gravel with flashing; perlite/EPS composite insulation		S.F. Roof	4.02	.67	0.8%
3020	Roof Openings	N/A		—	—	—	
C. INTERIORS							
1010	Partitions	Gypsum board and sound deadening board on metal studs	8 S.F. Floor/L.F. Partitions	S.F. Partition	4.70	4.70	
1020	Interior Doors	15% solid core wood, 85% hollow core wood	80 S.F Floor/Door	Each	472	5.91	
1030	Fittings	Kitchen cabinets		S.F. Floor	2.24	2.24	
2010	Stair Construction	Concrete filled metal pan		Flight	4100	1.09	29.4%
3010	Wall Finishes	70% paint, 25% vinyl wall covering, 5% ceramic tile		S.F. Surface	2.29	2.29	
3020	Floor Finishes	60% carpet, 30% vinyl composition tile, 10% ceramic tile		S.F. Floor	4.42	4.42	
3030	Ceiling Finishes	Painted gypsum board on resilient channels		S.F. Ceiling	2.83	2.83	
D. SERVICES							
	D10 Conveying						
1010	Elevators & Lifts	Two geared passenger elevators		Each	134,400	4.48	5.6%
1020	Escalators & Moving Walks	N/A		—	—	—	
	D20 Plumbing						
2010	Plumbing Fixtures	Kitchen, bathroom and service fixtures, supply and drainage	1 Fixture/215 S.F. Floor	Each	1425	6.63	
2020	Domestic Water Distribution	Gas fired water heater		S.F. Floor	2.03	2.03	11.0%
2040	Rain Water Drainage	Roof drains		S.F. Roof	.84	.14	
	D30 HVAC						
3010	Energy Supply	Oil fired hot water, baseboard radiation		S.F. Floor	4.26	4.26	
3020	Heat Generating Systems	N/A		—	—	—	
3030	Cooling Generating Systems	Chilled water, air cooled condenser system		S.F. Floor	6.12	6.12	13.0%
3050	Terminal & Package Units	N/A		—	—	—	
3090	Other HVAC Sys. & Equipment	N/A		—	—	—	
	D40 Fire Protection						
4010	Sprinklers	Wet pipe sprinkler system		S.F. Floor	1.51	1.51	2.3%
4020	Standpipes	Standpipe		S.F. Floor	.29	.29	
	D50 Electrical						
5010	Electrical Service/Distribution	1600 ampere service, panel board and feeders		S.F. Floor	1.51	1.51	
5020	Lighting & Branch Wiring	Incandescent fixtures, receptacles, switches, A.C. and misc. power		S.F. Floor	4.72	4.72	8.3%
5030	Communications & Security	Alarm systems, emergency lighting, and intercom		S.F. Floor	.22	.22	
5090	Other Electrical Systems	Emergency generator, 11.5 kW		S.F. Floor	.14	.14	
E. EQUIPMENT & FURNISHINGS							
1010	Commercial Equipment	N/A		—	—	—	
1020	Institutional Equipment	N/A		—	—	—	
1030	Vehicular Equipment	N/A		—	—	—	0.0%
1090	Other Equipment	N/A		—	—	—	
F. SPECIAL CONSTRUCTION & DEMOLITION							
1020	Integrated Construction	N/A		—	—	—	
1040	Special Facilities	N/A		—	—	—	0.0%
G. BUILDING SITEWORK	**N/A**						

			Sub-Total	79.78	100%
	CONTRACTOR FEES (General Requirements: 10%, Overhead: 5%, Profit: 10%)		25%	19.95	
	ARCHITECT FEES		7%	6.97	
		Total Building Cost		106.70	

Costs per square foot of floor area

Exterior Wall	S.F. Area	95000	112000	129000	145000	170000	200000	275000	400000	600000
	L.F. Perimeter	345	386	406	442	480	510	530	570	630
Ribbed Precast Concrete Panel	Steel Frame	124.15	121.90	119.40	118.20	116.20	114.15	110.05	106.65	104.15
	R/Conc. Frame	118.40	116.20	113.85	112.65	110.80	108.80	104.95	101.75	99.40
Face Brick with Concrete Block Back-up	Steel Frame	117.00	114.65	112.15	110.85	108.85	106.75	102.50	99.00	96.40
	R/Conc. Frame	120.40	118.10	115.60	114.30	112.35	110.25	106.00	102.55	100.00
Stucco on Concrete Block	Steel Frame	110.45	108.45	106.45	105.35	103.80	102.15	99.00	96.40	94.50
	R/Conc. Frame	113.85	111.85	109.90	108.80	107.25	105.65	102.55	99.95	98.10
Perimeter Adj., Add or Deduct	Per 100 L.F.	4.95	4.15	3.65	3.20	2.75	2.35	1.70	1.20	.80
Story Hgt. Adj., Add or Deduct	Per 1 Ft.	1.65	1.55	1.45	1.40	1.30	1.15	.90	.65	.45
For Basement, add $22.25 per square foot of basement area										

The above costs were calculated using the basic specifications shown on the facing page. These costs should be adjusted where necessary for design alternatives and owner's requirements. Reported completed project costs, for this type of structure, range from $55.45 to $127.75 per S.F.

Common additives

Description	Unit	$ Cost	Description	Unit	$ Cost
Appliances			Closed Circuit Surveillance, One station		
Cooking range, 30" free standing			Camera and monitor	Each	1375
1 oven	Each	340 - 1475	For additional camera stations, add	Each	750
2 oven	Each	1475 - 1650	Elevators, Electric passenger, 10 stops		
30" built-in			3000# capacity	Each	217,000
1 oven	Each	420 - 1550	4000# capacity	Each	219,000
2 oven	Each	1225 - 2100	5000# capacity	Each	223,000
Counter top cook tops, 4 burner	Each	281 - 625	Additional stop, add	Each	5675
Microwave oven	Each	203 - 650	Emergency Lighting, 25 watt, battery operated		
Combination range, refrig. & sink, 30" wide	Each	1225 - 2450	Lead battery	Each	227
72" wide	Each	3725	Nickel cadmium	Each	660
Combination range, refrigerator, sink,			Laundry Equipment		
microwave oven & icemaker	Each	5500	Dryer, gas, 16 lb. capacity	Each	725
Compactor, residential, 4-1 compaction	Each	465 - 550	30 lb. capacity	Each	2800
Dishwasher, built-in, 2 cycles	Each	495 - 740	Washer, 4 cycle	Each	810
4 cycles	Each	530 - 1000	Commercial	Each	1225
Garbage disposer, sink type	Each	128 - 287	Smoke Detectors		
Hood for range, 2 speed, vented, 30" wide	Each	199 - 760	Ceiling type	Each	151
42" wide	Each	350 - 1100	Duct type	Each	405
Refrigerator, no frost 10-12 C.F.	Each	555 - 860			
18-20 C.F.	Each	630 - 950			

Important: See the Reference Section for Location Factors

BUILDING TYPES

Model costs calculated for a 15 story building with 10'-6" story height and 145,000 square feet of floor area

				Unit	Unit Cost	Cost Per S.F.	% Of Sub-Total
A. SUBSTRUCTURE							
1010	Standard Foundations	Poured concrete; strip and spread footings		S.F. Ground	10.65	.71	
1030	Slab on Grade	4" reinforced concrete with vapor barrier and granular base		S.F. Slab	3.42	.23	
2010	Basement Excavation	Site preparation for slab and trench for foundation wall and footing		S.F. Ground	1.05	.07	1.3%
2020	Basement Walls	4' Foundation wall		L.F. Wall	49	.18	
B. SHELL							
	B10 Superstructure						
1010	Floor Construction	Open web steel joists, slab form, concrete, interior steel columns		S.F. Floor	13.04	12.17	13.9%
1020	Roof Construction	Open web steel joists with rib metal deck, interior steel columns		S.F. Roof	3.90	.26	
	B20 Exterior Enclosure						
2010	Exterior Walls	Ribbed precast concrete panel	87% of wall	S.F. Wall	16.81	7.02	
2020	Exterior Windows	Aluminum horizontal sliding	13% of wall	Each	314	1.24	10.8%
2030	Exterior Doors	Aluminum and glass		Each	1593	1.37	
	B30 Roofing						
3010	Roof Coverings	Built-up tar and gravel with flashing; perlite/EPS composite insulation		S.F. Roof	3.90	.26	0.3%
3020	Roof Openings	N/A		—	—	—	
C. INTERIORS							
1010	Partitions	Gypsum board on concrete block and metal studs	10 S.F. of Floor/L.F. Partition	S.F. Partition	9.31	9.31	
1020	Interior Doors	15% solid core wood, 85% hollow core wood	80 S.F. Floor/Door	Each	482	6.03	
1030	Fittings	Kitchen cabinets		S.F. Floor	2.32	2.32	
2010	Stair Construction	Concrete filled metal pan		Flight	4100	1.22	31.8%
3010	Wall Finishes	70% paint, 25% vinyl wall covering, 5% ceramic tile		S.F. Surface	2.27	2.27	
3020	Floor Finishes	60% carpet, 30% vinyl composition tile, 10% ceramic tile		S.F. Floor	4.42	4.42	
3030	Ceiling Finishes	Painted gypsum board on resilient channels		S.F. Ceiling	2.83	2.83	
D. SERVICES							
	D10 Conveying						
1010	Elevators & Lifts	Four geared passenger elevators		Each	235,625	6.50	7.3%
1020	Escalators & Moving Walks	N/A		—	—	—	
	D20 Plumbing						
2010	Plumbing Fixtures	Kitchen, bathroom and service fixtures, supply and drainage	1 Fixture/210 S.F. Floor	Each	1593	7.59	
2020	Domestic Water Distribution	Gas fired water heater		S.F. Floor	2.33	2.33	11.2%
2040	Rain Water Drainage	Roof drains		S.F Roof	1.65	.11	
	D30 HVAC						
3010	Energy Supply	Oil fired hot water, baseboard radiation		S.F.Floor	4.26	4.26	
3020	Heat Generating Systems	N/A		—	—	—	
3030	Cooling Generating Systems	Chilled water, air cooled condenser system		S.F. Floor	6.12	6.12	11.6%
3050	Terminal & Package Units	N/A		—	—	—	
3090	Other HVAC Sys. & Equipment	N/A		—	—	—	
	D40 Fire Protection						
4010	Sprinklers	Wet pipe sprinkler system		S.F. Floor	1.47	1.47	2.5%
4020	Standpipes	Standpipe		S.F. Floor	.77	.77	
	D50 Electrical						
5010	Electrical Service/Distribution	4000 ampere service, panel board and feeders		S.F. Floor	1.36	1.36	
5020	Lighting & Branch Wiring	Incandescent fixtures, receptacles, switches, A.C. and misc. power		S.F. Floor	5.09	5.09	
5030	Communications & Security	Alarm systems, emergency lighting, antenna, intercom and security television		S.F. Floor	1.52	1.52	9.3%
5090	Other Electrical Systems	Emergency generator, 80KW		S.F. Floor	.16	.16	
E. EQUIPMENT & FURNISHINGS							
1010	Commercial Equipment	N/A		—	—	—	
1020	Institutional Equipment	N/A		—	—	—	
1030	Vehicular Equipment	N/A		—	—	—	0.0%
1090	Other Equipment	N/A		—	—	—	
F. SPECIAL CONSTRUCTION & DEMOLITION							
1020	Integrated Construction	N/A		—	—	—	0.0%
1040	Special Facilities	N/A		—	—	—	
G. BUILDING SITEWORK	**N/A**						

		Sub-Total	89.19	100%
CONTRACTOR FEES (General Requirements: 10%, Overhead: 5%, Profit: 10%)		25%	22.30	
ARCHITECT FEES		6%	6.71	
		Total Building Cost	**118.20**	

BUILDING TYPES

BUILDING TYPES

Costs per square foot of floor area

Exterior Wall	S.F. Area	12000	15000	18000	21000	24000	27000	30000	33000	36000
	L.F. Perimeter	440	500	540	590	640	665	700	732	770
Face Brick with Concrete Block Back-up	Steel Frame	127.95	123.95	120.15	117.90	116.20	113.95	112.45	111.15	110.25
	Bearing Wall	123.75	120.10	116.70	114.65	113.15	111.10	109.80	108.60	107.80
Precast Concrete	Steel Frame	120.25	116.95	113.90	112.05	110.70	108.90	107.70	106.65	105.95
Decorative Concrete Block	Bearing Wall	112.45	109.85	107.40	105.95	104.85	103.40	102.50	101.65	101.05
Concrete Block	Steel Frame	116.35	114.25	112.35	111.15	110.30	109.20	108.50	107.80	107.35
	Bearing Wall	109.05	106.75	104.60	103.35	102.40	101.15	100.30	99.60	99.10
Perimeter Adj., Add or Deduct	Per 100 L.F.	8.80	7.10	5.85	5.05	4.40	3.90	3.55	3.20	2.95
Story Hgt. Adj., Add or Deduct	Per 1 Ft.	1.40	1.30	1.10	1.05	1.00	.90	.90	.85	.80

For Basement, add $17.75 per square foot of basement area

The above costs were calculated using the basic specifications shown on the facing page. These costs should be adjusted where necessary for design alternatives and owner's requirements. Reported completed project costs, for this type of structure, range from $53.30 to $139.65 per S.F.

Common additives

Description	Unit	$ Cost
Closed Circuit Surveillance, One station		
Camera and monitor	Each	1375
For additional camera stations, add	Each	750
Emergency Lighting, 25 watt, battery operated		
Lead battery	Each	227
Nickel cadmium	Each	660
Seating		
Auditorium chair, all veneer	Each	171
Veneer back, padded seat	Each	207
Upholstered, spring seat	Each	208
Classroom, movable chair & desk	Set	65 - 120
Lecture hall, pedestal type	Each	156 - 485
Smoke Detectors		
Ceiling type	Each	151
Duct type	Each	405
Sound System		
Amplifier, 250 watts	Each	1675
Speaker, ceiling or wall	Each	147
Trumpet	Each	275

Important: See the Reference Section for Location Factors

Model costs calculated for a 1 story building
with 24' story height and 24,000 square feet
of floor area

Auditorium

				Unit	Unit Cost	Cost Per S.F.	% Of Sub-Total
A. SUBSTRUCTURE							
1010	Standard Foundations	Poured concrete; strip and spread footings		S.F. Ground	.96	.96	
1030	Slab on Grade	6" reinforced concrete with vapor barrier and granular base		S.F. Slab	4.10	4.10	
2010	Basement Excavation	Site preparation for slab and trench for foundation wall and footing		S.F. Ground	1.09	1.09	9.3%
2020	Basement Walls	4" foundation wall		L.F. Wall	58	1.56	
B. SHELL							
	B10 Superstructure						
1010	Floor Construction	Open web steel joists, slab form, concrete (balcony)		S.F. Floor	14.64	1.83	9.0%
1020	Roof Construction	Metal deck on steel truss		S.F. Roof	5.58	5.58	
	B20 Exterior Enclosure						
2010	Exterior Walls	Precast concrete panel	80% of wall (adjusted for end walls)	S.F. Wall	16.93	8.67	
2020	Exterior Windows	Glass curtain wall	20% of wall	Each	30	3.85	16.7%
2030	Exterior Doors	Double aluminum and glass and hollow metal		Each	2557	1.28	
	B30 Roofing						
3010	Roof Coverings	Built-up tar and gravel with flashing; perlite/EPS composite insulation		S.F. Roof	3.75	3.75	4.7%
3020	Roof Openings	Roof hatches		S.F. Roof	.10	.10	
C. INTERIORS							
1010	Partitions	Concrete Block and toilet partitions	40 S.F. Floor/L.F. Partition	S.F. Partition	6.03	2.41	
1020	Interior Doors	Single leaf hollow metal	400 S.F. Floor/Door	Each	537	1.34	
1030	Fittings	Toilet partitions		S.F. Floor	—	—	
2010	Stair Construction	Concrete filled metal pan		Flight	6050	.76	21.3%
3010	Wall Finishes	70% paint, 30% epoxy coating		S.F. Surface	7.60	3.04	
3020	Floor Finishes	70% vinyl tile, 30% carpet		S.F. Floor	7.04	7.04	
3030	Ceiling Finishes	Fiberglass board, suspended		S.F. Ceiling	3.06	3.06	
D. SERVICES							
	D10 Conveying						
1010	Elevators & Lifts	One hydraulic passenger elevator		Each	58,320	2.43	2.9%
1020	Escalators & Moving Walks	N/A		—	—	—	
	D20 Plumbing						
2010	Plumbing Fixtures	Toilet and service fixtures, supply and drainage	1 Fixture/800 S.F. Floor	Each	2368	2.96	
2020	Domestic Water Distribution	Gas fired water heater		S.F. Floor	.24	.24	4.6%
2040	Rain Water Drainage	Roof drains		S.F. Roof	.60	.60	
	D30 HVAC						
3010	Energy Supply	N/A		—	—	—	
3020	Heat Generating Systems	Included in D3030		—	—	—	
3030	Cooling Generating Systems	Single zone rooftop unit, gas heating, electric cooling		S.F. Floor	11.44	11.44	13.8%
3050	Terminal & Package Units	N/A		—	—	—	
3090	Other HVAC Sys. & Equipment	N/A		—	—	—	
	D40 Fire Protection						
4010	Sprinklers	Wet pipe sprinkler system		S.F. Floor	1.79	1.79	2.2%
4020	Standpipes	N/A		—	—	—	
	D50 Electrical						
5010	Electrical Service/Distribution	800 ampere service, panel board and feeders		S.F. Floor	1.49	1.49	
5020	Lighting & Branch Wiring	Fluorescent fixtures, receptacles, switches, A.C. and misc. power		S.F. Floor	8.08	8.08	15.5%
5030	Communications & Security	Alarm systems and emergency lighting, and public address system		S.F. Floor	2.31	2.31	
5090	Other Electrical Systems	Emergency generator, 100KW		S.F. Floor	1.01	1.01	
E. EQUIPMENT & FURNISHINGS							
1010	Commercial Equipment	N/A		—	—	—	
1020	Institutional Equipment	N/A		—	—	—	
1030	Vehicular Equipment	N/A		—	—	—	0.0%
1090	Other Equipment	N/A		—	—	—	
F. SPECIAL CONSTRUCTION & DEMOLITION							
1020	Integrated Construction	N/A		—	—	—	0.0%
1040	Special Facilities	N/A		—	—	—	
G. BUILDING SITEWORK	**N/A**						

		Sub-Total	82.77	100%
CONTRACTOR FEES (General Requirements: 10%, Overhead: 5%, Profit: 10%)			25%	20.69
ARCHITECT FEES			7%	7.24

Total Building Cost	**110.70**

BUILDING TYPES

Costs per square foot of floor area

Exterior Wall	S.F. Area	2000	2700	3400	4100	4800	5500	6200	6900	7600
	L.F. Perimeter	180	208	236	256	280	303	317	337	357
Face Brick with Concrete Block Back-up	Steel Frame	169.20	158.65	152.45	147.20	143.95	141.35	138.50	136.80	135.35
	R/Conc. Frame	181.60	171.10	164.90	159.60	156.40	153.80	150.95	149.20	147.75
Precast Concrete Panel	Steel Frame	159.20	150.15	144.80	140.25	137.45	135.25	132.85	131.35	130.15
	R/Conc. Frame	171.65	162.55	157.20	152.70	149.90	147.70	145.30	143.80	142.55
Limestone with Concrete Block Back-up	Steel Frame	181.65	169.30	162.05	155.80	152.00	149.00	145.60	143.55	141.85
	R/Conc. Frame	194.05	181.75	174.50	168.25	164.45	161.45	158.00	155.95	154.25
Perimeter Adj., Add or Deduct	Per 100 L.F.	32.30	23.90	19.00	15.75	13.45	11.80	10.45	9.35	8.45
Story Hgt. Adj., Add or Deduct	Per 1 Ft.	3.10	2.65	2.40	2.15	2.00	1.90	1.75	1.65	1.60

For Basement, add $20.55 per square foot of basement area

The above costs were calculated using the basic specifications shown on the facing page. These costs should be adjusted where necessary for design alternatives and owner's requirements. Reported completed project costs, for this type of structure, range from $83.75 to $206.00 per S.F.

Common additives

Description	Unit	$ Cost	Description	Unit	$ Cost
Bulletproof Teller Window, 44" x 60"	Each	3450	Service Windows, Pass thru, steel		
60" x 48"	Each	4200	24" x 36"	Each	2350
Closed Circuit Surveillance, One station			48" x 48"	Each	3025
Camera and monitor	Each	1375	72" x 40"	Each	4275
For additional camera stations, add	Each	750	Smoke Detectors		
Counters, Complete	Station	4425	Ceiling type	Each	151
Door & Frame, 3' x 6'-8", bullet resistant steel			Duct type	Each	405
with vision panel	Each	3850 - 4900	Twenty-four Hour Teller		
Drive-up Window, Drawer & micr., not incl. glass	Each	5025 - 9900	Automatic deposit cash & memo	Each	42,400
Emergency Lighting, 25 watt, battery operated			Vault Front, Door & frame		
Lead battery	Each	227	1 hour test, 32"x 78"	Opening	3600
Nickel cadmium	Each	660	2 hour test, 32" door	Opening	4275
Night Depository	Each	6875 - 10,500	40" door	Opening	4700
Package Receiver, painted	Each	1425	4 hour test, 32" door	Opening	4375
stainless steel	Each	2200	40" door	Opening	5225
Partitions, Bullet resistant to 8' high	L.F.	238 - 385	Time lock, two movement, add	Each	1650
Pneumatic Tube Systems, 2 station	Each	24,400			
With TV viewer	Each	45,100			

BUILDING TYPES

Model costs calculated for a 1 story building with 14' story height and 4,100 square feet of floor area

Bank

				Unit	Unit Cost	Cost Per S.F.	% Of Sub-Total
A.	**SUBSTRUCTURE**						
1010	Standard Foundations	Poured concrete; strip and spread footings		S.F. Ground	2.80	2.80	
1030	Slab on Grade	4" reinforced concrete with vapor barrier and granular base		S.F. Slab	3.42	3.42	
2010	Basement Excavation	Site preparation for slab and trench for foundation wall and footing		S.F. Ground	1.18	1.18	9.9%
2020	Basement Walls	4' Foundation wall		L.F. Wall	52	3.96	
B.	**SHELL**						
	B10 Superstructure						
1010	Floor Construction	Cast-in-place columns		L.F. Column	75.65	3.69	
1020	Roof Construction	Cast-in-place concrete flat plate		S.F. Roof	10.75	10.75	12.6%
	B20 Exterior Enclosure						
2010	Exterior Walls	Face brick with concrete block backup	80% of wall	S.F. Wall	21	14.76	
2020	Exterior Windows	Horizontal aluminum sliding	20% of wall	Each	396	4.62	17.7%
2030	Exterior Doors	Double aluminum and glass and hollow metal		Each	2065	1.01	
	B30 Roofing						
3010	Roof Coverings	Built-up tar and gravel with flashing; perlite/EPS composite insulation		S.F. Roof	4.72	4.72	
3020	Roof Openings	N/A		—	—	—	4.1%
C.	**INTERIORS**						
1010	Partitions	Gypsum board on metal studs	20 S.F. of Floor/L.F. Partition	S.F. Partition	6.96	3.48	
1020	Interior Doors	Single leaf hollow metal	200 S.F. Floor/Door	Each	537	2.69	
1030	Fittings	N/A		—	—	—	
2010	Stair Construction	N/A		—	—	—	13.4%
3010	Wall Finishes	50% vinyl wall covering, 50% paint		S.F. Surface	2.02	1.01	
3020	Floor Finishes	50% carpet tile, 40% vinyl composition tile, 10% quarry tile		S.F. Floor	4.47	4.47	
3030	Ceiling Finishes	Mineral fiber tile on concealed zee bars		S.F. Ceiling	3.71	3.71	
D.	**SERVICES**						
	D10 Conveying						
1010	Elevators & Lifts	N/A		—	—	—	
1020	Escalators & Moving Walks	N/A		—	—	—	0.0%
	D20 Plumbing						
2010	Plumbing Fixtures	Toilet and service fixtures, supply and drainage	1 Fixture/580 S.F. Floor	Each	2743	4.73	
2020	Domestic Water Distribution	Gas fired water heater		S.F. Floor	.58	.58	5.2%
2040	Rain Water Drainage	Roof drains		S.F. Roof	.65	.65	
	D30 HVAC						
3010	Energy Supply	N/A		—	—	—	
3020	Heat Generating Systems	Included in D3030		—	—	—	
3030	Cooling Generating Systems	Single zone rooftop unit, gas heating, electric cooling		S.F. Floor	9.74	9.74	8.5%
3050	Terminal & Package Units	N/A		—	—	—	
3090	Other HVAC Sys. & Equipment	N/A		—	—	—	
	D40 Fire Protection						
4010	Sprinklers	Wet pipe sprinkler system		S.F. Floor	2.34	2.34	
4020	Standpipes	N/A		—	—	—	2.0%
	D50 Electrical						
5010	Electrical Service/Distribution	200 ampere service, panel board and feeders		S.F. Floor	1.65	1.65	
5020	Lighting & Branch Wiring	Fluorescent fixtures, receptacles, switches, A.C. and misc. power		S.F. Floor	7.15	7.15	
5030	Communications & Security	Alarm systems, emergency lighting, and security television		S.F. Floor	1.69	1.69	9.2%
5090	Other Electrical Systems	Emergency generator, 15KW		S.F. Floor	.16	.16	
E.	**EQUIPMENT & FURNISHINGS**						
1010	Commercial Equipment	Automatic teller, drive up window, night depository		S.F. Floor	5.80	5.80	
1020	Institutional Equipment	Closed circuit TV monitoring system		S.F. Floor	3.23	3.23	
1030	Vehicular Equipment	N/A		—	—	—	7.8%
1090	Other Equipment	N/A		—	—	—	
F.	**SPECIAL CONSTRUCTION & DEMOLITION**						
1020	Integrated Construction	N/A		—	—	—	
1040	Special Facilities	Security vault door		S.F. Floor	11.05	11.05	9.6%
G.	**BUILDING SITEWORK**	**N/A**					

		Sub-Total	115.04	**100%**
CONTRACTOR FEES (General Requirements: 10%, Overhead: 5%, Profit: 10%)		25%	28.76	
ARCHITECT FEES		11%	15.80	
		Total Building Cost	159.60	

BUILDING TYPES

87

Costs per square foot of floor area

Exterior Wall	S.F. Area	12000	14000	16000	18000	20000	22000	24000	26000	28000
	L.F. Perimeter	460	491	520	540	566	593	620	645	670
Concrete Block	Steel Roof Deck	76.00	73.85	72.20	70.70	69.65	68.80	68.10	67.45	66.90
Decorative Concrete Block	Steel Roof Deck	79.55	77.15	75.20	73.50	72.25	71.25	70.45	69.75	69.10
Face Brick on Concrete Block	Steel Roof Deck	84.40	81.60	79.35	77.30	75.85	74.70	73.75	72.90	72.15
Jumbo Brick on Concrete Block	Steel Roof Deck	81.85	79.25	77.20	75.30	74.00	72.90	72.05	71.25	70.55
Stucco on Concrete Block	Steel Roof Deck	78.50	76.15	74.30	72.70	71.50	70.55	69.75	69.10	68.45
Precast Concrete Panel	Steel Roof Deck	77.30	75.05	73.30	71.75	70.60	69.70	68.95	68.30	67.70
Perimeter Adj., Add or Deduct	Per 100 L.F.	3.50	3.00	2.60	2.35	2.10	1.95	1.75	1.60	1.50
Story Hgt. Adj., Add or Deduct	Per 1 Ft.	.80	.70	.65	.60	.60	.55	.55	.50	.50
For Basement, add $20.25 per square foot of basement area										

The above costs were calculated using the basic specifications shown on the facing page. These costs should be adjusted where necessary for design alternatives and owner's requirements. Reported completed project costs, for this type of structure, range from $37.10 to $90.95 per S.F.

Common additives

Description	Unit	$ Cost
Bowling Alleys, Incl. alley, pinsetter,		
scorer, counter & misc. supplies, average	Lane	50,500
For automatic scorer, add (maximum)	Lane	7475
Emergency Lighting, 25 watt, battery operated		
Lead battery	Each	227
Nickel cadmium	Each	660
Lockers, Steel, Single tier, 60" or 72"	Opening	131 - 227
2 tier, 60" or 72" total	Opening	74 - 125
5 tier, box lockers	Opening	42 - 62
Locker bench, lam., maple top only	L.F.	19.50
Pedestals, steel pipe	Each	63
Seating		
Auditorium chair, all veneer	Each	171
Veneer back, padded seat	Each	207
Upholstered, spring seat	Each	208
Sound System		
Amplifier, 250 watts	Each	1675
Speaker, ceiling or wall	Each	147
Trumpet	Each	275

BUILDING TYPES

Important: See the Reference Section for Location Factors

Model costs calculated for a 1 story building with 14' story height and 20,000 square feet of floor area

Bowling Alley

				Unit	Unit Cost	Cost Per S.F.	% Of Sub-Total
A. SUBSTRUCTURE							
1010	Standard Foundations	Poured concrete; strip and spread footings		S.F. Ground	.81	.81	
1030	Slab on Grade	4" reinforced concrete with vapor barrier and granular base		S.F. Slab	3.42	3.42	
2010	Basement Excavation	Site preparation for slab and trench for foundation wall and footing		S.F. Ground	1.31	1.31	13.5%
2020	Basement Walls	4' Foundation wall		L.F. Wall	.52	1.50	
B. SHELL							
B10 Superstructure							
1010	Floor Construction	N/A		—	—	—	
1020	Roof Construction	Metal deck on open web steel joists with columns and beams		S.F. Roof	5.65	5.65	10.9%
B20 Exterior Enclosure							
2010	Exterior Walls	Concrete block	90% of wall	S.F. Wall	8.05	2.87	
2020	Exterior Windows	Horizontal pivoted	10% of wall	Each	1489	2.46	10.9%
2030	Exterior Doors	Double aluminum and glass and hollow metal		Each	1459	.36	
B30 Roofing							
3010	Roof Coverings	Built-up tar and gravel with flashing; perlite/EPS composite insulation		S.F. Roof	3.61	3.61	
3020	Roof Openings	Roof hatches		S.F. Roof	.09	.09	7.1%
C. INTERIORS							
1010	Partitions	Concrete block	50 S.F. Floor/L.F. Partition	S.F. Partition	6.70	1.34	
1020	Interior Doors	Hollow metal single leaf	1000 S.F. Floor/Door	Each	537	.54	
1030	Fittings	N/A		—	—	—	
2010	Stair Construction	N/A		—	—	—	9.3%
3010	Wall Finishes	Paint and block filler		S.F. Surface	8.20	1.64	
3020	Floor Finishes	Vinyl tile	25% of floor	S.F. Floor	2.12	.53	
3030	Ceiling Finishes	Suspended fiberglass board	25% of area	S.F. Ceiling	3.06	.77	
D. SERVICES							
D10 Conveying							
1010	Elevators & Lifts	N/A		—	—	—	
1020	Escalators & Moving Walks	N/A		—	—	—	0.0%
D20 Plumbing							
2010	Plumbing Fixtures	Toilet and service fixtures, supply and drainage	1 Fixture/2200 S.F. Floor	Each	1804	.82	
2020	Domestic Water Distribution	Gas fired water heater		S.F. Floor	.11	.11	2.5%
2040	Rain Water Drainage	Roof drains		S.F. Roof	.38	.38	
D30 HVAC							
3010	Energy Supply	N/A		—	—	—	
3020	Heat Generating Systems	Included in D3030		—	—	—	
3030	Cooling Generating Systems	Single zone rooftop unit, gas heating, electric cooling		S.F. Floor	12.95	12.95	24.9%
3050	Terminal & Package Units	N/A		—	—	—	
3090	Other HVAC Sys. & Equipment	N/A		—	—	—	
D40 Fire Protection							
4010	Sprinklers	Sprinklers, light hazard		S.F. Floor	1.50	1.50	
4020	Standpipes	N/A		—	—	—	2.9%
D50 Electrical							
5010	Electrical Service/Distribution	800 ampere service, panel board and feeders		S.F. Floor	1.78	1.78	
5020	Lighting & Branch Wiring	Fluorescent fixtures, receptacles, switches, A.C. and misc. power		S.F. Floor	7.33	7.33	
5030	Communications & Security	Alarm systems and emergency lighting		S.F. Floor	.23	.23	18.0%
5090	Other Electrical Systems	Emergency generator, 7.5KW		S.F. Floor	.06	.06	
E. EQUIPMENT & FURNISHINGS							
1010	Commercial Equipment	N/A		—	—	—	
1020	Institutional Equipment	N/A		—	—	—	
1030	Vehicular Equipment	N/A		—	—	—	0.0%
1090	Other Equipment	N/A		—	—	—	
F. SPECIAL CONSTRUCTION & DEMOLITION							
1020	Integrated Construction	N/A		—	—	—	
1040	Special Facilities	N/A		—	—	—	0.0%
G. BUILDING SITEWORK	**N/A**						

			Sub-Total	52.06	**100%**
CONTRACTOR FEES (General Requirements: 10%, Overhead: 5%, Profit: 10%)			25%	13.02	
ARCHITECT FEES			7%	4.57	
			Total Building Cost	69.65	

BUILDING TYPES

Costs per square foot of floor area

Exterior Wall	S.F. Area	6000	8000	10000	12000	14000	16000	18000	20000	22000
	L.F. Perimeter	320	386	453	520	540	597	610	660	710
Face Brick with Concrete Block Back-up	Bearing Walls	105.75	101.65	99.30	97.70	94.70	93.70	91.60	90.95	90.40
	Steel Frame	107.30	103.25	100.90	99.35	96.40	95.45	93.35	92.70	92.20
Decorative Concrete Block	Bearing Walls	100.50	96.90	94.80	93.40	90.90	90.00	88.25	87.70	87.25
	Steel Frame	102.05	98.50	96.45	95.05	92.60	91.75	90.00	89.45	89.00
Precast Concrete Panels	Bearing Walls	99.40	95.90	93.90	92.50	90.10	89.25	87.55	87.05	86.55
	Steel Frame	100.95	97.50	95.50	94.15	91.80	91.00	89.30	88.80	88.35
Perimeter Adj., Add or Deduct	Per 100 L.F.	9.10	6.80	5.45	4.50	3.90	3.40	3.05	2.75	2.45
Story Hgt. Adj., Add or Deduct	Per 1 Ft.	1.50	1.35	1.30	1.20	1.10	1.05	.95	.95	.90
Basement—Not Applicable										

The above costs were calculated using the basic specifications shown on the facing page. These costs should be adjusted where necessary for design alternatives and owner's requirements. Reported completed project costs, for this type of structure, range from $44.75 to $107.85 per S.F.

Common additives

Description	Unit	$ Cost
Directory Boards, Plastic, glass covered		
30" x 20"	Each	575
36" x 48"	Each	1075
Aluminum, 24" x 18"	Each	440
36" x 24"	Each	555
48" x 32"	Each	780
48" x 60"	Each	1675
Emergency Lighting, 25 watt, battery operated		
Lead battery	Each	227
Nickel cadmium	Each	660
Benches, Hardwood	L.F.	84 - 157
Ticket Printer	Each	6475
Turnstiles, One way		
4 arm, 46" dia., manual	Each	430
Electric	Each	1550
High security, 3 arm		
65" dia., manual	Each	3175
Electric	Each	4400
Gate with horizontal bars		
65" dia., 7' high transit type	Each	4075

Important: See the Reference Section for Location Factors

Model costs calculated for a 1 story building with 14' story height and 12,000 square feet of floor area

Bus Terminal

				Unit	Unit Cost	Cost Per S.F.	% Of Sub-Total
A. SUBSTRUCTURE							
1010	Standard Foundations	Poured concrete; strip and spread footings		S.F. Ground	1.72	1.72	
1030	Slab on Grade	6" reinforced concrete with vapor barrier and granular base		S.F. Slab	5.09	5.09	
2010	Basement Excavation	Site preparation for slab and trench for foundation wall and footing		S.F. Ground	1.18	1.18	14.2%
2020	Basement Walls	4' Foundation wall		L.F. Wall	54	2.36	
B. SHELL							
	B10 Superstructure						
1010	Floor Construction	N/A		—	—	—	
1020	Roof Construction	Metal deck on open web steel joists with columns and beams		S.F. Roof	3.84	3.84	5.3%
	B20 Exterior Enclosure						
2010	Exterior Walls	Face brick with concrete block backup	70% of wall	S.F. Wall	21	8.96	
2020	Exterior Windows	Store front	30% of wall	Each	19.60	3.57	18.7%
2030	Exterior Doors	Double aluminum and glass		Each	3350	1.12	
	B30 Roofing						
3010	Roof Coverings	Built-up tar and gravel with flashing; perlite/EPS composite insulation		S.F. Roof	4.20	4.20	5.8%
3020	Roof Openings	N/A		—	—	—	
C. INTERIORS							
1010	Partitions	Lightweight concrete block	15 S.F. Floor/L.F. Partition	S.F. Partition	6.69	4.46	
1020	Interior Doors	Hollow metal	150 S.F. Floor/Door	Each	537	3.58	
1030	Fittings	N/A		—	—	—	
2010	Stair Construction	N/A		—	—	—	26.6%
3010	Wall Finishes	Glazed coating		S.F. Surface	2.87	1.91	
3020	Floor Finishes	Quarry tile and vinyl composition tile		S.F. Floor	5.80	5.80	
3030	Ceiling Finishes	Mineral fiber tile on concealed zee bars		S.F. Ceiling	3.71	3.71	
D. SERVICES							
	D10 Conveying						
1010	Elevators & Lifts	N/A		—	—	—	
1020	Escalators & Moving Walks	N/A		—	—	—	0.0%
	D20 Plumbing						
2010	Plumbing Fixtures	Toilet and service fixtures, supply and drainage	1 Fixture/850 S.F. Floor	Each	1759	2.07	
2020	Domestic Water Distribution	Electric water heater		S.F. Floor	.41	.41	4.0%
2040	Rain Water Drainage	Roof drains		S.F. Roof	.42	.42	
	D30 HVAC						
3010	Energy Supply	N/A		—	—	—	
3020	Heat Generating Systems	Included in D3030		—	—	—	
3030	Cooling Generating Systems	Single zone rooftop unit, gas heating, electric cooling		S.F. Floor	9.72	9.72	13.3%
3050	Terminal & Package Units	N/A		—	—	—	
3090	Other HVAC Sys. & Equipment	N/A		—	—	—	
	D40 Fire Protection						
4010	Sprinklers	Wet pipe sprinkler system		S.F. Floor	2.30	2.30	3.1%
4020	Standpipes	N/A		—	—	—	
	D50 Electrical						
5010	Electrical Service/Distribution	400 ampere service, panel board and feeders		S.F. Floor	1.15	1.15	
5020	Lighting & Branch Wiring	Fluorescent fixtures, receptacles, switches, A.C. and misc. power		S.F. Floor	4.96	4.96	9.0%
5030	Communications & Security	Alarm systems and emergency lighting		S.F. Floor	.19	.19	
5090	Other Electrical Systems	Emergency generator, 7.5 KW		S.F. Floor	.31	.31	
E. EQUIPMENT & FURNISHINGS							
1010	Commercial Equipment	N/A		—	—	—	
1020	Institutional Equipment	N/A		—	—	—	
1030	Vehicular Equipment	N/A		—	—	—	0.0%
1090	Other Equipment	N/A		—	—	—	
F. SPECIAL CONSTRUCTION & DEMOLITION							
1020	Integrated Construction	N/A		—	—	—	
1040	Special Facilities	N/A		—	—	—	0.0%
G. BUILDING SITEWORK	**N/A**						

				Sub-Total	73.03	100%
CONTRACTOR FEES (General Requirements: 10%, Overhead: 5%, Profit: 10%)				25%	18.26	
ARCHITECT FEES				7%	6.41	

Total Building Cost		**97.70**

BUILDING TYPES

91

Costs per square foot of floor area

Exterior Wall	S.F. Area	600	800	1000	1200	1600	2000	2400	3000	4000
	L.F. Perimeter	100	114	128	139	164	189	214	250	314
Brick with Concrete Block Back-up	Steel Frame	224.70	215.15	209.45	204.70	199.45	196.30	194.20	191.90	190.00
	Bearing Walls	221.65	212.10	206.40	201.60	196.40	193.20	191.15	188.85	186.90
Concrete Block	Steel Frame	197.20	191.65	188.35	185.60	182.55	180.70	179.50	178.15	177.05
	Bearing Walls	194.15	188.60	185.30	182.50	179.50	177.65	176.45	175.10	173.95
Galvanized Steel Siding	Steel Frame	197.45	192.65	189.75	187.35	184.70	183.10	182.05	180.90	179.90
Metal Sandwich Panel	Steel Frame	195.45	189.55	186.00	183.05	179.80	177.85	176.55	175.15	173.95
Perimeter Adj., Add or Deduct	Per 100 L.F.	65.55	49.20	39.35	32.80	24.55	19.70	16.40	13.10	9.85
Story Hgt. Adj., Add or Deduct	Per 1 Ft.	4.00	3.40	3.10	2.80	2.45	2.25	2.15	2.00	1.90
Basement—Not Applicable										

The above costs were calculated using the basic specifications shown on the facing page. These costs should be adjusted where necessary for design alternatives and owner's requirements. Reported completed project costs, for this type of structure, range from $80.25 to $227.95 per S.F.

Common additives

Description	Unit	$ Cost
Air Compressors, Electric		
1-1/2 H.P., standard controls	Each	1050
Dual controls	Each	1225
5 H.P., 115/230 volt, standard control	Each	2750
Dual controls	Each	2850
Emergency Lighting, 25 watt, battery operated		
Lead battery	Each	227
Nickel cadmium	Each	660
Fence, Chain link, 6' high		
9 ga. wire, galvanized	L.F.	16.40
6 ga. wire	L.F.	21
Gate	Each	297
Product Dispenser with vapor recovery		
for 6 nozzles	Each	16,500
Laundry Equipment		
Dryer, gas, 16 lb. capacity	Each	725
30 lb. capacity	Each	2800
Washer, 4 cycle	Each	810
Commercial	Each	1225

Description	Unit	$ Cost
Lockers, Steel, single tier, 60" or 72"	Opening	131 - 227
2 tier, 60" or 72" total	Opening	74 - 125
5 tier, box lockers	Opening	42 - 62
Locker bench, lam. maple top only	L.F.	19.50
Pedestals, steel pipe	Each	63
Paving, Bituminous		
Wearing course plus base course	Sq. Yard	4.24
Safe, Office type, 4 hour rating		
30" x 18" x 18"	Each	3200
62" x 33" x 20"	Each	6975
Sidewalks, Concrete 4" thick	S.F.	2.95
Yard Lighting,		
20' aluminum pole with 400 watt high pressure sodium fixture	Each	2195

Important: See the Reference Section for Location Factors

Model costs calculated for a 1 story building with 12' story height and 800 square feet of floor area

				Unit	Unit Cost	Cost Per S.F.	% Of Sub-Total
A.	**SUBSTRUCTURE**						
1010	Standard Foundations	Poured concrete; strip and spread footings		S.F. Ground	2.04	2.04	
1030	Slab on Grade	5" reinforced concrete with vapor barrier and granular base		S.F. Slab	4.47	4.47	9.3%
2010	Basement Excavation	Site preparation for slab and trench for foundation wall and footing		S.F. Ground	1.71	1.71	
2020	Basement Walls	4' Foundation wall		L.F. Wall	46	6.68	
B.	**SHELL**						
	B10 Superstructure						
1010	Floor Construction	N/A		–	–	–	3.0%
1020	Roof Construction	Metal deck, open web steel joists, beams, columns		S.F. Roof	4.80	4.80	
	B20 Exterior Enclosure						
2010	Exterior Walls	Face brick with concrete block backup	70% of wall	S.F. Wall	21	25.26	
2020	Exterior Windows	Horizontal pivoted steel	5% of wall	Each	578	5.49	24.3%
2030	Exterior Doors	Steel overhead hollow metal		Each	1676	8.39	
	B30 Roofing						
3010	Roof Coverings	Built-up tar and gravel with flashing; perlite/EPS composite insulation		S.F. Roof	5.47	5.47	3.4%
3020	Roof Openings	N/A		–	–	–	
C.	**INTERIORS**						
1010	Partitions	Concrete block	20 S.F. Floor/S.F. Partition	S.F. Partition	6.42	3.21	
1020	Interior Doors	Hollow metal	600 S.F. Floor/Door	Each	537	.90	
1030	Fittings	N/A		–	–	–	
2010	Stair Construction	N/A		–	–	–	3.1%
3010	Wall Finishes	Paint		S.F. Surface	1.74	.87	
3020	Floor Finishes	N/A		–	–	–	
3030	Ceiling Finishes	N/A		–	–	–	
D.	**SERVICES**						
	D10 Conveying						
1010	Elevators & Lifts	N/A		–	–	–	0.0%
1020	Escalators & Moving Walks	N/A		–	–	–	
	D20 Plumbing						
2010	Plumbing Fixtures	Toilet and service fixtures, supply and drainage	1 Fixture/160 S.F. Floor	Each	2014	12.59	
2020	Domestic Water Distribution	Gas fired water heater		S.F. Floor	16.53	16.53	19.6%
2040	Rain Water Drainage	Roof drains		S.F. Roof	2.45	2.45	
	D30 HVAC						
3010	Energy Supply	N/A		–	–	–	
3020	Heat Generating Systems	Unit heaters		Each	12.25	12.25	
3030	Cooling Generating Systems	N/A		–	–	–	7.6%
3050	Terminal & Package Units	N/A		–	–	–	
3090	Other HVAC Sys. & Equipment	N/A		–	–	–	
	D40 Fire Protection						
4010	Sprinklers	N/A		–	–	–	0.0%
4020	Standpipes	N/A		–	–	–	
	D50 Electrical						
5010	Electrical Service/Distribution	400 ampere service, panel board and feeders		S.F. Floor	15.42	15.42	
5020	Lighting & Branch Wiring	Fluorescent fixtures, receptacles, switches and misc. power		S.F. Floor	30	30.99	29.7%
5030	Communications & Security	N/A		–	–	–	
5090	Other Electrical Systems	Emergency generator, 7.5KW		–	1.35	1.35	
E.	**EQUIPMENT & FURNISHINGS**						
1010	Commercial Equipment	N/A		–	–	–	
1020	Institutional Equipment	N/A		–	–	–	0.0%
1030	Vehicular Equipment	N/A		–	–	–	
1090	Other Equipment	N/A		–	–	–	
F.	**SPECIAL CONSTRUCTION & DEMOLITION**						
1020	Integrated Construction	N/A		–	–	–	0.0%
1040	Special Facilities	N/A		–	–	–	
G.	**BUILDING SITEWORK**	**N/A**					

		Sub-Total	160.87	100%
	CONTRACTOR FEES (General Requirements: 10%, Overhead: 5%, Profit: 10%)	25%	40.22	
	ARCHITECT FEES	7%	14.06	
	Total Building Cost		**215.15**	

BUILDING TYPES

93

Costs per square foot of floor area

Exterior Wall	S.F. Area	2000	7000	12000	17000	22000	27000	32000	37000	42000
	L.F. Perimeter	180	340	470	540	640	740	762	793	860
Decorative Concrete Brick	Wood Arch	171.00	128.50	119.20	112.25	109.70	108.00	104.75	102.55	101.65
	Steel Truss	167.05	124.55	115.25	108.30	105.75	104.05	100.80	98.60	97.70
Stone with Concrete Block Back-up	Wood Arch	197.90	143.00	130.90	121.75	118.40	116.20	111.90	108.95	107.80
	Steel Truss	189.45	134.60	122.50	113.35	109.95	107.75	103.45	100.55	99.35
Face Brick with Concrete Block Back-up	Wood Arch	189.05	138.25	127.05	118.65	115.55	113.50	109.55	106.85	105.75
	Steel Truss	180.65	129.85	118.65	110.20	107.10	105.10	101.10	98.45	97.35
Perimeter Adj., Add or Deduct	Per 100 L.F.	44.35	12.65	7.40	5.20	4.00	3.30	2.75	2.40	2.10
Story Hgt. Adj., Add or Deduct	Per 1 Ft.	2.65	1.45	1.15	.95	.85	.80	.70	.60	.60

For Basement, add $21.15 per square foot of basement area

The above costs were calculated using the basic specifications shown on the facing page. These costs should be adjusted where necessary for design alternatives and owner's requirements. Reported completed project costs, for this type of structure, range from $51.30 to $209.50 per S.F.

Common additives

Description	Unit	$ Cost
Altar, Wood, custom design, plain	Each	2175
Deluxe	Each	11,000
Granite or marble, average	Each	8800
Deluxe	Each	27,700
Ark, Prefabricated, plain	Each	8125
Deluxe	Each	93,500
Baptistry, Fiberglass, incl. plumbing	Each	3725 - 6775
Bells & Carillons, 48 bells	Each	605,000
24 bells	Each	242,000
Confessional, Prefabricated wood		
Single, plain	Each	3025
Deluxe	Each	7525
Double, plain	Each	5725
Deluxe	Each	16,900
Emergency Lighting, 25 watt, battery operated		
Lead battery	Each	227
Nickel cadmium	Each	660
Lecterns, Wood, plain	Each	610
Deluxe	Each	2975

Description	Unit	$ Cost
Pews/Benches, Hardwood	L.F.	84 - 157
Pulpits, Prefabricated, hardwood	Each	1375 - 8150
Railing, Hardwood	L.F.	165
Steeples, translucent fiberglass		
30" square, 15' high	Each	4875
25' high	Each	5625
Painted fiberglass, 24" square, 14' high	Each	4275
28' high	Each	4875
Aluminum		
20' high, 3'- 6" base	Each	4725
35' high, 8'- 0" base	Each	17,900
60' high, 14'- 0" base	Each	40,300

Model costs calculated for a 1 story building with 24' story height and 17,000 square feet of floor area

Church

BUILDING TYPES

				Unit	Unit Cost	Cost Per S.F.	% Of Sub-Total
A.	**SUBSTRUCTURE**						
1010	Standard Foundations	Poured concrete; strip and spread footings		S.F. Ground	2.68	2.68	
1030	Slab on Grade	4" reinforced concrete with vapor barrier and granular base		S.F. Slab	3.42	3.42	
2010	Basement Excavation	Site preparation for slab and trench for foundation wall and footing		S.F. Ground	1.09	1.09	11.4%
2020	Basement Walls	4' Foundation wall		L.F. Wall	52	2.52	
B.	**SHELL**						
	B10 Superstructure						
1010	Floor Construction	N/A		–	–	–	
1020	Roof Construction	Wood deck on laminated wood arches		S.F. Roof	13.80	14.72	17.2%
	B20 Exterior Enclosure						
2010	Exterior Walls	Face brick with concrete block backup	80% of wall (adjusted for end walls)	S.F. Wall	25	15.28	
2020	Exterior Windows	Aluminum, top hinged, in-swinging and curtain wall panels	20% of wall	Each	20	3.16	22.0%
2030	Exterior Doors	Double hollow metal swinging, single hollow metal		Each	1088	.38	
	B30 Roofing						
3010	Roof Coverings	Asphalt shingles with flashing; polystyrene insulation		S.F. Roof	3.35	3.35	3.9%
3020	Roof Openings	N/A		–	–	–	
C.	**INTERIORS**						
1010	Partitions	Plaster on metal studs	40 S.F. Floor/L.F. Partitions	S.F. Partition	10.55	6.33	
1020	Interior Doors	Hollow metal	400 S.F. Floor/Door	Each	537	1.34	
1030	Fittings	N/A		–	–	–	
2010	Stair Construction	N/A		–	–	–	16.5%
3010	Wall Finishes	Paint		S.F. Surface	1.63	.98	
3020	Floor Finishes	Carpet		S.F. Floor	5.49	5.49	
3030	Ceiling Finishes	N/A		–	–	–	
D.	**SERVICES**						
	D10 Conveying						
1010	Elevators & Lifts	N/A		–	–	–	0.0%
1020	Escalators & Moving Walks	N/A		–	–	–	
	D20 Plumbing						
2010	Plumbing Fixtures	Kitchen, toilet and service fixtures, supply and drainage	1 Fixture/2430 S.F. Floor	Each	2284	.94	
2020	Domestic Water Distribution	Gas fired hot water heater		S.F. Floor	.15	.15	1.3%
2040	Rain Water Drainage	N/A		–	–	–	
	D30 HVAC						
3010	Energy Supply	Oil fired hot water, wall fin radiation		S.F. Floor	6.10	6.10	
3020	Heat Generating Systems	N/A		–	–	–	
3030	Cooling Generating Systems	Split systems with air cooled condensing units		S.F. Floor	7.28	7.28	15.6%
3050	Terminal & Package Units	N/A		–	–	–	
3090	Other HVAC Sys. & Equipment	N/A		–	–	–	
	D40 Fire Protection						
4010	Sprinklers	Wet pipe sprinkler system		S.F. Floor	1.79	1.79	2.1%
4020	Standpipes	N/A		–	–	–	
	D50 Electrical						
5010	Electrical Service/Distribution	400 ampere service, panel board and feeders		S.F. Floor	.53	.53	
5020	Lighting & Branch Wiring	Fluorescent fixtures, receptacles, switches, A.C. and misc. power		S.F. Floor	5.90	5.90	10.0%
5030	Communications & Security	Alarm systems, sound system and emergency lighting		S.F. Floor	1.98	1.98	
5090	Other Electrical Systems	Emergency generator, 7.5KW		S.F. Floor	.09	.09	
E.	**EQUIPMENT & FURNISHINGS**						
1010	Commercial Equipment	N/A		–	–	–	
1020	Institutional Equipment	N/A		–	–	–	0.0%
1030	Vehicular Equipment	N/A		–	–	–	
1090	Other Equipment	N/A		–	–	–	
F.	**SPECIAL CONSTRUCTION & DEMOLITION**						
1020	Integrated Construction	N/A		–	–	–	0.0%
1040	Special Facilities	N/A		–	–	–	
G.	**BUILDING SITEWORK**	**N/A**					

		Sub-Total	85.50	100%
CONTRACTOR FEES (General Requirements: 10%, Overhead: 5%, Profit: 10%)		25%	21.38	
ARCHITECT FEES		11%	11.77	

	Total Building Cost	**118.65**

95

Costs per square foot of floor area

Exterior Wall	S.F. Area	2000	4000	6000	8000	12000	15000	18000	20000	22000
	L.F. Perimeter	180	280	340	386	460	535	560	600	640
Stone Ashlar Veneer On Concrete Block	Wood Truss	175.60	152.90	141.55	134.85	127.35	125.05	121.95	121.10	120.35
	Steel Joists	172.75	150.10	138.75	132.05	124.55	122.25	119.15	118.25	117.55
Stucco on Concrete Block	Wood Truss	161.50	142.00	132.70	127.35	121.35	119.50	117.10	116.40	115.80
	Steel Joists	158.95	139.40	130.10	124.75	118.75	116.95	114.55	113.80	113.25
Brick Veneer	Wood Frame	168.05	147.00	136.70	130.70	123.95	121.90	119.15	118.35	117.70
Wood Shingles	Wood Frame	160.00	140.75	131.60	126.35	120.50	118.70	116.35	115.65	115.10
Perimeter Adj., Add or Deduct	Per 100 L.F.	28.60	14.30	9.50	7.15	4.75	3.80	3.15	2.85	2.60
Story Hgt. Adj., Add or Deduct	Per 1 Ft.	3.10	2.45	1.95	1.65	1.35	1.25	1.05	1.05	1.00

For Basement, add $18.55 per square foot of basement area

The above costs were calculated using the basic specifications shown on the facing page. These costs should be adjusted where necessary for design alternatives and owner's requirements. Reported completed project costs, for this type of structure, range from $50.70 to $168.60 per S.F.

Common additives

Description	Unit	$ Cost		Description	Unit	$ Cost
Bar, Front bar	L.F.	287		Sauna, Prefabricated, complete, 6' x 4'	Each	4825
Back bar	L.F.	229		6' x 9'	Each	5550
Booth, Upholstered, custom, straight	L.F.	145 - 265		8' x 8'	Each	6525
"L" or "U" shaped	L.F.	150 - 252		10' x 12'	Each	9475
Fireplaces, Brick not incl. chimney				Smoke Detectors		
or foundation, 30" x 24" opening	Each	2075		Ceiling type	Each	151
Chimney, standard brick				Duct type	Each	405
Single flue 16" x 20"	V.L.F.	66		Sound System		
20" x 20"	V.L.F.	81		Amplifier, 250 watts	Each	1675
2 flue, 20" x 32"	V.L.F.	125		Speaker, ceiling or wall	Each	147
Lockers, Steel, single tier, 60" or 72"	Opening	131 - 227		Trumpet	Each	275
2 tier, 60" or 72" total	Opening	74 - 125		Steam Bath, Complete, to 140 C.F.	Each	1325
5 tier, box lockers	Opening	42 - 62		To 300 C.F.	Each	1500
Locker bench, lam. maple top only	L.F.	19.50		To 800 C.F.	Each	4825
Pedestals, steel pipe	Each	63		To 2500 C.F.	Each	5125
Refrigerators, Prefabricated, walk-in				Swimming Pool Complete, gunite	S.F.	48 - 59
7'-6" High, 6' x 6'	S.F.	121		Tennis Court, Complete with fence		
10' x 10'	S.F.	95		Bituminous	Each	23,100 - 28,700
12' x 14'	S.F.	84		Clay	Each	25,400 - 30,900
12' x 20'	S.F.	74				

Important: See the Reference Section for Location Factors

Model costs calculated for a 1 story building with 12' story height and 6,000 square feet of floor area

Club, Country

				Unit	Unit Cost	Cost Per S.F.	% Of Sub-Total
A. SUBSTRUCTURE							
1010	Standard Foundations	Poured concrete; strip and spread footings		S.F. Ground	1.89	1.89	
1030	Slab on Grade	4" reinforced concrete with vapor barrier and granular base		S.F. Slab	3.42	3.42	
2010	Basement Excavation	Site preparation for slab and trench for foundation wall and footing		S.F. Ground	1.18	1.18	10.2%
2020	Basement Walls	4' Foundation wall		L.F. Wall	52	3.89	
B. SHELL							
	B10 Superstructure						
1010	Floor Construction	N/A		–	–	–	
1020	Roof Construction	Wood truss with plywood sheathing		S.F. Roof	4.90	4.90	4.8%
	B20 Exterior Enclosure						
2010	Exterior Walls	Stone ashlar veneer on concrete block	65% of wall	S.F. Wall	24	10.70	
2020	Exterior Windows	Aluminum horizontal sliding	35% of wall	Each	314	4.98	17.1%
2030	Exterior Doors	Double aluminum and glass, hollow metal		Each	1729	1.73	
	B30 Roofing						
3010	Roof Coverings	Asphalt shingles		S.F. Roof	10.74	1.88	
3020	Roof Openings	N/A		–	–	–	1.8%
C. INTERIORS							
1010	Partitions	Gypsum board on metal studs, load bearing	14 S.F. Floor/L.F. Partition	S.F. Partition	5.25	3.75	
1020	Interior Doors	Single leaf wood	140 S.F. Floor/Door	Each	432	3.09	
1030	Fittings	N/A		–	–	–	
2010	Stair Construction	N/A		–	–	–	18.2%
3010	Wall Finishes	40% vinyl wall covering, 40% paint, 20% ceramic tile		S.F. Surface	3.84	2.74	
3020	Floor Finishes	50% carpet, 30% hardwood tile, 20% ceramic tile		S.F. Floor	6.02	6.02	
3030	Ceiling Finishes	Gypsum plaster on wood furring		S.F. Ceiling	2.95	2.95	
D. SERVICES							
	D10 Conveying						
1010	Elevators & Lifts	N/A		–	–	–	
1020	Escalators & Moving Walks	N/A		–	–	–	0.0%
	D20 Plumbing						
2010	Plumbing Fixtures	Kitchen, toilet and service fixtures, supply and drainage	1 Fixture/125 S.F. Floor	Each	1663	13.31	
2020	Domestic Water Distribution	Gas fired hot water		S.F. Floor	3.12	3.12	16.1%
2040	Rain Water Drainage	N/A		–	–	–	
	D30 HVAC						
3010	Energy Supply	N/A		–	–	–	
3020	Heat Generating Systems	Included in D3030		–	–	–	
3030	Cooling Generating Systems	Multizone rooftop unit, gas heating, electric cooling		S.F. Floor	23	23.70	23.2%
3050	Terminal & Package Units	N/A		–	–	–	
3090	Other HVAC Sys. & Equipment	N/A		–	–	–	
	D40 Fire Protection						
4010	Sprinklers	Wet pipe sprinkler system		S.F. Floor	2.34	2.34	
4020	Standpipes	N/A		–	–	–	2.3%
	D50 Electrical						
5010	Electrical Service/Distribution	400 ampere service, panel board and feeders		S.F. Floor	2.26	2.26	
5020	Lighting & Branch Wiring	Fluorescent fixtures, receptacles, switches, A.C. and misc. power		S.F. Floor	3.72	3.72	6.3%
5030	Communications & Security	Alarm systems and emergency lighting		S.F. Floor	.27	.27	
5090	Other Electrical Systems	Emergency generator, 11.5KW		S.F. Floor	.17	.17	
E. EQUIPMENT & FURNISHINGS							
1010	Commercial Equipment	N/A		–	–	–	
1020	Institutional Equipment	N/A		–	–	–	
1030	Vehicular Equipment	N/A		–	–	–	0.0%
1090	Other Equipment	N/A		–	–	–	
F. SPECIAL CONSTRUCTION & DEMOLITION							
1020	Integrated Construction	N/A		–	–	–	
1040	Special Facilities	N/A		–	–	–	0.0%
G. BUILDING SITEWORK	**N/A**						

				Sub-Total	102.01	100%
CONTRACTOR FEES (General Requirements: 10%, Overhead: 5%, Profit: 10%)				25%	25.50	
ARCHITECT FEES				11%	14.04	

Total Building Cost 141.55

BUILDING TYPES

97

Costs per square foot of floor area

Exterior Wall	S.F. Area	4000	8000	12000	17000	22000	27000	32000	37000	42000
	L.F. Perimeter	280	386	520	585	640	740	840	940	940
Stone Ashlar on Concrete Block	Steel Joists	132.80	113.40	108.50	101.55	97.40	95.95	94.90	94.25	92.00
	Wood Joists	130.50	111.70	107.00	100.20	96.20	94.80	93.80	93.15	91.00
Face Brick on Concrete Block	Steel Joists	130.55	111.85	107.10	100.40	96.45	95.05	94.10	93.40	91.25
	Wood Joists	128.70	110.45	105.85	99.35	95.55	94.15	93.20	92.55	90.50
Decorative Concrete Block	Steel Joists	125.05	108.05	103.70	97.70	94.15	92.90	92.00	91.40	89.50
	Wood Joists	122.55	106.05	101.80	96.05	92.60	91.40	90.50	89.95	88.15
Perimeter Adj., Add or Deduct	Per 100 L.F.	17.15	8.60	5.75	4.05	3.10	2.50	2.15	1.80	1.60
Story Hgt. Adj., Add or Deduct	Per 1 Ft.	2.80	1.95	1.75	1.40	1.15	1.10	1.05	1.00	.90
For Basement, add $18.90 per square foot of basement area										

The above costs were calculated using the basic specifications shown on the facing page. These costs should be adjusted where necessary for design alternatives and owner's requirements. Reported completed project costs, for this type of structure, range from $51.80 to $148.90 per S.F.

Common additives

Description	Unit	$ Cost
Bar, Front bar	L.F.	287
Back bar	L.F.	229
Booth, Upholstered, custom, straight	L.F.	145 - 265
"L" or "U" shaped	L.F.	150 - 252
Emergency Lighting, 25 watt, battery operated		
Lead battery	Each	227
Nickel cadmium	Each	660
Flagpoles, Complete		
Aluminum, 20' high	Each	1100
40' High	Each	2700
70' High	Each	8350
Fiberglass, 23' High	Each	1425
39'-5" High	Each	3000
59' High	Each	7425
Kitchen Equipment		
Broiler	Each	4275
Coffee urn, twin 6 gallon	Each	7375
Cooler, 6 ft. long, reach-in	Each	3375
Dishwasher, 10-12 racks per hr.	Each	3225
Food warmer, counter 1.2 kw	Each	745

Description	Unit	$ Cost
Kitchen Equipment, cont.		
Freezer, 44 C.F., reach-in	Each	8775
Ice cube maker, 50 lb. per day	Each	1900
Lockers, Steel, single tier, 60" or 72"	Opening	131 - 227
2 tier, 60" or 72" total	Opening	74 - 125
5 tier, box lockers	Opening	42 - 62
Locker bench, lam. maple top only	L.F.	19.50
Pedestals, steel pipe	Each	63
Refrigerators, Prefabricated, walk-in		
7'-6" High, 6' x 6'	S.F.	121
10' x 10'	S.F.	95
12' x 14'	S.F.	84
12' x 20'	S.F.	74
Smoke Detectors		
Ceiling type	Each	151
Duct type	Each	405
Sound System		
Amplifier, 250 watts	Each	1675
Speaker, ceiling or wall	Each	147
Trumpet	Each	275

Important: See the Reference Section for Location Factors

Club, Social

Model costs calculated for a 1 story building with 12' story height and 22,000 square feet of floor area

		Unit	Unit Cost	Cost Per S.F.	% Of Sub-Total
A. SUBSTRUCTURE					
1010	Standard Foundations	Poured concrete; strip and spread footings	S.F. Ground	1.19	1.19	
1030	Slab on Grade	4" reinforced concrete with vapor barrier and granular base	S.F. Slab	3.42	3.42	
2010	Basement Excavation	Site preparation for slab and trench for foundation wall and footing	S.F. Ground	1.09	1.09	
2020	Basement Walls	4' foundation wall	L.F. Wall	52	2.46	11.2%
B. SHELL					
B10 Superstructure					
1010	Floor Construction	N/A				
1020	Roof Construction	Metal deck on open web steel joists	S.F. Roof	4.04	4.04	5.5%
B20 Exterior Enclosure					
2010	Exterior Walls	Stone, ashlar veneer on concrete block	65% of wall	S.F. Wall	24	5.49	
2020	Exterior Windows	Window wall	35% of wall	Each	34	4.18	
2030	Exterior Doors	Double aluminum and glass doors	Each	1729	.47	13.9%
B30 Roofing					
3010	Roof Coverings	Built-up tar and gravel with flashing; perlite/EPS composite insulation	S.F. Roof	3.82	3.82	
3020	Roof Openings	N/A				5.2%
C. INTERIORS					
1010	Partitions	Lightweight concrete block	14 S.F. Floor/L.F. Partition	S.F. Partition	7.55	5.39	
1020	Interior Doors	Single leaf wood	140 S.F. Floor/Door	Each	432	3.09	
1030	Fittings	N/A				
2010	Stair Construction	N/A				
3010	Wall Finishes	65% paint, 25% vinyl wall covering, 10% ceramic tile	S.F. Surface	2.58	1.84	
3020	Floor Finishes	60% carpet, 35% hardwood, 15% ceramic tile	S.F. Floor	5.71	5.71	
3030	Ceiling Finishes	Mineral fiber tile on concealed zee bars	S.F. Ceiling	3.71	3.71	27.1%
D. SERVICES					
D10 Conveying					
1010	Elevators & Lifts	N/A				
1020	Escalators & Moving Walks	N/A				0.0%
D20 Plumbing					
1010	Plumbing Fixtures	Kitchen, toilet and service fixtures, supply and drainage	1 Fixture/1050 S.F. Floor	Each	1900	1.81	
2020	Domestic Water Distribution	Gas fired hot water heater	S.F. Floor	.12	.12	
2040	Rain Water Drainage	Roof drains	S.F. Roof	.35	.35	3.1%
D30 HVAC					
3010	Energy Supply	N/A				
3020	Heat Generating Systems	Included in D3030				
3030	Cooling Generating Systems	Multizone rooftop unit, gas heating, electric cooling	S.F. Floor	18.05	18.05	
3050	Terminal & Package Units	N/A				
3090	Other HVAC Sys. & Equipment	N/A				24.8%
D40 Fire Protection					
4010	Sprinklers	Wet pipe sprinkler system	S.F. Floor	1.79	1.79	
4020	Standpipes	N/A				2.5%
D50 Electrical					
5010	Electrical Service/Distribution	400 ampere service, panel board and feeders	S.F. Floor	.76	.76	
5020	Lighting & Branch Wiring	Fluorescent fixtures, receptacles, switches, A.C. and misc. power	S.F. Floor	3.80	3.80	
5030	Communications & Security	Alarm systems and emergency lighting	S.F. Floor	.15	.15	
5090	Other Electrical Systems	Emergency generator, 11.5kW	S.F. Floor	.09	.09	6.7%
E. EQUIPMENT & FURNISHINGS					
1010	Commercial Equipment	N/A				
1020	Institutional Equipment	N/A				
1030	Vehicular Equipment	N/A				
1090	Other Equipment	N/A				0.0%
F. SPECIAL CONSTRUCTION & DEMOLITION					
1020	Integrated Construction	N/A				
1040	Special Facilities	N/A				0.0%
G. BUILDING SITEWORK	N/A				
	Sub-Total				72.82	100%
	CONTRACTOR FEES (General Requirements: 10%, Overhead: 5%, Profit: 10%)			25%	18.21	
	ARCHITECT FEES			7%	6.37	
	Total Building Cost				**97.40**	

BUILDING TYPES

Costs per square foot of floor area

Exterior Wall	S.F. Area	15000	20000	28000	38000	50000	65000	85000	100000	150000
	L.F. Perimeter	350	400	480	550	630	660	750	825	1035
Face Brick with Concrete Block Back-up	Steel Frame	140.75	133.05	126.40	121.25	117.75	114.10	111.85	110.85	108.60
	Bearing Walls	141.70	133.10	125.75	120.00	116.00	111.75	109.15	108.00	105.45
Decorative Concrete Block	Steel Frame	135.80	128.75	122.70	118.20	115.05	111.95	109.95	109.05	107.10
	Bearing Walls	136.70	128.85	122.10	116.90	**113.25**	109.60	107.25	106.20	103.95
Stucco on Concrete Block	Steel Frame	133.60	126.65	120.65	116.15	113.05	110.05	108.05	107.20	105.30
	Bearing Walls	136.20	128.40	121.70	116.55	113.00	109.35	107.05	106.05	103.80
Perimeter Adj., Add or Deduct	Per 100 L.F.	6.40	4.80	3.45	2.55	1.90	1.50	1.10	.95	.70
Story Hgt. Adj., Add or Deduct	Per 1 Ft.	1.60	1.35	1.15	.95	.85	.70	.60	.55	.50

For Basement, add $22.00 per square foot of basement area

The above costs were calculated using the basic specifications shown on the facing page. These costs should be adjusted where necessary for design alternatives and owner's requirements. Reported completed project costs, for this type of structure, range from $74.95 to $176.20 per S.F.

Common additives

Description	Unit	$ Cost
Carrels Hardwood	Each	750 - 980
Clock System		
20 Room	Each	13,000
50 Room	Each	31,500
Elevators, Hydraulic passenger, 2 stops		
1500# capacity	Each	43,425
2500# capacity	Each	44,625
3500# capacity	Each	48,525
Additional stop, add	Each	3800
Emergency Lighting, 25 watt, battery operated		
Lead battery	Each	227
Nickel cadmium	Each	660
Flagpoles, Complete		
Aluminum, 20' high	Each	1100
40' High	Each	2700
70' High	Each	8350
Fiberglass, 23' High		1425
39'-5" High	Each	3000
59' High	Each	7425

Description	Unit	$ Cost
Lockers, Steel, single tier, 60" or 72"	Opening	131 - 227
2 tier, 60" or 72" total	Opening	74 - 125
5 tier, box lockers	Opening	42 - 62
Locker bench, lam. maple top only	L.F.	19.50
Pedestals, steel pipe	Each	63
Seating		
Auditorium chair, all veneer	Each	171
Veneer back, padded seat	Each	207
Upholstered, spring seat	Each	208
Classroom, movable chair & desk	Set	65 - 120
Lecture hall, pedestal type	Each	156 - 485
Smoke Detectors		
Ceiling type	Each	151
Duct type	Each	405
Sound System		
Amplifier, 250 watts	Each	1675
Speaker, ceiling or wall	Each	147
Trumpet	Each	275
TV Antenna, Master system, 12 outlet	Outlet	238
30 outlet	Outlet	153
100 outlet	Outlet	145

Important: See the Reference Section for Location Factors

BUILDING TYPES

Model costs calculated for a 2 story building with 12' story height and 50,000 square feet of floor area

College, Classroom, 2-3 Story

				Unit	Unit Cost	Cost Per S.F.	% Of Sub-Total
A. SUBSTRUCTURE							
1010	Standard Foundations	Poured concrete; strip and spread footings		S.F. Ground	.86	.43	
1030	Slab on Grade	4" reinforced concrete with vapor barrier and granular base		S.F. Slab	3.42	1.71	4.3%
2010	Basement Excavation	Site preparation for slab and trench for foundation wall and footing		S.F. Ground	1.05	.52	
2020	Basement Walls	4' Foundation wall		L.F. Wall	49	1.02	
B. SHELL							
	B10 Superstructure						
1010	Floor Construction	Open web steel joists, slab form, concrete		S.F. Floor	10.34	5.17	9.4%
1020	Roof Construction	Metal deck on open web steel joists, columns		S.F. Roof	5.64	2.82	
	B20 Exterior Enclosure						
2010	Exterior Walls	Decorative concrete block	65% of wall	S.F. Wall	10.79	2.12	
2020	Exterior Windows	Window wall	35% of wall	Each	29	3.11	6.6%
2030	Exterior Doors	Double glass and aluminum with transom		Each	3350	.40	
	B30 Roofing						
3010	Roof Coverings	Built-up tar and gravel with flashing; perlite/EPS composite insulation		S.F. Roof	3.72	1.86	2.2%
3020	Roof Openings	N/A		—	—	—	
C. INTERIORS							
1010	Partitions	Concrete block	20 S.F. Floor/L.F. Partition	S.F. Partition	9.86	4.93	
1020	Interior Doors	Single leaf hollow metal	200 S.F. Floor/Door	Each	537	2.69	
1030	Fittings	Chalkboards, counters, cabinets		S.F. Floor	3.53	3.53	
2010	Stair Construction	Concrete filled metal pan		Flight	6050	1.21	26.4%
3010	Wall Finishes	95% paint, 5% ceramic tile		S.F. Surface	5.54	2.77	
3020	Floor Finishes	70% vinyl composition tile, 25% carpet, 5% ceramic tile		S.F. Floor	3.48	3.48	
3030	Ceiling Finishes	Mineral fiber tile on concealed zee bars		S.F. Ceiling	3.71	3.71	
D. SERVICES							
	D10 Conveying						
1010	Elevators & Lifts	Two hydraulic passenger elevators		Each	52,250	2.09	2.5%
1020	Escalators & Moving Walks	N/A		—	—	—	
	D20 Plumbing						
2010	Plumbing Fixtures	Toilet and service fixtures, supply and drainage	1 Fixture/455 S.F. Floor	Each	4317	9.49	
2020	Domestic Water Distribution	Oil fired hot water heater		S.F. Floor	.56	.56	12.5%
2040	Rain Water Drainage	Roof drains		S.F. Roof	1.00	.50	
	D30 HVAC						
3010	Energy Supply	N/A		—	—	—	
3020	Heat Generating Systems	Included in D3030		—	—	—	
3030	Cooling Generating Systems	Multizone unit, gas heating, electric cooling		S.F. Floor	13.85	13.85	16.4%
3050	Terminal & Package Units	N/A		—	—	—	
3090	Other HVAC Sys. & Equipment	N/A		—	—	—	
	D40 Fire Protection						
4010	Sprinklers	Sprinklers, light hazard		S.F. Floor	1.33	1.33	1.6%
4020	Standpipes	N/A		—	—	—	
	D50 Electrical						
5010	Electrical Service/Distribution	2000 ampere service, panel board and feeders		S.F. Floor	2.91	2.91	
5020	Lighting & Branch Wiring	Fluorescent fixtures, receptacles, switches, A.C. and misc. power		S.F. Floor	8.52	8.52	18.1%
5030	Communications & Security	Alarm systems, communications systems and emergency lighting		S.F. Floor	3.41	3.41	
5090	Other Electrical Systems	Emergency generator, 100KW		S.F. Floor	.54	.54	
E. EQUIPMENT & FURNISHINGS							
1010	Commercial Equipment	N/A		—	—	—	
1020	Institutional Equipment	N/A		—	—	—	0.0%
1030	Vehicular Equipment	N/A		—	—	—	
1090	Other Equipment	N/A		—	—	—	
F. SPECIAL CONSTRUCTION & DEMOLITION							
1020	Integrated Construction	N/A		—	—	—	0.0%
1040	Special Facilities	N/A		—	—	—	
G. BUILDING SITEWORK	**N/A**						

		Sub-Total	84.68	100%
CONTRACTOR FEES (General Requirements: 10%, Overhead: 5%, Profit: 10%)		25%	21.17	
ARCHITECT FEES		7%	7.40	
	Total Building Cost		**113.25**	

BUILDING TYPES

Costs per square foot of floor area

Exterior Wall	S.F. Area	10000	15000	25000	40000	55000	70000	80000	90000	100000
	L.F. Perimeter	260	320	400	476	575	628	684	721	772
Face Brick with Concrete Block Back-up	R/Conc. Frame	140.85	130.30	119.85	112.55	109.80	107.35	106.60	105.75	105.20
	Steel Frame	145.20	134.65	124.15	116.90	114.15	111.70	110.95	110.10	109.55
Decorative Concrete Block	R/Conc. Frame	131.10	122.40	114.05	108.40	106.25	104.40	103.80	103.15	102.70
	Steel Frame	134.90	126.20	117.85	112.20	110.00	108.15	107.55	106.95	106.50
Precast Concrete Panels	R/Conc. Frame	135.15	125.65	116.40	110.05	107.60	105.50	104.80	104.10	103.60
	Steel Frame	139.40	129.85	120.60	114.25	111.80	109.70	109.05	108.30	107.85
Perimeter Adj., Add or Deduct	Per 100 L.F.	12.50	8.35	5.00	3.15	2.25	1.80	1.55	1.35	1.25
Story Hgt. Adj., Add or Deduct	Per 1 Ft.	2.40	1.95	1.45	1.10	.95	.85	.80	.70	.70

For Basement, add $21.40 per square foot of basement area

The above costs were calculated using the basic specifications shown on the facing page. These costs should be adjusted where necessary for design alternatives and owner's requirements. Reported completed project costs, for this type of structure, range from $48.45 to $142.65 per S.F.

Common additives

Description	Unit	$ Cost
Carrels Hardwood	Each	750 - 980
Closed Circuit Surveillance, One station		
Camera and monitor	Each	1375
For additional camera stations, add	Each	750
Elevators, Hydraulic passenger, 2 stops		
2000# capacity	Each	43,925
2500# capacity	Each	44,625
3500# capacity	Each	48,525
Additional stop, add	Each	3800
Emergency Lighting, 25 watt, battery operated		
Lead battery	Each	227
Nickel cadmium	Each	660
Furniture	Student	1900 - 3600
Intercom System, 25 station capacity		
Master station	Each	2025
Intercom outlets	Each	128
Handset	Each	335

Description	Unit	$ Cost
Kitchen Equipment		
Broiler	Each	4275
Coffee urn, twin 6 gallon	Each	7375
Cooler, 6 ft. long	Each	3375
Dishwasher, 10-12 racks per hr.	Each	3225
Food warmer	Each	745
Freezer, 44 C.F., reach-in	Each	8775
Ice cube maker, 50 lb. per day	Each	1900
Range with 1 oven	Each	2500
Laundry Equipment		
Dryer, gas, 16 lb. capacity	Each	725
30 lb. capacity	Each	2800
Washer, 4 cycle	Each	810
Commercial	Each	1225
Smoke Detectors		
Ceiling type	Each	151
Duct type	Each	405
TV Antenna, Master system, 12 outlet	Outlet	238
30 outlet	Outlet	153
100 outlet	Outlet	145

Important: See the Reference Section for Location Factors

Model costs calculated for a 3 story building with 12' story height and 25,000 square feet of floor area

College, Dormitory, 2-3 Story

				Unit	Unit Cost	Cost Per S.F.	% Of Sub-Total
A. SUBSTRUCTURE							
1010	Standard Foundations	Poured concrete; strip and spread footings		S.F. Ground	4.20	1.40	
1030	Slab on Grade	4" reinforced concrete with vapor barrier and granular base		S.F. Slab	3.42	1.14	
2010	Basement Excavation	Site preparation for slab and trench for foundation wall and footing		S.F. Ground	1.09	.36	4.2%
2020	Basement Walls	4' Foundation wall		L.F. Wall	52	.85	
B. SHELL							
	B10 Superstructure						
1010	Floor Construction	Concrete flat plate		S.F. Floor	16.47	10.98	
1020	Roof Construction	Concrete flat plate		S.F. Roof	9.30	3.10	15.7%
	B20 Exterior Enclosure						
2010	Exterior Walls	Face brick with concrete block backup	80% of wall	S.F. Wall	21	9.72	
2020	Exterior Windows	Aluminum horizontal sliding	20% of wall	Each	509	2.55	14.7%
2030	Exterior Doors	Double glass & aluminum doors		Each	3900	.94	
	B30 Roofing						
3010	Roof Coverings	Single ply membrane, loose laid and ballasted; perlite/EPS composite insulation		S.F. Roof	3.81	1.27	1.4%
3020	Roof Openings	N/A		—	—	—	
C. INTERIORS							
1010	Partitions	Gypsum board on metal studs, concrete block	9 S.F. Floor/L.F. Partition	S.F. Partition	4.10	4.56	
1020	Interior Doors	Single leaf wood	90 S.F. Floor/Door	Each	432	4.80	
1030	Fittings	Closet shelving, mirrors, bathroom accessories		S.F. Floor	1.66	1.66	
2010	Stair Construction	Cast in place concrete		Flight	3400	1.77	24.9%
3010	Wall Finishes	95% paint, 5% ceramic tile		S.F. Surface	2.63	2.92	
3020	Floor Finishes	80% carpet, 10% vinyl composition tile, 10% ceramic tile		S.F. Floor	5.91	5.91	
3030	Ceiling Finishes	90% paint, 10% suspended fiberglass board		S.F. Ceiling	.71	.71	
D. SERVICES							
	D10 Conveying						
1010	Elevators & Lifts	One hydraulic passenger elevator		Each	68,750	2.75	3.1%
1020	Escalators & Moving Walks	N/A		—	—	—	
	D20 Plumbing						
2010	Plumbing Fixtures	Toilet and service fixtures, supply and drainage	1 Fixture/455 S.F. Floor	Each	2852	6.27	
2020	Domestic Water Distribution	Electric water heater		S.F. Floor	1.75	1.75	9.2%
2040	Rain Water Drainage	Roof drains		S.F. Roof	.69	.23	
	D30 HVAC						
3010	Energy Supply	N/A		—	—	—	
3020	Heat Generating Systems	Included in D3030		—	—	—	
3030	Cooling Generating Systems	Rooftop multizone unit system		S.F. Floor	8.70	8.70	9.7%
3050	Terminal & Package Units	N/A		—	—	—	
3090	Other HVAC Sys. & Equipment	N/A		—	—	—	
	D40 Fire Protection						
4010	Sprinklers	Wet pipe sprinkler system		S.F. Floor	1.56	1.56	1.7%
4020	Standpipes	N/A		—	—	—	
	D50 Electrical						
5010	Electrical Service/Distribution	800 ampere service, panel board and feeders		S.F. Floor	1.56	1.56	
5020	Lighting & Branch Wiring	Fluorescent fixtures, receptacles, switches, A.C. and misc. power		S.F. Floor	6.53	6.53	12.0%
5030	Communications & Security	Alarm systems, communications systems and emergency lighting		S.F. Floor	2.47	2.47	
5090	Other Electrical Systems	Emergency generator, 7.5KW		S.F. Floor	.06	.06	
E. EQUIPMENT & FURNISHINGS							
1010	Commercial Equipment	N/A		—	—	—	
1020	Institutional Equipment	N/A		—	—	—	
1030	Vehicular Equipment	N/A		—	—	—	3.4%
1090	Other Equipment	Dormitory furniture		S.F. Floor	3.08	3.08	
F. SPECIAL CONSTRUCTION & DEMOLITION							
1020	Integrated Construction	N/A		—	—	—	
1040	Special Facilities	N/A		—	—	—	0.0%
G. BUILDING SITEWORK	**N/A**						

			Sub-Total	89.60	100%
CONTRACTOR FEES (General Requirements: 10%, Overhead: 5%, Profit: 10%)			25%	22.40	
ARCHITECT FEES			7%	7.85	

Total Building Cost	**119.85**

BUILDING TYPES

BUILDING TYPES

Costs per square foot of floor area

Exterior Wall	S.F. Area	20000	35000	45000	65000	85000	110000	135000	160000	200000
	L.F. Perimeter	260	340	400	440	500	540	560	590	640
Face Brick with Concrete Block Back-up	R/Conc. Frame	134.40	123.10	120.15	114.05	111.30	108.60	106.50	105.20	103.90
	Steel Frame	141.40	130.15	127.20	121.05	118.35	115.60	113.55	112.25	110.90
Decorative Concrete Block	R/Conc. Frame	125.15	116.20	113.80	109.20	107.10	105.00	103.50	102.50	101.55
	Steel Frame	132.25	123.30	120.95	116.30	114.20	112.15	110.65	109.65	108.65
Precast Concrete Panels With Exposed Aggregate	R/Conc. Frame	133.60	122.50	119.60	113.60	110.95	108.25	106.25	104.95	103.65
	Steel Frame	140.55	129.50	126.60	120.60	117.95	115.30	113.25	112.00	110.70
Perimeter Adj., Add or Deduct	Per 100 L.F.	11.90	6.80	5.25	3.65	2.75	2.15	1.75	1.50	1.15
Story Hgt. Adj., Add or Deduct	Per 1 Ft.	2.40	1.80	1.60	1.25	1.05	.90	.75	.70	.55

For Basement, add $21.40 per square foot of basement area

The above costs were calculated using the basic specifications shown on the facing page. These costs should be adjusted where necessary for design alternatives and owner's requirements. Reported completed project costs, for this type of structure, range from $68.80 to $166.00 per S.F.

Common additives

Description	Unit	$ Cost
Carrels Hardwood	Each	750 - 980
Closed Circuit Surveillance, One station		
Camera and monitor	Each	1375
For additional camera stations, add	Each	750
Elevators, Electric passenger, 5 stops		
2000# capacity	Each	101,400
2500# capacity	Each	103,900
3500# capacity	Each	105,400
Additional stop, add	Each	5675
Emergency Lighting, 25 watt, battery operated		
Lead battery	Each	227
Nickel cadmium	Each	660
Furniture	Student	1900 - 3600
Intercom System, 25 station capacity		
Master station	Each	2025
Intercom outlets	Each	128
Handset	Each	335

Description	Unit	$ Cost
Kitchen Equipment		
Broiler	Each	4275
Coffee urn, twin, 6 gallon	Each	7375
Cooler, 6 ft. long	Each	3375
Dishwasher, 10-12 racks per hr.	Each	3225
Food warmer	Each	745
Freezer, 44 C.F., reach-in	Each	8775
Ice cube maker, 50 lb. per day	Each	1900
Range with 1 oven	Each	2500
Laundry Equipment		
Dryer, gas, 16 lb. capacity	Each	725
30 lb. capacity	Each	2800
Washer, 4 cycle	Each	810
Commercial	Each	1225
Smoke Detectors		
Ceiling type	Each	151
Duct type	Each	405
TV Antenna, Master system, 12 outlet	Outlet	238
30 outlet	Outlet	153
100 outlet	Outlet	145

Important: See the Reference Section for Location Factors

Model costs calculated for a 6 story building with 12' story height and 85,000 square feet of floor area

				Unit	Unit Cost	Cost Per S.F.	% Of Sub-Total

A. SUBSTRUCTURE

				Unit	Unit Cost	Cost Per S.F.	% Of Sub-Total
1010	Standard Foundations	Poured concrete; strip and spread footings		S.F. Ground	6.24	1.04	
1030	Slab on Grade	4" reinforced concrete with vapor barrier and granular base		S.F. Slab	3.42	.57	2.6%
2010	Basement Excavation	Site preparation for slab and trench for foundation wall and footing		S.F. Ground	1.05	.17	
2020	Basement Walls	4' foundation wall		L.F. Wall	49	.43	

B. SHELL

B10 Superstructure

				Unit	Unit Cost	Cost Per S.F.	% Of Sub-Total
1010	Floor Construction	Concrete slab with metal deck and beams		S.F. Floor	16.03	13.36	18.0%
1020	Roof Construction	Concrete slab with metal deck and beams		S.F. Roof	12.24	2.04	

B20 Exterior Enclosure

				Unit	Unit Cost	Cost Per S.F.	% Of Sub-Total
2010	Exterior Walls	Decorative concrete block	80% of wall	S.F. Wall	11.98	4.06	
2020	Exterior Windows	Aluminum horizontal sliding	20% of wall	Each	314	1.77	7.0%
2030	Exterior Doors	Double glass & aluminum doors		Each	2025	.19	

B30 Roofing

				Unit	Unit Cost	Cost Per S.F.	% Of Sub-Total
3010	Roof Coverings	Single ply membrane, loose laid and ballasted; perlite/EPS composite insulation		S.F. Roof	3.48	.58	0.7%
3020	Roof Openings	N/A		—	—	—	

C. INTERIORS

				Unit	Unit Cost	Cost Per S.F.	% Of Sub-Total
1010	Partitions	Concrete block	9 S.F. Floor/L.F. Partition	S.F. Partition	6.69	7.43	
1020	Interior Doors	Single leaf wood	90 S.F. Floor/Door	Each	432	4.80	
1030	Fittings	Closet shelving, mirrors, bathroom accessories		S.F. Floor	1.36	1.36	
2010	Stair Construction	Concrete filled metal pan		Flight	6050	1.28	29.8%
3010	Wall Finishes	95% paint, 5% ceramic tile		S.F. Surface	3.55	3.94	
3020	Floor Finishes	80% carpet, 10% vinyl composition tile, 10% ceramic tile		S.F. Floor	5.91	5.91	
3030	Ceiling Finishes	Mineral fiber tile on concealed zee bars, paint		S.F. Ceiling	.71	.71	

D. SERVICES

D10 Conveying

				Unit	Unit Cost	Cost Per S.F.	% Of Sub-Total
1010	Elevators & Lifts	Four geared passenger elevators		Each	137,487	6.47	7.6%
1020	Escalators & Moving Walks	N/A		—	—	—	

D20 Plumbing

				Unit	Unit Cost	Cost Per S.F.	% Of Sub-Total
2010	Plumbing Fixtures	Toilet and service fixtures, supply and drainage	1 Fixture/390 S.F. Floor	Each	2340	6.00	
2020	Domestic Water Distribution	Oil fired water heater		S.F. Floor	.43	.43	7.8%
2040	Rain Water Drainage	Roof drains		S.F. Roof	1.20	.20	

D30 HVAC

				Unit	Unit Cost	Cost Per S.F.	% Of Sub-Total
3010	Energy Supply	Oil fired hot water, wall fin radiation		S.F. Floor	2.78	2.78	
3020	Heat Generating Systems	N/A		—	—	—	
3030	Cooling Generating Systems	Chilled water, air cooled condenser system		S.F. Floor	6.12	6.12	10.4%
3050	Terminal & Package Units	N/A		—	—	—	
3090	Other HVAC Sys. & Equipment	N/A		—	—	—	

D40 Fire Protection

				Unit	Unit Cost	Cost Per S.F.	% Of Sub-Total
4010	Sprinklers	Sprinklers, light hazard		S.F. Floor	1.50	1.50	1.8%
4020	Standpipes	N/A		—	—	—	

D50 Electrical

				Unit	Unit Cost	Cost Per S.F.	% Of Sub-Total
5010	Electrical Service/Distribution	1200 ampere service, panel board and feeders		S.F. Floor	.72	.72	
5020	Lighting & Branch Wiring	Fluorescent fixtures, receptacles, switches, A.C. and misc. power		S.F. Floor	6.42	6.42	11.3%
5030	Communications & Security	Alarm systems, communications systems and emergency lighting		S.F. Floor	2.45	2.45	
5090	Other Electrical Systems	Emergency generator, 30 kW		S.F. Floor	.15	.15	

E. EQUIPMENT & FURNISHINGS

				Unit	Unit Cost	Cost Per S.F.	% Of Sub-Total
1010	Commercial Equipment	N/A		—	—	—	
1020	Institutional Equipment	N/A		—	—	—	3.0%
1030	Vehicular Equipment	N/A		—	—	—	
1090	Other Equipment	Built-in dormitory furnishings		S.F. Floor	2.52	2.52	

F. SPECIAL CONSTRUCTION & DEMOLITION

				Unit	Unit Cost	Cost Per S.F.	% Of Sub-Total
1020	Integrated Construction	N/A		—	—	—	0.0%
1040	Special Facilities	N/A		—	—	—	

G. BUILDING SITEWORK   N/A

	Sub-Total	85.40	**100%**

CONTRACTOR FEES (General Requirements: 10%, Overhead: 5%, Profit: 10%)	25%	21.35
ARCHITECT FEES	7%	7.45

Total Building Cost	**114.20**

BUILDING TYPES

Costs per square foot of floor area

Exterior Wall	S.F. Area	12000	20000	28000	37000	45000	57000	68000	80000	92000
	L.F. Perimeter	470	600	698	793	900	1060	1127	1200	1320
Face Brick with Concrete Brick Back-up	Steel Frame	185.75	152.35	137.00	127.35	122.45	117.65	113.70	110.70	108.80
	Bearing Walls	183.15	149.75	134.40	124.75	119.85	115.05	111.10	108.10	106.25
Decorative Concrete Block	Steel Frame	181.30	148.95	134.20	124.90	120.15	115.50	111.80	109.00	107.20
	Bearing Walls	178.85	146.50	131.75	122.50	117.75	113.10	109.40	106.55	104.80
Stucco on Concrete Block	Steel Frame	180.25	148.15	133.55	124.35	119.65	115.05	111.40	108.60	106.85
	Bearing Walls	177.85	145.75	131.10	121.95	117.20	112.65	108.95	106.20	104.40
Perimeter Adj., Add or Deduct	Per 100 L.F.	7.25	4.35	3.15	2.30	1.90	1.55	1.30	1.05	.95
Story Hgt. Adj., Add or Deduct	Per 1 Ft.	1.35	1.00	.85	.70	.65	.65	.60	.50	.50
For Basement, add $20.45 per square foot of basement area										

The above costs were calculated using the basic specifications shown on the facing page. These costs should be adjusted where necessary for design alternatives and owner's requirements. Reported completed project costs, for this type of structure, range from $99.15 to $185.20 per S.F.

Common additives

Description	Unit	$ Cost
Cabinets, Base, door units, metal	L.F.	189
Drawer units	L.F.	360
Tall storage cabinets, open	L.F.	360
With doors	L.F.	445
Wall, metal 12-1/2" deep, open	L.F.	140
With doors	L.F.	250
Carrels Hardwood	Each	750 - 980
Countertops, not incl. base cabinets, acid proof	S.F.	32 - 44
Stainless steel	S.F.	85
Fume Hood, Not incl. ductwork	L.F.	695 - 1825
Ductwork	Hood	1775 - 6375
Glassware Washer, Distilled water rinse	Each	7050 - 9300
Seating		
Auditorium chair, all veneer	Each	171
Veneer back, padded seat	Each	207
Upholstered, spring seat	Each	208
Classroom, movable chair & desk	Set	65 - 120
Lecture hall, pedestal type	Each	156 - 485

Description	Unit	$ Cost
Safety Equipment, Eye wash, hand held	Each	315
Deluge shower	Each	605
Sink, One piece plastic		
Flask wash, freestanding	Each	1725
Tables, acid resist. top, drawers	L.F.	145
Titration Unit, Four 2000 ml reservoirs	Each	9200

Important: See the Reference Section for Location Factors

Model costs calculated for a 1 story building with 12' story height and 45,000 square feet of floor area

College, Laboratory

				Unit	Unit Cost	Cost Per S.F.	% Of Sub-Total
A. SUBSTRUCTURE							
1010	Standard Foundations	Poured concrete; strip and spread footings		S.F. Ground	2.05	2.05	
1030	Slab on Grade	4" reinforced concrete with vapor barrier and granular base		S.F. Slab	3.42	3.42	
2010	Basement Excavation	Site preparation for slab and trench for foundation wall and footing		S.F. Ground	1.09	1.09	12.2%
2020	Basement Walls	4' foundation wall		L.F. Wall	49	4.11	
B. SHELL							
	B10 Superstructure						
1010	Floor Construction	Metal deck on open web steel joists	(5680 S.F.)	S.F. Floor	10.31	1.34	4.8%
1020	Roof Construction	Metal deck on open web steel joists		S.F. Roof	2.85	2.85	
	B20 Exterior Enclosure						
2010	Exterior Walls	Face brick with concrete block backup	75% of wall	S.F. Wall	21	3.80	
2020	Exterior Windows	Window wall	25% of wall	Each	29	1.77	7.8%
2030	Exterior Doors	Glass and metal doors and entrances with transom		Each	2802	1.25	
	B30 Roofing						
3010	Roof Coverings	Built-up tar and gravel with flashing; perlite/EPS composite insulation		S.F. Roof	3.57	3.57	4.3%
3020	Roof Openings	Skylight		S.F. Roof	.21	.21	
C. INTERIORS							
1010	Partitions	Concrete block	10 S.F. Floor/L.F. Partition	S.F. Partition	7.76	7.76	
1020	Interior Doors	Single leaf-kalamein fire doors	820 S.F. Floor/Door	Each	666	.81	
1030	Fittings	Lockers		S.F. Floor	.03	.03	
2010	Stair Construction	N/A		—	—	—	24.3%
3010	Wall Finishes	60% paint, 40% epoxy coating		S.F. Surface	4.27	4.27	
3020	Floor Finishes	60% epoxy, 20% carpet, 20% vinyl composition tile		S.F. Floor	4.59	4.59	
3030	Ceiling Finishes	Mineral fiber tile on concealed zee runners		S.F. Ceiling	3.71	3.71	
D. SERVICES							
	D10 Conveying						
1010	Elevators & Lifts	N/A		—	—	—	0.0%
1020	Escalators & Moving Walks	N/A		—	—	—	
	D20 Plumbing						
2010	Plumbing Fixtures	Toilet and service fixtures, supply and drainage	1 Fixture/260 S.F. Floor	Each	3595	13.83	
2020	Domestic Water Distribution	Gas fired water heater		S.F. Floor	.67	.67	17.3%
2040	Rain Water Drainage	Roof drains		S.F. Roof	.59	.59	
	D30 HVAC						
3010	Energy Supply	N/A		—	—	—	
3020	Heat Generating Systems	Included in D3030		—	—	—	
3030	Cooling Generating Systems	Multizone unit, gas heating, electric cooling		S.F. Floor	13.85	13.85	15.9%
3050	Terminal & Package Units	N/A		—	—	—	
3090	Other HVAC Sys. & Equipment	N/A		—	—	—	
	D40 Fire Protection						
4010	Sprinklers	Sprinklers, light hazard		S.F. Floor	1.50	1.50	1.7%
4020	Standpipes	N/A		—	—	—	
	D50 Electrical						
5010	Electrical Service/Distribution	1000 ampere service, panel board and feeders		S.F. Floor	1.28	1.28	
5020	Lighting & Branch Wiring	Fluorescent fixtures, receptacles, switches, A.C. and misc. power		S.F. Floor	7.24	7.24	10.6%
5030	Communications & Security	Alarm systems and emergency lighting		S.F. Floor	.50	.50	
5090	Other Electrical Systems	Emergency generator, 11.5 kW		S.F. Floor	.08	.08	
E. EQUIPMENT & FURNISHINGS							
1010	Commercial Equipment	N/A		—	—	—	
1020	Institutional Equipment	Cabinets, fume hoods, lockers, glassware washer		S.F. Floor	.99	.99	1.1%
1030	Vehicular Equipment	N/A		—	—	—	
1090	Other Equipment	N/A		—	—	—	
F. SPECIAL CONSTRUCTION & DEMOLITION							
1020	Integrated Construction	N/A		—	—	—	0.0%
1040	Special Facilities	N/A		—	—	—	
G. BUILDING SITEWORK	**N/A**						

		Sub-Total	87.16	100%
	CONTRACTOR FEES (General Requirements: 10%, Overhead: 5%, Profit: 10%)		25%	21.79
	ARCHITECT FEES		10%	10.90
		Total Building Cost	**119.85**	

BUILDING TYPES

Costs per square foot of floor area

Exterior Wall	S.F. Area	15000	20000	25000	30000	35000	40000	45000	50000	55000
	L.F. Perimeter	354	425	457	513	568	583	629	644	683
Brick Face with Concrete Block Back-up	Steel Frame	117.05	113.50	109.95	108.25	107.10	105.25	104.45	103.20	102.65
	R/Conc. Frame	113.85	110.25	106.75	105.05	103.90	102.05	101.25	100.00	99.45
Precast Concrete Panel	Steel Frame	114.30	111.05	107.90	106.35	105.35	103.65	102.90	101.80	101.30
	R/Conc. Frame	111.45	108.10	104.85	103.30	102.25	100.55	99.80	98.70	98.15
Limestone Face Concrete Block Back-up	Steel Frame	122.10	118.05	113.85	111.95	110.60	108.40	107.45	106.00	105.30
	R/Conc. Frame	118.90	114.85	110.65	108.70	107.40	105.15	104.25	102.75	102.10
Perimeter Adj., Add or Deduct	Per 100 L.F.	6.40	4.80	3.85	3.20	2.70	2.35	2.10	1.95	1.70
Story Hgt. Adj., Add or Deduct	Per 1 Ft.	1.60	1.45	1.25	1.20	1.10	1.00	.95	.90	.85

For Basement, add $23.00 per square foot of basement area

The above costs were calculated using the basic specifications shown on the facing page. These costs should be adjusted where necessary for design alternatives and owner's requirements. Reported completed project costs, for this type of structure, range from $81.20 to $165.70 per S.F.

Common additives

Description	Unit	$ Cost
Carrels Hardwood	Each	750 - 980
Elevators, Hydraulic passenger, 2 stops		
2000# capacity	Each	43,925
2500# capacity	Each	44,625
3500# capacity	Each	48,525
Emergency Lighting, 25 watt, battery operated		
Lead battery	Each	227
Nickel cadmium	Each	660
Escalators, Metal		
32" wide, 10' story height	Each	91,000
20' story height	Each	105,500
48" wide, 10' Story height	Each	95,500
20' story height	Each	110,000
Glass		
32" wide, 10' story height	Each	89,500
20' story height	Each	105,500
48" wide, 10' story height	Each	95,500
20' story height	Each	110,000

Description	Unit	$ Cost
Lockers, Steel, Single tier, 60" or 72"	Opening	131 - 227
2 tier, 60" or 72" total	Opening	74 - 125
5 tier, box lockers	Opening	42 - 62
Locker bench, lam. maple top only	L.F.	19.50
Pedestals, steel pipe	Each	63
Sound System		
Amplifier, 250 watts	Each	1675
Speaker, ceiling or wall	Each	147
Trumpet	Each	275

BUILDING TYPES

Model costs calculated for a 2 story building with 12' story height and 25,000 square feet of floor area

College, Student Union

				Unit	Unit Cost	Cost Per S.F.	% Of Sub-Total

A. SUBSTRUCTURE

				Unit	Unit Cost	Cost Per S.F.	% Of Sub-Total
1010	Standard Foundations	Poured concrete; strip and spread footings		S.F. Ground	3.12	1.56	
1030	Slab on Grade	4" reinforced concrete with vapor barrier and granular base		S.F. Slab	3.42	1.71	
2010	Basement Excavation	Site preparation for slab and trench for foundation wall and footing		S.F. Ground	1.09	.55	6.0%
2020	Basement Walls	4' foundation wall		L.F. Wall	52	.97	

B. SHELL

B10 Superstructure

				Unit	Unit Cost	Cost Per S.F.	% Of Sub-Total
1010	Floor Construction	Concrete flat plate		S.F. Floor	18.22	9.11	17.9%
1020	Roof Construction	Concrete flat plate		S.F. Roof	10.38	5.19	

B20 Exterior Enclosure

				Unit	Unit Cost	Cost Per S.F.	% Of Sub-Total
2010	Exterior Walls	Face brick with concrete block backup	75% of wall	S.F. Wall	21	6.94	
2020	Exterior Windows	Window wall	25% of wall	Each	29	3.22	13.1%
2030	Exterior Doors	Double aluminum and glass		Each	1730	.28	

B30 Roofing

				Unit	Unit Cost	Cost Per S.F.	% Of Sub-Total
3010	Roof Coverings	Built-up tar and gravel with flashing; perlite/EPS composite insulation		S.F. Roof	4.02	2.01	2.6%
3020	Roof Openings	Roof hatches		S.F. Roof	.08	.04	

C. INTERIORS

				Unit	Unit Cost	Cost Per S.F.	% Of Sub-Total
1010	Partitions	Gypsum board on metal studs	14 S.F. Floor/L.F. Partition	S.F. Partition	4.00	2.86	
1020	Interior Doors	Single leaf hollow metal	140 S.F. Floor/Door	Each	537	3.84	
1030	Fittings	N/A		—	—	—	
2010	Stair Construction	Cast in place concrete		Flight	5250	.84	22.1%
3010	Wall Finishes	50% paint, 50% vinyl wall covering		S.F. Surface	3.33	2.38	
3020	Floor Finishes	50% carpet, 50% vinyl composition tile		S.F. Floor	4.62	4.62	
3030	Ceiling Finishes	Suspended fiberglass board		S.F. Ceiling	3.06	3.06	

D. SERVICES

D10 Conveying

				Unit	Unit Cost	Cost Per S.F.	% Of Sub-Total
1010	Elevators & Lifts	One hydraulic passenger elevator		Each	52,250	2.09	2.6%
1020	Escalators & Moving Walks	N/A		—	—	—	

D20 Plumbing

				Unit	Unit Cost	Cost Per S.F.	% Of Sub-Total
2010	Plumbing Fixtures	Toilet and service fixtures, supply and drainage	1 Fixture/1040 S.F. Floor	Each	1445	1.39	
2020	Domestic Water Distribution	Gas fired water heater		S.F. Floor	.33	.33	2.4%
2040	Rain Water Drainage	Roof drains		S.F. Roof	.34	.17	

D30 HVAC

				Unit	Unit Cost	Cost Per S.F.	% Of Sub-Total
3010	Energy Supply	N/A		—	—	—	
3020	Heat Generating Systems	Included in D3030		—	—	—	
3030	Cooling Generating Systems	Multizone unit, gas heating, electric cooling		S.F. Floor	13.85	13.85	17.4%
3050	Terminal & Package Units	N/A		—	—	—	
3090	Other HVAC Sys. & Equipment	N/A		—	—	—	

D40 Fire Protection

				Unit	Unit Cost	Cost Per S.F.	% Of Sub-Total
4010	Sprinklers	Wet pipe sprinkler system		S.F. Floor	1.63	1.63	2.0%
4020	Standpipes	N/A		—	—	—	

D50 Electrical

				Unit	Unit Cost	Cost Per S.F.	% Of Sub-Total
5010	Electrical Service/Distribution	600 ampere service, panel board and feeders		S.F. Floor	1.41	1.41	
5020	Lighting & Branch Wiring	Fluorescent fixtures, receptacles, switches, A.C. and misc. power		S.F. Floor	8.45	8.45	13.9%
5030	Communications & Security	Alarm systems, communications systems and emergency lighting		S.F. Floor	1.18	1.18	
5090	Other Electrical Systems	Emergency generator, 11.5 kW		S.F. Floor	.12	.12	

E. EQUIPMENT & FURNISHINGS

				Unit	Unit Cost	Cost Per S.F.	% Of Sub-Total
1010	Commercial Equipment	N/A		—	—	—	
1020	Institutional Equipment	N/A		—	—	—	
1030	Vehicular Equipment	N/A		—	—	—	0.0%
1090	Other Equipment	N/A		—	—	—	

F. SPECIAL CONSTRUCTION & DEMOLITION

				Unit	Unit Cost	Cost Per S.F.	% Of Sub-Total
1020	Integrated Construction	N/A		—	—	—	0.0%
1040	Special Facilities	N/A		—	—	—	

G. BUILDING SITEWORK   N/A

Sub-Total	79.80	**100%**

CONTRACTOR FEES (General Requirements: 10%, Overhead: 5%, Profit: 10%)	25%	19.95	
ARCHITECT FEES	7%	7.00	

Total Building Cost	**106.75**

BUILDING TYPES

Costs per square foot of floor area

Exterior Wall		S.F. Area	4000	6000	8000	10000	12000	14000	16000	18000	20000
		L.F. Perimeter	260	340	420	453	460	510	560	610	600
Face Brick with Concrete Block Back-up	Bearing Walls		112.90	107.10	104.15	99.55	95.20	93.95	93.00	92.25	89.85
	Steel Frame		109.45	104.35	101.75	97.85	94.10	93.00	92.20	91.55	89.50
Decorative Concrete Block	Bearing Walls		99.70	95.10	92.75	89.15	85.80	84.75	84.00	83.45	81.55
	Steel Frame		100.70	96.70	94.70	91.75	89.00	88.10	87.50	87.00	85.45
Tilt Up Concrete Wall Panels	Bearing Walls		101.25	96.90	94.70	91.45	88.35	87.40	86.70	86.20	84.45
	Steel Frame		97.80	94.15	92.35	89.70	87.25	86.50	85.90	85.45	84.10
Perimeter Adj., Add or Deduct	Per 100 L.F.		15.05	10.05	7.50	6.00	5.00	4.30	3.75	3.35	3.00
Story Hgt. Adj., Add or Deduct	Per 1 Ft.		1.90	1.65	1.55	1.30	1.15	1.05	1.05	1.00	.90

For Basement, add $ 20.10 per square foot of basement area

The above costs were calculated using the basic specifications shown on the facing page. These costs should be adjusted where necessary for design alternatives and owner's requirements. Reported completed project costs, for this type of structure, range from $ 49.60 to $ 156.50 per S.F.

Common additives

Description	Unit	$ Cost	Description	Unit	$ Cost
Bar, Front bar	L.F.	287	Movie Equipment		
Back bar	L.F.	229	Projector, 35mm	Each	9700 - 13,300
Booth, Upholstered, custom straight	L.F.	145 - 265	Screen, wall or ceiling hung	S.F.	6.60 - 10.10
"L" or "U" shaped	L.F.	150 - 252	Partitions, Folding leaf, wood		
Bowling Alleys, incl. alley, pinsetter			Acoustic type	S.F.	55 - 91
Scorer, counter & misc. supplies, average	Lane	50,500	Seating		
For automatic scorer, add	Lane	7475	Auditorium chair, all veneer	Each	171
Emergency Lighting, 25 watt, battery operated			Veneer back, padded seat	Each	207
Lead battery	Each	227	Upholstered, spring seat	Each	208
Nickel cadmium	Each	660	Classroom, movable chair & desk	Set	65 - 120
Kitchen Equipment			Lecture hall, pedestal type	Each	156 - 485
Broiler	Each	4275	Sound System		
Coffee urn, twin 6 gallon	Each	7375	Amplifier, 250 watts	Each	1675
Cooler, 6 ft. long	Each	3375	Speaker, ceiling or wall	Each	147
Dishwasher, 10-12 racks per hr.	Each	3225	Trumpet	Each	275
Food warmer	Each	745	Stage Curtains, Medium weight	S.F.	8 - 28
Freezer, 44 C.F., reach-in	Each	8775	Curtain Track, Light duty	L.F.	58
Ice cube maker, 50 lb. per day	Each	1900	Swimming Pools, Complete, gunite	S.F.	48 - 59
Range with 1 oven	Each	2500			

Model costs calculated for a 1 story building with 12' story height and 10,000 square feet of floor area

Community Center

BUILDING TYPES

				Unit	Unit Cost	Cost Per S.F.	% Of Sub-Total
A.	**SUBSTRUCTURE**						
1010	Standard Foundations	Poured concrete; strip and spread footings		S.F. Ground	1.51	1.51	
1030	Slab on Grade	4" reinforced concrete with vapor barrier and granular base		S.F. Slab	3.42	3.42	
2010	Basement Excavation	Site preparation for slab and trench for foundation wall and footing		S.F. Ground	1.18	1.18	15.5%
2020	Basement Walls	4' foundation wall		L.F. Wall	49	5.23	
B.	**SHELL**						
	B10 Superstructure						
1010	Floor Construction	N/A		—	—	—	5.5%
1020	Roof Construction	Metal deck on open web steel joists		S.F. Roof	4.02	4.02	
	B20 Exterior Enclosure						
2010	Exterior Walls	Face brick with concrete block backup	80% of wall	S.F. Wall	21	9.18	
2020	Exterior Windows	Aluminum sliding	20% of wall	Each	459	1.56	15.6%
2030	Exterior Doors	Double aluminum and glass and hollow metal		Each	1560	.63	
	B30 Roofing						
3010	Roof Coverings	Built-up tar and gravel with flashing; perlite/EPS composite insulation		S.F. Roof	4.62	4.62	6.4%
3020	Roof Openings	Roof hatches		S.F. Roof	.05	.05	
C.	**INTERIORS**						
1010	Partitions	Gypsum board on metal studs	14 S.F. Floor/L.F. Partition	S.F. Partition	5.45	3.89	
1020	Interior Doors	Single leaf hollow metal	140 S.F. Floor/Door	Each	537	3.84	
1030	Fittings	Toilet partitions, directory board, mailboxes		S.F. Floor	1.43	1.43	
2010	Stair Construction	N/A		—	—	—	26.1%
3010	Wall Finishes	Paint		S.F. Surface	2.48	1.77	
3020	Floor Finishes	50% carpet, 50% vinyl tile		S.F. Floor	4.46	4.46	
3030	Ceiling Finishes	Mineral fiber tile on concealed zee bars		S.F. Ceiling	3.71	3.71	
D.	**SERVICES**						
	D10 Conveying						
1010	Elevators & Lifts	N/A		—	—	—	0.0%
1020	Escalators & Moving Walks	N/A		—	—	—	
	D20 Plumbing						
2010	Plumbing Fixtures	Kitchen, toilet and service fixtures, supply and drainage	1 Fixture/910 S.F. Floor	Each	1719	1.89	
2020	Domestic Water Distribution	Electric water heater		S.F. Floor	3.22	3.22	7.4%
2040	Rain Water Drainage	Roof drains		S.F. Roof	.29	.29	
	D30 HVAC						
3010	Energy Supply	N/A		—	—	—	
3020	Heat Generating Systems	Included in D3030		—	—	—	
3030	Cooling Generating Systems	Single zone rooftop unit, gas heating, electric cooling		S.F. Floor	8.96	8.96	12.3%
3050	Terminal & Package Units	N/A		—	—	—	
3090	Other HVAC Sys. & Equipment	N/A		—	—	—	
	D40 Fire Protection						
4010	Sprinklers	Wet pipe sprinkler system		S.F. Floor	1.79	1.79	2.4%
4020	Standpipes	N/A		—	—	—	
	D50 Electrical						
5010	Electrical Service/Distribution	200 ampere service, panel board and feeders		S.F. Floor	.68	.68	
5020	Lighting & Branch Wiring	Incandescent fixtures, receptacles, switches, A.C. and misc. power		S.F. Floor	3.38	3.38	6.1%
5030	Communications & Security	Alarm systems and emergency lighting		S.F. Floor	.27	.27	
5090	Other Electrical Systems	Emergency generator, 15 kW		S.F. Floor	.13	.13	
E.	**EQUIPMENT & FURNISHINGS**						
1010	Commercial Equipment	Freezer, chest type		S.F. Floor	.60	.60	
1020	Institutional Equipment	N/A		—	—	—	2.7%
1030	Vehicular Equipment	N/A		—	—	—	
1090	Other Equipment	Kitchen equipment, directory board, mailboxes, built-in coat racks		S.F. Floor	1.37	1.37	
F.	**SPECIAL CONSTRUCTION & DEMOLITION**						
1020	Integrated Construction	N/A		—	—	—	0.0%
1040	Special Facilities	N/A		—	—	—	
G.	**BUILDING SITEWORK**	**N/A**					

			Sub-Total	73.08	100%
CONTRACTOR FEES (General Requirements: 10%, Overhead: 5%, Profit: 10%)			25%	18.27	
ARCHITECT FEES			9%	8.20	
		Total Building Cost		99.55	

111

Costs per square foot of floor area

Exterior Wall	S.F. Area	16000	23000	30000	37000	44000	51000	58000	65000	72000
	L.F. Perimeter	597	763	821	968	954	1066	1090	1132	1220
Limestone with Concrete Block Back-up	R/Conc. Frame	145.80	140.85	135.35	133.90	129.95	129.05	127.20	125.90	125.40
	Steel Frame	143.05	138.10	132.60	131.15	127.20	126.30	124.45	123.15	122.65
Face Brick with Concrete Block Back-up	R/Conc. Frame	138.30	134.20	129.90	128.65	125.60	124.85	123.40	122.40	122.00
	Steel Frame	135.55	131.45	127.10	125.90	122.85	122.10	120.65	119.65	119.25
Stone with Concrete Block Back-up	R/Conc. Frame	139.85	135.55	131.00	129.70	126.45	125.70	124.15	123.10	122.70
	Steel Frame	137.15	132.90	128.30	127.00	123.80	123.00	121.50	120.40	120.00
Perimeter Adj., Add or Deduct	Per 100 L.F.	4.90	3.45	2.65	2.15	1.80	1.55	1.35	1.25	1.10
Story Hgt. Adj., Add or Deduct	Per 1 Ft.	1.75	1.55	1.25	1.20	1.00	.95	.85	.80	.80
For Basement, add $ 19.05 per square foot of basement area										

The above costs were calculated using the basic specifications shown on the facing page. These costs should be adjusted where necessary for design alternatives and owner's requirements. Reported completed project costs, for this type of structure, range from $ 87.20 to $ 162.50 per S.F.

Common additives

Description	Unit	$ Cost	Description	Unit	$ Cost
Benches, Hardwood	L.F.	84 - 157	Flagpoles, Complete		
Clock System			Aluminum, 20' high	Each	1100
20 room	Each	13,000	40' high	Each	2700
50 room	Each	31,500	70' high	Each	8350
Closed Circuit Surveillance, One station			Fiberglass, 23' high	Each	1425
Camera and monitor	Each	1375	39'-5" high	Each	3000
For additional camera stations, add	Each	750	59' high	Each	7425
Directory Boards, Plastic, glass covered			Intercom System, 25 station capacity		
30" x 20"	Each	575	Master station	Each	2025
36" x 48"	Each	1075	Intercom outlets	Each	128
Aluminum, 24" x 18"	Each	440	Handset	Each	335
36" x 24"	Each	555	Safe, Office type, 4 hour rating		
48" x 32"	Each	780	30" x 18" x 18"	Each	3200
48" x 60"	Each	1675	62" x 33" x 20"	Each	6975
Emergency Lighting, 25 watt, battery operated			Smoke Detectors		
Lead battery	Each	227	Ceiling type	Each	151
Nickel cadmium	Each	660	Duct type	Each	405

Important: See the Reference Section for Location Factors

Model costs calculated for a 1 story building with 14' story height and 30,000 square feet of floor area

				Unit	Unit Cost	Cost Per S.F.	% Of Sub-Total
A. SUBSTRUCTURE							
1010	Standard Foundations	Poured concrete; strip and spread footings		S.F. Ground	1.15	1.15	
1030	Slab on Grade	4" reinforced concrete with vapor barrier and granular base		S.F. Slab	3.42	3.42	7.1%
2010	Basement Excavation	Site preparation for slab and trench for foundation wall and footing		S.F. Ground	1.09	1.09	
2020	Basement Walls	4' foundation wall		L.F. Wall	52	1.55	
B. SHELL							
	B10 Superstructure						
1010	Floor Construction	Cast-in-place columns		L.F. Column	47.45	1.20	14.7%
1020	Roof Construction	Cast-in-place concrete waffle slab		S.F. Roof	13.65	13.65	
	B20 Exterior Enclosure						
2010	Exterior Walls	Limestone panels with concrete block backup	75% of wall	S.F. Wall	35	10.17	
2020	Exterior Windows	Aluminum with insulated glass	25% of wall	Each	509	2.12	12.5%
2030	Exterior Doors	Double wood		Each	1481	.35	
	B30 Roofing						
3010	Roof Coverings	Built-up tar and gravel with flashing; perlite/EPS composite insulation		S.F. Roof	3.60	3.60	3.6%
3020	Roof Openings	Roof hatches		S.F. Roof	.05	.05	
C. INTERIORS							
1010	Partitions	Plaster on metal studs	10 S.F. Floor/L.F. Partition	S.F. Partition	9.13	10.95	
1020	Interior Doors	Single leaf wood	100 S.F. Floor/Door	Each	432	4.32	
1030	Fittings	Toilet partitions		S.F. Floor	.43	.43	
2010	Stair Construction	N/A		—	—	—	36.2%
3010	Wall Finishes	70% paint, 20% wood paneling, 10% vinyl wall covering		S.F. Surface	3.71	4.45	
3020	Floor Finishes	60% hardwood, 20% carpet, 20% terrazzo		S.F. Floor	9.45	9.45	
3030	Ceiling Finishes	Gypsum plaster on metal lath, suspended		S.F. Ceiling	7.00	7.00	
D. SERVICES							
	D10 Conveying						
1010	Elevators & Lifts	N/A		—	—	—	0.0%
1020	Escalators & Moving Walks	N/A		—	—	—	
	D20 Plumbing						
2010	Plumbing Fixtures	Toilet and service fixtures, supply and drainage	1 Fixture/1110 S.F. Floor	Each	1986	1.79	
2020	Domestic Water Distribution	Electric hot water heater		S.F. Floor	1.16	1.16	3.3%
2040	Rain Water Drainage	Roof drains		S.F. Roof	.34	.34	
	D30 HVAC						
3010	Energy Supply	N/A		—	—	—	
3020	Heat Generating Systems	Included in D3030		—	—	—	
3030	Cooling Generating Systems	Multizone unit, gas heating, electric cooling		S.F. Floor	13.85	13.85	13.7%
3050	Terminal & Package Units	N/A		—	—	—	
3090	Other HVAC Sys. & Equipment	N/A		—	—	—	
	D40 Fire Protection						
4010	Sprinklers	Wet pipe sprinkler system		S.F. Floor	1.50	1.50	1.5%
4020	Standpipes	N/A		—	—	—	
	D50 Electrical						
5010	Electrical Service/Distribution	400 ampere service, panel board and feeders		S.F. Floor	.64	.64	
5020	Lighting & Branch Wiring	Fluorescent fixtures, receptacles, switches, A.C. and misc. power		S.F. Floor	6.57	6.57	7.4%
5030	Communications & Security	Alarm systems and emergency lighting		S.F. Floor	.31	.31	
5090	Other Electrical Systems	Emergency generator, 11.5 kW		S.F. Floor	.10	.10	
E. EQUIPMENT & FURNISHINGS							
1010	Commercial Equipment	N/A		—	—	—	
1020	Institutional Equipment	N/A		—	—	—	0.0%
1030	Vehicular Equipment	N/A		—	—	—	
1090	Other Equipment	N/A		—	—	—	
F. SPECIAL CONSTRUCTION & DEMOLITION							
1020	Integrated Construction	N/A		—	—	—	0.0%
1040	Special Facilities	N/A		—	—	—	
G. BUILDING SITEWORK	**N/A**						

	Sub-Total	101.21	**100%**
CONTRACTOR FEES (General Requirements: 10%, Overhead: 5%, Profit: 10%)	25%	25.30	
ARCHITECT FEES	7%	8.84	
Total Building Cost		135.35	

Costs per square foot of floor area

Exterior Wall	S.F. Area	30000	40000	45000	50000	60000	70000	80000	90000	100000
	L.F. Perimeter	410	493	535	533	600	666	733	800	795
Limestone with Concrete Block Back-up	R/Conc. Frame	155.10	149.15	147.20	143.60	140.95	139.00	137.60	136.50	134.00
	Steel Frame	152.95	147.00	145.05	141.45	138.80	136.85	135.45	134.35	131.85
Face Brick with Concrete Block Back-up	R/Conc. Frame	148.05	142.75	141.05	138.10	135.80	134.10	132.85	131.90	129.90
	Steel Frame	145.90	140.65	138.90	135.95	133.65	131.95	130.75	129.75	127.75
Stone with Concrete Block Back-up	R/Conc. Frame	149.55	144.15	142.35	139.30	136.90	135.15	133.90	132.90	130.75
	Steel Frame	147.40	142.00	140.25	137.15	134.75	133.05	131.75	130.75	128.60
Perimeter Adj., Add or Deduct	Per 100 L.F.	7.65	5.70	5.10	4.55	3.80	3.30	2.85	2.55	2.25
Story Hgt. Adj., Add or Deduct	Per 1 Ft.	2.45	2.20	2.15	1.90	1.80	1.70	1.65	1.60	1.40

For Basement, add $19.35 per square foot of basement area

The above costs were calculated using the basic specifications shown on the facing page. These costs should be adjusted where necessary for design alternatives and owner's requirements. Reported completed project costs, for this type of structure, range from $87.65 to $163.50 per S.F.

Common additives

Description	Unit	$ Cost
Benches, Hardwood	L.F.	84 - 157
Clock System		
20 room	Each	13,000
50 room	Each	31,500
Closed Circuit Surveillance, One station		
Camera and monitor	Each	1375
For additional camera stations, add	Each	750
Directory Boards, Plastic, glass covered		
30" x 20"	Each	575
36" x 48"	Each	1075
Aluminum, 24" x 18"	Each	440
36" x 24"	Each	555
48" x 32"	Each	780
48" x 60"	Each	1675
Elevators, Hydraulic passenger, 2 stops		
1500# capacity	Each	43,425
2500# capacity	Each	44,625
3500# capacity	Each	48,525
Additional stop, add	Each	3800

Description	Unit	$ Cost
Emergency Lighting, 25 watt, battery operated		
Lead battery	Each	227
Nickel cadmium	Each	660
Flagpoles, Complete		
Aluminum, 20' high	Each	1100
40' high	Each	2700
70' high	Each	8350
Fiberglass, 23' high	Each	1425
39'-5" high	Each	3000
59' high	Each	7425
Intercom System, 25 station capacity		
Master station	Each	2025
Intercom outlets	Each	128
Handset	Each	335
Safe, Office type, 4 hour rating		
30" x 18" x 18"	Each	3200
62" x 33" x 20"	Each	6975
Smoke Detectors		
Ceiling type	Each	151
Duct type	Each	405

Important: See the Reference Section for Location Factors

Model costs calculated for a 3 story building with 12' story height and 60,000 square feet of floor area

Courthouse, 2-3 Story

				Unit	Unit Cost	Cost Per S.F.	% Of Sub-Total
A. SUBSTRUCTURE							
1010	Standard Foundations	Poured concrete; strip and spread footings		S.F. Ground	2.64	.88	
1030	Slab on Grade	4" reinforced concrete with vapor barrier and granular base		S.F. Slab	3.42	1.14	
2010	Basement Excavation	Site preparation for slab and trench for foundation wall and footing		S.F. Ground	1.09	.36	3.0%
2020	Basement Walls	4' foundation wall		L.F. Wall	49	.61	
B. SHELL							
	B10 Superstructure						
1010	Floor Construction	Concrete slab with metal deck and beams		S.F. Floor	16.65	11.10	14.9%
1020	Roof Construction	Concrete slab with metal deck and beams		S.F. Roof	11.22	3.74	
	B20 Exterior Enclosure						
2010	Exterior Walls	Face brick with concrete block backup	75% of wall	S.F. Wall	22	6.11	
2020	Exterior Windows	Horizontal pivoted steel	25% of wall	Each	554	5.54	11.8%
2030	Exterior Doors	Double aluminum and glass and hollow metal		Each	1895	.19	
	B30 Roofing						
3010	Roof Coverings	Built-up tar and gravel with flashing; perlite/EPS composite insulation		S.F. Roof	3.84	1.28	1.3%
3020	Roof Openings	N/A		—	—	—	
C. INTERIORS							
1010	Partitions	Plaster on metal studs	10 S.F. Floor/L.F. Partition	S.F. Partition	9.23	9.23	
1020	Interior Doors	Single leaf wood	100 S.F. Floor/Door	Each	432	4.32	
1030	Fittings	Toilet partitions		S.F. Floor	.21	.21	
2010	Stair Construction	Concrete filled metal pan		Flight	7050	1.18	35.1%
3010	Wall Finishes	70% paint, 20% wood paneling, 10% vinyl wall covering		S.F. Surface	3.72	3.72	
3020	Floor Finishes	60% hardwood, 20% terrazzo, 20% carpet		S.F. Floor	9.45	9.45	
3030	Ceiling Finishes	Gypsum plaster on metal lath, suspended		S.F. Ceiling	7.00	7.00	
D. SERVICES							
	D10 Conveying						
1010	Elevators & Lifts	Five hydraulic passenger elevators		Each	79,680	6.64	6.6%
1020	Escalators & Moving Walks	N/A		—	—	—	
	D20 Plumbing						
2010	Plumbing Fixtures	Toilet and service fixtures, supply and drainage	1 Fixture/665 S.F. Floor	Each	1177	1.77	
2020	Domestic Water Distribution	Electric water heaterD...		S.F. Floor	1.98	1.98	4.0%
2040	Rain Water Drainage	Roof drains		S.F. Roof	.60	.20	
	D30 HVAC						
3010	Energy Supply	N/A		—	—	—	
3020	Heat Generating Systems	Included in D3030		—	—	—	
3030	Cooling Generating Systems	Multizone unit, gas heating, electric cooling		S.F. Floor	13.85	13.85	13.9%
3050	Terminal & Package Units	N/A		—	—	—	
3090	Other HVAC Sys. & Equipment	N/A		—	—	—	
	D40 Fire Protection						
4010	Sprinklers	Wet pipe sprinkler system		S.F. Floor	1.56	1.56	1.6%
4020	Standpipes	N/A		—	—	—	
	D50 Electrical						
5010	Electrical Service/Distribution	800 ampere service, panel board and feeders		S.F. Floor	.74	.74	
5020	Lighting & Branch Wiring	Fluorescent fixtures, receptacles, switches, A.C. and misc. power		S.F. Floor	6.66	6.66	7.8%
5030	Communications & Security	Alarm systems and emergency lighting		S.F. Floor	.31	.31	
5090	Other Electrical Systems	Emergency generator, 15 kW		S.F. Floor	.16	.16	
E. EQUIPMENT & FURNISHINGS							
1010	Commercial Equipment	N/A		—	—	—	
1020	Institutional Equipment	N/A		—	—	—	
1030	Vehicular Equipment	N/A		—	—	—	0.0%
1090	Other Equipment	N/A		—	—	—	
F. SPECIAL CONSTRUCTION & DEMOLITION							
1020	Integrated Construction	N/A		—	—	—	0.0%
1040	Special Facilities	N/A		—	—	—	
G. BUILDING SITEWORK	**N/A**						

			Sub-Total	99.93	100%
CONTRACTOR FEES (General Requirements: 10%, Overhead: 5%, Profit: 10%)		25%	24.98		
ARCHITECT FEES		7%	8.74		

Total Building Cost	**133.65**

115

Costs per square foot of floor area

Exterior Wall	S.F. Area	12000	18000	24000	30000	36000	42000	48000	54000	60000
	L.F. Perimeter	460	580	713	730	826	880	965	1006	1045
Concrete Block	Steel Frame	81.30	76.85	74.90	72.00	71.10	69.95	69.40	68.60	68.00
	Bearing Walls	79.65	75.20	73.25	70.40	69.50	68.30	67.75	66.95	66.35
Precast Concrete Panels	Steel Frame	85.25	80.20	78.00	74.55	73.50	72.15	71.50	70.50	69.75
Insulated Metal Panels	Steel Frame	82.45	77.85	75.80	72.75	71.80	70.60	70.00	69.15	68.50
Face Brick on Common Brick	Steel Frame	94.10	87.65	84.85	80.15	78.80	76.95	76.15	74.85	73.80
Tilt-up Concrete Panel	Steel Frame	82.90	78.25	76.20	73.05	72.10	70.85	70.30	69.40	68.75
Perimeter Adj., Add or Deduct	Per 100 L.F.	3.70	2.50	1.85	1.50	1.25	1.05	.95	.85	.75
Story Hgt. Adj., Add or Deduct	Per 1 Ft.	.55	.45	.45	.35	.35	.30	.30	.25	.25
For Basement, add $19.25 per square foot of basement area										

The above costs were calculated using the basic specifications shown on the facing page. These costs should be adjusted where necessary for design alternatives and owner's requirements. Reported completed project costs, for this type of structure, range from $28.05 to $108.60 per S.F.

Common additives

Description	Unit	$ Cost	Description	Unit	$ Cost
Clock System			Dock Levelers, Hinged 10 ton cap.		
20 room	Each	13,000	6' x 8'	Each	5300
50 room	Each	31,500	7' x 8'	Each	5000
Dock Bumpers, Rubber blocks			Partitions, Woven wire, 10 ga., 1-1/2" mesh		
4-1/2" thick, 10" high, 14" long	Each	61	4' wide x 7' high	Each	136
24" long	Each	98	8' high	Each	146
36" long	Each	107	10' High	Each	172
12" high, 14" long	Each	101	Platform Lifter, Portable, 6'x 6'		
24" long	Each	114	3000# cap.	Each	7850
36" long	Each	131	4000# cap.	Each	9675
6" thick, 10" high, 14" long	Each	91	Fixed, 6' x 8', 5000# cap.	Each	10,100
24" long	Each	125			
36" long	Each	149			
20" high, 11" long	Each	155			
Dock Boards, Heavy					
60" x 60" Aluminum, 5,000# cap.	Each	1225			
9000# cap.	Each	1375			
15,000# cap.	Each	1450			

Important: See the Reference Section for Location Factors

Model costs calculated for a 1 story building with 20' story height and 30,000 square feet of floor area

Factory, 1 Story

					Unit	Unit Cost	Cost Per S.F.	% Of Sub-Total
A. SUBSTRUCTURE								
1010	Standard Foundations		Poured concrete; strip and spread footings		S.F. Ground	.95	.95	
1030	Slab on Grade		4" reinforced concrete with vapor barrier and granular base		S.F. Slab	4.47	4.47	
2010	Basement Excavation		Site preparation for slab and trench for foundation wall and footing		S.F. Ground	1.09	1.09	14.9%
2020	Basement Walls		4' foundation wall		L.F. Wall	52	1.54	
B. SHELL								
	B10 Superstructure							
1010	Floor Construction		N/A		—	—	—	
1020	Roof Construction		Metal deck, open web steel joists, beams and columns		S.F. Roof	5.56	5.56	10.3%
	B20 Exterior Enclosure							
2010	Exterior Walls		Concrete block	75% of wall	S.F. Wall	5.26	1.92	
2020	Exterior Windows		Industrial horizontal pivoted steel	25% of wall	Each	640	2.43	9.4%
2030	Exterior Doors		Double aluminum and glass, hollow metal, steel overhead		Each	1387	.69	
	B30 Roofing							
3010	Roof Coverings		Built-up tar and gravel with flashing; perlite/EPS composite insulation		S.F. Roof	3.69	3.69	
3020	Roof Openings		Roof hatches		S.F. Roof	.20	.20	7.2%
C. INTERIORS								
1010	Partitions		Concrete block	60 S.F. Floor/L.F. Partition	S.F. Partition	5.70	1.14	
1020	Interior Doors		Single leaf hollow metal and fire doors	600 S.F. Floor/Door	Each	537	.90	
1030	Fittings		Toilet partitions		S.F. Floor	.86	.86	
2010	Stair Construction		N/A		—	—	—	9.0%
3010	Wall Finishes		Paint		S.F. Surface	6.40	1.28	
3020	Floor Finishes		Vinyl composition tile	10% of floor	S.F. Floor	2.70	.27	
3030	Ceiling Finishes		Fiberglass board on exposed grid system	10% of area	S.F. Ceiling	3.71	.37	
D. SERVICES								
	D10 Conveying							
1010	Elevators & Lifts		N/A		—	—	—	
1020	Escalators & Moving Walks		N/A		—	—	—	0.0%
	D20 Plumbing							
2010	Plumbing Fixtures		Toilet and service fixtures, supply and drainage	1 Fixture/1000 S.F. Floor	Each	2360	2.36	
2020	Domestic Water Distribution		Gas fired water heater		S.F. Floor	.22	.22	5.5%
2040	Rain Water Drainage		Roof drains		S.F. Roof	.39	.39	
	D30 HVAC							
3010	Energy Supply		Oil fired hot water, unit heaters		S.F. Floor	5.74	5.74	
3020	Heat Generating Systems		N/A		—	—	—	
3030	Cooling Generating Systems		Chilled water, air cooled condenser system		S.F. Floor	8.01	8.01	25.5%
3050	Terminal & Package Units		N/A		—	—	—	
3090	Other HVAC Sys. & Equipment		N/A		—	—	—	
	D40 Fire Protection							
4010	Sprinklers		Sprinklers, ordinary hazard		S.F. Floor	2.24	2.24	
4020	Standpipes		N/A		—	—	—	4.2%
	D50 Electrical							
5010	Electrical Service/Distribution		600 ampere service, panel board and feeders		S.F. Floor	.84	.84	
5020	Lighting & Branch Wiring		High intensity discharge fixtures, receptacles, switches, A.C. and misc. power		S.F. Floor	6.38	6.38	
5030	Communications & Security		Alarm systems and emergency lighting		S.F. Floor	.31	.31	14.0%
5090	Other Electrical Systems		N/A		—	—	—	
E. EQUIPMENT & FURNISHINGS								
1010	Commercial Equipment		N/A		—	—	—	
1020	Institutional Equipment		N/A		—	—	—	
1030	Vehicular Equipment		N/A		—	—	—	0.0%
1090	Other Equipment		N/A		—	—	—	
F. SPECIAL CONSTRUCTION & DEMOLITION								
1020	Integrated Construction		N/A		—	—	—	
1040	Special Facilities		N/A		—	—	—	0.0%
G. BUILDING SITEWORK	**N/A**							

		Sub-Total	53.85	100%
CONTRACTOR FEES (General Requirements: 10%, Overhead: 5%, Profit: 10%)		25%	13.46	
ARCHITECT FEES		7%	4.69	
	Total Building Cost		72.00	

BUILDING TYPES

117

Costs per square foot of floor area

Exterior Wall	S.F. Area	20000	30000	40000	50000	60000	70000	80000	90000	100000
	L.F. Perimeter	362	410	493	576	600	628	660	700	744
Face Brick Common Brick Back-up	Steel Frame	102.25	92.60	89.35	87.40	84.35	82.25	80.80	79.80	79.10
	Concrete Frame	102.35	92.70	89.45	87.50	84.45	82.35	80.90	79.90	79.20
Face Brick Concrete Block Back-up	Steel Frame	101.25	91.95	88.75	86.90	83.90	81.85	80.50	79.55	78.80
	Concrete Frame	101.15	91.85	88.65	86.75	83.80	81.75	80.35	79.40	78.70
Stucco on Concrete Block	Steel Frame	94.45	86.75	84.10	82.55	80.15	78.50	77.35	76.55	76.00
	Concrete Frame	94.30	86.65	84.00	82.40	80.00	78.35	77.25	76.45	75.90
Perimeter Adj., Add or Deduct	Per 100 L.F.	8.90	5.95	4.45	3.55	2.95	2.55	2.25	2.00	1.80
Story Hgt. Adj., Add or Deduct	Per 1 Ft.	2.50	1.85	1.70	1.60	1.35	1.25	1.15	1.05	1.00
For Basement, add $21.00 per square foot of basement area										

The above costs were calculated using the basic specifications shown on the facing page. These costs should be adjusted where necessary for design alternatives and owner's requirements. Reported completed project costs, for this type of structure, range from $28.05 to $108.60 per S.F.

Common additives

Description	Unit	$ Cost	Description	Unit	$ Cost
Clock System			Dock Levelers, Hinged 10 ton cap.		
20 room	Each	13,000	6' x 8'	Each	5300
50 room	Each	31,500	7' x 8'	Each	5000
Dock Bumpers, Rubber blocks			Elevator, Hydraulic freight, 2 stops		
4-1/2" thick, 10" high, 14" long	Each	61	3500# capacity	Each	65,100
24" long	Each	98	4000# capacity	Each	67,600
36" long	Each	107	Additional stop, add	Each	3800
12" high, 14" long	Each	101	Partitions, Woven wire, 10 ga., 1-1/2" mesh		
24" long	Each	114	4' Wide x 7" high	Each	136
36" long	Each	131	8' High	Each	146
6" thick, 10" high, 14" long	Each	91	10' High	Each	172
24" long	Each	125	Platform Lifter, Portable, 6' x 6'		
36" long	Each	149	3000# cap.	Each	7850
20" high, 11" long	Each	155	4000# cap.	Each	9675
Dock Boards, Heavy			Fixed, 6' x 8', 5000# cap.	Each	10,100
60" x 60" Aluminum, 5,000# cap.	Each	1225			
9000# cap.	Each	1375			
15,000# cap.	Each	1450			

Important: See the Reference Section for Location Factors

BUILDING TYPES

Model costs calculated for a 3 story building with 12' story height and 90,000 square feet of floor area

				Unit	Unit Cost	Cost Per S.F.	% Of Sub-Total
A. SUBSTRUCTURE							
1010	Standard Foundations	Poured concrete; strip and spread footings		S.F. Ground	4.44	1.48	
1030	Slab on Grade	4" reinforced concrete with vapor barrier and granular base		S.F. Slab	5.84	1.95	7.0%
2010	Basement Excavation	Site preparation for slab and trench for foundation wall and footing		S.F. Ground	1.09	.36	
2020	Basement Walls	4' foundation wall		L.F. Wall	52	.41	
B. SHELL							
	B10 Superstructure						
1010	Floor Construction	Concrete flat slab		S.F. Floor	16.29	10.86	24.0%
1020	Roof Construction	Concrete flat slab		S.F. Roof	10.80	3.60	
	B20 Exterior Enclosure						
2010	Exterior Walls	Face brick with common brick backup	70% of wall	S.F. Wall	23	4.52	
2020	Exterior Windows	Industrial, horizontal pivoted steel	30% of wall	Each	986	5.18	16.5%
2030	Exterior Doors	Double aluminum & glass, hollow metal, overhead doors		Each	1332	.25	
	B30 Roofing						
3010	Roof Coverings	Built-up tar and gravel with flashing; perlite/EPS composite insulation		S.F. Roof	3.57	1.19	2.1%
3020	Roof Openings	Roof hatches		S.F. Roof	.21	.07	
C. INTERIORS							
1010	Partitions	Gypsum board on metal studs	50 S.F. Floor/L.F. Partition	S.F. Partition	2.70	.54	
1020	Interior Doors	Single leaf fire doors	500 S.F. Floor/Door	Each	666	1.33	
1030	Fittings	Toilet partitions		S.F. Floor	.43	.43	
2010	Stair Construction	Concrete		Flight	5250	.47	9.0%
3010	Wall Finishes	Paint		S.F. Surface	1.15	.23	
3020	Floor Finishes	90% metallic hardener, 10% vinyl composition tile		S.F. Floor	2.11	2.11	
3030	Ceiling Finishes	Fiberglass board on exposed grid systems	10% of area	S.F. Ceiling	3.06	.31	
D. SERVICES							
	D10 Conveying						
1010	Elevators & Lifts	Two hydraulic freight elevators		Each	95,400	2.12	3.5%
1020	Escalators & Moving Walks	N/A		—	—	—	
	D20 Plumbing						
2010	Plumbing Fixtures	Toilet and service fixtures, supply and drainage	1 Fixture/1345 S.F. Floor	Each	2568	1.91	
2020	Domestic Water Distribution	Gas fired water heater		S.F. Floor	.13	.13	3.7%
2040	Rain Water Drainage	Roof drains		S.F. Roof	.51	.17	
	D30 HVAC						
3010	Energy Supply	Oil fired hot water, unit heaters		S.F. Floor	2.96	2.96	
3020	Heat Generating Systems	N/A		—	—	—	
3030	Cooling Generating Systems	Chilled water, air cooled condenser system		S.F. Floor	8.01	8.01	18.2%
3050	Terminal & Package Units	N/A		—	—	—	
3090	Other HVAC Sys. & Equipment	N/A		—	—	—	
	D40 Fire Protection						
4010	Sprinklers	Wet pipe sprinkler system		S.F. Floor	2.02	2.02	3.3%
4020	Standpipes	N/A		—	—	—	
	D50 Electrical						
5010	Electrical Service/Distribution	800 ampere service, panel board and feeders		S.F. Floor	.50	.50	
5020	Lighting & Branch Wiring	High intensity discharge fixtures, switches, A.C. and misc. power		S.F. Floor	6.74	6.74	12.7%
5030	Communications & Security	Alarm systems and emergency lighting		S.F. Floor	.31	.31	
5090	Other Electrical Systems	Emergency generator, 30 kW		S.F. Floor	.15	.15	
E. EQUIPMENT & FURNISHINGS							
1010	Commercial Equipment	N/A		—	—	—	
1020	Institutional Equipment	N/A		—	—	—	
1030	Vehicular Equipment	N/A		—	—	—	0.0%
1090	Other Equipment	N/A		—	—	—	
F. SPECIAL CONSTRUCTION & DEMOLITION							
1020	Integrated Construction	N/A		—	—	—	0.0%
1040	Special Facilities	N/A		—	—	—	
G. BUILDING SITEWORK	**N/A**						

		Sub-Total	60.31	100%
CONTRACTOR FEES (General Requirements: 10%, Overhead: 5%, Profit: 10%)		25%	15.08	
ARCHITECT FEES		6%	4.51	
		Total Building Cost	79.90	

BUILDING TYPES

BUILDING TYPES

Costs per square foot of floor area

Exterior Wall	S.F. Area	4000	4500	5000	5500	6000	6500	7000	7500	8000
	L.F. Perimeter	260	280	300	320	320	336	353	370	386
Face Brick Concrete Block Back-up	Steel Joists	114.95	112.85	111.20	109.80	106.80	105.70	104.75	104.00	103.15
	Bearing Walls	112.95	110.85	109.20	107.80	104.80	103.70	102.75	102.00	101.15
Decorative Concrete Block	Steel Joists	105.45	103.80	102.45	101.30	99.05	98.15	97.40	96.80	96.15
	Bearing Walls	103.65	102.00	100.65	99.50	97.25	96.35	95.60	95.00	94.35
Limestone with Concrete Block Back-up	Steel Joists	123.15	120.70	118.75	117.15	113.55	112.20	111.15	110.20	109.25
	Bearing Walls	121.35	118.90	116.95	115.35	111.75	110.40	109.35	108.40	107.45
Perimeter Adj., Add or Deduct	Per 100 L.F.	13.65	12.15	10.95	9.95	9.10	8.40	7.80	7.25	6.85
Story Hgt. Adj., Add or Deduct	Per 1 Ft.	1.80	1.70	1.65	1.60	1.45	1.40	1.40	1.35	1.35
For Basement, add $23.50 per square foot of basement area										

The above costs were calculated using the basic specifications shown on the facing page. These costs should be adjusted where necessary for design alternatives and owner's requirements. Reported completed project costs, for this type of structure, range from $45.35 to $134.35 per S.F.

Common additives

Description	Unit	$ Cost	Description	Unit	$ Cost
Appliances			Appliances, cont.		
Cooking range, 30" free standing			Refrigerator, no frost 10-12 C.F.	Each	555 - 860
1 oven	Each	340 - 1475	14-16 C.F.	Each	570 - 710
2 oven	Each	1475 - 1650	18-20 C.F.	Each	630 - 950
30" built-in			Lockers, Steel, single tier, 60" or 72"	Opening	131 - 227
1 oven	Each	420 - 1550	2 tier, 60" or 72" total	Opening	74 - 125
2 oven	Each	1225 - 2100	5 tier, box lockers	Opening	42 - 62
Counter top cook tops, 4 burner	Each	281 - 625	Locker bench, lam. maple top only	L.F.	19.50
Microwave oven	Each	203 - 650	Pedestals, steel pipe	Each	63
Combination range, refrig. & sink, 30" wide	Each	1225 - 2450	Sound System		
60" wide	Each	3250	Amplifier, 250 watts	Each	1675
72" wide	Each	3725	Speaker, ceiling or wall	Each	147
Combination range refrigerator, sink			Trumpet	Each	275
microwave oven & icemaker	Each	5500			
Compactor, residential, 4-1 compaction	Each	465 - 550			
Dishwasher, built-in, 2 cycles	Each	495 - 740			
4 cycles	Each	530 - 1000			
Garbage disposer, sink type	Each	128 - 287			
Hood for range, 2 speed, vented, 30" wide	Each	199 - 760			
42" wide	Each	350 - 1100			

Important: See the Reference Section for Location Factors

Model costs calculated for a 1 story building with 14' story height and 6,000 square feet of floor area

Fire Station, 1 Story

				Unit	Unit Cost	Cost Per S.F.	% Of Sub-Total
A. SUBSTRUCTURE							
1010	Standard Foundations	Poured concrete; strip and spread footings		S.F. Ground	2.18	2.18	
1030	Slab on Grade	6" reinforced concrete with vapor barrier and granular base		S.F. Slab	4.47	4.47	
2010	Basement Excavation	Site preparation for slab and trench for foundation wall and footing		S.F. Ground	1.31	1.31	13.6%
2020	Basement Walls	4' foundation wall		L.F. Wall	52	2.82	
B. SHELL							
	B10 Superstructure						
1010	Floor Construction	N/A		—	—	—	
1020	Roof Construction	Metal deck, open web steel joists, beams on columns		S.F. Roof	5.47	5.47	6.9%
	B20 Exterior Enclosure						
2010	Exterior Walls	Face brick with concrete block backup	75% of wall	S.F. Wall	21	11.82	
2020	Exterior Windows	Aluminum insulated glass	10% of wall	Each	640	1.49	20.1%
2030	Exterior Doors	Single aluminum and glass, overhead, hollow metal	15% of wall	Each	23	2.60	
	B30 Roofing						
3010	Roof Coverings	Built-up tar and gravel with flashing; perlite/EPS composite insulation		S.F. Roof	4.47	4.47	5.8%
3020	Roof Openings	Skylights, roof hatches		S.F. Roof	.09	.09	
C. INTERIORS							
1010	Partitions	Concrete block	17 S.F. Floor/L.F. Partition	S.F. Partition	6.03	3.55	
1020	Interior Doors	Single leaf hollow metal	500 S.F. Floor/Door	Each	537	1.07	
1030	Fittings	Toilet partitions		S.F. Floor	.43	.43	
2010	Stair Construction	N/A		—	—	—	13.4%
3010	Wall Finishes	Paint		S.F. Surface	2.89	1.70	
3020	Floor Finishes	50% vinyl tile, 50% paint		S.F. Floor	1.98	1.98	
3030	Ceiling Finishes	Fiberglass board on exposed grid, suspended	50% of area	S.F. Ceiling	3.71	1.86	
D. SERVICES							
	D10 Conveying						
1010	Elevators & Lifts	N/A		—	—	—	
1020	Escalators & Moving Walks	N/A		—	—	—	0.0%
	D20 Plumbing						
2010	Plumbing Fixtures	Kitchen, toilet and service fixtures, supply and drainage	1 Fixture/375 S.F. Floor	Each	2088	5.57	
2020	Domestic Water Distribution	Gast fired water heater		S.F. Floor	1.03	1.03	9.0%
2040	Rain Water Drainage	Roof drains		S.F. Roof	.56	.56	
	D30 HVAC						
3010	Energy Supply	N/A		—	—	—	
3020	Heat Generating Systems	Included in D3030		—	—	—	
3030	Cooling Generating Systems	Rooftop multizone unit system		S.F. Floor	16.50	16.50	20.9%
3050	Terminal & Package Units	N/A		—	—	—	
3090	Other HVAC Sys. & Equipment	N/A		—	—	—	
	D40 Fire Protection						
4010	Sprinklers	Wet pipe sprinkler system		S.F. Floor	2.34	2.34	3.0%
4020	Standpipes	N/A		—	—	—	
	D50 Electrical						
5010	Electrical Service/Distribution	200 ampere service, panel board and feeders		S.F. Floor	.99	.99	
5020	Lighting & Branch Wiring	Fluorescent fixtures, receptacles, switches, A.C. and misc. power		S.F. Floor	4.52	4.52	7.3%
5030	Communications & Security	Alarm systems		S.F. Floor	.31	.31	
5090	Other Electrical Systems	N/A		—	—	—	
E. EQUIPMENT & FURNISHINGS							
1010	Commercial Equipment	N/A		—	—	—	
1020	Institutional Equipment	N/A		—	—	—	
1030	Vehicular Equipment	N/A		—	—	—	0.0%
1090	Other Equipment	N/A		—	—	—	
F. SPECIAL CONSTRUCTION & DEMOLITION							
1020	Integrated Construction	N/A		—	—	—	
1040	Special Facilities	N/A		—	—	—	0.0%
G. BUILDING SITEWORK	**N/A**						

	Sub-Total	79.13	100%
CONTRACTOR FEES (General Requirements: 10%, Overhead: 5%, Profit: 10%)	25%	19.78	
ARCHITECT FEES	8%	7.89	
Total Building Cost		**106.80**	

BUILDING TYPES

121

BUILDING TYPES

Costs per square foot of floor area

Exterior Wall	S.F. Area	6000	7000	8000	9000	10000	11000	12000	13000	14000
	L.F. Perimeter	220	240	260	280	286	303	320	336	353
Face Brick with Concrete Block Back-up	Steel Joists	118.00	114.60	112.05	110.10	107.20	105.75	104.60	103.50	102.60
	Precast Conc.	123.45	120.05	117.55	115.60	112.75	111.30	110.15	109.05	108.20
Decorative Concrete Block	Steel Joists	110.15	107.30	105.10	103.40	101.10	99.85	98.85	97.95	97.20
	Precast Conc.	115.55	112.75	110.60	108.90	106.65	105.40	104.45	103.55	102.80
Limestone with Concrete Block Back-up	Steel Joists	127.25	123.25	120.25	117.95	114.45	112.70	111.30	110.00	109.00
	Precast Conc.	132.60	128.65	125.65	123.35	119.90	118.15	116.80	115.50	114.50
Perimeter Adj., Add or Deduct	Per 100 L.F.	15.65	13.40	11.70	10.45	9.35	8.50	7.80	7.20	6.75
Story Hgt. Adj., Add or Deduct	Per 1 Ft.	2.05	1.90	1.80	1.70	1.60	1.55	1.45	1.40	1.40

For Basement, add $22.80 per square foot of basement area

The above costs were calculated using the basic specifications shown on the facing page. These costs should be adjusted where necessary for design alternatives and owner's requirements. Reported completed project costs, for this type of structure, range from $45.35 to $134.35 per S.F.

Common additives

Description	Unit	$ Cost
Appliances		
Cooking range, 30" free standing		
1 oven	Each	340 - 1475
2 oven	Each	1475 - 1650
30" built-in		
1 oven	Each	420 - 1550
2 oven	Each	1225 - 2100
Counter top cook tops, 4 burner	Each	281 - 625
Microwave oven	Each	203 - 650
Combination range, refrig. & sink, 30" wide	Each	1225 - 2450
60" wide	Each	3250
72" wide	Each	3725
Combination range, refrigerator, sink, microwave oven & icemaker	Each	5500
Compactor, residential, 4-1 compaction	Each	465 - 550
Dishwasher, built-in, 2 cycles	Each	495 - 740
4 cycles	Each	530 - 1000
Garbage diposer, sink type	Each	128 - 287
Hood for range, 2 speed, vented, 30" wide	Each	199 - 760
42" wide	Each	350 - 1100

Description	Unit	$ Cost
Appliances, cont.		
Refrigerator, no frost 10-12 C.F.	Each	555 - 860
14-16 C.F.	Each	570 - 710
18-20 C.F.	Each	630 - 950
Elevators, Hydraulic passenger, 2 stops		
1500# capacity	Each	43,425
2500# capacity	Each	44,625
3500# capacity	Each	48,525
Lockers, Steel, single tier, 60" or 72"	Opening	131 - 227
2 tier, 60" or 72" total	Opening	74 - 125
5 tier, box lockers	Opening	42 - 62
Locker bench, lam. maple top only	L.F.	19.50
Pedestals, steel pipe	Each	63
Sound System		
Amplifier, 250 watts	Each	1675
Speaker, ceiling or wall	Each	147
Trumpet	Each	275

Important: See the Reference Section for Location Factors

Model costs calculated for a 2 story building with 14' story height and 10,000 square feet of floor area

				Unit	Unit Cost	Cost Per S.F.	% Of Sub-Total
A.	**SUBSTRUCTURE**						
1010	Standard Foundations	Poured concrete; strip and spread footings		S.F. Ground	2.06	1.03	
1030	Slab on Grade	6" reinforced concrete with vapor barrier and granular base		S.F. Slab	4.47	2.23	7.5%
2010	Basement Excavation	Site preparation for slab and trench for foundation wall and footing		S.F. Ground	1.31	.65	
2020	Basement Walls	4' foundation wall		L.F. Wall	58	1.67	
B.	**SHELL**						
	B10 Superstructure						
1010	Floor Construction	Open web steel joists, slab form, concrete		S.F. Floor	8.78	4.39	7.8%
1020	Roof Construction	Metal deck on open web steel joists		S.F. Roof	2.84	1.42	
	B20 Exterior Enclosure						
2010	Exterior Walls	Decorative concrete block	75% of wall	S.F. Wall	13.55	8.14	
2020	Exterior Windows	Aluminum insulated glass	10% of wall	Each	509	1.77	16.0%
2030	Exterior Doors	Single aluminum and glass, steel overhead, hollow metal	15% of wall	Each	17.48	2.10	
	B30 Roofing						
3010	Roof Coverings	Built-up tar and gravel with flashing; perlite/EPS composite insulation		S.F. Roof	4.60	2.30	3.1%
3020	Roof Openings	N/A		—	—	—	
C.	**INTERIORS**						
1010	Partitions	Concrete block	10 S.F. Floor/L.F. Partition	S.F. Partition	6.03	3.55	
1020	Interior Doors	Single leaf hollow metal	500 S.F. Floor/Door	Each	666	1.33	
1030	Fittings	Toilet partitions		S.F. Floor	.39	.39	
2010	Stair Construction	Concrete filled metal pan		Flight	7050	1.41	17.4%
3010	Wall Finishes	Paint		S.F. Surface	4.83	2.84	
3020	Floor Finishes	50% vinyl tile, 50% paint		S.F. Floor	1.98	1.98	
3030	Ceiling Finishes	Fiberglass board on exposed grid, suspended	50% of area	S.F. Ceiling	3.06	1.53	
D.	**SERVICES**						
	D10 Conveying						
1010	Elevators & Lifts	One hydraulic passenger elevator		Each	52,200	5.22	7.0%
1020	Escalators & Moving Walks	N/A		—	—	—	
	D20 Plumbing						
2010	Plumbing Fixtures	Kitchen toilet and service fixtures, supply and drainage	1 Fixture/400 S.F. Floor	Each	2044	5.11	
2020	Domestic Water Distribution	Gas fired water heater		S.F. Floor	.81	.81	8.3%
2040	Rain Water Drainage	Roof drains		S.F. Roof	.62	.31	
	D30 HVAC						
3010	Energy Supply	N/A		—	—	—	
3020	Heat Generating Systems	Included in D3030		—	—	—	
3030	Cooling Generating Systems	Rooftop multizone unit system		S.F. Floor	16.50	16.50	22.0%
3050	Terminal & Package Units	N/A		—	—	—	
3090	Other HVAC Sys. & Equipment	N/A		—	—	—	
	D40 Fire Protection						
4010	Sprinklers	Wet pipe sprinkler system		S.F. Floor	1.95	1.95	2.6%
4020	Standpipes	N/A		—	—	—	
	D50 Electrical						
5010	Electrical Service/Distribution	400 ampere service, panel board and feeders		S.F. Floor	1.08	1.08	
5020	Lighting & Branch Wiring	Fluorescent fixtures, receptacles, switches, A.C. and misc. power		S.F. Floor	4.68	4.68	8.3%
5030	Communications & Security	Alarm systems and emergency lighting		S.F. Floor	.31	.31	
5090	Other Electrical Systems	Emergency generator, 15 kW		S.F. Floor	.18	.18	
E.	**EQUIPMENT & FURNISHINGS**						
1010	Commercial Equipment	N/A		—	—	—	
1020	Institutional Equipment	N/A		—	—	—	0.0%
1030	Vehicular Equipment	N/A		—	—	—	
1090	Other Equipment	N/A		—	—	—	
F.	**SPECIAL CONSTRUCTION & DEMOLITION**						
1020	Integrated Construction	N/A		—	—	—	0.0%
1040	Special Facilities	N/A		—	—	—	
G.	**BUILDING SITEWORK**	**N/A**					

			Sub-Total	74.88	100%
CONTRACTOR FEES (General Requirements: 10%, Overhead: 5%, Profit: 10%)			25%	18.72	
ARCHITECT FEES			8%	7.50	
			Total Building Cost	**101.10**	

BUILDING TYPES

Costs per square foot of floor area

Exterior Wall	S.F. Area	4000	5000	6000	8000	10000	12000	14000	16000	18000
	L.F. Perimeter	180	205	230	260	300	340	353	386	420
Cedar Beveled Siding	Wood Frame	114.95	108.95	104.90	98.85	95.70	93.50	91.20	89.95	89.00
Aluminum Siding	Wood Frame	114.15	108.15	104.15	98.10	94.90	92.75	90.40	89.15	88.20
Board and Batten	Wood Frame	112.00	107.00	103.60	98.35	95.70	93.85	91.75	90.70	89.85
Face Brick on Block	Bearing Wall	133.85	126.75	121.95	114.20	110.30	107.70	104.45	102.90	101.70
Stucco on Block	Bearing Wall	121.95	115.80	111.70	105.40	102.15	99.90	97.45	96.15	95.20
Decorative Block	Wood Joists	118.80	112.50	108.25	101.65	98.25	95.95	93.35	92.05	91.00
Perimeter Adj., Add or Deduct	Per 100 L.F.	10.75	8.60	7.15	5.40	4.30	3.60	3.10	2.70	2.40
Story Hgt. Adj., Add or Deduct	Per 1 Ft.	1.35	1.25	1.20	1.00	.90	.85	.75	.75	.70

For Basement, add $14.50 per square foot of basement area

The above costs were calculated using the basic specifications shown on the facing page. These costs should be adjusted where necessary for design alternatives and owner's requirements. Reported completed project costs, for this type of structure, range from $65.80 to $144.00 per S.F.

Common additives

Description	Unit	$ Cost
Appliances		
Cooking range, 30" free standing		
1 oven	Each	340 - 1475
2 oven	Each	1475 - 1650
30" built-in		
1 oven	Each	420 - 1550
2 oven	Each	1225 - 2100
Counter top cook tops, 4 burner	Each	281 - 625
Microwave oven	Each	203 - 650
Combination range, refrig. & sink, 30" wide	Each	1225 - 2450
60" wide	Each	3250
72" wide	Each	3725
Combination range, refrigerator, sink,		
microwave oven & icemaker	Each	5500
Compactor, residential, 4-1 compaction	Each	465 - 550
Dishwasher, built-in, 2 cycles	Each	495 - 740
4 cycles	Each	530 - 1000
Garbage disposer, sink type	Each	128 - 287
Hood for range, 2 speed, vented, 30" wide	Each	199 - 760
42" wide	Each	350 - 1100

Description	Unit	$ Cost
Appliances, cont.		
Refrigerator, no frost 10-12 C.F.	Each	555 - 860
14-16 C.F.	Each	570 - 710
18-20 C.F.	Each	630 - 950
Elevators, Hydraulic passenger, 2 stops		
1500# capacity	Each	43,425
2500# capacity	Each	44,625
3500# capacity	Each	48,525
Laundry Equipment		
Dryer, gas, 16 lb. capacity	Each	725
30 lb. capacity	Each	2800
Washer, 4 cycle	Each	810
Commercial	Each	1225
Sound System		
Amplifier, 250 watts	Each	1675
Speaker, ceiling or wall	Each	147
Trumpet	Each	275

BUILDING TYPES

Model costs calculated for a 2 story building
with 10' story height and 10,000 square feet
of floor area

Fraternity/Sorority House

				Unit	Unit Cost	Cost Per S.F.	% Of Sub-Total
A. SUBSTRUCTURE							
1010	Standard Foundations	Poured concrete; strip and spread footings		S.F. Ground	2.00	1.00	
1030	Slab on Grade	4" reinforced concrete with vapor barrier and granular base		S.F. Slab	3.42	1.71	
2010	Basement Excavation	Site preparation for slab and trench for foundation wall and footing		S.F. Ground	1.31	.65	6.8%
2020	Basement Walls	4' foundation wall		L.F. Wall	46	1.41	
B. SHELL							
	B10 Superstructure						
1010	Floor Construction	Plywood on wood joists		S.F. Floor	3.34	1.67	
1020	Roof Construction	Plywood on wood rafters (pitched)		S.F. Roof	2.80	1.57	4.6%
	B20 Exterior Enclosure						
2010	Exterior Walls	Cedar bevel siding on wood studs, insulated	80% of wall	S.F. Wall	8.00	3.84	
2020	Exterior Windows	Double hung wood	20% of wall	Each	440	2.11	9.9%
2030	Exterior Doors	Solid core wood		Each	1453	1.02	
	B30 Roofing						
3010	Roof Coverings	Asphalt shingles with flashing (pitched); rigid fiberglass insulation		S.F. Roof	3.04	1.52	2.2%
3020	Roof Openings	N/A		—	—	—	
C. INTERIORS							
1010	Partitions	Gypsum board on wood studs	25 S.F. Floor/L.F. Partition	S.F. Partition	5.59	1.79	
1020	Interior Doors	Single leaf wood	200 S.F. Floor/Door	Each	432	2.16	
1030	Fittings	N/A		—	—	—	
2010	Stair Construction	Wood		Flight	1195	.48	20.5%
3010	Wall Finishes	Paint		S.F. Surface	1.16	.37	
3020	Floor Finishes	10% hardwood, 70% carpet, 20% ceramic tile		S.F. Floor	6.67	6.67	
3030	Ceiling Finishes	Gypsum board on wood furring		S.F. Ceiling	2.95	2.95	
D. SERVICES							
	D10 Conveying						
1010	Elevators & Lifts	One hydraulic passenger elevator		Each	52,200	5.22	7.4%
1020	Escalators & Moving Walks	N/A		—	—	—	
	D20 Plumbing						
2010	Plumbing Fixtures	Kitchen toilet and service fixtures, supply and drainage	1 Fixture/150 S.F. Floor	Each	670	4.47	
2020	Domestic Water Distribution	Gas fired water heater		S.F. Floor	.94	.94	7.7%
2040	Rain Water Drainage	N/A		—	—	—	
	D30 HVAC						
3010	Energy Supply	Oil fired hot water, baseboard radiation		S.F. Floor	4.38	4.38	
3020	Heat Generating Systems	N/A		—	—	—	
3030	Cooling Generating Systems	Split system with air cooled condensing unit		S.F. Floor	6.22	6.22	15.1%
3050	Terminal & Package Units	N/A		—	—	—	
3090	Other HVAC Sys. & Equipment	N/A		—	—	—	
	D40 Fire Protection						
4010	Sprinklers	Wet pipe sprinkler system		S.F. Floor	1.95	1.95	2.8%
4020	Standpipes	N/A		—	—	—	
	D50 Electrical						
5010	Electrical Service/Distribution	600 ampere service, panel board and feeders		S.F. Floor	3.07	3.07	
5020	Lighting & Branch Wiring	Fluorescent fixtures, receptacles, switches, A.C. and misc. power		S.F. Floor	6.74	6.74	23.0%
5030	Communications & Security	Alarm, communication system and generator set		S.F. Floor	6.16	6.16	
5090	Other Electrical Systems	Emergency generator, 7.5 kW		S.F. Floor	.15	.15	
E. EQUIPMENT & FURNISHINGS							
1010	Commercial Equipment	N/A		—	—	—	
1020	Institutional Equipment	N/A		—	—	—	
1030	Vehicular Equipment	N/A		—	—	—	0.0%
1090	Other Equipment	N/A		—	—	—	
F. SPECIAL CONSTRUCTION & DEMOLITION							
1020	Integrated Construction	N/A		—	—	—	0.0%
1040	Special Facilities	N/A		—	—	—	
G. BUILDING SITEWORK	**N/A**						

	Sub-Total	70.22	**100%**
CONTRACTOR FEES (General Requirements: 10%, Overhead: 5%, Profit: 10%)	25%	17.56	
ARCHITECT FEES	9%	7.92	
Total Building Cost		**95.70**	

BUILDING TYPES

Costs per square foot of floor area

Exterior Wall	S.F. Area	4000	6000	8000	10000	12000	14000	16000	18000	20000
	L.F. Perimeter	260	320	384	424	460	484	510	540	576
Vertical Redwood Siding	Wood Frame	111.85	101.25	96.10	92.20	89.50	87.25	85.65	84.40	83.60
Brick Veneer	Wood Frame	119.80	107.90	102.15	97.60	94.45	91.80	89.85	88.45	87.50
Aluminum Siding	Wood Frame	109.75	99.65	94.70	91.05	88.50	86.45	84.95	83.80	83.05
Brick on Block	Wood Truss	125.25	112.55	106.45	101.55	98.10	95.15	93.10	91.50	90.50
Limestone on Block	Wood Truss	137.05	122.20	115.10	109.15	104.95	101.35	98.75	96.85	95.60
Stucco on Block	Wood Truss	114.30	103.60	98.35	94.40	91.65	89.35	87.70	86.50	85.65
Perimeter Adj., Add or Deduct	Per 100 L.F.	8.45	5.60	4.20	3.35	2.80	2.45	2.10	1.90	1.65
Story Hgt. Adj., Add or Deduct	Per 1 Ft.	1.05	.85	.75	.70	.60	.55	.50	.50	.45

For Basement, add $19.35 per square foot of basement area

The above costs were calculated using the basic specifications shown on the facing page. These costs should be adjusted where necessary for design alternatives and owner's requirements. Reported completed project costs, for this type of structure, range from $64.46 to $198.20 per S.F.

Common additives

Description	Unit	$ Cost
Autopsy Table, Standard	Each	8200
Deluxe	Each	12,300
Directory Boards, Plastic, glass covered		
30" x 20"	Each	575
36" x 48"	Each	1075
Aluminum, 24" x 18"	Each	440
36" x 24"	Each	555
48" x 32"	Each	780
48" x 60"	Each	1675
Emergency Lighting, 25 watt, battery operated		
Lead battery	Each	227
Nickel cadmium	Each	660
Mortuary Refrigerator, End operated		
Two capacity	Each	12,000
Six capacity	Each	22,000

Description	Unit	$ Cost
Planters, Precast concrete		
48" diam., 24" high	Each	610
7" diam., 36" high	Each	1025
Fiberglass, 36" diam., 24" high	Each	425
60" diam., 24" high	Each	740
Smoke Detectors		
Ceiling type	Each	151
Duct type	Each	405

BUILDING TYPES

Model costs calculated for a 1 story building with 12' story height and 10,000 square feet of floor area

Funeral Home

				Unit	Unit Cost	Cost Per S.F.	% Of Sub-Total
A.	**SUBSTRUCTURE**						
1010	Standard Foundations	Poured concrete; strip and spread footings		S.F. Ground	1.19	1.19	
1030	Slab on Grade	4" reinforced concrete with vapor barrier and granular base		S.F. Slab	3.42	3.42	11.8%
2010	Basement Excavation	Site preparation for slab and trench for foundation wall and footing		S.F. Ground	1.18	1.18	
2020	Basement Walls	4' foundation wall		L.F. Wall	46	2.19	
B.	**SHELL**						
	B10 Superstructure						
1010	Floor Construction	N/A		–	–	–	5.9%
1020	Roof Construction	Plywood on wood truss		S.F. Roof	3.97	3.97	
	B20 Exterior Enclosure						
2010	Exterior Walls	1" x 4" vertical T & G redwood siding on wood studs	90% of wall	S.F. Wall	8.69	3.99	
2020	Exterior Windows	Double hung wood	10% of wall	Each	354	1.50	9.5%
2030	Exterior Doors	Wood swinging double doors, single leaf hollow metal		Each	1546	.93	
	B30 Roofing						
3010	Roof Coverings	Single ply membrane, fully adhered; polyisocyanurate sheets		S.F. Roof	3.83	3.83	5.7%
3020	Roof Openings	N/A		–	–	–	
C.	**INTERIORS**						
1010	Partitions	Gypsum board on wood studs with sound deadening board	15 S.F. Floor/L.F. Partition	S.F. Partition	6.41	3.42	
1020	Interior Doors	Single leaf wood	150 S.F. Floor/Door	Each	432	2.88	
1030	Fittings	N/A		–	–	–	
2010	Stair Construction	N/A		–	–	–	28.8%
3010	Wall Finishes	50% wallpaper, 25% wood paneling, 25% paint		S.F. Surface	4.72	2.52	
3020	Floor Finishes	70% carpet, 30% ceramic tile		S.F. Floor	7.63	7.63	
3030	Ceiling Finishes	Fiberglass board on exposed grid, suspended		S.F. Ceiling	3.06	3.06	
D.	**SERVICES**						
	D10 Conveying						
1010	Elevators & Lifts	N/A		–	–	–	0.0%
1020	Escalators & Moving Walks	N/A		–	–	–	
	D20 Plumbing						
2010	Plumbing Fixtures	Toilet and service fixtures, supply and drainage	1 Fixture/770 S.F. Floor	Each	1339	1.74	
2020	Domestic Water Distribution	Electric water heater		S.F. Floor	4.36	4.36	9.5%
2040	Rain Water Drainage	Roof drain		S.F. Roof	.32	.32	
	D30 HVAC						
3010	Energy Supply	N/A		–	–	–	
3020	Heat Generating Systems	Included in D3030		–	–	–	
3030	Cooling Generating Systems	Multizone rooftop unit, gas heating, electric cooling		S.F. Floor	12.60	12.60	18.6%
3050	Terminal & Package Units	N/A		–	–	–	
3090	Other HVAC Sys. & Equipment	N/A		–	–	–	
	D40 Fire Protection						
4010	Sprinklers	Wet pipe sprinkler system		S.F. Floor	1.79	1.79	2.6%
4020	Standpipes	N/A		–	–	–	
	D50 Electrical						
5010	Electrical Service/Distribution	400 ampere service, panel board and feeders		S.F. Floor	1.22	1.22	
5020	Lighting & Branch Wiring	Fluorescent fixtures, receptacles, switches, A.C. and misc. power		S.F. Floor	3.72	3.72	7.6%
5030	Communications & Security	Alarm systems and emergency lighting		S.F. Floor	.16	.16	
5090	Other Electrical Systems	Emergency generator, 15 kW		S.F. Floor	.06	.06	
E.	**EQUIPMENT & FURNISHINGS**						
1010	Commercial Equipment	N/A		–	–	–	
1020	Institutional Equipment	N/A		–	–	–	0.0%
1030	Vehicular Equipment	N/A		–	–	–	
1090	Other Equipment	N/A		–	–	–	
F.	**SPECIAL CONSTRUCTION & DEMOLITION**						
1020	Integrated Construction	N/A		–	–	–	0.0%
1040	Special Facilities	N/A		–	–	–	
G.	**BUILDING SITEWORK**	**N/A**					

		Sub-Total	67.68	100%
CONTRACTOR FEES (General Requirements: 10%, Overhead: 5%, Profit: 10%)		25%	16.92	
ARCHITECT FEES		9%	7.60	

Total Building Cost	**92.20**

Costs per square foot of floor area

Exterior Wall	S.F. Area	12000	14000	16000	19000	21000	23000	26000	28000	30000
	L.F. Perimeter	440	474	510	556	583	607	648	670	695
E.I.F.S. on Concrete Block	Steel Frame	78.75	76.55	75.00	73.00	71.90	70.90	69.80	69.05	68.55
Tilt-up Concrete Wall	Steel Frame	75.55	73.60	72.20	70.45	69.45	68.60	67.60	67.00	66.50
Face Brick with Concrete Block Back-up	Bearing Walls	79.35	76.85	75.10	72.90	71.60	70.45	69.25	68.45	67.80
	Steel Frame	82.15	79.65	77.90	75.70	74.40	73.25	72.05	71.20	70.60
Stucco on Concrete Block	Bearing Walls	73.70	71.65	70.20	68.35	67.30	66.40	65.40	64.70	64.20
	Steel Frame	76.85	74.75	73.30	71.45	70.40	69.50	68.50	67.80	67.30
Perimeter Adj., Add or Deduct	Per 100 L.F.	5.00	4.25	3.70	3.15	2.85	2.60	2.30	2.15	2.00
Story Hgt. Adj., Add or Deduct	Per 1 Ft.	1.20	1.10	1.05	.95	.90	.85	.80	.75	.75
For Basement, add $21.95 per square foot of basement area										

The above costs were calculated using the basic specifications shown on the facing page. These costs should be adjusted where necessary for design alternatives and owner's requirements. Reported completed project costs, for this type of structure, range from $34.30 to $99.80 per S.F.

Common additives

Description	Unit	$ Cost
Emergency Lighting, 25 watt, battery operated		
Lead battery	Each	227
Nickel cadmium	Each	660
Smoke Detectors		
Ceiling type	Each	151
Duct type	Each	405
Sound System		
Amplifier, 250 watts	Each	1675
Speaker, ceiling or wall	Each	147
Trumpet	Each	275

BUILDING TYPES

Model costs calculated for a 1 story building with 14' story height and 21,000 square feet of floor area

Garage, Auto Sales

				Unit	Unit Cost	Cost Per S.F.	% Of Sub-Total
A. SUBSTRUCTURE							
1010	Standard Foundations	Poured concrete; strip and spread footings		S.F. Ground	.96	.96	
1030	Slab on Grade	4" reinforced concrete with vapor barrier and granular base		S.F. Slab	5.09	5.09	
2010	Basement Excavation	Site preparation for slab and trench for foundation wall and footing		S.F. Ground	1.18	1.18	16.2%
2020	Basement Walls	4' foundation wall		L.F. Wall	52	1.47	
B. SHELL							
	B10 Superstructure						
1010	Floor Construction	N/A		—	—	—	
1020	Roof Construction	Metal deck, open web steel joists, beams, columns		S.F. Roof	5.63	5.63	10.5%
	B20 Exterior Enclosure						
2010	Exterior Walls	E.I.F.S.	70% of wall	S.F. Wall	18.82	5.12	
2020	Exterior Windows	Window wall	30% of wall	Each	37	4.39	21.7%
2030	Exterior Doors	Double aluminum and glass, hollow metal, steel overhead		Each	2497	2.14	
	B30 Roofing						
3010	Roof Coverings	Built-up tar and gravel with flashing; perlite/EPS composite insulation		S.F. Roof	3.69	3.69	6.9%
3020	Roof Openings	Skylight		S.F. Roof	—	—	
C. INTERIORS							
1010	Partitions	Gypsum board on metal studs	28 S.F. Floor/L.F. Partition	S.F. Partition	2.82	1.21	
1020	Interior Doors	Hollow metal	280 S.F. Floor/Door	Each	537	1.92	
1030	Fittings	N/A		—	—	—	
2010	Stair Construction	N/A		—	—	—	13.3%
3010	Wall Finishes	Paint		S.F. Surface	1.17	.50	
3020	Floor Finishes	50% vinyl tile, 50% paint		S.F. Floor	1.98	1.98	
3030	Ceiling Finishes	Fiberglass board on exposed grid, suspended	50% of area	S.F. Ceiling	3.06	1.53	
D. SERVICES							
	D10 Conveying						
1010	Elevators & Lifts	N/A		—	—	—	
1020	Escalators & Moving Walks	N/A		—	—	—	0.0%
	D20 Plumbing						
2010	Plumbing Fixtures	Toilet and service fixtures, supply and drainage	1 Fixture/1500 S.F. Floor	Each	1425	.95	
2020	Domestic Water Distribution	Gas fired water heater		S.F. Floor	.65	.65	4.1%
2040	Rain Water Drainage	Roof drains		S.F. Roof	.58	.58	
	D30 HVAC						
3010	Energy Supply	Gas fired hot water, unit heaters (service area)		S.F. Floor	2.61	2.61	
3020	Heat Generating Systems	N/A		—	—	—	
3030	Cooling Generating Systems	Single zone rooftop unit, gas heating, electric (office and showroom)		S.F. Floor	3.76	3.76	12.3%
3050	Terminal & Package Units	N/A		—	—	—	
3090	Other HVAC Sys. & Equipment	Underfloor garage exhaust system		S.F. Floor	.22	.22	
	D40 Fire Protection						
4010	Sprinklers	Wet pipe sprinkler system		S.F. Floor	2.30	2.30	4.3%
4020	Standpipes	N/A		—	—	—	
	D50 Electrical						
5010	Electrical Service/Distribution	200 ampere service, panel board and feeders		S.F. Floor	.30	.30	
5020	Lighting & Branch Wiring	Fluorescent fixtures, receptacles, switches, A.C. and misc. power		S.F. Floor	4.46	4.46	9.0%
5030	Communications & Security	Alarm systems and emergency lighting		S.F. Floor	.16	.16	
5090	Other Electrical Systems	Emergency generator, 7.5 kW		S.F. Floor	.05	.05	
E. EQUIPMENT & FURNISHINGS							
1010	Commercial Equipment	N/A		—	—	—	
1020	Institutional Equipment	N/A		—	—	—	
1030	Vehicular Equipment	Hoists, compressor, fuel pump		S.F. Floor	.89	.89	1.7%
1090	Other Equipment	N/A		—	—	—	
F. SPECIAL CONSTRUCTION & DEMOLITION							
1020	Integrated Construction	N/A		—	—	—	
1040	Special Facilities	N/A		—	—	—	0.0%
G. BUILDING SITEWORK	**N/A**						

	Sub-Total	53.74	100%
CONTRACTOR FEES (General Requirements: 10%, Overhead: 5%, Profit: 10%)	25%	13.44	
ARCHITECT FEES	7%	4.72	
Total Building Cost		**71.90**	

BUILDING TYPES

Costs per square foot of floor area

Exterior Wall	S.F. Area	85000	115000	145000	175000	205000	235000	265000	295000	325000
	L.F. Perimeter	529	638	723	823	923	951	1037	1057	1132
Face Brick with Concrete Block Back-up	Steel Frame	39.55	38.55	37.80	37.40	37.15	36.70	36.45	36.15	36.00
	R/Conc. Frame	33.45	32.45	31.70	31.30	31.05	30.60	30.35	30.05	29.90
Precast Concrete	Steel Frame	40.15	39.10	38.40	38.00	37.70	37.25	37.00	36.70	36.55
	R/Conc. Frame	33.55	32.55	31.80	31.40	31.10	30.65	30.40	30.15	30.00
Reinforced Concrete	Steel Frame	39.45	38.55	37.90	37.55	37.25	36.90	36.70	36.40	36.30
	R/Conc. Frame	32.65	31.75	31.05	30.70	30.45	30.05	29.85	29.60	29.45
Perimeter Adj., Add or Deduct	Per 100 L.F.	1.00	.75	.60	.50	.45	.35	.35	.30	.25
Story Hgt. Adj., Add or Deduct	Per 1 Ft.	.35	.30	.30	.25	.25	.20	.25	.20	.20
Basement—Not Applicable										

The above costs were calculated using the basic specifications shown on the facing page. These costs should be adjusted where necessary for design alternatives and owner's requirements. Reported completed project costs, for this type of structure, range from $19.60 to $81.40 per S.F.

Common additives

Description	Unit	$ Cost
Elevators, Electric passenger, 5 stops		
2000# capacity	Each	101,400
3500# capacity	Each	107,400
5000# capacity	Each	111,900
Barrier gate w/programmable controller	Each	3450
Booth for attendant, average	Each	10,800
Fee computer	Each	13,200
Ticket spitter with time/date stamp	Each	6475
Mag strip encoding	Each	18,600
Collection station, pay on foot	Each	111,500
Parking control software	Each	22,200 - 91,000
Painting, Parking stalls	Stall	8.05
Parking Barriers		
Timber with saddles, 4" x 4"	L.F.	5.80
Precast concrete, 6" x 10" x 6'	Each	45
Traffic Signs, directional, 12" x 18", high densit	Each	59

BUILDING TYPES

Model costs calculated for a 5 story building with 10' story height and 145,000 square feet of floor area

			Unit	Unit Cost	Cost Per S.F.	% Of Sub-Total
A. SUBSTRUCTURE						
1010	Standard Foundations	Poured concrete; strip and spread footings	S.F. Ground	4.70	.94	
1030	Slab on Grade	6" reinforced concrete with vapor barrier and granular base	S.F. Slab	4.41	.88	12.5%
2010	Basement Excavation	Site preparation for slab and trench for foundation wall and footing	S.F. Ground	1.09	.22	
2020	Basement Walls	4' foundation wall	L.F. Wall	46	.94	
B. SHELL						
	B10 Superstructure					
1010	Floor Construction	Double tee precast concrete slab, precast concrete columns	S.F. Floor	15.27	12.22	51.1%
1020	Roof Construction	N/A	—	—	—	
	B20 Exterior Enclosure					
2010	Exterior Walls	Face brick with concrete block backup 40% of story height	S.F. Wall	21	2.10	
2020	Exterior Windows	N/A	—	—	—	8.8%
2030	Exterior Doors	N/A	—	—	—	
	B30 Roofing					
3010	Roof Coverings	N/A	—	—	—	0.0%
3020	Roof Openings	N/A	—	—	—	
C. INTERIORS						
1010	Partitions	Concrete block	S.F. Partition	21	.81	
1020	Interior Doors	Hollow metal	Each	10,740	.07	
1030	Fittings	N/A	—	—	—	
2010	Stair Construction	Concrete	Flight	2800	.19	4.8%
3010	Wall Finishes	Paint	S.F. Surface	2.34	.09	
3020	Floor Finishes	N/A	—	—	—	
3030	Ceiling Finishes	N/A	—	—	—	
D. SERVICES						
	D10 Conveying					
1010	Elevators & Lifts	Two hydraulic passenger elevators	Each	93,525	1.29	5.4%
1020	Escalators & Moving Walks	N/A	—	—	—	
	D20 Plumbing					
2010	Plumbing Fixtures	Toilet and service fixtures, supply and drainage 1 Fixture/18,125 S.F. Floor	Each	1812	.10	
2020	Domestic Water Distribution	Electric water heater	S.F. Floor	.04	.04	3.6%
2040	Rain Water Drainage	Roof drains	S.F. Roof	3.55	.71	
	D30 HVAC					
3010	Energy Supply	N/A	—	—	—	
3020	Heat Generating Systems	N/A	—	—	—	
3030	Cooling Generating Systems	N/A	—	—	—	0.0%
3050	Terminal & Package Units	N/A	—	—	—	
3090	Other HVAC Sys. & Equipment	N/A	—	—	—	
	D40 Fire Protection					
4010	Sprinklers	N/A	—	—	—	0.2%
4020	Standpipes	Standpipes and hose systems	S.F. Floor	.05	.05	
	D50 Electrical					
5010	Electrical Service/Distribution	400 ampere service, panel board and feeders	S.F. Floor	.15	.15	
5020	Lighting & Branch Wiring	Fluorescent fixtures, receptacles, switches and misc. power	S.F. Floor	2.03	2.03	9.5%
5030	Communications & Security	Alarm systems and emergency lighting	S.F. Floor	.09	.09	
5090	Other Electrical Systems	Emergency generator, 7.5 kW	S.F. Floor	.04	.04	
E. EQUIPMENT & FURNISHINGS						
1010	Commercial Equipment	N/A	—	—	—	
1020	Institutional Equipment	N/A	—	—	—	4.1%
1030	Vehicular Equipment	Ticket dispensers, booths, automatic gates	S.F. Floor	.97	.97	
1090	Other Equipment	N/A	—	—	—	
F. SPECIAL CONSTRUCTION & DEMOLITION						
1020	Integrated Construction	N/A	—	—	—	0.0%
1040	Special Facilities	N/A	—	—	—	
G. BUILDING SITEWORK	**N/A**					

				Sub-Total	23.93	100%
	CONTRACTOR FEES (General Requirements: 10%, Overhead: 5%, Profit: 10%)			25%	5.98	
	ARCHITECT FEES			6%	1.79	

Total Building Cost	**31.70**	

BUILDING TYPES

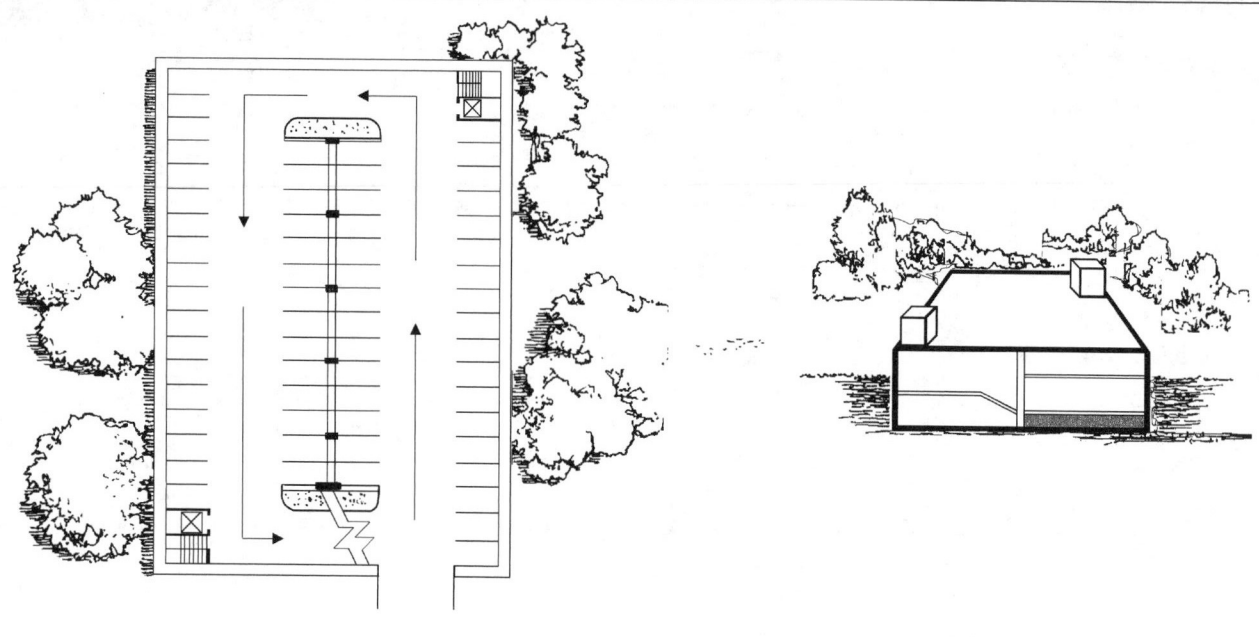

Costs per square foot of floor area

Exterior Wall	S.F. Area	20000	30000	40000	50000	75000	100000	125000	150000	175000
	L.F. Perimeter	400	500	600	650	775	900	1000	1100	1185
Reinforced Concrete	R/Conc. Frame	62.40	57.90	55.65	53.50	50.65	49.20	48.20	47.55	46.95
Perimeter Adj., Add or Deduct	Per 100 L.F.	3.95	2.65	2.00	1.55	1.05	.80	.65	.55	.45
Story Hgt. Adj., Add or Deduct	Per 1 Ft.	1.50	1.25	1.15	1.00	.75	.70	.60	.55	.50
Basement—Not Applicable										

The above costs were calculated using the basic specifications shown on the facing page. These costs should be adjusted where necessary for design alternatives and owner's requirements. Reported completed project costs, for this type of structure, range from $30.60 to $72.95 per S.F.

Common additives

Description	Unit	$ Cost
Elevators, Hydraulic passenger, 2 stops		
1500# capacity	Each	43,425
2500# capacity	Each	44,625
3500# capacity	Each	48,525
Barrier gate w/programmable controller	Each	3450
Booth for attendant, average	Each	10,800
Fee computer	Each	13,200
Ticket spitter with time/date stamp	Each	6475
Mag strip encoding	Each	18,600
Collection station, pay on foot	Each	111,500
Parking control software	Each	22,200 - 91,000
Painting, Parking stalls	Stall	8.05
Parking Barriers		
Timber with saddles, 4" x 4"	L.F.	5.80
Precast concrete, 6" x 10" x 6'	Each	45
Traffic Signs, directional, 12" x 18"	Each	59

Garage, Underground Parking

				Unit	Unit Cost	Cost Per S.F.	% Of Sub-Total
A. SUBSTRUCTURE							
1010	Standard Foundations	Poured concrete; strip and spread footings and waterproofing		S.F. Ground	5.26	2.63	
1030	Slab on Grade	5" reinforced concrete with vapor barrier and granular base		S.F. Slab	4.47	2.23	23.2%
2010	Basement Excavation	Excavation 24' deep		S.F. Ground	7.20	3.60	
2020	Basement Walls	N/A		—	—	—	
B. SHELL							
	B10 Superstructure						
1010	Floor Construction	Cast-in-place concrete beam and slab, concrete columns		S.F. Floor	18.58	9.29	49.1%
1020	Roof Construction	Cast-in-place concrete beam and slab, concrete columns		S.F. Roof	17.20	8.60	
	B20 Exterior Enclosure						
2010	Exterior Walls	Cast-in place concrete		S.F. Wall	16.33	2.94	
2020	Exterior Windows	N/A		—	—	—	8.4%
2030	Exterior Doors	Steel overhead, hollow metal		Each	3078	.12	
	B30 Roofing						
3010	Roof Coverings	Neoprene membrane traffic deck		S.F. Roof	3.30	1.65	4.5%
3020	Roof Openings	N/A		—	—	—	
C. INTERIORS							
1010	Partitions	Concrete block		S.F. Partition	27	.53	
1020	Interior Doors	Hollow metal		Each	4296	.04	
1030	Fittings	N/A		—	—	—	
2010	Stair Construction	Concrete		Flight	4325	.22	2.4%
3010	Wall Finishes	Paint		S.F. Surface	3.64	.07	
3020	Floor Finishes	N/A		—	—	—	
3030	Ceiling Finishes	N/A		—	—	—	
D. SERVICES							
	D10 Conveying						
1010	Elevators & Lifts	Two hydraulic passenger elevators		Each	52,000	1.04	2.9%
1020	Escalators & Moving Walks	N/A		—	—	—	
	D20 Plumbing						
2010	Plumbing Fixtures	Drainage in parking areas, toilets, & service fixtures	1 Fixture/5000 S.F. Floor	Each	.23	.23	
2020	Domestic Water Distribution	Electric water heater		S.F. Floor	.06	.06	2.4%
2040	Rain Water Drainage	Roof drains		S.F. Roof	1.14	.57	
	D30 HVAC						
3010	Energy Supply	N/A		—	—	—	
3020	Heat Generating Systems	N/A		—	—	—	
3030	Cooling Generating Systems	N/A		—	—	—	0.3%
3050	Terminal & Package Units	Exhaust fans		S.F. Floor	.10	.10	
3090	Other HVAC Sys. & Equipment	N/A		—	—	—	
	D40 Fire Protection						
4010	Sprinklers	N/A		—	—	—	0.2%
4020	Standpipes	Dry standpipe system, class 1		S.F. Floor	.08	.08	
	D50 Electrical						
5010	Electrical Service/Distribution	200 ampere service, panel board and feeders		S.F. Floor	.08	.08	
5020	Lighting & Branch Wiring	Fluorescent fixtures, receptacles, switches and misc. power		S.F. Floor	1.73	1.73	5.8%
5030	Communications & Security	Alarm systems and emergency lighting		S.F. Floor	.30	.30	
5090	Other Electrical Systems	Emergency generator, 11.5 kW		S.F. Floor	.05	.05	
E. EQUIPMENT & FURNISHINGS							
1010	Commercial Equipment	N/A		—	—	—	
1020	Institutional Equipment	N/A		—	—	—	0.8%
1030	Vehicular Equipment	Ticket dispensers, booths, automatic gates		S.F. Floor	.29	.29	
1090	Other Equipment	N/A		—	—	—	
F. SPECIAL CONSTRUCTION & DEMOLITION							
1020	Integrated Construction	N/A		—	—	—	0.0%
1040	Special Facilities	N/A		—	—	—	
G. BUILDING SITEWORK	**N/A**						

		Sub-Total	36.45	100%
CONTRACTOR FEES (General Requirements: 10%, Overhead: 5%, Profit: 10%)		25%	9.11	
ARCHITECT FEES		8%	3.64	

Total Building Cost	**49.20**

BUILDING TYPES

Costs per square foot of floor area

Exterior Wall	S.F. Area	2000	4000	6000	8000	10000	12000	14000	16000	18000
	L.F. Perimeter	180	260	340	420	500	580	586	600	610
Concrete Block	Wood Joists	108.15	92.90	87.80	85.30	83.80	82.75	80.20	78.45	77.05
	Steel Joists	107.75	92.95	88.00	85.55	84.10	83.10	80.55	78.85	77.45
Poured Concrete	Wood Joists	118.35	100.45	94.50	91.55	89.75	88.55	85.20	82.95	81.10
	Steel Joists	118.75	100.85	94.90	91.90	90.15	88.95	85.60	83.35	81.50
Stucco	Wood Frame	117.30	99.70	93.85	90.95	89.20	88.00	84.75	82.50	80.75
Insulated Metal Panels	Steel Frame	105.30	91.85	87.40	85.15	83.85	82.95	80.75	79.25	78.05
Perimeter Adj., Add or Deduct	Per 100 L.F.	17.10	8.60	5.70	4.25	3.40	2.85	2.45	2.15	1.90
Story Hgt. Adj., Add or Deduct	Per 1 Ft.	1.25	.95	.80	.75	.70	.70	.60	.55	.45

For Basement, add $20.70 per square foot of basement area

The above costs were calculated using the basic specifications shown on the facing page. These costs should be adjusted where necessary for design alternatives and owner's requirements. Reported completed project costs, for this type of structure, range from $43.90 to $131.80 per S.F.

Common additives

Description	Unit	$ Cost
Air Compressors		
Electric 1-1/2 H.P., standard controls	Each	1050
Dual controls	Each	1225
5 H.P. 115/230 Volt, standard controls	Each	2750
Dual controls	Each	2850
Product Dispenser		
with vapor recovery for 6 nozzles	Each	16,500
Hoists, Single post		
8000# cap., swivel arm	Each	6500
Two post, adjustable frames, 11,000# cap.	Each	9175
24,000# cap.	Each	13,700
7500# Frame support	Each	7625
Four post, roll on ramp	Each	7075
Lockers, Steel, single tier, 60" or 72"	Opening	131 - 227
2 tier, 60" or 72" total	Opening	74 - 125
5 tier, box lockers	Opening	42 - 62
Locker bench, lam. maple top only	L.F.	19.50
Pedestals, steel pipe	Each	63
Lube Equipment		
3 reel type, with pumps, no piping	Each	8725
Spray Painting Booth, 26' long, complete	Each	16,300

Model costs calculated for a 1 story building with 14' story height and 10,000 square feet of floor area

					Unit	Unit Cost	Cost Per S.F.	% Of Sub-Total
A. SUBSTRUCTURE								
1010	Standard Foundations	Poured concrete; strip and spread footings			S.F. Ground	1.47	1.47	
1030	Slab on Grade	6" reinforced concrete with vapor barrier and granular base			S.F. Slab	5.09	5.09	16.7%
2010	Basement Excavation	Site preparation for slab and trench for foundation wall and footing			S.F. Ground	1.18	1.18	
2020	Basement Walls	4' foundation wall			L.F. Wall	52	2.64	
B. SHELL								
	B10 Superstructure							
1010	Floor Construction	N/A			—	—	—	5.1%
1020	Roof Construction	Metal deck on open web steel joists			S.F. Roof	3.19	3.19	
	B20 Exterior Enclosure							
2010	Exterior Walls	Concrete block		80% of wall	S.F. Wall	8.05	4.51	
2020	Exterior Windows	Hopper type commercial steel		5% of wall	Each	314	.73	10.4%
2030	Exterior Doors	Steel overhead and hollow metal		15% of wall	Each	12.00	1.26	
	B30 Roofing							
3010	Roof Coverings	Built-up tar and gravel; perlite/EPS composite insulation			S.F. Roof	4.22	4.22	6.8%
3020	Roof Openings	Skylight			S.F. Roof	.01	.01	
C. INTERIORS								
1010	Partitions	Concrete block		50 S.F. Floor/L.F. Partition	S.F. Partition	5.05	1.01	
1020	Interior Doors	Single leaf hollow metal		3000 S.F. Floor/Door	Each	537	.18	
1030	Fittings	Toilet partitions			S.F. Floor	.13	.13	
2010	Stair Construction	N/A			—	—	—	7.1%
3010	Wall Finishes	Paint			S.F. Surface	10.45	2.09	
3020	Floor Finishes	90% metallic floor hardener, 10% vinyl composition tile			S.F. Floor	.80	.80	
3030	Ceiling Finishes	Gypsum board on wood joists in office and washrooms		10% of area	S.F. Ceiling	2.37	.24	
D. SERVICES								
	D10 Conveying							
1010	Elevators & Lifts	N/A			—	—	—	0.0%
1020	Escalators & Moving Walks	N/A			—	—	—	
	D20 Plumbing							
2010	Plumbing Fixtures	Toilet and service fixtures, supply and drainage		1 Fixture/500 S.F. Floor	Each	755	1.51	
2020	Domestic Water Distribution	Gas fired water heater			S.F. Floor	.24	.24	4.1%
2040	Rain Water Drainage	Roof drains			S.F. Roof	.80	.80	
	D30 HVAC							
3010	Energy Supply	Oil fired hot water, unit heaters			S.F. Floor	5.22	5.22	
3020	Heat Generating Systems	N/A			—	—	—	
3030	Cooling Generating Systems	Split systems with air cooled condensing units			S.F. Floor	8.07	8.07	21.9%
3050	Terminal & Package Units	N/A			—	—	—	
3090	Other HVAC Sys. & Equipment	Garage exhaust system			S.F. Floor	.34	.34	
	D40 Fire Protection							
4010	Sprinklers	Sprinklers, ordinary hazard			S.F. Floor	2.54	2.54	4.1%
4020	Standpipes	N/A			—	—	—	
	D50 Electrical							
5010	Electrical Service/Distribution	200 ampere service, panel board and feeders			S.F. Floor	.34	.34	
5020	Lighting & Branch Wiring	Fluorescent fixtures, receptacles, switches, A.C. and misc. power			S.F. Floor	4.58	4.58	8.2%
5030	Communications & Security	Alarm systems and emergency lighting			S.F. Floor	.11	.11	
5090	Other Electrical Systems	Emergency generator, 15 kW			S.F. Floor	.06	.06	
E. EQUIPMENT & FURNISHINGS								
1010	Commercial Equipment	N/A			—	—	—	
1020	Institutional Equipment	N/A			—	—	—	15.6%
1030	Vehicular Equipment	Hoists			S.F. Floor	9.75	9.75	
1090	Other Equipment	N/A			—	—	—	
F. SPECIAL CONSTRUCTION & DEMOLITION								
1020	Integrated Construction	N/A			—	—	—	0.0%
1040	Special Facilities	N/A			—	—	—	
G. BUILDING SITEWORK	**N/A**							

	Sub-Total	62.31	**100%**
CONTRACTOR FEES (General Requirements: 10%, Overhead: 5%, Profit: 10%)	25%	15.58	
ARCHITECT FEES	8%	6.21	
Total Building Cost		**84.10**	

BUILDING TYPES

135

Costs per square foot of floor area

Exterior Wall	S.F. Area	600	800	1000	1200	1400	1600	1800	2000	2200
	L.F. Perimeter	100	120	126	140	153	160	170	180	190
Face Brick with Concrete Block Back-up	Wood Truss	160.80	147.85	133.55	127.10	122.20	116.75	113.30	110.55	108.25
	Steel Joists	154.80	141.80	127.55	121.10	116.15	110.70	107.25	104.50	102.20
Enameled Sandwich Panel	Steel Frame	140.05	128.60	116.55	111.00	106.75	102.15	99.20	96.90	94.90
Tile on Concrete Block	Steel Joists	166.45	152.45	136.70	129.70	124.30	118.25	114.50	111.45	108.95
Aluminum Siding	Wood Frame	136.50	125.95	115.15	110.10	106.25	102.15	99.50	97.40	95.65
Wood Siding	Wood Frame	138.85	128.05	116.90	111.75	107.80	103.55	100.80	98.65	96.85
Perimeter Adj., Add or Deduct	Per 100 L.F.	77.40	58.05	46.45	38.70	33.15	29.05	25.80	23.25	21.10
Story Hgt. Adj., Add or Deduct	Per 1 Ft.	5.10	4.60	3.85	3.55	3.35	3.05	2.90	2.75	2.65
Basement—Not Applicable										

The above costs were calculated using the basic specifications shown on the facing page. These costs should be adjusted where necessary for design alternatives and owner's requirements. Reported completed project costs, for this type of structure, range from $31.60 to $159.65 per S.F.

Common additives

Description	Unit	$ Cost
Air Compressors		
Electric 1-1/2 H.P., standard controls	Each	1050
Dual controls	Each	1225
5 H.P. 115/230 volt, standard controls	Each	2750
Dual controls	Each	2850
Product Dispenser		
with vapor recovery for 6 nozzles	Each	16,500
Hoists, Single post		
8000# cap. swivel arm	Each	6500
Two post, adjustable frames, 11,000# cap.	Each	9175
24,000# cap.	Each	13,700
7500# cap.	Each	7625
Four post, roll on ramp	Each	7075
Lockers, Steel, single tier, 60" or 72"	Opening	131 - 227
2 tier, 60" or 72" total	Opening	74 - 125
5 tier, box lockers	Each	42 - 62
Locker bench, lam. maple top only	L.F.	19.50
Pedestals, steel pipe	Each	63
Lube Equipment		
3 reel type, with pumps, no piping	Each	8725

BUILDING TYPES

Important: See the Reference Section for Location Factors

Model costs calculated for a 1 story building with 12' story height and 1,400 square feet of floor area

				Unit	Unit Cost	Cost Per S.F.	% Of Sub-Total
A. SUBSTRUCTURE							
1010	Standard Foundations	Poured concrete; strip and spread footings		S.F. Ground	2.80	2.80	
1030	Slab on Grade	6" reinforced concrete with vapor barrier and granular base		S.F. Slab	5.09	5.09	
2010	Basement Excavation	Site preparation for slab and trench for foundation wall and footing		S.F. Ground	1.71	1.71	16.3%
2020	Basement Walls	4' foundation wall		L.F. Wall	46	5.13	
B. SHELL							
	B10 Superstructure						
1010	Floor Construction	N/A		—	—	—	5.4%
1020	Roof Construction	Plywood on wood trusses		S.F. Roof	4.90	4.90	
	B20 Exterior Enclosure						
2010	Exterior Walls	Face brick with concrete block backup	60% of wall	S.F. Wall	21	16.60	
2020	Exterior Windows	Store front and metal top hinged outswinging	20% of wall	Each	29	7.71	34.3%
2030	Exterior Doors	Steel overhead, aluminum & glass and hollow metal	20% of wall	Each	25	6.73	
	B30 Roofing						
3010	Roof Coverings	Asphalt shingles with flashing; perlite/EPS composite insulation		S.F. Roof	2.29	2.29	2.5%
3020	Roof Openings	N/A		—	—	—	
C. INTERIORS							
1010	Partitions	Concrete block	25 S.F. Floor/L.F. Partition	S.F. Partition	5.03	1.61	
1020	Interior Doors	Single leaf hollow metal	700 S.F. Floor/Door	Each	537	.77	
1030	Fittings	Toilet partitions		S.F. Floor	1.84	1.84	
2010	Stair Construction	N/A		—	—	—	9.6%
3010	Wall Finishes	Paint		S.F. Surface	7.78	2.49	
3020	Floor Finishes	Vinyl composition tile	35% of floor area	S.F. Floor	2.74	.96	
3030	Ceiling Finishes	Painted gypsum board on furring in sales area & washrooms	35% of floor area	S.F. Ceiling	2.95	1.03	
D. SERVICES							
	D10 Conveying						
1010	Elevators & Lifts	N/A		—	—	—	0.0%
1020	Escalators & Moving Walks	N/A		—	—	—	
	D20 Plumbing						
2010	Plumbing Fixtures	Toilet and service fixtures, supply and drainage	1 Fixture/235 S.F. Floor	Each	1527	6.50	
2020	Domestic Water Distribution	Gas fired water heater		S.F. Floor	1.68	1.68	9.0%
2040	Rain Water Drainage	N/A		—	—	—	
	D30 HVAC						
3010	Energy Supply	Oil fired hot water, wall fin radiation		S.F. Floor	5.22	5.22	
3020	Heat Generating Systems	N/A		—	—	—	
3030	Cooling Generating Systems	Split systems with air cooled condensing units		S.F. Floor	6.56	6.56	13.0%
3050	Terminal & Package Units	N/A		—	—	—	
3090	Other HVAC Sys. & Equipment	N/A		—	—	—	
	D40 Fire Protection						
4010	Sprinklers	N/A		—	—	—	0.0%
4020	Standpipes	N/A		—	—	—	
	D50 Electrical						
5010	Electrical Service/Distribution	200 ampere service, panel board and feeders		S.F. Floor	3.14	3.14	
5020	Lighting & Branch Wiring	Fluorescent fixtures, receptacles, switches, A.C. and misc. power		S.F. Floor	5.00	5.00	9.9%
5030	Communications & Security	Alarm systems and emergency lighting		S.F. Floor	.29	.29	
5090	Other Electrical Systems	Emergency generator		S.F. Floor	.46	.46	
E. EQUIPMENT & FURNISHINGS							
1010	Commercial Equipment	N/A		—	—	—	
1020	Institutional Equipment	N/A		—	—	—	0.0%
1030	Vehicular Equipment	N/A		—	—	—	
1090	Other Equipment	N/A		—	—	—	
F. SPECIAL CONSTRUCTION & DEMOLITION							
1020	Integrated Construction	N/A		—	—	—	0.0%
1040	Special Facilities	N/A		—	—	—	
G. BUILDING SITEWORK	**N/A**						

			Sub-Total	90.51	100%
CONTRACTOR FEES (General Requirements: 10%, Overhead: 5%, Profit: 10%)			25%	22.63	
ARCHITECT FEES			8%	9.06	
		Total Building Cost		**122.20**	

BUILDING TYPES

Costs per square foot of floor area

Exterior Wall	S.F. Area	12000	16000	20000	25000	30000	35000	40000	45000	50000
	L.F. Perimeter	440	520	600	700	708	780	841	910	979
Reinforced Concrete Block	Lam. Wood Arches	107.95	103.80	101.30	99.30	96.30	95.15	94.10	93.45	92.90
	Rigid Steel Frame	106.75	102.55	100.10	98.10	95.10	93.95	92.90	92.25	91.70
Face Brick with Concrete Block Back-up	Lam. Wood Arches	125.20	119.05	115.40	112.45	107.40	105.60	104.00	102.95	102.10
	Rigid Steel Frame	123.95	117.85	114.20	111.25	106.20	104.40	102.80	101.75	100.90
Metal Sandwich Panels	Lam. Wood Arches	104.25	100.55	98.30	96.50	93.95	92.90	92.00	91.40	90.95
	Rigid Steel Frame	103.05	99.35	97.10	95.30	92.75	91.70	90.80	90.20	89.70
Perimeter Adj., Add or Deduct	Per 100 L.F.	4.55	3.40	2.75	2.20	1.85	1.55	1.35	1.20	1.10
Story Hgt. Adj., Add or Deduct	Per 1 Ft.	.65	.55	.50	.50	.40	.35	.35	.35	.35
Basement—Not Applicable										

The above costs were calculated using the basic specifications shown on the facing page. These costs should be adjusted where necessary for design alternatives and owner's requirements. Reported completed project costs, for this type of structure, range from $47.95 to $143.20 per S.F.

Common additives

Description	Unit	$ Cost
Bleachers, Telescoping, manual		
To 15 tier	Seat	79 - 110
16-20 tier	Seat	161 - 196
21-30 tier	Seat	171 - 206
For power operation, add	Seat	31 - 48
Gym Divider Curtain, Mesh top		
Manual roll-up	S.F.	9.45
Gym Mats		
2" naugahyde covered	S.F.	3.45
2" nylon	S.F.	5.15
1-1/2" wall pads	S.F.	6.80
1" wrestling mats	S.F.	4.62
Scoreboard		
Basketball, one side	Each	2775 - 17,500
Basketball Backstop		
Wall mtd., 6' extended, fixed	Each	1675 - 2150
Swing up, wall mtd.	Each	1875 - 2825

Description	Unit	$ Cost
Lockers, Steel, single tier, 60" or 72"	Opening	131 - 227
2 tier, 60" or 72" total	Opening	74 - 125
5 tier, box lockers	Opening	42 - 62
Locker bench, lam. maple top only	L.F.	19.50
Pedestals, steel pipe	Each	63
Sound System		
Amplifier, 250 watts	Each	1675
Speaker, ceiling or wall	Each	147
Trumpet	Each	275
Emergency Lighting, 25 watt, battery operated		
Lead battery	Each	227
Nickel cadmium	Each	660

Model costs calculated for a 1 story building with 25' story height and 20,000 square feet of floor area

				Unit	Unit Cost	Cost Per S.F.	% Of Sub-Total
A. SUBSTRUCTURE							
1010	Standard Foundations	Poured concrete; strip and spread footings		S.F. Ground	1.17	1.17	
1030	Slab on Grade	4" reinforced concrete with vapor barrier and granular base		S.F. Slab	3.42	3.42	9.7%
2010	Basement Excavation	Site preparation for slab and trench for foundation wall and footing		S.F. Ground	1.09	1.09	
2020	Basement Walls	4' foundation wall		L.F. Wall	46	1.65	
B. SHELL							
B10 Superstructure							
1010	Floor Construction	N/A		—	—	—	
1020	Roof Construction	Wood deck on laminated wood arches		S.F. Roof	13.63	13.63	18.0%
B20 Exterior Enclosure							
2010	Exterior Walls	Reinforced concrete block (end walls included)	90% of wall	S.F. Wall	8.77	5.92	
2020	Exterior Windows	Metal horizontal pivoted	10% of wall	Each	312	2.34	11.3%
2030	Exterior Doors	Aluminum and glass, hollow metal, steel overhead		Each	1085	.33	
B30 Roofing							
3010	Roof Coverings	EPDM, 60 mils, fully adhered; polyisocyanurate insulation		S.F. Roof	3.02	3.02	4.0%
3020	Roof Openings	N/A		—	—	—	
C. INTERIORS							
1010	Partitions	Concrete block	50 S.F. Floor/L.F. Partition	S.F. Partition	6.05	1.21	
1020	Interior Doors	Single leaf hollow metal	500 S.F. Floor/Door	Each	537	1.07	
1030	Fittings	Toilet partitions		S.F. Floor	.26	.26	
2010	Stair Construction	N/A		—	—	—	20.8%
3010	Wall Finishes	50% paint, 50% ceramic tile		S.F. Surface	14.05	2.81	
3020	Floor Finishes	90% hardwood, 10% ceramic tile		S.F. Floor	9.83	9.83	
3030	Ceiling Finishes	Mineral fiber tile on concealed zee bars	15% of area	S.F. Ceiling	3.71	.56	
D. SERVICES							
D10 Conveying							
1010	Elevators & Lifts	N/A		—	—	—	0.0%
1020	Escalators & Moving Walks	N/A		—	—	—	
D20 Plumbing							
2010	Plumbing Fixtures	Toilet and service fixtures, supply and drainage	1 Fixture/515 S.F. Floor	Each	2065	4.01	
2020	Domestic Water Distribution	Electric water heater		S.F. Floor	2.43	2.43	8.5%
2040	Rain Water Drainage	N/A		—	—	—	
D30 HVAC							
3010	Energy Supply	N/A		—	—	—	
3020	Heat Generating Systems	Included in D3030		—	—	—	
3030	Cooling Generating Systems	Single zone rooftop unit, gas heating, electric cooling		S.F. Floor	9.72	9.72	12.8%
3050	Terminal & Package Units	N/A		—	—	—	
3090	Other HVAC Sys. & Equipment	N/A		—	—	—	
D40 Fire Protection							
4010	Sprinklers	Wet pipe sprinkler system		S.F. Floor	1.79	1.79	2.4%
4020	Standpipes	N/A		—	—	—	
D50 Electrical							
5010	Electrical Service/Distribution	400 ampere service, panel board and feeders		S.F. Floor	.69	.69	
5020	Lighting & Branch Wiring	Fluorescent fixtures, receptacles, switches, A.C. and misc. power		S.F. Floor	6.15	6.15	11.0%
5030	Communications & Security	Alarm systems, sound system and emergency lighting		S.F. Floor	1.33	1.33	
5090	Other Electrical Systems	Emergency generator, 7.5 kW		S.F. Floor	.15	.15	
E. EQUIPMENT & FURNISHINGS							
1010	Commercial Equipment	N/A		—	—	—	
1020	Institutional Equipment	N/A		—	—	—	1.5%
1030	Vehicular Equipment	N/A		—	—	—	
1090	Other Equipment	Bleachers, sauna, weight room		S.F. Floor	1.15	1.15	
F. SPECIAL CONSTRUCTION & DEMOLITION							
1020	Integrated Construction	N/A		—	—	—	0.0%
1040	Special Facilities	N/A		—	—	—	
G. BUILDING SITEWORK	**N/A**						

	Sub-Total	75.73	**100%**
CONTRACTOR FEES (General Requirements: 10%, Overhead: 5%, Profit: 10%)		25%	18.93
ARCHITECT FEES		7%	6.64
	Total Building Cost	101.30	

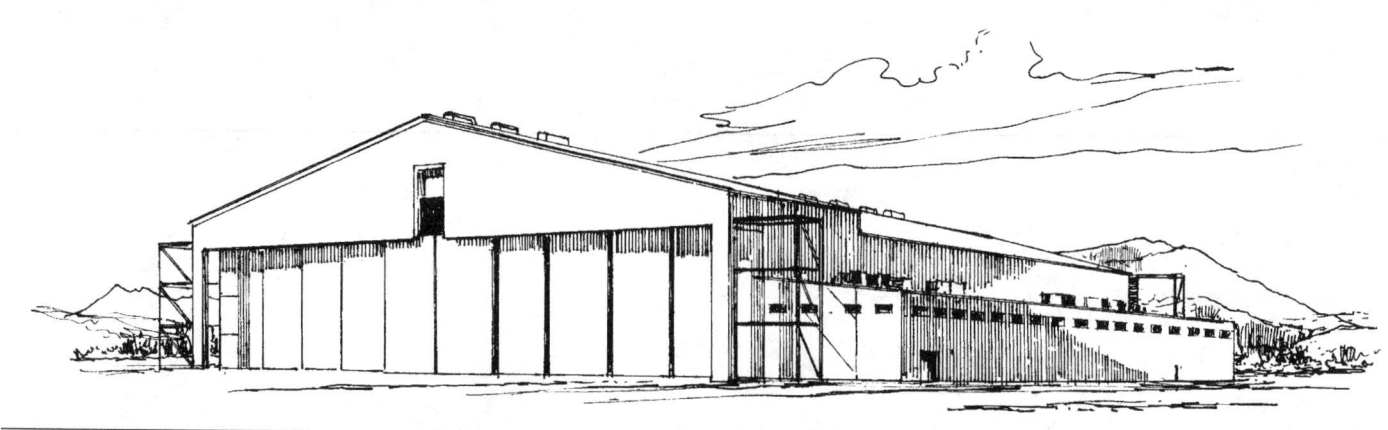

Costs per square foot of floor area

Exterior Wall	S.F. Area	5000	10000	15000	20000	30000	40000	50000	75000	100000
	L.F. Perimeter	300	410	500	580	710	830	930	1150	1300
Concrete Block Reinforced	Steel Frame	94.50	82.60	77.75	75.05	71.70	69.90	68.50	66.50	65.00
	Bearing Walls	97.95	85.90	81.05	78.30	74.95	73.10	71.70	69.65	68.15
Precast Concrete	Steel Frame	102.30	87.90	82.10	78.80	74.75	72.55	70.90	68.45	66.70
	Bearing Walls	105.45	91.05	85.25	81.95	77.90	75.70	74.05	71.60	69.85
Galv. Steel Siding	Steel Frame	93.90	82.20	77.45	74.75	71.50	69.65	68.30	66.30	64.85
Metal Sandwich Panel	Steel Frame	98.05	85.00	79.75	76.75	73.10	71.10	69.65	67.40	65.80
Perimeter Adj., Add or Deduct	Per 100 L.F.	12.55	6.25	4.15	3.15	2.05	1.55	1.25	.80	.60
Story Hgt. Adj., Add or Deduct	Per 1 Ft.	1.30	.85	.70	.60	.50	.45	.40	.30	.30
Basement—Not Applicable										

The above costs were calculated using the basic specifications shown on the facing page. These costs should be adjusted where necessary for design alternatives and owner's requirements. Reported completed project costs, for this type of structure, range from $27.50 to $125.50 per S.F.

Common additives

Description	Unit	$ Cost
Closed Circuit Surveillance, One station		
Camera and monitor	Each	1375
For additional camera stations, add	Each	750
Emergency Lighting, 25 watt, battery operated		
Lead battery	Each	227
Nickel cadmium	Each	660
Lockers, Steel, single tier, 60" or 72"	Opening	131 - 227
2 tier, 60" or 72" total	Opening	74 - 125
5 tier, box lockers	Opening	42 - 62
Locker bench, lam. maple top only	L.F.	19.50
Pedestals, steel pipe	Each	63
Safe, Office type, 4 hour rating		
30" x 18" x 18"	Each	3200
62" x 33" x 20"	Each	6975
Sound System		
Amplifier, 250 watts	Each	1675
Speaker, ceiling or wall	Each	147
Trumpet	Each	275

BUILDING TYPES

Model costs calculated for a 1 story building with 24' story height and 20,000 square feet of floor area

Hangar, Aircraft

				Unit	Unit Cost	Cost Per S.F.	% Of Sub-Total
A. SUBSTRUCTURE							
1010	Standard Foundations	Poured concrete; strip and spread footings and		S.F. Ground	1.29	1.29	
1030	Slab on Grade	6" reinforced concrete with vapor barrier and granular base		S.F. Slab	5.09	5.09	16.3%
2010	Basement Excavation	Site preparation for slab and trench for foundation wall and footing		S.F. Ground	1.18	1.18	
2020	Basement Walls	4' foundation wall		L.F. Wall	52	1.53	
B. SHELL							
	B10 Superstructure						
1010	Floor Construction	N/A		—	—	—	11.4%
1020	Roof Construction	Metal deck, open web steel joists, beams, columns		S.F. Roof	6.38	6.38	
	B20 Exterior Enclosure						
2010	Exterior Walls	Galvanized steel siding		S.F. Wall	6.58	2.29	
2020	Exterior Windows	Industrial horizontal pivoted steel	20% of wall	Each	1489	8.64	31.1%
2030	Exterior Doors	Steel overhead and sliding	30% of wall	Each	30	6.44	
	B30 Roofing						
3010	Roof Coverings	Elastomeric membrane; fiberboard insulation		S.F. Roof	3.16	3.16	5.8%
3020	Roof Openings	Roof hatches		S.F. Roof	.06	.06	
C. INTERIORS							
1010	Partitions	Concrete block	200 S.F. Floor/L.F. Partition	S.F. Partition	6.40	.32	
1020	Interior Doors	Single leaf hollow metal	5000 S.F. Floor/Door	Each	537	.11	
1030	Fittings	Toilet partitions		S.F. Floor	.22	.22	
2010	Stair Construction	N/A		—	—	—	1.3%
3010	Wall Finishes	Paint		S.F. Surface	1.80	.09	
3020	Floor Finishes	N/A		—	—	—	
3030	Ceiling Finishes	N/A		—	—	—	
D. SERVICES							
	D10 Conveying						
1010	Elevators & Lifts	N/A		—	—	—	0.0%
1020	Escalators & Moving Walks	N/A		—	—	—	
	D20 Plumbing						
2010	Plumbing Fixtures	Toilet and service fixtures, supply and drainage	1 Fixture/1000 S.F. Floor	Each	2010	2.01	
2020	Domestic Water Distribution	Gas fired water heater		S.F. Floor	.31	.31	4.2%
2040	Rain Water Drainage	N/A		—	—	—	
	D30 HVAC						
3010	Energy Supply	N/A		—	—	—	
3020	Heat Generating Systems	Unit heaters		Each	4.75	4.75	
3030	Cooling Generating Systems	N/A		—	—	—	8.8%
3050	Terminal & Package Units	Exhaust fan		S.F. Floor	.16	.16	
3090	Other HVAC Sys. & Equipment	N/A		—	—	—	
	D40 Fire Protection						
4010	Sprinklers	Sprinklers, extra hazard		S.F. Floor	3.58	3.58	6.4%
4020	Standpipes	N/A		—	—	—	
	D50 Electrical						
5010	Electrical Service/Distribution	200 ampere service, panel board and feeders		S.F. Floor	.59	.59	
5020	Lighting & Branch Wiring	High intensity discharge fixtures, receptacles, switches and misc. power		S.F. Floor	7.29	7.29	14.7%
5030	Communications & Security	Alarm systems and emergency lighting		S.F. Floor	.34	.34	
5090	Other Electrical Systems	Emergency generator, 7.5 kW		S.F. Floor	.05	.05	
E. EQUIPMENT & FURNISHINGS							
1010	Commercial Equipment	N/A		—	—	—	
1020	Institutional Equipment	N/A		—	—	—	0.0%
1030	Vehicular Equipment	N/A		—	—	—	
1090	Other Equipment	N/A		—	—	—	
F. SPECIAL CONSTRUCTION & DEMOLITION							
1020	Integrated Construction	N/A		—	—	—	0.0%
1040	Special Facilities	N/A		—	—	—	
G. BUILDING SITEWORK	**N/A**						

		Sub-Total	55.88	100%
CONTRACTOR FEES (General Requirements: 10%, Overhead: 5%, Profit: 10%)		25%	13.97	
ARCHITECT FEES		7%	4.90	
	Total Building Cost		**74.75**	

BUILDING TYPES

141

Costs per square foot of floor area

Exterior Wall	S.F. Area	25000	40000	55000	70000	85000	100000	115000	130000	145000
	L.F. Perimeter	388	520	566	666	766	866	878	962	1045
Face Brick with Structural Facing Tile	Steel Frame	160.30	150.25	143.25	140.40	138.65	137.35	135.25	134.45	133.80
	R/Conc. Frame	167.75	157.70	150.70	147.85	146.05	144.80	142.70	141.85	141.25
Face Brick with Concrete Block Back-up	Steel Frame	155.00	145.80	139.75	137.15	135.55	134.40	132.65	131.90	131.35
	R/Conc. Frame	162.45	153.25	147.15	144.60	143.00	141.80	140.05	139.35	138.80
Precast Concrete Panels	Steel Frame	154.85	145.70	139.65	137.10	135.45	134.30	132.55	131.85	131.30
	R/Conc. Frame	162.30	153.15	147.10	144.50	142.90	141.75	140.00	139.30	138.75
Perimeter Adj., Add or Deduct	Per 100 L.F.	6.20	3.90	2.85	2.25	1.80	1.55	1.35	1.20	1.10
Story Hgt. Adj., Add or Deduct	Per 1 Ft.	1.80	1.50	1.20	1.10	1.05	1.00	.90	.90	.85

For Basement, add $21.60 per square foot of basement area

The above costs were calculated using the basic specifications shown on the facing page. These costs should be adjusted where necessary for design alternatives and owner's requirements. Reported completed project costs, for this type of structure, range from $105.80 to $267.25 per S.F.

Common additives

Description	Unit	$ Cost
Cabinet, Base, door units, metal	L.F.	189
Drawer units	L.F.	360
Tall storage cabinets, 7' high, open	L.F.	360
With doors	L.F.	445
Wall, metal 12-1/2" deep, open	L.F.	140
With doors	L.F.	250
Closed Circuit TV (Patient monitoring)		
One station camera & monitor	Each	1375
For additional camera, add	Each	750
For automatic iris for low light, add	Each	1925
Doctors In-Out Register, 200 names	Each	13,600
Comb. control & recall, 200 names	Each	17,000
Recording register	Each	6425
Transformers	Each	293
Pocket pages	Each	965
Hubbard Tank, with accessories		
Stainless steel, 125 GPM 45 psi	Each	23,100
For electric hoist, add	Each	2525
Mortuary Refrigerator, End operated		
2 capacity	Each	12,000
6 capacity	Each	22,000

Description	Unit	$ Cost
Nurses Call Station		
Single bedside call station	Each	245
Ceiling speaker station	Each	110
Emergency call station	Each	149
Pillow speaker	Each	230
Double bedside call station	Each	400
Duty station	Each	247
Standard call button	Each	118
Master control station for 20 stations	Each	4725
Sound System		
Amplifier, 250 watts	Each	1675
Speaker, ceiling or wall	Each	147
Trumpet	Each	275
Station, Dietary with ice	Each	14,300
Sterilizers		
Single door, steam	Each	141,500
Double door, steam	Each	181,500
Portable, countertop, steam	Each	3400 - 5300
Gas	Each	35,000
Automatic washer/sterilizer	Each	48,200

Important: See the Reference Section for Location Factors

BUILDING TYPES

Model costs calculated for a 3 story building with 12' story height and 55,000 square feet of floor area

Hospital, 2-3 Story

BUILDING TYPES

				Unit	Unit Cost	Cost Per S.F.	% Of Sub-Total
A. SUBSTRUCTURE							
1010	Standard Foundations	Poured concrete; strip and spread footings		S.F. Ground	3.72	1.24	
1030	Slab on Grade	4" reinforced concrete with vapor barrier and granular base		S.F. Slab	3.42	1.14	3.0%
2010	Basement Excavation	Site preparation for slab and trench for foundation wall and footing		S.F. Ground	1.09	.36	
2020	Basement Walls	4' foundation wall		L.F. Wall	52	.54	
B. SHELL							
	B10 Superstructure						
1010	Floor Construction	Cast-in-place concrete beam and slab		S.F. Floor	15.94	10.63	13.6%
1020	Roof Construction	Cast-in-place concrete beam and slab		S.F. Roof	13.14	4.38	
	B20 Exterior Enclosure						
2010	Exterior Walls	Face brick and structural facing tile	85% of wall	S.F. Wall	29	9.23	
2020	Exterior Windows	Aluminum sliding	15% of wall	Each	509	1.23	9.6%
2030	Exterior Doors	Double aluminum and glass and sliding doors		Each	1878	.20	
	B30 Roofing						
3010	Roof Coverings	Built-up tar and gravel with flashing; perlite/EPS composite insulation		S.F. Roof	3.87	1.29	1.2%
3020	Roof Openings	Roof hatches		S.F. Roof	.03	.01	
C. INTERIORS							
1010	Partitions	Concrete block, gypsum board on metal studs	9 S.F. Floor/L.F. Partition	S.F. Partition	4.45	4.94	
1020	Interior Doors	Single leaf hollow metal	90 S.F. Floor/Door	Each	537	5.97	
1030	Fittings	Hospital Curtains		S.F. Floor	.69	.69	
2010	Stair Construction	Concrete filled metal pan		Flight	4225	1.00	26.4%
3010	Wall Finishes	40% vinyl wall covering, 35% ceramic tile, 25% epoxy coating		S.F. Surface	5.42	6.02	
3020	Floor Finishes	60% vinyl tile, 20% ceramic, 20% terrazzo		S.F. Floor	6.83	6.83	
3030	Ceiling Finishes	Plaster on suspended metal lath		S.F. Ceiling	3.71	3.71	
D. SERVICES							
	D10 Conveying						
1010	Elevators & Lifts	Two hydraulic hospital elevators		Each	75,900	2.76	2.5%
1020	Escalators & Moving Walks	N/A		—	—	—	
	D20 Plumbing						
2010	Plumbing Fixtures	Kitchen, toilet and service fixtures, supply and drainage	1 Fixture/265 S.F. Floor	Each	3339	12.60	
2020	Domestic Water Distribution	Electric water heater		S.F. Floor	3.93	3.93	15.2%
2040	Rain Water Drainage	Roof drains		S.F. Roof	.81	.27	
	D30 HVAC						
3010	Energy Supply	Oil fired hot water, wall fin radiation		S.F. Floor	3.06	3.06	
3020	Heat Generating Systems	N/A		—	—	—	
3030	Cooling Generating Systems	Chilled water, fan coil units		S.F. Floor	9.15	9.15	11.0%
3050	Terminal & Package Units	N/A		—	—	—	
3090	Other HVAC Sys. & Equipment	N/A		—	—	—	
	D40 Fire Protection						
4010	Sprinklers	Wet pipe sprinkler system		S.F. Floor	1.56	1.56	1.4%
4020	Standpipes	N/A		—	—	—	
	D50 Electrical						
5010	Electrical Service/Distribution	1200 ampere service, panel board and feeders		S.F. Floor	1.12	1.12	
5020	Lighting & Branch Wiring	Fluorescent fixtures, receptacles, switches, A.C. and misc. power		S.F. Floor	8.32	8.32	10.8%
5030	Communications & Security	Alarm systems, communications systems, emergency lighting, emergency generator		S.F. Floor	1.67	1.67	
5090	Other Electrical Systems	Emergency generator, 125 kW		S.F. Floor	.92	.92	
E. EQUIPMENT & FURNISHINGS							
1010	Commercial Equipment	N/A		—	—	—	
1020	Institutional Equipment	Oxygen piping, curtain partitions		S.F. Floor	.72	.72	4.8%
1030	Vehicular Equipment	N/A		—	—	—	
1090	Other Equipment	Patient wall systems		S.F. Floor	4.57	4.57	
F. SPECIAL CONSTRUCTION & DEMOLITION							
1020	Integrated Construction	Conductive flooring		S.F. Floor	.54	.54	0.5%
1040	Special Facilities	N/A		—	—	—	
G. BUILDING SITEWORK	**N/A**						

Sub-Total		110.60	**100%**
CONTRACTOR FEES (General Requirements: 10%, Overhead: 5%, Profit: 10%)		25%	27.65
ARCHITECT FEES		9%	12.45
Total Building Cost		150.70	

143

Costs per square foot of floor area

Exterior Wall	S.F. Area	100000	125000	150000	175000	200000	225000	250000	275000	300000
	L.F. Perimeter	594	705	816	783	866	950	1033	1116	1200
Face Brick with Structural Facing Tile	Steel Frame	132.85	130.25	128.50	124.80	123.70	122.95	122.30	121.75	121.35
	R/Conc. Frame	140.55	137.90	136.15	132.50	131.40	130.65	129.95	129.40	129.00
Face Brick with Concrete Block Back-up	Steel Frame	129.55	127.10	125.45	122.30	121.30	120.60	120.00	119.45	119.10
	R/Conc. Frame	137.20	134.75	133.10	129.95	128.95	128.25	127.65	127.15	126.75
Precast Concrete Panels With Exposed Aggregate	Steel Frame	128.40	126.05	124.45	121.65	120.65	120.00	119.40	118.90	118.55
	R/Conc. Frame	135.30	132.95	131.40	128.55	127.60	126.90	126.35	125.85	125.45
Perimeter Adj., Add or Deduct	Per 100 L.F.	3.00	2.35	1.95	1.70	1.50	1.30	1.20	1.10	1.00
Story Hgt. Adj., Add or Deduct	Per 1 Ft.	1.40	1.30	1.30	1.05	1.00	1.00	.95	.95	.95
For Basement, add $22.30 per square foot of basement area										

The above costs were calculated using the basic specifications shown on the facing page. These costs should be adjusted where necessary for design alternatives and owner's requirements. Reported completed project costs, for this type of structure, range from $108.20 to $263.90 per S.F.

Common additives

Description	Unit	$ Cost	Description	Unit	$ Cost
Cabinets, Base, door units, metal	L.F.	189	Nurses Call Station		
Drawer units	L.F.	360	Single bedside call station	Each	245
Tall storage cabinets, 7' high, open	L.F.	360	Ceiling speaker station	Each	110
With doors	L.F.	445	Emergency call station	Each	149
Wall, metal 12-1/2" deep, open	L.F.	140	Pillow speaker	Each	230
With doors	L.F.	250	Double bedside call station	Each	400
Closed Circuit TV (Patient monitoring)			Duty station	Each	247
One station camera & monitor	Each	1375	Standard call button	Each	118
For additional camera add	Each	750	Master control station for 20 stations	Each	4725
For automatic iris for low light add	Each	1925	Sound System		
Doctors In-Out Register, 200 names	Each	13,600	Amplifier, 250 watts	Each	1675
Comb. control & recall, 200 names	Each	17,000	Speaker, ceiling or wall	Each	147
Recording register	Each	6425	Trumpet	Each	275
Transformers	Each	293	Station, Dietary with ice	Each	14,300
Pocket pages	Each	965	Sterilizers		
Hubbard Tank, with accessories			Single door, steam	Each	141,500
Stainless steel, 125 GPM 45 psi	Each	23,100	Double door, steam	Each	181,500
For electric hoist, add	Each	2525	Portable, counter top, steam	Each	3400 - 5300
Mortuary Refrigerator, End operated			Gas	Each	35,000
2 capacity	Each	12,000	Automatic washer/sterilizer	Each	48,200
6 capacity	Each	22,000			

Important: See the Reference Section for Location Factors

Model costs calculated for a 6 story building with 12' story height and 200,000 square feet of floor area

				Unit	Unit Cost	Cost Per S.F.	% Of Sub-Total

A. SUBSTRUCTURE

1010	Standard Foundations	Poured concrete; strip and spread footings		S.F. Ground	3.66	.61	
1030	Slab on Grade	4" reinforced concrete with vapor barrier and granular base		S.F. Slab	3.42	.57	1.8%
2010	Basement Excavation	Site preparation for slab and trench for foundation wall and footing		S.F. Ground	1.09	.18	
2020	Basement Walls	4' foundation wall		L.F. Wall	52	.23	

B. SHELL

B10 Superstructure

| 1010 | Floor Construction | Concrete slab with metal deck and beams, steel columns | | S.F. Floor | 12.23 | 10.19 | 12.1% |
| 1020 | Roof Construction | Metal deck, open web steel joists, beams, interior columns | | S.F. Roof | 4.92 | .82 | |

B20 Exterior Enclosure

2010	Exterior Walls	Face brick and structural facing tile	70% of wall	S.F. Wall	29	6.39	
2020	Exterior Windows	Aluminum sliding	30% of wall	Each	396	2.47	10.2%
2030	Exterior Doors	Double aluminum and glass and sliding doors		Each	3091	.43	

B30 Roofing

| 3010 | Roof Coverings | Built-up tar and gravel with flashing; perlite/EPS composite insulation | | S.F. Roof | 3.54 | .59 | 0.7% |
| 3020 | Roof Openings | Roof hatches | | S.F. Roof | .12 | .02 | |

C. INTERIORS

1010	Partitions	Gypsum board on metal studs with sound deadening board	9 S.F. Floor/L.F. Partition	S.F. Partition	3.91	4.34	
1020	Interior Doors	Single leaf hollow metal	90 S.F. Floor/Door	Each	537	5.97	
1030	Fittings	Hospital Curtains		S.F. Floor	.76	.76	
2010	Stair Construction	Concrete filled metal pan		Flight	4100	.53	31.0%
3010	Wall Finishes	40% vinyl wall covering, 35% ceramic tile, 25% epoxy coating		S.F. Surface	5.41	6.01	
3020	Floor Finishes	60% vinyl tile, 20% ceramic, 20% terrazzo		S.F. Floor	6.83	6.83	
3030	Ceiling Finishes	Plaster on suspended metal lath		S.F. Ceiling	3.71	3.71	

D. SERVICES

D10 Conveying

| 1010 | Elevators & Lifts | Six geared hospital elevators | | Each | 147,000 | 4.41 | 4.9% |
| 1020 | Escalators & Moving Walks | N/A | | — | — | — | |

D20 Plumbing

2010	Plumbing Fixtures	Kitchen, toilet and service fixtures, supply and drainage	1 Fixture/275 S.F. Floor	Each	3495	12.71	
2020	Domestic Water Distribution	Electric water heater		S.F. Floor	1.48	1.48	16.0%
2040	Rain Water Drainage	Roof drains		S.F. Floor	2.22	.37	

D30 HVAC

3010	Energy Supply	Oil fired hot water, wall fin radiation		S.F. Floor	3.06	3.06	
3020	Heat Generating Systems	N/A		—	—	—	
3030	Cooling Generating Systems	Chilled water, fan coil units		S.F. Floor	2.08	2.08	5.7%
3050	Terminal & Package Units	N/A		—	—	—	
3090	Other HVAC Sys. & Equipment	N/A		—	—	—	

D40 Fire Protection

| 4010 | Sprinklers | Wet pipe sprinkler system | | S.F. Floor | 1.22 | 1.22 | 1.5% |
| 4020 | Standpipes | Standpipe | | S.F. Floor | .10 | .10 | |

D50 Electrical

5010	Electrical Service/Distribution	4000 ampere service, panel board and feeders		S.F. Floor	1.15	1.15	
5020	Lighting & Branch Wiring	Fluorescent fixtures, receptacles, switches, A.C. and misc. power		S.F. Floor	7.61	7.61	12.4%
5030	Communications & Security	Alarm systems, communications system, emergency lighting, and emergency generator		S.F. Floor	2.08	2.08	
5090	Other Electrical Systems	Emergency generator, 500 kW		S.F. Floor	.51	.51	

E. EQUIPMENT & FURNISHINGS

1010	Commercial Equipment	N/A		—	—	—	
1020	Institutional Equipment	Conductive flooring, oxygen piping, curtain partitions		S.F. Floor	.35	.35	3.7%
1030	Vehicular Equipment	N/A		—	—	—	
1090	Other Equipment	Patient wall systems		S.F. Floor	3.02	3.02	

F. SPECIAL CONSTRUCTION & DEMOLITION

| 1020 | Integrated Construction | N/A | | — | — | — | 0.0% |
| 1040 | Special Facilities | N/A | | — | — | — | |

G. BUILDING SITEWORK   N/A

| | | **Sub-Total** | | | | 90.80 | 100% |

| CONTRACTOR FEES (General Requirements: 10%, Overhead: 5%, Profit: 10%) | | | | | 25% | 22.70 | |
| ARCHITECT FEES | | | | | 9% | 10.20 | |

| | | **Total Building Cost** | | | | **123.70** | |

BUILDING TYPES

Costs per square foot of floor area

Exterior Wall	S.F. Area	35000	55000	75000	95000	115000	135000	155000	175000	195000
	L.F. Perimeter	314	401	497	555	639	722	716	783	850
Face Brick with Concrete Block Back-up	Steel Frame	118.10	111.00	107.95	105.30	104.05	103.15	101.20	100.55	100.15
	R/Conc. Frame	121.45	114.35	111.30	108.60	107.40	106.50	104.50	103.90	103.45
Glass and Metal Curtain Walls	Steel Frame	113.05	106.90	104.20	102.00	100.90	100.15	98.60	98.05	97.70
	R/Conc. Frame	116.55	110.40	107.70	105.50	104.40	103.65	102.10	101.55	101.20
Precast Concrete Panels	Steel Frame	121.60	113.85	110.50	107.55	106.20	105.20	102.95	102.30	101.80
	R/Conc. Frame	125.80	117.95	114.60	111.50	110.20	109.15	106.85	106.15	105.70
Perimeter Adj., Add or Deduct	Per 100 L.F.	6.70	4.30	3.15	2.45	2.00	1.70	1.50	1.35	1.20
Story Hgt. Adj., Add or Deduct	Per 1 Ft.	1.85	1.50	1.40	1.20	1.15	1.10	.95	.95	.95
For Basement, add $21.25 per square foot of basement area										

The above costs were calculated using the basic specifications shown on the facing page. These costs should be adjusted where necessary for design alternatives and owner's requirements. Reported completed project costs, for this type of structure, range from $71.05 to $137.00 per S.F.

Common additives

Description	Unit	$ Cost
Bar, Front bar	L.F.	287
Back bar	L.F.	229
Booth, Upholstered, custom, straight	L.F.	145 - 265
"L" or "U" shaped	L.F.	150 - 252
Closed Circuit Surveillance, One station		
Camera and monitor	Each	1375
For additional camera stations, add	Each	750
Directory Boards, Plastic, glass covered		
30" x 20"	Each	575
36" x 48"	Each	1075
Aluminum, 24" x 18"	Each	440
48" x 32"	Each	780
48" x 60"	Each	1675
Elevators, Electric passenger, 5 stops		
3500# capacity	Each	107,400
5000# capacity	Each	111,900
Additional stop, add	Each	5675
Emergency Lighting, 25 watt, battery operated		
Lead battery	Each	227
Nickel cadmium	Each	660

Description	Unit	$ Cost
Laundry Equipment		
Folders, blankets & sheets, king size	Each	53,500
Ironers, 110" single roll	Each	29,100
Combination washer extractor 50#	Each	10,200
125#	Each	27,000
Sauna, Prefabricated, complete		
6' x 4'	Each	4825
6' x 6'	Each	5625
6' x 9'	Each	5550
8' x 8'	Each	6525
10' x 12'	Each	9475
Smoke Detectors		
Ceiling type	Each	151
Duct type	Each	405
Sound System		
Amplifier, 250 watts	Each	1675
Speaker, ceiling or wall	Each	147
Trumpet	Each	275
TV Antenna, Master system, 12 outlet	Outlet	238
30 outlet	Outlet	153
100 outlet	Outlet	145

Important: See the Reference Section for Location Factors

Model costs calculated for a 6 story building with 10' story height and 135,000 square feet of floor area

	Unit	Unit Cost	Cost Per S.F.	% Of Sub-Total

A. SUBSTRUCTURE

				Unit	Unit Cost	Cost Per S.F.	% Of Sub-Total
1010	Standard Foundations	Poured concrete; strip and spread footings		S.F. Ground	4.14	.69	
1030	Slab on Grade	4" reinforced concrete with vapor barrier and granular base		S.F. Slab	3.42	.57	
2010	Basement Excavation	Site preparation for slab and trench for foundation wall and footing		S.F. Ground	1.09	.18	2.6%
2020	Basement Walls	4' foundation wall		L.F. Wall	52	.57	

B. SHELL

B10 Superstructure

				Unit	Unit Cost	Cost Per S.F.	% Of Sub-Total
1010	Floor Construction	Concrete slab with metal deck, beams, columns		S.F. Floor	12.02	10.02	14.1%
1020	Roof Construction	Metal deck, open web steel joists, beams, columns		S.F. Roof	5.70	.95	

B20 Exterior Enclosure

				Unit	Unit Cost	Cost Per S.F.	% Of Sub-Total
2010	Exterior Walls	Face brick with concrete block backup	80% of wall	S.F. Wall	21	5.42	
2020	Exterior Windows	Window wall	20% of wall	Each	35	2.30	10.1%
2030	Exterior Doors	Glass amd metal doors and entrances		Each	2267	.17	

B30 Roofing

			Unit	Unit Cost	Cost Per S.F.	% Of Sub-Total
3010	Roof Coverings	Built-up tar and gravel with flashing; perlite/EPS composite insulation	S.F. Roof	3.84	.64	0.8%
3020	Roof Openings	Roof hatches	S.F. Roof	.06	.01	

C. INTERIORS

				Unit	Unit Cost	Cost Per S.F.	% Of Sub-Total
1010	Partitions	Gypsum board and sound deadening board, steel studs	9 S.F. Floor/L.F. Partition	S.F. Partition	4.79	4.26	
1020	Interior Doors	Single leaf hollow metal	90 S.F. Floor/Door	Each	537	5.97	
1030	Fittings	N/A		—	—	—	
2010	Stair Construction	Concrete filled metal pan		Flight	5075	.53	25.2%
3010	Wall Finishes	20% paint, 75% vinyl cover, 5% ceramic tile		S.F. Surface	2.95	2.62	
3020	Floor Finishes	80% carpet, 10% vinyl composition tile, 10% ceramic tile		S.F. Floor	4.62	4.62	
3030	Ceiling Finishes	Mineral fiber tile applied with adhesive		S.F. Ceiling	1.60	1.60	

D. SERVICES

D10 Conveying

			Unit	Unit Cost	Cost Per S.F.	% Of Sub-Total
1010	Elevators & Lifts	Four geared passenger elevators	Each	134,662	3.99	5.1%
1020	Escalators & Moving Walks	N/A	—	—	—	

D20 Plumbing

				Unit	Unit Cost	Cost Per S.F.	% Of Sub-Total
2010	Plumbing Fixtures	Kitchen, toilet and service fixtures, supply and drainage	1 Fixture/155 S.F. Floor	Each	1621	10.46	
2020	Domestic Water Distribution	Oil fired water heater		S.F. Floor	.61	.61	14.4%
2040	Rain Water Drainage	Roof drains		S.F. Roof	.72	.12	

D30 HVAC

			Unit	Unit Cost	Cost Per S.F.	% Of Sub-Total
3010	Energy Supply	Oil fired hot water, wall fin radiation	S.F. Floor	3.06	3.06	
3020	Heat Generating Systems	N/A	—	—	—	
3030	Cooling Generating Systems	Chilled water, fan coil units	S.F. Floor	8.32	8.32	14.6%
3050	Terminal & Package Units	N/A	—	—	—	
3090	Other HVAC Sys. & Equipment	N/A	—	—	—	

D40 Fire Protection

			Unit	Unit Cost	Cost Per S.F.	% Of Sub-Total
4010	Sprinklers	Sprinkler system, light hazard	S.F. Floor	1.50	1.50	2.0%
4020	Standpipes	Standpipes and hose systems	S.F. Floor	.07	.07	

D50 Electrical

			Unit	Unit Cost	Cost Per S.F.	% Of Sub-Total
5010	Electrical Service/Distribution	2000 ampere service, panel board and feeders	S.F. Floor	.83	.83	
5020	Lighting & Branch Wiring	Fluorescent fixtures, receptacles, switches, A.C. and misc. power	S.F. Floor	5.78	5.78	
5030	Communications & Security	Alarm systems, communications systems and emergency lighting	S.F. Floor	1.58	1.58	11.1%
5090	Other Electrical Systems	Emergency generator, 250 kW	S.F. Floor	.41	.41	

E. EQUIPMENT & FURNISHINGS

			Unit	Unit Cost	Cost Per S.F.	% Of Sub-Total
1010	Commercial Equipment	N/A	—	—	—	
1020	Institutional Equipment	N/A	—	—	—	
1030	Vehicular Equipment	N/A	—	—	—	0.0%
1090	Other Equipment	N/A	—	—	—	

F. SPECIAL CONSTRUCTION & DEMOLITION

			Unit	Unit Cost	Cost Per S.F.	% Of Sub-Total
1020	Integrated Construction	N/A	—	—	—	0.0%
1040	Special Facilities	N/A	—	—	—	

G. BUILDING SITEWORK   N/A

		Cost Per S.F.	%
	Sub-Total	77.85	100%
CONTRACTOR FEES (General Requirements: 10%, Overhead: 5%, Profit: 10%)	25%	19.46	
ARCHITECT FEES	6%	5.84	
Total Building Cost		103.15	

BUILDING TYPES

Costs per square foot of floor area

Exterior Wall	S.F. Area	140000	243000	346000	450000	552000	655000	760000	860000	965000
	L.F. Perimeter	403	587	672	800	936	1073	1213	1195	1312
Face Brick with Concrete Block Back-up	Steel Frame	100.15	95.25	91.75	90.30	89.55	89.05	88.65	87.40	87.10
	R/Conc. Frame	101.85	96.95	93.40	92.00	91.25	90.75	90.30	89.05	88.80
Face Brick Veneer On Steel Studs	Steel Frame	98.15	93.60	90.40	89.10	88.40	87.90	87.55	86.40	86.15
	R/Conc. Frame	100.30	95.70	92.45	91.15	90.50	90.00	89.60	88.50	88.25
Glass and Metal Curtain Walls	Steel Frame	110.30	103.65	99.80	**98.10**	97.10	96.45	95.90	94.85	94.50
	R/Conc. Frame	112.40	105.75	101.85	100.15	99.20	98.55	98.00	96.95	96.60
Perimeter Adj., Add or Deduct	Per 100 L.F.	3.90	2.25	1.60	1.20	1.00	.85	.70	.65	.60
Story Hgt. Adj., Add or Deduct	Per 1 Ft.	1.50	1.30	1.05	.95	.90	.85	.85	.70	.70
For Basement, add $22.00 per square foot of basement area										

The above costs were calculated using the basic specifications shown on the facing page. These costs should be adjusted where necessary for design alternatives and owner's requirements. Reported completed project costs, for this type of structure, range from $79.20 to $138.55 per S.F.

Common additives

Description	Unit	$ Cost	Description	Unit	$ Cost
Bar, Front bar	L.F.	287	Laundry Equipment		
Back bar	L.F.	229	Folders, blankets & sheets, king size	Each	53,500
Booth, Upholstered, custom, straight	L.F.	145 - 265	Ironers, 110" single roll	Each	29,100
"L" or "U" shaped	L.F.	150 - 252	Combination washer & extractor 50#	Each	10,200
Closed Circuit Surveillance, One station			125#	Each	27,000
Camera and monitor	Each	1375	Sauna, Prefabricated, complete		
For additional camera stations, add	Each	750	6' x 4'	Each	4825
Directory Boards, Plastic, glass covered			6' x 6'	Each	5625
30" x 20"	Each	575	6' x 9'	Each	5550
36" x 48"	Each	1075	8' x 8'	Each	6525
Aluminum, 24" x 18"	Each	440	10' x 12'	Each	9475
48" x 32"	Each	780	Smoke Detectors		
48" x 60"	Each	1675	Ceiling type	Each	151
Elevators, Electric passenger, 10 stops			Duct type	Each	405
3500# capacity	Each	217,000	Sound System	Each	
5000# capacity	Each	223,000	Amplifier, 250 watts	Each	1675
Additional stop, add	Each	5675	Speaker, ceiling or wall	Each	147
Emergency Lighting, 25 watt, battery operated			Trumpet	Each	275
Lead battery	Each	227	TV Antenna, Master system, 12 outlet	Outlet	238
Nickel cadmium	Each	660	30 outlet	Outlet	153
			100 outlet	Outlet	145

Important: See the Reference Section for Location Factors

Model costs calculated for a 15 story building with 10' story height and 450,000 square feet of floor area

Hotel, 8-24 Story

				Unit	Unit Cost	Cost Per S.F.	% Of Sub-Total
A.	**SUBSTRUCTURE**						
1010	Standard Foundations	Poured concrete; strip and spread footings		S.F. Ground	12.15	.81	
1030	Slab on Grade	4" reinforced concrete with vapor barrier and granular base		S.F. Slab	3.42	.23	
2010	Basement Excavation	Site preparation for slab and trench for foundation wall and footing		S.F. Ground	1.09	.07	1.7%
2020	Basement Walls	4' foundation wall		L.F. Wall	52	.12	
B.	**SHELL**						
	B10 Superstructure						
1010	Floor Construction	Open web steel joists, slab form, concrete, columns		S.F. Floor	12.15	11.34	15.8%
1020	Roof Construction	Metal deck, open web steel joists, beams, columns		S.F. Roof	4.95	.33	
	B20 Exterior Enclosure						
2010	Exterior Walls	N/A		—	—	—	
2020	Exterior Windows	Glass and metal curtain walls	100% of wall	Each	16.95	4.52	6.3%
2030	Exterior Doors	Glass and metal doors and entrances		Each	2036	.14	
	B30 Roofing						
3010	Roof Coverings	Built-up tar and gravel with flashing; perlite/EPS composite insulation		S.F. Roof	3.75	.25	0.3%
3020	Roof Openings	N/A		—	—	—	
C.	**INTERIORS**						
1010	Partitions	Gypsum board and sound deadening board, steel studs	9 S.F. Floor/L.F. Partition	S.F. Partition	4.65	4.13	
1020	Interior Doors	Single leaf hollow metal	90 S.F. Floor/Door	Each	537	5.97	
1030	Fittings	N/A		—	—	—	
2010	Stair Construction	Concrete filled metal pan		Flight	5075	1.03	28.6%
3010	Wall Finishes	20% paint, 75% vinyl cover, 5% ceramic tile		S.F. Surface	2.94	2.61	
3020	Floor Finishes	80% carpet tile, 10% vinyl composition tile, 10% ceramic tile		S.F. Floor	4.62	4.62	
3030	Ceiling Finishes	Gypsum board on resilient channel		S.F. Ceiling	2.83	2.83	
D.	**SERVICES**						
	D10 Conveying						
1010	Elevators & Lifts	One geared freight, six geared passenger elevators		Each	241,500	3.22	4.3%
1020	Escalators & Moving Walks	N/A		—	—	—	
	D20 Plumbing						
2010	Plumbing Fixtures	Kitchen, toilet and service fixtures, supply and drainage	1 Fixture/165 S.F. Floor	Each	1587	9.62	
2020	Domestic Water Distribution	Electric water heater		S.F. Floor	1.37	1.37	15.0%
2040	Rain Water Drainage	Roof drains		S.F. Roof	1.35	.09	
	D30 HVAC						
3010	Energy Supply	Oil fired hot water, wall fin radiation		S.F. Floor	1.55	1.55	
3020	Heat Generating Systems	N/A		—	—	—	
3030	Cooling Generating Systems	Chilled water, fan coil units		S.F. Floor	8.32	8.32	13.3%
3050	Terminal & Package Units	N/A		—	—	—	
3090	Other HVAC Sys. & Equipment	N/A		—	—	—	
	D40 Fire Protection						
4010	Sprinklers	Sprinkler system, light hazard		S.F. Floor	2.13	2.13	3.2%
4020	Standpipes	Standpipes and hose systems		S.F. Floor	.23	.23	
	D50 Electrical						
5010	Electrical Service/Distribution	6000 ampere service, panel board and feeders		S.F. Floor	.91	.91	
5020	Lighting & Branch Wiring	Fluorescent fixtures, receptacles, switches, A.C. and misc. power		S.F. Floor	5.69	5.69	11.5%
5030	Communications & Security	Alarm systems, communications systems and emergency lighting		S.F. Floor	1.59	1.59	
5090	Other Electrical Systems	Emergency generator, 500 kW		S.F. Floor	.31	.31	
E.	**EQUIPMENT & FURNISHINGS**						
1010	Commercial Equipment	N/A		—	—	—	
1020	Institutional Equipment	N/A		—	—	—	
1030	Vehicular Equipment	N/A		—	—	—	0.0%
1090	Other Equipment	N/A		—	—	—	
F.	**SPECIAL CONSTRUCTION & DEMOLITION**						
1020	Integrated Construction	N/A		—	—	—	0.0%
1040	Special Facilities	N/A		—	—	—	
G.	**BUILDING SITEWORK**	**N/A**					

		Sub-Total	74.03	100%
CONTRACTOR FEES (General Requirements: 10%, Overhead: 5%, Profit: 10%)		25%	18.51	
ARCHITECT FEES		6%	5.56	
	Total Building Cost		**98.10**	

BUILDING TYPES

149

Costs per square foot of floor area

Exterior Wall	S.F. Area	5500	12000	20000	30000	40000	60000	80000	100000	145000
	L.F. Perimeter	310	330	400	490	530	590	680	760	920
Face Brick with Concrete Block Back-up	Steel Frame	299.25	221.50	199.45	188.60	180.75	172.40	168.90	166.60	163.50
	R/Conc. Frame	304.30	226.55	204.50	193.60	185.80	177.40	173.90	171.60	168.55
Stucco on Concrete Block	Steel Frame	279.40	211.85	192.50	182.90	176.15	169.00	166.00	164.00	161.35
	R/Conc. Frame	284.35	216.80	197.40	187.80	181.10	173.90	170.90	168.90	166.30
Reinforced Concrete	Steel Frame	288.10	216.10	195.55	185.40	178.20	170.50	167.30	165.15	162.35
	R/Conc. Frame	293.00	221.00	200.50	190.35	183.15	175.45	172.20	170.10	167.25
Perimeter Adj., Add or Deduct	Per 100 L.F.	34.10	15.60	9.35	6.25	4.70	3.10	2.35	1.90	1.30
Story Hgt. Adj., Add or Deduct	Per 1 Ft.	8.15	4.00	2.90	2.35	1.90	1.40	1.25	1.10	.90

For Basement, add $30.50 per square foot of basement area

The above costs were calculated using the basic specifications shown on the facing page. These costs should be adjusted where necessary for design alternatives and owner's requirements. Reported completed project costs, for this type of structure, range from $115.30 to $306.20 per S.F.

Common additives

Description	Unit	$ Cost	Description	Unit	$ Cost
Clock System			Emergency Lighting, 25 watt, battery operated		
20 room	Each	13,000	Nickel cadmium	Each	660
50 room	Each	31,500	Flagpoles, Complete		
Closed Circuit Surveillance, One station			Aluminum, 20' high	Each	1100
Camera and monitor	Each	1375	40' High	Each	2700
For additional camera stations, add	Each	750	70' High	Each	8350
Elevators, Hydraulic passenger, 2 stops			Fiberglass, 23' High	Each	1425
1500# capacity	Each	43,425	39'-5" High	Each	3000
2500# capacity	Each	44,625	59' High	Each	7425
3500# capacity	Each	48,525	Laundry Equipment	Each	
Additional stop, add	Each	3800	Folders, blankets, & sheets	Each	53,500
Emergency Generators, Complete system, gas			Ironers, 110" single roll	Each	29,100
15 KW	Each	13,000	Combination washer & extractor, 50#	Each	10,200
70 KW	Each	25,700	125#	Each	27,000
85 KW	Each	28,500	Safe, Office type, 4 hour rating		
115 KW	Each	55,000	30" x 18" x 18"	Each	3200
170 KW	Each	72,000	62" x 33" x 20"	Each	6975
Diesel, 50 KW	Each	24,600	Sound System		
100 KW	Each	35,300	Amplifier, 250 watts	Each	1675
150 KW	Each	42,500	Speaker, ceiling or wall	Each	147
350 KW	Each	67,500	Trumpet	Each	275

BUILDING TYPES

Model costs calculated for a 3 story building with 12' story height and 40,000 square feet of floor area

				Unit	Unit Cost	Cost Per S.F.	% Of Sub-Total
A. SUBSTRUCTURE							
1010	Standard Foundations	Poured concrete; strip and spread footings		S.F. Ground	1.41	.47	
1030	Slab on Grade	4" reinforced concrete with vapor barrier and granular base		S.F. Slab	3.42	1.14	1.9%
2010	Basement Excavation	Site preparation for slab and trench for foundation wall and footing		S.F. Ground	1.18	.39	
2020	Basement Walls	4' foundation wall		L.F. Wall	46	.62	
B. SHELL							
	B10 Superstructure						
1010	Floor Construction	Concrete slab with metal deck and beams		S.F. Floor	17.10	11.40	11.8%
1020	Roof Construction	Concrete slab with metal deck and beams		S.F. Roof	13.74	4.58	
	B20 Exterior Enclosure						
2010	Exterior Walls	Face brick with reinforced concrete block backup	85% of wall	S.F. Wall	21	8.55	
2020	Exterior Windows	Bullet resisting and metal horizontal pivoted	15% of wall	Each	108	7.76	12.1%
2030	Exterior Doors	Metal		Each	3800	.10	
	B30 Roofing						
3010	Roof Coverings	Built-up tar and gravel with flashing; perlite/EPS composite insulation		S.F. Roof	3.99	1.33	1.0%
3020	Roof Openings	Roof hatches		S.F. Roof	.03	.01	
C. INTERIORS							
1010	Partitions	Concrete block	45 S.F. of Floor/L.F. Partition	S.F. Partition	6.67	1.78	
1020	Interior Doors	Hollow metal fire doors	930 S.F. Floor/Door	Each	537	.58	
1030	Fittings	N/A		—	—	—	
2010	Stair Construction	Concrete filled metal pan		Flight	6050	.76	6.2%
3010	Wall Finishes	Paint		S.F. Surface	5.74	1.53	
3020	Floor Finishes	70% vinyl composition tile, 20% carpet, 10% ceramic tile		S.F. Floor	2.59	2.59	
3030	Ceiling Finishes	Mineral tile on zee runners	30% of area	S.F. Ceiling	3.71	1.11	
D. SERVICES							
	D10 Conveying						
1010	Elevators & Lifts	One hydraulic passenger elevator		Each	102,400	2.56	1.9%
1020	Escalators & Moving Walks	N/A		—	—	—	
	D20 Plumbing						
2010	Plumbing Fixtures	Toilet and service fixtures, supply and drainage	1 Fixture/78 S.F. Floor	Each	1517	19.45	
2020	Domestic Water Distribution	Electric water heater		S.F. Floor	3.02	3.02	16.9%
2040	Rain Water Drainage	Roof drains		S.F. Roof	.93	.31	
	D30 HVAC						
3010	Energy Supply	N/A		—	—	—	
3020	Heat Generating Systems	Included in D3030		—	—	—	
3030	Cooling Generating Systems	Rooftop multizone unit systems		S.F. Floor	10.88	10.88	8.1%
3050	Terminal & Package Units	N/A		—	—	—	
3090	Other HVAC Sys. & Equipment	N/A		—	—	—	
	D40 Fire Protection						
4010	Sprinklers	N/A		—	—	—	0.2%
4020	Standpipes	Standpipes and hose systems		S.F. Floor	.31	.31	
	D50 Electrical						
5010	Electrical Service/Distribution	600 ampere service, panel board and feeders		S.F. Floor	.76	.76	
5020	Lighting & Branch Wiring	Fluorescent fixtures, receptacles, switches, A.C. and misc. power		S.F. Floor	6.68	6.68	7.8%
5030	Communications & Security	Alarm systems, intercom, and emergency lighting		S.F. Floor	2.57	2.57	
5090	Other Electrical Systems	Emergency generator, 80 kW		S.F. Floor	.56	.56	
E. EQUIPMENT & FURNISHINGS							
1010	Commercial Equipment	N/A		—	—	—	
1020	Institutional Equipment	Prefabricated Cells, visitor cubicles		S.F. Floor	43	43.35	32.1%
1030	Vehicular Equipment	N/A		—	—	—	
1090	Other Equipment	N/A		—	—	—	
F. SPECIAL CONSTRUCTION & DEMOLITION							
1020	Integrated Construction	N/A		—	—	—	0.0%
1040	Special Facilities	N/A		—	—	—	
G. BUILDING SITEWORK	**N/A**						

	Sub-Total	135.15	**100%**
CONTRACTOR FEES (General Requirements: 10%, Overhead: 5%, Profit: 10%)	25%	33.79	
ARCHITECT FEES	7%	11.81	
	Total Building Cost	**180.75**	

Costs per square foot of floor area

Exterior Wall	S.F. Area	1000	2000	3000	4000	5000	10000	15000	20000	25000
	L.F. Perimeter	126	179	219	253	283	400	490	568	632
Decorative Concrete Block	Steel Frame	164.70	140.45	130.75	125.35	121.80	113.50	110.05	108.10	106.70
	Bearing Walls	165.80	141.25	131.50	125.95	122.40	113.95	110.45	108.45	107.05
Face Brick with Concrete Block Back-up	Steel Frame	185.35	155.70	143.55	136.70	132.15	121.40	116.85	114.25	112.40
	Bearing Walls	184.50	154.55	142.35	135.35	130.80	119.90	115.30	112.65	110.80
Metal Sandwich Panel	Steel Frame	164.70	141.15	131.80	126.55	123.10	115.15	111.85	109.95	108.65
Precast Concrete Panel	Bearing Walls	170.00	144.25	133.90	128.05	124.25	115.25	111.55	109.40	107.90
Perimeter Adj., Add or Deduct	Per 100 L.F.	37.45	18.70	12.50	9.35	7.50	3.75	2.50	1.90	1.50
Story Hgt. Adj., Add or Deduct	Per 1 Ft.	2.60	1.85	1.50	1.30	1.15	.80	.65	.60	.50

For Basement, add $20.20 per square foot of basement area

The above costs were calculated using the basic specifications shown on the facing page. These costs should be adjusted where necessary for design alternatives and owner's requirements. Reported completed project costs, for this type of structure, range from $64.20 to $156.90 per S.F.

Common additives

Description	Unit	$ Cost
Closed Circuit Surveillance, One station		
Camera and monitor	Each	1375
For additional camera stations, add	Each	750
Emergency Lighting, 25 watt, battery operated		
Lead battery	Each	227
Nickel cadmium	Each	660
Laundry Equipment		
Dryers, coin operated 30 lb.	Each	2800
Double stacked	Each	5925
50 lb.	Each	3225
Dry cleaner 20 lb.	Each	35,500
30 lb.	Each	48,400
Washers, coin operated	Each	1225
Washer/extractor 20 lb.	Each	4250
30 lb.	Each	8175
50 lb.	Each	10,200
75 lb.	Each	21,200
Smoke Detectors		
Ceiling type	Each	151
Duct type	Each	405

Model costs calculated for a 1 story building with 12' story height and 3,000 square feet of floor area

Laundromat

				Unit	Unit Cost	Cost Per S.F.	% Of Sub-Total
A. SUBSTRUCTURE							
1010	Standard Foundations	Poured concrete; strip and spread footings		S.F. Ground	2.20	2.20	
1030	Slab on Grade	5" reinforced concrete with vapor barrier and granular base		S.F. Slab	3.70	3.70	
2010	Basement Excavation	Site preparation for slab and trench for foundation wall and footing		S.F. Ground	1.31	1.31	10.9%
2020	Basement Walls	4' Foundation wall		L.F. Wall	46	3.42	
B. SHELL							
	B10 Superstructure						
1010	Floor Construction	N/A		—	—	—	
1020	Roof Construction	Metal deck, open web steel joists, beams, columns		S.F. Roof	5.48	5.48	5.6%
	B20 Exterior Enclosure						
2010	Exterior Walls	Decorative concrete block	90% of wall	S.F. Wall	10.54	8.31	
2020	Exterior Windows	Store front	10% of wall	Each	30	2.69	12.3%
2030	Exterior Doors	Double aluminum and glass		Each	3075	1.03	
	B30 Roofing						
3010	Roof Coverings	Built-up tar and gravel with flashing; perlite/EPS composite insulation		S.F. Roof	5.99	5.99	6.1%
3020	Roof Openings	N/A		—	—	—	
C. INTERIORS							
1010	Partitions	Gypsum board on metal studs	60 S.F. Floor/L.F. Partition	S.F. Partition	17.22	2.87	
1020	Interior Doors	Single leaf wood	750 S.F. Floor/Door	Each	432	.58	
1030	Fittings	N/A		—	—	—	
2010	Stair Construction	N/A		—	—	—	7.7%
3010	Wall Finishes	Paint		S.F. Surface	1.14	.19	
3020	Floor Finishes	Vinyl composition tile		S.F. Floor	1.49	1.49	
3030	Ceiling Finishes	Fiberglass board on exposed grid system		S.F. Ceiling	2.37	2.37	
D. SERVICES							
	D10 Conveying						
1010	Elevators & Lifts	N/A		—	—	—	
1020	Escalators & Moving Walks	N/A		—	—	—	0.0%
	D20 Plumbing						
2010	Plumbing Fixtures	Toilet and service fixtures, supply and drainage	1 Fixture/600 S.F. Floor	Each	3156	5.26	
2020	Domestic Water Distribution	Gas fired hot water heater		S.F. Floor	20	20.13	26.6%
2040	Rain Water Drainage	Roof drains		S.F. Roof	.64	.64	
	D30 HVAC						
3010	Energy Supply	N/A		—	—	—	
3020	Heat Generating Systems	Included in D3030		—	—	—	
3030	Cooling Generating Systems	Rooftop single zone unit systems		S.F. Floor	8.71	8.71	8.9%
3050	Terminal & Package Units	N/A		—	—	—	
3090	Other HVAC Sys. & Equipment	N/A		—	—	—	
	D40 Fire Protection						
4010	Sprinklers	Sprinkler, ordinary hazard		S.F. Floor	2.54	2.54	
4020	Standpipes	N/A		—	—	—	2.6%
	D50 Electrical						
5010	Electrical Service/Distribution	200 ampere service, panel board and feeders		S.F. Floor	2.47	2.47	
5020	Lighting & Branch Wiring	Fluorescent fixtures, receptacles, switches, A.C. and misc. power		S.F. Floor	16.03	16.03	19.3%
5030	Communications & Security	Alarm systems and emergency lighting		S.F. Floor	—	—	
5090	Other Electrical Systems	Emergency generator, 7.5 kW		S.F. Floor	.36	.36	
E. EQUIPMENT & FURNISHINGS							
1010	Commercial Equipment	N/A		—	—	—	
1020	Institutional Equipment	N/A		—	—	—	
1030	Vehicular Equipment	N/A		—	—	—	0.0%
1090	Other Equipment	N/A		—	—	—	
F. SPECIAL CONSTRUCTION & DEMOLITION							
1020	Integrated Construction	N/A		—	—	—	
1040	Special Facilities	N/A		—	—	—	0.0%
G. BUILDING SITEWORK	N/A						
				Sub-Total		97.77	100%
	CONTRACTOR FEES (General Requirements: 10%, Overhead: 5%, Profit: 10%)				25%	24.44	
	ARCHITECT FEES				7%	8.54	
				Total Building Cost		**130.75**	

Costs per square foot of floor area

Exterior Wall	S.F. Area	7000	10000	13000	16000	19000	22000	25000	28000	31000
	L.F. Perimeter	240	300	336	386	411	435	472	510	524
Face Brick with Concrete Block Back-up	R/Conc. Frame	125.05	118.45	112.85	110.30	107.15	104.80	103.60	102.65	101.05
	Steel Frame	120.35	113.75	108.15	105.60	102.45	100.05	98.85	97.90	96.35
Limestone with Concrete Block	R/Conc. Frame	139.80	131.35	123.95	120.70	116.45	113.30	111.70	110.45	108.30
	Steel Frame	135.10	126.65	119.25	116.00	111.75	108.60	107.00	105.75	103.60
Precast Concrete Panels	R/Conc. Frame	118.15	112.40	107.65	105.45	102.75	100.80	99.80	98.95	97.65
	Steel Frame	113.45	107.70	102.95	100.75	98.05	96.10	95.05	94.25	92.90
Perimeter Adj., Add or Deduct	Per 100 L.F.	15.65	10.90	8.45	6.85	5.75	5.00	4.40	3.90	3.55
Story Hgt. Adj., Add or Deduct	Per 1 Ft.	2.40	2.05	1.75	1.65	1.50	1.35	1.30	1.25	1.20

For Basement, add $28.90 per square foot of basement area

The above costs were calculated using the basic specifications shown on the facing page. These costs should be adjusted where necessary for design alternatives and owner's requirements. Reported completed project costs, for this type of structure, range from $61.70 to $158.30 per S.F.

Common additives

Description	Unit	$ Cost
Carrels Hardwood	Each	750 - 980
Closed Circuit Surveillance, One station		
Camera and monitor	Each	1375
For additional camera stations, add	Each	750
Elevators, Hydraulic passenger, 2 stops		
1500# capacity	Each	43,425
2500# capacity	Each	44,625
3500# capacity	Each	48,525
Emergency Lighting, 25 watt, battery operated		
Lead battery	Each	227
Nickel cadmium	Each	660
Flagpoles, Complete		
Aluminum, 20' high	Each	1100
40' high	Each	2700
70' high	Each	8350
Fiberglass, 23' high	Each	1425
39'-5" high	Each	3000
59' high	Each	7425

Description	Unit	$ Cost
Library Furnishings		
Bookshelf, 90" high, 10" shelf double face	L.F.	140
single face	L.F.	203
Charging desk, built-in with counter		
Plastic laminated top	L.F.	530
Reading table, laminated		
top 60" x 36"	Each	720

Important: See the Reference Section for Location Factors

Model costs calculated for a 2 story building with 14' story height and 22,000 square feet of floor area

Library

BUILDING TYPES

				Unit	Unit Cost	Cost Per S.F.	% Of Sub-Total
A.	**SUBSTRUCTURE**						
1010	Standard Foundations	Poured concrete; strip and spread footings		S.F. Ground	2.62	1.31	
1030	Slab on Grade	4" reinforced concrete with vapor barrier and granular base		S.F. Slab	3.42	1.71	
2010	Basement Excavation	Site preparation for slab and trench for foundation wall and footing		S.F. Ground	1.18	.59	6.0%
2020	Basement Walls	4' foundation wall		L.F. Wall	52	1.04	
B.	**SHELL**						
	B10 Superstructure						
1010	Floor Construction	Concrete waffle slab		S.F. Floor	18.04	9.02	20.2%
1020	Roof Construction	Concrete waffle slab		S.F. Roof	13.36	6.68	
	B20 Exterior Enclosure						
2010	Exterior Walls	Face brick with concrete block backup	90% of wall	S.F. Wall	21	10.56	
2020	Exterior Windows	Window wall	10% of wall	Each	37	2.06	16.6%
2030	Exterior Doors	Double aluminum and glass, single leaf hollow metal		Each	3350	.30	
	B30 Roofing						
3010	Roof Coverings	Single ply membrane, EPDM, fully adhered; perlite/EPS composite insulation		S.F. Roof	3.76	1.88	2.4%
3020	Roof Openings	Roof hatches		S.F. Roof	.04	.02	
C.	**INTERIORS**						
1010	Partitions	Gypsum board on metal studs	30 S.F. Floor/L.F. Partition	S.F. Partition	7.80	3.12	
1020	Interior Doors	Single leaf wood	300 S.F. Floor/Door	Each	432	1.44	
1030	Fittings	N/A		—	—	—	
2010	Stair Construction	Concrete filled metal pan		Flight	6125	.56	16.6%
3010	Wall Finishes	Paint		S.F. Surface	1.15	.46	
3020	Floor Finishes	50% carpet, 50% vinyl tile		S.F. Floor	3.61	3.61	
3030	Ceiling Finishes	Mineral fiber on concealed zee bars		S.F. Ceiling	3.71	3.71	
D.	**SERVICES**						
	D10 Conveying						
1010	Elevators & Lifts	One hydraulic passenger elevator		Each	55,000	2.50	3.2%
1020	Escalators & Moving Walks	N/A		—	—	—	
	D20 Plumbing						
2010	Plumbing Fixtures	Toilet and service fixtures, supply and drainage	1 Fixture/1835 S.F. Floor	Each	1633	.89	
2020	Domestic Water Distribution	Gas fired water heater		S.F. Floor	.44	.44	2.0%
2040	Rain Water Drainage	Roof drains		S.F. Roof	.46	.23	
	D30 HVAC						
3010	Energy Supply	N/A		—	—	—	
3020	Heat Generating Systems	Included in D3030		—	—	—	
3030	Cooling Generating Systems	Multizone unit, gas heating, electric cooling		S.F. Floor	15.00	15.00	19.3%
3050	Terminal & Package Units	N/A		—	—	—	
3090	Other HVAC Sys. & Equipment	N/A		—	—	—	
	D40 Fire Protection						
4010	Sprinklers	Wet pipe sprinkler system		S.F. Floor	1.63	1.63	2.1%
4020	Standpipes	N/A		—	—	—	
	D50 Electrical						
5010	Electrical Service/Distribution	400 ampere service, panel board and feeders		S.F. Floor	.74	.74	
5020	Lighting & Branch Wiring	Fluorescent fixtures, receptacles, switches, A.C. and misc. power		S.F. Floor	7.54	7.54	11.6%
5030	Communications & Security	Alarm systems and emergency lighting		S.F. Floor	.51	.51	
5090	Other Electrical Systems	Emergency generator, 7.5 kW		S.F. Floor	.07	.07	
E.	**EQUIPMENT & FURNISHINGS**						
1010	Commercial Equipment	N/A		—	—	—	
1020	Institutional Equipment	N/A		—	—	—	
1030	Vehicular Equipment	N/A		—	—	—	0.0%
1090	Other Equipment	N/A		—	—	—	
F.	**SPECIAL CONSTRUCTION & DEMOLITION**						
1020	Integrated Construction	N/A		—	—	—	0.0%
1040	Special Facilities	N/A		—	—	—	
G.	**BUILDING SITEWORK**	**N/A**					

		Sub-Total	77.62	100%
CONTRACTOR FEES (General Requirements: 10%, Overhead: 5%, Profit: 10%)		25%	19.41	
ARCHITECT FEES		8%	7.77	
	Total Building Cost		**104.80**	

BUILDING TYPES

Costs per square foot of floor area

Exterior Wall	S.F. Area	4000	5500	7000	8500	10000	11500	13000	14500	16000
	L.F. Perimeter	280	320	380	440	453	503	510	522	560
Face Brick with Concrete Block Back-up	Steel Joists	127.50	120.70	118.25	116.60	113.30	112.30	110.05	108.40	107.85
	Wood Truss	132.65	125.85	123.40	121.75	118.45	117.45	115.20	113.55	113.00
Stucco on Concrete Block	Steel Joists	119.85	114.40	112.30	110.95	108.35	107.55	105.75	104.50	104.00
	Wood Truss	125.00	119.55	117.45	116.10	113.50	112.70	110.90	109.65	109.15
Brick Veneer	Wood Truss	129.55	123.30	121.00	119.45	116.45	115.55	113.45	112.00	111.45
Wood Siding	Wood Frame	122.65	117.60	115.65	114.35	112.00	111.25	109.60	108.45	108.00
Perimeter Adj., Add or Deduct	Per 100 L.F.	11.50	8.40	6.55	5.40	4.60	4.00	3.50	3.20	2.85
Story Hgt. Adj., Add or Deduct	Per 1 Ft.	2.35	1.95	1.80	1.75	1.50	1.50	1.30	1.20	1.15

For Basement, add $18.80 per square foot of basement area

The above costs were calculated using the basic specifications shown on the facing page. These costs should be adjusted where necessary for design alternatives and owner's requirements. Reported completed project costs, for this type of structure, range from $52.00 to $133.85 per S.F.

Common additives

Description	Unit	$ Cost
Cabinets, Hospital, base		
Laminated plastic	L.F.	283
Stainless steel	L.F.	335
Counter top, laminated plastic	L.F.	54
Stainless steel	L.F.	118
For drop-in sink, add	Each	505
Nurses station, door type		
Laminated plastic	L.F.	315
Enameled steel	L.F.	284
Stainless steel	L.F.	355
Wall cabinets, laminated plastic	L.F.	206
Enameled steel	L.F.	224
Stainless steel	L.F.	320

Description	Unit	$ Cost
Directory Boards, Plastic, glass covered		
30" x 20"	Each	575
36" x 48"	Each	1075
Aluminum, 24" x 18"	Each	440
36" x 24"	Each	555
48" x 32"	Each	780
48" x 60"	Each	1675
Heat Therapy Unit		
Humidified, 26" x 78" x 28"	Each	2975
Smoke Detectors		
Ceiling type	Each	151
Duct type	Each	405
Tables, Examining, vinyl top		
with base cabinets	Each	1400 - 3800
Utensil Washer, Sanitizer	Each	9950
X-Ray, Mobile	Each	11,900 - 67,000

Important: See the Reference Section for Location Factors

Model costs calculated for a 1 story building with 10' story height and 7,000 square feet of floor area

				Unit	Unit Cost	Cost Per S.F.	% Of Sub-Total

A. SUBSTRUCTURE

				Unit	Unit Cost	Cost Per S.F.	% Of Sub-Total
1010	Standard Foundations	Poured concrete; strip and spread footings		S.F. Ground	1.39	1.39	
1030	Slab on Grade	4" reinforced concrete with vapor barrier and granular base		S.F. Slab	3.42	3.42	9.8%
2010	Basement Excavation	Site preparation for slab and trench for foundation wall and footing		S.F. Ground	1.18	1.18	
2020	Basement Walls	4' foundation wall		L.F. Wall	52	2.87	

B. SHELL

B10 Superstructure

				Unit	Unit Cost	Cost Per S.F.	% Of Sub-Total
1010	Floor Construction	N/A		—	—	—	5.4%
1020	Roof Construction	Plywood on wood trusses		S.F. Roof	4.90	4.90	

B20 Exterior Enclosure

				Unit	Unit Cost	Cost Per S.F.	% Of Sub-Total
2010	Exterior Walls	Face brick with concrete block backup	70% of wall	S.F. Wall	21	8.06	
2020	Exterior Windows	Wood double hung	30% of wall	Each	440	4.22	15.6%
2030	Exterior Doors	Aluminum and glass doors and entrance with transoms		Each	1088	1.87	

B30 Roofing

				Unit	Unit Cost	Cost Per S.F.	% Of Sub-Total
3010	Roof Coverings	Asphalt shingles with flashing (Pitched); rigid fiber glass insulation, gutters		S.F. Roof	2.79	2.79	3.1%
3020	Roof Openings	N/A		—	—	—	

C. INTERIORS

				Unit	Unit Cost	Cost Per S.F.	% Of Sub-Total
1010	Partitions	Gypsum bd. & sound deadening bd. on wood studs w/insul.	6 S.F. Floor/L.F. Partition	S.F. Partition	6.21	8.28	
1020	Interior Doors	Single leaf wood	60 S.F. Floor/Door	Each	432	7.20	
1030	Fittings	N/A		—	—	—	
2010	Stair Construction	N/A		—	—	—	27.9%
3010	Wall Finishes	50% paint, 50% vinyl wall covering		S.F. Surface	2.01	2.68	
3020	Floor Finishes	50% carpet, 50% vinyl composition tile		S.F. Floor	4.62	4.62	
3030	Ceiling Finishes	Mineral fiber tile on concealed zee bars		S.F. Ceiling	2.52	2.52	

D. SERVICES

D10 Conveying

				Unit	Unit Cost	Cost Per S.F.	% Of Sub-Total
1010	Elevators & Lifts	N/A		—	—	—	0.0%
1020	Escalators & Moving Walks	N/A		—	—	—	

D20 Plumbing

				Unit	Unit Cost	Cost Per S.F.	% Of Sub-Total
2010	Plumbing Fixtures	Toilet and service fixtures, supply and drainage	1 Fixture/195 S.F. Floor	Each	2152	11.04	
2020	Domestic Water Distribution	Gas fired water heater		S.F. Floor	1.01	1.01	13.3%
2040	Rain Water Drainage	N/A		—	—	—	

D30 HVAC

				Unit	Unit Cost	Cost Per S.F.	% Of Sub-Total
3010	Energy Supply	N/A		—	—	—	
3020	Heat Generating Systems	Included in D3030		—	—	—	
3030	Cooling Generating Systems	Multizone unit, gas heating, electric cooling		S.F. Floor	11.37	11.37	12.6%
3050	Terminal & Package Units	N/A		—	—	—	
3090	Other HVAC Sys. & Equipment	N/A		—	—	—	

D40 Fire Protection

				Unit	Unit Cost	Cost Per S.F.	% Of Sub-Total
4010	Sprinklers	Wet pipe sprinkler system		S.F. Floor	2.34	2.34	2.6%
4020	Standpipes	N/A		—	—	—	

D50 Electrical

				Unit	Unit Cost	Cost Per S.F.	% Of Sub-Total
5010	Electrical Service/Distribution	200 ampere service, panel board and feeders		S.F. Floor	1.00	1.00	
5020	Lighting & Branch Wiring	Fluorescent fixtures, receptacles, switches, A.C. and misc. power		S.F. Floor	5.07	5.07	9.7%
5030	Communications & Security	Alarm systems, intercom system, and emergency lighting		S.F. Floor	2.08	2.08	
5090	Other Electrical Systems	Emergency generator, 7.5 kW		S.F. Floor	.65	.65	

E. EQUIPMENT & FURNISHINGS

				Unit	Unit Cost	Cost Per S.F.	% Of Sub-Total
1010	Commercial Equipment	N/A		—	—	—	
1020	Institutional Equipment	N/A		—	—	—	0.0%
1030	Vehicular Equipment	N/A		—	—	—	
1090	Other Equipment	N/A		—	—	—	

F. SPECIAL CONSTRUCTION & DEMOLITION

				Unit	Unit Cost	Cost Per S.F.	% Of Sub-Total
1020	Integrated Construction	N/A		—	—	—	0.0%
1040	Special Facilities	N/A		—	—	—	

G. BUILDING SITEWORK   N/A

			Sub-Total			90.56	100%
	CONTRACTOR FEES (General Requirements: 10%, Overhead: 5%, Profit: 10%)				25%	22.64	
	ARCHITECT FEES				9%	10.20	
			Total Building Cost			**123.40**	

BUILDING TYPES

157

Costs per square foot of floor area

Exterior Wall	S.F. Area	4000	5500	7000	8500	10000	11500	13000	14500	16000
	L.F. Perimeter	180	210	240	270	286	311	336	361	386
Face Brick with Concrete Block Back-up	Steel Joists	141.15	134.05	130.00	127.40	124.50	122.95	121.75	120.80	120.05
	Wood Joists	146.25	139.20	135.15	132.55	129.65	128.10	126.90	125.95	125.20
Stucco on Concrete Block	Steel Joists	133.85	127.85	124.40	122.25	119.85	118.50	117.50	116.75	116.10
	Wood Joists	138.80	132.90	129.45	127.30	124.90	123.60	122.60	121.85	121.20
Brick Veneer	Wood Frame	142.40	135.90	132.20	129.80	127.20	125.75	124.65	123.80	123.10
Wood Siding	Wood Frame	136.50	130.90	127.70	125.65	123.45	122.20	121.25	120.55	119.95
Perimeter Adj., Add or Deduct	Per 100 L.F.	19.50	14.15	11.10	9.15	7.80	6.75	6.00	5.35	4.85
Story Hgt. Adj., Add or Deduct	Per 1 Ft.	2.85	2.40	2.15	2.00	1.80	1.70	1.65	1.55	1.50
For Basement, add $20.85 per square foot of basement area										

The above costs were calculated using the basic specifications shown on the facing page. These costs should be adjusted where necessary for design alternatives and owner's requirements. Reported completed project costs, for this type of structure, range from $53.35 to $182.25 per S.F.

Common additives

Description	Unit	$ Cost
Cabinets, Hospital, base		
Laminated plastic	L.F.	283
Stainless steel	L.F.	335
Counter top, laminated plastic	L.F.	54
Stainless steel	L.F.	118
For drop-in sink, add	Each	505
Nurses station, door type		
Laminated plastic	L.F.	315
Enameled steel	L.F.	284
Stainless steel	L.F.	355
Wall cabinets, laminated plastic	L.F.	206
Enameled steel	L.F.	224
Stainless steel	L.F.	320
Elevators, Hydraulic passenger, 2 stops		
1500# capacity	Each	43,425
2500# capacity	Each	44,625
3500# capacity	Each	48,525

Description	Unit	$ Cost
Directory Boards, Plastic, glass covered		
30" x 20"	Each	575
36" x 48"	Each	1075
Aluminum, 24" x 18"	Each	440
36" x 24"	Each	555
48" x 32"	Each	780
48" x 60"	Each	1675
Emergency Lighting, 25 watt, battery operated		
Lead battery	Each	227
Nickel cadmium	Each	660
Heat Therapy Unit		
Humidified, 26" x 78" x 28"	Each	2975
Smoke Detectors		
Ceiling type	Each	151
Duct type	Each	405
Tables, Examining, vinyl top		
with base cabinets	Each	1400 - 3800
Utensil Washer, Sanitizer	Each	9950
X-Ray, Mobile	Each	11,900 - 67,000

Important: See the Reference Section for Location Factors

Model costs calculated for a 2 story building with 10' story height and 7,000 square feet of floor area

Medical Office, 2 Story

				Unit	Unit Cost	Cost Per S.F.	% Of Sub-Total
A. SUBSTRUCTURE							
1010	Standard Foundations	Poured concrete; strip and spread footings		S.F. Ground	2.22	1.11	
1030	Slab on Grade	4" reinforced concrete with vapor barrier and granular base		S.F. Slab	3.42	1.71	
2010	Basement Excavation	Site preparation for slab and trench for foundation wall and footing		S.F. Ground	1.09	.55	5.7%
2020	Basement Walls	4' foundation wall		L.F. Wall	52	1.81	
B. SHELL							
	B10 Superstructure						
1010	Floor Construction	Open web steel joists, slab form, concrete, columns		S.F. Floor	8.52	4.26	6.5%
1020	Roof Construction	Metal deck, open web steel joists, beams, columns		S.F. Roof	3.32	1.66	
	B20 Exterior Enclosure						
2010	Exterior Walls	Stucco on concrete block		S.F. Wall	12.77	6.13	
2020	Exterior Windows	Outward projecting metal	30% of wall	Each	314	4.31	12.5%
2030	Exterior Doors	Aluminum and glass doors with transoms		Each	3350	.96	
	B30 Roofing						
3010	Roof Coverings	Built-up tar and gravel with flashing; perlite/EPS composite insulation		S.F. Roof	4.66	2.33	2.6%
3020	Roof Openings	Roof hatches		S.F. Roof	.14	.07	
C. INTERIORS							
1010	Partitions	Gypsum bd. & acous. insul. on metal studs	6 S.F. Floor/L.F. Partition	S.F. Partition	5.00	6.67	
1020	Interior Doors	Single leaf wood	60 S.F. Floor/Door	Each	432	7.20	
1030	Fittings	N/A		—	—	—	
2010	Stair Construction	Concrete filled metal pan		Flight	5075	1.45	27.5%
3010	Wall Finishes	45% paint, 50% vinyl wall coating, 5% ceramic tile		S.F. Surface	2.01	2.68	
3020	Floor Finishes	50% carpet, 50% vinyl asbestos tile		S.F. Floor	4.62	4.62	
3030	Ceiling Finishes	Mineral fiber tile on concealed zee bars		S.F. Ceiling	2.52	2.52	
D. SERVICES							
	D10 Conveying						
1010	Elevators & Lifts	One hydraulic hospital elevator		Each	65,030	9.29	10.2%
1020	Escalators & Moving Walks	N/A		—	—	—	
	D20 Plumbing						
2010	Plumbing Fixtures	Toilet and service fixtures, supply and drainage	1 Fixture/160 S.F. Floor	Each	1318	8.24	
2020	Domestic Water Distribution	Gas fired water heater		S.F. Floor	.89	.89	10.2%
2040	Rain Water Drainage	Roof drains		S.F. Roof	.34	.17	
	D30 HVAC						
3010	Energy Supply	N/A		—	—	—	
3020	Heat Generating Systems	Included in D3030		—	—	—	
3030	Cooling Generating Systems	Multizone unit, gas heating, electric cooling		S.F. Floor	11.37	11.37	12.5%
3050	Terminal & Package Units	N/A		—	—	—	
3090	Other HVAC Sys. & Equipment	N/A		—	—	—	
	D40 Fire Protection						
4010	Sprinklers	Wet pipe sprinkler system		S.F. Floor	1.95	1.95	2.1%
4020	Standpipes	N/A		—	—	—	
	D50 Electrical						
5010	Electrical Service/Distribution	400 ampere service, panel board and feeders		S.F. Floor	1.57	1.57	
5020	Lighting & Branch Wiring	Fluorescent fixtures, receptacles, switches, A.C. and misc. power		S.F. Floor	5.07	5.07	10.2%
5030	Communications & Security	Alarm system, intercom system, and emergency lighting		S.F. Floor	2.08	2.08	
5090	Other Electrical Systems	Emergency generator, 7.5 kW		S.F. Floor	.65	.65	
E. EQUIPMENT & FURNISHINGS							
1010	Commercial Equipment	N/A		—	—	—	
1020	Institutional Equipment	N/A		—	—	—	
1030	Vehicular Equipment	N/A		—	—	—	0.0%
1090	Other Equipment	N/A		—	—	—	
F. SPECIAL CONSTRUCTION & DEMOLITION							
1020	Integrated Construction	N/A		—	—	—	0.0%
1040	Special Facilities	N/A		—	—	—	
G. BUILDING SITEWORK	**N/A**						

Sub-Total		91.32	100%
CONTRACTOR FEES (General Requirements: 10%, Overhead: 5%, Profit: 10%)	25%	22.83	
ARCHITECT FEES	9%	10.25	
Total Building Cost		**124.40**	

BUILDING TYPES

159

Costs per square foot of floor area

Exterior Wall	S.F. Area	2000	3000	4000	6000	8000	10000	12000	14000	16000
	L.F. Perimeter	240	260	280	380	480	560	580	660	740
Brick Veneer	Wood Frame	127.65	111.65	103.70	99.60	97.60	95.60	92.30	91.60	91.10
Aluminum Siding	Wood Frame	115.50	102.90	96.60	93.20	91.50	89.95	87.40	86.85	86.45
Wood Siding	Wood Frame	115.70	103.05	96.75	93.30	91.60	90.00	87.45	86.90	86.50
Wood Shingles	Wood Frame	118.05	104.75	98.10	94.55	92.80	91.10	88.45	87.85	87.40
Precast Concrete Block	Wood Truss	114.55	102.20	96.05	92.70	91.05	89.50	87.00	86.45	86.05
Brick on Concrete Block	Wood Truss	132.85	115.45	106.70	102.35	100.20	98.00	94.35	93.65	93.10
Perimeter Adj., Add or Deduct	Per 100 L.F.	19.75	13.15	9.85	6.60	4.95	3.95	3.30	2.80	2.45
Story Hgt. Adj., Add or Deduct	Per 1 Ft.	3.30	2.35	1.90	1.70	1.65	1.55	1.30	1.30	1.25

For Basement, add $14.40 per square foot of basement area

The above costs were calculated using the basic specifications shown on the facing page. These costs should be adjusted where necessary for design alternatives and owner's requirements. Reported completed project costs, for this type of structure, range from $43.10 to $203.85 per S.F.

Common additives

Description	Unit	$ Cost
Closed Circuit Surveillance, One station		
Camera and monitor	Each	1375
For additional camera stations, add	Each	750
Emergency Lighting, 25 watt, battery operated		
Lead battery	Each	227
Nickel cadmium	Each	660
Laundry Equipment		
Dryer, gas, 16 lb. capacity	Each	725
30 lb. capacity	Each	2800
Washer, 4 cycle	Each	810
Commercial	Each	1225
Sauna, Prefabricated, complete		
6' x 4'	Each	4825
6' x 6'	Each	5625
6' x 9'	Each	5550
8' x 8'	Each	6525
8' x 10'	Each	7225
10' x 12'	Each	9475
Smoke Detectors		
Ceiling type	Each	151
Duct type	Each	405

Description	Unit	$ Cost
Swimming Pools, Complete, gunite	S.F.	48 - 59
TV Antenna, Master system, 12 outlet	Outlet	238
30 outlet	Outlet	153
100 outlet	Outlet	145

BUILDING TYPES

Model costs calculated for a 1 story building with 9' story height and 8,000 square feet of floor area

Motel, 1 Story

					Unit	Unit Cost	Cost Per S.F.	% Of Sub-Total
A. SUBSTRUCTURE								
1010	Standard Foundations	Poured concrete; strip and spread footings			S.F. Ground	2.10	2.10	
1030	Slab on Grade	4" reinforced concrete with vapor barrier and granular base			S.F. Slab	3.42	3.42	
2010	Basement Excavation	Site preparation for slab and trench for foundation wall and footing			S.F. Ground	1.18	1.18	14.8%
2020	Basement Walls	4' foundation wall			L.F. Wall	49	4.08	
B. SHELL								
	B10 Superstructure							
1010	Floor Construction	N/A			—	—	—	
1020	Roof Construction	Plywood on wood trusses			S.F. Roof	4.90	4.90	6.7%
	B20 Exterior Enclosure							
2010	Exterior Walls	Face brick on wood studs with sheathing, insulation and paper	80% of wall		S.F. Wall	16.60	7.17	
2020	Exterior Windows	Wood double hung	20% of wall		Each	354	2.73	17.7%
2030	Exterior Doors	Wood solid core			Each	1088	2.99	
	B30 Roofing							
3010	Roof Coverings	Asphalt shingles with flashing (pitched); rigid fiberglass insulation			S.F. Roof	2.66	2.66	3.6%
3020	Roof Openings	N/A			—	—	—	
C. INTERIORS								
1010	Partitions	Gypsum bd. and sound deadening bd. on wood studs	9 S.F. Floor/L.F. Partition		S.F. Partition	6.64	5.90	
1020	Interior Doors	Single leaf hollow core wood	300 S.F. Floor/Door		Each	388	1.29	
1030	Fittings	N/A			—	—	—	
2010	Stair Construction	N/A			—	—	—	25.3%
3010	Wall Finishes	90% paint, 10% ceramic tile			S.F. Surface	2.16	1.92	
3020	Floor Finishes	85% carpet, 15% ceramic tile			S.F. Floor	6.40	6.40	
3030	Ceiling Finishes	Painted gypsum board on furring			S.F. Ceiling	2.95	2.95	
D. SERVICES								
	D10 Conveying							
1010	Elevators & Lifts	N/A			—	—	—	
1020	Escalators & Moving Walks	N/A			—	—	—	0.0%
	D20 Plumbing							
2010	Plumbing Fixtures	Toilet and service fixtures, supply and drainage	1 Fixture/90 S.F. Floor		Each	967	10.75	
2020	Domestic Water Distribution	Gas fired water heater...			S.F. Floor	.75	.75	15.8%
2040	Rain Water Drainage	N/A			—	—	—	
	D30 HVAC							
3010	Energy Supply	N/A			—	—	—	
3020	Heat Generating Systems	Included in D3050			—	—	—	
3030	Cooling Generating Systems				—	—	—	4.9%
3050	Terminal & Package Units	Through the wall electric heating and cooling units			S.F. Floor	3.59	3.59	
3090	Other HVAC Sys. & Equipment	N/A			—	—	—	
	D40 Fire Protection							
4010	Sprinklers	Wet pipe sprinkler system			S.F. Floor	2.34	2.34	3.2%
4020	Standpipes	N/A			—	—	—	
	D50 Electrical							
5010	Electrical Service/Distribution	200 ampere service, panel board and feeders			S.F. Floor	.83	.83	
5020	Lighting & Branch Wiring	Fluorescent fixtures, receptacles, switches and misc. power			S.F. Floor	4.71	4.71	8.0%
5030	Communications & Security	Alarm systems			S.F. Floor	.30	.30	
5090	Other Electrical Systems	N/A			—	—	—	
E. EQUIPMENT & FURNISHINGS								
1010	Commercial Equipment	N/A			—	—	—	
1020	Institutional Equipment	N/A			—	—	—	
1030	Vehicular Equipment	N/A			—	—	—	0.0%
1090	Other Equipment	N/A			—	—	—	
F. SPECIAL CONSTRUCTION & DEMOLITION								
1020	Integrated Construction	N/A			—	—	—	
1040	Special Facilities	N/A			—	—	—	0.0%
G. BUILDING SITEWORK	**N/A**							

			Sub-Total	72.96	100%
CONTRACTOR FEES (General Requirements: 10%, Overhead: 5%, Profit: 10%)			25%	18.24	
ARCHITECT FEES			7%	6.40	

	Total Building Cost	**97.60**

BUILDING TYPES

Costs per square foot of floor area

Exterior Wall	S.F. Area	25000	37000	49000	61000	73000	81000	88000	96000	104000
	L.F. Perimeter	433	593	606	720	835	911	978	1054	1074
Decorative Concrete Block	Wood Joists	114.45	111.60	107.80	106.75	106.10	105.75	105.50	105.25	104.65
	Precast Conc.	121.50	118.70	114.90	113.85	113.20	112.85	112.55	112.35	111.75
Stucco on Concrete Block	Wood Joists	114.00	111.20	107.30	106.30	105.60	105.25	105.00	104.75	104.15
	Precast Conc.	121.75	118.95	115.05	114.05	113.35	113.00	112.75	112.50	111.90
Wood Siding	Wood Frame	112.55	109.85	106.25	105.25	104.65	104.30	104.05	103.85	103.25
Brick Veneer	Wood Frame	116.20	113.20	108.85	107.80	107.05	106.65	106.40	106.15	105.45
Perimeter Adj., Add or Deduct	Per 100 L.F.	3.30	2.20	1.65	1.30	1.10	1.00	.95	.90	.75
Story Hgt. Adj., Add or Deduct	Per 1 Ft.	1.15	1.05	.80	.80	.75	.75	.75	.70	.70

For Basement, add $18.85 per square foot of basement area

The above costs were calculated using the basic specifications shown on the facing page. These costs should be adjusted where necessary for design alternatives and owner's requirements. Reported completed project costs, for this type of structure, range from $43.10 to $203.85 per S.F.

Common additives

Description	Unit	$ Cost
Closed Circuit Surveillance, One station		
Camera and monitor	Each	1375
For additional camera station, add	Each	750
Elevators, Hydraulic passenger, 2 stops		
1500# capacity	Each	43,425
2500# capacity	Each	44,625
3500# capacity	Each	48,525
Additional stop, add	Each	3800
Emergency Lighting, 25 watt, battery operated		
Lead battery	Each	227
Nickel cadmium	Each	660
Laundry Equipment		
Dryer, gas, 16 lb. capacity	Each	725
30 lb. capacity	Each	2800
Washer, 4 cycle	Each	810
Commercial	Each	1225

Description	Unit	$ Cost
Sauna, Prefabricated, complete		
6' x 4'	Each	4825
6' x 6'	Each	5625
6' x 9'	Each	5550
8' x 8'	Each	6525
8' x 10'	Each	7225
10' x 12'	Each	9475
Smoke Detectors		
Ceiling type	Each	151
Duct type	Each	405
Swimming Pools, Complete, gunite	S.F.	48 - 59
TV Antenna, Master system, 12 outlet	Outlet	238
30 outlet	Outlet	153
100 outlet	Outlet	145

BUILDING TYPES

Model costs calculated for a 3 story building with 9' story height and 49,000 square feet of floor area

Motel, 2-3 Story

				Unit	Unit Cost	Cost Per S.F.	% Of Sub-Total
A. SUBSTRUCTURE							
1010	Standard Foundations	Poured concrete; strip and spread footings		S.F. Ground	.99	.33	
1030	Slab on Grade	4" reinforced concrete with vapor barrier and granular base		S.F. Slab	3.42	1.14	3.6%
2010	Basement Excavation	Site preparation for slab and trench for foundation wall and footing		S.F. Ground	1.09	.36	
2020	Basement Walls	4' foundation wall		L.F. Wall	52	1.31	
B. SHELL							
	B10 Superstructure						
1010	Floor Construction	Precast concrete plank		S.F. Floor	7.39	4.93	8.5%
1020	Roof Construction	Precast concrete plank		S.F. Roof	7.20	2.40	
	B20 Exterior Enclosure						
2010	Exterior Walls	Decorative concrete block	85% of wall	S.F. Wall	11.98	3.40	
2020	Exterior Windows	Aluminum sliding	15% of wall	Each	396	1.32	9.7%
2030	Exterior Doors	Aluminum and glass doors and entrance with transom		Each	1124	3.72	
	B30 Roofing						
3010	Roof Coverings	Built-up tar and gravel with flashing; perlite/EPS composite insulation		S.F. Roof	4.20	1.40	1.6%
3020	Roof Openings	Roof hatches		S.F. Roof	.06	.02	
C. INTERIORS							
1010	Partitions	Concrete block	7 S.F. Floor/L.F. Partition	S.F. Partition	12.77	14.59	
1020	Interior Doors	Wood hollow core	70 S.F. Floor/Door	Each	388	5.54	
1030	Fittings	N/A	—	—	—	—	
2010	Stair Construction	Concrete filled metal pan		Flight	5075	1.24	38.1%
3010	Wall Finishes	90% paint, 10% ceramic tile		S.F. Surface	2.15	2.46	
3020	Floor Finishes	85% carpet, 5% vinyl composition tile, 10% ceramic tile		S.F. Floor	6.40	6.40	
3030	Ceiling Finishes	Textured finish		S.F. Ceiling	2.79	2.79	
D. SERVICES							
	D10 Conveying						
1010	Elevators & Lifts	Two hydraulic passenger elevators		Each	66,150	2.70	3.1%
1020	Escalators & Moving Walks	N/A		—	—	—	
	D20 Plumbing						
2010	Plumbing Fixtures	Toilet and service fixtures, supply and drainage	1 Fixture/180 S.F. Floor	Each	2892	16.07	
2020	Domestic Water Distribution	Gas fired water heater		S.F. Floor	.46	.46	19.5%
2040	Rain Water Drainage	Roof drains		S.F. Roof	1.05	.35	
	D30 HVAC						
3010	Energy Supply	N/A		—	—	—	
3020	Heat Generating Systems	Included in D3050		—	—	—	
3030	Cooling Generating Systems	N/A		—	—	—	6.5%
3050	Terminal & Package Units	Through the wall electric heating and cooling units		S.F. Floor	5.60	5.60	
3090	Other HVAC Sys. & Equipment	N/A		—	—	—	
	D40 Fire Protection						
4010	Sprinklers	Sprinklers, light hazard		S.F. Floor	1.26	1.26	1.5%
4020	Standpipes	N/A		—	—	—	
	D50 Electrical						
5010	Electrical Service/Distribution	800 ampere service, panel board and feeders		S.F. Floor	1.18	1.18	
5020	Lighting & Branch Wiring	Fluorescent fixtures, receptacles, switches and misc. power		S.F. Floor	5.18	5.18	7.9%
5030	Communications & Security	Alarm systems and emergency lighting		S.F. Floor	.46	.46	
5090	Other Electrical Systems	Emergency generator, 7.5 kW		S.F. Floor	.09	.09	
E. EQUIPMENT & FURNISHINGS							
1010	Commercial Equipment	N/A		—	—	—	
1020	Institutional Equipment	N/A		—	—	—	0.0%
1030	Vehicular Equipment	N/A		—	—	—	
1090	Other Equipment	N/A		—	—	—	
F. SPECIAL CONSTRUCTION & DEMOLITION							
1020	Integrated Construction	N/A		—	—	—	0.0%
1040	Special Facilities	N/A		—	—	—	
G. BUILDING SITEWORK	**N/A**						

		Sub-Total	86.70	100%
CONTRACTOR FEES (General Requirements: 10%, Overhead: 5%, Profit: 10%)		25%	21.68	
ARCHITECT FEES		6%	6.52	
	Total Building Cost		**114.90**	

BUILDING TYPES

Costs per square foot of floor area

Exterior Wall	S.F. Area	9000	10000	12000	13000	14000	15000	16000	18000	20000
	L.F. Perimeter	385	410	460	460	480	500	510	547	583
Decorative Concrete Block	Steel Joists	108.60	105.80	101.55	98.85	97.30	96.00	94.50	92.55	90.95
Painted Concrete Block	Steel Joists	102.35	99.80	95.90	93.65	92.25	91.10	89.80	88.05	86.65
Face Brick on Conc. Block	Steel Joists	117.25	114.10	109.30	106.00	104.25	102.70	100.95	98.65	96.80
Precast Concrete Panels	Steel Joists	112.00	109.05	104.55	101.65	100.00	98.60	97.00	94.90	93.20
Tilt-up Panels	Steel Joists	112.90	109.95	105.40	102.40	100.75	99.35	97.70	95.55	93.85
Metal Sandwich Panels	Steel Joists	104.00	101.40	97.40	95.05	93.65	92.40	91.05	89.25	87.80
Perimeter Adj., Add or Deduct	Per 100 L.F.	6.20	5.55	4.65	4.25	4.00	3.70	3.45	3.05	2.75
Story Hgt. Adj., Add or Deduct	Per 1 Ft.	.90	.85	.80	.75	.70	.70	.65	.65	.60
Basement—Not Applicable										

The above costs were calculated using the basic specifications shown on the facing page. These costs should be adjusted where necessary for design alternatives and owner's requirements. Reported completed project costs, for this type of structure, range from $50.80 to $132.70 per S.F.

Common additives

Description	Unit	$ Cost
Emergency Lighting, 25 watt, battery operated		
Lead battery	Each	227
Nickel cadmium	Each	660
Seating		
Auditorium chair, all veneer	Each	171
Veneer back, padded seat	Each	207
Upholstered, spring seat	Each	208
Classroom, movable chair & desk	Set	65 - 120
Lecture hall, pedestal type	Each	156 - 485
Smoke Detectors		
Ceiling type	Each	151
Duct type	Each	405
Sound System		
Amplifier, 250 watts	Each	1675
Speaker, ceiling or wall	Each	147
Trumpet	Each	275

Important: See the Reference Section for Location Factors

BUILDING TYPES

Model costs calculated for a 1 story building with 20' story height and 12,000 square feet of floor area

Movie Theater

				Unit	Unit Cost	Cost Per S.F.	% Of Sub-Total
A. SUBSTRUCTURE							
1010	Standard Foundations	Poured concrete; strip and spread footings		S.F. Ground	.98	.98	
1030	Slab on Grade	4" reinforced concrete with vapor barrier and granular base		S.F. Slab	3.42	3.42	
2010	Basement Excavation	Site preparation for slab and trench for foundation wall and footing		S.F. Ground	2.49	2.49	11.7%
2020	Basement Walls	4' foundation wall		L.F. Wall	52	2.03	
B. SHELL							
	B10 Superstructure						
1010	Floor Construction	Open web steel joists, slab form, concrete	mezzanine 2250 S.F.	S.F. Floor	6.84	1.28	10.5%
1020	Roof Construction	Metal deck on open web steel joists		S.F. Roof	6.72	6.72	
	B20 Exterior Enclosure						
2010	Exterior Walls	Decorative concrete block	100% of wall	S.F. Wall	13.55	10.39	
2020	Exterior Windows	N/A		—	—	—	14.6%
2030	Exterior Doors	Sliding mallfront aluminum and glass and hollow metal		Each	1438	.72	
	B30 Roofing						
3010	Roof Coverings	Built-up tar and gravel with flashing; perlite/EPS composite insulation		S.F. Roof	3.94	3.94	5.2%
3020	Roof Openings	N/A		—	—	—	
C. INTERIORS							
1010	Partitions	Concrete block	40 S.F. Floor/L.F. Partition	S.F. Partition	6.03	2.41	
1020	Interior Doors	Single leaf hollow metal	705 S.F. Floor/Door	Each	537	.76	
1030	Fittings	Toilet partitions		S.F. Floor	.54	.54	
2010	Stair Construction	Concrete filled metal pan		Flight	6050	1.01	19.1%
3010	Wall Finishes	Paint		S.F. Surface	6.60	2.64	
3020	Floor Finishes	Carpet 50%, ceramic tile 5%	50% of area	S.F. Floor	6.94	3.47	
3030	Ceiling Finishes	Mineral fiber tile on concealed zee runners suspended		S.F. Ceiling	3.71	3.71	
D. SERVICES							
	D10 Conveying						
1010	Elevators & Lifts	N/A		—	—	—	0.0%
1020	Escalators & Moving Walks	N/A		—	—	—	
	D20 Plumbing						
2010	Plumbing Fixtures	Toilet and service fixtures, supply and drainage	1 Fixture/500 S.F. Floor	Each	1280	2.56	
2020	Domestic Water Distribution	Gas fired water heater		S.F. Floor	.15	.15	3.9%
2040	Rain Water Drainage	Roof drains		S.F. Roof	.22	.22	
	D30 HVAC						
3010	Energy Supply	N/A		—	—	—	
3020	Heat Generating Systems	Included in D3030		—	—	—	
3030	Cooling Generating Systems	Single zone rooftop unit, gas heating, electric cooling		S.F. Floor	8.96	8.96	11.8%
3050	Terminal & Package Units	N/A		—	—	—	
3090	Other HVAC Sys. & Equipment	N/A		—	—	—	
	D40 Fire Protection						
4010	Sprinklers	Wet pipe sprinkler system		S.F. Floor	2.17	2.17	2.9%
4020	Standpipes	N/A		—	—	—	
	D50 Electrical						
5010	Electrical Service/Distribution	400 ampere service, panel board and feeders		S.F. Floor	1.14	1.14	
5020	Lighting & Branch Wiring	Fluorescent fixtures, receptacles, switches, A.C. and misc. power		S.F. Floor	3.25	3.25	6.5%
5030	Communications & Security	Alarm systems, sound system and emergency lighting		S.F. Floor	.38	.38	
5090	Other Electrical Systems	Emergency generator, 7.5 kW		S.F. Floor	.13	.13	
E. EQUIPMENT & FURNISHINGS							
1010	Commercial Equipment	N/A		—	—	—	
1020	Institutional Equipment	Projection equipment, screen		S.F. Floor	7.02	7.02	13.8%
1030	Vehicular Equipment	N/A		—	—	—	
1090	Other Equipment	Movie theater seating		S.F. Floor	3.44	3.44	
F. SPECIAL CONSTRUCTION & DEMOLITION							
1020	Integrated Construction	N/A		—	—	—	0.0%
1040	Special Facilities	N/A		—	—	—	
G. BUILDING SITEWORK	**N/A**						

	Sub-Total	75.93	100%
CONTRACTOR FEES (General Requirements: 10%, Overhead: 5%, Profit: 10%)	25%	18.98	
ARCHITECT FEES	7%	6.64	
	Total Building Cost	**101.55**	

Costs per square foot of floor area

Exterior Wall	S.F. Area	10000	15000	20000	25000	30000	35000	40000	45000	50000
	L.F. Perimeter	286	370	453	457	513	568	624	680	735
Precast Concrete Panels	Bearing Walls	115.05	109.85	107.30	103.35	102.00	101.00	100.30	99.75	99.30
	Steel Frame	117.80	112.60	110.00	106.05	104.70	103.75	103.00	102.50	102.00
Face Brick with Concrete Block Back-up	Bearing Walls	117.60	112.05	109.30	104.95	103.50	102.45	101.70	101.10	100.60
	Steel Joists	120.65	115.10	112.35	108.00	106.55	105.50	104.75	104.15	103.65
Stucco on Concrete Block	Bearing Walls	111.70	106.95	104.60	101.15	99.95	99.10	98.45	97.95	97.55
	Steel Joists	114.75	110.00	107.65	104.20	103.00	102.15	101.50	101.00	100.60
Perimeter Adj., Add or Deduct	Per 100 L.F.	7.60	5.05	3.80	3.00	2.55	2.15	1.90	1.65	1.55
Story Hgt. Adj., Add or Deduct	Per 1 Ft.	1.70	1.45	1.35	1.05	1.00	.95	.90	.90	.90
For Basement, add $19.50 per square foot of basement area										

The above costs were calculated using the basic specifications shown on the facing page. These costs should be adjusted where necessary for design alternatives and owner's requirements. Reported completed project costs, for this type of structure, range from $54.50 to $136.50 per S.F.

Common additives

Description	Unit	$ Cost
Beds, Manual	Each	910 - 1575
Doctors In-Out Register, 200 names	Each	13,600
Elevators, Hydraulic passenger, 2 stops		
1500# capacity	Each	43,425
2500# capacity	Each	44,625
3500# capacity	Each	48,525
Emergency Lighting, 25 watt, battery operated		
Lead battery	Each	227
Nickel cadmium	Each	660
Intercom System, 25 station capacity		
Master station	Each	2025
Intercom outlets	Each	128
Handset	Each	335
Kitchen Equipment		
Broiler	Each	4275
Coffee urn, twin 6 gallon	Each	7375
Cooler, 6 ft. long	Each	3375
Dishwasher, 10-12 racks per hr.	Each	3225
Food warmer	Each	745
Freezer, 44 C.F., reach-in	Each	8775

Description	Unit	$ Cost
Kitchen Equipment, cont.		
Ice cube maker, 50 lb. per day	Each	1900
Range with 1 oven	Each	2500
Laundry Equipment		
Dryer, gas, 16 lb. capacity	Each	725
30 lb. capacity	Each	2800
Washer, 4 cycle	Each	810
Commercial	Each	1225
Nurses Call System		
Single bedside call station	Each	245
Pillow speaker	Each	230
Refrigerator, Prefabricated, walk-in		
7'-6" high, 6' x 6'	S.F.	121
10' x 10'	S.F.	95
12' x 14'	S.F.	84
12' x 20'	S.F.	74
TV Antenna, Master system, 12 outlet	Outlet	238
30 outlet	Outlet	153
100 outlet	Outlet	145
Whirlpool Bath, Mobile, 18" x 24" x 60"	Each	4025
X-Ray, Mobile	Each	11,900 - 67,000

Nursing Home

Model costs calculated for a 2 story building with 10' story height and 25,000 square feet of floor area

			Unit	Unit Cost	Cost Per S.F.	% Of Sub-Total
A. SUBSTRUCTURE						
1010	Standard Foundations	Poured concrete; strip and spread footings	S.F. Ground	.94	.47	
1030	Slab on Grade	4" reinforced concrete with vapor barrier and granular base	S.F. Slab	3.42	1.71	5.2%
2010	Basement Excavation	Site preparation for slab and trench for foundation wall and footing	S.F. Ground	1.18	.59	
2020	Basement Walls	4' foundation wall	L.F. Wall	46	1.11	
B. SHELL						
B10 Superstructure						
1010	Floor Construction	Pre-cast double tees with concrete topping	S.F. Floor	8.40	4.20	10.2%
1020	Roof Construction	Pre-Cast double tees	S.F. Roof	6.80	3.40	
B20 Exterior Enclosure						
2010	Exterior Walls	Precast concrete panels	85% of wall	S.F. Wall	17.47	5.43
2020	Exterior Windows	Wood double hung	15% of wall	Each	396	1.45	9.6%
2030	Exterior Doors	Double aluminum & glass doors, single leaf hollow metal	Each	1540	.30	
B30 Roofing						
3010	Roof Coverings	Built-up tar and gravel with flashing; perlite/EPS composite insulation	S.F. Roof	4.20	2.10	2.8%
3020	Roof Openings	Roof hatches	S.F. Roof	.04	.02	
C. INTERIORS						
1010	Partitions	Gypsum board on metal studs	8 S.F. Floor/L.F. Partition	S.F. Partition	4.87	4.87
1020	Interior Doors	Single leaf wood	80 S.F. Floor/Door	Each	432	5.40
1030	Fittings	N/A		—	—	—
2010	Stair Construction	Concrete stair	Flight	4325	1.04	26.4%
3010	Wall Finishes	50% vinyl wall coverings, 45% paint, 5% ceramic tile	S.F. Surface	2.51	2.51	
3020	Floor Finishes	95% vinyl tile, 5% ceramic tile	S.F. Floor	2.98	2.98	
3030	Ceiling Finishes	Painted gypsum board	S.F. Ceiling	2.83	2.83	
D. SERVICES						
D10 Conveying						
1010	Elevators & Lifts	One hydraulic hospital elevator	Each	65,000	2.60	3.5%
1020	Escalators & Moving Walks	N/A		—	—	—
D20 Plumbing						
2010	Plumbing Fixtures	Kitchen, toilet and service fixtures, supply and drainage	1 Fixture/230 S.F. Floor	Each	2021	8.79
2020	Domestic Water Distribution	Oil fired water heater	S.F. Floor	.28	.28	12.4%
2040	Rain Water Drainage	Roof drains	S.F. Roof	.36	.18	
D30 HVAC						
3010	Energy Supply	Oil fired hot water, wall fin radiation	S.F. Floor	4.68	4.68	
3020	Heat Generating Systems	N/A		—	—	—
3030	Cooling Generating Systems	Split systems with air cooled condensing units	S.F. Floor	5.64	5.64	13.9%
3050	Terminal & Package Units	N/A		—	—	—
3090	Other HVAC Sys. & Equipment	N/A		—	—	—
D40 Fire Protection						
4010	Sprinklers	Sprinkler, light hazard	S.F. Floor	3.24	3.24	4.4%
4020	Standpipes	N/A		—	—	—
D50 Electrical						
5010	Electrical Service/Distribution	800 ampere service, panel board and feeders	S.F. Floor	1.30	1.30	
5020	Lighting & Branch Wiring	Incandescent fixtures, receptacles, switches, A.C. and misc. power	S.F. Floor	6.45	6.45	11.6%
5030	Communications & Security	Alarm systems and emergency lighting	S.F. Floor	.51	.51	
5090	Other Electrical Systems	Emergency generator, 15 kW	S.F. Floor	.39	.39	
E. EQUIPMENT & FURNISHINGS						
1010	Commercial Equipment	N/A		—	—	—
1020	Institutional Equipment	N/A		—	—	—
1030	Vehicular Equipment	N/A		—	—	0.0%
1090	Other Equipment	N/A		—	—	—
F. SPECIAL CONSTRUCTION & DEMOLITION						
1020	Integrated Construction	N/A		—	—	—
1040	Special Facilities	N/A		—	—	0.0%
G. BUILDING SITEWORK	N/A					N/A
Sub-Total					74.47	100%
CONTRACTOR FEES (General Requirements: 10%, Overhead: 5%, Profit: 10%)				25%	18.62	
ARCHITECT FEES				11%	10.26	
Total Building Cost					**103.35**	

Costs per square foot of floor area

Exterior Wall	S.F. Area	5000	8000	12000	16000	20000	35000	50000	65000	80000
	L.F. Perimeter	220	260	310	330	360	440	490	548	580
Face Brick with Concrete Block Back-up	Wood Joists	156.00	133.05	120.00	111.00	106.30	96.95	92.50	90.20	88.40
	Steel Joists	156.40	133.45	120.35	111.40	106.65	97.30	92.85	90.55	88.75
Glass and Metal Curtain Wall	Steel Frame	152.40	131.25	119.25	111.10	106.80	98.30	94.30	92.25	90.60
	R/Conc. Frame	156.40	135.30	123.25	115.15	110.80	102.35	98.30	96.25	94.65
Wood Siding	Wood Frame	128.75	110.85	100.65	94.00	90.40	83.45	80.20	78.55	77.25
Brick Veneer	Wood Frame	140.45	119.45	107.50	99.45	95.20	86.80	82.80	80.75	79.15
Perimeter Adj., Add or Deduct	Per 100 L.F.	25.70	16.05	10.70	8.05	6.40	3.70	2.55	2.00	1.60
Story Hgt. Adj., Add or Deduct	Per 1 Ft.	4.20	3.10	2.45	1.95	1.70	1.20	.90	.80	.65

For Basement, add $22.10 per square foot of basement area

The above costs were calculated using the basic specifications shown on the facing page. These costs should be adjusted where necessary for design alternatives and owner's requirements. Reported completed project costs, for this type of structure, range from $43.55 to $168.50 per S.F.

Common additives

Description	Unit	$ Cost	Description	Unit	$ Cost
Clock System			Smoke Detectors		
20 room	Each	13,000	Ceiling type	Each	151
50 room	Each	31,500	Duct type	Each	405
Closed Circuit Surveillance, One station			Sound System		
Camera and monitor	Each	1375	Amplifier, 250 watts	Each	1675
For additional camera stations, add	Each	750	Speaker, ceiling or wall	Each	147
Directory Boards, Plastic, glass covered			Trumpet	Each	275
30" x 20"	Each	575	TV Antenna, Master system, 12 outlet	Outlet	238
36" x 48"	Each	1075	30 outlet	Outlet	153
Aluminum, 24" x 18"	Each	440	100 outlet	Outlet	145
36" x 24"	Each	555			
48" x 32"	Each	780			
48" x 60"	Each	1675			
Elevators, Hydraulic passenger, 2 stops					
1500# capacity	Each	43,425			
2500# capacity	Each	44,625			
3500# capacity	Each	48,525			
Additional stop, add	Each	3800			
Emergency Lighting, 25 watt, battery operated					
Lead battery	Each	227			
Nickel cadmium	Each	660			

Important: See the Reference Section for Location Factors

Model costs calculated for a 3 story building with 12' story height and 20,000 square feet of floor area

BUILDING TYPES

				Unit	Unit Cost	Cost Per S.F.	% Of Sub-Total
A. SUBSTRUCTURE							
1010	Standard Foundations	Poured concrete; strip and spread footings		S.F. Ground	5.04	1.68	
1030	Slab on Grade	4" reinforced concrete with vapor barrier and granular base		S.F. Slab	3.42	1.14	
2010	Basement Excavation	Site preparation for slab and trench for foundation wall and footing		S.F. Ground	1.09	.36	5.4%
2020	Basement Walls	4' foundation wall		L.F. Wall	49	1.11	
B. SHELL							
	B10 Superstructure						
1010	Floor Construction	Open web steel joists, slab form, concrete, columns		S.F. Floor	11.41	7.61	11.2%
1020	Roof Construction	Metal deck, open web steel joists, columns		S.F. Roof	3.96	1.32	
	B20 Exterior Enclosure						
2010	Exterior Walls	Face brick with concrete block backup	80% of wall	S.F. Wall	21	10.94	
2020	Exterior Windows	Aluminum outward projecting	20% of wall	Each	509	2.87	18.1%
2030	Exterior Doors	Aluminum and glass, hollow metal		Each	2088	.64	
	B30 Roofing						
3010	Roof Coverings	Built-up tar and gravel with flashing; perlite/EPS composite		S.F. Roof	4.53	1.51	1.9%
3020	Roof Openings	N/A		—	—	—	
C. INTERIORS							
1010	Partitions	Gypsum board on metal studs	20 S.F. Floor/L.F. Partition	S.F. Partition	6.78	2.71	
1020	Interior Doors	Single leaf hollow metal	200 S.F. Floor/Door	Each	537	2.69	
1030	Fittings	Toilet partitions		S.F. Floor	.93	.93	
2010	Stair Construction	Concrete filled metal pan		Flight	5075	1.78	22.6%
3010	Wall Finishes	60% vinyl wall covering, 40% paint		S.F. Surface	2.20	.88	
3020	Floor Finishes	60% carpet, 30% vinyl composition tile, 10% ceramic tile		S.F. Floor	5.29	5.29	
3030	Ceiling Finishes	Mineral fiber tile on concealed zee bars		S.F. Ceiling	3.71	3.71	
D. SERVICES							
	D10 Conveying						
1010	Elevators & Lifts	Two hydraulic passenger elevators		Each	69,600	6.96	8.7%
1020	Escalators & Moving Walks	N/A		—	—	—	
	D20 Plumbing						
2010	Plumbing Fixtures	Toilet and service fixtures, supply and drainage	1 Fixture/1320 S.F. Floor	Each	1716	1.30	
2020	Domestic Water Distribution	Gas fired water heater		S.F. Floor	.15	.15	2.1%
2040	Rain Water Drainage	Roof drains		S.F. Roof	.72	.24	
	D30 HVAC						
3010	Energy Supply	N/A		—	—	—	
3020	Heat Generating Systems	Included in D3030		—	—	—	
3030	Cooling Generating Systems	Multizone unit gas heating, electric cooling		S.F. Floor	11.80	11.80	14.8%
3050	Terminal & Package Units	N/A		—	—	—	
3090	Other HVAC Sys. & Equipment	N/A		—	—	—	
	D40 Fire Protection						
4010	Sprinklers	N/A		—	—	—	0.8%
4020	Standpipes	Standpipes and hose systems		S.F. Floor	.61	.61	
	D50 Electrical						
5010	Electrical Service/Distribution	1000 ampere service, panel board and feeders		S.F. Floor	2.89	2.89	
5020	Lighting & Branch Wiring	Fluorescent fixtures, receptacles, switches, A.C. and misc. power		S.F. Floor	7.99	7.99	14.4%
5030	Communications & Security	Alarm systems and emergency lighting		S.F. Floor	.57	.57	
5090	Other Electrical Systems	Emergency generator, 7.5 kW		S.F. Floor	.05	.05	
E. EQUIPMENT & FURNISHINGS							
1010	Commercial Equipment	N/A		—	—	—	
1020	Institutional Equipment	N/A		—	—	—	0.0%
1030	Vehicular Equipment	N/A		—	—	—	
1090	Other Equipment	N/A		—	—	—	
F. SPECIAL CONSTRUCTION & DEMOLITION							
1020	Integrated Construction	N/A		—	—	—	0.0%
1040	Special Facilities	N/A		—	—	—	
G. BUILDING SITEWORK	**N/A**						

		Sub-Total	79.73	100%
CONTRACTOR FEES (General Requirements: 10%, Overhead: 5%, Profit: 10%)		25%	19.93	
ARCHITECT FEES		7%	6.99	
	Total Building Cost		**106.65**	

Costs per square foot of floor area

Exterior Wall	S.F. Area	20000	40000	60000	80000	100000	150000	200000	250000	300000
	L.F. Perimeter	260	360	400	420	460	520	600	640	700
Precast Concrete Panel	Steel Frame	124.05	107.70	99.25	94.25	91.90	87.90	86.20	84.70	83.95
	R/Conc. Frame	126.55	110.05	101.50	96.50	94.15	90.10	88.40	86.85	86.10
Face Brick with Concrete Block Back-up	Steel Frame	128.20	110.45	101.25	95.85	93.30	88.95	87.10	85.50	84.65
	R/Conc. Frame	130.30	112.60	103.40	98.00	95.45	91.05	89.25	87.60	86.75
Limestone Panel Concrete Block Back-up	Steel Frame	147.10	123.55	110.95	103.50	100.00	93.95	91.50	89.20	88.00
	R/Conc. Frame	149.25	125.70	113.10	105.65	102.15	96.10	93.60	91.35	90.15
Perimeter Adj., Add or Deduct	Per 100 L.F.	15.00	7.50	5.00	3.75	3.00	2.00	1.50	1.25	1.00
Story Hgt. Adj., Add or Deduct	Per 1 Ft.	3.05	2.10	1.60	1.20	1.10	.80	.70	.60	.55

For Basement, add $23.85 per square foot of basement area

The above costs were calculated using the basic specifications shown on the facing page. These costs should be adjusted where necessary for design alternatives and owner's requirements. Reported completed project costs, for this type of structure, range from $48.95 to $143.95 per S.F.

Common additives

Description	Unit	$ Cost
Clock System		
20 room	Each	13,000
50 room	Each	31,500
Closed Circuit Surveillance, One station		
Camera and monitor	Each	1375
For additional camera stations, add	Each	750
Directory Boards, Plastic, glass covered		
30" x 20"	Each	575
36" x 48"	Each	1075
Aluminum, 24" x 18"	Each	440
36" x 24"	Each	555
48" x 32"	Each	780
48" x 60"	Each	1675
Elevators, Electric passenger, 5 stops		
2000# capacity	Each	101,400
3500# capacity	Each	107,400
5000# capacity	Each	111,900
Additional stop, add	Each	5675
Emergency Lighting, 25 watt, battery operated		
Lead battery	Each	227
Nickel cadmium	Each	660

Description	Unit	$ Cost
Intercom System, 25 station capacity		
Master station	Each	2025
Intercom outlets	Each	128
Handset	Each	335
Smoke Detectors		
Ceiling type	Each	151
Duct type	Each	405
Sound System		
Amplifier, 250 watts	Each	1675
Speaker, ceiling or wall	Each	147
Trumpet	Each	275
TV Antenna, Master system, 12 oulet	Outlet	238
30 outlet	Outlet	153
100 outlet	Outlet	145

Important: See the Reference Section for Location Factors

Model costs calculated for an 8 story building with 12' story height and 80,000 square feet of floor area.

					Unit	Unit Cost	Cost Per S.F.	% Of Sub-Total

A. SUBSTRUCTURE

				Unit	Unit Cost	Cost Per S.F.	% Of Sub-Total
1010	Standard Foundations	Poured concrete; strip and spread footings		S.F. Ground	8.16	1.02	
1030	Slab on Grade	4" reinforced concrete with vapor barrier and granular base		S.F. Slab	3.42	.43	
2010	Basement Excavation	Site preparation for slab and trench for foundation wall and footing		S.F. Ground	1.18	.15	2.8%
2020	Basement Walls	4' foundation wall		L.F. Wall	52	.36	

B. SHELL

B10 Superstructure

				Unit	Unit Cost	Cost Per S.F.	% Of Sub-Total
1010	Floor Construction	Concrete slab with metal deck and beams		S.F. Floor	13.41	11.73	17.2%
1020	Roof Construction	Metal deck, open web steel joists, interior columns		S.F. Roof	3.92	.49	

B20 Exterior Enclosure

				Unit	Unit Cost	Cost Per S.F.	% Of Sub-Total
2010	Exterior Walls	Precast concrete panels	80% of wall	S.F. Wall	18.28	7.37	
2020	Exterior Windows	Vertical pivoted steel	20% of wall	Each	396	2.66	14.3%
2030	Exterior Doors	Double aluminum and glass doors and entrance with transoms		Each	2432	.16	

B30 Roofing

				Unit	Unit Cost	Cost Per S.F.	% Of Sub-Total
3010	Roof Coverings	Built-up tar and gravel with flashing; perlite/EPS composite insulation		S.F. Roof	3.92	.49	0.7%
3020	Roof Openings	N/A		—	—	—	

C. INTERIORS

				Unit	Unit Cost	Cost Per S.F.	% Of Sub-Total
1010	Partitions	Gypsum board on metal studs	30 S.F. Floor/L.F. Partition	S.F. Partition	6.51	2.17	
1020	Interior Doors	Single leaf hollow metal	400 S.F. Floor/Door	Each	537	1.34	
1030	Fittings	Toilet Partitions		S.F. Floor	.62	.62	
2010	Stair Construction	Concrete filled metal pan		Flight	5075	1.08	21.0%
3010	Wall Finishes	60% vinyl wall covering, 40% paint		S.F. Surface	2.16	.72	
3020	Floor Finishes	60% carpet, 30% vinyl composition tile, 10% ceramic tile		S.F. Floor	5.29	5.29	
3030	Ceiling Finishes	Mineral fiber tile on concealed zee bars		S.F. Ceiling	3.71	3.71	

D. SERVICES

D10 Conveying

				Unit	Unit Cost	Cost Per S.F.	% Of Sub-Total
1010	Elevators & Lifts	Four geared passenger elevators		Each	159,200	7.96	11.2%
1020	Escalators & Moving Walks	N/A		—	—	—	

D20 Plumbing

				Unit	Unit Cost	Cost Per S.F.	% Of Sub-Total
2010	Plumbing Fixtures	Toilet and service fixtures, supply and drainage	1 Fixture/1370 S.F. Floor	Each	1712	1.25	
2020	Domestic Water Distribution	Gas fired water heater		S.F. Floor	.16	.16	2.1%
2040	Rain Water Drainage	Roof drains		S.F. Roof	.80	.10	

D30 HVAC

				Unit	Unit Cost	Cost Per S.F.	% Of Sub-Total
3010	Energy Supply	N/A		—	—	—	
3020	Heat Generating Systems	Included in D3030		—	—	—	
3030	Cooling Generating Systems	Multizone unit gas heating, electric cooling		S.F. Floor	11.80	11.80	16.6%
3050	Terminal & Package Units	N/A		—	—	—	
3090	Other HVAC Sys. & Equipment	N/A		—	—	—	

D40 Fire Protection

				Unit	Unit Cost	Cost Per S.F.	% Of Sub-Total
4010	Sprinklers	N/A		—	—	—	
4020	Standpipes	Standpipes and hose systems		S.F. Floor	.17	.17	0.2%

D50 Electrical

				Unit	Unit Cost	Cost Per S.F.	% Of Sub-Total
5010	Electrical Service/Distribution	1600 ampere service, panel board and feeders		S.F. Floor	1.08	1.08	
5020	Lighting & Branch Wiring	Fluorescent fixtures, receptacles, switches, A.C. and misc. power		S.F. Floor	7.91	7.91	13.9%
5030	Communications & Security	Alarm systems and emergency lighting		S.F. Floor	.48	.48	
5090	Other Electrical Systems	Emergency generator, 100 kW		S.F. Floor	.44	.44	

E. EQUIPMENT & FURNISHINGS

				Unit	Unit Cost	Cost Per S.F.	% Of Sub-Total
1010	Commercial Equipment	N/A		—	—	—	
1020	Institutional Equipment	N/A		—	—	—	
1030	Vehicular Equipment	N/A		—	—	—	0.0%
1090	Other Equipment	N/A		—	—	—	

F. SPECIAL CONSTRUCTION & DEMOLITION

				Unit	Unit Cost	Cost Per S.F.	% Of Sub-Total
1020	Integrated Construction	N/A		—	—	—	0.0%
1040	Special Facilities	N/A		—	—	—	

G. BUILDING SITEWORK   N/A

		Sub-Total	71.14	100%
CONTRACTOR FEES (General Requirements: 10%, Overhead: 5%, Profit: 10%)		25%	17.79	
ARCHITECT FEES		6%	5.32	

	Total Building Cost	**94.25**

BUILDING TYPES

171

Costs per square foot of floor area

Exterior Wall	S.F. Area	120000	145000	170000	200000	230000	260000	400000	600000	800000
	L.F. Perimeter	420	450	470	490	510	530	600	730	820
Double Glazed Heat Absorbing Tinted Plate Glass Panels	Steel Frame	104.80	100.90	97.70	94.75	92.65	90.95	86.05	83.35	81.65
	R/Conc. Frame	106.25	101.55	97.70	94.25	91.70	89.75	84.00	80.80	78.85
Face Brick with Concrete Block Back-up	Steel Frame	103.75	99.70	96.45	93.55	91.35	89.75	84.95	82.20	80.60
	R/Conc. Frame	99.60	95.55	92.40	89.50	87.40	85.75	81.00	78.35	76.75
Precast Concrete Panel With Exposed Aggregate	Steel Frame	101.45	97.65	94.65	91.90	89.90	88.40	83.95	81.40	79.90
	R/Conc. Frame	97.55	93.75	90.80	88.15	86.15	84.65	80.25	77.75	76.30
Perimeter Adj., Add or Deduct	Per 100 L.F.	6.25	5.20	4.40	3.75	3.25	2.90	1.85	1.25	.95
Story Hgt. Adj., Add or Deduct	Per 1 Ft.	2.55	2.25	2.00	1.75	1.60	1.50	1.10	.90	.75

For Basement, add $23.85 per square foot of basement area

The above costs were calculated using the basic specifications shown on the facing page. These costs should be adjusted where necessary for design alternatives and owner's requirements. Reported completed project costs, for this type of structure, range from $61.50 to $149.90 per S.F.

Common additives

Description	Unit	$ Cost		Description	Unit	$ Cost
Clock System				Escalators, Metal		
20 room	Each	13,000		32" wide, 10' story height	Each	91,000
50 room	Each	31,500		20' story height	Each	105,500
Directory Boards, Plastic, glass covered				48" wide, 10' story height	Each	95,500
30" x 20"	Each	575		20' story height	Each	110,000
36" x 48"	Each	1075		Glass		
Aluminum, 24" x 18"	Each	440		32" wide, 10' story height	Each	89,500
36" x 24"	Each	555		20' story height	Each	105,500
48" x 32"	Each	780		48" wide, 10' story height	Each	95,500
48" x 60"	Each	1675		20' story height	Each	110,000
Elevators, Electric passenger, 10 stops				Smoke Detectors		
3000# capacity	Each	217,000		Ceiling type	Each	151
4000# capacity	Each	219,000		Duct type	Each	405
5000# capacity	Each	223,000		Sound System		
Additional stop, add	Each	5675		Amplifier, 250 watts	Each	1675
Emergency Lighting, 25 watt, battery operated				Speaker, ceiling or wall	Each	147
Lead battery	Each	227		Trumpet	Each	275
Nickel cadmium	Each	660		TV Antenna, Master system, 12 oulet	Outlet	238
				30 oulet	Outlet	153
				100 outlet	Outlet	145

Model costs calculated for a 16 story building with 10' story height and 260,000 square feet of floor area

				Unit	Unit Cost	Cost Per S.F.	% Of Sub-Total
A. SUBSTRUCTURE							
1010	Standard Foundations	Poured concrete; strip and spread footings		S.F. Ground	9.92	.62	
1030	Slab on Grade	4" reinforced concrete with vapor barrier and granular base		S.F. Slab	3.42	.21	1.7%
2010	Basement Excavation	Site preparation for slab and trench for foundation wall and footing		S.F. Ground	1.18	.07	
2020	Basement Walls	4' foundation wall		L.F. Wall	52	.26	
B. SHELL							
	B10 Superstructure						
1010	Floor Construction	Concrete slab, metal deck, beams		S.F. Floor	16.25	15.23	22.6%
1020	Roof Construction	Metal deck, open web steel joists, beams, columns		S.F. Roof	4.64	.29	
	B20 Exterior Enclosure						
2010	Exterior Walls	N/A		—	—	—	
2020	Exterior Windows	Double glazed heat absorbing, tinted plate glass wall panels	100% of wall	Each	31	10.32	15.6%
2030	Exterior Doors	Double aluminum & glass doors		Each	3760	.41	
	B30 Roofing						
3010	Roof Coverings	Single ply membrane, fully adhered; perlite/EPS composite insulation		S.F. Roof	3.84	.24	0.3%
3020	Roof Openings	N/A		—	—	—	
C. INTERIORS							
1010	Partitions	Gypsum board on metal studs	30 S.F. Floor/L.F. Partition	S.F. Partition	5.81	1.55	
1020	Interior Doors	Single leaf hollow metal	400 S.F. Floor/Door	Each	537	1.34	
1030	Fittings	Toilet partitions		S.F. Floor	.36	.36	
2010	Stair Construction	Concrete filled metal pan		Flight	6050	.81	18.4%
3010	Wall Finishes	60% vinyl wall covering, 40% paint		S.F. Surface	2.17	.58	
3020	Floor Finishes	60% carpet tile, 30% vinyl composition tile, 10% ceramic tile		S.F. Floor	4.28	4.28	
3030	Ceiling Finishes	Mineral fiber tile on concealed zee bars		S.F. Ceiling	3.71	3.71	
D. SERVICES							
	D10 Conveying						
1010	Elevators & Lifts	Four geared passenger elevators		Each	235,300	3.62	5.3%
1020	Escalators & Moving Walks	N/A		—	—	—	
	D20 Plumbing						
2010	Plumbing Fixtures	Toilet and service fixtures, supply and drainage	1 Fixture/1345 S.F. Floor	Each	995	.74	
2020	Domestic Water Distribution	Oil fired water heater...		S.F. Floor	.06	.06	1.3%
2040	Rain Water Drainage	Roof drains		S.F. Roof	.96	.06	
	D30 HVAC						
3010	Energy Supply	Oil fired hot water		S.F. Floor	3.06	3.06	
3020	Heat Generating Systems	N/A		—	—	—	
3030	Cooling Generating Systems	Chilled water, fan coil units		S.F. Floor	9.07	9.07	17.7%
3050	Terminal & Package Units	N/A		—	—	—	
3090	Other HVAC Sys. & Equipment	N/A		—	—	—	
	D40 Fire Protection						
4010	Sprinklers	Sprinkler system, light hazard		S.F. Floor	2.60	2.60	4.2%
4020	Standpipes	Standpipes and hose systems		S.F. Floor	.27	.27	
	D50 Electrical						
5010	Electrical Service/Distribution	2400 ampere service, panel board and feeders		S.F. Floor	.66	.66	
5020	Lighting & Branch Wiring	Fluorescent fixtures, receptacles, switches, A.C. and misc. power		S.F. Floor	7.84	7.84	12.9%
5030	Communications & Security	Alarm systems and emergency lighting		S.F. Floor	.21	.21	
5090	Other Electrical Systems	Emergency generator, 200 kW		S.F. Floor	.18	.18	
E. EQUIPMENT & FURNISHINGS							
1010	Commercial Equipment	N/A		—	—	—	
1020	Institutional Equipment	N/A		—	—	—	0.0%
1030	Vehicular Equipment	N/A		—	—	—	
1090	Other Equipment	N/A		—	—	—	
F. SPECIAL CONSTRUCTION & DEMOLITION							
1020	Integrated Construction	N/A		—	—	—	0.0%
1040	Special Facilities	N/A		—	—	—	
G. BUILDING SITEWORK	**N/A**						

		Sub-Total	68.65	100%
CONTRACTOR FEES (General Requirements: 10%, Overhead: 5%, Profit: 10%)		25%	17.16	
ARCHITECT FEES		6%	5.14	
	Total Building Cost		**90.95**	

BUILDING TYPES

173

Costs per square foot of floor area

Exterior Wall	S.F. Area	7000	9000	11000	13000	15000	17000	19000	21000	23000
	L.F. Perimeter	240	280	303	325	354	372	397	422	447
Limestone with Concrete Block Back-up	Bearing Walls	161.15	150.60	141.55	135.15	131.15	127.20	124.60	122.45	120.75
	R/Conc. Frame	170.00	159.95	151.55	145.60	141.80	138.15	135.70	133.75	132.10
Face Brick with Concrete Block Back-up	Bearing Walls	141.85	133.00	125.95	120.95	117.70	114.65	112.60	110.90	109.50
	R/Conc. Frame	157.20	148.30	141.25	136.25	133.00	129.95	127.90	126.20	124.85
Decorative Concrete Block	Bearing Walls	135.10	126.85	120.50	116.00	113.00	110.35	108.45	106.95	105.70
	R/Conc. Frame	150.40	142.15	135.80	131.30	128.35	125.65	123.75	122.25	121.00
Perimeter Adj., Add or Deduct	Per 100 L.F.	21.30	16.60	13.60	11.50	9.95	8.80	7.85	7.15	6.50
Story Hgt. Adj., Add or Deduct	Per 1 Ft.	3.75	3.40	3.00	2.75	2.60	2.40	2.30	2.20	2.15
For Basement, add $17.45 per square foot of basement area										

The above costs were calculated using the basic specifications shown on the facing page. These costs should be adjusted where necessary for design alternatives and owner's requirements. Reported completed project costs, for this type of structure, range from $70.55 to $184.10 per S.F.

Common additives

Description	Unit	$ Cost
Cells Prefabricated, 5'-6' wide, 7'-8' high, 7'-8' deep	Each	9550
Elevators, Hydraulic passenger, 2 stops		
1500# capacity	Each	43,425
2500# capacity	Each	44,625
3500# capacity	Each	48,525
Emergency Lighting, 25 watt, battery operated		
Lead battery	Each	227
Nickel cadmium	Each	660
Flagpoles, Complete		
Aluminum, 20' high	Each	1100
40' high	Each	2700
70' high	Each	8350
Fiberglass, 23' high	Each	1425
39'-5" high	Each	3000
59' high	Each	7425

Description	Unit	$ Cost
Lockers, Steel, Single tier, 60" to 72"	Opening	131 - 227
2 tier, 60" or 72" total	Opening	74 - 125
5 tier, box lockers	Opening	42 - 62
Locker bench, lam. maple top only	L.F.	19.50
Pedestals, steel pipe	Each	63
Safe, Office type, 4 hour rating		
30" x 18" x 18"	Each	3200
62" x 33" x 20"	Each	6975
Shooting Range, Incl. bullet traps, target provisions, and contols, not incl. structural shell	Each	24,600
Smoke Detectors		
Ceiling type	Each	151
Duct type	Each	405
Sound System		
Amplifier, 250 watts	Each	1675
Speaker, ceiling or wall	Each	147
Trumpet	Each	275

Important: See the Reference Section for Location Factors

BUILDING TYPES

Model costs calculated for a 2 story building with 12' story height and 11,000 square feet of floor area

BUILDING TYPES

			Unit	Unit Cost	Cost Per S.F.	% Of Sub-Total
A. SUBSTRUCTURE						
1010	Standard Foundations	Poured concrete; strip and spread footings	S.F. Ground	1.88	.94	
1030	Slab on Grade	4" reinforced concrete with vapor barrier and granular base	S.F. Slab	3.42	1.71	5.0%
2010	Basement Excavation	Site preparation for slab and trench for foundation wall and footing	S.F. Ground	1.18	.59	
2020	Basement Walls	4' foundation wall	L.F. Wall	52	1.94	
B. SHELL						
	B10 Superstructure					
1010	Floor Construction	Open web steel joists, slab form, concrete	S.F. Floor	7.22	3.61	4.8%
1020	Roof Construction	Metal deck on open web steel joists	S.F. Roof	2.76	1.38	
	B20 Exterior Enclosure					
2010	Exterior Walls	Limestone with concrete block backup 80% of wall	S.F. Wall	35	18.72	
2020	Exterior Windows	Metal horizontal sliding 20% of wall	Each	758	6.68	25.5%
2030	Exterior Doors	Hollow metal	Each	1516	1.09	
	B30 Roofing					
3010	Roof Coverings	Built-up tar and gravel with flashing; perlite/EPS composite insulation	S.F. Roof	4.52	2.26	2.2%
3020	Roof Openings	N/A	—	—	—	
C. INTERIORS						
1010	Partitions	Concrete block 20 S.F. Floor/L.F. Partition	S.F. Partition	6.04	3.02	
1020	Interior Doors	Single leaf kalamein fire door 200 S.F. Floor/Door	Each	537	2.69	
1030	Fittings	Toilet partitions	S.F. Floor	.71	.71	
2010	Stair Construction	Concrete filled metal pan	Flight	6050	1.10	16.8%
3010	Wall Finishes	90% paint, 10% ceramic tile	S.F. Surface	5.26	2.63	
3020	Floor Finishes	70% vinyl composition tile, 20% carpet tile, 10% ceramic tile	S.F. Floor	3.59	3.59	
3030	Ceiling Finishes	Mineral fiber tile on concealed zee bars	S.F. Ceiling	3.71	3.71	
D. SERVICES						
	D10 Conveying					
1010	Elevators & Lifts	One hydraulic passenger elevator	Each	52,250	4.75	4.6%
1020	Escalators & Moving Walks	N/A	—	—	—	
	D20 Plumbing					
2010	Plumbing Fixtures	Toilet and service fixtures, supply and drainage 1 Fixture/580 S.F. Floor	Each	2047	3.53	
2020	Domestic Water Distribution	Oil fired water heater	S.F. Floor	.32	.32	4.3%
2040	Rain Water Drainage	Roof drains	S.F. Roof	1.16	.58	
	D30 HVAC					
3010	Energy Supply	Oil fired hot water, wall fin radiation	S.F. Floor	6.71	6.71	
3020	Heat Generating Systems	N/A	—	—	—	
3030	Cooling Generating Systems	Split systems with air cooled condensing units	S.F. Floor	8.43	8.43	14.6%
3050	Terminal & Package Units	N/A	—	—	—	
3090	Other HVAC Sys. & Equipment	N/A	—	—	—	
	D40 Fire Protection					
4010	Sprinklers	Wet pipe sprinkler system	S.F. Floor	1.95	1.95	1.9%
4020	Standpipes	N/A	—	—	—	
	D50 Electrical					
5010	Electrical Service/Distribution	400 ampere service, panel board and feeders	S.F. Floor	1.25	1.25	
5020	Lighting & Branch Wiring	Fluorescent fixtures, receptacles, switches, A.C. and misc. power	S.F. Floor	7.40	7.40	8.8%
5030	Communications & Security	Alarm systems and emergency lighting	S.F. Floor	.51	.51	
5090	Other Electrical Systems	Emergency generator, 15 kW	S.F. Floor	.17	.17	
E. EQUIPMENT & FURNISHINGS						
1010	Commercial Equipment	N/A	—	—	—	
1020	Institutional Equipment	Lockers, detention rooms, cells, gasoline dispensers	S.F. Floor	10.12	10.12	11.5%
1030	Vehicular Equipment	Gasoline dispenser system	S.F. Floor	1.50	1.50	
1090	Other Equipment	Flag pole	S.F. Floor	.30	.30	
F. SPECIAL CONSTRUCTION & DEMOLITION						
1020	Integrated Construction	N/A	—	—	—	0.0%
1040	Special Facilities	N/A	—	—	—	
G. BUILDING SITEWORK	**N/A**					

				Sub-Total	103.89	100%
	CONTRACTOR FEES (General Requirements: 10%, Overhead: 5%, Profit: 10%)			25%	25.97	
	ARCHITECT FEES			9%	11.69	

Total Building Cost	**141.55**

Costs per square foot of floor area

Exterior Wall	S.F. Area	5000	7000	9000	11000	13000	15000	17000	19000	21000
	L.F. Perimeter	300	380	420	486	468	513	540	580	620
Face Brick with Concrete Block Back-up	Steel Frame	106.80	100.65	94.55	92.10	86.50	84.95	83.10	82.00	81.20
	Bearing Walls	105.25	99.05	92.95	90.50	84.90	83.35	81.50	80.45	79.65
Limestone with Concrete Block Back-up	Steel Frame	119.90	112.50	104.75	101.75	94.35	92.40	90.00	88.70	87.65
	Bearing Walls	117.75	110.35	102.55	99.60	92.15	90.25	87.85	86.55	85.50
Decorative Concrete Block	Steel Frame	99.90	94.40	89.15	87.05	82.35	81.00	79.40	78.50	77.80
	Bearing Walls	98.30	92.80	87.60	85.45	80.75	79.40	77.85	76.95	76.25
Perimeter Adj., Add or Deduct	Per 100 L.F.	12.10	8.65	6.70	5.45	4.65	4.05	3.55	3.20	2.85
Story Hgt. Adj., Add or Deduct	Per 1 Ft.	1.85	1.70	1.50	1.40	1.15	1.05	1.00	.95	.95
For Basement, add $18.30 per square foot of basement area										

The above costs were calculated using the basic specifications shown on the facing page. These costs should be adjusted where necessary for design alternatives and owner's requirements. Reported completed project costs, for this type of structure, range from $58.55 to $151.95 per S.F.

Common additives

Description	Unit	$ Cost	Description	Unit	$ Cost
Closed Circuit Surveillance, One station			Mail Boxes, Horizontal, key lock, 15" x 6" x 5"	Each	45
Camera and monitor	Each	1375	Double 15" x 12" x 5"	Each	80
For additional camera stations, add	Each	750	Quadruple 15" x 12" x 10"	Each	142
Emergency Lighting, 25 watt, battery operated			Vertical, 6" x 5" x 15", aluminum	Each	37
Lead battery	Each	227	Bronze	Each	59
Nickel cadmium	Each	660	Steel, enameled	Each	43
Flagpoles, Complete			Scales, Dial type, 5 ton cap.		
Aluminum, 20' high	Each	1100	8' x 6' platform	Each	8575
40' high	Each	2700	9' x 7' platform	Each	11,300
70' high	Each	8350	Smoke Detectors		
Fiberglass, 23' high	Each	1425	Ceiling type	Each	151
39'-5" high	Each	3000	Duct type	Each	405
59' high	Each	7425			

BUILDING TYPES

Model costs calculated for a 1 story building with 14' story height and 13,000 square feet of floor area

				Unit	Unit Cost	Cost Per S.F.	% Of Sub-Total
A.	**SUBSTRUCTURE**						
1010	Standard Foundations	Poured concrete; strip and spread footings		S.F. Ground	1.30	1.30	
1030	Slab on Grade	4" reinforced concrete with vapor barrier and granular base		S.F. Slab	3.42	3.42	13.2%
2010	Basement Excavation	Site preparation for slab and trench for foundation wall and footing		S.F. Ground	1.18	1.18	
2020	Basement Walls	4' foundation wall		L.F. Wall	52	2.47	
B.	**SHELL**						
	B10 Superstructure						
1010	Floor Construction	Steel column fireproofing		L.F. Column	23.20	.07	8.6%
1020	Roof Construction	Metal deck, open web steel joists, columns		S.F. Roof	5.38	5.38	
	B20 Exterior Enclosure						
2010	Exterior Walls	Face brick with concrete block backup	80% of wall	S.F. Wall	21	8.51	
2020	Exterior Windows	Double strength window glass	20% of wall	Each	509	2.23	18.1%
2030	Exterior Doors	Double aluminum & glass, single aluminum, hollow metal, steel overhead		Each	1572	.72	
	B30 Roofing						
3010	Roof Coverings	Built-up tar and gravel with flashing; perlite/EPS composite insulation		S.F. Roof	4.00	4.00	6.3%
3020	Roof Openings	N/A		—	—	—	
C.	**INTERIORS**						
1010	Partitions	Concrete block	15 S.F. Floor/L.F. Partition	S.F. Partition	6.03	4.82	
1020	Interior Doors	Single leaf hollow metal	150 S.F. Floor/Door	Each	537	3.58	
1030	Fittings	Toilet partitions, cabinets, shelving, lockers		S.F. Floor	1.00	1.00	
2010	Stair Construction	N/A		—	—	—	23.3%
3010	Wall Finishes	Paint		S.F. Surface	3.13	2.50	
3020	Floor Finishes	50% vinyl tile, 50% paint		S.F. Floor	1.98	1.98	
3030	Ceiling Finishes	Mineral fiber tile on concealed zee bars	25% of area	S.F. Ceiling	3.71	.93	
D.	**SERVICES**						
	D10 Conveying						
1010	Elevators & Lifts	N/A		—	—	—	0.0%
1020	Escalators & Moving Walks	N/A		—	—	—	
	D20 Plumbing						
2010	Plumbing Fixtures	Toilet and service fixtures, supply and drainage	1 Fixture/1180 S.F. Floor	Each	1215	1.03	
2020	Domestic Water Distribution	Gas fired water heater		S.F. Floor	.18	.18	2.7%
2040	Rain Water Drainage	Roof drains		S.F. Roof	.52	.52	
	D30 HVAC						
3010	Energy Supply	N/A		—	—	—	
3020	Heat Generating Systems	Included in D3030		—	—	—	
3030	Cooling Generating Systems	Single zone, gas heating, electric cooling		S.F. Floor	7.52	7.52	11.8%
3050	Terminal & Package Units	N/A		—	—	—	
3090	Other HVAC Sys. & Equipment	N/A		—	—	—	
	D40 Fire Protection						
4010	Sprinklers	Wet pipe sprinkler system		S.F. Floor	1.79	1.79	2.8%
4020	Standpipes	N/A		—	—	—	
	D50 Electrical						
5010	Electrical Service/Distribution	600 ampere service, panel board and feeders		S.F. Floor	1.95	1.95	
5020	Lighting & Branch Wiring	Fluorescent fixtures, receptacles, switches, A.C. and misc. power		S.F. Floor	5.99	5.99	13.2%
5030	Communications & Security	Alarm systems and emergency lighting		S.F. Floor	.30	.30	
5090	Other Electrical Systems	Emergency generator, 15 kW		S.F. Floor	.10	.10	
E.	**EQUIPMENT & FURNISHINGS**						
1010	Commercial Equipment	N/A		—	—	—	
1020	Institutional Equipment	N/A		—	—	—	0.0%
1030	Vehicular Equipment	N/A		—	—	—	
1090	Other Equipment	N/A		—	—	—	
F.	**SPECIAL CONSTRUCTION & DEMOLITION**						
1020	Integrated Construction	N/A		—	—	—	0.0%
1040	Special Facilities	N/A		—	—	—	
G.	**BUILDING SITEWORK**	**N/A**					

			Sub-Total	63.47	100%
CONTRACTOR FEES (General Requirements: 10%, Overhead: 5%, Profit: 10%)		25%	15.87		
ARCHITECT FEES		9%	7.16		

Total Building Cost	**86.50**

Costs per square foot of floor area

Exterior Wall	S.F. Area	5000	10000	15000	21000	25000	30000	40000	50000	60000
	L.F. Perimeter	287	400	500	600	700	700	834	900	1000
Face Brick with Concrete Block Back-up	Steel Frame	165.90	139.95	130.45	124.05	122.85	117.55	114.35	111.00	109.40
	Bearing Walls	161.70	137.05	127.95	121.75	120.70	115.50	112.40	109.10	107.55
Concrete Block	Steel Frame	144.30	124.60	117.50	112.85	111.85	108.40	106.10	103.90	102.80
Brick Veneer	Steel Frame	160.70	136.05	127.05	121.00	119.85	115.05	112.10	109.05	107.60
Galvanized Steel Siding	Steel Frame	136.40	118.95	112.75	108.70	107.75	104.90	102.90	101.10	100.15
Metal Sandwich Panel	Steel Frame	138.35	120.45	114.10	109.90	108.95	106.00	104.00	102.05	101.10
Perimeter Adj., Add or Deduct	Per 100 L.F.	20.30	10.15	6.80	4.85	4.05	3.40	2.50	2.05	1.70
Story Hgt. Adj., Add or Deduct	Per 1 Ft.	4.15	2.85	2.40	2.05	2.00	1.65	1.50	1.30	1.20

For Basement, add $17.65 per square foot of basement area

The above costs were calculated using the basic specifications shown on the facing page. These costs should be adjusted where necessary for design alternatives and owner's requirements. Reported completed project costs, for this type of structure, range from $55.50 to $173.00 per S.F.

Common additives

Description	Unit	$ Cost
Bar, Front Bar	L.F.	287
Back Bar	L.F.	229
Booth, Upholstered, custom straight	L.F.	145 - 265
"L" or "U" shaped	L.F.	150 - 252
Bleachers, Telescoping, manual		
To 15 tier	Seat	79 - 110
21-30 tier	Seat	171 - 206
Courts		
Ceiling	Court	5750
Floor	Court	10,100
Walls	Court	20,300
Emergency Lighting, 25 watt, battery operated		
Lead battery	Each	227
Nickel cadmium	Each	660
Kitchen Equipment		
Broiler	Each	4275
Cooler, 6 ft. long, reach-in	Each	3375
Dishwasher, 10-12 racks per hr.	Each	3225
Food warmer, counter 1.2 KW	Each	745
Freezer, reach-in, 44 C.F.	Each	8775
Ice cube maker, 50 lb. per day	Each	1900

Description	Unit	$ Cost
Lockers, Steel, single tier, 60" or 72"	Opening	131 - 227
2 tier, 60" or 72" total	Opening	74 - 125
5 tier, box lockers	Opening	42 - 62
Locker bench, lam. maple top only	L.F.	19.50
Pedestals, steel pipe	Each	63
Sauna, Prefabricated, complete		
6' x 4'	Each	4825
6' x 9'	Each	5550
8' x 8'	Each	6525
8' x 10'	Each	7225
10' x 12'	Each	9475
Sound System		
Amplifier, 250 watts	Each	1675
Speaker, ceiling or wall	Each	147
Trumpet	Each	275
Steam Bath, Complete, to 140 C.F.	Each	1325
To 300 C.F.	Each	1500
To 800 C.F.	Each	4825
To 2500 C.F.	Each	5125

Model costs calculated for a 2 story building with 12' story height and 30,000 square feet of floor area

				Unit	Unit Cost	Cost Per S.F.	% Of Sub-Total
A. SUBSTRUCTURE							
1010	Standard Foundations	Poured concrete; strip and spread footings		S.F. Ground	2.32	1.16	
1030	Slab on Grade	5" reinforced concrete with vapor barrier and granular base		S.F. Slab	4.47	2.91	7.0%
2010	Basement Excavation	Site preparation for slab and trench for foundation wall and footing		S.F. Ground	1.18	.77	
2020	Basement Walls	4' foundation wall		L.F. Wall	52	1.23	
B. SHELL							
	B10 Superstructure						
1010	Floor Construction	Open web steel joists, slab form, concrete, columns	50% of area	S.F. Floor	10.62	5.31	9.3%
1020	Roof Construction	Metal deck on open web steel joists, columns		S.F. Roof	5.64	2.82	
	B20 Exterior Enclosure						
2010	Exterior Walls	Face brick with concrete block backup	95% of wall	S.F. Wall	21	11.28	
2020	Exterior Windows	Storefront	5% of wall	Each	69	1.94	15.5%
2030	Exterior Doors	Aluminum and glass and hollow metal		Each	2004	.27	
	B30 Roofing						
3010	Roof Coverings	Built-up tar and gravel with flashing; perlite/EPS composite insulation		S.F. Roof	5.04	2.52	2.9%
3020	Roof Openings	Roof hatches		S.F. Roof	.08	.04	
C. INTERIORS							
1010	Partitions	Concrete block, gypsum board on metal studs	25 S.F. Floor/L.F. Partition	S.F. Partition	9.50	3.80	
1020	Interior Doors	Single leaf hollow metal	810 S.F. Floor/Door	Each	537	.66	
1030	Fittings	Toilet partitions		S.F. Floor	.34	.34	
2010	Stair Construction	Concrete filled metal pan		Flight	5075	.51	12.5%
3010	Wall Finishes	Paint		S.F. Surface	1.95	.78	
3020	Floor Finishes	80% carpet tile, 20% ceramic tile	50% of floor area	S.F. Floor	5.12	2.56	
3030	Ceiling Finishes	Mineral fiber tile on concealed zee bars	60% of area	S.F. Ceiling	3.71	2.23	
D. SERVICES							
	D10 Conveying						
1010	Elevators & Lifts	N/A		—	—	—	0.0%
1020	Escalators & Moving Walks	N/A		—	—	—	
	D20 Plumbing						
2010	Plumbing Fixtures	Kitchen, bathroom and service fixtures, supply and drainage	1 Fixture/1000 S.F. Floor	Each	2130	2.13	
2020	Domestic Water Distribution	Gas fired water heater		S.F. Floor	1.31	1.31	4.3%
2040	Rain Water Drainage	Roof drains		S.F. Roof	.56	.28	
	D30 HVAC						
3010	Energy Supply	N/A		—	—	—	
3020	Heat Generating Systems	Included in D3030		—	—	—	
3030	Cooling Generating Systems	Multizone unit, gas heating, electric cooling		S.F. Floor	20	20.35	23.4%
3050	Terminal & Package Units	N/A		—	—	—	
3090	Other HVAC Sys. & Equipment	N/A		—	—	—	
	D40 Fire Protection						
4010	Sprinklers	Sprinklers, light hazard		S.F. Floor	.27	.27	0.3%
4020	Standpipes	N/A		—	—	—	
	D50 Electrical						
5010	Electrical Service/Distribution	400 ampere service, panel board and feeders		S.F. Floor	.56	.56	
5020	Lighting & Branch Wiring	Fluorescent and high intensity discharge fixtures, receptacles, switches, A.C. and misc. power		S.F. Floor	9.08	9.08	11.3%
5030	Communications & Security	Alarm systems and emergency lighting		S.F. Floor	.21	.21	
5090	Other Electrical Systems	Emergency generator, 15 kW		S.F. Floor	.04	.04	
E. EQUIPMENT & FURNISHINGS							
1010	Commercial Equipment	N/A		—	—	—	
1020	Institutional Equipment	N/A		—	—	—	13.5%
1030	Vehicular Equipment	N/A		—	—	—	
1090	Other Equipment	Courts, sauna baths		S.F. Floor	11.72	11.72	
F. SPECIAL CONSTRUCTION & DEMOLITION							
1020	Integrated Construction	N/A		—	—	—	0.0%
1040	Special Facilities	N/A		—	—	—	
G. BUILDING SITEWORK	**N/A**						

			Sub-Total	87.08	100%
CONTRACTOR FEES (General Requirements: 10%, Overhead: 5%, Profit: 10%)			25%	21.77	
ARCHITECT FEES			8%	8.70	
		Total Building Cost		**117.55**	

Costs per square foot of floor area

Exterior Wall	S.F. Area	5000	6000	7000	8000	9000	10000	11000	12000	13000
	L.F. Perimeter	286	320	353	386	397	425	454	460	486
Face Brick with Concrete Block Back-up	Steel Joists	122.25	119.35	117.20	115.55	112.95	111.80	110.90	109.15	108.45
	Wood Joists	119.75	116.75	114.50	112.80	110.10	108.90	108.00	106.10	105.40
Stucco on Concrete Block	Steel Joists	118.30	115.65	113.70	112.20	109.90	108.85	108.10	106.50	105.85
	Wood Joists	112.90	110.30	108.45	107.00	104.80	103.80	103.05	101.50	100.90
Limestone with Concrete Block Back-up	Steel Joists	136.55	132.65	129.80	127.60	124.00	122.45	121.25	118.70	117.80
	Wood Joists	131.15	127.35	124.55	122.40	118.85	117.35	116.20	113.75	112.85
Perimeter Adj., Add or Deduct	Per 100 L.F.	11.15	9.30	7.95	7.00	6.20	5.55	5.05	4.60	4.25
Story Hgt. Adj., Add or Deduct	Per 1 Ft.	1.95	1.80	1.70	1.65	1.50	1.45	1.40	1.30	1.25

For Basement, add $18.55 per square foot of basement area

The above costs were calculated using the basic specifications shown on the facing page. These costs should be adjusted where necessary for design alternatives and owner's requirements. Reported completed project costs, for this type of structure, range from $48.45 to $150.00 per S.F.

Common additives

Description	Unit	$ Cost
Carrels Hardwood	Each	750 - 980
Emergency Lighting, 25 watt, battery operated		
Lead battery	Each	227
Nickel cadmium	Each	660
Flagpoles, Complete		
Aluminum, 20' high	Each	1100
40' high	Each	2700
70' high	Each	8350
Fiberglass, 23' high	Each	1425
39'-5" high	Each	3000
59' high	Each	7425
Gym Floor, Incl. sleepers and finish, maple	S.F.	9.85
Intercom System, 25 Station capacity		
Master station	Each	2025
Intercom outlets	Each	128
Handset	Each	335

Description	Unit	$ Cost
Lockers, Steel, single tier, 60" to 72"	Opening	131 - 227
2 tier, 60" to 72" total	Opening	74 - 125
5 tier, box lockers	Opening	42 - 62
Locker bench, lam. maple top only	L.F.	19.50
Pedestals, steel pipe	Each	63
Seating		
Auditorium chair, all veneer	Each	171
Veneer back, padded seat	Each	207
Upholstered, spring seat	Each	208
Classroom, movable chair & desk	Set	65 - 120
Lecture hall, pedestal type	Each	156 - 485
Smoke Detectors		
Ceiling type	Each	151
Duct type	Each	405
Sound System		
Amplifier, 250 watts	Each	1675
Speaker, ceiling or wall	Each	147
Trumpet	Each	275
Swimming Pools, Complete, gunite	S.F.	48 - 59

Important: See the Reference Section for Location Factors

BUILDING TYPES

Model costs calculated for a 1 story building with 12' story height and 10,000 square feet of floor area

Religious Education

					Unit	Unit Cost	Cost Per S.F.	% Of Sub-Total
A.	**SUBSTRUCTURE**							
1010	Standard Foundations	Poured concrete; strip and spread footings			S.F. Ground	1.60	1.60	
1030	Slab on Grade	4" reinforced concrete with vapor barrier and granular base			S.F. Slab	3.42	3.42	10.3%
2010	Basement Excavation	Site preparation for slab and trench for foundation wall and footing			S.F. Ground	1.18	1.18	
2020	Basement Walls	4' foundation wall			L.F. Wall	52	2.25	
B.	**SHELL**							
	B10 Superstructure							
1010	Floor Construction	N/A			—	—	—	6.1%
1020	Roof Construction	Metal deck, open web steel joists, beams, interior columns			S.F. Roof	5.01	5.01	
	B20 Exterior Enclosure							
2010	Exterior Walls	Face brick with concrete block backup		85% of wall	S.F. Wall	21	9.15	
2020	Exterior Windows	Window wall		15% of wall	Each	34	2.62	16.2%
2030	Exterior Doors	Double aluminum and glass			Each	3737	1.50	
	B30 Roofing							
3010	Roof Coverings	Built-up tar and gravel with flashing; perlite/EPS composite insulation			S.F. Roof	4.04	4.04	4.9%
3020	Roof Openings	N/A			—	—	—	
C.	**INTERIORS**							
1010	Partitions	Concrete block		8 S.F. Floor/S.F. Partition	S.F. Partition	6.03	7.54	
1020	Interior Doors	Single leaf hollow metal		700 S.F. Floor/Door	Each	537	.77	
1030	Fittings	Toilet partitions			S.F. Floor	1.03	1.03	
2010	Stair Construction	N/A			—	—	—	26.2%
3010	Wall Finishes	Paint			S.F. Surface	3.19	3.99	
3020	Floor Finishes	50% vinyl tile, 50% carpet			S.F. Floor	4.46	4.46	
3030	Ceiling Finishes	Mineral fiber tile on concealed zee bars			S.F. Ceiling	3.71	3.71	
D.	**SERVICES**							
	D10 Conveying							
1010	Elevators & Lifts	N/A			—	—	—	0.0%
1020	Escalators & Moving Walks	N/A			—	—	—	
	D20 Plumbing							
2010	Plumbing Fixtures	Toilet and service fixtures, supply and drainage		1 Fixture/455 S.F. Floor	Each	1474	3.24	
2020	Domestic Water Distribution	Gas fired water heater			S.F. Floor	.59	.59	5.1%
2040	Rain Water Drainage	Roof drains			S.F. Roof	.36	.36	
	D30 HVAC							
3010	Energy Supply	Oil fired hot water, wall fin radiation			S.F. Floor	6.71	6.71	
3020	Heat Generating Systems	N/A			—	—	—	
3030	Cooling Generating Systems	Split systems with air cooled condensing units			S.F. Floor	9.71	9.71	20.0%
3050	Terminal & Package Units	N/A			—	—	—	
3090	Other HVAC Sys. & Equipment	N/A			—	—	—	
	D40 Fire Protection							
4010	Sprinklers	Sprinkler, light hazard			S.F. Floor	1.79	1.79	2.2%
4020	Standpipes	N/A			—	—	—	
	D50 Electrical							
5010	Electrical Service/Distribution	200 ampere service, panel board and feeders			S.F. Floor	.68	.68	
5020	Lighting & Branch Wiring	Fluorescent fixtures, receptacles, switches, A.C. and misc. power			S.F. Floor	6.35	6.35	9.0%
5030	Communications & Security	Alarm systems and emergency lighting			S.F. Floor	.30	.30	
5090	Other Electrical Systems	Emergency generator, 15 kW			S.F. Floor	.06	.06	
E.	**EQUIPMENT & FURNISHINGS**							
1010	Commercial Equipment	N/A			—	—	—	
1020	Institutional Equipment	N/A			—	—	—	0.0%
1030	Vehicular Equipment	N/A			—	—	—	
1090	Other Equipment	N/A			—	—	—	
F.	**SPECIAL CONSTRUCTION & DEMOLITION**							
1020	Integrated Construction	N/A			—	—	—	0.0%
1040	Special Facilities	N/A			—	—	—	
G.	**BUILDING SITEWORK**	**N/A**						

		Sub-Total	82.06	100%
CONTRACTOR FEES (General Requirements: 10%, Overhead: 5%, Profit: 10%)		25%	20.52	
ARCHITECT FEES		9%	9.22	

Total Building Cost	**111.80**

181

BUILDING TYPES

Costs per square foot of floor area

Exterior Wall	S.F. Area	2000	2800	3500	4200	5000	5800	6500	7200	8000
	L.F. Perimeter	180	212	240	268	300	314	336	344	368
Wood Siding	Wood Frame	157.40	146.45	141.05	137.35	134.50	131.20	129.50	127.45	126.25
Brick Veneer	Wood Frame	166.00	153.70	147.55	143.40	140.15	136.30	134.40	131.95	130.55
Face Brick with Concrete Block Back-up	Wood Joists	170.75	157.70	151.20	146.85	143.40	139.25	137.20	134.55	133.10
	Steel Joists	168.80	154.50	147.35	142.60	138.80	134.15	131.90	128.90	127.30
Stucco on Concrete Block	Wood Joists	162.00	150.35	144.55	140.65	137.55	133.95	132.20	129.90	128.60
	Steel Joists	152.15	140.50	134.70	130.80	127.70	124.15	122.35	120.05	118.80
Perimeter Adj., Add or Deduct	Per 100 L.F.	19.15	13.70	10.95	9.15	7.65	6.60	5.95	5.35	4.80
Story Hgt. Adj., Add or Deduct	Per 1 Ft.	2.00	1.70	1.50	1.40	1.35	1.20	1.15	1.10	1.05
For Basement, add $21.40 per square foot of basement area										

The above costs were calculated using the basic specifications shown on the facing page. These costs should be adjusted where necessary for design alternatives and owner's requirements. Reported completed project costs, for this type of structure, range from $75.10 to $175.00 per S.F.

Common additives

Description	Unit	$ Cost
Bar, Front Bar	L.F.	287
Back bar	L.F.	229
Booth, Upholstered, custom straight	L.F.	145 - 265
"L" or "U" shaped	L.F.	150 - 252
Cupola, Stock unit, redwood		
30" square, 37" high, aluminum roof	Each	490
Copper roof	Each	254
Fiberglass, 5'-0" base, 63" high	Each	2975 - 3600
6'-0" base, 63" high	Each	4475 - 4950
Decorative Wood Beams, Non load bearing		
Rough sawn, 4" x 6"	L.F.	7.15
4" x 8"	L.F.	8.20
4" x 10"	L.F.	9.90
4" x 12"	L.F.	11.30
8" x 8"	L.F.	14.05
Emergency Lighting, 25 watt, battery operated		
Lead battery	Each	227
Nickel cadmium	Each	660

Description	Unit	$ Cost
Fireplace, Brick, not incl. chimney or foundation		
30" x 29" opening	Each	2075
Chimney, standard brick		
Single flue, 16" x 20"	V.L.F.	66
20" x 20"	V.L.F.	81
2 Flue, 20" x 24"	V.L.F.	95
20" x 32"	V.L.F.	125
Kitchen Equipment		
Broiler	Each	4275
Coffee urn, twin 6 gallon	Each	7375
Cooler, 6 ft. long	Each	3375
Dishwasher, 10-12 racks per hr.	Each	3225
Food warmer, counter, 1.2 KW	Each	745
Freezer, 44 C.F., reach-in	Each	8775
Ice cube maker, 50 lb. per day	Each	1900
Range with 1 oven	Each	2500
Refrigerators, Prefabricated, walk-in		
7'-6" high, 6' x 6'	S.F.	121
10' x 10'	S.F.	95
12' x 14'	S.F.	84
12' x 20'	S.F.	74

Important: See the Reference Section for Location Factors

Model costs calculated for a 1 story building with 12' story height and 5,000 square feet of floor area

Restaurant

				Unit	Unit Cost	Cost Per S.F.	% Of Sub-Total
A.	**SUBSTRUCTURE**						
1010	Standard Foundations	Poured concrete; strip and spread footings		S.F. Ground	1.74	1.74	
1030	Slab on Grade	4" reinforced concrete with vapor barrier and granular base		S.F. Slab	3.42	3.42	
2010	Basement Excavation	Site preparation for slab and trench for foundation wall and footing		S.F. Ground	1.31	1.31	9.6%
2020	Basement Walls	4' foundation wall		L.F. Wall	52	3.17	
B.	**SHELL**						
	B10 Superstructure						
1010	Floor Construction	Wood columns		S.F. Floor	.36	.90	
1020	Roof Construction	Plywood on wood truss (pitched)		S.F. Roof	4.89	5.48	6.3%
	B20 Exterior Enclosure						
2010	Exterior Walls	Cedar siding on wood studs with insulation	70% of wall	S.F. Wall	7.86	3.96	
2020	Exterior Windows	Storefront windows	30% of wall	Each	30	6.54	13.7%
2030	Exterior Doors	Aluminum and glass doors and entrance with transom		Each	3266	3.27	
	B30 Roofing						
3010	Roof Coverings	Cedar shingles with flashing (pitched); rigid fiberglass insulation		S.F. Roof	12.25	5.34	5.4%
3020	Roof Openings	Skylights		S.F. Roof	.06	.06	
C.	**INTERIORS**						
1010	Partitions	Gypsum board on wood studs,	25 S.F. Floor/L.F. Partition	S.F. Partition	7.05	2.82	
1020	Interior Doors	Hollow core wood	250 S.F. Floor/Door	Each	388	1.55	
1030	Fittings	Toilet partitions		S.F. Floor	.62	.62	
2010	Stair Construction	N/A		—	—	—	16.3%
3010	Wall Finishes	75% paint, 25% ceramic tile		S.F. Surface	3.65	1.46	
3020	Floor Finishes	65% carpet, 35% quarry tile		S.F. Floor	6.22	6.22	
3030	Ceiling Finishes	Mineral fiber tile on concealed zee bars		S.F. Ceiling	3.71	3.71	
D.	**SERVICES**						
	D10 Conveying						
1010	Elevators & Lifts	N/A		—	—	—	0.0%
1020	Escalators & Moving Walks	N/A		—	—	—	
	D20 Plumbing						
2010	Plumbing Fixtures	Kitchen, bathroom and service fixtures, supply and drainage	1 Fixture/355 S.F. Floor	Each	1785	5.03	
2020	Domestic Water Distribution	Gas fired water heater		S.F. Floor	2.74	2.74	8.3%
2040	Rain Water Drainage	Roof drains		S.F. Roof	.60	.60	
	D30 HVAC						
3010	Energy Supply	N/A		—	—	—	
3020	Heat Generating Systems	Included in D3030		—	—	—	
3030	Cooling Generating Systems	Multizone unit, gas heating, electric cooling		S.F. Floor	28	28.45	28.3%
3050	Terminal & Package Units	N/A		—	—	—	
3090	Other HVAC Sys. & Equipment	N/A		—	—	—	
	D40 Fire Protection						
4010	Sprinklers	Sprinklers, light hazard		S.F. Floor	1.79	1.79	1.8%
4020	Standpipes	N/A		—	—	—	
	D50 Electrical						
5010	Electrical Service/Distribution	400 ampere service, panel board and feeders		S.F. Floor	2.87	2.87	
5020	Lighting & Branch Wiring	Fluorescent fixtures, receptacles, switches, A.C. and misc. power		S.F. Floor	6.82	6.82	10.3%
5030	Communications & Security	Alarm systems and emergency lighting		S.F. Floor	.41	.41	
5090	Other Electrical Systems	Emergency generator, 15 kW		S.F. Floor	.26	.26	
E.	**EQUIPMENT & FURNISHINGS**						
1010	Commercial Equipment	N/A		—	—	—	
1020	Institutional Equipment	N/A		—	—	—	
1030	Vehicular Equipment	N/A		—	—	—	0.0%
1090	Other Equipment	N/A		—	—	—	
F.	**SPECIAL CONSTRUCTION & DEMOLITION**						
1020	Integrated Construction	N/A		—	—	—	0.0%
1040	Special Facilities	N/A		—	—	—	
G.	**BUILDING SITEWORK**	**N/A**					

		Sub-Total	100.54	100%
CONTRACTOR FEES (General Requirements: 10%, Overhead: 5%, Profit: 10%)		25%	25.14	
ARCHITECT FEES		7%	8.82	

Total Building Cost	**134.50**	

BUILDING TYPES

Costs per square foot of floor area

Exterior Wall	S.F. Area	2000	2800	3500	4000	5000	5800	6500	7200	8000
	L.F. Perimeter	180	212	240	260	300	314	336	344	368
Face Brick with Concrete Block Back-up	Bearing Walls	140.60	130.80	125.90	123.45	120.05	116.60	115.05	112.85	111.70
	Steel Frame	143.60	133.75	128.90	126.40	123.00	119.55	118.00	115.80	114.65
Concrete Block With Stucco	Bearing Walls	133.55	124.85	120.50	118.35	115.30	112.35	111.00	109.10	108.05
	Steel Frame	135.75	127.20	122.95	120.80	117.75	114.85	113.50	111.65	110.65
Wood Siding	Wood Frame	130.75	122.70	118.65	116.60	113.80	111.05	109.80	108.05	107.15
Brick Veneer	Steel Frame	139.30	130.20	125.65	123.35	120.15	117.00	115.55	113.50	112.45
Perimeter Adj., Add or Deduct	Per 100 L.F.	24.60	17.60	14.05	12.30	9.85	8.45	7.55	6.85	6.15
Story Hgt. Adj., Add or Deduct	Per 1 Ft.	3.20	2.75	2.45	2.35	2.15	1.95	1.85	1.70	1.65
Basement—Not Applicable										

The above costs were calculated using the basic specifications shown on the facing page. These costs should be adjusted where necessary for design alternatives and owner's requirements. Reported completed project costs, for this type of structure, range from $74.90 to $146.15 per S.F.

Common additives

Description	Unit	$ Cost
Bar, Front Bar	L.F.	287
Back bar	L.F.	229
Booth, Upholstered, custom straight	L.F.	145 - 265
"L" or "U" shaped	L.F.	150 - 252
Drive-up Window	Each	5025 - 9900
Emergency Lighting, 25 watt, battery operated		
Lead battery	Each	227
Nickel cadmium	Each	660
Kitchen Equipment		
Broiler	Each	4275
Coffee urn, twin 6 gallon	Each	7375
Cooler, 6 ft. long	Each	3375
Dishwasher, 10-12 racks per hr.	Each	3225
Food warmer, counter, 1.2 KW	Each	745
Freezer, 44 C.F., reach-in	Each	8775
Ice cube maker, 50 lb. per day	Each	1900
Range with 1 oven	Each	2500

Description	Unit	$ Cost
Refrigerators, Prefabricated, walk-in		
7'-6" High, 6' x 6'	S.F.	121
10' x 10'	S.F.	95
12' x 14'	S.F.	84
12' x 20'	S.F.	74
Serving		
Counter top (Stainless steel)	L.F.	118
Base cabinets	L.F.	283 - 335
Sound System		
Amplifier, 250 watts	Each	1675
Speaker, ceiling or wall	Each	147
Trumpet	Each	275
Storage		
Shelving	S.F.	9.00
Washing		
Stainless steel counter	L.F.	118

Important: See the Reference Section for Location Factors

Model costs calculated for a 1 story building with 10' story height and 4,000 square feet of floor area

Restaurant, Fast Food

				Unit	Unit Cost	Cost Per S.F.	% Of Sub-Total
A. SUBSTRUCTURE							
1010	Standard Foundations	Poured concrete; strip and spread footings		S.F. Ground	1.66	1.66	
1030	Slab on Grade	4" reinforced concrete with vapor barrier and granular base		S.F. Slab	3.42	3.42	10.7%
2010	Basement Excavation	Site preparation for slab and trench for foundation wall and footing		S.F. Ground	1.31	1.31	
2020	Basement Walls	4' foundation wall		L.F. Wall	52	3.44	
B. SHELL							
	B10 Superstructure						
1010	Floor Construction	N/A		—	—	—	4.6%
1020	Roof Construction	Metal deck on open web steel joists		S.F. Roof	4.17	4.17	
	B20 Exterior Enclosure						
2010	Exterior Walls	Face brick with concrete block backup	70% of wall	S.F. Wall	17.81	9.60	
2020	Exterior Windows	Window wall	30% of wall	Each	34	6.67	23.5%
2030	Exterior Doors	Aluminum and glass		Each	2590	5.19	
	B30 Roofing						
3010	Roof Coverings	Built-up tar and gravel with flashing; perlite/EPS composite insulation		S.F. Roof	4.36	4.36	5.1%
3020	Roof Openings	Roof hatches		S.F. Roof	.29	.29	
C. INTERIORS							
1010	Partitions	Gypsum board on metal studs	25 S.F. Floor/L.F. Partition	S.F. Partition	4.00	1.44	
1020	Interior Doors	Hollow core wood	1000 S.F. Floor/Door	Each	388	.39	
1030	Fittings	N/A		—	—	—	
2010	Stair Construction	N/A		—	—	—	17.9%
3010	Wall Finishes	Paint		S.F. Surface	3.92	1.41	
3020	Floor Finishes	Quarry tile		S.F. Floor	9.46	9.46	
3030	Ceiling Finishes	Mineral fiber tile on concealed zee bars		S.F. Ceiling	3.71	3.71	
D. SERVICES							
	D10 Conveying						
1010	Elevators & Lifts	N/A		—	—	—	0.0%
1020	Escalators & Moving Walks	N/A		—	—	—	
	D20 Plumbing						
2010	Plumbing Fixtures	Kitchen, bathroom and service fixtures, supply and drainage	1 Fixture/400 S.F. Floor	Each	1796	4.49	
2020	Domestic Water Distribution	Gas fired water heater		S.F. Floor	2.04	2.04	8.0%
2040	Rain Water Drainage	Roof drains		S.F. Roof	.78	.78	
	D30 HVAC						
3010	Energy Supply	N/A		—	—	—	
3020	Heat Generating Systems	Included in D3030		—	—	—	
3030	Cooling Generating Systems	Multizone unit, gas heating, electric cooling		S.F. Floor	9.74	9.74	10.7%
3050	Terminal & Package Units	N/A		—	—	—	
3090	Other HVAC Sys. & Equipment	N/A		—	—	—	
	D40 Fire Protection						
4010	Sprinklers	Sprinklers, light hazard		S.F. Floor	2.34	2.34	2.6%
4020	Standpipes	N/A		—	—	—	
	D50 Electrical						
5010	Electrical Service/Distribution	400 ampere service, panel board and feeders		S.F. Floor	3.02	3.02	
5020	Lighting & Branch Wiring	Fluorescent fixtures, receptacles, switches, A.C. and misc. power		S.F. Floor	7.45	7.45	12.1%
5030	Communications & Security	Alarm systems and emergency lighting		S.F. Floor	.34	.34	
5090	Other Electrical Systems	Emergency generator, 7.5 kW		S.F. Floor	.30	.30	
E. EQUIPMENT & FURNISHINGS							
1010	Commercial Equipment	N/A		—	—	—	
1020	Institutional Equipment	N/A		—	—	—	4.8%
1030	Vehicular Equipment	N/A		—	—	—	
1090	Other Equipment	Walk-in refrigerator		S.F. Floor	4.43	4.43	
F. SPECIAL CONSTRUCTION & DEMOLITION							
1020	Integrated Construction	N/A		—	—	—	0.0%
1040	Special Facilities	N/A		—	—	—	
G. BUILDING SITEWORK	**N/A**						

			Sub-Total	91.45	100%
	CONTRACTOR FEES (General Requirements: 10%, Overhead: 5%, Profit: 10%)		25%	22.86	
	ARCHITECT FEES		8%	9.14	

Total Building Cost 123.45

Costs per square foot of floor area

Exterior Wall	S.F. Area	10000	15000	20000	25000	30000	35000	40000	45000	50000
	L.F. Perimeter	450	500	600	700	740	822	890	920	966
Face Brick with Concrete Block Back-up	Steel Frame	135.00	122.55	118.60	116.25	112.85	111.50	110.20	108.40	107.20
	Lam. Wood Truss	137.60	125.65	121.85	119.65	116.35	115.05	113.80	112.10	110.95
Concrete Block	Steel Frame	133.30	125.90	123.45	122.00	120.00	119.20	118.45	117.40	116.75
	Lam. Wood Truss	121.70	114.30	111.85	110.40	108.45	107.60	106.85	105.80	105.15
Galvanized Steel Siding	Steel Frame	128.30	120.90	118.45	117.00	115.00	114.15	113.40	112.40	111.70
Metal Sandwich Panel	Steel Joists	134.15	125.25	122.35	120.60	118.15	117.20	116.25	115.05	114.15
Perimeter Adj., Add or Deduct	Per 100 L.F.	9.25	6.15	4.65	3.70	3.10	2.65	2.30	2.05	1.85
Story Hgt. Adj., Add or Deduct	Per 1 Ft.	1.40	1.05	.95	.85	.80	.75	.70	.65	.65
Basement—Not Applicable										

The above costs were calculated using the basic specifications shown on the facing page. These costs should be adjusted where necessary for design alternatives and owner's requirements. Reported completed project costs, for this type of structure, range from $44.35 to $137.90 per S.F.

Common additives

Description	Unit	$ Cost		Description	Unit	$ Cost
Bar, Front Bar	L.F.	287		Rink		
Back bar	L.F.	229		Dasher boards & top guard	Each	152,000
Booth, Upholstered, custom straight	L.F.	145 - 265		Mats, rubber	S.F.	16.70
"L" or "U" shaped	L.F.	150 - 252		Score Board	Each	11,000 - 32,100
Bleachers, Telescoping, manual						
To 15 tier	Seat	79 - 110				
16-20 tier	Seat	161 - 196				
21-30 tier	Seat	171 - 206				
For power operation, add	Seat	31 - 48				
Emergency Lighting, 25 watt, battery operated						
Lead battery	Each	227				
Nickel cadmium	Each	660				
Lockers, Steel, single tier, 60" or 72"	Opening	131 - 227				
2 tier, 60" or 72" total	Opening	74 - 125				
5 tier, box lockers	Opening	42 - 62				
Locker bench, lam. maple top only	L.F.	19.50				
Pedestals, steel pipe	Each	63				

Important: See the Reference Section for Location Factors

Model costs calculated for a 1 story building with 24' story height and 30,000 square feet of floor area

				Unit	Unit Cost	Cost Per S.F.	% Of Sub-Total
A. SUBSTRUCTURE							
1010	Standard Foundations	Poured concrete; strip and spread footings		S.F. Ground	1.18	1.18	
1030	Slab on Grade	6" reinforced concrete with vapor barrier and granular base		S.F. Slab	4.10	4.10	9.0%
2010	Basement Excavation	Site preparation for slab and trench for foundation wall and footing		S.F. Ground	1.09	1.09	
2020	Basement Walls	4' foundation wall		L.F. Wall	49	1.70	
B. SHELL							
	B10 Superstructure						
1010	Floor Construction	Wide flange beams and columns		S.F. Floor	20.01	14.01	26.6%
1020	Roof Construction	Metal deck on steel joist		S.F. Roof	9.82	9.82	
	B20 Exterior Enclosure						
2010	Exterior Walls	Concrete block	95% of wall	S.F. Wall	8.23	4.63	
2020	Exterior Windows	Store front	5% of wall	Each	33	.99	6.9%
2030	Exterior Doors	Aluminum and glass, hollow metal, overhead		Each	1990	.54	
	B30 Roofing						
3010	Roof Coverings	Elastomeric neoprene membrane with flashing; perlite/EPS composite insulation		S.F. Roof	3.44	3.44	4.0%
3020	Roof Openings	Roof hatches		S.F. Roof	.13	.13	
C. INTERIORS							
1010	Partitions	Concrete block	140 S.F. Floor/L.F. Partition	S.F. Partition	6.07	.52	
1020	Interior Doors	Hollow metal	2500 S.F. Floor/Door	Each	537	.21	
1030	Fittings	N/A		—	—	—	
2010	Stair Construction	N/A		—	—	—	5.5%
3010	Wall Finishes	Paint		S.F. Surface	16.68	1.43	
3020	Floor Finishes	80% rubber mat, 20% paint	50% of floor area	S.F. Floor	4.76	2.38	
3030	Ceiling Finishes	Mineral fiber tile on concealed zee bar	10% of area	S.F. Ceiling	3.71	.37	
D. SERVICES							
	D10 Conveying						
1010	Elevators & Lifts	N/A		—	—	—	0.0%
1020	Escalators & Moving Walks	N/A		—	—	—	
	D20 Plumbing						
2010	Plumbing Fixtures	Toilet and service fixtures, supply and drainage	1 Fixture/1070 S.F. Floor	Each	2332	2.18	
2020	Domestic Water Distribution	Oil fired water heater		S.F. Floor	1.24	1.24	3.8%
2040	Rain Water Drainage	N/A		—	—	—	
	D30 HVAC						
3010	Energy Supply	Oil fired hot water, unit heaters	10% of area	S.F. Floor	.52	.52	
3020	Heat Generating Systems	N/A		—	—	—	
3030	Cooling Generating Systems	Single zone, electric cooling	90% of area	S.F. Floor	11.66	11.66	13.6%
3050	Terminal & Package Units	N/A		—	—	—	
3090	Other HVAC Sys. & Equipment	N/A		—	—	—	
	D40 Fire Protection						
4010	Sprinklers	N/A		—	—	—	0.0%
4020	Standpipes	Standpipes and hose systems		S.F. Floor	—	—	
	D50 Electrical						
5010	Electrical Service/Distribution	600 ampere service, panel board and feeders		S.F. Floor	1.17	1.17	
5020	Lighting & Branch Wiring	High intensity discharge and fluorescent fixtures, receptacles, switches, A.C. and misc. power		S.F. Floor	5.42	5.42	8.2%
5030	Communications & Security	Alarm systems, emergency lighting and public address		S.F. Floor	.76	.76	
5090	Other Electrical Systems	Emergency generator		S.F. Floor	.15	.15	
E. EQUIPMENT & FURNISHINGS							
1010	Commercial Equipment	N/A		—	—	—	
1020	Institutional Equipment	N/A		—	—	—	0.0%
1030	Vehicular Equipment	N/A		—	—	—	
1090	Other Equipment	N/A		—	—	—	
F. SPECIAL CONSTRUCTION & DEMOLITION							
1020	Integrated Construction	N/A		—	—	—	22.4%
1040	Special Facilities	Dasher boards and rink (including ice making system)		S.F. Floor	20	20.09	
G. BUILDING SITEWORK	**N/A**						
				Sub-Total		89.73	100%
	CONTRACTOR FEES (General Requirements: 10%, Overhead: 5%, Profit: 10%)				25%	22.43	
	ARCHITECT FEES				7%	7.84	
				Total Building Cost		**120.00**	

187

Costs per square foot of floor area

Exterior Wall	S.F. Area	25000	30000	35000	40000	45000	50000	55000	60000	65000
	L.F. Perimeter	700	740	823	906	922	994	1033	1100	1166
Face Brick with Concrete Block Back-up	Steel Frame	99.75	97.00	95.85	94.95	93.30	92.75	91.90	91.45	91.15
	Bearing Walls	96.00	93.30	92.15	91.25	89.60	89.00	88.15	87.75	87.40
Stucco on Concrete Block	Steel Frame	97.05	94.65	93.60	92.80	91.35	90.85	90.10	89.70	89.40
	Bearing Walls	93.30	90.90	89.85	89.05	87.60	87.10	86.35	85.95	85.65
Decorative Concrete Block	Steel Frame	97.40	94.95	93.90	93.10	91.60	91.05	90.30	89.90	89.60
	Bearing Walls	93.65	91.20	90.15	89.35	87.85	87.30	86.55	86.20	85.90
Perimeter Adj., Add or Deduct	Per 100 L.F.	2.65	2.15	1.85	1.65	1.50	1.30	1.20	1.10	1.00
Story Hgt. Adj., Add or Deduct	Per 1 Ft.	.90	.75	.70	.70	.65	.60	.55	.55	.55

For Basement, add $15.85 per square foot of basement area

The above costs were calculated using the basic specifications shown on the facing page. These costs should be adjusted where necessary for design alternatives and owner's requirements. Reported completed project costs, for this type of structure, range from $51.75 to $131.74 per S.F.

Common additives

Description	Unit	$ Cost
Bleachers, Telescoping, manual		
To 15 tier	Seat	79 - 110
16-20 tier	Seat	161 - 196
21-30 tier	Seat	171 - 206
For power operation, add	Seat	31 - 48
Carrels Hardwood	Each	750 - 980
Clock System		
20 room	Each	13,000
50 room	Each	31,500
Emergency Lighting, 25 watt, battery operated		
Lead battery	Each	227
Nickel cadmium	Each	660
Flagpoles, Complete		
Aluminum, 20' high	Each	1100
40' high	Each	2700
Fiberglass, 23' high	Each	1425
39'-5" high	Each	3000
Kitchen Equipment		
Broiler	Each	4275
Cooler, 6 ft. long, reach-in	Each	3375

Description	Unit	$ Cost
Kitchen Equipment, cont.		
Dishwasher, 10-12 racks per hr.	Each	3225
Food warmer, counter, 1.2 KW	Each	745
Freezer, 44 C.F., reach-in	Each	8775
Ice cube maker, 50 lb. per day	Each	1900
Range with 1 oven	Each	2500
Lockers, Steel, single tier, 60" to 72"	Opening	131 - 227
2 tier, 60" to 72" total	Opening	74 - 125
5 tier, box lockers	Opening	42 - 62
Locker bench, lam. maple top only	L.F.	19.50
Pedestals, steel pipe	Each	63
Seating		
Auditorium chair, all veneer	Each	171
Veneer back, padded seat	Each	207
Upholstered, spring seat	Each	208
Classroom, movable chair & desk	Set	65 - 120
Lecture hall, pedestal type	Each	156 - 485
Sound System		
Amplifier, 250 watts	Each	1675
Speaker, ceiling or wall	Each	147
Trumpet	Each	275

Important: See the Reference Section for Location Factors

Model costs calculated for a 1 story building with 12' story height and 45,000 square feet of floor area

School, Elementary

				Unit	Unit Cost	Cost Per S.F.	% Of Sub-Total
A.	**SUBSTRUCTURE**						
1010	Standard Foundations	Poured concrete; strip and spread footings		S.F. Ground	1.99	1.99	
1030	Slab on Grade	4" reinforced concrete with vapor barrier and granular base		S.F. Slab	3.42	3.42	12.5%
2010	Basement Excavation	Site preparation for slab and trench for foundation wall and footing		S.F. Ground	1.09	1.09	
2020	Basement Walls	4' foundation wall		L.F. Wall	52	1.84	
B.	**SHELL**						
	B10 Superstructure						
1010	Floor Construction	N/A		—	—	—	3.7%
1020	Roof Construction	Metal deck on open web steel joists		S.F. Roof	2.51	2.51	
	B20 Exterior Enclosure						
2010	Exterior Walls	Face brick with concrete block backup	70% of wall	S.F. Wall	21	3.63	
2020	Exterior Windows	Steel outward projecting	25% of wall	Each	509	1.63	8.4%
2030	Exterior Doors	Metal and glass	5% of wall	Each	2203	.39	
	B30 Roofing						
3010	Roof Coverings	Built-up tar and gravel with flashing; perlite/EPS composite insulation		S.F. Roof	3.58	3.58	5.3%
3020	Roof Openings	N/A		—	—	—	
C.	**INTERIORS**						
1010	Partitions	Concrete block	20 S.F. Floor/L.F. Partition	S.F. Partition	6.42	3.21	
1020	Interior Doors	Single leaf kalamein fire doors	700 S.F. Floor/Door	Each	537	.77	
1030	Fittings	Toilet partitions		S.F. Floor	1.54	1.54	
2010	Stair Construction	N/A		—	—	—	25.7%
3010	Wall Finishes	75% paint, 15% glazed coating, 10% ceramic tile		S.F. Surface	5.62	2.81	
3020	Floor Finishes	65% vinyl composition tile, 25% carpet, 10% terrazzo		S.F. Floor	5.15	5.15	
3030	Ceiling Finishes	Mineral fiber tile on concealed zee bars		S.F. Ceiling	3.71	3.71	
D.	**SERVICES**						
	D10 Conveying						
1010	Elevators & Lifts	N/A		—	—	—	0.0%
1020	Escalators & Moving Walks	N/A		—	—	—	
	D20 Plumbing						
2010	Plumbing Fixtures	Kitchen, bathroom and service fixtures, supply and drainage	1 Fixture/625 S.F. Floor	Each	2193	3.51	
2020	Domestic Water Distribution	Gas fired water heater		S.F. Floor	.21	.21	6.0%
2040	Rain Water Drainage	Roof drains		S.F. Roof	.28	.28	
	D30 HVAC						
3010	Energy Supply	Oil fired hot water, wall fin radiation		S.F. Floor	6.10	6.10	
3020	Heat Generating Systems	N/A		—	—	—	
3030	Cooling Generating Systems	Split systems with air cooled condensing units		S.F. Floor	8.82	8.82	22.3%
3050	Terminal & Package Units	N/A		—	—	—	
3090	Other HVAC Sys. & Equipment	N/A		—	—	—	
	D40 Fire Protection						
4010	Sprinklers	Sprinklers, light hazard		S.F. Floor	1.50	1.50	2.2%
4020	Standpipes	N/A		—	—	—	
	D50 Electrical						
5010	Electrical Service/Distribution	800 ampere service, panel board and feeders		S.F. Floor	.87	.87	
5020	Lighting & Branch Wiring	Fluorescent fixtures, receptacles, switches, A.C. and misc. power		S.F. Floor	7.05	7.05	13.8%
5030	Communications & Security	Alarm systems, communications systems and emergency lighting		S.F. Floor	1.20	1.20	
5090	Other Electrical Systems	Emergency generator, 15 kW		S.F. Floor	.07	.07	
E.	**EQUIPMENT & FURNISHINGS**						
1010	Commercial Equipment	N/A		—	—	—	
1020	Institutional Equipment	Chalkboards		S.F. Floor	.10	.10	0.1%
1030	Vehicular Equipment	N/A		—	—	—	
1090	Other Equipment	N/A		—	—	—	
F.	**SPECIAL CONSTRUCTION & DEMOLITION**						
1020	Integrated Construction	N/A		—	—	—	0.0%
1040	Special Facilities	N/A		—	—	—	
G.	**BUILDING SITEWORK**	**N/A**					

			Sub-Total	66.98	100%
	CONTRACTOR FEES (General Requirements: 10%, Overhead: 5%, Profit: 10%)		25%	16.75	
	ARCHITECT FEES		7%	5.87	

Total Building Cost	**89.60**

Costs per square foot of floor area

Exterior Wall	S.F. Area	50000	70000	90000	110000	130000	150000	170000	190000	210000
	L.F. Perimeter	816	1083	1100	1300	1290	1450	1433	1566	1700
Face Brick with Concrete Block Back-up	Steel Frame	101.20	98.40	93.75	92.70	90.15	89.55	87.95	87.55	87.20
	R/Conc. Frame	104.85	102.40	97.90	96.95	94.45	93.90	92.30	91.95	91.65
Decorative Concrete Block	Steel Frame	97.25	94.90	91.05	90.15	87.95	87.50	86.10	85.80	85.55
	R/Conc. Frame	101.90	99.55	95.70	94.80	92.60	92.15	90.80	90.45	90.20
Limestone with Concrete Block Back-up	Steel Frame	105.85	103.05	97.45	96.35	93.20	92.55	90.55	90.15	89.80
	R/Conc. Frame	111.10	108.30	102.70	101.60	98.45	97.80	95.80	95.35	95.05
Perimeter Adj., Add or Deduct	Per 100 L.F.	2.15	1.50	1.20	1.00	.85	.70	.65	.55	.55
Story Hgt. Adj., Add or Deduct	Per 1 Ft.	1.25	1.20	1.00	.90	.80	.75	.70	.65	.65
For Basement, add $20.75 per square foot of basement area										

The above costs were calculated using the basic specifications shown on the facing page. These costs should be adjusted where necessary for design alternatives and owner's requirements. Reported completed project costs, for this type of structure, range from $60.20 to $141.75 per S.F.

Common additives

Description	Unit	$ Cost
Bleachers, Telescoping, manual		
To 15 tier	Seat	79 - 110
16-20 tier	Seat	161 - 196
21-30 tier	Seat	171 - 206
For power operation, add	Seat	31 - 48
Carrels Hardwood	Each	750 - 980
Clock System		
20 room	Each	13,000
50 room	Each	31,500
Elevators, Hydraulic passenger, 2 stops		
1500# capacity	Each	43,425
2500# capacity	Each	44,625
Emergency Lighting, 25 watt, battery operated		
Lead battery	Each	227
Nickel cadmium	Each	660
Flagpoles, Complete		
Aluminum, 20' high	Each	1100
40' high	Each	2700
Fiberglass, 23' high	Each	1425
39'-5" high	Each	3000

Description	Unit	$ Cost
Kitchen Equipment		
Broiler	Each	4275
Cooler, 6 ft. long, reach-in	Each	3375
Dishwasher, 10-12 racks per hr.	Each	3225
Food warmer, counter, 1.2 KW	Each	745
Freezer, 44 C.F., reach-in	Each	8775
Lockers, Steel, single tier, 60" or 72"	Opening	131 - 227
2 tier, 60" or 72" total	Opening	74 - 125
5 tier, box lockers	Opening	42 - 62
Locker bench, lam. maple top only	L.F.	19.50
Pedestals, steel pipe	Each	63
Seating		
Auditorium chair, all veneer	Each	171
Veneer back, padded seat	Each	207
Upholstered, spring seat	Each	208
Classroom, movable chair & desk	Set	65 - 120
Lecture hall, pedestal type	Each	156 - 485
Sound System		
Amplifier, 250 watts	Each	1675
Speaker, ceiling or wall	Each	147
Trumpet	Each	275

Important: See the Reference Section for Location Factors

BUILDING TYPES

Model costs calculated for a 2 story building with 12' story height and 130,000 square feet of floor area

School, High, 2-3 Story

				Unit	Unit Cost	Cost Per S.F.	% Of Sub-Total
A. SUBSTRUCTURE							
1010	Standard Foundations	Poured concrete; strip and spread footings		S.F. Ground	1.42	.71	
1030	Slab on Grade	4" reinforced concrete with vapor barrier and granular base		S.F. Slab	3.42	1.71	
2010	Basement Excavation	Site preparation for slab and trench for foundation wall and footing		S.F. Ground	1.09	.55	4.9%
2020	Basement Walls	4' foundation wall		L.F. Wall	52	.52	
B. SHELL							
	B10 Superstructure						
1010	Floor Construction	Concrete slab without drop panel, concrete columns		S.F. Floor	14.78	7.39	17.8%
1020	Roof Construction	Concrete slab without drop panel		S.F. Roof	10.38	5.19	
	B20 Exterior Enclosure						
2010	Exterior Walls	Face brick with concrete block backup	75% of wall	S.F. Wall	21	3.77	
2020	Exterior Windows	Window wall	25% of wall	Each	45	2.71	9.6%
2030	Exterior Doors	Metal and glass		Each	1275	.28	
	B30 Roofing						
3010	Roof Coverings	Built-up tar and gravel with flashing; perlite/EPS composite insulation		S.F. Roof	3.50	1.75	2.5%
3020	Roof Openings	Roof hatches		S.F. Roof	.06	.03	
C. INTERIORS							
1010	Partitions	Concrete block	25 S.F. Floor/L.F. Partition	S.F. Partition	7.38	2.95	
1020	Interior Doors	Single leaf kalamein fire doors	700 S.F. Floor/Door	Each	537	.77	
1030	Fittings	Toilet partitions, chalkboards		S.F. Floor	1.09	1.09	
2010	Stair Construction	Concrete filled metal pan		Flight	5075	.23	22.2%
3010	Wall Finishes	75% paint, 15% glazed coating, 10% ceramic tile		S.F. Surface	4.85	1.94	
3020	Floor Finishes	70% vinyl composition tile, 20% carpet, 10% terrazzo		S.F. Floor	4.99	4.99	
3030	Ceiling Finishes	Mineral fiber tile on concealed zee bars		S.F. Ceiling	3.71	3.71	
D. SERVICES							
	D10 Conveying						
1010	Elevators & Lifts	One hydraulic passenger elevator		Each	52,000	.40	0.6%
1020	Escalators & Moving Walks	N/A		—	—	—	
	D20 Plumbing						
2010	Plumbing Fixtures	Kitchen, bathroom and service fixtures, supply and drainage	1 Fixture/860 S.F. Floor	Each	2468	2.87	
2020	Domestic Water Distribution	Gas fired water heater		S.F. Floor	.35	.35	5.1%
2040	Rain Water Drainage	Roof drains		S.F. Roof	.74	.37	
	D30 HVAC						
3010	Energy Supply	Oil fired hot water, wall fin radiation		S.F. Floor	3.20	3.20	
3020	Heat Generating Systems	N/A		—	—	—	
3030	Cooling Generating Systems	Chilled water, cooling tower systems		S.F. Floor	11.01	11.01	20.1%
3050	Terminal & Package Units	N/A		—	—	—	
3090	Other HVAC Sys. & Equipment	N/A		—	—	—	
	D40 Fire Protection						
4010	Sprinklers	Sprinklers, light hazard		S.F. Floor	1.33	1.33	1.9%
4020	Standpipes	N/A		—	—	—	
	D50 Electrical						
5010	Electrical Service/Distribution	2000 ampere service, panel board and feeders		S.F. Floor	.86	.86	
5020	Lighting & Branch Wiring	Fluorescent fixtures, receptacles, switches, A.C. and misc. power		S.F. Floor	7.00	7.00	15.0%
5030	Communications & Security	Alarm systems, communications systems and emergency lighting		S.F. Floor	2.28	2.28	
5090	Other Electrical Systems	Emergency generator, 250 kW		S.F. Floor	.43	.43	
E. EQUIPMENT & FURNISHINGS							
1010	Commercial Equipment	N/A		—	—	—	
1020	Institutional Equipment	Laboratory counters		S.F. Floor	.07	.07	0.3%
1030	Vehicular Equipment	N/A		—	—	—	
1090	Other Equipment	Built-in athletic equipment		S.F. Floor	.14	.14	
F. SPECIAL CONSTRUCTION & DEMOLITION							
1020	Integrated Construction	N/A		—	—	—	0.0%
1040	Special Facilities	N/A		—	—	—	
G. BUILDING SITEWORK	**N/A**						

		Sub-Total	70.60	100%
CONTRACTOR FEES (General Requirements: 10%, Overhead: 5%, Profit: 10%)		25%	17.65	
ARCHITECT FEES		7%	6.20	
	Total Building Cost		**94.45**	

BUILDING TYPES

Costs per square foot of floor area

Exterior Wall	S.F. Area	50000	65000	80000	95000	110000	125000	140000	155000	170000
	L.F. Perimeter	816	1016	1000	1150	1890	2116	2287	2438	2552
Face Brick with Concrete Block Back-up	Steel Frame	99.45	98.50	95.25	94.70	99.55	99.25	98.60	97.95	97.25
	Bearing Walls	95.55	94.60	91.35	90.80	95.65	95.35	94.70	94.05	93.35
Concrete Block Stucco Face	Steel Frame	95.85	95.05	92.50	92.05	95.80	95.50	95.00	94.50	93.95
	Bearing Walls	91.95	91.15	88.60	88.15	91.90	91.60	91.10	90.60	90.05
Decorative Concrete Block	Steel Frame	96.40	95.55	92.90	92.40	96.35	96.10	95.55	95.00	94.40
	Bearing Walls	92.05	91.15	88.50	88.05	92.00	91.70	91.15	90.65	90.05
Perimeter Adj., Add or Deduct	Per 100 L.F.	1.95	1.50	1.25	1.05	.90	.80	.70	.65	.55
Story Hgt. Adj., Add or Deduct	Per 1 Ft.	1.10	1.05	.85	.85	1.20	1.20	1.15	1.10	1.05

For Basement, add $21.60 per square foot of basement area

The above costs were calculated using the basic specifications shown on the facing page. These costs should be adjusted where necessary for design alternatives and owner's requirements. Reported completed project costs, for this type of structure, range from $56.85 to $133.45 per S.F.

Common additives

Description	Unit	$ Cost	Description	Unit	$ Cost
Bleachers, Telescoping, manual			Kitchen Equipment		
To 15 tier	Seat	79 - 110	Broiler	Each	4275
16-20 tier	Seat	161 - 196	Cooler, 6 ft. long, reach-in	Each	3375
21-30 tier	Seat	171 - 206	Dishwasher, 10-12 racks per hr.	Each	3225
For power operation, add	Seat	31 - 48	Food warmer, counter, 1.2 KW	Each	745
Carrels Hardwood	Each	750 - 980	Freezer, 44 C.F., reach-in	Each	8775
Clock System			Lockers, Steel, single tier, 60" to 72"	Opening	131 - 227
20 room	Each	13,000	2 tier, 60" to 72" total	Opening	74 - 125
50 room	Each	31,500	5 tier, box lockers	Opening	42 - 62
Elevators, Hydraulic passenger, 2 stops			Locker bench, lam. maple top only	L.F.	19.50
1500# capacity	Each	43,425	Pedestals, steel pipe	Each	63
2500# capacity	Each	44,625	Seating		
Emergency Lighting, 25 watt, battery operated			Auditorium chair, all veneer	Each	171
Lead battery	Each	227	Veneer back, padded seat	Each	207
Nickel cadmium	Each	660	Upholstered, spring seat	Each	208
Flagpoles, Complete			Classroom, movable chair & desk	Set	65 - 120
Aluminum, 20' high	Each	1100	Lecture hall, pedestal type	Each	156 - 485
40' high	Each	2700	Sound System		
Fiberglass, 23' high	Each	1425	Amplifier, 250 watts	Each	1675
39'-5" high	Each	3000	Speaker, ceiling or wall	Each	147
			Trumpet	Each	275

BUILDING TYPES

Model costs calculated for a 2 story building with 12' story height and 110,000 square feet of floor area

					Unit	Unit Cost	Cost Per S.F.	% Of Sub-Total

A. SUBSTRUCTURE

				Unit	Unit Cost	Cost Per S.F.	% Of Sub-Total
1010	Standard Foundations	Poured concrete; strip and spread footings		S.F. Ground	1.66	.83	
1030	Slab on Grade	4" reinforced concrete with vapor barrier and granular base		S.F. Slab	3.42	1.71	5.4%
2010	Basement Excavation	Site preparation for slab and trench for foundation wall and footing		S.F. Ground	1.09	.55	
2020	Basement Walls	4' foundation wall		L.F. Wall	52	.91	

B. SHELL

B10 Superstructure

				Unit	Unit Cost	Cost Per S.F.	% Of Sub-Total
1010	Floor Construction	Open web steel joists, slab form, concrete, columns		S.F. Floor	16.30	8.15	15.2%
1020	Roof Construction	Metal deck, open web steel joists, columns		S.F. Roof	6.34	3.17	

B20 Exterior Enclosure

				Unit	Unit Cost	Cost Per S.F.	% Of Sub-Total
2010	Exterior Walls	Face brick with concrete block backup	75% of wall	S.F. Wall	21	6.53	
2020	Exterior Windows	Window wall	25% of wall	Each	34	3.53	14.0%
2030	Exterior Doors	Double aluminum & glass		Each	1405	.37	

B30 Roofing

			Unit	Unit Cost	Cost Per S.F.	% Of Sub-Total
3010	Roof Coverings	Built-up tar and gravel with flashing; perlite/EPS composite insulation	S.F. Roof	3.96	1.98	2.7%
3020	Roof Openings	Roof hatches	S.F. Roof	.04	.02	

C. INTERIORS

				Unit	Unit Cost	Cost Per S.F.	% Of Sub-Total
1010	Partitions	Concrete block	20 S.F. Floor/L.F. Partition	S.F. Partition	6.04	3.02	
1020	Interior Doors	Single leaf kalamein fire doors	750 S.F. Floor/Door	Each	537	.72	
1030	Fittings	Toilet partitions, chalkboards		S.F. Floor	1.17	1.17	
2010	Stair Construction	Concrete filled metal pan		Flight	5075	.28	25.2%
3010	Wall Finishes	50% paint, 40% glazed coatings, 10% ceramic tile		S.F. Surface	5.96	2.98	
3020	Floor Finishes	50% vinyl composition tile, 30% carpet, 20% terrrazzo		S.F. Floor	6.88	6.88	
3030	Ceiling Finishes	Mineral fiberboard on concealed zee bars		S.F. Ceiling	3.71	3.71	

D. SERVICES

D10 Conveying

				Unit	Unit Cost	Cost Per S.F.	% Of Sub-Total
1010	Elevators & Lifts	One hydraulic passenger elevator		Each	51,700	.47	0.6%
1020	Escalators & Moving Walks	N/A		—	—	—	

D20 Plumbing

				Unit	Unit Cost	Cost Per S.F.	% Of Sub-Total
2010	Plumbing Fixtures	Kitchen, toilet and service fixtures, supply and drainage	1 Fixture/1170 S.F. Floor	Each	2480	2.12	
2020	Domestic Water Distribution	Gas fired water heater...		S.F. Floor	.22	.22	3.5%
2040	Rain Water Drainage	Roof drains		S.F. Roof	.46	.23	

D30 HVAC

			Unit	Unit Cost	Cost Per S.F.	% Of Sub-Total
3010	Energy Supply	N/A	—	—	—	
3020	Heat Generating Systems	Included in D3030	—	—	—	
3030	Cooling Generating Systems	Multizone unit, gas heating, electric cooling	S.F. Floor	13.85	13.85	18.6%
3050	Terminal & Package Units	N/A	—	—	—	
3090	Other HVAC Sys. & Equipment	N/A	—	—	—	

D40 Fire Protection

				Unit	Unit Cost	Cost Per S.F.	% Of Sub-Total
4010	Sprinklers	Sprinklers, light hazard	10% of area	S.F. Floor	.27	.27	0.4%
4020	Standpipes	N/A		—	—	—	

D50 Electrical

			Unit	Unit Cost	Cost Per S.F.	% Of Sub-Total
5010	Electrical Service/Distribution	1600 ampere service, panel board and feeders	S.F. Floor	.84	.84	
5020	Lighting & Branch Wiring	Fluorescent fixtures, receptacles, switches, A.C. and misc. power	S.F. Floor	7.08	7.08	14.3%
5030	Communications & Security	Alarm systems, communications systems and emergency lighting	S.F. Floor	2.46	2.46	
5090	Other Electrical Systems	Emergency generator, 100 kW	S.F. Floor	.32	.32	

E. EQUIPMENT & FURNISHINGS

			Unit	Unit Cost	Cost Per S.F.	% Of Sub-Total
1010	Commercial Equipment	N/A	—	—	—	
1020	Institutional Equipment	Laboratory counters	S.F. Floor	.08	.08	0.1%
1030	Vehicular Equipment	N/A	—	—	—	
1090	Other Equipment	N/A	—	—	—	

F. SPECIAL CONSTRUCTION & DEMOLITION

			Unit	Unit Cost	Cost Per S.F.	% Of Sub-Total
1020	Integrated Construction	N/A	—	—	—	0.0%
1040	Special Facilities	N/A	—	—	—	

G. BUILDING SITEWORK   N/A

	Sub-Total	74.45	**100%**
CONTRACTOR FEES (General Requirements: 10%, Overhead: 5%, Profit: 10%)	25%	18.61	
ARCHITECT FEES	7%	6.49	
	Total Building Cost	**99.55**	

BUILDING TYPES

Costs per square foot of floor area

Exterior Wall	S.F. Area	20000	30000	40000	50000	60000	70000	80000	90000	100000
	L.F. Perimeter	400	500	685	700	800	900	1000	1100	1200
Face Brick with Concrete Block Back-up	Steel Frame	102.60	98.80	98.90	95.80	95.05	94.50	94.05	93.75	93.55
	Bearing Walls	99.70	95.95	96.05	92.95	92.20	91.65	91.20	90.90	90.70
Decorative Concrete Block	Steel Frame	96.95	94.15	94.10	91.90	91.30	90.90	90.55	90.35	90.20
	Bearing Walls	94.10	91.25	91.25	89.00	88.45	88.05	87.70	87.50	87.30
Steel Siding on Steel Studs	Steel Frame	94.15	91.80	91.75	89.95	89.45	89.10	88.80	88.65	88.50
Metal Sandwich Panel	Steel Frame	94.10	91.75	91.70	89.90	89.40	89.05	88.80	88.60	88.45
Perimeter Adj., Add or Deduct	Per 100 L.F.	4.70	3.15	2.35	1.85	1.55	1.35	1.20	1.05	.90
Story Hgt. Adj., Add or Deduct	Per 1 Ft.	1.30	1.10	1.15	.90	.90	.85	.85	.85	.80

For Basement, add $20.95 per square foot of basement area

The above costs were calculated using the basic specifications shown on the facing page. These costs should be adjusted where necessary for design alternatives and owner's requirements. Reported completed project costs, for this type of structure, range from $51.00 to $145.65 per S.F.

Common additives

Description	Unit	$ Cost
Carrels Hardwood	Each	750 - 980
Clock System		
20 room	Each	13,000
50 room	Each	31,500
Directory Boards, Plastic, glass covered		
30" x 20"	Each	575
36" x 48"	Each	1075
Aluminum, 24" x 18"	Each	440
36" x 24"	Each	555
48" x 32"	Each	780
48" x 60"	Each	1675
Elevators, Hydraulic passenger, 2 stops		
1500# capacity	Each	43,425
2500# capacity	Each	44,625
3500# capacity	Each	48,525
Emergency Lighting, 25 watt, battery operated		
Lead battery	Each	227
Nickel cadmium	Each	660

Description	Unit	$ Cost
Flagpoles, Complete		
Aluminum, 20' high	Each	1100
40' high	Each	2700
Fiberglass, 23' high	Each	1425
39'-5" high	Each	3000
Seating		
Auditorium chair, all veneer	Each	171
Veneer back, padded seat	Each	207
Upholstered, spring seat	Each	208
Classroom, movable chair & desk	Set	65 - 120
Lecture hall, pedestal type	Each	156 - 485
Shops & Workroom:		
Benches, metal	Each	405
Parts bins 6'-3" high, 3' wide, 12" deep, 72 bins	Each	505
Shelving, metal 1' x 3'	S.F.	13.85
Wide span 6' wide x 24" deep	S.F.	9.00
Sound System		
Amplifier, 250 watts	Each	1675
Speaker, ceiling or wall	Each	147
Trumpet	Each	275

Model costs calculated for a 2 story building with 12' story height and 40,000 square feet of floor area

					Unit	Unit Cost	Cost Per S.F.	% Of Sub-Total

A. SUBSTRUCTURE

				Unit	Unit Cost	Cost Per S.F.	% Of Sub-Total
1010	Standard Foundations	Poured concrete; strip and spread footings		S.F. Ground	1.80	.90	
1030	Slab on Grade	5" reinforced concrete with vapor barrier and granular base		S.F. Slab	6.96	3.48	
2010	Basement Excavation	Site preparation for slab and trench for foundation wall and footing		S.F. Ground	1.09	.55	
2020	Basement Walls	4' foundation wall		L.F. Wall	52	.91	8.1%

B. SHELL

B10 Superstructure

				Unit	Unit Cost	Cost Per S.F.	% Of Sub-Total
1010	Floor Construction	Open web steel joists, slab form, concrete, beams, columns		S.F. Floor	10.76	5.38	
1020	Roof Construction	Metal deck, open web steel joists, beams, columns		S.F. Roof	3.64	1.82	10.0%

B20 Exterior Enclosure

				Unit	Unit Cost	Cost Per S.F.	% Of Sub-Total
2010	Exterior Walls	Face brick with concrete block backup	85% of wall	S.F. Wall	21	7.37	
2020	Exterior Windows	Steel outward projecting	15% of wall	Each	34	2.11	
2030	Exterior Doors	Metal and glass doors		Each	1683	.34	13.7%

B30 Roofing

				Unit	Unit Cost	Cost Per S.F.	% Of Sub-Total
3010	Roof Coverings	Built-up tar and gravel with flashing; perlite/EPS composite insulation		S.F. Roof	3.96	1.98	
3020	Roof Openings	Roof hatches		S.F. Roof	.18	.09	2.9%

C. INTERIORS

				Unit	Unit Cost	Cost Per S.F.	% Of Sub-Total
1010	Partitions	Concrete block	20 S.F. Floor/L.F. Partition	S.F. Partition	6.04	3.02	
1020	Interior Doors	Single leaf kalamein fire doors	600 S.F. Floor/Door	Each	537	.90	
1030	Fittings	Toilet partitions, chalkboards		S.F. Floor	1.28	1.28	
2010	Stair Construction	Concrete filled metal pan		Flight	5075	.51	22.3%
3010	Wall Finishes	50% paint, 40% glazed coating, 10% ceramic tile		S.F. Surface	6.14	3.07	
3020	Floor Finishes	70% vinyl composition tile, 20% carpet, 10% terrazzo		S.F. Floor	4.99	4.99	
3030	Ceiling Finishes	Mineral fiber tile on concealed zee bars	60% of area	S.F. Ceiling	3.71	2.23	

D. SERVICES

D10 Conveying

				Unit	Unit Cost	Cost Per S.F.	% Of Sub-Total
1010	Elevators & Lifts	One hydraulic passenger elevator		Each	52,400	1.31	
1020	Escalators & Moving Walks	N/A		—	—	—	1.8%

D20 Plumbing

				Unit	Unit Cost	Cost Per S.F.	% Of Sub-Total
2010	Plumbing Fixtures	Toilet and service fixtures, supply and drainage	1 Fixture/700 S.F. Floor	Each	2037	2.91	
2020	Domestic Water Distribution	Gas fired water heater		S.F. Floor	.37	.37	
2040	Rain Water Drainage	Roof drains		S.F. Roof	.66	.33	5.0%

D30 HVAC

				Unit	Unit Cost	Cost Per S.F.	% Of Sub-Total
3010	Energy Supply	Oil fired hot water, wall fin radiation		S.F. Floor	5.75	5.75	
3020	Heat Generating Systems	N/A		—	—	—	
3030	Cooling Generating Systems	Chilled water, cooling tower systems		S.F. Floor	9.58	9.58	
3050	Terminal & Package Units	N/A		—	—	—	
3090	Other HVAC Sys. & Equipment	N/A		—	—	—	21.3%

D40 Fire Protection

				Unit	Unit Cost	Cost Per S.F.	% Of Sub-Total
4010	Sprinklers	Sprinklers, light hazard		S.F. Floor	.90	.90	
4020	Standpipes	N/A		—	—	—	1.3%

D50 Electrical

				Unit	Unit Cost	Cost Per S.F.	% Of Sub-Total
5010	Electrical Service/Distribution	800 ampere service, panel board and feeders		S.F. Floor	.98	.98	
5020	Lighting & Branch Wiring	Fluorescent fixtures, receptacles, switches, A.C. and misc. power		S.F. Floor	7.09	7.09	
5030	Communications & Security	Alarm systems, communications systems and emergency lighting		S.F. Floor	1.50	1.50	
5090	Other Electrical Systems	Emergency generator, 11/5 kW		S.F. Floor	.08	.08	13.5%

E. EQUIPMENT & FURNISHINGS

				Unit	Unit Cost	Cost Per S.F.	% Of Sub-Total
1010	Commercial Equipment	N/A		—	—	—	
1020	Institutional Equipment	Stainless steel countertops		S.F. Floor	.09	.09	
1030	Vehicular Equipment	N/A		—	—	—	
1090	Other Equipment	N/A		—	—	—	0.1%

F. SPECIAL CONSTRUCTION & DEMOLITION

				Unit	Unit Cost	Cost Per S.F.	% Of Sub-Total
1020	Integrated Construction	N/A		—	—	—	
1040	Special Facilities	N/A		—	—	—	0.0%

G. BUILDING SITEWORK   N/A

Sub-Total	71.82	**100%**
CONTRACTOR FEES (General Requirements: 10%, Overhead: 5%, Profit: 10%) 25%	17.96	
ARCHITECT FEES 7%	6.27	
Total Building Cost	**96.05**	

Costs per square foot of floor area

Exterior Wall	S.F. Area	1000	2000	3000	4000	6000	8000	10000	12000	15000
	L.F. Perimeter	126	179	219	253	310	358	400	438	490
Wood Siding	Wood Frame	91.25	81.75	77.45	74.90	71.85	70.10	68.85	67.95	66.95
Face Brick Veneer	Wood Frame	111.00	95.60	88.65	84.50	79.60	76.70	74.70	73.25	71.60
Stucco on Concrete Block	Steel Frame	106.60	93.30	87.30	83.75	79.50	77.00	75.25	74.00	72.60
	Bearing Walls	104.35	91.05	85.05	81.50	77.30	74.75	73.00	71.75	70.35
Metal Sandwich Panel	Steel Frame	113.45	98.30	91.45	87.40	82.60	79.70	77.75	76.30	74.70
Precast Concrete	Steel Frame	118.70	101.70	94.05	89.50	84.10	80.90	78.65	77.05	75.25
Perimeter Adj., Add or Deduct	Per 100 L.F.	26.05	13.00	8.65	6.50	4.35	3.25	2.60	2.15	1.70
Story Hgt. Adj., Add or Deduct	Per 1 Ft.	1.90	1.35	1.05	.95	.80	.65	.60	.55	.50

For Basement, add $15.80 per square foot of basement area

The above costs were calculated using the basic specifications shown on the facing page. These costs should be adjusted where necessary for design alternatives and owner's requirements. Reported completed project costs, for this type of structure, range from $42.25 to $146.40 per S.F.

Common additives

Description	Unit	$ Cost
Check Out Counter		
Single belt	Each	2275
Double belt	Each	3900
Emergency Lighting, 25 watt, battery operated		
Lead battery	Each	227
Nickel cadmium	Each	660
Refrigerators, Prefabricated, walk-in		
7'-6" high, 6' x 6'	S.F.	121
10' x 10'	S.F.	95
12' x 14'	S.F.	84
12' x 20'	S.F.	74
Refrigerated Food Cases		
Dairy, multi deck, 12' long	Each	8825
Delicatessen case, single deck, 12' long	Each	5975
Multi deck, 18 S.F. shelf display	Each	7325
Freezer, self-contained chest type, 30 C.F.	Each	4400
Glass door upright, 78 C.F.	Each	8275

Description	Unit	$ Cost
Refrigerated Food Cases, cont.		
Frozen food, chest type, 12' long	Each	6050
Glass door reach-in, 5 door	Each	11,400
Island case 12' long, single deck	Each	6800
Multi deck	Each	14,200
Meat cases, 12' long, single deck	Each	5025
Multi deck	Each	8450
Produce, 12' long single deck	Each	6550
Multi deck	Each	7275
Safe, Office type, 4 hour rating		
30" x 18" x 18"	Each	3200
62" x 33" x 20"	Each	6975
Smoke Detectors		
Ceiling type	Each	151
Duct type	Each	405
Sound System		
Amplifier, 250 watts	Each	1675
Speaker, ceiling or wall	Each	147
Trumpet	Each	275

Important: See the Reference Section for Location Factors

Model costs calculated for a 1 story building with 12' story height and 4,000 square feet of floor area

Store, Convenience

				Unit	Unit Cost	Cost Per S.F.	% Of Sub-Total
A. SUBSTRUCTURE							
1010	Standard Foundations	Poured concrete; strip and spread footings		S.F. Ground	1.09	1.09	
1030	Slab on Grade	4" reinforced concrete with vapor barrier and granular base		S.F. Slab	3.42	3.42	
2010	Basement Excavation	Site preparation for slab and trench for foundation wall and footing		S.F. Ground	1.31	1.31	15.7%
2020	Basement Walls	4' foundation wall		L.F. Wall	46	2.97	
B. SHELL							
B10 Superstructure							
1010	Floor Construction	N/A		—	—	—	
1020	Roof Construction	Wood truss with plywood sheathing		S.F. Roof	4.90	4.90	8.8%
B20 Exterior Enclosure							
2010	Exterior Walls	Wood siding on wood studs, insulated	80% of wall	S.F. Wall	6.75	4.10	
2020	Exterior Windows	Storefront	20% of wall	Each	33	3.37	16.1%
2030	Exterior Doors	Double aluminum and glass, solid core wood		Each	2100	1.57	
B30 Roofing							
3010	Roof Coverings	Asphalt shingles; rigid fiberglass insulation		S.F. Roof	2.93	2.93	5.2%
3020	Roof Openings	N/A		—	—	—	
C. INTERIORS							
1010	Partitions	Gypsum board on wood studs	60 S.F. Floor/L.F. Partition	S.F. Partition	9.00	1.50	
1020	Interior Doors	Single leaf wood, hollow metal	1300 S.F. Floor/Door	Each	660	.51	
1030	Fittings	N/A		—	—	—	
2010	Stair Construction	N/A		—	—	—	13.5%
3010	Wall Finishes	Paint		S.F. Surface	1.62	.27	
3020	Floor Finishes	Vinyl composition tile		S.F. Floor	2.74	2.74	
3030	Ceiling Finishes	Mineral fiber tile on wood furring		S.F. Ceiling	2.52	2.52	
D. SERVICES							
D10 Conveying							
1010	Elevators & Lifts	N/A		—	—	—	
1020	Escalators & Moving Walks	N/A		—	—	—	0.0%
D20 Plumbing							
2010	Plumbing Fixtures	Toilet and service fixtures, supply and drainage	1 Fixture/1000 S.F. Floor	Each	1380	1.38	
2020	Domestic Water Distribution	Gas fired water heater		S.F. Floor	.68	.68	3.7%
2040	Rain Water Drainage	N/A		—	—	—	
D30 HVAC							
3010	Energy Supply	N/A		—	—	—	
3020	Heat Generating Systems	Included in D3030		—	—	—	
3030	Cooling Generating Systems	Single zone rooftop unit, gas heating, electric cooling		S.F. Floor	6.25	6.25	11.2%
3050	Terminal & Package Units	N/A		—	—	—	
3090	Other HVAC Sys. & Equipment	N/A		—	—	—	
D40 Fire Protection							
4010	Sprinklers	Sprinkler, ordinary hazard		S.F. Floor	2.34	2.34	4.2%
4020	Standpipes	N/A		—	—	—	
D50 Electrical							
5010	Electrical Service/Distribution	200 ampere service, panel board and feeders		S.F. Floor	1.85	1.85	
5020	Lighting & Branch Wiring	Fluorescent fixtures, receptacles, switches, A.C. and misc. power		S.F. Floor	7.17	7.17	21.6%
5030	Communications & Security	Alarm systems and emergency lighting		S.F. Floor	2.86	2.86	
5090	Other Electrical Systems	Emergency generator, 7.5 kW		S.F. Floor	.27	.27	
E. EQUIPMENT & FURNISHINGS							
1010	Commercial Equipment	N/A		—	—	—	
1020	Institutional Equipment	N/A		—	—	—	
1030	Vehicular Equipment	N/A		—	—	—	0.0%
1090	Other Equipment	N/A		—	—	—	
F. SPECIAL CONSTRUCTION & DEMOLITION							
1020	Integrated Construction	N/A		—	—	—	
1040	Special Facilities	N/A		—	—	—	0.0%
G. BUILDING SITEWORK	**N/A**						
				Sub-Total		56.00	**100%**
	CONTRACTOR FEES (General Requirements: 10%, Overhead: 5%, Profit: 10%)				25%	14.00	
	ARCHITECT FEES				7%	4.90	
				Total Building Cost		**74.90**	

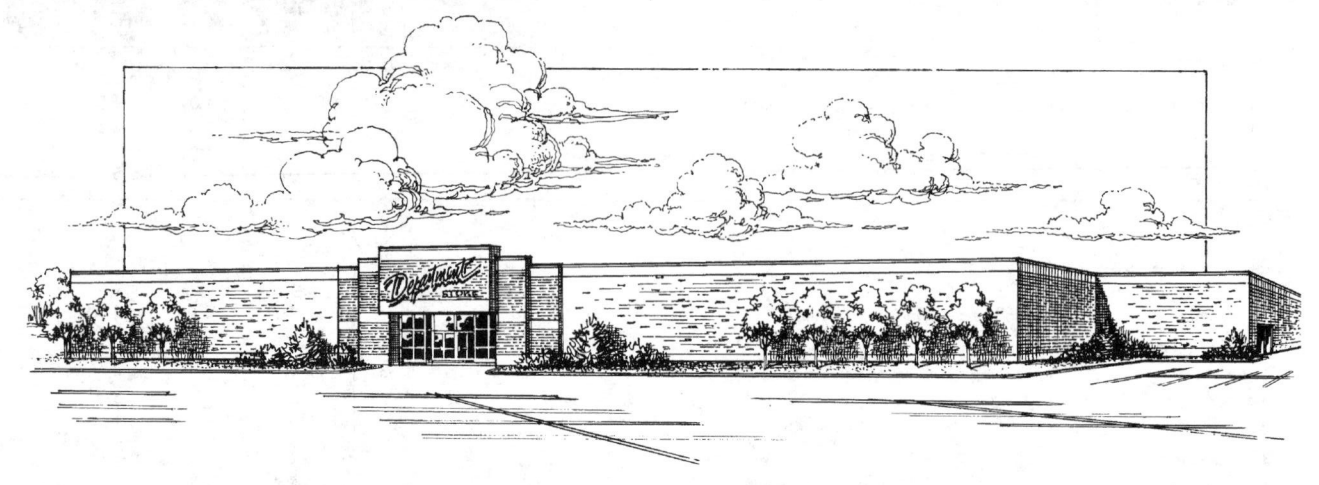

Costs per square foot of floor area

Exterior Wall	S.F. Area	50000	65000	80000	95000	110000	125000	140000	155000	170000
	L.F. Perimeter	920	1065	1167	1303	1333	1433	1533	1633	1733
Face Brick with Concrete Block Back-up	R/Conc. Frame	85.00	83.15	81.60	80.85	79.65	79.05	78.65	78.30	78.00
	Steel Frame	71.55	69.75	68.20	67.40	66.20	65.65	65.20	64.90	64.60
Decorative Concrete Block	R/Conc. Frame	81.50	80.05	78.85	78.25	77.35	76.90	76.55	76.30	76.10
	Steel Joists	68.75	67.25	65.95	65.30	64.35	63.90	63.55	63.30	63.05
Precast Concrete Panels	R/Conc. Frame	81.90	80.45	79.25	78.65	77.75	77.30	76.95	76.70	76.45
	Steel Joists	68.15	66.65	65.45	64.85	63.95	63.50	63.20	62.90	62.70
Perimeter Adj., Add or Deduct	Per 100 L.F.	1.25	.95	.75	.65	.55	.50	.50	.40	.35
Story Hgt. Adj., Add or Deduct	Per 1 Ft.	.65	.55	.50	.45	.40	.40	.40	.40	.35

For Basement, add $15.60 per square foot of basement area

The above costs were calculated using the basic specifications shown on the facing page. These costs should be adjusted where necessary for design alternatives and owner's requirements. Reported completed project costs, for this type of structure, range from $36.40 to $90.70 per S.F.

Common additives

Description	Unit	$ Cost
Closed Circuit Surveillance, One station		
Camera and monitor	Each	1375
For additional camera stations, add	Each	750
Directory Boards, Plastic, glass covered		
30" x 20"	Each	575
36" x 48"	Each	1075
Aluminum, 24" x 18"	Each	440
36" x 24"	Each	555
48" x 32"	Each	780
48" x 60"	Each	1675
Emergency Lighting, 25 watt, battery operated		
Lead battery	Each	227
Nickel cadmium	Each	660
Safe, Office type, 4 hour rating		
30" x 18" x 18"	Each	3200
62" x 33" x 20"	Each	6975
Sound System		
Amplifier, 250 watts	Each	1675
Speaker, ceiling or wall	Each	147
Trumpet	Each	275

BUILDING TYPES

Important: See the Reference Section for Location Factors

Model costs calculated for a 1 story building with 14' story height and 110,000 square feet of floor area

				Unit	Unit Cost	Cost Per S.F.	% Of Sub-Total
A. SUBSTRUCTURE							
1010	Standard Foundations	Poured concrete; strip and spread footings		S.F. Ground	.62	.62	
1030	Slab on Grade	4" reinforced concrete with vapor barrier and granular base		S.F. Slab	3.42	3.42	9.5%
2010	Basement Excavation	Site preparation for slab and trench for foundation wall and footing		S.F. Ground	1.05	1.05	
2020	Basement Walls	4' foundation wall		L.F. Wall	52	.64	
B. SHELL							
	B10 Superstructure						
1010	Floor Construction	Cast-in-place concrete columns		L.F. Column	47.45	.60	26.8%
1020	Roof Construction	Precast concrete beam and plank, concrete columns		S.F. Roof	15.49	15.49	
	B20 Exterior Enclosure						
2010	Exterior Walls	Face brick with concrete block backup	90% of wall	S.F. Wall	21	3.24	
2020	Exterior Windows	Storefront	10% of wall	Each	44	.74	7.1%
2030	Exterior Doors	Sliding electric operated entrance, hollow metal		Each	5145	.28	
	B30 Roofing						
3010	Roof Coverings	Built-up tar and gravel with flashing; perlite/EPS composite insulation		S.F. Roof	3.32	3.32	5.6%
3020	Roof Openings	Roof hatches		S.F. Roof	.03	.03	
C. INTERIORS							
1010	Partitions	Gypsum board on metal studs	60 S.F. Floor/L.F. Partition	S.F. Partition	6.84	1.14	
1020	Interior Doors	Single leaf hollow metal	600 S.F. Floor/Door	Each	537	.90	
1030	Fittings	N/A		—	—	—	
2010	Stair Construction	N/A		—	—	—	24.8%
3010	Wall Finishes	Paint		S.F. Surface	1.14	.19	
3020	Floor Finishes	50% ceramic tile, 50% carpet tile		S.F. Floor	11.07	11.07	
3030	Ceiling Finishes	Mineral fiber board on exposed grid system, suspended		S.F. Ceiling	1.60	1.60	
D. SERVICES							
	D10 Conveying						
1010	Elevators & Lifts	N/A		—	—	—	0.0%
1020	Escalators & Moving Walks	N/A		—	—	—	
	D20 Plumbing						
2010	Plumbing Fixtures	Toilet and service fixtures, supply and drainage	1 Fixture/4075 S.F. Floor	Each	1670	.41	
2020	Domestic Water Distribution	Gas fired water heater		S.F. Floor	.07	.07	1.2%
2040	Rain Water Drainage	Roof drains		S.F. Roof	.27	.27	
	D30 HVAC						
3010	Energy Supply	N/A		—	—	—	
3020	Heat Generating Systems	Included in D3030		—	—	—	
3030	Cooling Generating Systems	Single zone unit gas heating, electric cooling		S.F. Floor	6.93	6.93	11.5%
3050	Terminal & Package Units	N/A		—	—	—	
3090	Other HVAC Sys. & Equipment	N/A		—	—	—	
	D40 Fire Protection						
4010	Sprinklers	Sprinklers, light hazard		S.F. Floor	1.50	1.50	2.5%
4020	Standpipes	N/A		—	—	—	
	D50 Electrical						
5010	Electrical Service/Distribution	1200 ampere service, panel board and feeders		S.F. Floor	.56	.56	
5020	Lighting & Branch Wiring	Fluorescent fixtures, receptacles, switches, A.C. and misc. power		S.F. Floor	5.69	5.69	11.0%
5030	Communications & Security	Alarm systems and emergency lighting		S.F. Floor	.31	.31	
5090	Other Electrical Systems	Emergency generator, 7.5 kW		S.F. Floor	.03	.03	
E. EQUIPMENT & FURNISHINGS							
1010	Commercial Equipment	N/A		—	—	—	
1020	Institutional Equipment	N/A		—	—	—	0.0%
1030	Vehicular Equipment	N/A		—	—	—	
1090	Other Equipment	N/A		—	—	—	
F. SPECIAL CONSTRUCTION & DEMOLITION							
1020	Integrated Construction	N/A		—	—	—	0.0%
1040	Special Facilities	N/A		—	—	—	
G. BUILDING SITEWORK	**N/A**						

		Sub-Total	60.10	100%
	CONTRACTOR FEES (General Requirements: 10%, Overhead: 5%, Profit: 10%)	25%	15.03	
	ARCHITECT FEES	6%	4.52	
	Total Building Cost		**79.65**	

BUILDING TYPES

199

BUILDING TYPES

Costs per square foot of floor area

Exterior Wall	S.F. Area	50000	65000	80000	95000	110000	125000	140000	155000	170000
	L.F. Perimeter	533	593	670	715	778	840	871	923	976
Face Brick with Concrete Block Back-up	Steel Frame	95.50	91.75	89.80	87.95	86.80	85.95	84.90	84.25	83.80
	R/Conc. Frame	99.25	95.50	93.55	91.65	90.55	89.70	88.65	88.00	87.55
Face Brick on Steel Studs	Steel Frame	92.45	89.20	87.45	85.80	84.80	84.05	83.15	82.55	82.20
	R/Conc. Frame	97.45	94.15	92.40	90.80	89.80	89.05	88.15	87.55	87.15
Precast Concrete Panels Exposed Aggregate	Steel Frame	92.45	89.20	87.45	85.80	84.85	84.10	83.20	82.60	82.25
	R/Conc. Frame	97.05	93.80	92.05	90.40	89.45	88.70	87.80	87.20	86.80
Perimeter Adj., Add or Deduct	Per 100 L.F.	3.35	2.60	2.10	1.75	1.55	1.35	1.20	1.10	.95
Story Hgt. Adj., Add or Deduct	Per 1 Ft.	1.00	.85	.80	.75	.65	.65	.60	.55	.55

For Basement, add $26.40 per square foot of basement area

The above costs were calculated using the basic specifications shown on the facing page. These costs should be adjusted where necessary for design alternatives and owner's requirements. Reported completed project costs, for this type of structure, range from $37.65 to $100.75 per S.F.

Common additives

Description	Unit	$ Cost
Closed Circuit Surveillance, One station		
Camera and monitor	Each	1375
For additional camera stations, add	Each	750
Directory Boards, Plastic, glass covered		
30" x 20"	Each	575
36" x 48"	Each	1075
Aluminum, 24" x 18"	Each	440
36" x 24"	Each	555
48" x 32"	Each	780
48" x 60"	Each	1675
Elevators, Hydraulic passenger, 2 stops		
1500# capacity	Each	43,425
2500# capacity	Each	44,625
3500# capacity	Each	48,525
Additional stop, add	Each	3800
Emergency Lighting, 25 watt, battery operated		
Lead battery	Each	227
Nickel cadmium	Each	660

Description	Unit	$ Cost
Escalators, Metal		
32" wide, 10' story height	Each	91,000
20' story height	Each	105,500
48" wide, 10' story height	Each	95,500
20' story height	Each	110,000
Glass		
32" wide, 10' story height	Each	89,500
20' story height	Each	105,500
48" wide, 10' story height	Each	95,500
20' story height	Each	110,000
Safe, Office type, 4 hour rating		
30" x 18" x 18"	Each	3200
62" x 33" x 20"	Each	6975
Sound System		
Amplifier, 250 watts	Each	1675
Speaker, ceiling or wall	Each	147
Trumpet	Each	275

Important: See the Reference Section for Location Factors

Model costs calculated for a 3 story building with 16' story height and 95,000 square feet of floor area

Store, Department, 3 Story

				Unit	Unit Cost	Cost Per S.F.	% Of Sub-Total
A. SUBSTRUCTURE							
1010	Standard Foundations	Poured concrete; strip and spread footings		S.F. Ground	2.49	.83	
1030	Slab on Grade	4" reinforced concrete with vapor barrier and granular base		S.F. Slab	3.42	1.14	4.1%
2010	Basement Excavation	Site preparation for slab and trench for foundation wall and footing		S.F. Ground	1.09	.36	
2020	Basement Walls	4' foundation wall		L.F. Wall	49	.42	
B. SHELL							
	B10 Superstructure						
1010	Floor Construction	Concrete slab with metal deck and beams, steel columns		S.F. Floor	16.36	10.91	19.3%
1020	Roof Construction	Metal deck, open web steel joists, beams, columns		S.F. Roof	5.64	1.88	
	B20 Exterior Enclosure						
2010	Exterior Walls	Face brick with concrete block backup	90% of wall	S.F. Wall	21	6.89	
2020	Exterior Windows	Storefront	10% of wall	Each	32	1.19	14.0%
2030	Exterior Doors	Revolving and sliding panel, mall-front		Each	8162	1.20	
	B30 Roofing						
3010	Roof Coverings	Built-up tar and gravel with flashing; perlite/EPS composite insulation		S.F. Roof	3.66	1.22	1.9%
3020	Roof Openings	Roof hatches		S.F. Roof	.12	.04	
C. INTERIORS							
1010	Partitions	Gypsum board on metal studs	60 S.F. Floor/L.F. Partition	S.F. Partition	4.02	.67	
1020	Interior Doors	Single leaf hollow metal	600 S.F. Floor/Door	Each	537	.90	
1030	Fittings	N/A		—	—	—	
2010	Stair Construction	Concrete filled metal pan		Flight	7050	.74	27.6%
3010	Wall Finishes	70% paint, 20% vinyl wall covering, 10% ceramic tile		S.F. Surface	6.84	1.14	
3020	Floor Finishes	50% carpet tile, 40% marble tile, 10% terrazzo		S.F. Floor	11.13	11.13	
3030	Ceiling Finishes	Mineral fiber tile on concealed zee bars		S.F. Ceiling	3.71	3.71	
D. SERVICES							
	D10 Conveying						
1010	Elevators & Lifts	One hydraulic passenger, one hydraulic freight		Each	156,750	1.65	8.6%
1020	Escalators & Moving Walks	Four escalators		Each	96,662	4.07	
	D20 Plumbing						
2010	Plumbing Fixtures	Toilet and service fixtures, supply and drainage	1 Fixture/2570 S.F. Floor	Each	1002	.39	
2020	Domestic Water Distribution	Gas fired water heater		S.F. Floor	.06	.06	1.0%
2040	Rain Water Drainage	Roof drains		S.F. Roof	.60	.20	
	D30 HVAC						
3010	Energy Supply	N/A		—	—	—	
3020	Heat Generating Systems	Included in D3030		—	—	—	
3030	Cooling Generating Systems	Multizone rooftop unit, gas heating, electric cooling		S.F. Floor	6.93	6.93	10.4%
3050	Terminal & Package Units	N/A		—	—	—	
3090	Other HVAC Sys. & Equipment	N/A		—	—	—	
	D40 Fire Protection						
4010	Sprinklers	Sprinklers, light hazard		S.F. Floor	1.26	1.26	1.9%
4020	Standpipes	N/A		—	—	—	
	D50 Electrical						
5010	Electrical Service/Distribution	1200 ampere service, panel board and feeders		S.F. Floor	.84	.84	
5020	Lighting & Branch Wiring	Fluorescent fixtures, receptacles, switches, A.C. and misc. power		S.F. Floor	6.02	6.02	11.2%
5030	Communications & Security	Alarm systems and emergency lighting		S.F. Floor	.31	.31	
5090	Other Electrical Systems	Emergency generator, 50 kW		S.F. Floor	.26	.26	
E. EQUIPMENT & FURNISHINGS							
1010	Commercial Equipment	N/A		—	—	—	
1020	Institutional Equipment	N/A		—	—	—	0.0%
1030	Vehicular Equipment	N/A		—	—	—	
1090	Other Equipment	N/A		—	—	—	
F. SPECIAL CONSTRUCTION & DEMOLITION							
1020	Integrated Construction	N/A		—	—	—	0.0%
1040	Special Facilities	N/A		—	—	—	
G. BUILDING SITEWORK	**N/A**						

			Sub-Total	66.36	100%
CONTRACTOR FEES (General Requirements: 10%, Overhead: 5%, Profit: 10%)			25%	16.59	
ARCHITECT FEES			6%	5.00	
		Total Building Cost		**87.95**	

BUILDING TYPES

Costs per square foot of floor area

Exterior Wall		4000	6000	8000	10000	12000	15000	18000	20000	22000
	S.F. Area									
	L.F. Perimeter	260	340	360	410	440	490	540	565	594
Split Face Concrete Block	Steel Joists	93.25	85.45	78.20	75.20	72.45	69.90	68.15	67.10	66.35
Stucco on Concrete Block	Steel Joists	91.40	83.80	76.95	74.05	71.45	68.95	67.30	66.30	65.60
Painted Concrete Block	Steel Joists	86.85	79.55	73.05	70.35	67.85	65.55	64.00	63.05	62.35
Face Brick on Concrete Block	Steel Joists	103.50	94.35	85.30	81.70	78.25	75.05	72.90	71.55	70.60
Painted Reinforced Concrete	Steel Joists	100.15	91.45	83.00	79.60	76.35	73.40	71.35	70.10	69.20
Tilt-up Concrete Panels	Steel Joists	94.20	86.25	78.85	75.80	73.00	70.35	68.60	67.50	66.75
Perimeter Adj., Add or Deduct	Per 100 L.F.	11.00	7.35	5.50	4.40	3.65	2.90	2.45	2.20	2.00
Story Hgt. Adj., Add or Deduct	Per 1 Ft.	1.40	1.25	1.00	.90	.80	.70	.65	.60	.60
For Basement, add $21.85 per square foot of basement area										

The above costs were calculated using the basic specifications shown on the facing page. These costs should be adjusted where necessary for design alternatives and owner's requirements. Reported completed project costs, for this type of structure, range from $36.75 to $127.80 per S.F.

Common additives

Description	Unit	$ Cost
Emergency Lighting, 25 watt, battery operated		
Lead battery	Each	227
Nickel cadmium	Each	660
Safe, Office type, 4 hour rating		
30" x 18" x 18"	Each	3200
62" x 33" x 20"	Each	6975
Smoke Detectors		
Ceiling type	Each	151
Duct type	Each	405
Sound System		
Amplifier, 250 watts	Each	1675
Speaker, ceiling or wall	Each	147
Trumpet	Each	275

Important: See the Reference Section for Location Factors

BUILDING TYPES

Model costs calculated for a 1 story building with 14' story height and 8,000 square feet of floor area

				Unit	Unit Cost	Cost Per S.F.	% Of Sub-Total
A.	**SUBSTRUCTURE**						
1010	Standard Foundations	Poured concrete; strip and spread footings		S.F. Ground	1.24	1.24	
1030	Slab on Grade	4" reinforced concrete with vapor barrier and granular base		S.F. Slab	3.42	3.42	
2010	Basement Excavation	Site preparation for slab and trench for foundation wall and footing		S.F. Ground	1.18	1.18	14.2%
2020	Basement Walls	4' foundation wall		L.F. Wall	52	2.38	
B.	**SHELL**						
	B10 Superstructure						
1010	Floor Construction	N/A		—	—	—	
1020	Roof Construction	Metal deck, open web steel joists, beams, interior columns		S.F. Roof	4.03	4.03	7.0%
	B20 Exterior Enclosure						
2010	Exterior Walls	Decorative concrete block	90% of wall	S.F. Wall	11.98	6.79	
2020	Exterior Windows	Storefront windows	10% of wall	Each	32	2.07	15.9%
2030	Exterior Doors	Sliding entrance door and hollow metal service doors		Each	1393	.35	
	B30 Roofing						
3010	Roof Coverings	Built-up tar and gravel with flashing; perlite/EPS composite insulation		S.F. Roof	4.09	4.09	7.2%
3020	Roof Openings	Roof hatches		S.F. Roof	.07	.07	
C.	**INTERIORS**						
1010	Partitions	Gypsum board on metal studs	60 S.F. Floor/L.F. Partition	S.F. Partition	4.02	.67	
1020	Interior Doors	Single leaf hollow metal	600 S.F. Floor/Door	Each	537	.90	
1030	Fittings	N/A		—	—	—	
2010	Stair Construction	N/A		—	—	—	16.3%
3010	Wall Finishes	Paint		S.F. Surface	8.52	1.42	
3020	Floor Finishes	Vinyl tile		S.F. Floor	2.74	2.74	
3030	Ceiling Finishes	Mineral fiber tile on concealed zee bars		S.F. Ceiling	3.71	3.71	
D.	**SERVICES**						
	D10 Conveying						
1010	Elevators & Lifts	N/A		—	—	—	0.0%
1020	Escalators & Moving Walks	N/A		—	—	—	
	D20 Plumbing						
2010	Plumbing Fixtures	Toilet and service fixtures, supply and drainage	1 Fixture/890 S.F. Floor	Each	1602	1.80	
2020	Domestic Water Distribution	Gas fired water heater		S.F. Floor	1.85	1.85	7.6%
2040	Rain Water Drainage	Roof drains		S.F. Roof	.78	.78	
	D30 HVAC						
3010	Energy Supply	N/A		—	—	—	
3020	Heat Generating Systems	Included in D3030		—	—	—	
3030	Cooling Generating Systems	Single zone unit, gas heating, electric cooling		S.F. Floor	6.93	6.93	12.0%
3050	Terminal & Package Units	N/A		—	—	—	
3090	Other HVAC Sys. & Equipment	N/A		—	—	—	
	D40 Fire Protection						
4010	Sprinklers	Wet pipe sprinkler system		S.F. Floor	2.30	2.30	4.0%
4020	Standpipes	N/A		—	—	—	
	D50 Electrical						
5010	Electrical Service/Distribution	400 ampere service, panel board and feeders		S.F. Floor	1.72	1.72	
5020	Lighting & Branch Wiring	Fluorescent fixtures, receptacles, switches, A.C. and misc. power		S.F. Floor	7.04	7.04	15.8%
5030	Communications & Security	Alarm systems and emergency lighting		S.F. Floor	.30	.30	
5090	Other Electrical Systems	Emergency generator, 15 kW		S.F. Floor	.16	.16	
E.	**EQUIPMENT & FURNISHINGS**						
1010	Commercial Equipment	N/A		—	—	—	
1020	Institutional Equipment	N/A		—	—	—	
1030	Vehicular Equipment	N/A		—	—	—	0.0%
1090	Other Equipment	N/A		—	—	—	
F.	**SPECIAL CONSTRUCTION & DEMOLITION**						
1020	Integrated Construction	N/A		—	—	—	0.0%
1040	Special Facilities	N/A		—	—	—	
G.	**BUILDING SITEWORK**	**N/A**					

			Sub-Total	57.94	100%
	CONTRACTOR FEES (General Requirements: 10%, Overhead: 5%, Profit: 10%)		25%	14.49	
	ARCHITECT FEES		8%	5.77	

			Total Building Cost	**78.20**

BUILDING TYPES

203

Costs per square foot of floor area

Exterior Wall	S.F. Area	12000	16000	20000	26000	32000	38000	44000	52000	60000
	L.F. Perimeter	450	510	570	650	716	780	840	920	1020
Face Brick with Concrete Block Back-up	Steel Frame	91.20	85.15	81.55	77.90	75.25	73.40	72.00	70.65	69.90
	Bearing Walls	88.55	82.50	78.90	75.20	72.60	70.75	69.35	68.00	67.20
Stucco on Concrete Block	Steel Frame	84.55	79.55	76.50	73.45	71.30	69.75	68.65	67.55	66.90
	Bearing Walls	84.65	79.60	76.60	73.55	71.40	69.85	68.75	67.60	66.95
Precast Concrete Panels	Steel Frame	87.65	82.15	78.85	75.50	73.10	71.45	70.20	68.95	68.25
Metal Sandwich Panels	Steel Frame	78.65	74.50	72.00	69.50	67.75	66.55	65.60	64.70	64.20
Perimeter Adj., Add or Deduct	Per 100 L.F.	6.55	4.95	3.90	3.00	2.50	2.10	1.80	1.50	1.35
Story Hgt. Adj., Add or Deduct	Per 1 Ft.	1.35	1.15	1.00	.90	.80	.75	.70	.65	.65
For Basement, add $15.90 per square foot of basement area										

The above costs were calculated using the basic specifications shown on the facing page. These costs should be adjusted where necessary for design alternatives and owner's requirements. Reported completed project costs, for this type of structure, range from $42.90 to $122.90 per S.F.

Common additives

Description	Unit	$ Cost
Check Out Counter		
Single belt	Each	2275
Double belt	Each	3900
Scanner, registers, guns & memory 2 lanes	Each	13,900
10 lanes	Each	131,500
Power take away	Each	4825
Emergency Lighting, 25 watt, battery operated		
Lead battery	Each	227
Nickel cadmium	Each	660
Refrigerators, Prefabricated, walk-in		
7'-6" High, 6' x 6'	S.F.	121
10' x 10'	S.F.	95
12' x 14'	S.F.	84
12' x 20'	S.F.	74
Refrigerated Food Cases		
Dairy, multi deck, 12' long	Each	8825
Delicatessen case, single deck, 12' long	Each	5975
Multi deck, 18 S.F. shelf display	Each	7325
Freezer, self-contained chest type, 30 C.F.	Each	4400
Glass door upright, 78 C.F.	Each	8275

Description	Unit	$ Cost
Refrigerated Food Cases, cont.		
Frozen food, chest type, 12' long	Each	6050
Glass door reach in, 5 door	Each	11,400
Island case 12' long, single deck	Each	6800
Multi deck	Each	14,200
Meat case 12' long, single deck	Each	5025
Multi deck	Each	8450
Produce, 12' long single deck	Each	6550
Multi deck	Each	7275
Safe, Office type, 4 hour rating		
30" x 18" x 18"	Each	3200
62" x 33" x 20"	Each	6975
Smoke Detectors		
Ceiling type	Each	151
Duct type	Each	405
Sound System		
Amplifier, 250 watts	Each	1675
Speaker, ceiling or wall	Each	147
Trumpet	Each	275

Model costs calculated for a 1 story building with 18' story height and 44,000 square feet of floor area

				Unit	Unit Cost	Cost Per S.F.	% Of Sub-Total
A. SUBSTRUCTURE							
1010	Standard Foundations	Poured concrete; strip and spread footings		S.F. Ground	.75	.75	
1030	Slab on Grade	4" reinforced concrete with vapor barrier and granular base		S.F. Slab	3.42	3.42	12.1%
2010	Basement Excavation	Site preparation for slab and trench for foundation wall and footing		S.F. Ground	1.09	1.09	
2020	Basement Walls	4' foundation wall		L.F. Wall	52	1.01	
B. SHELL							
	B10 Superstructure						
1010	Floor Construction	N/A		–	–	–	8.4%
1020	Roof Construction	Metal deck, open web steel joists, beams, interior columns		S.F. Roof	4.38	4.38	
	B20 Exterior Enclosure						
2010	Exterior Walls	Face brick with concrete block backup	85% of wall	S.F. Wall	21	6.16	
2020	Exterior Windows	Storefront windows	15% of wall	Each	48	2.51	18.4%
2030	Exterior Doors	Sliding entrance doors with electrical operator, hollow metal		Each	3544	.89	
	B30 Roofing						
3010	Roof Coverings	Built-up tar and gravel with flashing; perlite/EPS composite insulation		S.F. Roof	3.48	3.48	6.8%
3020	Roof Openings	Roof hatches		S.F. Roof	.07	.07	
C. INTERIORS							
1010	Partitions	50% concrete block, 50% gypsum board on metal studs	50 S.F. Floor/L.F. Partition	S.F. Partition	5.00	1.20	
1020	Interior Doors	Single leaf hollow metal	2000 S.F. Floor/Door	Each	537	.27	
1030	Fittings	N/A		–	–	–	
2010	Stair Construction	N/A		–	–	–	18.5%
3010	Wall Finishes	Paint		S.F. Surface	7.00	1.68	
3020	Floor Finishes	Vinyl composition tile		S.F. Floor	2.74	2.74	
3030	Ceiling Finishes	Mineral fiber tile on concealed zee bars		S.F. Ceiling	3.71	3.71	
D. SERVICES							
	D10 Conveying						
1010	Elevators & Lifts	N/A		–	–	–	0.0%
1020	Escalators & Moving Walks	N/A		–	–	–	
	D20 Plumbing						
2010	Plumbing Fixtures	Toilet and service fixtures, supply and drainage	1 Fixture/1820 S.F. Floor	Each	837	.46	
2020	Domestic Water Distribution	Gas fired water heater		S.F. Floor	.16	.16	1.8%
2040	Rain Water Drainage	Roof drains		S.F. Roof	.29	.29	
	D30 HVAC						
3010	Energy Supply	N/A		–	–	–	
3020	Heat Generating Systems	Included in D3030		–	–	–	
3030	Cooling Generating Systems	Single zone rooftop unit, gas heating, electric cooling		S.F. Floor	6.73	6.73	13.0%
3050	Terminal & Package Units	N/A		–	–	–	
3090	Other HVAC Sys. & Equipment	N/A		–	–	–	
	D40 Fire Protection						
4010	Sprinklers	Sprinklers, light hazard		S.F. Floor	1.79	1.79	3.5%
4020	Standpipes	N/A		–	–	–	
	D50 Electrical						
5010	Electrical Service/Distribution	600 ampere service, panel board and feeders		S.F. Floor	.80	.80	
5020	Lighting & Branch Wiring	Fluorescent fixtures, receptacles, switches, A.C. and misc. power		S.F. Floor	8.08	8.08	17.5%
5030	Communications & Security	Alarm systems and emergency lighting		S.F. Floor	.14	.14	
5090	Other Electrical Systems	Emergency generator, 15 kW		S.F. Floor	.03	.03	
E. EQUIPMENT & FURNISHINGS							
1010	Commercial Equipment	N/A		–	–	–	
1020	Institutional Equipment	N/A		–	–	–	0.0%
1030	Vehicular Equipment	N/A		–	–	–	
1090	Other Equipment	N/A		–	–	–	
F. SPECIAL CONSTRUCTION & DEMOLITION							
1020	Integrated Construction	N/A		–	–	–	0.0%
1040	Special Facilities	N/A		–	–	–	
G. BUILDING SITEWORK	**N/A**						

		Sub-Total	51.84	100%
CONTRACTOR FEES (General Requirements: 10%, Overhead: 5%, Profit: 10%)			25%	12.96	
ARCHITECT FEES			7%	4.55	

	Total Building Cost	**69.35**

BUILDING TYPES

205

Costs per square foot of floor area

Exterior Wall	S.F. Area	10000	16000	20000	22000	24000	26000	28000	30000	32000
	L.F. Perimeter	420	510	600	640	680	673	706	740	737
Face Brick with Concrete Block Back-up	Wood Truss	173.90	160.20	157.30	156.05	154.95	152.00	151.05	150.30	148.30
	Precast Conc.	202.95	186.50	183.10	181.60	180.30	176.65	175.55	174.65	172.20
Metal Sandwich Panel	Wood Truss	175.80	165.55	163.30	162.30	161.50	159.35	158.70	158.10	156.65
Precast Concrete Panel	Precast Conc.	185.65	173.30	170.65	169.45	168.50	165.85	165.00	164.30	162.55
Painted Concrete Block	Wood Frame	176.45	166.35	164.10	163.15	162.35	160.25	159.55	159.00	157.60
	Precast Conc.	188.00	175.15	172.40	171.20	170.20	167.45	166.60	165.85	164.00
Perimeter Adj., Add or Deduct	Per 100 L.F.	11.45	7.15	5.75	5.20	4.75	4.40	4.10	3.85	3.60
Story Hgt. Adj., Add or Deduct	Per 1 Ft.	1.45	1.10	1.05	1.00	.95	.85	.85	.85	.80
For Basement, add $23.55 per square foot of basement area										

The above costs were calculated using the basic specifications shown on the facing page. These costs should be adjusted where necessary for design alternatives and owner's requirements. Reported completed project costs, for this type of structure, range from $67.80 to $199.35 per S.F.

Common additives

Description	Unit	$ Cost
Bleachers, Telescoping, manual		
To 15 tier	Seat	79 - 110
16-20 tier	Seat	161 - 196
21-30 tier	Seat	171 - 206
For power operation, add	Seat	31 - 48
Emergency Lighting, 25 watt, battery operated		
Lead battery	Each	227
Nickel cadmium	Each	660
Lockers, Steel, single tier, 60" or 72"	Opening	131 - 227
2 tier, 60" or 72" total	Opening	74 - 125
5 tier, box lockers	Opening	42 - 62
Locker bench, lam. maple top only	L.F.	19.50
Pedestal, steel pipe	Each	63
Pool Equipment		
Diving stand, 3 meter	Each	6975
1 meter	Each	4425
Diving board, 16' aluminum	Each	1875
Fiberglass	Each	1475
Lifeguard chair, fixed	Each	1675
Portable	Each	1900
Lights, underwater, 12 volt, 300 watt	Each	1175

Description	Unit	$ Cost
Sauna, Prefabricated, complete		
6' x 4'	Each	4825
6' x 6'	Each	5625
6' x 9'	Each	5550
8' x 8'	Each	6525
8' x 10'	Each	7225
10' x 12'	Each	9475
Sound System		
Amplifier, 250 watts	Each	1675
Speaker, ceiling or wall	Each	147
Trumpet	Each	275
Steam Bath, Complete, to 140 C.F.	Each	1325
To 300 C.F.	Each	1500
To 800 C.F.	Each	4825
To 2500 C.F.	Each	5125

Important: See the Reference Section for Location Factors

BUILDING TYPES

Model costs calculated for a 1 story building with 24' story height and 20,000 square feet of floor area

Swimming Pool, Enclosed

				Unit	Unit Cost	Cost Per S.F.	% Of Sub-Total
A.	**SUBSTRUCTURE**						
1010	Standard Foundations	Poured concrete; strip and spread footings		S.F. Ground	1.08	1.08	
1030	Slab on Grade	4" reinforced concrete with vapor barrier and granular base		S.F. Slab	3.42	3.42	15.5%
2010	Basement Excavation	Site preparation for slab and trench for foundation wall and footing		S.F. Ground	7.20	7.20	
2020	Basement Walls	4' foundation wall		L.F. Wall	52	6.32	
B.	**SHELL**						
	B10 Superstructure						
1010	Floor Construction	N/A		—	—	—	10.8%
1020	Roof Construction	Wood deck on laminated wood truss		S.F. Roof	12.63	12.63	
	B20 Exterior Enclosure						
2010	Exterior Walls	Face brick with concrete block backup	80% of wall	S.F. Wall	21	12.21	
2020	Exterior Windows	Outward projecting steel	20% of wall	Each	32	4.74	14.9%
2030	Exterior Doors	Double aluminum and glass, hollow metal		Each	1514	.38	
	B30 Roofing						
3010	Roof Coverings	Asphalt shingles with flashing; perlite/EPS composite insulation		S.F. Roof	1.30	1.30	1.1%
3020	Roof Openings	N/A		—	—	—	
C.	**INTERIORS**						
1010	Partitions	Concrete block	100 S.F. Floor/L.F. Partition	S.F. Partition	6.00	.60	
1020	Interior Doors	Single leaf hollow metal	1000 S.F. Floor/Door	Each	537	.54	
1030	Fittings	Toilet partitions		S.F. Floor	.15	.15	
2010	Stair Construction	N/A		—	—	—	15.7%
3010	Wall Finishes	Acrylic glazed coating		S.F. Surface	15.80	1.58	
3020	Floor Finishes	70% terrazzo, 30% ceramic tile		S.F. Floor	15.11	15.11	
3030	Ceiling Finishes	Mineral fiber tile on concealed zee bars	20% of area	S.F. Ceiling	1.60	.32	
D.	**SERVICES**						
	D10 Conveying						
1010	Elevators & Lifts	N/A		—	—	—	0.0%
1020	Escalators & Moving Walks	N/A		—	—	—	
	D20 Plumbing						
2010	Plumbing Fixtures	Toilet and service fixtures, supply and drainage	1 Fixture/210 S.F. Floor	Each	1816	8.65	
2020	Domestic Water Distribution	Gas fired water heater		S.F. Floor	1.37	1.37	8.6%
2040	Rain Water Drainage	N/A		—	—	—	
	D30 HVAC						
3010	Energy Supply	Terminal unit heaters	10% of area	S.F. Floor	.52	.52	
3020	Heat Generating Systems	N/A		—	—	—	
3030	Cooling Generating Systems	Single zone unit gas heating, electric cooling		S.F. Floor	11.66	11.66	10.5%
3050	Terminal & Package Units	N/A		—	—	—	
3090	Other HVAC Sys. & Equipment	N/A		—	—	—	
	D40 Fire Protection						
4010	Sprinklers	Sprinklers, light hazard		S.F. Floor	.90	.90	0.8%
4020	Standpipes	N/A		—	—	—	
	D50 Electrical						
5010	Electrical Service/Distribution	400 ampere service, panel board and feeders		S.F. Floor	.72	.72	
5020	Lighting & Branch Wiring	Fluorescent fixtures, receptacles, switches, A.C. and misc. power		S.F. Floor	6.26	6.26	6.2%
5030	Communications & Security	Alarm systems and emergency lighting		S.F. Floor	.24	.24	
5090	Other Electrical Systems	Emergency generator, 15 kW		S.F. Floor	.06	.06	
E.	**EQUIPMENT & FURNISHINGS**						
1010	Commercial Equipment	N/A		—	—	—	
1020	Institutional Equipment	N/A		—	—	—	15.9%
1030	Vehicular Equipment	N/A		—	—	—	
1090	Other Equipment	Swimming pool		S.F. Floor	18.57	18.57	
F.	**SPECIAL CONSTRUCTION & DEMOLITION**						
1020	Integrated Construction	N/A		—	—	—	0.0%
1040	Special Facilities	N/A		—	—	—	
G.	**BUILDING SITEWORK**	**N/A**					

			Sub-Total	116.53	100%
CONTRACTOR FEES (General Requirements: 10%, Overhead: 5%, Profit: 10%)			25%	29.13	
ARCHITECT FEES			8%	11.64	

Total Building Cost	**157.30**

BUILDING TYPES

Costs per square foot of floor area

Exterior Wall	S.F. Area	2000	3000	4000	5000	6000	7000	8000	9000	10000
	L.F. Perimeter	180	220	260	286	320	353	368	397	425
Face Brick with Concrete Block Back-up	Steel Frame	138.75	123.95	116.55	110.25	106.95	104.45	101.10	99.60	98.30
	Bearing Walls	136.20	121.40	114.00	107.70	104.40	101.95	98.60	97.05	95.75
Limestone with Concrete Block Back-up	Steel Frame	155.90	137.95	128.90	121.15	117.10	114.10	109.90	108.00	106.40
	Bearing Walls	153.35	135.40	126.35	118.60	114.55	111.55	107.35	105.45	103.85
Decorative Concrete Block	Steel Frame	129.70	116.60	110.00	104.50	101.60	99.40	96.50	95.15	94.05
	Bearing Walls	127.15	114.05	107.45	101.95	99.05	96.85	93.95	92.60	91.50
Perimeter Adj., Add or Deduct	Per 100 L.F.	33.05	22.05	16.50	13.25	11.00	9.45	8.25	7.30	6.55
Story Hgt. Adj., Add or Deduct	Per 1 Ft.	3.90	3.15	2.80	2.50	2.30	2.20	2.00	1.90	1.85
For Basement, add $22.90 per square foot of basement area										

The above costs were calculated using the basic specifications shown on the facing page. These costs should be adjusted where necessary for design alternatives and owner's requirements. Reported completed project costs, for this type of structure, range from $59.95 to $216.25 per S.F.

Common additives

Description	Unit	$ Cost
Emergency Lighting, 25 watt, battery operated		
Lead battery	Each	227
Nickel cadmium	Each	660
Smoke Detectors		
Ceiling type	Each	151
Duct type	Each	405
Emergency Generators, complete system, gas		
15 kw	Each	13,000
85 kw	Each	28,500
170 kw	Each	72,000
Diesel, 50 kw	Each	24,600
150 kw	Each	42,500
350 kw	Each	67,500

BUILDING TYPES

Important: See the Reference Section for Location Factors

Telephone Exchange

BUILDING TYPES

Model costs calculated for a 1 story building with 12' story height and 5,000 square feet of floor area

			Unit	Unit Cost	Cost Per S.F.	% of Sub-Total
A. SUBSTRUCTURE						
1010	Standard Foundations	Poured concrete; strip and spread footings	S.F. Ground		1.83	
1030	Slab on Grade	4" reinforced concrete with vapor barrier and granular base	S.F. Slab		3.42	
2010	Basement Excavation	Site preparation for slab and trench for foundation wall and footing	S.F. Ground		1.31	12.1%
2020	Basement Walls	4' foundation wall	L.F. Wall	52	3.02	
B. SHELL						
B10 Superstructure						
1010	Floor Construction	N/A	—		—	
1020	Roof Construction	Metal deck, open web steel joists, beams, columns	S.F. Roof		5.48	6.9%
B20 Exterior Enclosure						
2010	Exterior Walls	Face brick with concrete block backup	80% of wall	S.F. Wall	21	11.59	
2020	Exterior Windows	Outward projecting steel	20% of wall	Each	1256	8.62	26.7%
2030	Exterior Doors	Single aluminum glass with transom	Each	2478	.99	
B30 Roofing						
3010	Roof Coverings	Built-up tar and gravel with flashing; perlite/EPS composite insulation	S.F. Roof		4.38	5.5%
3020	Roof Openings	Roof hatches	S.F. Roof		—	
C. INTERIORS						
1010	Partitions	Double layer gypsum board on metal studs	15 S.F. Floor/L.F. Partition	S.F. Partition	4.01	2.67	
1020	Interior Doors	Single leaf hollow metal	150 S.F. Floor/Door	Each	537	3.58	
1030	Fittings	Toilet partitions	S.F. Floor		1.03	
2010	Stair Construction	N/A	—		—	23.2%
3010	Wall Finishes	Paint	S.F. Surface		2.94	1.96	
3020	Floor Finishes	90% carpet, 10% terrazzo	S.F. Floor		6.10	
3030	Ceiling Finishes	Fiberglass board on exposed grid system, suspended	S.F. Ceiling		3.06	
D. SERVICES						
D10 Conveying						
1010	Elevators & Lifts	N/A	—		—	0.0%
1020	Escalators & Moving Walks	N/A	—		—	
D20 Plumbing						
2010	Plumbing Fixtures	Kitchen, toilet and service fixtures, supply and drainage	1 Fixture/715 S.F. Floor	Each	1716	2.40	
2020	Domestic Water Distribution	Gas fired water heater	S.F. Floor		.95	5.7%
2040	Rain Water Drainage	Roof drains	S.F. Roof		1.14	
D30 HVAC						
3010	Energy Supply	N/A	—		—	
3020	Heat Generating Systems	Included in D3030	—		—	
3030	Cooling Generating Systems	Single zone unit, gas heating, electric cooling	S.F. Floor		6.97	8.8%
3050	Terminal & Package Units	N/A	—		—	
3090	Other HVAC Sys. & Equipment	N/A	—		—	
D40 Fire Protection						
4010	Sprinklers	Wet pipe sprinkler system	S.F. Floor		2.54	3.2%
4020	Standpipes	N/A	—		—	
D50 Electrical						
5010	Electrical Service/Distribution	200 ampere service, panel board and feeders	S.F. Floor		1.19	
5020	Lighting & Branch Wiring	Fluorescent fixtures, receptacles, switches, A.C. and misc. power	S.F. Floor		4.13	7.9%
5030	Communications & Security	Alarm systems and emergency lighting	S.F. Floor		.96	
5090	Other Electrical Systems	Emergency generator, 15 kW	S.F. Floor		.13	
E. EQUIPMENT & FURNISHINGS						
1010	Commercial Equipment	N/A	—		—	
1020	Institutional Equipment	N/A	—		—	0.0%
1030	Vehicular Equipment	N/A	—		—	
1090	Other Equipment	N/A	—		—	
F. SPECIAL CONSTRUCTION & DEMOLITION						
1020	Integrated Construction	N/A	—		—	0.0%
1040	Special Facilities	N/A	—		—	
G. BUILDING SITEWORK	N/A					
			Sub-Total		79.45	**100%**
CONTRACTOR FEES (General Requirements: 10%, Overhead: 5%, Profit: 10%)				25%	19.86	
ARCHITECT FEES				11%	10.94	
			Total Building Cost		**110.25**	

Costs per square foot of floor area

Exterior Wall	S.F. Area	5000	6500	8000	9500	11000	14000	17500	21000	24000
	L.F. Perimeter	300	360	386	396	435	510	550	620	680
Face Brick with Concrete Block Back-up	Steel Joists	105.65	101.50	96.65	92.45	90.80	88.45	85.25	83.95	83.05
	Wood Joists	107.65	103.45	98.45	94.15	92.45	90.05	86.80	85.40	84.55
Stone with Concrete Block Back-up	Steel Joists	107.80	103.45	98.35	93.90	92.20	89.70	86.35	85.00	84.10
	Wood Joists	109.80	105.40	100.20	95.65	93.85	91.35	87.90	86.45	85.55
Brick Veneer	Wood Frame	101.50	97.60	93.15	89.35	87.80	85.65	82.75	81.50	80.75
E.I.F.S.	Wood Frame	96.75	93.25	89.35	86.05	84.70	82.75	80.30	79.20	78.50
Perimeter Adj., Add or Deduct	Per 100 L.F.	10.60	8.10	6.60	5.60	4.80	3.80	3.00	2.50	2.20
Story Hgt. Adj., Add or Deduct	Per 1 Ft.	1.95	1.75	1.55	1.35	1.25	1.15	1.00	.95	.90

For Basement, add $18.85 per square foot of basement area

The above costs were calculated using the basic specifications shown on the facing page. These costs should be adjusted where necessary for design alternatives and owner's requirements. Reported completed project costs, for this type of structure, range from $52.90 to $149.00 per S.F.

Common additives

Description	Unit	$ Cost		Description	Unit	$ Cost
Directory Boards, Plastic, glass covered				Smoke Detectors		
30" x 20"	Each	575		Ceiling type	Each	151
36" x 48"	Each	1075		Duct type	Each	405
Aluminum, 24" x 18"	Each	440		Vault Front, Door & frame		
36" x 24"	Each	555		1 Hour test, 32" x 78"	Opening	3600
48" x 32"	Each	780		2 Hour test, 32" door	Opening	4275
48" x 60"	Each	1675		40" door	Opening	4700
Emergency Lighting, 25 watt, battery operated				4 Hour test, 32" door	Opening	4375
Lead battery	Each	227		40" door	Opening	5225
Nickel cadmium	Each	660		Time lock movement; two movement	Each	1650
Flagpoles, Complete						
Aluminum, 20' high	Each	1100				
40' high	Each	2700				
70' high	Each	8350				
Fiberglass, 23' high	Each	1425				
39'-5" high	Each	3000				
59' high	Each	7425				
Safe, Office type, 4 hour rating						
30" x 18" x 18"	Each	3200				
62" x 33" x 20"	Each	6975				

Important: See the Reference Section for Location Factors

Model costs calculated for a 1 story building with 12' story height and 11,000 square feet of floor area

				Unit	Unit Cost	Cost Per S.F.	% Of Sub-Total
A. SUBSTRUCTURE							
1010	Standard Foundations	Poured concrete; strip and spread footings		S.F. Ground	1.11	1.11	
1030	Slab on Grade	4" reinforced concrete with vapor barrier and granular base		S.F. Slab	3.42	3.42	11.7%
2010	Basement Excavation	Site preparation for slab and trench for foundation wall and footing		S.F. Ground	1.18	1.18	
2020	Basement Walls	4' foundation wall		L.F. Wall	52	2.09	
B. SHELL							
B10 Superstructure							
1010	Floor Construction	N/A		—	—	—	6.3%
1020	Roof Construction	Metal deck, open web steel joists, beams, interior columns		S.F. Roof	4.17	4.17	
B20 Exterior Enclosure							
2010	Exterior Walls	Face brick with concrete block backup	70% of wall	S.F. Wall	21	7.01	
2020	Exterior Windows	Metal outward projecting	30% of wall	Each	509	3.15	16.1%
2030	Exterior Doors	Metal and glass with transom		Each	1629	.59	
B30 Roofing							
3010	Roof Coverings	Built-up tar and gravel with flashing; perlite/EPS composite insulation		S.F. Roof	4.10	4.10	6.2%
3020	Roof Openings	Roof hatches		S.F. Roof	—	—	
C. INTERIORS							
1010	Partitions	Gypsum board on metal studs	20 S.F. Floor/L.F. Partition	S.F. Partition	6.02	3.01	
1020	Interior Doors	Wood solid core	200 S.F. Floor/Door	Each	432	2.16	
1030	Fittings	Toilet partitions		S.F. Floor	.42	.42	
2010	Stair Construction	N/A		—	—	—	25.0%
3010	Wall Finishes	90% paint, 10% ceramic tile		S.F. Surface	2.16	1.08	
3020	Floor Finishes	70% carpet tile, 15% terrazzo, 15% vinyl composition tile		S.F. Floor	6.30	6.30	
3030	Ceiling Finishes	Mineral fiber tile on concealed zee bars		S.F. Ceiling	3.71	3.71	
D. SERVICES							
D10 Conveying							
1010	Elevators & Lifts	N/A		—	—	—	0.0%
1020	Escalators & Moving Walks	N/A		—	—	—	
D20 Plumbing							
2010	Plumbing Fixtures	Kitchen, toilet and service fixtures, supply and drainage	1 Fixture/500 S.F. Floor	Each	1950	3.90	
2020	Domestic Water Distribution	Gas fired water heater		S.F. Floor	.68	.68	7.7%
2040	Rain Water Drainage	Roof drains		S.F. Roof	.52	.52	
D30 HVAC							
3010	Energy Supply	N/A		—	—	—	
3020	Heat Generating Systems	Included in D3030					
3030	Cooling Generating Systems	Multizone unit, gas heating, electric cooling		S.F. Floor	7.52	7.52	11.3%
3050	Terminal & Package Units	N/A		—	—	—	
3090	Other HVAC Sys. & Equipment	N/A		—	—	—	
D40 Fire Protection							
4010	Sprinklers	Wet pipe sprinkler system		S.F. Floor	1.79	1.79	2.7%
4020	Standpipes	N/A		—	—	—	
D50 Electrical							
5010	Electrical Service/Distribution	400 ampere service, panel board and feeders		S.F. Floor	1.25	1.25	
5020	Lighting & Branch Wiring	Fluorescent fixtures, receptacles, switches, A.C. and misc. power		S.F. Floor	6.89	6.89	13.0%
5030	Communications & Security	Alarm systems and emergency lighting		S.F. Floor	.46	.46	
5090	Other Electrical Systems	Emergency generator, 15 kW		S.F. Floor	.12	.12	
E. EQUIPMENT & FURNISHINGS							
1010	Commercial Equipment	N/A		—	—	—	
1020	Institutional Equipment	N/A		—	—	—	0.0%
1030	Vehicular Equipment	N/A		—	—	—	
1090	Other Equipment	N/A		—	—	—	
F. SPECIAL CONSTRUCTION & DEMOLITION							
1020	Integrated Construction	N/A		—	—	—	0.0%
1040	Special Facilities	N/A		—	—	—	
G. BUILDING SITEWORK	**N/A**						

Sub-Total		66.63	100%
CONTRACTOR FEES (General Requirements: 10%, Overhead: 5%, Profit: 10%)	25%	16.66	
ARCHITECT FEES	9%	7.51	
Total Building Cost		**90.80**	

BUILDING TYPES

211

Costs per square foot of floor area

Exterior Wall	S.F. Area	8000	10000	12000	15000	18000	24000	28000	35000	40000
	L.F. Perimeter	206	233	260	300	320	360	393	451	493
Face Brick with Concrete Block Back-up	Steel Frame	124.00	118.05	114.10	110.10	106.00	100.85	98.95	96.65	95.55
	R/Conc. Frame	132.25	126.35	122.40	118.35	114.30	109.15	107.25	104.95	103.85
Stone with Concrete Block Back-up	Steel Frame	119.60	114.35	110.80	107.20	103.50	98.80	97.10	95.05	94.05
	R/Conc. Frame	135.00	128.85	124.70	120.50	116.15	110.75	108.75	106.30	105.15
Limestone with Concrete Block Back-up	Steel Frame	136.60	129.50	124.75	119.90	114.70	108.25	105.85	103.00	101.60
	R/Conc. Frame	144.90	137.80	133.00	128.20	123.00	116.50	114.10	111.30	109.90
Perimeter Adj., Add or Deduct	Per 100 L.F.	16.20	12.95	10.80	8.65	7.20	5.40	4.65	3.70	3.20
Story Hgt. Adj., Add or Deduct	Per 1 Ft.	2.45	2.25	2.05	1.95	1.70	1.45	1.35	1.25	1.20

For Basement, add $18.35 per square foot of basement area

The above costs were calculated using the basic specifications shown on the facing page. These costs should be adjusted where necessary for design alternatives and owner's requirements. Reported completed project costs, for this type of structure, range from $52.95 to $149.00 per S.F.

Common additives

Description	Unit	$ Cost
Directory Boards, Plastic, glass covered		
30" x 20"	Each	575
36" x 48"	Each	1075
Aluminum, 24" x 18"	Each	440
36" x 24"	Each	555
48" x 32"	Each	780
48" x 60"	Each	1675
Elevators, Hydraulic passenger, 2 stops		
1500# capacity	Each	43,425
2500# capacity	Each	44,625
3500# capacity	Each	48,525
Additional stop, add	Each	3800
Emergency Lighting, 25 watt, battery operated		
Lead battery	Each	227
Nickel cadmium	Each	660

Description	Unit	$ Cost
Flagpoles, Complete		
Aluminum, 20' high	Each	1100
40' high	Each	2700
70' high	Each	8350
Fiberglass, 23' high	Each	1425
39'-5" high	Each	3000
59' high	Each	7425
Safe, Office type, 4 hour rating		
30" x 18" x 18"	Each	3200
62" x 33" x 20"	Each	6975
Smoke Detectors		
Ceiling type	Each	151
Duct type	Each	405
Vault Front, Door & frame		
1 Hour test, 32" x 78"	Opening	3600
2 Hour test, 32" door	Opening	4275
40" door	Opening	4700
4 Hour test, 32" door	Opening	4375
40" door	Opening	5225
Time lock movement; two movement	Each	1650

BUILDING TYPES

Model costs calculated for a 3 story building with 12' story height and 18,000 square feet of floor area

				Unit	Unit Cost	Cost Per S.F.	% Of Sub-Total
A.	**SUBSTRUCTURE**						
1010	Standard Foundations	Poured concrete; strip and spread footings		S.F. Ground	2.73	.91	
1030	Slab on Grade	4" reinforced concrete with vapor barrier and granular base		S.F. Slab	3.42	1.14	
2010	Basement Excavation	Site preparation for slab and trench for foundation wall and footing		S.F. Ground	1.09	.36	4.6%
2020	Basement Walls	4' foundation wall		L.F. Wall	49	1.05	
B.	**SHELL**						
	B10 Superstructure						
1010	Floor Construction	Open web steel joists, slab form, concrete, wide flange steel columns		S.F. Floor	14.64	9.76	14.9%
1020	Roof Construction	Metal deck, open web steel joists, beams, interior columns		S.F. Roof	4.59	1.53	
	B20 Exterior Enclosure						
2010	Exterior Walls	Stone with concrete block backup	70% of wall	S.F. Wall	24	10.84	
2020	Exterior Windows	Metal outward projecting	10% of wall	Each	732	1.69	17.1%
2030	Exterior Doors	Metal and glass with transoms		Each	1514	.42	
	B30 Roofing						
3010	Roof Coverings	Built-up tar and gravel with flashing; perlite/EPS composite insulation		S.F. Roof	4.11	1.37	1.8%
3020	Roof Openings	N/A		—	—	—	
C.	**INTERIORS**						
1010	Partitions	Gypsum board on metal studs	20 S.F. Floor/L.F. Partition	S.F. Partition	6.74	3.37	
1020	Interior Doors	Wood solid core	200 S.F. Floor/Door	Each	432	2.16	
1030	Fittings	Toilet partitions		S.F. Floor	.26	.26	
2010	Stair Construction	Concrete filled metal pan		Flight	6050	2.02	24.9%
3010	Wall Finishes	90% paint, 10% ceramic tile		S.F. Surface	2.16	1.08	
3020	Floor Finishes	70% carpet tile, 15% terrazzo, 15% vinyl composition tile		S.F. Floor	6.30	6.30	
3030	Ceiling Finishes	Mineral fiber tile on concealed zee bars		S.F. Ceiling	3.71	3.71	
D.	**SERVICES**						
	D10 Conveying						
1010	Elevators & Lifts	Two hydraulic elevators		Each	69,570	7.73	10.2%
1020	Escalators & Moving Walks	N/A		—	—	—	
	D20 Plumbing						
2010	Plumbing Fixtures	Toilet and service fixtures, supply and drainage	1 Fixture/1385 S.F. Floor	Each	1551	1.12	
2020	Domestic Water Distribution	Gas fired water heater		S.F. Floor	.26	.26	2.2%
2040	Rain Water Drainage	Roof drains		S.F. Roof	.81	.27	
	D30 HVAC						
3010	Energy Supply	N/A		—	—	—	
3020	Heat Generating Systems	Included in D3030		—	—	—	
3030	Cooling Generating Systems	Multizone unit, gas heating, electric cooling		S.F. Floor	7.52	7.52	9.9%
3050	Terminal & Package Units	N/A		—	—	—	
3090	Other HVAC Sys. & Equipment	N/A		—	—	—	
	D40 Fire Protection						
4010	Sprinklers	Sprinklers, light hazard		S.F. Floor	1.72	1.72	2.3%
4020	Standpipes	N/A		—	—	—	
	D50 Electrical						
5010	Electrical Service/Distribution	400 ampere service, panel board and feeders		S.F. Floor	1.06	1.06	
5020	Lighting & Branch Wiring	Fluorescent fixtures, receptacles, switches, A.C. and misc. power		S.F. Floor	7.70	7.70	12.1%
5030	Communications & Security	Alarm systems and emergency lighting		S.F. Floor	.46	.46	
5090	Other Electrical Systems	Emergency generator, 15 kW		S.F. Floor	.14	.14	
E.	**EQUIPMENT & FURNISHINGS**						
1010	Commercial Equipment	N/A		—	—	—	
1020	Institutional Equipment	N/A		—	—	—	
1030	Vehicular Equipment	N/A		—	—	—	0.0%
1090	Other Equipment	N/A		—	—	—	
F.	**SPECIAL CONSTRUCTION & DEMOLITION**						
1020	Integrated Construction	N/A		—	—	—	0.0%
1040	Special Facilities	N/A		—	—	—	
G.	**BUILDING SITEWORK**	N/A					

Sub-Total		75.95	**100%**
CONTRACTOR FEES (General Requirements: 10%, Overhead: 5%, Profit: 10%)	25%	18.99	
ARCHITECT FEES	9%	8.56	
Total Building Cost		103.50	

BUILDING TYPES

213

Costs per square foot of floor area

Exterior Wall	S.F. Area	10000	15000	20000	25000	30000	35000	40000	50000	60000
	L.F. Perimeter	410	500	600	700	700	766	833	966	1000
Brick with Concrete Block Back-up	Steel Frame	87.30	78.50	74.55	72.15	67.70	66.10	65.00	63.35	60.80
	Bearing Walls	85.80	76.85	72.75	70.30	65.75	64.15	63.00	61.30	58.75
Concrete Block	Steel Frame	71.40	65.50	62.80	61.20	58.55	57.50	56.75	55.75	54.25
	Bearing Walls	69.60	63.65	60.90	59.25	56.55	55.55	54.75	53.70	52.20
Galvanized Steel Siding	Steel Frame	75.00	68.95	66.20	64.55	61.80	60.80	60.00	58.95	57.40
Metal Sandwich Panels	Steel Frame	75.85	69.15	66.05	64.25	61.10	59.90	59.05	57.85	56.05
Perimeter Adj., Add or Deduct	Per 100 L.F.	8.70	5.75	4.35	3.45	2.90	2.45	2.15	1.75	1.45
Story Hgt. Adj., Add or Deduct	Per 1 Ft.	1.20	.95	.90	.80	.70	.65	.60	.60	.50
For Basement, add $17.65 per square foot of basement area										

The above costs were calculated using the basic specifications shown on the facing page. These costs should be adjusted where necessary for design alternatives and owner's requirements. Reported completed project costs, for this type of structure, range from $22.60 to $90.00 per S.F.

Common additives

Description	Unit	$ Cost
Dock Leveler, 10 ton cap.		
6' x 8'	Each	5300
7' x 8'	Each	5000
Emergency Lighting, 25 watt, battery operated		
Lead battery	Each	227
Nickel cadmium	Each	660
Fence, Chain link, 6' high		
9 ga. wire	L.F.	16.40
6 ga. wire	L.F.	21
Gate	Each	297
Flagpoles, Complete		
Aluminum, 20' high	Each	1100
40' high	Each	2700
70' high	Each	8350
Fiberglass, 23' high	Each	1425
39'-5" high	Each	3000
59' high	Each	7425
Paving, Bituminous		
Wearing course plus base course	S.Y.	4.24
Sidewalks, Concrete 4" thick	S.F.	2.95

Description	Unit	$ Cost
Sound System		
Amplifier, 250 watts	Each	1675
Speaker, ceiling or wall	Each	147
Trumpet	Each	275
Yard Lighting, 20' aluminum pole with 400 watt high pressure sodium fixture.	Each	2195

BUILDING TYPES

Important: See the Reference Section for Location Factors

Model costs calculated for a 1 story building with 24' story height and 30,000 square feet of floor area

					Unit	Unit Cost	Cost Per S.F.	% Of Sub-Total

A. SUBSTRUCTURE

				Unit	Unit Cost	Cost Per S.F.	% Of Sub-Total
1010	Standard Foundations	Poured concrete; strip and spread footings		S.F. Ground	.95	.95	
1030	Slab on Grade	5" reinforced concrete with vapor barrier and granular base		S.F. Slab	6.96	6.96	
2010	Basement Excavation	Site preparation for slab and trench for foundation wall and footing		S.F. Ground	1.09	1.09	25.1%
2020	Basement Walls	4' foundation wall		L.F. Wall	52	1.60	

B. SHELL

B10 Superstructure

				Unit	Unit Cost	Cost Per S.F.	% Of Sub-Total
1010	Floor Construction	Mezzanine: open web steel joists, slab form, concrete beams, columns	10% of area	S.F. Floor	12.60	1.26	12.3%
1020	Roof Construction	Metal deck, open web steel joists, beams, columns		S.F. Roof	3.95	3.95	

B20 Exterior Enclosure

				Unit	Unit Cost	Cost Per S.F.	% Of Sub-Total
2010	Exterior Walls	Concrete block	95% of wall	S.F. Wall	8.05	4.28	
2020	Exterior Windows	N/A		—	—	—	11.7%
2030	Exterior Doors	Steel overhead, hollow metal	5% of wall	Each	1784	.65	

B30 Roofing

			Unit	Unit Cost	Cost Per S.F.	% Of Sub-Total
3010	Roof Coverings	Built-up tar and gravel with flashing; perlite/EPS composite insulation	S.F. Roof	3.58	3.58	8.9%
3020	Roof Openings	Roof hatches and skylight	S.F. Roof	.19	.19	

C. INTERIORS

				Unit	Unit Cost	Cost Per S.F.	% Of Sub-Total
1010	Partitions	Concrete block (office and washrooms)	100 S.F. Floor/L.F. Partition	S.F. Partition	6.00	.48	
1020	Interior Doors	Single leaf hollow metal	5000 S.F. Floor/Door	Each	537	.11	
1030	Fittings	N/A		—	—	—	
2010	Stair Construction	Steel gate with rails		Flight	4825	.32	9.4%
3010	Wall Finishes	Paint		S.F. Surface	18.75	1.50	
3020	Floor Finishes	90% hardener, 10% vinyl composition tile		S.F. Floor	1.18	1.18	
3030	Ceiling Finishes	Suspended mineral tile on zee channels in office area	10% of area	S.F. Ceiling	3.71	.37	

D. SERVICES

D10 Conveying

			Unit	Unit Cost	Cost Per S.F.	% Of Sub-Total
1010	Elevators & Lifts	N/A	—	—	—	0.0%
1020	Escalators & Moving Walks	N/A	—	—	—	

D20 Plumbing

				Unit	Unit Cost	Cost Per S.F.	% Of Sub-Total
2010	Plumbing Fixtures	Toilet and service fixtures, supply and drainage	1 Fixture/2500 S.F. Floor	Each	1700	.68	
2020	Domestic Water Distribution	Gas fired water heater		S.F. Floor	.16	.16	3.7%
2040	Rain Water Drainage	Roof drains		S.F. Roof	.71	.71	

D30 HVAC

				Unit	Unit Cost	Cost Per S.F.	% Of Sub-Total
3010	Energy Supply	Oil fired hot water, unit heaters	90% of area	S.F. Floor	3.36	3.36	
3020	Heat Generating Systems	N/A		—	—	—	
3030	Cooling Generating Systems	Single zone unit gas, heating, electric cooling	10% of area	S.F. Floor	.83	.83	9.9%
3050	Terminal & Package Units	N/A		—	—	—	
3090	Other HVAC Sys. & Equipment	N/A		—	—	—	

D40 Fire Protection

			Unit	Unit Cost	Cost Per S.F.	% Of Sub-Total
4010	Sprinklers	Sprinklers, ordinary hazard	S.F. Floor	2.02	2.02	4.8%
4020	Standpipes	N/A	—	—	—	

D50 Electrical

			Unit	Unit Cost	Cost Per S.F.	% Of Sub-Total
5010	Electrical Service/Distribution	200 ampere service, panel board and feeders	S.F. Floor	.34	.34	
5020	Lighting & Branch Wiring	Fluorescent fixtures, receptacles, switches, A.C. and misc. power	S.F. Floor	3.34	3.34	9.3%
5030	Communications & Security	Alarm systems	S.F. Floor	.31	.31	
5090	Other Electrical Systems	N/A	—	—	—	

E. EQUIPMENT & FURNISHINGS

			Unit	Unit Cost	Cost Per S.F.	% Of Sub-Total
1010	Commercial Equipment	N/A	—	—	—	
1020	Institutional Equipment	N/A	—	—	—	
1030	Vehicular Equipment	Dock boards, dock levelers	S.F. Floor	2.06	2.06	4.9%
1090	Other Equipment	N/A	—	—	—	

F. SPECIAL CONSTRUCTION & DEMOLITION

			Unit	Unit Cost	Cost Per S.F.	% Of Sub-Total
1020	Integrated Construction	N/A	—	—	—	0.0%
1040	Special Facilities	N/A	—	—	—	

G. BUILDING SITEWORK   N/A

	Sub-Total	42.28	**100%**
CONTRACTOR FEES (General Requirements: 10%, Overhead: 5%, Profit: 10%)	25%	10.57	
ARCHITECT FEES	7%	3.70	
Total Building Cost		56.55	

BUILDING TYPES

Costs per square foot of floor area

Exterior Wall	S.F. Area	2000	3000	5000	8000	12000	20000	30000	50000	100000
	L.F. Perimeter	180	220	300	420	580	900	1300	2100	4100
Concrete Block	Steel Frame	127.30	110.50	97.05	89.45	85.25	81.90	80.20	78.90	77.85
	R/Conc. Frame	105.45	96.30	89.00	84.85	82.60	80.75	79.85	79.10	78.60
Metal Sandwich Panel	Steel Frame	99.65	89.85	82.00	77.55	75.10	73.15	72.15	71.40	70.80
Tilt-up Concrete Panel	R/Conc. Frame	111.75	102.30	94.70	90.40	88.05	86.15	85.20	84.45	83.90
Precast Concrete Panel	Steel Frame	98.25	88.60	80.85	76.50	74.10	72.15	71.15	70.40	69.85
	R/Conc. Frame	115.50	105.20	96.95	92.30	89.75	87.70	86.65	85.85	85.20
Perimeter Adj., Add or Deduct	Per 100 L.F.	25.25	16.80	10.10	6.35	4.20	2.50	1.70	1.00	.55
Story Hgt. Adj., Add or Deduct	Per 1 Ft.	1.95	1.55	1.30	1.15	1.05	.95	.95	.90	.90
Basement—Not Applicable										

The above costs were calculated using the basic specifications shown on the facing page. These costs should be adjusted where necessary for design alternatives and owner's requirements. Reported completed project costs, for this type of structure, range from $21.60 to $123.95 per S.F.

Common additives

Description	Unit	$ Cost
Dock Leveler, 10 ton cap.		
6' x 8'	Each	5300
7' x 8'	Each	5000
Emergency Lighting, 25 watt, battery operated		
Lead battery	Each	227
Nickel cadmium	Each	660
Fence, Chain link, 6' high		
9 ga. wire	L.F.	16.40
6 ga. wire	L.F.	21
Gate	Each	297
Flagpoles, Complete		
Aluminum, 20' high	Each	1100
40' high	Each	2700
70' high	Each	8350
Fiberglass, 23' high	Each	1425
39'-5" high	Each	3000
59' high	Each	7425
Paving, Bituminous		
Wearing course plus base course	S.Y.	4.24
Sidewalks, Concrete 4" thick	S.F.	2.95

Description	Unit	$ Cost
Sound System		
Amplifier, 250 watts	Each	1675
Speaker, ceiling or wall	Each	147
Trumpet	Each	275
Yard Lighting, 20' aluminum pole with 400 watt high pressure sodium fixture	Each	2195

Important: See the Reference Section for Location Factors

Model costs calculated for a 1 story building with 12' story height and 20,000 square feet of floor area

Warehouse, Mini

				Unit	Unit Cost	Cost Per S.F.	% Of Sub-Total
A. SUBSTRUCTURE							
1010	Standard Foundations	Poured concrete; strip and spread footings		S.F. Ground	1.26	1.26	
1030	Slab on Grade	4" reinforced concrete with vapor barrier and granular base		S.F. Slab	5.84	5.84	
2010	Basement Excavation	Site preparation for slab and trench for foundation wall and footing		S.F. Ground	1.18	1.18	17.0%
2020	Basement Walls	4' foundation wall		L.F. Wall	46	2.11	
B. SHELL							
	B10 Superstructure						
1010	Floor Construction	Precast concrete beams		S.F. Floor	4.35	4.35	
1020	Roof Construction	Metal deck, open web steel joists, beams, columns		S.F. Roof	5.01	5.01	15.3%
	B20 Exterior Enclosure						
2010	Exterior Walls	Concrete block	70% of wall	S.F. Wall	9.07	3.43	
2020	Exterior Windows	Aluminum projecting	5% of wall	Each	583	5.25	17.5%
2030	Exterior Doors	Steel overhead, hollow metal	25% of wall	Each	933	2.05	
	B30 Roofing						
3010	Roof Coverings	Built-up tar and gravel with flashing; perlite/EPS composite insulation		S.F. Roof	3.23	3.23	
3020	Roof Openings	N/A		—	—	—	5.3%
C. INTERIORS							
1010	Partitions	Concrete block, gypsum board on metal studs	10.65 S.F. Floor/L.F. Partition	S.F. Partition	5.99	10.27	
1020	Interior Doors	Single leaf hollow metal	300 S.F. Floor/Door	Each	867	2.89	
1030	Fittings	N/A		—	—	—	
2010	Stair Construction	N/A		—	—	—	21.5%
3010	Wall Finishes	N/A		—	—	—	
3020	Floor Finishes	N/A		—	—	—	
3030	Ceiling Finishes	N/A		—	—	—	
D. SERVICES							
	D10 Conveying						
1010	Elevators & Lifts	N/A		—	—	—	
1020	Escalators & Moving Walks	N/A		—	—	—	0.0%
	D20 Plumbing						
2010	Plumbing Fixtures	Toilet and service fixtures, supply and drainage	1 Fixture/5000 S.F. Floor	Each	1400	.28	
2020	Domestic Water Distribution	Gas fired water heater		S.F. Floor	.12	.12	1.8%
2040	Rain Water Drainage	Roof drains		S.F. Roof	.72	.72	
	D30 HVAC						
3010	Energy Supply	N/A		—	—	—	
3020	Heat Generating Systems	Oil fired hot water, unit heaters		Each	5.62	5.62	
3030	Cooling Generating Systems	N/A		—	—	—	9.2%
3050	Terminal & Package Units	N/A		—	—	—	
3090	Other HVAC Sys. & Equipment	N/A		—	—	—	
	D40 Fire Protection						
4010	Sprinklers	Wet pipe sprinkler system		S.F. Floor	2.30	2.30	
4020	Standpipes	N/A		—	—	—	3.8%
	D50 Electrical						
5010	Electrical Service/Distribution	200 ampere service, panel board and feeders		S.F. Floor	1.05	1.05	
5020	Lighting & Branch Wiring	Fluorescent fixtures, receptacles, switches and misc. power		S.F. Floor	3.90	3.90	
5030	Communications & Security	Alarm systems		S.F. Floor	.29	.29	8.6%
5090	Other Electrical Systems	Emergency generator, 7.5 kW		S.F. Floor	.08	.08	
E. EQUIPMENT & FURNISHINGS							
1010	Commercial Equipment	N/A		—	—	—	
1020	Institutional Equipment	N/A		—	—	—	
1030	Vehicular Equipment	N/A		—	—	—	0.0%
1090	Other Equipment	N/A		—	—	—	
F. SPECIAL CONSTRUCTION & DEMOLITION							
1020	Integrated Construction	N/A		—	—	—	
1040	Special Facilities	N/A		—	—	—	0.0%
G. BUILDING SITEWORK	**N/A**						

			Sub-Total	61.23	100%
CONTRACTOR FEES (General Requirements: 10%, Overhead: 5%, Profit: 10%)			25%	15.31	
ARCHITECT FEES			7%	5.36	

Total Building Cost	**81.90**

BUILDING TYPES

217

Deterioration occurs within the structure itself and is determined by the observation of both materials and equipment.

Curable deterioration can be remedied either by maintenance, repair or replacement, within prudent economic limits.

Incurable deterioration that has progressed to the point of actually affecting the structural integrity of the structure, making repair or replacement not economically feasible.

Actual Versus Observed Age

The observed age of a structure refers to the age that the structure appears to be. Periodic maintenance, remodeling and renovation all tend to reduce the amount of deterioration that has taken place, thereby decreasing the observed age. Actual age on the other hand relates solely to the year that the structure was built.

The Depreciation Table shown here relates to the observed age.

Obsolescence arises from conditions either occurring within the structure (functional) or caused by factors outside the limits of the structure (economic).

Functional obsolescence is any inadequacy caused by outmoded design, dated construction materials or oversized or undersized areas, all of which cause excessive operation costs.

Incurable is so costly as not to economically justify the capital expenditure required to correct the deficiency.

Economic obsolescence is caused by factors outside the limits of the structure. The prime causes of economic obsolescence are:
- zoning and environmental laws
- government legislation
- negative neighborhood influences
- business climate
- proximity to transportation facilities

Depreciation Table
Commercial/Industrial/Institutional

Building Material			
Observed Age (Years)	Frame	Masonry On Wood	Masonry On Masonry or Steel
1	1%	0%	0%
2	2	1	0
3	3	2	1
4	4	3	2
5	6	5	3
10	20	15	8
15	25	20	15
20	30	25	20
25	35	30	25
30	40	35	30
35	45	40	35
40	50	45	40
45	55	50	45
50	60	55	50
55	65	60	55
60	70	65	60

Assemblies Section

Table of Contents

Table of Contents

Introduction to the Assemblies Section

This section of the manual contains the technical description and installed cost of all components used in the commercial/industrial/institutional section. In most cases, there is an illustration of the component to aid in identification.

All Costs Include:

Materials purchased in lot sizes typical of normal construction projects with discounts appropriate to established contractors.

Installation work performed by union labor working under standard conditions at a normal pace.

Installing Contractor's Overhead & Profit

Standard installing contractor's overhead & profit are included in the assemblies costs.

These assemblies costs contain no provisions for General Contractor's overhead and profit or Architect's fees, nor do they include premiums for labor or materials or savings which may be realized under certain economic situations.

The assemblies tables are arranged by physical size (dimensions) of the component wherever possible.

How to Use the Assemblies Cost Tables

The following is a detailed explanation of a sample Assemblies Cost Table. Most Assembly Tables are separated into two parts: 1) an illustration of the system to be estimated with descriptive notes; and 2) the costs for similar systems with dimensional and/or size variations. Next to each bold number below is the item being described with the appropriate component of the sample entry following in parenthesis. In most cases, if the work is to be subcontracted, the general contractor will need to add an additional markup (R.S. Means suggests using 10%) to the "Total" figures.

1 System/Line Numbers (A1030 120 2240)

Each Assemblies Cost Line has been assigned a unique identification number based on the UNIFORMAT II classification sstem.

UNIFORMAT II Major Group

A1030   120   2240

UNIFORMAT II Level 3

Means Major Classification

Means Individual Line Number

A10 Foundations

A1030   Slab on Grade

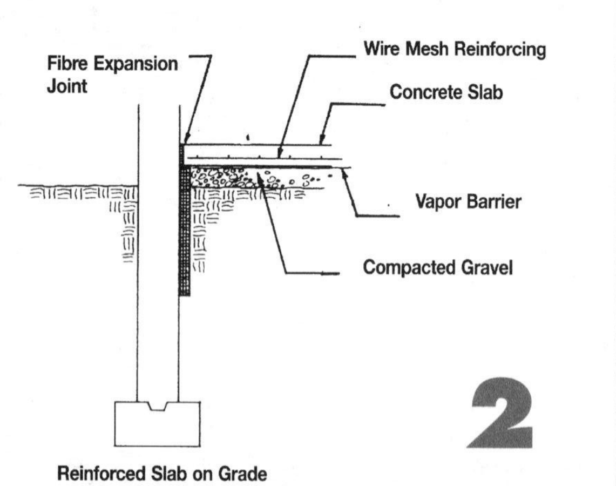

Reinforced Slab on Grade

- Fibre Expansion Joint
- Wire Mesh Reinforcing
- Concrete Slab
- Vapor Barrier
- Compacted Gravel

A Slab on Grade system includes fine grading; 6″ of compacted gravel; vapor barrier; 3500 p.s.i. concrete; bituminous fiber expansion joint; all necessary edge forms 4 uses; steel trowel finish; and sprayed on membrane curing compound. Wire mesh reinforcing used in all reinforced slabs.

Non-industrial slabs are for foot traffic only with negligible abrasion. Light industrial slabs are for pneumatic wheels and light abrasion. Industrial slabs are for solid rubber wheels and moderate abrasion. Heavy industrial slabs are for steel wheels and severe abrasion.

A1030 120	Plain & Reinforced	MAT.	INST.	TOTAL
2220	Slab on grade, 4″ thick, non industrial, non reinforced	1.29	1.78	3.07
2240	Reinforced	1.37	2.05	3.42
2260	Light industrial, non reinforced	1.65	2.18	
2280	Reinforced	1.73	.45	.18
2300	Industrial, non reinforced	2.01	.48	5.49
2320	Reinforced	2.09	3.75	5.84
3340	5″ thick, non industrial, non reinforced	1.52	1.83	3.35
3360	Reinforced	1.60	2.10	3.70
3380	Light industrial, non reinforced	1.89	2.23	4.12
3400	Reinforced	1.97	2.50	4.47
3420	Heavy industrial, non reinforced	2.60	4.04	6.64
3440	Reinforced	2.62	4.34	6.96
4460	6″ thick, non industrial, non reinforced	1.83	1.79	3.62
4480	Reinforced	1.95	2.15	4.10
4500	Light industrial, non reinforced	2.22	2.19	4.41
4520	Reinforced	2.41	2.68	5.09
4540	Heavy industrial, non reinforced	2.93	4.11	7.04
4560	Reinforced	3.05	4.47	7.52

COST PER S.F.

2 *Illustration*

At the top of most assembly pages is an illustration, a brief description, and the design criteria used to develop the cost.

3 *System Description*

The components of a typical system are listed in the description to show what has been included in the development of the total system price. The rest of the table contains prices for other similar systems with dimensional and/or size variations.

4 *Unit of Measure for Each System (S.F.)*

Costs shown in the three right hand columns have been adjusted by the component quantity and unit of measure for the entire system. In this example, "Cost per S.F." is the unit of measure for this system or "assembly."

5 *Materials (1.37)*

This column contains the Materials Cost of each component. These cost figures are bare costs plus 10% for profit.

6 *Installation (2.05)*

Installation includes labor and equipment plus the installing contractor's overhead and profit. Equipment costs are the bare rental costs plus 10% for profit. The labor overhead and profit is defined on the inside back cover of this book.

7 *Total (3.42)*

The figure in this column is the sum of the material and installation costs.

Material Cost	+	Installation Cost	=	Total
$1.37	+	$2.05	=	$3.42

Example

Assemblies costs can be used to calculate the cost for each component of a building and, accumulated with appropriate markups, produce a cost for the complete structure. Components from a model building with similar characteristics in conjunction with Assemblies components may be used to calculate costs for a complete structure.

Example:

Outline Specifications

General	Building size—60′ x 100′, 8 suspended floors, 12′ floor to floor, 4′ high parapet above roof, full basement 11′-8″ floor to floor, bay size 25′ x 30′, ceiling heights - 9′ in office area and 8′ in core area.
A Substructure	Concrete, spread and strip footings, concrete walls waterproofed, 4″ slab on grade.
B Shell	
B10 Superstructure	Columns, wide flange; 3 hr. fire rated; Floors, composite steel frame & deck with concrete slab; Roof, steel beams, open web joists and deck.
B20 Exterior Enclosure	Walls; North, East & West, brick & lightweight concrete block with 2″ cavity insulation, 25% window; South, 8″ lightweight concrete block insulated, 10% window. Doors, aluminum & glass. Windows, aluminum, 3′-0″ x 5′-4″, insulating glass.
B30 Roofing	Tar & gravel, 2″ rigid insulation, R 12.5.
C Interiors	Core—6″ lightweight concrete block, full height with plaster on exposed faces to ceiling height.
	Corridors—1st & 2nd floor; 3-5/8″ steel studs with F.R. gypsum board, full height.
	Exterior Wall—plaster, ceiling height.
	Fittings—Toilet accessories, directory boards.
	Doors—hollow metal.
	Wall Finishes—lobby, mahogany paneling on furring, remainder plaster & gypsum board, paint.
	Floor Finishes—1st floor lobby, corridors & toilet rooms, terrazzo; remainder, concrete, tenant developed. 2nd thru 8th, toilet rooms, ceramic tile; office & corridor, carpet.
	Ceiling Finishes—24″ x 48″ fiberglass board on Tee grid.
D Services	
D10 Conveying	2-2500 lb. capacity, 200 F.P.M. geared elevators, 9 stops.
D20 Plumbing	Fixtures, see sketch.
	Roof Drains—2-4″ C.I. pipe.
D30 HVAC	Heating—fin tube radiation, forced hot water.
	Air Conditioning—chilled water with water cooled condenser.
D40 Fire Protection	Fire Protection—4″ standpipe, 9 hose cabinets.
D50 Electrical	Lighting, 1st thru 8th, 15 fluorescent fixtures/1000 S.F., 3 Watts/S.F.
	Basement, 10 fluorescent fixtures/1000 S.F., 2 Watts/S.F.
	Receptacles, 1st thru 8th, 16.5/1000 S.F., 2 Watts/S.F.
	Basement, 10 receptacles/1000 S.F., 1.2 Watts/S.F.
	Air Conditioning, 4 Watts/S.F.
	Miscellaneous Connections, 1.2 Watts/S.F.
	Elevator Power, 2-10 H.P., 230 volt motors.
	Wall Switches, 2/1000 S.F.
	Service, panel board & feeder, 2000 Amp.
	Fire Detection System, pull stations, signals, smoke and heat detectors.
	Emergency Lighting Generator, 30 KW.
E Equipment & Furnishings	NA.
F Special Construction & Demolition	NA.
G Building Sitework	NA.

Front Elevation

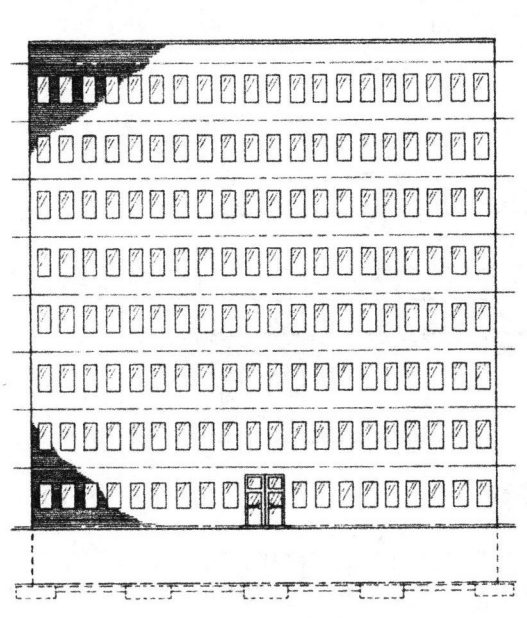

Basement Plan

100'-0"

60'-0"

Ground Floor Plan

100'-0"

60'-0"

Lobby

Typical Floor Plan

100'-0"

40'-0"

60'-0"

32'-0"

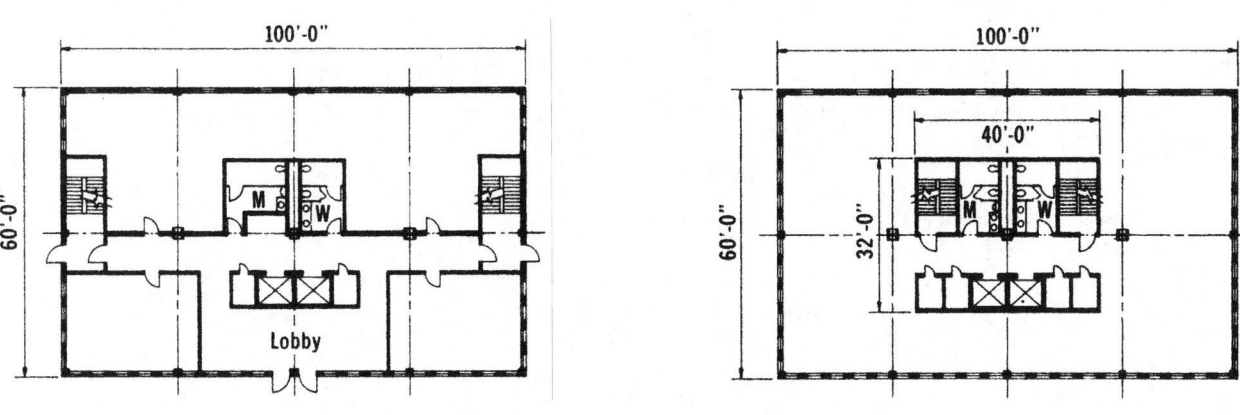

225

Example

If spread footing & column sizes are unknown, develop approximate loads as follows. Enter tables with these loads to determine costs.

Superimposed Load Ranges

Apartments & Residential Structures	65	to	75 psf
Assembly Areas & Retail Stores	110	to	125 psf
Commercial & Manufacturing	150	to	250 psf
Offices	75	to	100 psf

Approximate loads/S.F. for roof & floors.
Roof. Assume 40 psf superimposed load.
Steel joists, beams & deck.
Table B1020 112—Line 3900

B1020 112	**Steel Joists, Beams, & Deck on Columns**							
	BAY SIZE (FT.)	SUPERIMPOSED LOAD (P.S.F.)	DEPTH (IN.)	TOTAL LOAD (P.S.F.)	COLUMN ADD	COST PER S.F.		
						MAT.	INST.	TOTAL
3500	25x30	20	22	40		2.67	1.15	3.82
3600					columns	.55	.20	.75
3900		40	25	60		3.25	1.35	4.60
4000					columns	.66	.24	.90

B1010 256	**Composite Beams, Deck & Slab**							
	BAY SIZE (FT.)	SUPERIMPOSED LOAD (P.S.F.)	SLAB THICKNESS (IN.)	TOTAL DEPTH (FT. - IN.)	TOTAL LOAD (P.S.F.)	COST PER S.F.		
						MAT.	INST.	TOTAL
3400	25x30	40	5-1/2	1 - 11-1/2	83	6.15	4.06	10.21
3600		75	5-1/2	1 - 11-1/2	119	6.60	4.10	10.70
3900		125	5-1/2	1 - 11-1/2	170	7.65	4.65	12.30
4000		200	6-1/4	2 - 6-1/4	252	9.60	5.30	14.90

Floors–Total load, 119 psf.
Interior foundation load.
Roof

[(25' x 30' x 60 psf) + 8 floors x (25' x 30' x 119 psf)] x 1/1000 lb./Kip =	759 Kips
Approximate Footing Loads, Interior footing =	759 Kips
Exterior footing (1/2 bay) 759 k x .6 =	455 Kips
Corner footing (1/4 bay) 759 k x .45 =	342 Kips
[Factors to convert Interior load to Exterior & Corner loads]	
Approximate average Column load 759 k/2 =	379 Kips

PRELIMINARY ESTIMATE

PROJECT	**Office Building**	TOTAL SITE AREA	
BUILDING TYPE		OWNER	
LOCATION		ARCHITECT	
DATE OF CONSTRUCTION		ESTIMATED CONSTRUCTION PERIOD	

BRIEF DESCRIPTION	**Building size: 60' x 100' - 8 structural floors - 12' floor to floor**
	4' high parapet above roof, full basement 11' - 8" floor to floor, bay size
	25' x 30', ceiling heights - 9' in office area + 8' in core area

TYPE OF PLAN					TYPE OF CONSTRUCTION			
QUALITY					BUILDING CAPACITY			

Floor				Wall Area			
Below Grade Levels			S.F.	Foundation Walls	L.F.	HT.	S.F.
Area			S.F.	Frost Walls	L.F.	HT.	S.F.
Area			S.F.	Exterior Closure	Total		S.F.
Total Area			S.F.	Comment			
Ground Floor				Fenestration	%		S.F.
Area			S.F.		%		S.F.
Area			S.F.	Exterior Wall	%		S.F.
Total Area			S.F.		%		S.F.

Supported Levels				Site Work			
Area				Parking	S.F. (For		Cars)
Area			S.F.	Access Roads	L.F. (X		Ft. Wide)
Area			S.F.	Sidewalk	L.F. (X		Ft. Wide)
Area			S.F.	Landscaping	S.F.		S.F. (% Unbuilt Site)
Total Area			S.F.	Building Codes			
			S.F.	City		Country	
Miscellaneous				National		Other	
Area			S.F.	Loading			
Area			S.F.	Roof	psf	Ground Floor	psf
Area			S.F.	Supported Floors	psf	Corridor	psf
Area			S.F.	Balcony	psf	Partition, allow.	psf
Total Area			S.F.	Miscellaneous			psf

Net Finished Area	S.F.			Live Load Reduction		
Net Floor Area	S.F.			Wind		
Gross Floor Area	**54,000**	S.F.		Earthquake	Zone	
				Comment		
Roof						
Total Area	S.F.			Soil Type		
Comments				Bearing Capacity		K.S.F.
				Frost Depth		Ft.
Volume				**Frame**		
Depth of Floor System				Type	Bay Spacing	
Minimum	In.			Foundation		
Maximum	In.			Special		
Foundation Wall Height	Ft.			Substructure		
Floor to Floor Height	Ft.			Comment		
Floor to Ceiling Height	Ft.			Superstructure, Vertical		
Subgrade Volume	C.F.			Fireproofing	☐ Columns	Hrs.
Above Grade Volume	C.F.			☐ Girders	Hrs.	☐ Beams	Hrs.
Total Building Volume	**648,000**	C.F.		☐ Floor	Hrs.	☐ None

NUMBER		QTY.	UNIT	TOTAL COST		COST
				UNIT	TOTAL	PER S.F.
A	**SUBSTRUCTURE**					
A1010 210 7900	Corner Footings 8'-6" SQ. x 27"	4	Ea.	1295	5,180	
8010	Exterior 9' -6" SQ. x 30"	8		1725	13,800	
8300	Interior 12" SQ.	3		3175	9,525	
A1010 110 2700	Strip ↓ 2' Wide x 1' Thick					
	320 L.F. [(4 x 8.5) + 8 x 9.5)] =	210	L.F.	25.60	5,376	
A1010 210 7262	Foundation Wall 12' High, 1' Thick	210	↓	162.50	34,125	
A1010 320 2800	Foundation Waterproofing	210	↓	12.87	2,703	
A1030 120 2240	4", non-industrial, reinf. slab on grade	6000	S.F.	3.42	20,520	
A2010 110 3440	Building Excavation + Backfill	6000	S.F.	4.64	27,840	
3500	(Interpolated ; 12' Between					
4620	8' and 16' ; 6,000 Between					
4680	4,000 and 10,000 S.F.					
	Total				119,069	2.20
B	**SHELL**					
B10	**Superstructure**					
B1010 208 5800	Columns: Load 379K Use 400K					
	Exterior 12 x 96' + Interior 3 x 108'	1476	V.L.F.	72	106,272	
B1010 274 3650	Column Fireproofing - Interior 4 sides					
3700	Exterior 1/2 (Interpolated for 12")	900	V.L.F	23.45	21,105	
B1010 256 3600	Floors: Composite steel + lt. wt. conc.	48000	S.F.	10.70	513,600	
B1020 112 3900	Roof: Open web joists, beams + decks	6000	S.F.	4.60	27,600	
	Total				668,577	12.38
B20	**Exterior Enclosure**					
B2010 132 1200	4" Brick + 6" Block - Insulated					
	75% NE + W Walls 220' x 100' x .75	16500	S.F.	19.85	327,525	
B2010 109 3410	8" Block - Insulated 90%					
	100' x 100' x .9	9000	S.F.	7.59	68,310	
B2020 105 6950	Double Aluminum + Glass Door	1	Pr.	3350	3,350	
6300	Single	2	Ea.	1770	3,540	
B2020 102 8800	Aluminum windows - Insul. glass					
	(5,500 S.F. + 1,000 S.F.)/(3 x 5.33)	406	Ea.	684.00	277,704	
B2010 103 6776	Precast concrete coping	320	L.F.	26.55	8,496	
	Total				688,925	12.76
B30	**Roofing**					
B3010 320 4020	Insulation: Perlite/Polyisocyanurate	6000	S.F.	1.40	8,400	
B3010 105 1400	Roof: 4 Ply T & G composite	6000	↓	1.68	10,080	
B3010 430 0300	Flashing: Aluminum - fabric backed	640	↓	2.12	1,357	
B3020 210 0500	Roof hatch	1	Ea.	801	801	
	Total				20,638	0.38
	Page Total					

ASSEMBLY NUMBER		QTY.	UNIT	TOTAL COST		COST PER S.F.
				UNIT	TOTAL	
C	**INTERIORS**					
C1010 102 5500	Core partitions: 6" Lt. Wt. concrete					
	block 288 L.F. x 11.5' x 8 Floors	26,449	S.F.	6.03	159,487	
C1010 144 0920	Plaster: [(196 L.F. x 8') + (144 L.F. x 9')] 8					
	+ Ext. Wall [(320 x 9 x 8 Floors) - 6500]	39,452	S.F.	2.17	85,611	
C1010 124 5400	Corridor partitions 1st + 2nd floors					
	steel studs + F.R. gypsum board	2,530	S.F.	2.70	6,831	
C1020 102 2600	Interior doors	66	Ea.	666	43,956	
	Wall finishes - Lobby: Mahogany w/furring					
C1010 128 0652	Furring: 150 L.F. x 9' + Ht.	1,350	S.F.	0.96	1,296	
C1030 210 0120	Towel Dispenser	16	Ea.	65.50	1,048	
C1030 210 0120	Grab Bar	32	Ea.	39.05	1,250	
C1030 210 0120	Mirror	31	Ea.	127.70	3,959	
C1020 210 0120	Toilet Tissue Dispenser	31	Ea.	25.40	787	
C1030 510 0140	Directory Board	8	Ea.	246	1,968	
C2010 110 0780	Stairs: Steel w/conc. fill	18	Flt.	6050	108,900	
C3010 105 0080	Mahogany paneling	1,350	S.F.	4.18	5,643	
	Paint, plaster + gypsum board					
C3020 410 1100	39,452 + [(220' x 9' x 2) - 1350] + 72	42,134	S.F.	0.82	34,550	
	Floor Finishes					
C3020 410 1720	1st floor lobby + terrazzo	2,175	S.F.	8.83	19,205	
	Remainder: Concrete - Tenant finished					
C3020 410 0060	2nd - 8th: Carpet - 4725 S.F. x 7	33,075	S.F.	3.92	129,654	
C3020 410 1720	Ceramic Tile - 300 S.F. x 7	2,100	S.F.	7.67	16,107	
C3030 130 2780	Suspended Ceiling Tiles- 5200 S.F. x 8	41,600	S.F.	1.46	60,736	
C3030 130 3260	Suspended Ceiling Grid- 5200 S.F. x 8	41,600	S.F.	1.18	49,088	
C1030 110 0700	Toilet partitions	31	Ea.	652	20,212	
0760	Handicapped addition	16	Ea.	286	4,576	
	Total				754,864	13.98
D	**SERVICES**					
D10	**Conveying**					
	2 Elevators - 2500# 200'/min. Geared					
D1010 140 1600	5 Floors $103,900					
1800	15 Floors $ 194,000					
	Diff. $90,100/10 = 9010/Flr.					
	$103,900 + (4 x 9,010) = $139,940/Ea.	2	Ea.	139,940	279,880	
	Total				279,880	5.18
D20	**Plumbing**					
D2010 310 1600	Plumbing - Lavatories	31	Ea.	970	30,070	
D2010 430 4340	-Service Sink	8	Ea.	1600	12,800	
D2010 210 2000	-Urinals	8	Ea.	930	7,440	

ASSEMBLY NUMBER		QTY.	UNIT	TOTAL COST		COST PER S.F.
				UNIT	TOTAL	
D2010 110 2080	-Water Closets	31	Ea.	1210	37,510	
	Water Control, Waste Vent Piping	45%			39,519	
D2020 210 4200	- Roof Drains, 4" C.I.	2	Ea.	975	1,950	
4240	- Pipe, 9 Flr. x 12' x 12' Ea.	216	L.F.	23.50	5,076	
	Total				134,365	2.49
D30	HVAC					
D3010 520 2000	10,000 S.F. @ $6.10/S.F. Interpolate					
2040	10,000 S.F. @ $2.78/S.F.	48,000	S.F.	4.44	213,120	
	Cooling - Chilled Water, Air Cool, Cond.					
D3030 115 4000	40,000 S.F. @8.89/S.F. Interpolate	48,000	S.F.	9	432,000	
4040	60,000 S.F. @9.07/S.F.					
	Total				645,120	11.95
D40	Fire Protection					
D4020 310 0560	Wet Stand Pipe: 4" x 10' 1st floor	12/10	Ea.	4200	5,040	
0580	Additional Floors: (12'/10 x 8)	9.6	Ea.	1155	11,088	
D4020 410 8400	Cabinet Assembly	9	Ea.	682	6,138	
	Total				22,266	0.41
D50	Electrical					
D5020 210 0280	Office Lighting, 15/1000 S.F. - 3 Watts/S.F.	48,000	S.F.	5.38	258,240	
0240	Basement Lighting, 10/1000 S.F. - 2 Watts/S.F.	6,000	S.F.	3.56	21,360	
D5020 110 0640	Office Receptacles 16.5/1000 S.F. - 2 Watts/S.F.	48,000	S.F.	3.09	148,320	
0560	Basement Receptacles 10/1000 S.F. - 1.2 Watts/S.F.	6,000	S.F.	2.35	14,100	
D5020 140 0280	Central A.C. - 4 Watts/S.F.	48,000	S.F.	0.40	19,200	
D5020 135 0320	Misc. Connections - 1.2 Watts/S.F.	48,000	S.F.	0.22	10,560	
D5020 145 0680	Elevator Motor Power - 10 H.P.	2	Ea.	2,125	4,250	
D5020 130 0280	Wall Switches - 2/1000 S.F.	54,000	S.F.	0.29	15,660	
D5010 120 0560	2000 Amp Service	1	Ea.	27,250	27,250	
D5010 240 0400	Switchgear	1	Ea.	38,300	38,300	
D5010 230 0560	Feeder	50	L.F.	328	16,400	
D5030 810 0400	Fire Detection System - 50 Detectors	1	Ea.	22,375	22,375	
D5090 210 0320	Emergency Generator - 30 kW	30	kW	438.50	13,155	
	Total				609,170	11.28
E	EQUIPMENT & FURNISHINGS					
G	BUILDING SITEWORK					
	Miscellaneous					
			Page Total			

Preliminary Estimate Cost Summary

PROJECT: Office Building	TOTAL AREA	54,000 S.F.	SHEET NO.
LOCATION	TOTAL VOLUME	648,000 C.F.	ESTIMATE NO.
ARCHITECT	COST PER S.F.		DATE
OWNER	COST PER C.F.		NO OF STORIES
QUANTITIES BY	EXTENSIONS BY		CHECKED BY

DIV	DESCRIPTION	SUBTOTAL COST	COST/S.F.	PERCENTAGE
A	SUBSTRUCTURE	$ 119,069	$ 2.20	
B10	SHELL: SUPERSTRUCTURE	$ 668,577	$ 12.38	
B20	SHELL: EXTERIOR ENCLOSURE	$ 688,925	$ 12.76	
B30	SHELL: ROOFING	$ 20,638	$ 0.38	
C	INTERIORS	$ 754,864	$ 13.98	
D10	SERVICES: CONVEYING	$ 279,880	$ 5.18	
D20	SERVICES: PLUMBING	$ 134,365	$ 2.49	
D30	SERVICES: HVAC	$ 645,120	$ 11.95	
D40	SERVICES: FIRE PROTECTION	$ 22,266	$ 0.41	
D50	SERVICES: ELECTRICAL	$ 609,170	$ 11.28	
E	EQUIPMENT & FURNISHINGS			
F	SPECIAL CONSTRUCTION & DEMO.			
G	BUILDING SITEWORK			
	BUILDING SUBTOTAL	$ 3,942,874		$ 3,942,874

Sales Tax   % x Subtotal /2   N/A $ -

General Conditions 10 % x Subtotal $ 394,287

                                        Subtotal "A" $ 4,337,161

Overhead   5 % x Subtotal "A" $ 216,858

                                        Subtotal "B" $ 4,554,019

Profit   10% x Subtotal "B" $ 455,402

                                        Subtotal "C" $ 5,009,421

Location Factor   % x Subtotal "C"   N/A   Localized Cost
    (Boston, MA)

Architects Fee 6.5% x Localized Cost = $ 325,612
Contingency   x Localized Cost =   N/A $ -
                                        Project Total Cost $ 5,335,034

Square Foot Cost $5,335,034/ 54,000 S.F. =   S.F. Cost $ 98.80
Cubic Foot Cost $5,335,034/ 648,000 C.F. =   C.F. Cost $ 8.23

A SUBSTRUCTURE

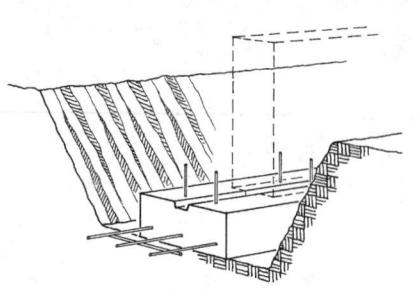

The Strip Footing System includes: excavation; hand trim; all forms needed for footing placement; forms for 2″ x 6″ keyway (four uses); dowels; and 3,000 p.s.i. concrete.

The footing size required varies for different soils. Soil bearing capacities are listed for 3 KSF and 6 KSF. Depths of the system range from 8″ to 24″. Widths range from 16″ to 96″. Smaller strip footings may not require reinforcement.

Please see the reference section for further design and cost information.

SUBSTRUCTURE **A**

A1010 110	Strip Footings	COST PER L.F.		
		MAT.	INST.	TOTAL
2100	Strip footing, load 2.6KLF, soil capacity 3KSF, 16″wide x 8″deep plain	4.68	9.65	14.33
2300	Load 3.9 KLF, soil capacity, 3 KSF, 24″wide x 8″deep, plain	5.95	10.70	16.65
2500	Load 5.1KLF, soil capacity 3 KSF, 24″wide x 12″deep, reinf.	9.40	16.20	25.60
2700	Load 11.1KLF, soil capacity 6 KSF, 24″wide x 12″deep, reinf.	9.40	16.20	25.60
2900	Load 6.8 KLF, soil capacity 3 KSF, 32″wide x 12″deep, reinf.	11.65	17.75	29.40
3100	Load 14.8 KLF, soil capacity 6 KSF, 32″wide x 12″deep, reinf.	11.65	17.75	29.40
3300	Load 9.3 KLF, soil capacity 3 KSF, 40″wide x 12″deep, reinf.	13.85	19.30	33.15
3500	Load 18.4 KLF, soil capacity 6 KSF, 40″wide x 12″deep, reinf.	13.95	19.45	33.40
4500	Load 10KLF, soil capacity 3 KSF, 48″wide x 16″deep, reinf.	19.95	24	43.95
4700	Load 22KLF, soil capacity 6 KSF, 48″wide, 16″deep, reinf.	20.50	25	45.50
5700	Load 15KLF, soil capacity 3 KSF, 72″wide x 20″deep, reinf.	35	34.50	69.50
5900	Load 33KLF, soil capacity 6 KSF, 72″wide x 20″deep, reinf.	37	37	74

A1010 Standard Foundations

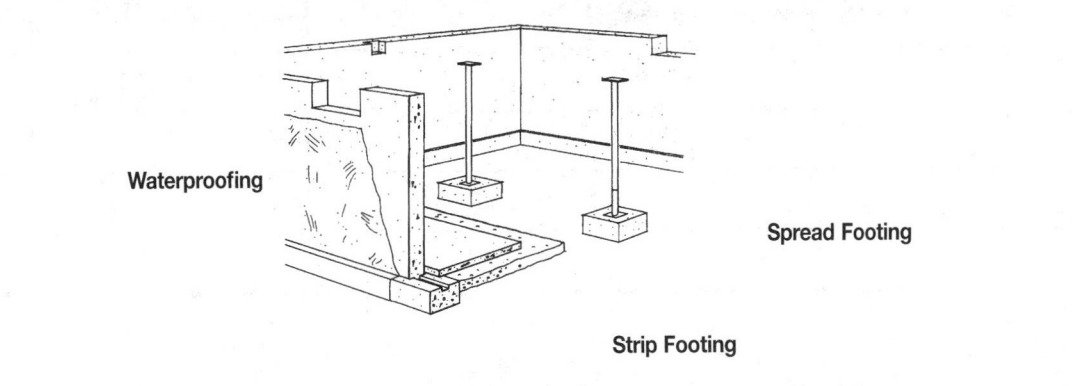

Waterproofing

Spread Footing

Strip Footing

A1010 210	Spread Footings	COST EACH		
		MAT.	INST.	TOTAL
7090	Spread footings, 3000 psi concrete, chute delivered			
7100	Load 25K, soil capacity 3 KSF, 3'-0" sq. x 12" deep	41	82	123
7150	Load 50K, soil capacity 3 KSF, 4'-6" sq. x 12" deep	86	143	229
7200	Load 50K, soil capacity 6 KSF, 3'-0" sq. x 12" deep	41	82	123
7250	Load 75K, soil capacity 3 KSF, 5'-6" sq. x 13" deep	135	202	337
7300	Load 75K, soil capacity 6 KSF, 4'-0" sq. x 12" deep	70	122	192
7350	Load 100K, soil capacity 3 KSF, 6'-0" sq. x 14" deep	170	242	412
7410	Load 100K, soil capacity 6 KSF, 4'-6" sq. x 15" deep	105	168	273
7450	Load 125K, soil capacity 3 KSF, 7'-0" sq. x 17" deep	269	345	614
7500	Load 125K, soil capacity 6 KSF, 5'-0" sq. x 16" deep	135	202	337
7550	Load 150K, soil capacity 3 KSF 7'-6" sq. x 18" deep	325	405	730
7610	Load 150K, soil capacity 6 KSF, 5'-6" sq. x 18" deep	179	253	432
7650	Load 200K, soil capacity 3 KSF, 8'-6" sq. x 20" deep	460	535	995
7700	Load 200K, soil capacity 6 KSF, 6'-0" sq. x 20" deep	234	310	544
7750	Load 300K, soil capacity 3 KSF, 10'-6" sq. x 25" deep	840	875	1,715
7810	Load 300K, soil capacity 6 KSF, 7'-6" sq. x 25" deep	440	530	970
7850	Load 400K, soil capacity 3 KSF, 12'-6" sq. x 28" deep	1,325	1,300	2,625
7900	Load 400K, soil capacity 6 KSF, 8'-6" sq. x 27" deep	610	685	1,295
8010	Load 500K, soil capacity 6 KSF, 9'-6" sq. x 30" deep	835	890	1,725
8100	Load 600K, soil capacity 6 KSF, 10'-6" sq. x 33" deep	1,125	1,150	2,275
8200	Load 700K, soil capacity 6 KSF, 11'-6" sq. x 36" deep	1,450	1,400	2,850
8300	Load 800K, soil capacity 6 KSF, 12'-0" sq. x 37" deep	1,625	1,550	3,175
8400	Load 900K, soil capacity 6 KSF, 13'-0" sq. x 39" deep	2,000	1,850	3,850
8500	Load 1000K, soil capacity 6 KSF, 13'-6" sq. x 41" deep	2,250	2,050	4,300

A1010 320	Foundation Dampproofing	COST PER L.F.		
		MAT.	INST.	TOTAL
1000	Foundation dampproofing, bituminous, 1 coat, 4' high	.28	3.04	3.32
1400	8' high	.56	6.05	6.61
1800	12' high	.84	9.40	10.24
2000	2 coats, 4' high	.44	3.76	4.20
2400	8' high	.88	7.50	8.38
2800	12' high	1.32	11.55	12.87
3000	Asphalt with fibers, 1/16" thick, 4' high	.68	3.76	4.44
3400	8' high	1.36	7.50	8.86
3800	12' high	2.04	11.55	13.59
4000	1/8" thick, 4' high	1.28	4.52	5.80
4400	8' high	2.56	9.05	11.61
4800	12' high	3.84	13.85	17.69
5000	Asphalt coated board and mastic, 1/4" thick, 4' high	2.56	4.08	6.64
5400	8' high	5.10	8.15	13.25
5800	12' high	7.70	12.50	20.20
6000	1/2" thick, 4' high	3.84	5.70	9.54

SUBSTRUCTURE

A

A1010 Standard Foundations

A1010 320	Foundation Dampproofing	COST PER L.F.		
		MAT.	INST.	TOTAL
6400	8' high	7.70	11.35	19.05
6800	12' high	11.50	17.30	28.80
7000	Cementitious coating, on walls, 1/8" thick coating, 4' high	6.60	4.20	10.80
7400	8' high	13.20	8.40	21.60
7800	12' high	19.80	12.60	32.40
8000	Cementitious/metallic slurry, 2 coat, 1/4" thick, 2' high	38.50	137	175.50
8400	4' high	77	274	351
8800	6' high	116	410	526

SUBSTRUCTURE A

Important: See the Reference Section for critical supporting data - Location Factors & Historical Cost Indexes

A1030 Slab on Grade

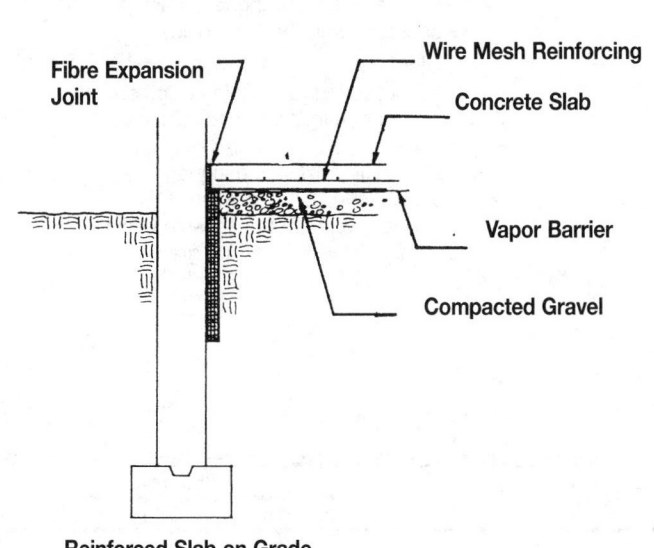

Fibre Expansion Joint

Wire Mesh Reinforcing

Concrete Slab

Vapor Barrier

Compacted Gravel

Reinforced Slab on Grade

A Slab on Grade system includes fine grading; 6″ of compacted gravel; vapor barrier; 3500 p.s.i. concrete; bituminous fiber expansion joint; all necessary edge forms
4 uses; steel trowel finish; and sprayed on membrane curing compound. Wire mesh reinforcing used in all reinforced slabs.

Non-industrial slabs are for foot traffic only with negligible abrasion. Light industrial slabs are for pneumatic wheels and light abrasion. Industrial slabs are for solid rubber wheels and moderate abrasion. Heavy industrial slabs are for steel wheels and severe abrasion.

A1030 120	Plain & Reinforced	COST PER S.F.		
		MAT.	INST.	TOTAL
2220	Slab on grade, 4″ thick, non industrial, non reinforced	1.29	1.78	3.07
2240	Reinforced	1.37	2.05	3.42
2260	Light industrial, non reinforced	1.65	2.18	3.83
2280	Reinforced	1.73	2.45	4.18
2300	Industrial, non reinforced	2.01	3.48	5.49
2320	Reinforced	2.09	3.75	5.84
3340	5″ thick, non industrial, non reinforced	1.52	1.83	3.35
3360	Reinforced	1.60	2.10	3.70
3380	Light industrial, non reinforced	1.89	2.23	4.12
3400	Reinforced	1.97	2.50	4.47
3420	Heavy industrial, non reinforced	2.60	4.04	6.64
3440	Reinforced	2.62	4.34	6.96
4460	6″ thick, non industrial, non reinforced	1.83	1.79	3.62
4480	Reinforced	1.95	2.15	4.10
4500	Light industrial, non reinforced	2.22	2.19	4.41
4520	Reinforced	2.41	2.68	5.09
4540	Heavy industrial, non reinforced	2.93	4.11	7.04
4560	Reinforced	3.05	4.47	7.52
5580	7″ thick, non industrial, non reinforced	2.09	1.85	3.94
5600	Reinforced	2.25	2.23	4.48
5620	Light industrial, non reinforced	2.48	2.25	4.73
5640	Reinforced	2.64	2.63	5.27
5660	Heavy industrial, non reinforced	3.18	4.03	7.21
5680	Reinforced	3.30	4.39	7.69
6700	8″ thick, non industrial, non reinforced	2.32	1.89	4.21
6720	Reinforced	2.45	2.21	4.66
6740	Light industrial, non reinforced	2.73	2.29	5.02
6760	Reinforced	2.86	2.61	5.47
6780	Heavy industrial, non reinforced	3.45	4.10	7.55
6800	Reinforced	3.64	4.45	8.09

SUBSTRUCTURE

A

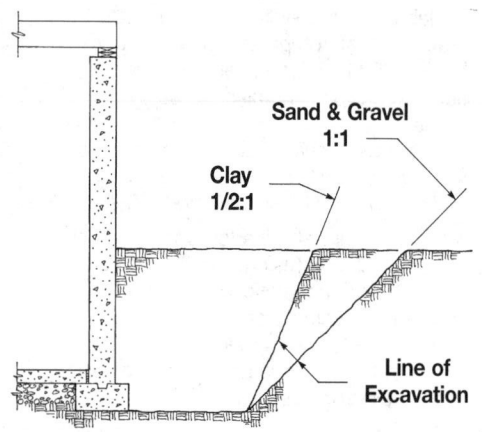

In general, the following items are accounted for in the table below.

Costs:
1) Excavation for building or other structure to depth and extent indicated.
2) Backfill compacted in place
3) Haul of excavated waste
4) Replacement of unsuitable backfill material with bank run gravel.

SUBSTRUCTURE A

A2010 110	Building Excavation & Backfill	COST PER S.F.		
		MAT.	INST.	TOTAL
2220	Excav & fill, 1000 S.F. 4' sand, gravel, or common earth, on site storage		1.71	1.71
2240	Off site storage		3.56	3.56
2260	Clay excavation, bank run gravel borrow for backfill	2.66	3.21	5.87
2280	8' deep, sand, gravel, or common earth, on site storage		3.96	3.96
2300	Off site storage		9.10	9.10
2320	Clay excavation, bank run gravel borrow for backfill	6.45	7.35	13.80
2340	16' deep, sand, gravel, or common earth, on site storage		10.95	10.95
2350	Off site storage		22	22
2360	Clay excavation, bank run gravel borrow for backfill	17.80	18.75	36.55
3380	4000 S.F.,4' deep, sand, gravel, or common earth, on site storage		1.31	1.31
3400	Off site storage		2.16	2.16
3420	Clay excavation, bank run gravel borrow for backfill	1.24	2.01	3.25
3440	8' deep, sand, gravel, or common earth, on site storage		2.87	2.87
3460	Off site storage		5.15	5.15
3480	Clay excavation, bank run gravel borrow for backfill	2.93	4.42	7.35
3500	16' deep, sand, gravel, or common earth, on site storage		6.90	6.90
3520	Off site storage		14.05	14.05
3540	Clay, excavation, bank run gravel borrow for backfill	7.85	10.50	18.35
4560	10,000 S.F., 4' deep, sand gravel, or common earth, on site storage		1.18	1.18
4580	Off site storage		1.70	1.70
4600	Clay excavation, bank run gravel borrow for backfill	.76	1.64	2.40
4620	8' deep, sand, gravel, or common earth, on site storage		2.49	2.49
4640	Off site storage		3.84	3.84
4660	Clay excavation, bank run gravel borrow for backfill	1.78	3.44	5.22
4680	16' deep, sand, gravel, or common earth, on site storage		5.70	5.70
4700	Off site storage		9.85	9.85
4720	Clay excavation, bank run gravel borrow for backfill	4.71	7.85	12.56
5740	30,000 S.F., 4' deep, sand, gravel, or common earth, on site storage		1.09	1.09
5760	Off site storage		1.38	1.38
5780	Clay excavation, bank run gravel borrow for backfill	.43	1.34	1.77
5860	16' deep, sand, gravel, or common earth, on site storage		4.94	4.94
5880	Off site storage		7.20	7.20
5900	Clay excavation, bank run gravel borrow for backfill	2.62	6.10	8.72
6910	100,000 S.F., 4' deep, sand, gravel, or common earth, on site storage		1.05	1.05
6940	8' deep, sand, gravel, or common earth, on site storage		2.12	2.12
6970	16' deep, sand, gravel, or common earth, on site storage		4.51	4.51
6980	Off site storage		5.70	5.70
6990	Clay excavation, bank run gravel borrow for backfill	1.41	5.10	6.51

Important: See the Reference Section for critical supporting data - Location Factors & Historical Cost Indexes

A2020 Basement Walls

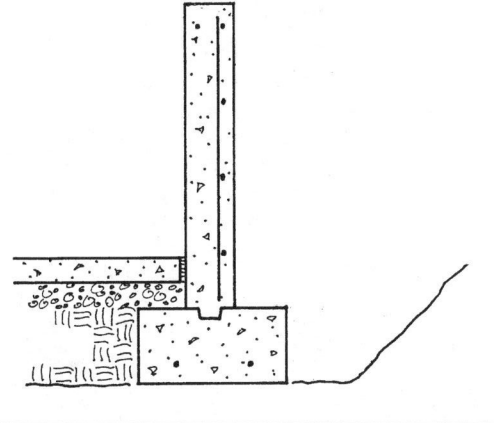

The Foundation Bearing Wall System includes: forms up to 16' high (four uses); 3,000 p.s.i. concrete placed and vibrated; and form removal with breaking form ties and patching walls. The wall systems list walls from 6" to 16" thick and are designed with minimum reinforcement.

Excavation and backfill are not included.

Please see the reference section for further design and cost information.

A2020 110			Walls, Cast in Place					
	WALL HEIGHT (FT.)	PLACING METHOD	CONCRETE (C.Y./L.F.)	REINFORCING (LBS./L.F.)	WALL THICKNESS (IN.)	COST PER L.F.		
						MAT.	INST.	TOTAL
1500	4'	direct chute	.074	3.3	6	12.05	31.50	43.55
1520			.099	4.8	8	14.40	32.50	46.90
1540			.123	6.0	10	16.60	33	49.60
1561			.148	7.2	12	18.85	34	52.85
1580			.173	8.1	14	21	34.50	55.50
1600			.197	9.44	16	23	35.50	58.50
3000	6'	direct chute	.111	4.95	6	18.10	47	65.10
3020			.149	7.20	8	21.50	48.50	70
3040			.184	9.00	10	25	49.50	74.50
3061			.222	10.8	12	28.50	51	79.50
5000	8'	direct chute	.148	6.6	6	24	63	87
5020			.199	9.6	8	29	65	94
5040			.250	12	10	33.50	66.50	100
5061			.296	14.39	12	37.50	68	105.50
6020	10'	direct chute	.248	12	8	36	81.50	117.50
6040			.307	14.99	10	41.50	83	124.50
6061			.370	17.99	12	47	85	132
7220	12'	pumped	.298	14.39	8	43.50	101	144.50
7240			.369	17.99	10	49.50	104	153.50
7262			.444	21.59	12	56.50	106	162.50
9220	16'	pumped	.397	19.19	8	57.50	134	191.50
9240			.492	23.99	10	66.50	138	204.50
9260			.593	28.79	12	75.50	143	218.50

For information about Means Estimating Seminars, see yellow pages 11 and 12 in back of book

SUBSTRUCTURE

A

B1010 Floor Construction

General: It is desirable for purposes of consistency and simplicity to maintain constant column sizes throughout building height. To do this, concrete strength may be varied (higher strength concrete at lower stories and lower strength concrete at upper stories), as well as varying the amount of reinforcing.

The table provides probably minimum column sizes with related costs and weight per lineal foot of story height.

B1010 203		C.I.P. Column, Square Tied						
	LOAD (KIPS)	STORY HEIGHT (FT.)	COLUMN SIZE (IN.)	COLUMN WEIGHT (P.L.F.)	CONCRETE STRENGTH (PSI)	COST PER V.L.F.		
						MAT.	INST.	TOTAL
0640	100	10	10	96	4000	7.45	31.50	38.95
0680		12	10	97	4000	7.25	31.50	38.75
0720		14	12	142	4000	9.45	38	47.45
0840	200	10	12	140	4000	9.55	38.50	48.05
0860		12	12	142	4000	9.50	38.50	48
0900		14	14	196	4000	11.90	44	55.90
0920	300	10	14	192	4000	12.20	44.50	56.70
0960		12	14	194	4000	12.10	44	56.10
0980		14	16	253	4000	13.70	48	61.70
1020	400	10	16	248	4000	14.90	50.50	65.40
1060		12	16	251	4000	14.75	50	64.75
1080		14	16	253	4000	14.65	50	64.65
1200	500	10	18	315	4000	19.95	61	80.95
1250		12	20	394	4000	20.50	63	83.50
1300		14	20	397	4000	20.50	63	83.50
1350	600	10	20	388	4000	24	69.50	93.50
1400		12	20	394	4000	23.50	69	92.50
1600		14	20	397	4000	23.50	68.50	92
3400	900	10	24	560	4000	34	90.50	124.50
3800		12	24	567	4000	33.50	89.50	123
4000		14	24	571	4000	33	89	122
7300	300	10	14	192	6000	12.05	44	56.05
7500		12	14	194	6000	12	44	56
7600		14	14	196	6000	11.90	44	55.90
8000	500	10	16	248	6000	14.90	50.50	65.40
8050		12	16	251	6000	14.75	50.50	65.25
8100		14	16	253	6000	14.65	50	64.65
8200	600	10	18	315	6000	17.95	58.50	76.45
8300		12	18	319	6000	17.80	58	75.80
8400		14	18	321	6000	17.65	58	75.65
8800	800	10	20	388	6000	20.50	63.50	84
8900		12	20	394	6000	20.50	63	83.50
9000		14	20	397	6000	20.50	63	83.50

B1010 Floor Construction

B1010 203	C.I.P. Column, Square Tied						
	LOAD (KIPS)	STORY HEIGHT (FT.)	COLUMN SIZE (IN.)	COLUMN WEIGHT (P.L.F.)	CONCRETE STRENGTH (PSI)	COST PER V.L.F.		
						MAT.	INST.	TOTAL
9100	900	10	20	388	6000	29	79.50	108.50
9300		12	20	394	6000	28.50	78.50	107
9600		14	20	397	6000	28	78	106

B1010 203	C.I.P. Column, Square Tied-Minimum Reinforcing						
	LOAD (KIPS)	STORY HEIGHT (FT.)	COLUMN SIZE (IN.)	COLUMN WEIGHT (P.L.F.)	CONCRETE STRENGTH (PSI)	COST PER V.L.F.		
						MAT.	INST.	TOTAL
9913	150	10-14	12	135	4000	9.30	38	47.30
9918	300	10-14	16	240	4000	13.60	48	61.60
9924	500	10-14	20	375	4000	20	62.50	82.50
9930	700	10-14	24	540	4000	29	80.50	109.50
9936	1000	10-14	28	740	4000	36.50	96	132.50
9942	1400	10-14	32	965	4000	47	111	158
9948	1800	10-14	36	1220	4000	57	128	185

B SHELL

B1010 Floor Construction

Concentric Load

Eccentric Load

General: Data presented here is for plant produced members transported 50 miles to 100 miles to the site and erected.

Design and pricing assumptions:
Normal wt. concrete, f'c = 5 KSI

Main reinforcement, fy = 60 KSI
Ties, fy = 40 KSI

Minimum design eccentricity, 0.1t.

Concrete encased structural steel haunches are assumed where practical; otherwise galvanized rebar haunches are assumed.

Base plates are integral with columns.

Foundation anchor bolts, nuts and washers are included in price.

B1010 207			Tied, Concentric Loaded Precast Concrete Columns					
	LOAD (KIPS)	STORY HEIGHT (FT.)	COLUMN SIZE (IN.)	COLUMN WEIGHT (P.L.F.)	LOAD LEVELS	COST PER V.L.F.		
						MAT.	INST.	TOTAL
0560	100	10	12x12	164	2	36	9.10	45.10
0570		12	12x12	162	2	35.50	7.55	43.05
0580		14	12x12	161	2	34.50	7.55	42.05
0590	150	10	12x12	166	3	35	7.55	42.55
0600		12	12x12	169	3	33	6.80	39.80
0610		14	12x12	162	3	33.50	6.80	40.30
0620	200	10	12x12	168	4	36.50	8.35	44.85
0630		12	12x12	170	4	37	7.55	44.55
0640		14	14x14	220	4	39	7.55	46.55

B1010 207			Tied, Eccentric Loaded Precast Concrete Columns					
	LOAD (KIPS)	STORY HEIGHT (FT.)	COLUMN SIZE (IN.)	COLUMN WEIGHT (P.L.F.)	LOAD LEVELS	COST PER V.L.F.		
						MAT.	INST.	TOTAL
1130	100	10	12x12	161	2	33	9.10	42.10
1140		12	12x12	159	2	35	7.55	42.55
1150		14	12x12	159	2	34.50	7.55	42.05
1390	600	10	18x18	385	4	65.50	8.35	73.85
1400		12	18x18	380	4	63.50	7.55	71.05
1410		14	18x18	375	4	63	7.55	70.55
1480	800	10	20x20	490	4	83.50	8.35	91.85
1490		12	20x20	480	4	82	7.55	89.55
1500		14	20x20	475	4	82	7.55	89.55

(A) Wide Flange

(B) Pipe

(C) Pipe, Concrete Filled

(G) Rectangular Tube, Concrete Filled

B1010 208					Steel Columns			
	LOAD (KIPS)	UNSUPPORTED HEIGHT (FT.)	WEIGHT (P.L.F.)	SIZE (IN.)	TYPE	COST PER V.L.F.		
						MAT.	INST.	TOTAL
1000	25	10	13	4	A	10.55	8	18.55
1020			7.58	3	B	6.15	8	14.15
1040			15	3-1/2	C	7.55	8	15.55
1120			20	4x3	G	8.70	8	16.70
1200		16	16	5	A	12	5.95	17.95
1220			10.79	4	B	8.10	5.95	14.05
1240			36	5-1/2	C	11.40	5.95	17.35
1320			64	8x6	G	17.70	5.95	23.65
1600	50	10	16	5	A	12.95	8	20.95
1620			14.62	5	B	11.85	8	19.85
1640			24	4-1/2	C	9.05	8	17.05
1720			28	6x3	G	11.60	8	19.60
1800		16	24	8	A	18	5.95	23.95
1840			36	5-1/2	C	11.40	5.95	17.35
1920			64	8x6	G	17.70	5.95	23.65
2000		20	28	8	A	19.90	5.95	25.85
2040			49	6-5/8	C	14.10	5.95	20.05
2120			64	8x6	G	16.80	5.95	22.75
2200	75	10	20	6	A	16.20	8	24.20
2240			36	4-1/2	C	22.50	8	30.50
2320			35	6x4	G	13.10	8	21.10
2400		16	31	8	A	23.50	5.95	29.45
2440			49	6-5/8	C	14.85	5.95	20.80
2520			64	8x6	G	17.70	5.95	23.65
2600		20	31	8	A	22	5.95	27.95
2640			81	8-5/8	C	21.50	5.95	27.45
2720			64	8x6	G	16.80	5.95	22.75
2800	100	10	24	8	A	19.45	8	27.45
2840			35	4-1/2	C	22.50	8	30.50
2920			46	8x4	G	16	8	24

SHELL

B

B1010 208	Steel Columns

	LOAD (KIPS)	UNSUPPORTED HEIGHT (FT.)	WEIGHT (P.L.F.)	SIZE (IN.)	TYPE	COST PER V.L.F.		
						MAT.	INST.	TOTAL
3000	100	16	31	8	A	23.50	5.95	29.45
3040			56	6-5/8	C	22	5.95	27.95
3120			64	8x6	G	17.70	5.95	23.65
3200		20	40	8	A	28.50	5.95	34.45
3240			81	8-5/8	C	21.50	5.95	27.45
3320			70	8x6	G	24	5.95	29.95
3400	125	10	31	8	A	25	8	33
3440			81	8	C	24	8	32
3520			64	8x6	G	19.10	8	27.10
3600		16	40	8	A	30	5.95	35.95
3640			81	8	C	22.50	5.95	28.45
3720			64	8x6	G	17.70	5.95	23.65
3800		20	48	8	A	34	5.95	39.95
3840			81	8	C	21.50	5.95	27.45
3920			60	8x6	G	24	5.95	29.95
4000	150	10	35	8	A	28.50	8	36.50
4040			81	8-5/8	C	24	8	32
4120			64	8x6	G	19.10	8	27.10
4200		16	45	10	A	34	5.95	39.95
4240			81	8-5/8	C	22.50	5.95	28.45
4320			70	8x6	G	25.50	5.95	31.45
4400		20	49	10	A	35	5.95	40.95
4440			123	10-3/4	C	30.50	5.95	36.45
4520			86	10x6	G	23.50	5.95	29.45
4600	200	10	45	10	A	36.50	8	44.50
4640			81	8-5/8	C	24	8	32
4720			70	8x6	G	27	8	35
4800		16	49	10	A	37	5.95	42.95
4840			123	10-3/4	C	32	5.95	37.95
4920			85	10x6	G	29.50	5.95	35.45
5200	300	10	61	14	A	49.50	8	57.50
5240			169	12-3/4	C	42.50	8	50.50
5320			86	10x6	G	40.50	8	48.50
5400		16	72	12	A	54	5.95	59.95
5440			169	12-3/4	C	39.50	5.95	45.45
5600		20	79	12	A	56	5.95	61.95
5640			169	12-3/4	C	37.50	5.95	43.45
5800	400	10	79	12	A	64	8	72
5840			178	12-3/4	C	55.50	8	63.50
6000		16	87	12	A	65.50	5.95	71.45
6040			178	12-3/4	C	51.50	5.95	57.45
6400	500	10	99	14	A	80.50	8	88.50
6600		16	109	14	A	82	5.95	87.95
6800		20	120	12	A	85	5.95	90.95
7000	600	10	120	12	A	97.50	8	105.50
7200		16	132	14	A	99	5.95	104.95
7400		20	132	14	A	94	5.95	99.95
7600	700	10	136	12	A	110	8	118
7800		16	145	14	A	109	5.95	114.95
8000		20	145	14	A	103	5.95	108.95
8200	800	10	145	14	A	118	8	126
8300		16	159	14	A	119	5.95	124.95
8400		20	176	14	A	125	5.95	130.95

Important: See the Reference Section for critical supporting data - Location Factors & Historical Cost Indexes

SHELL B

B **SHELL**

B10 Superstructure

B1010 Floor Construction

B1010 208	Steel Columns							
	LOAD (KIPS)	UNSUPPORTED HEIGHT (FT.)	WEIGHT (P.L.F.)	SIZE (IN.)	TYPE	COST PER V.L.F.		
						MAT.	INST.	TOTAL
8800	900	10	159	14	A	129	8	137
8900		16	176	14	A	132	5.95	137.95
9000		20	193	14	A	137	5.95	142.95
9100	1000	10	176	14	A	143	8	151
9200		16	193	14	A	145	5.95	150.95
9300		20	211	14	A	150	5.95	155.95

Description: Table below lists costs per S.F. of bay size for wood columns of various sizes and unsupported heights and the maximum allowable total load per S.F. per bay size.

B1010 210		Wood Columns						
	NOMINAL COLUMN SIZE (IN.)	BAY SIZE (FT.)	UNSUPPORTED HEIGHT (FT.)	MATERIAL (BF/M.S.F.)	TOTAL LOAD (P.S.F.)	COST PER S.F.		
						MAT.	INST.	TOTAL
1000	4 x 4	10 x 8	8	133	100	.24	.17	.41
1050			10	167	60	.30	.21	.51
1200		10 x 10	8	106	80	.19	.13	.32
1250			10	133	50	.24	.17	.41
1400		10 x 15	8	71	50	.13	.09	.22
1450			10	88	30	.16	.11	.27
1600		15 x 15	8	47	30	.08	.06	.14
1650			10	59	15	.10	.07	.17
2000	6 x 6	10 x 15	8	160	230	.34	.18	.52
2050			10	200	210	.43	.23	.66
2200		15 x 15	8	107	150	.23	.12	.35
2250			10	133	140	.28	.15	.43
2400		15 x 20	8	80	110	.17	.09	.26
2450			10	100	100	.21	.12	.33
2600		20 x 20	8	60	80	.13	.07	.20
2650			10	75	70	.16	.09	.25
2800		20 x 25	8	48	60	.10	.06	.16
2850			10	60	50	.13	.07	.20
3400	8 x 8	20 x 20	8	107	160	.24	.12	.36
3450			10	133	160	.30	.14	.44
3600		20 x 25	8	85	130	.15	.11	.26
3650			10	107	130	.19	.13	.32
3800		25 x 25	8	68	100	.16	.07	.23
3850			10	85	100	.21	.08	.29
4200	10 x 10	20 x 25	8	133	210	.32	.13	.45
4250			10	167	210	.40	.17	.57
4400		25 x 25	8	107	160	.26	.11	.37
4450			10	133	160	.32	.13	.45
4700	12 x 12	20 x 25	8	192	310	.47	.18	.65
4750			10	240	310	.59	.22	.81
4900		25 x 25	8	154	240	.38	.14	.52
4950			10	192	240	.47	.18	.65

Important: See the Reference Section for critical supporting data - Location Factors & Historical Cost Indexes

B1010 Floor Construction

"T" Shaped
Precast Beams

"L" Shaped
Precast Beams

B1010 214			"T" Shaped Precast Beams					
	SPAN (FT.)	SUPERIMPOSED LOAD (K.L.F.)	SIZE W X D (IN.)	BEAM WEIGHT (P.L.F.)	TOTAL LOAD (K.L.F.)	COST PER L.F.		
						MAT.	INST.	TOTAL
2300	15	2.8	12x16	260	3.06	106	16.10	122.10
2500		8.37	12x28	515	8.89	137	15.90	152.90
8900	45	3.34	12x60	1165	4.51	245	15.15	260.15
9900		6.5	24x60	1915	8.42	320	15.15	335.15

B1010 215			"L" Shaped Precast Beams					
	SPAN (FT.)	SUPERIMPOSED LOAD (K.L.F.)	SIZE W X D (IN.)	BEAM WEIGHT (P.L.F.)	TOTAL LOAD (K.L.F.)	COST PER L.F.		
						MAT.	INST.	TOTAL
2250	15	2.58	12x16	230	2.81	80.50	16.10	96.60
2400		5.92	12x24	370	6.29	98.50	15.15	113.65
4000	25	2.64	12x28	435	3.08	109	9.10	118.10
4450		6.44	18x36	790	7.23	149	11.40	160.40
5300	30	2.80	12x36	565	3.37	130	9.10	139.10
6400		8.66	24x44	1245	9.90	207	16.65	223.65

SHELL

B

B1010 Floor Construction

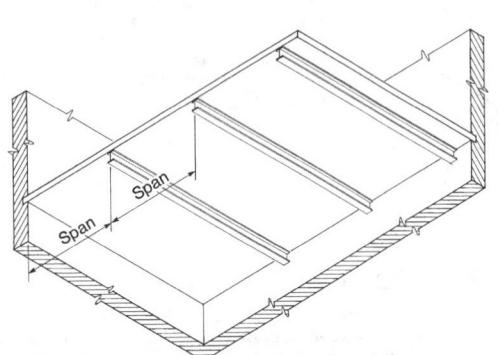

Cast in Place Floor Slab, One Way
General: Solid concrete slabs of uniform depth reinforced for flexure in one direction and for temperature and shrinkage in the other direction.

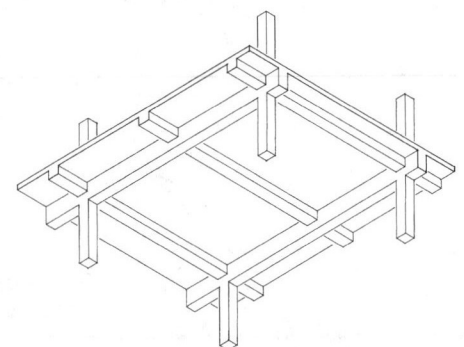

Cast in Place Beam & Slab, One Way
General: Solid concrete one way slab cast monolithically with reinforced concrete beams and girders.

B1010 217 — Cast in Place Slabs, One Way

	SLAB DESIGN & SPAN (FT.)	SUPERIMPOSED LOAD (P.S.F.)	THICKNESS (IN.)	TOTAL LOAD (P.S.F.)	COST PER S.F. MAT.	COST PER S.F. INST.	COST PER S.F. TOTAL
2500	Single 8	40	4	90	2.37	5.50	7.87
2600		75	4	125	2.45	5.95	8.40
2700		125	4-1/2	181	2.56	6	8.56
2800		200	5	262	2.85	6.20	9.05
3000	Single 10	40	4	90	2.51	5.90	8.41
3100		75	4	125	2.51	5.90	8.41
3200		125	5	188	2.81	6.05	8.86
3300		200	7-1/2	293	3.74	6.85	10.59
3500	Single 15	40	5-1/2	90	3.02	6.05	9.07
3600		75	6-1/2	156	3.35	6.35	9.70
3700		125	7-1/2	219	3.67	6.50	10.17
3800		200	8-1/2	306	3.99	6.65	10.64
4000	Single 20	40	7-1/2	115	3.64	6.35	9.99
4100		75	9	200	4.04	6.40	10.44
4200		125	10	250	4.46	6.65	11.11
4300		200	10	324	4.77	6.95	11.72

B1010 219 — Cast in Place Beam & Slab, One Way

	BAY SIZE (FT.)	SUPERIMPOSED LOAD (P.S.F.)	MINIMUM COL. SIZE (IN.)	SLAB THICKNESS (IN.)	TOTAL LOAD (P.S.F.)	COST PER S.F. MAT.	COST PER S.F. INST.	COST PER S.F. TOTAL
5000	20x25	40	12	5-1/2	121	3.60	7.55	11.15
5200		125	16	5-1/2	215	4.20	8.65	12.85
5500	25x25	40	12	6	129	3.78	7.45	11.23
5700		125	18	6	227	4.74	9.15	13.89
7500	30x35	40	16	8	158	4.68	8.45	13.13
7600		75	18	8	196	4.97	8.70	13.67
7700		125	22	8	254	5.50	9.65	15.15
7800		200	26	8	332	5.95	10.05	16
8000	35x35	40	16	9	169	5.20	8.70	13.90
8200		75	20	9	213	5.60	9.45	15.05
8400		125	24	9	272	6.10	9.90	16
8600		200	26	9	355	6.65	10.55	17.20

Important: See the Reference Section for critical supporting data - Location Factors & Historical Cost Indexes

B1010 Floor Construction

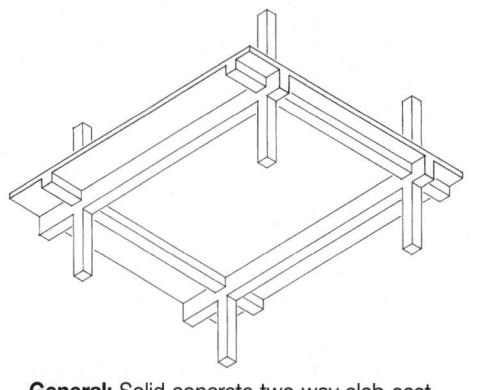

General: Solid concrete two way slab cast monolithically with reinforced concrete support beams and girders.

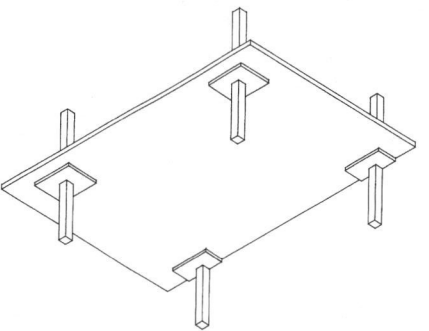

General: Flat Slab: Solid uniform depth concrete two way slabs with drop panels at columns and no column capitals.

B1010 220 — Cast in Place Beam & Slab, Two Way

	BAY SIZE (FT.)	SUPERIMPOSED LOAD (P.S.F.)	MINIMUM COL. SIZE (IN.)	SLAB THICKNESS (IN.)	TOTAL LOAD (P.S.F.)	COST PER S.F.		
						MAT.	INST.	TOTAL
4000	20 x 25	40	12	7	141	4.12	7.65	11.77
4300		75	14	7	181	4.67	8.35	13.02
4500		125	16	7	236	4.76	8.65	13.41
5100	25 x 25	40	12	7-1/2	149	4.30	7.70	12
5200		75	16	7-1/2	185	4.65	8.35	13
5300		125	18	7-1/2	250	5.05	9	14.05
7600	30 x 35	40	16	10	188	5.65	8.80	14.45
7700		75	18	10	225	5.95	9.15	15.10
8000		125	22	10	282	6.55	9.85	16.40
8500	35 x 35	40	16	10-1/2	193	6	9	15
8600		75	20	10-1/2	233	6.25	9.35	15.60
9000		125	24	10-1/2	287	6.95	10.05	17

B1010 222 — Cast in Place Flat Slab with Drop Panels

	BAY SIZE (FT.)	SUPERIMPOSED LOAD (P.S.F.)	MINIMUM COL. SIZE (IN.)	SLAB & DROP (IN.)	TOTAL LOAD (P.S.F.)	COST PER S.F.		
						MAT.	INST.	TOTAL
1960	20 x 20	40	12	7 - 3	132	3.74	6.10	9.84
1980		75	16	7 - 4	168	3.93	6.25	10.18
2000		125	18	7 - 6	221	4.34	6.50	10.84
3200	25 x 25	40	12	8-1/2 - 5-1/2	154	4.35	6.45	10.80
4000		125	20	8-1/2 - 8-1/2	243	4.85	6.90	11.75
4400		200	24	9 - 8-1/2	329	5.10	7.05	12.15
5000	25 x 30	40	14	9-1/2 - 7	168	4.71	6.65	11.36
5200		75	18	9-1/2 - 7	203	4.98	6.90	11.88
5600		125	22	9-1/2 - 8	256	5.20	7.10	12.30
6400	30 x 30	40	14	10-1/2 - 7-1/2	182	5.10	6.90	12
6600		75	18	10-1/2 - 7-1/2	217	5.35	7.15	12.50
6800		125	22	10-1/2 - 9	269	5.60	7.35	12.95
7400	30 x 35	40	16	11-1/2 - 9	196	5.50	7.15	12.65
7900		75	20	11-1/2 - 9	231	5.85	7.45	13.30
8000		125	24	11-1/2 - 11	284	6.05	7.60	13.65
9000	35 x 35	40	16	12 - 9	202	5.65	7.20	12.85
9400		75	20	12 - 11	240	6.05	7.55	13.60
9600		125	24	12 - 11	290	6.20	7.70	13.90

SHELL

B

B1010 Floor Construction

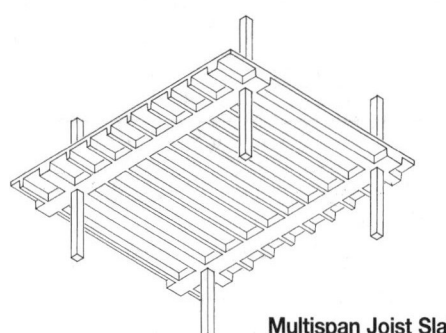

Multispan Joist Slab

General: Flat Plates: Solid uniform depth concrete two way slab without drops or interior beams. Primary design limit is shear at columns.

General: Combination of thin concrete slab and monolithic ribs at uniform spacing to reduce dead weight and increase rigidity.

B1010 223				Cast in Place Flat Plate				
	BAY SIZE (FT.)	SUPERIMPOSED LOAD (P.S.F.)	MINIMUM COL. SIZE (IN.)	SLAB THICKNESS (IN.)	TOTAL LOAD (P.S.F.)	COST PER S.F.		
						MAT.	INST.	TOTAL
3000	15 x 20	40	14	7	127	3.45	5.90	9.35
3400		75	16	7-1/2	169	3.67	6	9.67
3600		125	22	8-1/2	231	4.01	6.15	10.16
3800		175	24	8-1/2	281	4.03	6.15	10.18
4200	20 x 20	40	16	7	127	3.45	5.85	9.30
4400		75	20	7-1/2	175	3.69	6	9.69
4600		125	24	8-1/2	231	4.01	6.15	10.16
5000		175	24	8-1/2	281	4.04	6.20	10.24
5600	20 x 25	40	18	8-1/2	146	3.99	6.15	10.14
6000		75	20	9	188	4.12	6.25	10.37
6400		125	26	9-1/2	244	4.42	6.45	10.87
6600		175	30	10	300	4.60	6.50	11.10
7000	25 x 25	40	20	9	152	4.13	6.25	10.38
7400		75	24	9-1/2	194	4.35	6.40	10.75
7600		125	30	10	250	4.60	6.55	11.15

B1010 226				Cast in Place Multispan Joist Slab				
	BAY SIZE (FT.)	SUPERIMPOSED LOAD (P.S.F.)	MINIMUM COL. SIZE (IN.)	RIB DEPTH (IN.)	TOTAL LOAD (P.S.F.)	COST PER S.F.		
						MAT.	INST.	TOTAL
2000	15 x 15	40	12	8	115	3.48	7.15	10.63
2100		75	12	8	150	3.50	7.20	10.70
2200		125	12	8	200	3.58	7.25	10.83
2300		200	14	8	275	3.70	7.55	11.25
2600	15 x 20	40	12	8	115	3.53	7.20	10.73
2800		75	12	8	150	3.62	7.35	10.97
3000		125	14	8	200	3.77	7.70	11.47
3300		200	16	8	275	3.96	7.80	11.76
3600	20 x 20	40	12	10	120	3.62	7.05	10.67
3900		75	14	10	155	3.80	7.45	11.25
4000		125	16	10	205	3.84	7.60	11.44
4100		200	18	10	280	4.04	7.95	11.99
6200	30 x 30	40	14	14	131	4.09	7.40	11.49
6400		75	18	14	166	4.24	7.65	11.89
6600		125	20	14	216	4.49	8.05	12.54
6700		200	24	16	297	4.84	8.40	13.24

Important: See the Reference Section for critical supporting data - Location Factors & Historical Cost Indexes

SHELL B

B1010 Floor Construction

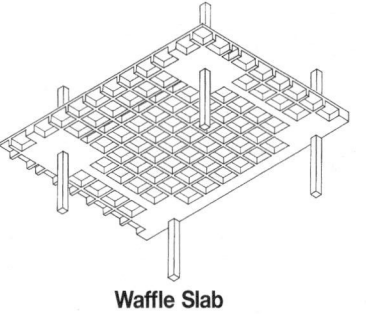

Waffle Slab

B1010 227	Cast in Place Waffle Slab						
	BAY SIZE (FT.)	SUPERIMPOSED LOAD (P.S.F.)	MINIMUM COL. SIZE (IN.)	RIB DEPTH (IN.)	TOTAL LOAD (P.S.F.)	COST PER S.F.		
						MAT.	INST.	TOTAL
3900	20 x 20	40	12	8	144	5.30	7.70	13
4000		75	12	8	179	5.40	7.80	13.20
4100		125	16	8	229	5.50	7.90	13.40
4200		200	18	8	304	5.75	8.15	13.90
4400	20 x 25	40	12	8	146	5.40	7.75	13.15
4500		75	14	8	181	5.55	7.85	13.40
4600		125	16	8	231	5.65	7.95	13.60
4700		200	18	8	306	5.85	8.20	14.05
4900	25 x 25	40	12	10	150	5.55	7.80	13.35
5000		75	16	10	185	5.70	7.95	13.65
5300		125	18	10	235	5.85	8.10	13.95
5500		200	20	10	310	6	8.25	14.25
5700	25 x 30	40	14	10	154	5.65	7.85	13.50
5800		75	16	10	189	5.80	8	13.80
5900		125	18	10	239	5.95	8.15	14.10
6000		200	20	12	329	6.55	8.55	15.10
6400	30 x 30	40	14	12	169	6.05	8.05	14.10
6500		75	18	12	204	6.20	8.20	14.40
6600		125	20	12	254	6.25	8.30	14.55
6700		200	24	12	329	6.80	8.75	15.55
6900	30 x 35	40	16	12	169	6.20	8.15	14.35
7000		75	18	12	204	6.20	8.15	14.35
7100		125	22	12	254	6.45	8.40	14.85
7200		200	26	14	334	7.15	8.95	16.10
7400	35 x 35	40	16	14	174	6.55	8.35	14.90
7500		75	20	14	209	6.65	8.50	15.15
7600		125	24	14	259	6.80	8.60	15.40
7700		200	26	16	346	7.25	9.05	16.30
8000	35 x 40	40	18	14	176	6.70	8.50	15.20
8300		75	22	14	211	6.90	8.70	15.60
8500		125	26	16	271	7.25	8.85	16.10
8750		200	30	20	372	7.90	9.40	17.30
9200	40 x 40	40	18	14	176	6.90	8.70	15.60
9400		75	24	14	211	7.15	8.95	16.10
9500		125	26	16	271	7.35	8.95	16.30
9700	40 x 45	40	20	16	186	7.15	8.80	15.95
9800		75	24	16	221	7.40	9.05	16.45
9900		125	28	16	271	7.50	9.15	16.65

SHELL

B

B1010   Floor Construction

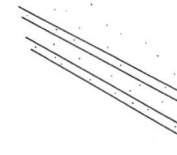

Precast Plank with No Topping

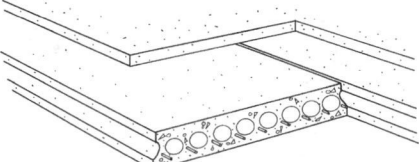

Precast Plank with 2″ Concrete Topping

B1010 229		**Precast Plank with No Topping**						
	SPAN (FT.)	SUPERIMPOSED LOAD (P.S.F.)	TOTAL DEPTH (IN.)	DEAD LOAD (P.S.F.)	TOTAL LOAD (P.S.F.)	COST PER S.F.		
						MAT.	INST.	TOTAL
0720	10	40	4	50	90	3.74	2.52	6.26
0750		75	6	50	125	5.05	2.16	7.21
0770		100	6	50	150	5.05	2.16	7.21
0800	15	40	6	50	90	5.05	2.16	7.21
0820		75	6	50	125	5.05	2.16	7.21
0850		100	6	50	150	5.05	2.16	7.21
0950	25	40	6	50	90	5.05	2.16	7.21
0970		75	8	55	130	5.50	1.89	7.39
1000		100	8	55	155	5.50	1.89	7.39
1200	30	40	8	55	95	5.50	1.89	7.39
1300		75	8	55	130	5.50	1.89	7.39
1400		100	10	70	170	4.15	1.69	5.84
1500	40	40	10	70	110	4.15	1.69	5.84
1600		75	12	70	145	6.20	1.51	7.71
1700	45	40	12	70	110	6.20	1.51	7.71

B1010 229		**Precast Plank with 2″ Concrete Topping**						
	SPAN (FT.)	SUPERIMPOSED LOAD (P.S.F.)	TOTAL DEPTH (IN.)	DEAD LOAD (P.S.F.)	TOTAL LOAD (P.S.F.)	COST PER S.F.		
						MAT.	INST.	TOTAL
2000	10	40	6	75	115	4.39	3.89	8.28
2100		75	8	75	150	5.70	3.53	9.23
2200		100	8	75	175	5.70	3.53	9.23
2500	15	40	8	75	115	5.70	3.53	9.23
2600		75	8	75	150	5.70	3.53	9.23
2700		100	8	75	175	5.70	3.53	9.23
3100	25	40	8	75	115	5.70	3.53	9.23
3200		75	8	75	150	5.70	3.53	9.23
3300		100	10	80	180	6.15	3.26	9.41
3400	30	40	10	80	120	6.15	3.26	9.41
3500		75	10	80	155	6.15	3.26	9.41
3600		100	10	80	180	6.15	3.26	9.41
4000	40	40	12	95	135	4.80	3.06	7.86
4500		75	14	95	170	6.85	2.88	9.73
5000	45	40	14	95	135	6.85	2.88	9.73

SHELL B

B1010 Floor Construction

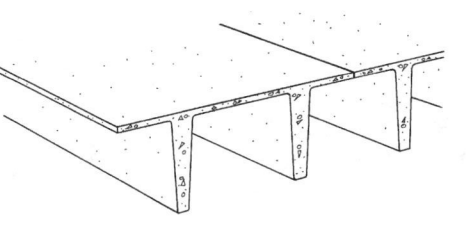

Precast Double "T" Beams with No Topping

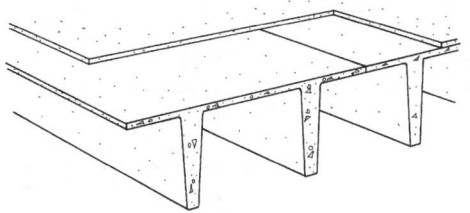

Precast Double "T" Beams with 2" Topping

Most widely used for moderate span floors and moderate and long span roofs. At shorter spans, they tend to be competitive with hollow core slabs. They are also used as wall panels.

B1010 234		Precast Double "T" Beams with No Topping						
	SPAN (FT.)	SUPERIMPOSED LOAD (P.S.F.)	DBL. "T" SIZE D (IN.) W (FT.)	CONCRETE "T" TYPE	TOTAL LOAD (P.S.F.)	COST PER S.F.		
						MAT.	INST.	TOTAL
4300	50	30	20x8	Lt. Wt.	66	5.35	1.16	6.51
4400		40	20x8	Lt. Wt.	76	5.75	1.05	6.80
4500		50	20x8	Lt. Wt.	86	6	1.27	7.27
4600		75	20x8	Lt. Wt.	111	6.30	1.44	7.74
5600	70	30	32x10	Lt. Wt.	78	6.20	.80	7
5750		40	32x10	Lt. Wt.	88	6.40	.96	7.36
5900		50	32x10	Lt. Wt.	98	6.60	1.11	7.71
6000		75	32x10	Lt. Wt.	123	6.95	1.35	8.30
6100		100	32x10	Lt. Wt.	148	7.60	1.83	9.43
6200	80	30	32x10	Lt. Wt.	78	6.60	1.11	7.71
6300		40	32x10	Lt. Wt.	88	6.95	1.35	8.30
6400		50	32x10	Lt. Wt.	98	7.30	1.59	8.89

B1010 234		Precast Double "T" Beams With 2" Topping						
	SPAN (FT.)	SUPERIMPOSED LOAD (P.S.F.)	DBL. "T" SIZE D (IN.) W (FT.)	CONCRETE "T" TYPE	TOTAL LOAD (P.S.F.)	COST PER S.F.		
						MAT.	INST.	TOTAL
7100	40	30	18x8	Reg. Wt.	120	5.05	2.26	7.31
7200		40	20x8	Reg. Wt.	130	4.83	2.15	6.98
7300		50	20x8	Reg. Wt.	140	5.10	2.36	7.46
7400		75	20x8	Reg. Wt.	165	5.25	2.45	7.70
7500		100	20x8	Reg. Wt.	190	5.65	2.76	8.41
7550	50	30	24x8	Reg. Wt.	120	5.35	2.26	7.61
7600		40	24x8	Reg. Wt.	130	5.45	2.33	7.78
7750		50	24x8	Reg. Wt.	140	5.50	2.35	7.85
7800		75	24x8	Reg. Wt.	165	5.90	2.65	8.55
7900		100	32x10	Reg. Wt.	189	6.25	2.15	8.40

SHELL

B

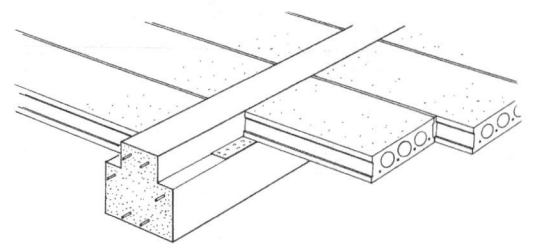

Precast Beam and Plank with No Topping

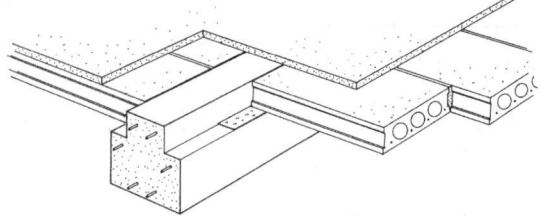

Precast Beam and Plank with 2″ Topping

B1010 236		Precast Beam & Plank with No Topping						
	BAY SIZE (FT.)	SUPERIMPOSED LOAD (P.S.F.)	PLANK THICKNESS (IN.)	TOTAL DEPTH (IN.)	TOTAL LOAD (P.S.F.)	COST PER S.F.		
						MAT.	INST.	TOTAL
6200	25x25	40	8	28	118	11.85	3.40	15.25
6400		75	8	36	158	12.45	3.40	15.85
6500		100	8	36	183	12.45	3.40	15.85
7000	25x30	40	8	36	110	11.35	3.40	14.75
7200		75	10	36	159	10.25	3.20	13.45
7400		100	12	36	188	12.90	3.02	15.92
7600	30x30	40	8	36	121	11.75	3.40	15.15
8000		75	10	44	140	12.40	3.40	15.80
8250		100	12	52	206	14.10	3.02	17.12
8500	30x35	40	12	44	135	11.95	3.02	14.97
8750		75	12	52	176	12.80	3.02	15.82
9000	35x35	40	12	52	141	12.80	3.02	15.82
9250		75	12	60	181	13.55	3.02	16.57
9500	35x40	40	12	52	137	12.20	3.79	15.99

B1010 238		Precast Beam & Plank with 2″ Topping						
	BAY SIZE (FT.)	SUPERIMPOSED LOAD (P.S.F.)	PLANK THICKNESS (IN.)	TOTAL DEPTH (IN.)	TOTAL LOAD (P.S.F.)	COST PER S.F.		
						MAT.	INST.	TOTAL
4300	20x20	40	6	22	135	12	5	17
4400		75	6	24	173	12.70	5	17.70
4500		100	6	28	200	13.15	5	18.15
4600	20x25	40	6	26	134	11.25	5	16.25
5000		75	8	30	177	12.25	4.73	16.98
5200		100	8	30	202	12.25	4.73	16.98
5400	25x25	40	6	38	143	12.65	4.96	17.61
5600		75	8	38	183	12.65	4.96	17.61
6000		100	8	46	216	14.15	4.69	18.84
6200	25x30	40	8	38	144	11.95	4.67	16.62
6400		75	10	46	200	11.50	4.47	15.97
6600		100	10	46	225	11.50	4.47	15.97
7000	30x30	40	8	46	150	13	4.66	17.66
7200		75	10	54	181	14	4.66	18.66
7600		100	10	54	231	14	4.66	18.66
7800	30x35	40	10	54	166	11.35	4.46	15.81
8000		75	12	54	200	13.15	4.28	17.43

SHELL B

Important: See the Reference Section for critical supporting data - Location Factors & Historical Cost Indexes

B1010 Floor Construction

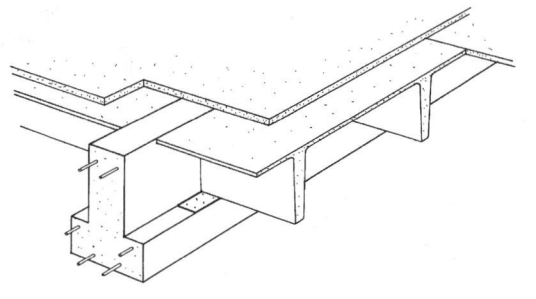

General: Beams and double tees priced here are for plant produced prestressed members transported to the site and erected.

The 2″ structural topping is applied after the beams and double tees are in place and is reinforced with W.W.F.

Note: Deduct from prices 20% for Southern States. Add to prices 10% for Western States.

With Topping

B1010 239		Precast Double "T" & 2″ Topping on Precast Beams						
	BAY SIZE (FT.)	SUPERIMPOSED LOAD (P.S.F.)	DEPTH (IN.)		TOTAL LOAD (P.S.F.)	COST PER S.F.		
						MAT.	INST.	TOTAL
3000	25x30	40	38		130	11.10	4.05	15.15
3100		75	38		168	11.10	4.04	15.14
3300		100	46		196	11.70	4.04	15.74
3600	30x30	40	46		150	12.15	4.02	16.17
3750		75	46		174	12.15	4.02	16.17
4000		100	54		203	12.85	4.02	16.87
4100	30x40	40	46		136	9.90	3.80	13.70
4300		75	54		173	10.40	3.80	14.20
4400		100	62		204	11.05	3.80	14.85
4600	30x50	40	54		138	9.40	3.69	13.09
4800		75	54		181	10	3.69	13.69
5000		100	54		219	11	3.47	14.47
5200	30x60	40	62		151	9.90	3.47	13.37
5400		75	62		192	10.55	3.47	14.02
5600		100	62		215	10.50	3.47	13.97
5800	35x40	40	54		139	10.65	3.80	14.45
6000		75	62		179	11.75	3.70	15.45
6250		100	62		212	12.10	3.70	15.80
6500	35x50	40	62		142	10.15	3.69	13.84
6750		75	62		186	10.80	3.69	14.49
7300		100	62		231	12.60	3.47	16.07
7600	35x60	40	54		154	11.35	3.37	14.72
7750		75	54		179	11.75	3.37	15.12
8000		100	62		224	12.45	3.37	15.82
8250	40x40	40	62		145	11.90	4.57	16.47
8400		75	62		187	12.55	4.57	17.12
8750		100	62		223	13.70	4.57	18.27
9000	40x50	40	62		151	11.15	4.44	15.59
9300		75	62		193	12	3.82	15.82
9800	40x60	40	62		164	14.20	3.97	18.17

SHELL

B

General: The following table is based upon structural wide flange (WF) beam and girder framing. Non-composite action is assumed between WF framing and decking. Deck costs not included.

The deck spans the short direction. The steel beams and girders are fireproofed with sprayed fiber fireproofing.

No columns included in price.

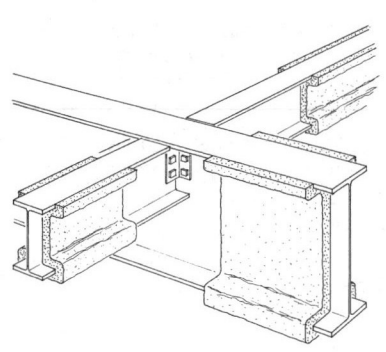

B1010 241	W Shape Beams & Girders						
	BAY SIZE (FT.) BEAM X GIRD	SUPERIMPOSED LOAD (P.S.F.)	STEEL FRAMING DEPTH (IN.)	FIREPROOFING (S.F./S.F.)	TOTAL LOAD (P.S.F.)	COST PER S.F.		
						MAT.	INST.	TOTAL
1350	15x20	40	12	.535	50	2.45	1.37	3.82
1400		40	16	.65	90	3.19	1.76	4.95
1450		75	18	.694	125	4.17	2.23	6.40
1500		125	24	.796	175	5.75	3.05	8.80
2000	20x20	40	12	.55	50	2.72	1.50	4.22
2050		40	14	.579	90	3.70	1.96	5.66
2100		75	16	.672	125	4.44	2.33	6.77
2150		125	16	.714	175	5.30	2.80	8.10
2900	20x25	40	16	.53	50	2.99	1.61	4.60
2950		40	18	.621	96	4.69	2.42	7.11
3000		75	18	.651	131	5.40	2.74	8.14
3050		125	24	.77	200	7.10	3.68	10.78
5200	25x25	40	18	.486	50	3.25	1.70	4.95
5300		40	18	.592	96	4.81	2.47	7.28
5400		75	21	.668	131	5.80	2.94	8.74
5450		125	24	.738	191	7.65	3.92	11.57
6300	25x30	40	21	.568	50	3.70	1.95	5.65
6350		40	21	.694	90	5	2.59	7.59
6400		75	24	.776	125	6.40	3.26	9.66
6450		125	30	.904	175	7.70	4.03	11.73
7700	30x30	40	21	.52	50	3.96	2.03	5.99
7750		40	24	.629	103	6.10	3.04	9.14
7800		75	30	.715	138	7.20	3.60	10.80
7850		125	36	.822	206	9.45	4.80	14.25
8250	30x35	40	21	.508	50	4.50	2.28	6.78
8300		40	24	.651	109	6.65	3.31	9.96
8350		75	33	.732	150	8.05	3.99	12.04
8400		125	36	.802	225	10	5.05	15.05
9550	35x35	40	27	.560	50	4.67	2.37	7.04
9600		40	36	.706	109	8.45	4.15	12.60
9650		75	36	.750	150	8.75	4.32	13.07
9820		125	36	.797	225	11.65	5.80	17.45

SHELL B

B1010   Floor Construction

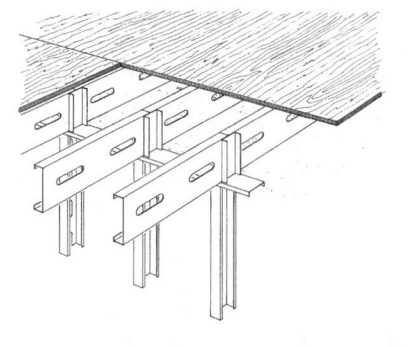

Description: Table below lists costs for light gage CEE or PUNCHED DOUBLE joists to suit the span and loading with the minimum thickness subfloor required by the joist spacing.

B1010 244	Light Gauge Steel Floor Systems							
	SPAN (FT.)	SUPERIMPOSED LOAD (P.S.F.)	FRAMING DEPTH (IN.)	FRAMING SPAC. (IN.)	TOTAL LOAD (P.S.F.)	COST PER S.F.		
						MAT.	INST.	TOTAL
1500	15	40	8	16	54	1.68	1.48	3.16
1550			8	24	54	1.78	1.52	3.30
1600		65	10	16	80	2.13	1.79	3.92
1650			10	24	80	2.14	1.75	3.89
1700		75	10	16	90	2.13	1.79	3.92
1750			10	24	90	2.14	1.75	3.89
1800		100	10	16	116	2.71	2.27	4.98
1850			10	24	116	2.14	1.75	3.89
1900		125	10	16	141	2.71	2.27	4.98
1950			10	24	141	2.14	1.75	3.89
2500	20	40	8	16	55	1.90	1.67	3.57
2550			8	24	55	1.78	1.52	3.30
2600		65	8	16	80	2.17	1.90	4.07
2650			10	24	80	2.15	1.76	3.91
2700		75	10	16	90	2.02	1.70	3.72
2750			12	24	90	2.37	1.56	3.93
2800		100	10	16	115	2.02	1.70	3.72
2850			10	24	116	2.67	2.01	4.68
2900		125	12	16	142	3.05	1.98	5.03
2950			12	24	141	2.88	2.13	5.01
3500	25	40	10	16	55	2.13	1.79	3.92
3550			10	24	55	2.14	1.75	3.89
3600		65	10	16	81	2.71	2.27	4.98
3650			12	24	81	2.60	1.70	4.30
3700		75	12	16	92	3.04	1.98	5.02
3750			12	24	91	2.88	2.13	5.01
3800		100	12	16	117	3.40	2.18	5.58
3850		125	12	16	143	3.85	2.84	6.69
4500	30	40	12	16	57	3.04	1.98	5.02
4550			12	24	56	2.59	1.69	4.28
4600		65	12	16	82	3.38	2.17	5.55
4650		75	12	16	92	3.38	2.17	5.55

B1010 Floor Construction

Description: The table below lists costs per S.F. for a floor system on bearing walls using open web steel joists, galvanized steel slab form and 2-1/2″ concrete slab reinforced with welded wire fabric. Costs of the bearing walls are not included.

B1010 246		Deck & Joists on Bearing Walls						
	SPAN (FT.)	SUPERIMPOSED LOAD (P.S.F.)	JOIST SPACING FT./IN.	DEPTH (IN.)	TOTAL LOAD (P.S.F.)	COST PER S.F.		
						MAT.	INST.	TOTAL
1050	20	40	2-0	14-1/2	83	3.02	2.73	5.75
1070		65	2-0	16-1/2	109	3.28	2.88	6.16
1100		75	2-0	16-1/2	119	3.28	2.88	6.16
1120		100	2-0	18-1/2	145	3.30	2.90	6.20
1150		125	1-9	18-1/2	170	4.04	3.44	7.48
1170	25	40	2-0	18-1/2	84	3.64	3.20	6.84
1200		65	2-0	20-1/2	109	3.77	3.28	7.05
1220		75	2-0	20-1/2	119	3.67	3.10	6.77
1250		100	2-0	22-1/2	145	3.73	3.12	6.85
1270		125	1-9	22-1/2	170	4.29	3.42	7.71
1300	30	40	2-0	22-1/2	84	3.73	2.80	6.53
1320		65	2-0	24-1/2	110	4.06	2.93	6.99
1350		75	2-0	26-1/2	121	4.22	2.99	7.21
1370		100	2-0	26-1/2	146	4.52	3.10	7.62
1400		125	2-0	24-1/2	172	4.92	4.07	8.99
1420	35	40	2-0	26-1/2	85	4.64	3.89	8.53
1450		65	2-0	28-1/2	111	4.80	3.99	8.79
1470		75	2-0	28-1/2	121	4.80	3.99	8.79
1500		100	1-11	28-1/2	147	5.25	4.26	9.51
1520		125	1-8	28-1/2	172	5.75	4.59	10.34
1550	5/8″ gyp. fireproof.							
1560	On metal furring, add					.46	2.02	2.48

Important: See the Reference Section for critical supporting data - Location Factors & Historical Cost Indexes

B1010 Floor Construction

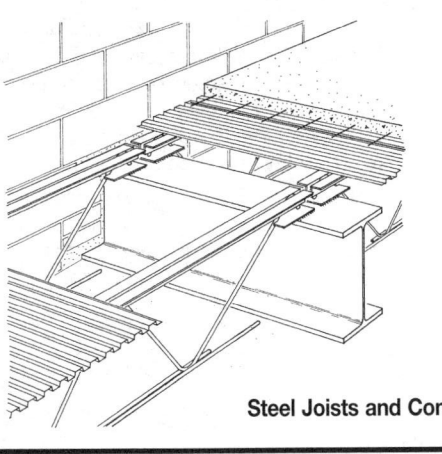

The table below lists costs for a floor system on exterior bearing walls and interior columns and beams using open web steel joists, galvanized steel slab form, 2-1/2″ concrete slab reinforced with welded wire fabric.

Columns are sized to accommodate one floor plus roof loading but costs for columns are from suspended floor to slab on grade only. Costs of the bearing walls are not included.

Steel Joists and Concrete Slab on Walls and Beams

B1010 248			Steel Joists on Beam & Wall					
	BAY SIZE (FT.)	SUPERIMPOSED LOAD (P.S.F.)	DEPTH (IN.)	TOTAL LOAD (P.S.F.)	COLUMN ADD	COST PER S.F.		
						MAT.	INST.	TOTAL
1720	25x25	40	23	84		4.43	3.44	7.87
1730					columns	.19	.09	.28
1750	25x25	65	29	110		4.67	3.57	8.24
1760					columns	.19	.09	.28
1770	25x25	75	26	120		4.81	3.50	8.31
1780					columns	.22	.11	.33
1800	25x25	100	29	145		5.45	3.84	9.29
1810					columns	.22	.11	.33
1820	25x25	125	29	170		5.70	3.96	9.66
1830					columns	.25	.11	.36
2020	30x30	40	29	84		4.74	3.15	7.89
2030					columns	.17	.08	.25
2050	30x30	65	29	110		5.35	3.40	8.75
2060					columns	.17	.08	.25
2070	30x30	75	32	120		5.55	3.47	9.02
2080					columns	.20	.09	.29
2100	30x30	100	35	145		6.10	3.71	9.81
2110					columns	.23	.11	.34
2120	30x30	125	35	172		6.70	4.76	11.46
2130					columns	.28	.12	.40
2170	30x35	40	29	85		5.40	3.40	8.80
2180					columns	.17	.08	.25
2200	30x35	65	29	111		6.05	4.45	10.50
2210					columns	.19	.08	.27
2220	30x35	75	32	121		6.05	4.45	10.50
2230					columns	.19	.09	.28
2250	30x35	100	35	148		6.55	3.87	10.42
2260					columns	.24	.11	.35
2320	35x35	40	32	85		5.55	3.47	9.02
2330					columns	.17	.08	.25
2350	35x35	65	35	111		6.35	4.61	10.96
2360					columns	.20	.09	.29
2370	35x35	75	35	121		6.50	4.68	11.18
2380					columns	.20	.09	.29
2400	35x35	100	38	148		6.75	4.85	11.60
2410					columns	.25	.12	.37
2460	5/8 gyp. fireproof.							
2475	On metal furring, add					.46	2.02	2.48

B1010   Floor Construction

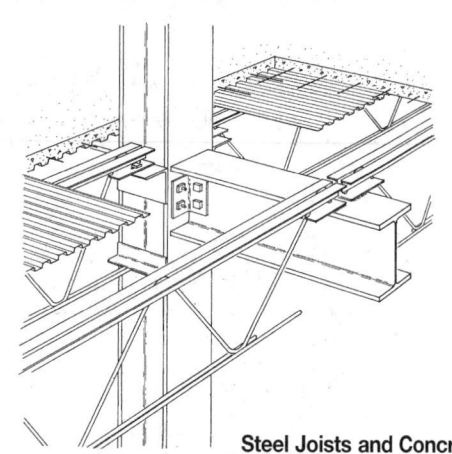

The table below lists costs per S.F. for a floor system on steel columns and beams using open web steel joists, galvanized steel slab form, 2-1/2″ concrete slab reinforced with welded wire fabric.

Columns are sized to accommodate one floor plus roof loading but costs for columns are from floor to grade only.

Steel Joists and Concrete Slab on Steel Columns and Beams

B1010 250	Steel Joists, Beams & Slab on Columns							
	BAY SIZE (FT.)	SUPERIMPOSED LOAD (P.S.F.)	DEPTH (IN.)	TOTAL LOAD (P.S.F.)	COLUMN ADD	COST PER S.F.		
						MAT.	INST.	TOTAL
2350	15x20	40	17	83		4.23	3.18	7.41
2400					column	.63	.30	.93
2450	15x20	65	19	108		4.66	3.38	8.04
2500					column	.63	.30	.93
2550	15x20	75	19	119		4.87	3.49	8.36
2600					column	.69	.32	1.01
2650	15x20	100	19	144		5.15	3.64	8.79
2700					column	.69	.32	1.01
2750	15x20	125	19	170		5.80	4.09	9.89
2800					column	.92	.43	1.35
2850	20x20	40	19	83		4.57	3.32	7.89
2900					column	.52	.24	.76
2950	20x20	65	23	109		5.05	3.56	8.61
3000					column	.69	.32	1.01
3100	20x20	75	26	119		5.30	3.68	8.98
3200					column	.69	.32	1.01
3400	20x20	100	23	144		5.50	3.77	9.27
3450					column	.69	.32	1.01
3500	20x20	125	23	170		6.10	4.08	10.18
3600					column	.83	.38	1.21
3700	20x25	40	44	83		5.25	3.80	9.05
3800					column	.55	.26	.81
3900	20x25	65	26	110		5.70	4.03	9.73
4000					column	.55	.26	.81
4100	20x25	75	26	120		5.60	3.85	9.45
4200					column	.66	.31	.97
4300	20x25	100	26	145		5.95	4.01	9.96
4400					column	.66	.31	.97
4500	20x25	125	29	170		6.65	4.38	11.03
4600					column	.77	.36	1.13
4700	25x25	40	23	84		5.60	3.96	9.56
4800					column	.53	.24	.77
4900	25x25	65	29	110		5.95	4.13	10.08
5000					column	.53	.24	.77
5100	25x25	75	26	120		6.20	4.12	10.32
5200					column	.62	.28	.90
5300	25x25	100	29	145		6.90	4.49	11.39
5400					column	.62	.28	.90

Important: See the Reference Section for critical supporting data - Location Factors & Historical Cost Indexes

B1010 Floor Construction

| B1010 250 | | Steel Joists, Beams & Slab on Columns | | | | | | |

	BAY SIZE (FT.)	SUPERIMPOSED LOAD (P.S.F.)	DEPTH (IN.)	TOTAL LOAD (P.S.F.)	COLUMN ADD	COST PER S.F.		
						MAT.	INST.	TOTAL
5500	25x25	125	32	170		7.30	4.67	11.97
5600					column	.69	.32	1.01
5700	25x30	40	29	84		5.85	4.16	10.01
5800					column	.52	.24	.76
5900	25x30	65	29	110		6.10	4.34	10.44
6000					column	.52	.24	.76
6050	25x30	75	29	120		6.55	3.93	10.48
6100					column	.57	.27	.84
6150	25x30	100	29	145		7.10	4.16	11.26
6200					column	.57	.27	.84
6250	25x30	125	32	170		7.60	5.15	12.75
6300					column	.66	.31	.97
6350	30x30	40	29	84		6.05	3.75	9.80
6400					column	.48	.22	.70
6500	30x30	65	29	110		6.90	4.10	11
6600					column	.48	.22	.70
6700	30x30	75	32	120		7.05	4.16	11.21
6800					column	.55	.26	.81
6900	30x30	100	35	145		7.80	4.48	12.28
7000					column	.64	.30	.94
7100	30x30	125	35	172		8.55	5.60	14.15
7200					column	.71	.32	1.03
7300	30x35	40	29	85		6.90	4.07	10.97
7400					column	.41	.19	.60
7500	30x35	65	29	111		7.65	5.15	12.80
7600					column	.53	.24	.77
7700	30x35	75	32	121		7.65	5.15	12.80
7800					column	.54	.24	.78
7900	30x35	100	35	148		8.30	4.66	12.96
8000					column	.66	.31	.97
8100	30x35	125	38	173		9.25	5.05	14.30
8200					column	.67	.31	.98
8300	35x35	40	32	85		7.05	4.15	11.20
8400					column	.47	.22	.69
8500	35x35	65	35	111		8.05	5.35	13.40
8600					column	.56	.26	.82
9300	35x35	75	38	121		8.25	5.45	13.70
9400					column	.56	.26	.82
9500	35x35	100	38	148		8.85	5.80	14.65
9600					column	.70	.32	1.02
9750	35x35	125	41	173		9.60	5.25	14.85
9800					column	.71	.32	1.03
9810	5/8 gyp. fireproof.							
9815	On metal furring, add					.46	2.02	2.48

SHELL

B

B1010 Floor Construction

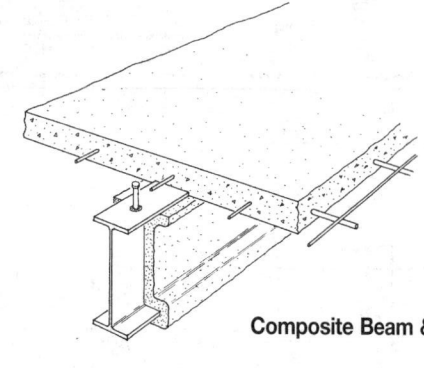

General: Composite construction of wide flange beams and concrete slabs is most efficiently used when loads are heavy and spans are moderately long. It is stiffer with less deflection than non-composite construction of similar depth and spans.

Composite Beam & Cast in Place Slab

B1010 252		Composite Beam & Cast in Place Slab						
	BAY SIZE (FT.)	SUPERIMPOSED LOAD (P.S.F.)	SLAB THICKNESS (IN.)	TOTAL DEPTH (FT. - IN.)	TOTAL LOAD (P.S.F.)	COST PER S.F.		
						MAT.	INST.	TOTAL
3800	20x25	40	4	1 - 8	94	5.30	7.15	12.45
3900		75	4	1 - 8	130	6.15	7.65	13.80
4000		125	4	1 - 10	181	7.10	8.20	15.30
4100		200	5	2 - 2	272	9.15	9.35	18.50
4200	25x25	40	4-1/2	1 - 8-1/2	99	5.55	7.25	12.80
4300		75	4-1/2	1 - 10-1/2	136	6.65	7.90	14.55
4400		125	5-1/2	2 - 0-1/2	200	7.85	8.60	16.45
4500		200	5-1/2	2 - 2-1/2	278	9.65	9.60	19.25
4600	25x30	40	4-1/2	1 - 8-1/2	100	6.15	7.60	13.75
4700		75	4-1/2	1 - 10-1/2	136	7.15	8.15	15.30
4800		125	5-1/2	2 - 2-1/2	202	8.75	9	17.75
4900		200	5-1/2	2 - 5-1/2	279	10.35	9.90	20.25
5000	30x30	40	4	1 - 8	95	6.15	7.55	13.70
5200		75	4	2 - 1	131	7.10	8.05	15.15
5400		125	4	2 - 4	183	8.65	8.95	17.60
5600		200	5	2 - 10	274	11	10.20	21.20
5800	30x35	40	4	2 - 1	95	6.50	7.70	14.20
6000		75	4	2 - 4	139	7.75	8.35	16.10
6250		125	4	2 - 4	185	9.95	9.55	19.50
6500		200	5	2 - 11	276	12.15	10.75	22.90
8000	35x35	40	4	2 - 1	96	6.80	7.90	14.70
8250		75	4	2 - 4	133	8.05	8.55	16.60
8500		125	4	2 - 10	185	9.60	9.40	19
8750		200	5	3 - 5	276	12.45	10.85	23.30
9000	35x40	40	4	2 - 4	97	7.55	8.30	15.85
9250		75	4	2 - 4	134	8.75	8.80	17.55
9500		125	4	2 - 7	186	10.60	9.85	20.45
9750		200	5	3 - 5	278	13.65	11.45	25.10

SHELL B

B1010 Floor Construction

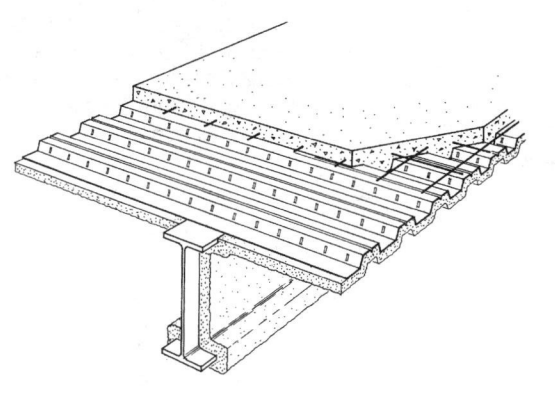

The table below lists costs per S.F. for floors using steel beams and girders, composite steel deck, concrete slab reinforced with W.W.F. and sprayed fiber fireproofing (non- asbestos) on the steel beams and girders and on the steel deck.

Wide Flange, Composite Deck & Slab

B1010 254	W Shape, Composite Deck, & Slab							
	BAY SIZE (FT.) BEAM X GIRD	SUPERIMPOSED LOAD (P.S.F.)	SLAB THICKNESS (IN.)	TOTAL DEPTH (FT.-IN.)	TOTAL LOAD (P.S.F.)	COST PER S.F.		
						MAT.	INST.	TOTAL
0540	15x20	40	5	1-7	89	6.10	4.58	10.68
0560		75	5	1-9	125	6.65	4.86	11.51
0580		125	5	1-11	176	7.65	5.35	13
0600		200	5	2-2	254	9.40	6.10	15.50
0700	20x20	40	5	1-9	90	6.55	4.79	11.34
0720		75	5	1-9	126	7.35	5.20	12.55
0740		125	5	1-11	177	8.30	5.65	13.95
0760		200	5	2-5	255	10	6.35	16.35
0780	20x25	40	5	1-9	90	6.75	4.76	11.51
0800		75	5	1-9	127	8.25	5.50	13.75
0820		125	5	1-11	180	9.95	6	15.95
0840		200	5	2-5	256	10.75	6.80	17.55
0940	25x25	40	5	1-11	91	7.55	5.10	12.65
0960		75	5	2-5	178	8.60	5.65	14.25
0980		125	5	2-5	181	10.60	6.30	16.90
1000		200	5-1/2	2-8-1/2	263	11.65	7.30	18.95
1400	25x30	40	5	2-5	91	7.45	5.25	12.70
1500		75	5	2-5	128	9.05	5.85	14.90
1600		125	5	2-8	180	10.15	6.50	16.65
1700		200	5	2-11	259	12.60	7.50	20.10
2200	30x30	40	5	2-2	92	8.25	5.60	13.85
2300		75	5	2-5	129	9.55	6.25	15.80
2400		125	5	2-11	182	11.25	7.05	18.30
2500		200	5	3-2	263	15.10	8.80	23.90
2600	30x35	40	5	2-5	94	8.95	5.90	14.85
2700		75	5	2-11	131	10.35	6.55	16.90
2800		125	5	3-2	183	12.05	7.35	19.40
2900		200	5-1/2	3-5-1/2	268	14.65	8.50	23.15
3800	35x35	40	5	2-8	94	9.20	5.90	15.10
3900		75	5	2-11	131	10.60	6.55	17.15
4000		125	5	3-5	184	12.65	7.55	20.20
4100		200	5-1/2	3-5-1/2	270	16.35	8.95	25.30

SHELL

B

B1010 Floor Construction

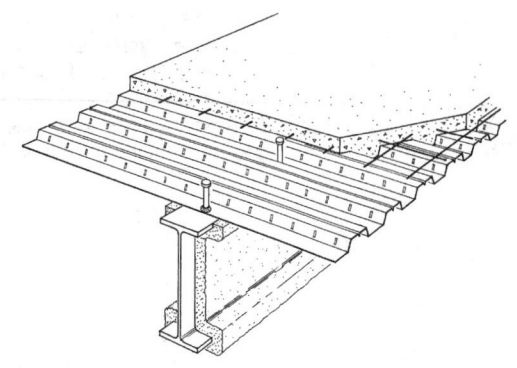

The table below lists costs per S.F. for floors using composite steel beams with welded shear studs, composite steel deck and light weight concrete slab reinforced with W.W.F.

Composite Beam, Deck & Slab

B1010 256	Composite Beams, Deck & Slab							
	BAY SIZE (FT.)	SUPERIMPOSED LOAD (P.S.F.)	SLAB THICKNESS (IN.)	TOTAL DEPTH (FT. - IN.)	TOTAL LOAD (P.S.F.)	COST PER S.F.		
						MAT.	INST.	TOTAL
2400	20x25	40	5-1/2	1 - 5-1/2	80	6	4.31	10.31
2500		75	5-1/2	1 - 9-1/2	115	6.25	4.30	10.55
2750		125	5-1/2	1 - 9-1/2	167	7.70	5.10	12.80
2900		200	6-1/4	1 - 11-1/2	251	8.65	5.50	14.15
3000	25x25	40	5-1/2	1 - 9-1/2	82	6	4.10	10.10
3100		75	5-1/2	1 - 11-1/2	118	6.65	4.17	10.82
3200		125	5-1/2	2 - 2-1/2	169	7	4.53	11.53
3300		200	6-1/4	2 - 6-1/4	252	9.55	5.30	14.85
3400	25x30	40	5-1/2	1 - 11-1/2	83	6.15	4.06	10.21
3600		75	5-1/2	1 - 11-1/2	119	6.60	4.10	10.70
3900		125	5-1/2	1 - 11-1/2	170	7.65	4.65	12.30
4000		200	6-1/4	2 - 6-1/4	252	9.60	5.30	14.90
4200	30x30	40	5-1/2	1 - 11-1/2	81	6.05	4.19	10.24
4400		75	5-1/2	2 - 2-1/2	116	6.60	4.38	10.98
4500		125	5-1/2	2 - 5-1/2	168	7.95	4.93	12.88
4700		200	6-1/4	2 - 9-1/4	252	9.55	5.75	15.30
4900	30x35	40	5-1/2	2 - 2-1/2	82	6.35	4.35	10.70
5100		75	5-1/2	2 - 5-1/2	117	6.95	4.45	11.40
5300		125	5-1/2	2 - 5-1/2	169	8.20	5.05	13.25
5500		200	6-1/4	2 - 9-1/4	254	9.60	5.75	15.35
5750	35x35	40	5-1/2	2 - 5-1/2	84	6.85	4.36	11.21
6000		75	5-1/2	2 - 5-1/2	121	7.80	4.67	12.47
7000		125	5-1/2	2 - 8-1/2	170	9.15	5.35	14.50
7200		200	5-1/2	2 - 11-1/2	254	10.65	5.95	16.60
7400	35x40	40	5-1/2	2 - 5-1/2	85	7.55	4.69	12.24
7600		75	5-1/2	2 - 5-1/2	121	8.20	4.86	13.06
8000		125	5-1/2	2 - 5-1/2	171	9.35	5.45	14.80
9000		200	5-1/2	2 - 11-1/2	255	11.45	6.20	17.65

SHELL B

Important: See the Reference Section for critical supporting data - Location Factors & Historical Cost Indexes

B1010 Floor Construction

Description: Table B1010 258 lists S.F. costs for steel deck and concrete slabs for various spans.

Description: Table B1010 261 lists the S.F. costs for wood joists and a minimum thickness plywood subfloor.

Description: Table B1010 264 lists the S.F. costs, total load, and member sizes, for various bay sizes and loading conditions.

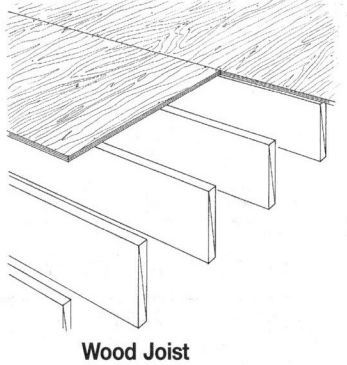

Metal Deck/Concrete Fill

Wood Joist

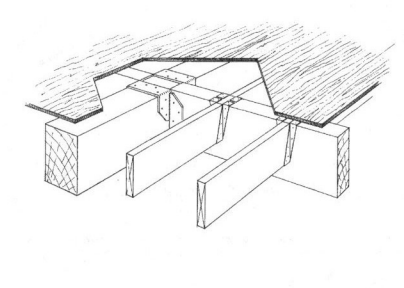

Wood Beam and Joist

B1010 258	Metal Deck/Concrete Fill							
	SUPERIMPOSED LOAD (P.S.F.)	DECK SPAN (FT.)	DECK GAGE DEPTH	SLAB THICKNESS (IN.)	TOTAL LOAD (P.S.F.)	COST PER S.F.		
						MAT.	INST.	TOTAL
0900	125	6	22 1-1/2	4	164	1.58	1.73	3.31
0920		7	20 1-1/2	4	164	1.71	1.83	3.54
0950		8	20 1-1/2	4	165	1.71	1.83	3.54
0970		9	18 1-1/2	4	165	1.97	1.83	3.80
1000		10	18 2	4	165	1.93	1.94	3.87
1020		11	18 3	5	169	2.39	2.09	4.48

B1010 261	Wood Joist	COST PER S.F.		
		MAT.	INST.	TOTAL
2902	Wood joist, 2" x 8",12" O.C.	1.61	1.27	2.88
2950	16" O.C.	1.36	1.09	2.45
3000	24" O.C.	1.30	1.01	2.31
3300	2"x10", 12" O.C.	2.08	1.45	3.53
3350	16" O.C.	1.70	1.22	2.92
3400	24" O.C.	1.52	1.09	2.61
3700	2"x12", 12" O.C.	2.63	1.47	4.10
3750	16" O.C.	2.11	1.23	3.34
3800	24" O.C.	1.80	1.10	2.90
4100	2"x14", 12" O.C.	2.87	1.59	4.46
4150	16" O.C.	2.31	1.33	3.64
4200	24" O.C.	1.93	1.17	3.10
7101	Note: Subfloor cost is included in these prices			

B1010 264	Wood Beam & Joist						
	BAY SIZE (FT.)	SUPERIMPOSED LOAD (P.S.F.)	GIRDER BEAM (IN.)	JOISTS (IN.)	TOTAL LOAD (P.S.F.)	COST PER S.F.		
						MAT.	INST.	TOTAL
2000	15x15	40	8 x 12 4 x 12	2 x 6 @ 16	53	4.98	2.93	7.91
2050		75	8 x 16 4 x 16	2 x 8 @ 16	90	6.60	3.22	9.82
2100		125	12 x 16 6 x 16	2 x 8 @ 12	144	10.50	4.13	14.63
2150		200	14 x 22 12 x 16	2 x 10 @ 12	227	19.65	6.30	25.95
3000	20x20	40	10 x 14 10 x 12	2 x 8 @ 16	63	7.40	2.96	10.36
3050		75	12 x 16 8 x 16	2 x 10 @ 16	102	9.30	3.31	12.61
3100		125	14 x 22 12 x 16	2 x 10 @ 12	163	15.85	5.30	21.15

SHELL

B

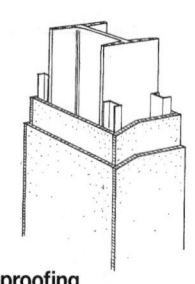

Steel Column Fireproofing

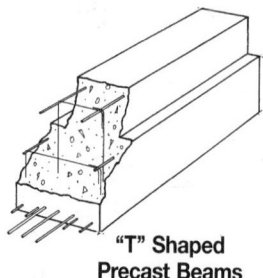

"T" Shaped
Precast Beams

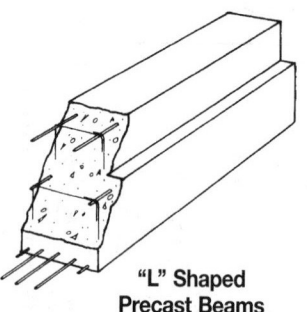

"L" Shaped
Precast Beams

B1010 274	Steel Column Fireproofing						
	ENCASEMENT SYSTEM	COLUMN SIZE (IN.)	THICKNESS (IN.)	FIRE RATING (HRS.)	WEIGHT (P.L.F.)	COST PER V.L.F.		
						MAT.	INST.	TOTAL
3000	Concrete	8	1	1	110	4.37	24	28.37
3300		14	1	1	258	7.20	35	42.20
3400			2	3	325	8.65	39.50	48.15
3450	Gypsum board	8	1/2	2	8	2.28	14.50	16.78
3550	1 layer	14	1/2	2	18	2.50	15.45	17.95
3600	Gypsum board	8	1	3	14	3.29	18.50	21.79
3650	1/2" fire rated	10	1	3	17	3.55	19.65	23.20
3700	2 layers	14	1	3	22	3.69	20	23.69
3750	Gypsum board	8	1-1/2	3	23	4.48	23.50	27.98
3800	1/2" fire rated	10	1-1/2	3	27	5.05	26	31.05
3850	3 layers	14	1-1/2	3	35	5.65	28.50	34.15
3900	Sprayed fiber	8	1-1/2	2	6.3	2.98	4.74	7.72
3950	Direct application		2	3	8.3	4.11	6.55	10.66
4050		10	1-1/2	2	7.9	3.60	5.75	9.35
4200		14	1-1/2	2	10.8	4.48	7.10	11.58

Important: See the Reference Section for critical supporting data - Location Factors & Historical Cost Indexes

B1020 Roof Construction

The table below lists prices per S.F. for roof rafters and sheathing by nominal size and spacing. Sheathing is 5/16″ CDX for 12″ and 16″ spacing and 3/8″ CDX for 24″ spacing.

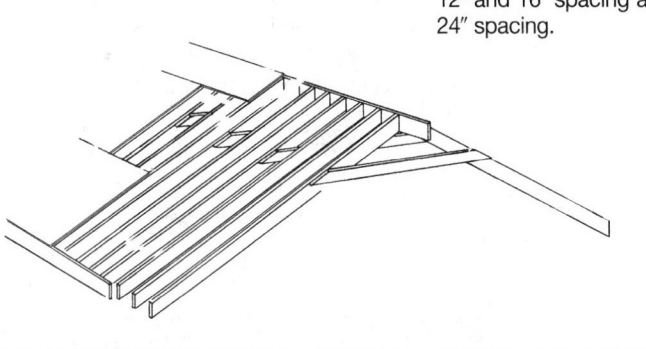

Factors for Converting Inclined to Horizontal

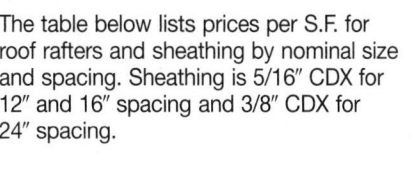

Roof Slope	Approx. Angle	Factor	Roof Slope	Approx. Angle	Factor
Flat	0°	1.000	12 in 12	45.0°	1.414
1 in 12	4.8°	1.003	13 in 12	47.3°	1.474
2 in 12	9.5°	1.014	14 in 12	49.4°	1.537
3 in 12	14.0°	1.031	15 in 12	51.3°	1.601
4 in 12	18.4°	1.054	16 in 12	53.1°	1.667
5 in 12	22.6°	1.083	17 in 12	54.8°	1.734
6 in 12	26.6°	1.118	18 in 12	56.3°	1.803
7 in 12	30.3°	1.158	19 in 12	57.7°	1.873
8 in 12	33.7°	1.202	20 in 12	59.0°	1.943
9 in 12	36.9°	1.250	21 in 12	60.3°	2.015
10 in 12	39.8°	1.302	22 in 12	61.4°	2.088
11 in 12	42.5°	1.357	23 in 12	62.4°	2.162

B1020 102	Wood/Flat or Pitched	MAT.	INST.	TOTAL
2500	Flat rafter, 2″x4″, 12″ O.C.	.93	1.15	2.08
2550	16″ O.C.	.82	.99	1.81
2600	24″ O.C.	.72	.84	1.56
2900	2″x6″, 12″ O.C.	1.18	1.15	2.33
2950	16″ O.C.	1.01	.99	2
3000	24″ O.C.	.84	.84	1.68
3300	2″x8″, 12″ O.C.	1.54	1.23	2.77
3350	16″ O.C.	1.29	1.05	2.34
3400	24″ O.C.	1.03	.89	1.92
3700	2″x10″, 12″ O.C.	2.01	1.41	3.42
3750	16″ O.C.	1.63	1.18	2.81
3800	24″ O.C.	1.25	.97	2.22
4100	2″x12″, 12″ O.C.	2.56	1.43	3.99
4150	16″ O.C.	2.04	1.19	3.23
4200	24″ O.C.	1.53	.98	2.51
4500	2″x14″, 12″ O.C.	2.80	2.09	4.89
4550	16″ O.C.	2.24	1.70	3.94
4600	24″ O.C.	1.66	1.32	2.98
4900	3″x6″, 12″ O.C.	2.52	1.22	3.74
4950	16″ O.C.	2.02	1.04	3.06
5000	24″ O.C.	1.52	.88	2.40
5300	3″x8″, 12″ O.C.	3.20	1.36	4.56
5350	16″ O.C.	2.52	1.14	3.66
5400	24″ O.C.	1.85	.94	2.79
5700	3″x10″, 12″ O.C.	3.87	1.55	5.42
5750	16″ O.C.	3.04	1.29	4.33
5800	24″ O.C.	2.19	1.04	3.23
6100	3″x12″, 12″ O.C.	4.54	1.86	6.40
6150	16″ O.C.	3.54	1.52	5.06
6200	24″ O.C.	2.52	1.19	3.71
7001	Wood truss, 4 in 12 slope, 24″ O.C., 24′ to 29′ span	2.96	1.78	4.74
7100	30′ to 43′ span	3.12	1.78	4.90
7200	44′ to 60′ span	3.30	1.78	5.08

SHELL

B

B1020 Roof Construction

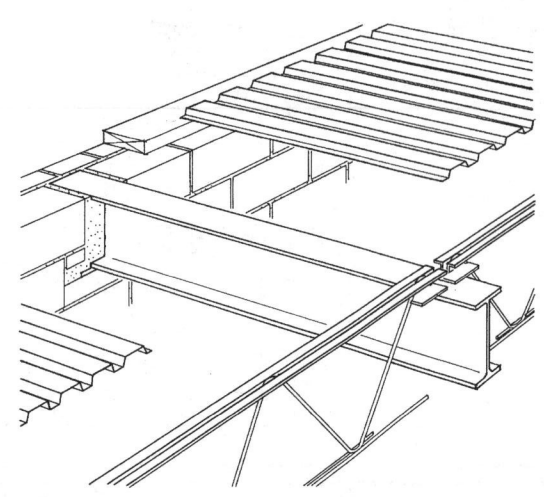

**Steel Joists, Beams and Deck
on Bearing Walls**

The table below lists the cost per S.F. for a roof system with steel columns, beams and deck using open web steel joists and 1-1/2″ galvanized metal deck. Perimeter of system is supported on bearing walls.

Fireproofing is not included. Costs/S.F. are based on a building 4 bays long and 4 bays wide.

Column costs are additive. Costs for the bearing walls are not included.

B1020 108		Steel Joists, Beams & Deck on Columns & Walls						
	BAY SIZE (FT.)	SUPERIMPOSED LOAD (P.S.F.)	DEPTH (IN.)	TOTAL LOAD (P.S.F.)	COLUMN ADD	COST PER S.F.		
						MAT.	INST.	TOTAL
3000	25x25	20	18	40		2.05	1.01	3.06
3100					columns	.18	.07	.25
3200		30	22	50		2.33	1.19	3.52
3300					columns	.24	.08	.32
3400		40	20	60		2.45	1.18	3.63
3500					columns	.24	.08	.32
3600	25x30	20	22	40		2.19	.96	3.15
3700					columns	.20	.07	.27
3800		30	20	50		2.39	1.26	3.65
3900					columns	.20	.07	.27
4000		40	25	60		2.60	1.11	3.71
4100					columns	.24	.08	.32
4200	30x30	20	25	42		2.41	1.04	3.45
4300					columns	.17	.05	.22
4400		30	22	52		2.64	1.13	3.77
4500					columns	.20	.07	.27
4600		40	28	62		2.74	1.16	3.90
4700					columns	.20	.07	.27
4800	30x35	20	22	42		2.52	1.08	3.60
4900					columns	.17	.07	.24
5000		30	28	52		2.66	1.13	3.79
5100					columns	.17	.07	.24
5200		40	25	62		2.89	1.22	4.11
5300					columns	.20	.07	.27
5400	35x35	20	28	42		2.54	1.09	3.63
5500					columns	.15	.05	.20

Important: See the Reference Section for critical supporting data - Location Factors & Historical Cost Indexes

B10 Superstructure

B1020 Roof Construction

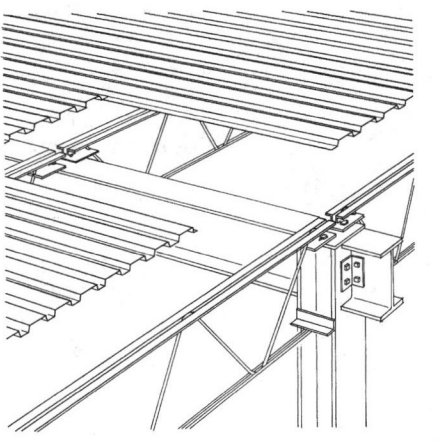

Description: The table below lists the cost per S.F. for a roof system with steel columns, beams, and deck, using open web steel joists and 1-1/2″ galvanized metal deck.

Column costs are additive. Fireproofing is not included.

Steel Joists, Beams and Deck on Columns

B1020 112 — Steel Joists, Beams, & Deck on Columns

	BAY SIZE (FT.)	SUPERIMPOSED LOAD (P.S.F.)	DEPTH (IN.)	TOTAL LOAD (P.S.F.)	COLUMN ADD	COST PER S.F. MAT.	INST.	TOTAL
1100	15x20	20	16	40		2.03	1	3.03
1200					columns	1.04	.37	1.41
1500		40	18	60		2.28	1.11	3.39
1600					columns	1.04	.37	1.41
2300	20x25	20	18	40		2.32	1.12	3.44
2400					columns	.62	.23	.85
2700		40	20	60		2.62	1.27	3.89
2800					columns	.83	.29	1.12
2900	25x25	20	18	40		2.70	1.25	3.95
3000					columns	.50	.17	.67
3100		30	22	50		2.94	1.43	4.37
3200					columns	.66	.24	.90
3300		40	20	60		3.14	1.44	4.58
3400					columns	.66	.24	.90
3500	25x30	20	22	40		2.67	1.15	3.82
3600					columns	.55	.20	.75
3900		40	25	60		3.25	1.35	4.60
4000					columns	.66	.24	.90
4100	30x30	20	25	42		3.02	1.26	4.28
4200					columns	.46	.16	.62
4300		30	22	52		3.34	1.38	4.72
4400					columns	.55	.20	.75
4500		40	28	62		3.49	1.44	4.93
4600					columns	.55	.20	.75
5300	35x35	20	28	42		3.31	1.37	4.68
5400					columns	.41	.15	.56
5500		30	25	52		3.70	1.51	5.21
5600					columns	.47	.17	.64
5700		40	28	62		4	1.63	5.63
5800					columns	.52	.19	.71

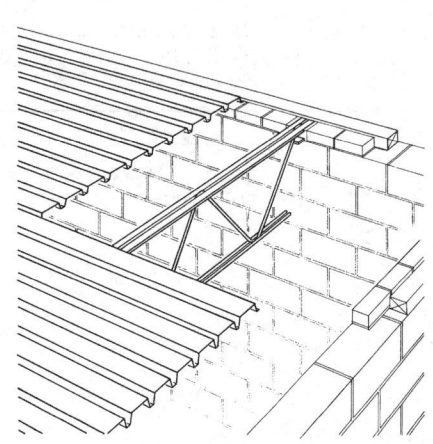

Description: The table below lists the cost per S.F. for a roof system using open web steel joists and 1-1/2″ galvanized metal deck. The system is assumed supported on bearing walls or other suitable support. Costs for the supports are not included.

SHELL B

B1020 116		Steel Joists & Deck on Bearing Walls						
	BAY SIZE (FT.)	SUPERIMPOSED LOAD (P.S.F.)	DEPTH (IN.)	TOTAL LOAD (P.S.F.)		COST PER S.F.		
						MAT.	INST.	TOTAL
1100	20	20	13-1/2	40		1.32	.73	2.05
1200		30	15-1/2	50		1.35	.75	2.10
1300		40	15-1/2	60		1.46	.81	2.27
1400	25	20	17-1/2	40		1.46	.81	2.27
1500		30	17-1/2	50		1.59	.88	2.47
1600		40	19-1/2	60		1.62	.89	2.51
1700	30	20	19-1/2	40		1.61	.89	2.50
1800		30	21-1/2	50		1.65	.90	2.55
1900		40	23-1/2	60		1.79	.98	2.77
2000	35	20	23-1/2	40		1.77	.82	2.59
2100		30	25-1/2	50		1.83	.84	2.67
2200		40	25-1/2	60		1.96	.89	2.85
2300	40	20	25-1/2	41		1.99	.90	2.89
2400		30	25-1/2	51		2.12	.95	3.07
2500		40	25-1/2	61		2.20	.99	3.19
2600	45	20	27-1/2	41		2.23	1.35	3.58
2700		30	31-1/2	51		2.37	1.43	3.80
2800		40	31-1/2	61		2.50	1.52	4.02
2900	50	20	29-1/2	42		2.51	1.53	4.04
3000		30	31-1/2	52		2.75	1.68	4.43
3100		40	31-1/2	62		2.93	1.79	4.72
3200	60	20	37-1/2	42		3.21	1.54	4.75
3300		30	37-1/2	52		3.58	1.72	5.30
3400		40	37-1/2	62		3.58	1.72	5.30
3500	70	20	41-1/2	42		3.57	1.71	5.28
3600		30	41-1/2	52		3.82	1.82	5.64
3700		40	41-1/2	64		4.68	2.22	6.90
3800	80	20	45-1/2	44		5.25	2.43	7.68
3900		30	45-1/2	54		5.25	2.43	7.68
4000		40	45-1/2	64		5.80	2.69	8.49
4400	100	20	57-1/2	44		4.55	2.04	6.59
4500		30	57-1/2	54		5.45	2.43	7.88
4600		40	57-1/2	65		6.05	2.68	8.73
4700	125	20	69-1/2	44		6.40	2.75	9.15
4800		30	69-1/2	56		7.50	3.19	10.69
4900		40	69-1/2	67		8.60	3.65	12.25

B1020 Roof Construction

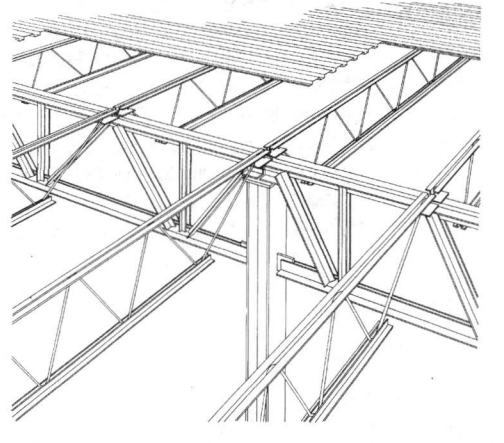

Description: The table below lists costs for a roof system supported on exterior bearing walls and interior columns. Costs include bracing, joist girders, open web steel joists and 1-1/2″ galvanized metal deck.

Column costs are additive. Fireproofing is not included.

Costs/S.F. are based on a building 4 bays long and 4 bays wide. Costs for bearing walls are not included.

B1020 120	Steel Joists, Joist Girders & Deck on Columns & Walls							
	BAY SIZE (FT.) GIRD X JOISTS	SUPERIMPOSED LOAD (P.S.F.)	DEPTH (IN.)	TOTAL LOAD (P.S.F.)	COLUMN ADD	COST PER S.F.		
						MAT.	INST.	TOTAL
2350	30x35	20	32-1/2	40		1.96	1.04	3
2400					columns	.17	.07	.24
2550		40	36-1/2	60		2.21	1.16	3.37
2600					columns	.20	.07	.27
3000	35x35	20	36-1/2	40		2.13	1.28	3.41
3050					columns	.15	.05	.20
3200		40	36-1/2	60		2.42	1.41	3.83
3250					columns	.19	.07	.26
3300	35x40	20	36-1/2	40		2.18	1.17	3.35
3350					columns	.15	.05	.20
3500		40	36-1/2	60		2.48	1.30	3.78
3550					columns	.17	.06	.23
3900	40x40	20	40-1/2	41		2.44	1.51	3.95
3950					columns	.15	.05	.20
4100		40	40-1/2	61		2.75	1.66	4.41
4150					columns	.15	.05	.20
5100	45x50	20	52-1/2	41		2.75	1.73	4.48
5150					columns	.10	.04	.14
5300		40	52-1/2	61		3.36	2.09	5.45
5350					columns	.15	.05	.20
5400	50x45	20	56-1/2	41		2.66	1.66	4.32
5450					columns	.10	.04	.14
5600		40	56-1/2	61		3.46	2.13	5.59
5650					columns	.14	.05	.19
5700	50x50	20	56-1/2	42		3	1.89	4.89
5750					columns	.11	.04	.15
5900		40	59	64		4.12	2.13	6.25
5950					columns	.15	.05	.20
6300	60x50	20	62-1/2	43		3.03	1.91	4.94
6350					columns	.17	.07	.24
6500		40	71	65		4.23	2.19	6.42
6550					columns	.22	.08	.30

SHELL

B

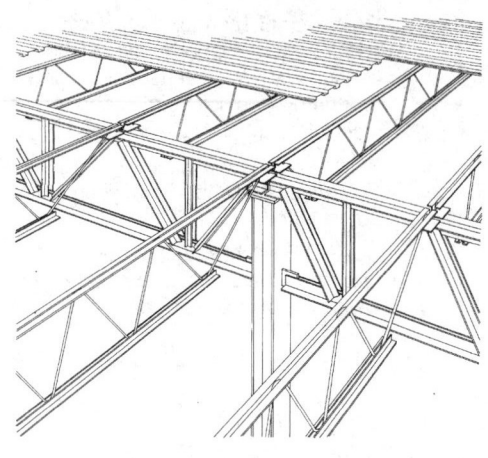

Description: The table below lists the cost per S.F. for a roof system supported on columns. Costs include joist girders, open web steel joists and 1-1/2″ galvanized metal deck. Column costs are additive.

Fireproofing is not included. Costs/S.F. are based on a building 4 bays long and 4 bays wide.

Costs for wind columns are not included.

SHELL B

B1020 124	Steel Joists & Joist Girders on Columns							
	BAY SIZE (FT.) **GIRD X JOISTS**	**SUPERIMPOSED** **LOAD (P.S.F.)**	**DEPTH** **(IN.)**	**TOTAL LOAD** **(P.S.F.)**	**COLUMN** **ADD**	**COST PER S.F.**		
						MAT.	**INST.**	**TOTAL**
2000	30x30	20	17-1/2	40		1.99	1.21	3.20
2050					columns	.56	.20	.76
2100		30	17-1/2	50		2.18	1.30	3.48
2150					columns	.56	.20	.76
2200		40	21-1/2	60		2.24	1.34	3.58
2250					columns	.56	.20	.76
2300	30x35	20	32-1/2	40		2.12	1.16	3.28
2350					columns	.48	.18	.66
2500		40	36-1/2	60		2.42	1.28	3.70
2550					columns	.56	.20	.76
3200	35x40	20	36-1/2	40		2.66	1.78	4.44
3250					columns	.42	.15	.57
3400		40	36-1/2	60		3.01	1.97	4.98
3450					columns	.46	.16	.62
3800	40x40	20	40-1/2	41		2.70	1.77	4.47
3850					columns	.40	.15	.55
4000		40	40-1/2	61		3.04	1.97	5.01
4050					columns	.40	.15	.55
4100	40x45	20	40-1/2	41		2.89	2.01	4.90
4150					columns	.36	.13	.49
4200		30	40-1/2	51		3.16	2.16	5.32
4250					columns	.36	.13	.49
4300		40	40-1/2	61		3.52	2.37	5.89
4350					columns	.41	.15	.56
5000	45x50	20	52-1/2	41		3.06	2.06	5.12
5050					columns	.29	.11	.40
5200		40	52-1/2	61		3.77	2.47	6.24
5250					columns	.37	.14	.51
5300	50x45	20	56-1/2	41		2.98	2	4.98
5350					columns	.29	.11	.40
5500		40	56-1/2	61		3.89	2.53	6.42
5550					columns	.37	.14	.51
5600	50x50	20	56-1/2	42		3.35	2.22	5.57
5650					columns	.29	.11	.40
5800		40	59	64		3.78	2.04	5.82
5850					columns	.41	.15	.56

Important: See the Reference Section for critical supporting data - Location Factors & Historical Cost Indexes

B2010 Exterior Walls

Table B2010 101 describes a concrete wall system for exterior closure. There are several types of wall finishes priced from plain finish to a finish with 3/4" rustication strip.

In Table B2010 103, precast concrete wall panels are either solid or insulated. Prices below are based on delivery within 50 miles of a plant. Small jobs can double the prices below. For large, highly repetitive jobs, deduct up to 15%.

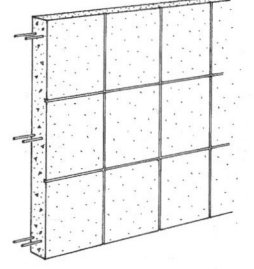

Cast in Place Concrete Wall

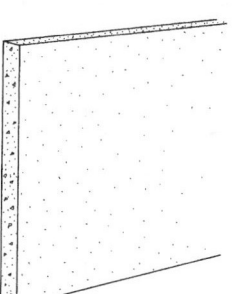

Flat Precast Concrete Wall

B2010 101	Cast In Place Concrete	COST PER S.F.		
		MAT.	INST.	TOTAL
2100	Conc wall reinforced, 8' high, 6" thick, plain finish, 3000 PSI	3.25	12.10	15.35
2700	Aged wood liner, 3000 PSI	5.10	12.60	17.70
3000	Sand blast light 1 side, 3000 PSI	3.49	12.95	16.44
3700	3/4" bevel rustication strip, 3000 PSI	3.44	14.50	17.94
4000	8" thick, plain finish, 3000 PSI	3.80	12.40	16.20
4100	4000 PSI	3.94	12.40	16.34
4300	Rub concrete 1 side, 3000 PSI	3.83	14.35	18.18
4550	8" thick, aged wood liner, 3000 PSI	5.65	12.90	18.55
4750	Sand blast light 1 side, 3000 PSI	4.04	13.25	17.29
5300	3/4" bevel rustication strip, 3000 PSI	3.99	14.80	18.79
5600	10" thick, plain finish, 3000 PSI	4.35	12.65	17
5900	Rub concrete 1 side, 3000 PSI	4.38	14.60	18.98
6200	Aged wood liner, 3000 PSI	6.20	13.15	19.35
6500	Sand blast light 1 side, 3000 PSI	4.59	13.55	18.14
7100	3/4" bevel rustication strip, 3000 PSI	4.54	15.10	19.64
7400	12" thick, plain finish, 3000 PSI	4.98	13	17.98
7700	Rub concrete 1 side, 3000 PSI	5	14.95	19.95
8000	Aged wood liner, 3000 PSI	6.85	13.50	20.35
8300	Sand blast light 1 side, 3000 PSI	5.20	13.90	19.10
8900	3/4" bevel rustication strip, 3000 PSI	5.15	15.40	20.55

B2010 103	Flat Precast Concrete							
	THICKNESS (IN.)	PANEL SIZE (FT.)	FINISHES	RIGID INSULATION (IN)	TYPE	COST PER S.F.		
						MAT.	INST.	TOTAL
3000	4	5x18	smooth gray	none	low rise	5.20	5.80	11
3050		6x18				4.35	4.85	9.20
3100		8x20				6.70	1.82	8.52
3150		12x20				6.35	1.72	8.07
3200	6	5x18	smooth gray	2	low rise	6.05	6.15	12.20
3250		6x18				5.20	5.20	10.40
3300		8x20				7.70	2.25	9.95
3350		12x20				7	2.06	9.06
3400	8	5x18	smooth gray	2	low rise	9.80	2.83	12.63
3450		6x18				9.30	2.70	12
3500		8x20				8.55	2.48	11.03
3550		12x20				7.80	2.28	10.08

SHELL

B

B2010 Exterior Walls

B2010 103 — Flat Precast Concrete

	THICKNESS (IN.)	PANEL SIZE (FT.)	FINISHES	RIGID INSULATION (IN)	TYPE	COST PER S.F.		
						MAT.	INST.	TOTAL
3600	4	4x8	white face	none	low rise	16.40	3.35	19.75
3650		8x8				12.30	2.51	14.81
3700		10x10				10.80	2.21	13.01
3750		20x10				9.80	1.99	11.79
3800	5	4x8	white face	none	low rise	16.75	3.42	20.17
3850		8x8				12.65	2.59	15.24
3900		10x10				11.20	2.28	13.48
3950		20x20				10.25	2.10	12.35
4000	6	4x8	white face	none	low rise	17.35	3.54	20.89
4050		8x8				13.20	2.69	15.89
4100		10x10				11.60	2.37	13.97
4150		20x10				10.65	2.18	12.83
4200	6	4x8	white face	2	low rise	18.25	3.96	22.21
4250		8x8				14.10	3.11	17.21
4300		10x10				12.50	2.79	15.29
4350		20x10				10.65	2.18	12.83
4400	7	4x8	white face	none	low rise	17.75	3.63	21.38
4450		8x8				13.65	2.79	16.44
4500		10x10				12.25	2.50	14.75
4550		20x10				11.20	2.28	13.48
4600	7	4x8	white face	2	low rise	18.65	4.05	22.70
4650		8x8				14.55	3.21	17.76
4700		10x10				13.15	2.92	16.07
4750		20x10				12.10	2.70	14.80
4800	8	4x8	white face	none	low rise	18.20	3.71	21.91
4850		8x8				14	2.86	16.86
4900		10x10				12.60	2.57	15.17
4950		20x10				11.60	2.37	13.97
5000	8	4x8	white face	2	low rise	19.05	4.13	23.18
5050		8x8				14.90	3.28	18.18
5100		10x10				13.50	2.99	16.49
5150		20x10				12.50	2.79	15.29

B2010 103 — Fluted Window or Mullion Precast Concrete

	THICKNESS (IN.)	PANEL SIZE (FT.)	FINISHES	RIGID INSULATION (IN)	TYPE	COST PER S.F.		
						MAT.	INST.	TOTAL
5200	4	4x8	smooth gray	none	high rise	11.90	13.25	25.15
5250		8x8				8.50	9.45	17.95
5300		10x10				12.85	3.46	16.31
5350		20x10				11.25	3.04	14.29
5400	5	4x8	smooth gray	none	high rise	12.10	13.50	25.60
5450		8x8				8.80	9.80	18.60
5500		10x10				13.40	3.61	17.01
5550		20x10				11.85	3.20	15.05
5600	6	4x8	smooth gray	none	high rise	12.45	13.85	26.30
5650		8x8				9.10	10.15	19.25
5700		10x10				13.75	3.72	17.47
5750		20x10				12.25	3.30	15.55
5800	6	4x8	smooth gray	2	high rise	13.30	14.25	27.55
5850		8x8				10	10.55	20.55
5900		10x10				14.65	4.14	18.79
5950		20x10				13.10	3.72	16.82

Important: See the Reference Section for critical supporting data - Location Factors & Historical Cost Indexes

B20 Exterior Enclosure

B2010 Exterior Walls

B2010 103 — Fluted Window or Mullion Precast Concrete

	THICKNESS (IN.)	PANEL SIZE (FT.)	FINISHES	RIGID INSULATION (IN)	TYPE	COST PER S.F.		
						MAT.	INST.	TOTAL
6000	7	4x8	smooth gray	none	high rise	12.70	14.15	26.85
6050		8x8				9.35	10.45	19.80
6100		10x10				14.35	3.88	18.23
6150		20x10				12.60	3.40	16
6200	7	4x8	smooth gray	2	high rise	13.60	14.55	28.15
6250		8x8				10.25	10.85	21.10
6300		10x10				15.25	4.30	19.55
6350		20x10				13.50	3.82	17.32
6400	8	4x8	smooth gray	none	high rise	12.85	14.35	27.20
6450		8x8				9.60	10.70	20.30
6500		10x10				14.80	4	18.80
6550		20x10				13.25	3.57	16.82
6600	8	4x8	smooth gray	2	high rise	13.75	14.75	28.50
6650		8x8				10.50	11.15	21.65
6700		10x10				15.70	4.42	20.12
6750		20x10				14.10	3.99	18.09

B2010 104 — Precast Concrete Specialties

	TYPE	SIZE				COST PER L.F.		
						MAT.	INST.	TOTAL
6773	Coping, precast	6"wide				18.70	8.85	27.55
6774	Stock units	10"wide				11.20	9.45	20.65
6775		12"wide				22.50	10.20	32.70
6776		14" wide				15.45	11.05	26.50
6777	Window sills	6" wide				10.10	9.45	19.55
6778	Precast	10" wide				14.15	11.05	25.20
6779		14" wide				13.50	13.25	26.75

B SHELL

B

277

Ribbed Precast Panel

SHELL B

B2010 105		Ribbed Precast Concrete						
	THICKNESS (IN.)	PANEL SIZE (FT.)	FINISHES	RIGID INSULATION(IN.)	TYPE	COST PER S.F.		
						MAT.	INST.	TOTAL
6800	4	4x8	aggregate	none	high rise	13.15	13.25	26.40
6850		8x8				9.60	9.65	19.25
6900		10x10				14.15	3.49	17.64
6950		20x10				12.50	3.09	15.59
7000	5	4x8	aggregate	none	high rise	13.40	13.50	26.90
7050		8x8				9.90	9.95	19.85
7100		10x10				14.70	3.63	18.33
7150		20x10				13.10	3.24	16.34
7200	6	4x8	aggregate	none	high rise	13.65	13.75	27.40
7250		8x8				10.15	10.25	20.40
7300		10x10				15.15	3.75	18.90
7350		20x10				13.60	3.36	16.96
7400	6	4x8	aggregate	2	high rise	14.55	14.15	28.70
7450		8x8				11.05	10.65	21.70
7500		10x10				16.05	4.17	20.22
7550		20x10				14.50	3.78	18.28
7600	7	4x8	aggregate	none	high rise	13.95	14.05	28
7650		8x8				10.40	10.45	20.85
7700		10x10				15.65	3.86	19.51
7750		20x10				14	3.46	17.46
7800	7	4x8	aggregate	2	high rise	14.85	14.45	29.30
7850		8x8				11.30	10.90	22.20
7900		10x10				16.55	4.28	20.83
7950		20x10				14.85	3.88	18.73
8000	8	4x8	aggregate	none	high rise	14.20	14.30	28.50
8050		8x8				10.70	10.80	21.50
8100		10x10				16.20	4	20.20
8150		20x10				14.60	3.61	18.21
8200	8	4x8	aggregate	2	high rise	15.10	14.75	29.85
8250		8x8				11.60	11.20	22.80
8300		10x10				17.05	4.42	21.47
8350		20x10				15.50	4.03	19.53

Important: See the Reference Section for critical supporting data - Location Factors & Historical Cost Indexes

B2010 Exterior Walls

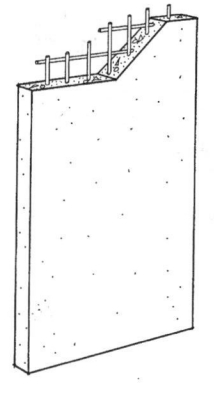

The advantage of tilt up construction is in the low cost of forms and placing of concrete and reinforcing. Tilt up has been used for several types of buildings, including warehouses, stores, offices, and schools. The panels are cast in forms on the ground, or floor slab. Most jobs use 5-1/2" thick solid reinforced concrete panels.

Tilt Up Concrete Panel

B2010 106	Tilt-Up Concrete Panel	COST PER S.F.		
		MAT.	INST.	TOTAL
3200	Tilt up conc panels, broom finish, 5-1/2" thick, 3000 PSI	2.80	4.20	7
3250	5000 PSI	2.88	4.11	6.99
3300	6" thick, 3000 PSI	3.08	4.30	7.38
3350	5000 PSI	3.18	4.22	7.40
3400	7-1/2" thick, 3000 PSI	3.91	4.47	8.38
3450	5000 PSI	4.05	4.39	8.44
3500	8" thick, 3000 PSI	4.20	4.58	8.78
3550	5000 PSI	4.36	4.49	8.85
3700	Steel trowel finish, 5-1/2" thick, 3000 PSI	2.80	4.27	7.07
3750	5000 PSI	2.88	4.19	7.07
3800	6" thick, 3000 PSI	3.08	4.38	7.46
3850	5000 PSI	3.18	4.30	7.48
3900	7-1/2" thick, 3000 PSI	3.91	4.55	8.46
3950	5000 PSI	4.05	4.47	8.52
4000	8" thick, 3000 PSI	4.20	4.66	8.86
4050	5000 PSI	4.36	4.57	8.93
4200	Exp. aggregate finish, 5-1/2" thick, 3000 PSI	7.10	4.27	11.37
4250	5000 PSI	7.15	4.19	11.34
4300	6" thick, 3000 PSI	7.35	4.38	11.73
4350	5000 PSI	7.45	4.30	11.75
4400	7-1/2" thick, 3000 PSI	8.20	4.56	12.76
4450	5000 PSI	8.35	4.47	12.82
4500	8" thick, 3000 PSI	8.45	4.66	13.11
4550	5000 PSI	8.65	4.58	13.23
4600	Exposed aggregate & vert. rustication 5-1/2" thick, 3000 PSI	8.50	7	15.50
4650	5000 PSI	8.60	6.90	15.50
4700	6" thick, 3000 PSI	8.80	7.10	15.90
4750	5000 PSI	8.90	7.05	15.95
4800	7-1/2" thick, 3000 PSI	9.60	7.30	16.90
4850	5000 PSI	9.75	7.20	16.95
4900	8" thick, 3000 PSI	9.90	7.40	17.30
4950	5000 PSI	10.05	7.30	17.35
5000	Vertical rib & light sandblast, 5-1/2" thick, 3000 PSI	4.98	5.10	10.08
5100	6" thick, 3000 PSI	5.25	5.20	10.45
5200	7-1/2" thick, 3000 PSI	6.10	5.35	11.45
5300	8" thick, 3000 PSI	6.35	5.45	11.80
6000	Broom finish w/2" urethane insulation, 6" thick, 3000 PSI	2.67	5.15	7.82
6100	Broom finish 2" fiberplank insulation, 6" thick, 3000 PSI	3.19	5.10	8.29
6200	Exposed aggregate w/2" urethane insulation, 6" thick, 3000 PSI	6.90	5.25	12.15
6300	Exposed aggregate 2" fiberplank insulation, 6" thick, 3000 PSI	7.40	5.20	12.60

SHELL

B

B2010 Exterior Walls

Exterior concrete block walls are defined in the following terms; structural reinforcement, weight, percent solid, size, strength and insulation. Within each of these categories, two to four variations are shown. No costs are included for brick shelf or relieving angles.

B2010 109		Concrete Block Wall - Regular Weight							
	TYPE	SIZE (IN.)	STRENGTH (P.S.I.)	CORE FILL		COST PER S.F.			
						MAT.	INST.	TOTAL	
1200	Hollow	4x8x16	2,000	none		1.11	4.10	5.21	
1250			4,500	none		1.02	4.10	5.12	
1400		8x8x16	2,000	perlite		2.19	4.98	7.17	
1410				styrofoam		2.62	4.70	7.32	
1440				none		1.72	4.70	6.42	
1450			4,500	perlite		2.02	4.98	7	
1460				styrofoam		2.45	4.70	7.15	
1490				none		1.55	4.70	6.25	
2000	75% solid	4x8x16	2,000	none		1.10	4.15	5.25	
2100		6x8x16	2,000	perlite		1.63	4.56	6.19	
2500	Solid	4x8x16	2,000	none		1.49	4.25	5.74	
2700		8x8x16	2,000	none		2.56	4.89	7.45	

B2010 109		Concrete Block Wall - Lightweight							
	TYPE	SIZE (IN.)	WEIGHT (P.C.F.)	CORE FILL		COST PER S.F.			
						MAT.	INST.	TOTAL	
3100	Hollow	8x4x16	105	perlite		1.50	4.70	6.20	
3110				styrofoam		1.93	4.42	6.35	
3200		4x8x16	105	none		1.29	4.01	5.30	
3250			85	none		1.32	3.93	5.25	
3300		6x8x16	105	perlite		2.05	4.52	6.57	
3310				styrofoam		2.63	4.30	6.93	
3340				none		1.73	4.30	6.03	
3400		8x8x16	105	perlite		2.57	4.87	7.44	
3410				styrofoam		3	4.59	7.59	
3440				none		2.10	4.59	6.69	
3450			85	perlite		2.32	4.75	7.07	
4000	75% solid	4x8x16	105	none		1.12	4.06	5.18	
4050			85	none		2.25	3.97	6.22	
4100		6x8x16	105	perlite		1.71	4.46	6.17	
4500	Solid	4x8x16	105	none		1.40	4.20	5.60	
4700		8x8x16	105	none		2.61	4.83	7.44	

SHELL B

Important: See the Reference Section for critical supporting data - Location Factors & Historical Cost Indexes

B2010 112	Reinforced Concrete Block Wall - Regular Weight							

	TYPE	SIZE (IN.)	STRENGTH (P.S.I.)	VERT. REINF & GROUT SPACING		COST PER S.F.		
						MAT.	INST.	TOTAL
5200	Hollow	4x8x16	2,000	#4 @ 48"		1.21	4.55	5.76
5300		6x8x16	2,000	#4 @ 48"		1.77	4.86	6.63
5330				#5 @ 32"		1.90	5.05	6.95
5340				#5 @ 16"		2.20	5.70	7.90
5350			4,500	#4 @ 28"		1.51	4.86	6.37
5390				#5 @ 16"		1.94	5.70	7.64
5400		8x8x16	2,000	#4 @ 48"		1.95	5.20	7.15
5430				#5 @ 32"		2.09	5.50	7.59
5440				#5 @ 16"		2.45	6.25	8.70
5450		8x8x16	4,500	#4 @ 48"		1.77	5.25	7.02
5490				#5 @ 16"		2.28	6.25	8.53
5500		12x8x16	2,000	#4 @ 48"		2.78	6.65	9.43
5540				#5 @ 16"		3.49	7.70	11.19
6100	75% solid	6x8x16	2,000	#4 @ 48"		1.56	4.83	6.39
6140				#5 @ 16"		1.82	5.50	7.32
6150			4,500	#4 @ 48"		1.91	4.83	6.74
6190				#5 @ 16"		2.17	5.50	7.67
6200		8x8x16	2,000	#4 @ 48"		2.29	5.20	7.49
6230				#5 @ 32"		2.38	5.40	7.78
6240				#5 @ 16"		2.57	6.05	8.62
6250			4,500	#4 @ 48"		2.56	5.20	7.76
6280				#5 @ 32"		2.65	5.40	8.05
6290				#5 @ 16"		2.84	6.05	8.89
6500	Solid-double	2-4x8x16	2,000	#4 @ 48" E.W.		3.61	9.65	13.26
6530	Wythe			#5 @ 16" E.W.		4.03	10.25	14.28
6550			4,500	#4 @ 48" E.W.		3.07	9.55	12.62
6580				#5 @ 16" E.W.		3.49	10.15	13.64

B2010 112	Reinforced Concrete Block Wall - Lightweight							

	TYPE	SIZE (IN.)	WEIGHT (P.C.F.)	VERT REINF. & GROUT SPACING		COST PER S.F.		
						MAT.	INST.	TOTAL
7100	Hollow	8x4x16	105	#4 @ 48"		1.25	4.97	6.22
7140				#5 @ 16"		1.76	6	7.76
7150			85	#4 @ 48"		3.23	4.87	8.10
7190				#5 @ 16"		3.74	5.90	9.64
7400		8x8x16	105	#4 @ 48"		2.32	5.15	7.47
7440				#5 @ 16"		2.83	6.15	8.98
7450		8x8x16	85	#4 @ 48"		2.07	5	7.07
7490				#5 @ 16"		2.58	6.05	8.63
7800		8x8x24	105	#4 @ 48"		1.75	5.50	7.25
7840				#5 @ 16"		2.26	6.50	8.76
7850			85	#4 @ 48"		3.53	4.72	8.25
7890				#5 @ 16"		4.04	5.75	9.79
8100	75% solid	6x8x16	105	#4 @ 48"		1.64	4.73	6.37
8130				#5 @ 32"		1.73	4.89	6.62
8150			85	#4 @ 48"		2.70	4.63	7.33
8180				#5 @ 32"		2.79	4.79	7.58
8200		8x8x16	105	#4 @ 48"		2.19	5.10	7.29
8230				#5 @ 32"		2.28	5.30	7.58
8250			85	#4 @ 48"		3.66	4.98	8.64
8280				#5 @ 32"		3.75	5.15	8.90

B SHELL

B2010 Exterior Walls

B2010 112	Reinforced Concrete Block Wall - Lightweight							

	TYPE	SIZE (IN.)	WEIGHT (P.C.F.)	VERT REINF. & GROUT SPACING		COST PER S.F.		
						MAT.	INST.	TOTAL
8500	Solid-double	2-4x8x16	105	#4 @ 48"		3.43	9.55	12.98
8530	Wythe		105	#5 @ 16"		3.85	10.15	14
8600		2-6x8x16	105	#4 @ 48"		4.50	10.20	14.70
8630			105	#5 @ 16"		4.92	10.80	15.72
8650			85	#4 @ 48"		6.30	9.75	16.05

Important: See the Reference Section for critical supporting data - Location Factors & Historical Cost Indexes

B2010 Exterior Walls

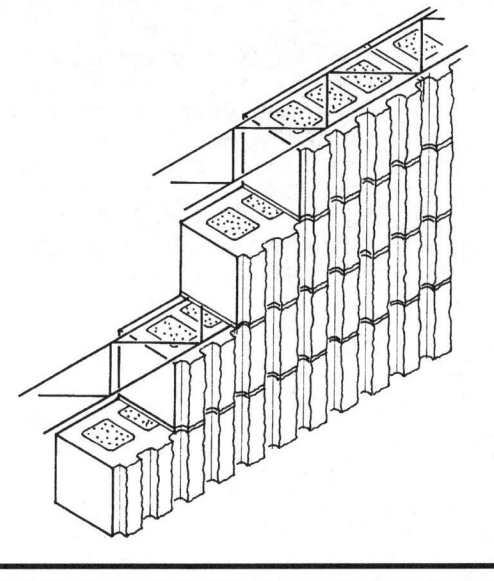

Exterior split ribbed block walls are defined in the following terms; structural reinforcement, weight, percent solid, size, number of ribs and insulation. Within each of these categories two to four variations are shown. No costs are included for brick shelf or relieving angles. Costs include control joints every 20′ and horizontal reinforcing.

B **SHELL**

B2010 113		Split Ribbed Block Wall - Regular Weight				COST PER S.F.		
	TYPE	SIZE (IN.)	RIBS	CORE FILL		MAT.	INST.	TOTAL
1220	Hollow	4x8x16	4	none		2.17	5.10	7.27
1250			8	none		2.53	5.10	7.63
1280			16	none		2.06	5.15	7.21
1430		8x8x16	8	perlite		4.41	6	10.41
1440				styrofoam		4.84	5.70	10.54
1450				none		3.94	5.70	9.64
1530		12x8x16	8	perlite		5.55	8	13.55
1540				styrofoam		5.90	7.45	13.35
1550				none		4.78	7.45	12.23
2120	75% solid	4x8x16	4	none		2.72	5.15	7.87
2150			8	none		3.21	5.15	8.36
2180			16	none		2.61	5.20	7.81
2520	Solid	4x8x16	4	none		3.31	5.20	8.51
2550			8	none		3.88	5.20	9.08
2580			16	none		3.13	5.30	8.43

B2010 113		Reinforced Split Ribbed Block Wall - Regular Weight				COST PER S.F.		
	TYPE	SIZE (IN.)	RIBS	VERT. REINF. & GROUT SPACING		MAT.	INST.	TOTAL
5200	Hollow	4x8x16	4	#4 @ 48″		2.27	5.55	7.82
5230			8	#4 @ 48″		2.63	5.55	8.18
5260			16	#4 @ 48″		2.16	5.60	7.76
5430		8x8x16	8	#4 @ 48″		4.16	6.30	10.46
5440				#5 @ 32″		4.31	6.50	10.81
5450				#5 @ 16″		4.67	7.30	11.97
5530		12x8x16	8	#4 @ 48″		5.10	8.05	13.15
5540				#5 @ 32″		5.30	8.25	13.55
5550				#5 @ 16″		5.80	9.10	14.90
6230	75% solid	6x8x16	8	#4 @ 48″		4.09	5.85	9.94
6240				#5 @ 32″		4.18	6	10.18
6250				#5 @ 16″		4.35	6.50	10.85
6330		8x8x16	8	#4 @ 48″		5.15	6.25	11.40
6340				#5 @ 32″		5.25	6.45	11.70
6350				#5 @ 16″		5.40	7.10	12.50

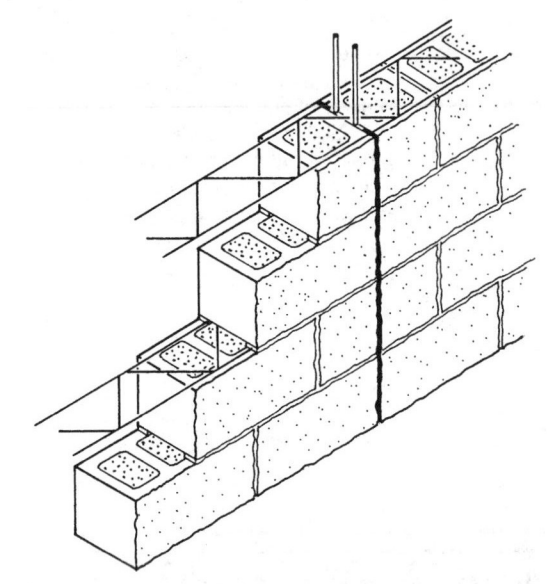

Exterior split face block walls are defined in the following terms; structural reinforcement, weight, percent solid, size, scores and insulation. Within each of these categories two to four variations are shown. No costs are included for brick shelf or relieving angles. Costs include control joints every 20′ and horizontal reinforcing.

SHELL B

B2010 115			Split Face Block Wall - Regular Weight					
	TYPE	SIZE (IN.)	SCORES	CORE FILL		COST PER S.F.		
						MAT.	INST.	TOTAL
1200	Hollow	8x4x16	0	perlite		6.45	6.65	13.10
1210				styrofoam		6.90	6.35	13.25
1240				none		6	6.35	12.35
1250			1	perlite		6.90	6.65	13.55
1260				styrofoam		7.35	6.35	13.70
1290				none		6.45	6.35	12.80
1300		12x4x16	0	perlite		8.40	7.55	15.95
1310				styrofoam		8.75	7	15.75
1340				none		7.65	7	14.65
1350			1	perlite		8.85	7.55	16.40
1360				styrofoam		9.20	7	16.20
1390				none		8.10	7	15.10
1400		4x8x16	0	none		2.17	5	7.17
1450			1	none		2.44	5.10	7.54
1500		6x8x16	0	perlite		3.13	5.75	8.88
1510				styrofoam		3.71	5.55	9.26
1540				none		2.81	5.55	8.36
1550			1	perlite		3.42	5.85	9.27
1560				styrofoam		4	5.65	9.65
1590				none		3.10	5.65	8.75
1600		8x8x16	0	perlite		3.80	6.20	10
1610				styrofoam		4.23	5.90	10.13
1640				none		3.33	5.90	9.23
1650	Hollow	8x8x16	1	perlite		4.03	6.30	10.33
1660				styrofoam		4.46	6	10.46
1690				none		3.56	6	9.56
1700		12x8x16	0	perlite		4.89	8.10	12.99
1710				styrofoam		5.25	7.55	12.80
1740				none		4.13	7.55	11.68
1750			1	perlite		5.10	8.25	13.35
1760				styrofoam		5.50	7.70	13.20
1790				none		4.36	7.70	12.06

B2010 Exterior Walls

| B2010 115 | Split Face Block Wall - Regular Weight | | | | | | | |

	TYPE	SIZE (IN.)	SCORES	CORE FILL		COST PER S.F.		
						MAT.	INST.	TOTAL
1800	75% solid	8x4x16	0	perlite		7.75	6.60	14.35
1840				none		7.55	6.45	14
1852	75% solid	8x4x16	1	perlite		8.25	6.60	14.85
1890				none		8.05	6.45	14.50
2000		4x8x16	0	none		2.72	5.10	7.82
2050			1	none		2.96	5.15	8.11
2400	Solid	8x4x16	0	none		9.25	6.55	15.80
2450			1	none		9.70	6.55	16.25
2900		12x8x16	0	none		6.30	7.85	14.15
2950			1	none		6.55	8	14.55

| B2010 115 | Reinforced Split Face Block Wall - Regular Weight | | | | | | | |

	TYPE	SIZE (IN.)	SCORES	VERT. REINF. & GROUT SPACING		COST PER S.F.		
						MAT.	INST.	TOTAL
5200	Hollow	8x4x16	0	#4 @ 48"		6.20	6.95	13.15
5210				#5 @ 32"		6.35	7.15	13.50
5240				#5 @ 16"		6.70	7.95	14.65
5250			1	#4 @ 48"		6.65	6.95	13.60
5260				#5 @ 32"		6.80	7.15	13.95
5290				#5 @ 16"		7.15	7.95	15.10
5700		12x8x16	0	#4 @ 48"		4.45	8.15	12.60
5710				#5 @ 32"		4.64	8.35	12.99
5740				#5 @ 16"		5.15	9.20	14.35
5750			1	#4 @ 48"		4.68	8.30	12.98
5760				#5 @ 32"		4.87	8.50	13.37
5790				#5 @ 16"		5.40	9.35	14.75
6000	75% solid	8x4x16	0	#4 @ 48"		7.65	6.90	14.55
6010				#5 @ 32"		7.70	7.10	14.80
6040				#5 @ 16"		7.90	7.75	15.65
6050			1	#4 @ 48"		8.15	6.90	15.05
6060				#5 @ 32"		8.20	7.10	15.30
6090				#5 @ 16"		8.40	7.75	16.15
6100		12x4x16	0	#4 @ 48"		9.85	7.65	17.50
6110				#5 @ 32"		10	7.80	17.80
6140				#5 @ 16"		10.25	8.30	18.55
6150			1	#4 @ 48"		10.25	7.65	17.90
6160				#5 @ 32"		10.40	7.80	18.20
6190				#5 @ 16"		10.70	8.45	19.15
6700	Solid-double Wythe	2-4x8x16	0	#4 @ 48" E.W.		7.25	11.45	18.70
6710				#5 @ 32" E.W.		7.40	11.55	18.95
6740				#5 @ 16" E.W.		7.65	12.05	19.70
6750			1	#4 @ 48" E.W.		7.70	11.55	19.25
6760				#5 @ 32" E.W.		7.85	11.65	19.50
6790				#5 @ 16" E.W.		8.15	12.15	20.30
6800		2-6x8x16	0	#4 @ 48" E.W.		9.45	12.55	22
6810				#5 @ 32" E.W.		9.60	12.65	22.25
6840				#5 @ 16" E.W.		9.85	13.15	23
6850			1	#4 @ 48" E.W.		9.90	12.75	22.65
6860				#5 @ 32" E.W.		10.05	12.85	22.90
6890				#5 @ 16" E.W.		10.30	13.35	23.65

B2010 Exterior Walls

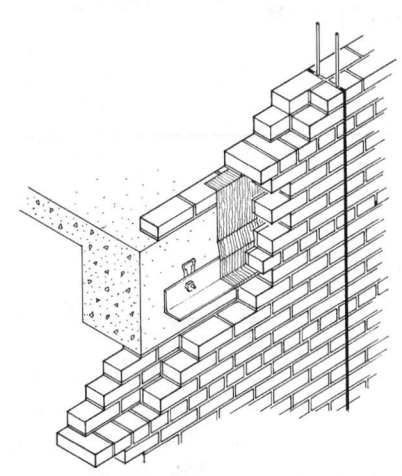

Exterior solid brick walls are defined in the following terms; structural reinforcement, type of brick, thickness, and bond. Nine different types of face bricks are presented, with single wythes shown in four different bonds. Shelf angles are included in system components. These walls do not include ties and as such are not tied to a backup wall.

SHELL B

B2010 119			Solid Brick Walls - Single Wythe					
	TYPE	THICKNESS (IN.)	BOND			COST PER S.F.		
						MAT.	INST.	TOTAL
1000	Common	4	running			2.90	9.30	12.20
1010			common			3.48	10.60	14.08
1050			Flemish			3.84	12.85	16.69
1100			English			4.29	13.60	17.89
1150	Standard	4	running			3.57	9.65	13.22
1160			common			4.21	11.10	15.31
1200			Flemish			4.66	13.25	17.91
1250			English			5.20	14.05	19.25
1300	Glazed	4	running			8.15	10	18.15
1310			common			9.70	11.60	21.30
1350			Flemish			10.75	14.05	24.80
1400			English			12.05	15	27.05
1600	Economy	4	running			3.69	7.40	11.09
1610			common			4.34	8.45	12.79
1650			Flemish			4.78	10	14.78
1700			English			9	10.60	19.60
1900	Fire	4-1/2	running			7.40	8.45	15.85
1910			common			8.80	9.65	18.45
1950			Flemish			9.75	11.60	21.35
2000			English			10.95	12.20	23.15
2050	King	3-1/2	running			3.11	7.55	10.66
2060			common			3.64	8.70	12.34
2100			Flemish			4.43	10.40	14.83
2150			English			4.89	10.85	15.74
2200	Roman	4	running			5.35	8.70	14.05
2210			common			6.35	10	16.35
2250			Flemish			7.05	11.90	18.95
2300			English			7.90	12.85	20.75
2350	Norman	4	running			4.02	7.20	11.22
2360			common			4.74	8.20	12.94
2400			Flemish			8.80	9.80	18.60
2450			English			5.85	10.40	16.25

B2010 Exterior Walls

B2010 119 — Solid Brick Walls - Single Wythe

	TYPE	THICKNESS (IN.)	BOND			COST PER S.F.		
						MAT.	INST.	TOTAL
2500	Norwegian	4	running			3.71	6.40	10.11
2510			common			4.37	7.30	11.67
2550			Flemish			4.82	8.70	13.52
2600			English			5.40	9.15	14.55

B2010 121 — Solid Brick Walls - Double Wythe

	TYPE	THICKNESS (IN.)	COLLAR JOINT THICKNESS (IN.)			COST PER S.F.		
						MAT.	INST.	TOTAL
4100	Common	8	3/4			5.75	15.60	21.35
4150	Standard	8	1/2			6.95	16.10	23.05
4200	Glazed	8	1/2			16.10	16.65	32.75
4250	Engineer	8	1/2			7.40	14.20	21.60
4300	Economy	8	3/4			7.20	12.20	19.40
4350	Double	8	3/4			9.35	9.90	19.25
4400	Fire	8	3/4			14.65	14.20	28.85
4450	King	7	3/4			6	12.50	18.50
4500	Roman	8	1			10.50	14.65	25.15
4550	Norman	8	3/4			7.85	11.95	19.80
4600	Norwegian	8	3/4			7.20	10.60	17.80
4650	Utility	8	3/4			6.70	9.15	15.85
4700	Triple	8	3/4			5.25	8.25	13.50
4750	SCR	12	3/4			9.95	12.20	22.15
4800	Norwegian	12	3/4			8.65	10.80	19.45
4850	Jumbo	12	3/4			9.40	9.45	18.85

B2010 121 — Reinforced Brick Walls

	TYPE	THICKNESS (IN.)	WYTHE	REINF. & SPACING		COST PER S.F.		
						MAT.	INST.	TOTAL
7200	Common	4"	1	#4 @ 48"vert		3.04	9.65	12.69
7220				#5 @ 32"vert		3.12	9.75	12.87
7230				#5 @ 16"vert		3.28	10.20	13.48
7700	King	9"	2	#4 @ 48" E.W.		6.35	13.10	19.45
7720				#5 @ 32" E.W.		6.50	13.20	19.70
7730				#5 @ 16" E.W.		6.75	13.60	20.35
7800	Common	10"	2	#4 @ 48" E.W.		6.05	16.20	22.25
7820				#5 @ 32" E.W.		6.20	16.30	22.50
7830				#5 @ 16" E.W.		6.45	16.70	23.15
7900	Standard	10"	2	#4 @ 48" E.W.		7.30	16.70	24
7920				#5 @ 32" E.W.		7.45	16.80	24.25
7930				#5 @ 16" E.W.		7.70	17.20	24.90
8100	Economy	10"	2	#4 @ 48" E.W.		7.55	12.80	20.35
8120				#5 @ 32" E.W.		7.70	12.90	20.60
8130				#5 @ 16" E.W.		7.95	13.30	21.25
8200	Double	10"	2	#4 @ 48" E.W.		9.70	10.50	20.20
8220				#5 @ 32" E.W.		9.85	10.60	20.45
8230				#5 @ 16" E.W.		10.10	11	21.10
8300	Fire	10"	2	#4 @ 48" E.W.		15	14.80	29.80
8320				#5 @ 32" E.W.		15.15	14.90	30.05
8330				#5 @ 16" E.W.		15.40	15.30	30.70
8400	Roman	10"	2	#4 @ 48" E.W.		10.85	15.25	26.10
8420				#5 @ 32" E.W.		11	15.35	26.35
8430				#5 @ 16" E.W.		11.25	15.75	27

SHELL

B

B2010 Exterior Walls

| B2010 121 | | | Reinforced Brick Walls | | | | | |

	TYPE	THICKNESS (IN.)	WYTHE	REINF. & SPACING		COST PER S.F.		
						MAT.	INST.	TOTAL
8500	Norman	10"	2	#4 @ 48" E.W.		8.20	12.55	20.75
8520				#5 @ 32" E.W.		8.35	12.65	21
8530				#5 @ 16" E.W.		8.60	13.05	21.65
8600	Norwegian	10"	2	#4 @ 48" E.W.		7.55	11.20	18.75
8620				#5 @ 32" E.W.		7.70	11.30	19
8630				#5 @ 16" E.W.		7.95	11.70	19.65
8800	Triple	10"	2	#4 @ 48" E.W.		5.55	8.85	14.40
8820				#5 @ 32" E.W.		5.70	8.95	14.65
8830				#5 @ 16" E.W.		5.95	9.35	15.30

SHELL B

Important: See the Reference Section for critical supporting data - Location Factors & Historical Cost Indexes

B2010 Exterior Walls

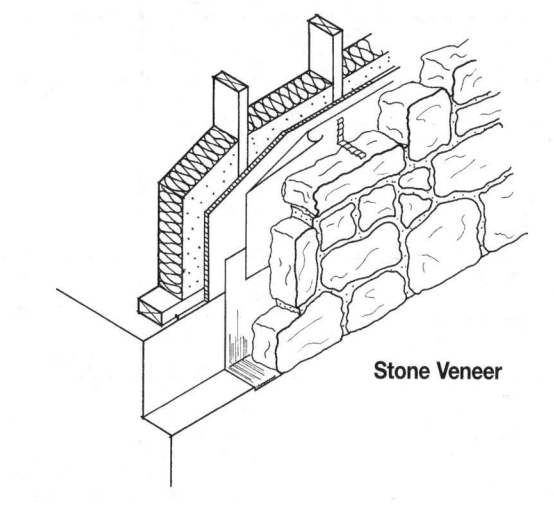

The table below lists costs per S.F. for stone veneer walls on various backup using different stone.

Stone Veneer

B2010 123	Stone Veneer	COST PER S.F.		
		MAT.	INST.	TOTAL
2000	Ashlar veneer,4″, 2″x4″ stud backup,16″ O.C., 8′ high, low priced stone	7.05	12.60	19.65
2100	Metal stud backup, 8′ high, 16″ O.C.	7.30	12.80	20.10
2150	24″ O.C.	7.10	12.50	19.60
2200	Conc. block backup, 4″ thick	7.65	15.40	23.05
2300	6″ thick	8.10	15.55	23.65
2350	8″ thick	8.25	15.95	24.20
2400	10″ thick	8.90	16.85	25.75
2500	12″ thick	8.95	18.10	27.05
3100	High priced stone, wood stud backup, 10′ high, 16″ O.C.	12.75	14.20	26.95
3200	Metal stud backup, 10′ high, 16″ O.C.	12.95	14.40	27.35
3250	24″ O.C.	12.75	14.10	26.85
3300	Conc. block backup, 10′ high, 4″ thick	13.30	17	30.30
3350	6″ thick	13.75	17	30.75
3400	8″ thick	13.90	17.55	31.45
3450	10″ thick	14.55	18.45	33
3500	12″ thick	14.60	19.70	34.30
4000	Indiana limestone 2″ thick,sawn finish,wood stud backup,10′ high,16″ O.C.	14.90	10.55	25.45
4100	Metal stud backup, 10′ high, 16″ O.C.	15.15	10.75	25.90
4150	24″ O.C.	14.95	10.45	25.40
4200	Conc. block backup, 4″ thick	15.50	13.35	28.85
4250	6″ thick	15.95	13.50	29.45
4300	8″ thick	16.10	13.90	30
4350	10″ thick	16.75	14.80	31.55
4400	12″ thick	16.80	16.05	32.85
4450	2″ thick, smooth finish, wood stud backup, 8′ high, 16″ O.C.	14.90	10.55	25.45
4550	Metal stud backup, 8′ high, 16″ O.C.	15.15	10.65	25.80
4600	24″ O.C.	14.95	10.30	25.25
4650	Conc. block backup, 4″ thick	15.50	13.20	28.70
4700	6″ thick	15.95	13.35	29.30
4750	8″ thick	16.10	13.75	29.85
4800	10″ thick	16.75	14.80	31.55
4850	12″ thick	16.80	16.05	32.85
5350	4″ thick, smooth finish, wood stud backup, 8′ high, 16″ O.C.	20	10.55	30.55
5450	Metal stud backup, 8′ high, 16″ O.C.	20.50	10.75	31.25
5500	24″ O.C.	20	10.45	30.45
5550	Conc. block backup, 4″ thick	20.50	13.35	33.85
5600	6″ thick	21	13.50	34.50
5650	8″ thick	21.50	13.90	35.40

SHELL

B

B2010 Exterior Walls

B2010 123	Stone Veneer	COST PER S.F.		
		MAT.	INST.	TOTAL
5700	10" thick	22	14.80	36.80
5750	12" thick	22	16.05	38.05
6000	Granite, grey or pink, 2" thick, wood stud backup, 8' high, 16" O.C.	19.40	17.90	37.30
6100	Metal studs, 8' high, 16" O.C.	19.65	18.10	37.75
6150	24" O.C.	19.45	17.80	37.25
6200	Conc. block backup, 4" thick	20	20.50	40.50
6250	6" thick	20.50	21	41.50
6300	8" thick	20.50	21	41.50
6350	10" thick	21.50	22	43.50
6400	12" thick	21.50	23.50	45
6900	4" thick, wood stud backup, 8' high, 16" O.C.	31.50	20.50	52
7000	Metal studs, 8' high, 16" O.C.	32	20.50	52.50
7050	24" O.C.	31.50	20	51.50
7100	Conc. block backup, 4" thick	32	23	55
7150	6" thick	32.50	23	55.50
7200	8" thick	32.50	23.50	56
7250	10" thick	33.50	24.50	58
7300	12" thick	33.50	25.50	59

SHELL B

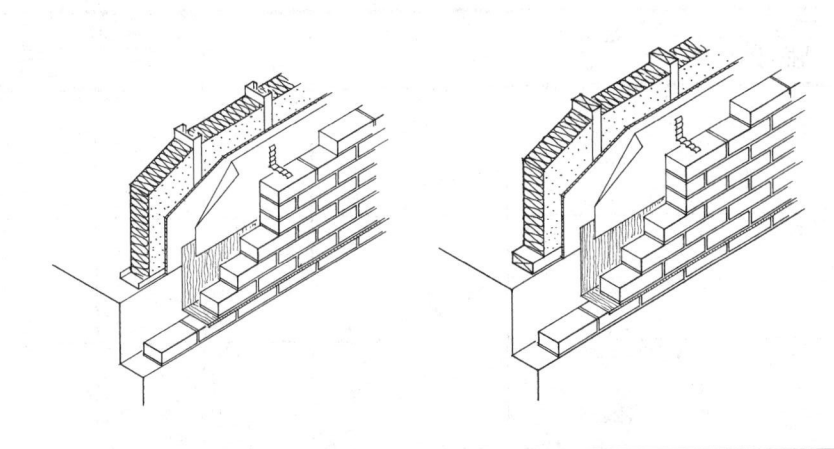

Exterior brick veneer/stud backup walls are defined in the following terms; type of brick and studs, stud spacing and bond. All systems include a brick shelf, ties to the backup and all necessary dampproofing and insulation.

B2010 126	Brick Veneer/Wood Stud Backup							
	FACE BRICK	STUD BACKUP	STUD SPACING (IN.)	BOND		COST PER S.F.		
						MAT.	INST.	TOTAL
1100	Standard	2x4-wood	16	running		4.92	11.35	16.27
1120				common		5.55	12.80	18.35
1140				Flemish		6	14.95	20.95
1160				English		6.55	15.75	22.30
1400		2x6-wood	16	running		5.15	11.45	16.60
1420				common		5.80	12.90	18.70
1440				Flemish		6.25	15.05	21.30
1460				English		6.80	15.85	22.65
1700	Glazed	2x4-wood	16	running		9.50	11.70	21.20
1720				common		11.05	13.30	24.35
1740				Flemish		12.10	15.75	27.85
1760				English		13.40	16.70	30.10
2300	Engineer	2x4-wood	16	running		5.15	10.15	15.30
2320				common		5.80	11.35	17.15
2340				Flemish		6.30	13.30	19.60
2360				English		6.85	13.90	20.75
2900	Roman	2x4-wood	16	running		6.70	10.40	17.10
2920				common		7.70	11.70	19.40
2940				Flemish		8.40	13.60	22
2960				English		9.25	14.55	23.80
4100	Norwegian	2x4-wood	16	running		5.05	8.15	13.20
4120				common		5.70	9	14.70
4140				Flemish		6.15	10.40	16.55
4160				English		6.75	10.85	17.60

B2010 126	Brick Veneer/Metal Stud Backup							
	FACE BRICK	STUD BACKUP	STUD SPACING (IN.)	BOND		COST PER S.F.		
						MAT.	INST.	TOTAL
5100	Standard	25ga.x6"NLB	24	running		4.60	11.10	15.70
5120				common		5.25	12.55	17.80
5140				Flemish		5.45	14.25	19.70
5160				English		6.25	15.50	21.75
5200		20ga.x3-5/8"NLB	16	running		4.68	11.55	16.23
5220				common		5.30	13	18.30
5240				Flemish		5.75	15.15	20.90
5260				English		6.30	15.95	22.25

SHELL

B

| B2010 126 | | | Brick Veneer/Metal Stud Backup | | | | |

SHELL B

	FACE BRICK	STUD BACKUP	STUD SPACING (IN.)	BOND		COST PER S.F.		
						MAT.	INST.	TOTAL
5400	Standard	16ga.x3-5/8"LB	16	running		5.05	11.70	16.75
5420				common		5.65	13.15	18.80
5440				Flemish		6.10	15.30	21.40
5460				English		6.65	16.10	22.75
5700	Glazed	25ga.x6"NLB	24	running		9.20	11.45	20.65
5720				common		10.75	13.05	23.80
5740				Flemish		11.80	15.50	27.30
5760				English		13.10	16.45	29.55
5800		20ga.x3-5/8"NLB	24	running		9.15	11.55	20.70
5820				common		10.70	13.15	23.85
5840				Flemish		11.75	15.60	27.35
5860				English		13.05	16.55	29.60
6000		16ga.x3-5/8"LB	16	running		9.60	12.05	21.65
6020				common		11.15	13.65	24.80
6040				Flemish		12.20	16.10	28.30
6060				English		13.50	17.05	30.55
6300	Engineer	25ga.x6"NLB	24	running		4.82	9.90	14.72
6320				common		5.50	11.10	16.60
6340				Flemish		5.95	13.05	19
6360				English		6.55	13.65	20.20
6400		20ga.x3-5/8"NLB	16	running		4.90	10.35	15.25
6420				common		5.60	11.55	17.15
6440				Flemish		6.05	13.50	19.55
6460				English		6.65	14.10	20.75
6900	Roman	25ga.x6"NLB	24	running		6.40	10.15	16.55
6920				common		7.40	11.45	18.85
6940				Flemish		8.10	13.35	21.45
6960				English		8.95	14.30	23.25
7000		20ga.x3-5/8"NLB	16	running		6.45	10.60	17.05
7020				common		7.45	11.90	19.35
7040				Flemish		8.15	13.80	21.95
7060				English		9	14.75	23.75
7500	Norman	25ga.x6"NLB	24	running		5.05	8.65	13.70
7520				common		5.75	9.65	15.40
7540				Flemish		9.85	11.25	21.10
7560				English		6.90	11.85	18.75
7600		20ga.x3-5/8"NLB	24	running		5.05	8.75	13.80
7620				common		5.75	9.75	15.50
7640				Flemish		9.80	11.35	21.15
7660				English		6.85	11.95	18.80
8100	Norwegian	25ga.x6"NLB	24	running		4.74	7.90	12.64
8120				common		5.40	8.75	14.15
8140				Flemish		5.85	10.15	16
8160				English		6.40	10.60	17
8400	Norwegian	16ga.x3-5/8"LB	16	running		5.15	8.50	13.65
8420				common		5.85	9.35	15.20
8440				Flemish		6.30	10.75	17.05
8460				English		6.85	11.20	18.05

B2010 Exterior Walls

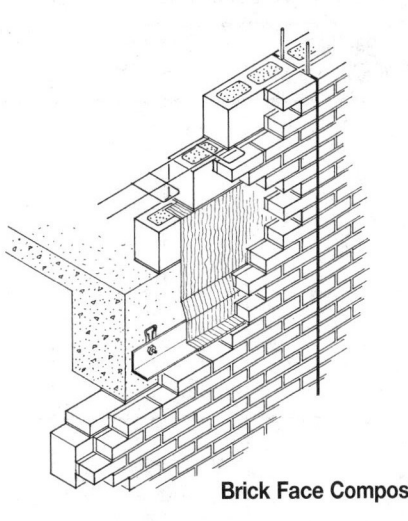

Brick Face Composite Wall

| B2010 130 | Brick Face Composite Wall - Double Wythe | | | | | | | |

	FACE BRICK	BACKUP MASONRY	BACKUP THICKNESS (IN.)	BACKUP CORE FILL		COST PER S.F.		
						MAT.	INST.	TOTAL
1000	Standard	common brick	4	none		6.25	17.65	23.90
1040		SCR brick	6	none		8.40	15.75	24.15
1080		conc. block	4	none		4.97	14.20	19.17
1120			6	perlite		5.75	14.70	20.45
1160				styrofoam		6.35	14.50	20.85
1200			8	perlite		6.05	15.05	21.10
1240				styrofoam		6.45	14.75	21.20
1520		glazed block	4	none		10.95	15.15	26.10
1560			6	perlite		11.65	15.60	27.25
1600				styrofoam		12.25	15.40	27.65
1640			8	perlite		12.15	16	28.15
1680				styrofoam		12.60	15.75	28.35
1720	Standard	clay tile	4	none		8.10	13.65	21.75
1760			6	none		8.15	14	22.15
1800			8	none		9.50	14.50	24
1840		glazed tile	4	none		11.45	18	29.45
1880								
2000	Glazed	common brick	4	none		10.80	18	28.80
2040		SCR brick	6	none		13	16.10	29.10
2080		conc. block	4	none		9.55	14.55	24.10
2200			8	perlite		10.60	15.40	26
2240				styrofoam		11.05	15.10	26.15
2280		L.W. block	4	none		9.75	14.45	24.20
2400			8	perlite		11	15.30	26.30
2520		glazed block	4	none		15.50	15.50	31
2560			6	perlite		16.25	15.95	32.20
2600				styrofoam		16.80	15.75	32.55
2640			8	perlite		16.75	16.35	33.10
2680				styrofoam		17.15	16.10	33.25

SHELL

B

B2010 Exterior Walls

B2010 130	Brick Face Composite Wall - Double Wythe

	FACE BRICK	BACKUP MASONRY	BACKUP THICKNESS (IN.)	BACKUP CORE FILL		COST PER S.F.		
						MAT.	INST.	TOTAL
2720		clay tile	4	none		12.70	14	26.70
2760			6	none		12.75	14.35	27.10
2800			8	none		14.05	14.85	28.90
2840		glazed tile	4	none		16	18.35	34.35
3000	Engineer	common brick	4	none		6.45	16.45	22.90
3040		SCR brick	6	none		8.65	14.55	23.20
3080		conc. block	4	none		5.20	13	18.20
3200			8	perlite		6.25	13.85	20.10
3280		L.W. block	4	none		4.26	9.20	13.46
3322	Engineer	L.W. block	6	perlite		6.10	13.40	19.50
3520		glazed block	4	none		11.15	13.95	25.10
3560			6	perlite		11.85	14.40	26.25
3600				styrofoam		12.45	14.20	26.65
3640			8	perlite		12.35	14.80	27.15
3680				styrofoam		12.80	14.55	27.35
3720		clay tile	4	none		8.35	12.45	20.80
3800			8	none		9.70	13.30	23
3840		glazed tile	4	none		11.65	16.80	28.45
4000	Roman	common brick	4	none		8	16.70	24.70
4040		SCR brick	6	none		10.20	14.80	25
4080		conc. block	4	none		6.75	13.25	20
4120			6	perlite		7.55	13.75	21.30
4200			8	perlite		7.80	14.10	21.90
5000	Norman	common brick	4	none		6.70	15.20	21.90
5040		SCR brick	6	none		8.85	13.30	22.15
5280		L.W. block	4	none		5.60	11.65	17.25
5320			6	perlite		6.35	12.30	18.65
5720		clay tile	4	none		8.55	11.20	19.75
5760			6	none		8.60	11.55	20.15
5840		glazed tile	4	none		11.90	15.55	27.45
6000	Norwegian	common brick	4	none		6.35	14.40	20.75
6040		SCR brick	6	none		8.55	12.50	21.05
6080		conc. block	4	none		5.10	10.95	16.05
6120			6	perlite		5.90	11.50	17.40
6160				styrofoam		6.50	11.25	17.75
6200			8	perlite		6.20	11.80	18
6520		glazed block	4	none		11.05	11.95	23
6560			6	perlite		11.80	12.40	24.20
7000	Utility	common brick	4	none		6.10	13.65	19.75
7040		SCR brick	6	none		8.30	11.75	20.05
7080		conc. block	4	none		4.85	10.20	15.05
7120			6	perlite		5.65	10.75	16.40
7160				styrofoam		6.20	10.50	16.70
7200			8	perlite		5.90	11.05	16.95
7240				styrofoam		6.35	10.80	17.15
7280		L.W. block	4	none		5.05	10.10	15.15
7320			6	perlite		5.80	10.65	16.45
7520		glazed block	4	none		10.80	11.20	22
7560			6	perlite		11.55	11.65	23.20
7720		clay tile	4	none		8	9.65	17.65
7760			6	none		8.05	10.05	18.10
7840		glazed tile	4	none		11.30	14	25.30

Important: See the Reference Section for critical supporting data - Location Factors & Historical Cost Indexes

B2010 Exterior Walls

B2010 130	Brick Face Composite Wall - Triple Wythe

	FACE BRICK	MIDDLE WYTHE	INSIDE MASONRY	TOTAL THICKNESS (IN.)		COST PER S.F.		
						MAT.	INST.	TOTAL
8000	Standard	common brick	standard brick	12		9.55	25	34.55
8100		4" conc. brick	standard brick	12		9.30	26	35.30
8120		4" conc. brick	common brick	12		8.65	25.50	34.15
8200	Glazed	common brick	standard brick	12		14.15	25.50	39.65
8300		4" conc. brick	standard brick	12		13.90	26	39.90
8320		4" conc. brick	glazed brick	12		13.20	26	39.20

SHELL

B

295

B2010 Exterior Walls

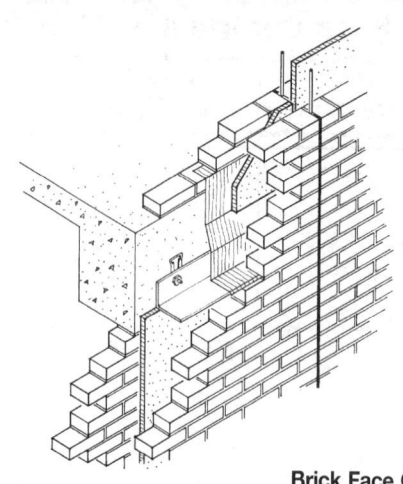

Brick Face Cavity Wall

B2010 132					Brick Face Cavity Wall				
	FACE BRICK	**BACKUP MASONRY**	**TOTAL THICKNESS (IN.)**	**CAVITY INSULATION**		**COST PER S.F.**			
						MAT.	**INST.**	**TOTAL**	
1000	Standard	4"common brick	10	polystyrene		6.10	17.65	23.75	
1020				none		5.95	17.20	23.15	
1040		6"SCR brick	12	polystyrene		8.30	15.75	24.05	
1060				none		8.15	15.30	23.45	
1080		4"conc. block	10	polystyrene		4.84	14.20	19.04	
1100				none		4.69	13.75	18.44	
1120		6"conc. block	12	polystyrene		5.30	14.50	19.80	
1140				none		5.15	14.05	19.20	
1160		4"L.W. block	10	polystyrene		5	14.10	19.10	
1180				none		4.87	13.65	18.52	
1200		6"L.W. block	12	polystyrene		5.45	14.40	19.85	
1220				none		5.30	13.95	19.25	
1240		4"glazed block	10	polystyrene		10.80	15.20	26	
1260				none		10.65	14.70	25.35	
1280		6"glazed block	12	polystyrene		10.80	15.20	26	
1300				none		10.65	14.70	25.35	
1320		4"clay tile	10	polystyrene		8	13.65	21.65	
1340				none		7.85	13.20	21.05	
1360		4"glazed tile	10	polystyrene		11.30	18	29.30	
1380				none		11.15	17.55	28.70	
1500	Glazed	4" common brick	10	polystyrene		10.70	18	28.70	
1520				none		10.55	17.55	28.10	
1580		4" conc. block	10	polystyrene		9.40	14.55	23.95	
1600				none		9.25	14.10	23.35	
1660		4" L.W. block	10	polystyrene		9.60	14.45	24.05	
1680				none		9.45	14	23.45	
1740		4" glazed block	10	polystyrene		15.40	15.55	30.95	
1760				none		15.25	15.05	30.30	
1820		4" clay tile	10	polystyrene		12.55	14	26.55	
1840				none		12.40	13.55	25.95	
1860		4" glazed tile	10	polystyrene		12.60	14.40	27	
1880				none		12.45	13.90	26.35	
2000	Engineer	4" common brick	10	polystyrene		6.30	16.45	22.75	
2020				none		6.15	16	22.15	
2080		4" conc. block	10	polystyrene		5.05	13	18.05	
2100				none		4.91	12.55	17.46	

Important: See the Reference Section for critical supporting data - Location Factors & Historical Cost Indexes

SHELL B

B2010 Exterior Walls

B2010 132	Brick Face Cavity Wall

	FACE BRICK	BACKUP MASONRY	TOTAL THICKNESS (IN.)	CAVITY INSULATION		COST PER S.F.		
						MAT.	INST.	TOTAL
2162	Engineer	4" L.W. block	10	polystyrene		5.25	12.90	18.15
2180				none		5.10	12.45	17.55
2240		4" glazed block	10	polystyrene		11	14	25
2260				none		10.85	13.50	24.35
2320		4" clay tile	10	polystyrene		8.20	12.45	20.65
2340				none		8.05	12	20.05
2360		4" glazed tile	10	polystyrene		11.50	16.80	28.30
2380				none		11.35	16.35	27.70
2500	Roman	4" common brick	10	polystyrene		7.90	16.70	24.60
2520				none		7.75	16.25	24
2580		4" conc. block	10	polystyrene		6.60	13.25	19.85
2600				none		6.45	12.80	19.25
2660		4" L.W. block	10	polystyrene		6.80	13.15	19.95
2680				none		6.65	12.70	19.35
2740		4" glazed block	10	polystyrene		12.60	14.25	26.85
2760				none		12.45	13.75	26.20
2820		4"clay tile	10	polystyrene		9.75	12.70	22.45
2840				none		9.60	12.25	21.85
2860		4"glazed tile	10	polystyrene		13.10	17.05	30.15
2880				none		12.95	16.60	29.55
3000	Norman	4" common brick	10	polystyrene		6.55	15.20	21.75
3020				none		6.40	14.75	21.15
3080		4" conc. block	10	polystyrene		5.30	11.75	17.05
3100				none		5.15	11.30	16.45
3160		4" L.W. block	10	polystyrene		5.45	11.65	17.10
3180				none		5.30	11.20	16.50
3240		4" glazed block	10	polystyrene		11.25	12.75	24
3260				none		11.10	12.25	23.35
3320		4" clay tile	10	polystyrene		8.45	11.20	19.65
3340				none		8.30	10.75	19.05
3360		4" glazed tile	10	polystyrene		11.75	15.55	27.30
3380				none		11.60	15.10	26.70
3500	Norwegian	4" common brick	10	polystyrene		6.25	14.45	20.70
3520				none		6.10	13.95	20.05
3580		4" conc. block	10	polystyrene		4.98	11	15.98
3600				none		4.83	10.50	15.33
3660		4" L.W. block	10	polystyrene		5.15	10.90	16.05
3680				none		5	10.45	15.45
3740		4" glazed block	10	polystyrene		10.95	11.95	22.90
3760				none		10.80	11.50	22.30
3820		4" clay tile	10	polystyrene		8.10	10.45	18.55
3840				none		7.95	9.95	17.90
3860		4" glazed tile	10	polystyrene		11.45	14.80	26.25
3880				none		11.30	14.30	25.60
4000	Utility	4" common brick	10	polystyrene		6	13.70	19.70
4020				none		5.85	13.20	19.05
4080		4" conc. block	10	polystyrene		4.72	10.25	14.97
4100				none		4.57	9.75	14.32
4160		4" L.W. block	10	polystyrene		4.90	10.15	15.05
4180				none		4.75	9.70	14.45
4240		4" glazed block	10	polystyrene		10.70	11.20	21.90
4260				none		10.55	10.75	21.30

SHELL

B

B2010 Exterior Walls

| B2010 132 | Brick Face Cavity Wall | | | | | | | |

	FACE BRICK	BACKUP MASONRY	TOTAL THICKNESS (IN.)	CAVITY INSULATION		COST PER S.F.		
						MAT.	INST.	TOTAL
4320	Utility	4" clay tile	10	polystyrene		7.85	9.70	17.55
4340				none		7.70	9.20	16.90
4360		4" glazed tile	10	polystyrene		11.20	14.05	25.25
4380				none		11.05	13.55	24.60

| B2010 132 | Brick Face Cavity Wall - Insulated Backup | | | | | | | |

	FACE BRICK	BACKUP MASONRY	TOTAL THICKNESS (IN.)	BACKUP CORE FILL		COST PER S.F.		
						MAT.	INST.	TOTAL
5100	Standard	6" conc. block	10	perlite		5.50	14.25	19.75
5120				styrofoam		6.05	14.05	20.10
5180		6" L.W. block	10	perlite		5.60	14.15	19.75
5200				styrofoam		6.20	13.95	20.15
5260		6" glazed block	10	perlite		11.35	15.15	26.50
5280				styrofoam		11.95	14.95	26.90
5340		6" clay tile	10	none		7.90	13.55	21.45
5360		8" clay tile	12	none		9.20	14.05	23.25
5600	Glazed	6" conc. block	10	perlite		10.05	14.60	24.65
5620				styrofoam		10.65	14.40	25.05
5680		6" L.W. block	10	perlite		10.20	14.50	24.70
5700				styrofoam		10.80	14.30	25.10
5760		6" glazed block	10	perlite		15.95	15.50	31.45
5780				styrofoam		16.55	15.30	31.85
5840		6" clay tile	10	none		12.45	13.90	26.35
5860		8"clay tile	8	none		13.80	14.40	28.20
6100	Engineer	6" conc. block	10	perlite		5.70	13.05	18.75
6120				styrofoam		6.30	12.85	19.15
6180		6" L.W. block	10	perlite		5.85	12.95	18.80
6200				styrofoam		6.40	12.75	19.15
6260		6" glazed block	10	perlite		11.60	13.95	25.55
6280				styrofoam		12.15	13.75	25.90
6340		6" clay tile	10	none		8.10	12.35	20.45
6360		8" clay tile	12	none		9.45	12.85	22.30
6600	Roman	6" conc. block	10	perlite		7.25	13.30	20.55
6620				styrofoam		7.85	13.10	20.95
6680		6" L.W. block	10	perlite		7.40	13.20	20.60
6700				styrofoam		8	13	21
6760		6" glazed block	10	perlite		13.15	14.20	27.35
6780				styrofoam		13.75	14	27.75
6840		6" clay tile	10	none		9.65	12.60	22.25
6860		8" clay tile	12	none		11	13.10	24.10
7100	Norman	6" conc. block	10	perlite		5.95	11.80	17.75
7120				styrofoam		6.50	11.60	18.10
7180		6" L.W. block	10	perlite		6.05	11.70	17.75
7200				styrofoam		6.65	11.50	18.15
7260		6" glazed block	10	perlite		11.80	12.70	24.50
7280				styrofoam		12.40	12.50	24.90
7340		6" clay tile	10	none		8.35	11.10	19.45
7360		8" clay tile	12	none		9.65	11.60	21.25
7600	Norwegian	6" conc. block	10	perlite		5.60	11.05	16.65
7620				styrofoam		6.20	10.80	17
7680		6" L.W. block	10	perlite		5.75	10.95	16.70
7700				styrofoam		6.35	10.70	17.05

B2010 Exterior Walls

B2010 132	Brick Face Cavity Wall - Insulated Backup							

	FACE BRICK	BACKUP MASONRY	TOTAL THICKNESS (IN.)	BACKUP CORE FILL		COST PER S.F.		
						MAT.	INST.	TOTAL
7760	Norwegian	6" glazed block	10	perlite		11.50	11.95	23.45
7780				styrofoam		12.10	11.70	23.80
7840		6" clay tile	10	none		8	10.35	18.35
7860		8" clay tile	12	none		9.35	10.80	20.15
8100	Utility	6" conc. block	10	perlite		5.35	10.30	15.65
8120				styrofoam		5.95	10.05	16
8180		6" L.W. block	10	perlite		5.50	10.20	15.70
8200				styrofoam		6.10	9.95	16.05
8260		6" glazed block	10	perlite		11.25	11.20	22.45
8280				styrofoam		11.85	10.95	22.80
8340		6" clay tile	10	none		7.75	9.60	17.35
8360		8" clay tile	12	none		9.10	10.05	19.15

SHELL

B

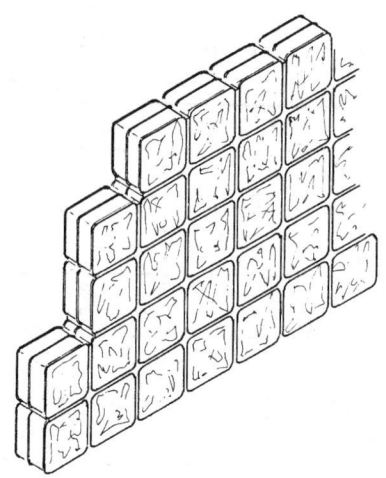

Glass Block

B2010 140	Glass Block	COST PER S.F.		
		MAT.	INST.	TOTAL
2300	Glass block 4" thick, 6"x6" plain, under 1,000 S.F.	15.65	16.50	32.15
2400	1,000 to 5,000 S.F.	15.40	14.30	29.70
2500	Over 5,000 S.F.	15.30	13.40	28.70
2600	Solar reflective, under 1,000 S.F.	22	22.50	44.50
2700	1,000 to 5,000 S.F.	21.50	19.35	40.85
2800	Over 5,000 S.F.	21.50	18.10	39.60
3500	8"x8" plain, under 1,000 S.F.	10.50	12.35	22.85
3600	1,000 to 5,000 S.F.	10.50	10.65	21.15
3700	Over 5,000 S.F.	10.05	9.60	19.65
3800	Solar reflective, under 1,000 S.F.	14.65	16.60	31.25
3900	1,000 to 5,000 S.F.	14.65	14.25	28.90
4000	Over 5,000 S.F.	14.05	12.75	26.80
5000	12"x12" plain, under 1,000 S.F.	12.40	11.40	23.80
5100	1,000 to 5,000 S.F.	13.10	9.60	22.70
5200	Over 5,000 S.F.	12.60	8.80	21.40
5300	Solar reflective, under 1,000 S.F.	17.35	15.30	32.65
5400	1,000 to 5,000 S.F.	18.30	12.75	31.05
5600	Over 5,000 S.F.	17.60	11.65	29.25
5800	3" thinline, 6"x6" plain, under 1,000 S.F.	14.45	16.50	30.95
5900	Over 5,000 S.F.	14.45	16.50	30.95
6000	Solar reflective, under 1,000 S.F.	20	22.50	42.50
6100	Over 5,000 S.F.	18.05	18.10	36.15
6200	8"x8" plain, under 1,000 S.F.	8.55	12.35	20.90
6300	Over 5,000 S.F.	7.65	9.60	17.25
6400	Solar reflective, under 1,000 S.F.	11.95	16.60	28.55
6500	Over 5,000 S.F.	10.70	12.75	23.45

SHELL B

Important: See the Reference Section for critical supporting data - Location Factors & Historical Cost Indexes

B2010 Exterior Walls

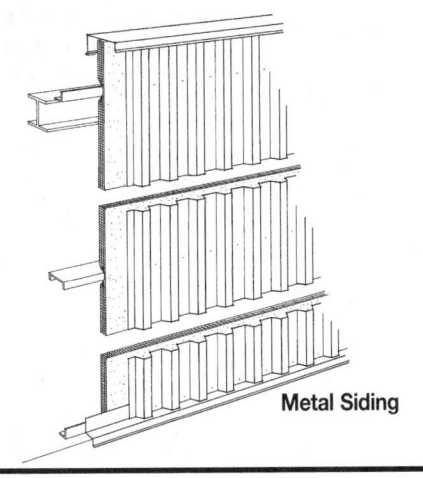

Metal Siding

The table below lists costs for metal siding of various descriptions, not including the steel frame, or the structural steel, of a building. Costs are per S.F. including all accessories and insulation.

B2010 146	Metal Siding Panel	COST PER S.F.		
		MAT.	INST.	TOTAL
1400	Metal siding aluminum panel, corrugated, .024" thick, natural	1.52	2.68	4.20
1450	Painted	1.74	2.68	4.42
1500	.032" thick, natural	1.91	2.68	4.59
1550	Painted	2.29	2.68	4.97
1600	Ribbed 4" pitch, .032" thick, natural	1.97	2.68	4.65
1650	Painted	2.30	2.68	4.98
1700	.040" thick, natural	2.29	2.68	4.97
1750	Painted	2.63	2.68	5.31
1800	.050" thick, natural	2.61	2.68	5.29
1850	Painted	2.92	2.68	5.60
1900	8" pitch panel, .032" thick, natural	1.88	2.57	4.45
1950	Painted	2.20	2.58	4.78
2000	.040" thick, natural	2.20	2.59	4.79
2050	Painted	2.52	2.61	5.13
2100	.050" thick, natural	2.52	2.60	5.12
2150	Painted	2.84	2.62	5.46
3000	Steel, corrugated or ribbed, 29 Ga. .0135" thick, galvanized	1.29	2.44	3.73
3050	Colored	1.70	2.47	4.17
3100	26 Ga. .0179" thick, galvanized	1.33	2.45	3.78
3150	Colored	1.65	2.48	4.13
3200	24 Ga. .0239" thick, galvanized	1.41	2.47	3.88
3250	Colored	1.84	2.50	4.34
3300	22 Ga. .0299" thick, galvanized	1.61	2.48	4.09
3350	Colored	2.05	2.51	4.56
3400	20 Ga. .0359" thick, galvanized	1.61	2.48	4.09
3450	Colored	2.17	2.65	4.82
4100	Sandwich panels, factory fab., 1" polystyrene, steel core, 26 Ga., galv.	2.93	3.81	6.74
4200	Colored, 1 side	3.97	3.81	7.78
4300	2 sides	5.10	3.81	8.91
4400	2" polystyrene, steel core, 26 Ga., galvanized	3.52	3.81	7.33
4500	Colored, 1 side	4.56	3.81	8.37
4600	2 sides	5.70	3.81	9.51
4700	22 Ga., baked enamel exterior	7.55	4.03	11.58
4800	Polyvinyl chloride exterior	7.95	4.03	11.98
5100	Textured aluminum, 4' x 8' x 5/16" plywood backing, single face	2.40	2.31	4.71
5200	Double face	3.62	2.31	5.93

B SHELL

B2010 Exterior Walls

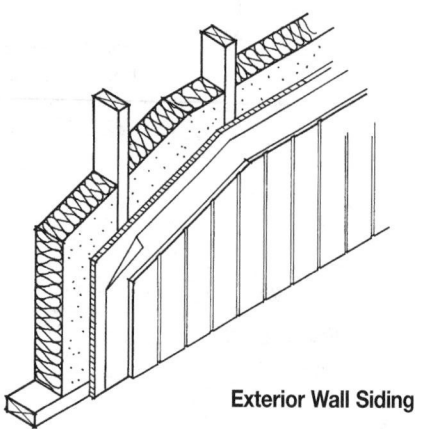

The table below lists costs per S.F. for exterior walls with wood siding. A variety of systems are presented using both wood and metal studs at 16″ and 24″ O.C.

Exterior Wall Siding

SHELL B

B2010 148	Wood & Other Siding	COST PER S.F.		
		MAT.	INST.	TOTAL
1400	Wood siding w/2″x4″studs, 16″O.C., insul. wall, 5/8″text 1-11 fir plywood	2.46	3.80	6.26
1450	5/8″ text 1-11 cedar plywood	3.86	3.67	7.53
1500	1″ x 4″ vert T.&G. redwood	4.44	5.20	9.64
1600	1″ x 8″ vert T.&G. redwood	4.03	4.68	8.71
1650	1″ x 5″ rabbetted cedar bev. siding	3.33	4.18	7.51
1700	1″ x 6″ cedar drop siding	3.37	4.21	7.58
1750	1″ x 12″ rough sawn cedar	2.74	3.83	6.57
1800	1″ x 12″ sawn cedar, 1″ x 4″ battens	3.73	4.13	7.86
1850	1″ x 10″ redwood shiplap siding	3.96	4	7.96
1900	18″ no. 1 red cedar shingles, 5-1/2″ exposed	3.32	5	8.32
1950	6″ exposed	3.19	4.85	8.04
2000	6-1/2″ exposed	3.06	4.68	7.74
2100	7″ exposed	2.93	4.52	7.45
2150	7-1/2″ exposed	2.80	4.35	7.15
3000	8″ wide aluminum siding	2.72	3.37	6.09
3150	8″ plain vinyl siding	2.13	3.39	5.52
3250	8″ insulated vinyl siding	2.21	3.39	5.60
3300				
3400	2″ x 6″ studs, 16″ O.C., insul. wall, w/ 5/8″ text 1-11 fir plywood	2.80	3.96	6.76
3500	5/8″ text 1-11 cedar plywood	4.20	3.96	8.16
3600	1″ x 4″ vert T.&G. redwood	4.78	5.35	10.13
3700	1″ x 8″ vert T.&G. redwood	4.37	4.84	9.21
3800	1″ x 5″ rabbetted cedar bev siding	3.67	4.34	8.01
3900	1″ x 6″ cedar drop siding	3.71	4.37	8.08
4000	1″ x 12″ rough sawn cedar	3.08	3.99	7.07
4200	1″ x 12″ sawn cedar, 1″ x 4″ battens	4.07	4.29	8.36
4500	1″ x 10″ redwood shiplap siding	4.30	4.16	8.46
4550	18″ no. 1 red cedar shingles, 5-1/2″ exposed	3.66	5.15	8.81
4600	6″ exposed	3.53	5	8.53
4650	6-1/2″ exposed	3.40	4.84	8.24
4700	7″ exposed	3.27	4.68	7.95
4750	7-1/2″ exposed	3.14	4.51	7.65
4800	8″ wide aluminum siding	3.06	3.53	6.59
4850	8″ plain vinyl siding	2.47	3.55	6.02
4900	8″ insulated vinyl siding	2.55	3.55	6.10
4910				
5000	2″ x 6″ studs, 24″ O.C., insul. wall, 5/8″ text 1-11, fir plywood	2.64	3.73	6.37
5050	5/8″ text 1-11 cedar plywood	4.04	3.73	7.77
5100	1″ x 4″ vert T.&G. redwood	4.62	5.10	9.72
5150	1″ x 8″ vert T.&G. redwood	4.21	4.61	8.82

Important: See the Reference Section for critical supporting data - Location Factors & Historical Cost Indexes

B20 Exterior Enclosure

B2010 Exterior Walls

B2010 148	Wood & Other Siding	COST PER S.F.		
		MAT.	INST.	TOTAL
5200	1" x 5" rabbetted cedar bev siding	3.51	4.11	7.62
5250	1" x 6" cedar drop siding	3.55	4.14	7.69
5300	1" x 12" rough sawn cedar	2.92	3.76	6.68
5400	1" x 12" sawn cedar, 1" x 4" battens	3.91	4.06	7.97
5450	1" x 10" redwood shiplap siding	4.14	3.93	8.07
5500	18" no. 1 red cedar shingles, 5-1/2" exposed	3.50	4.94	8.44
5550	6" exposed	3.37	4.78	8.15
5650	7" exposed	3.11	4.45	7.56
5700	7-1/2" exposed	2.98	4.28	7.26
5750	8" wide aluminum siding	2.90	3.30	6.20
5800	8" plain vinyl siding	2.31	3.32	5.63
5850	8" insulated vinyl siding	2.39	3.32	5.71
5900	3-5/8" metal studs, 16 Ga., 16" OC insul.wall, 5/8"text 1-11 fir plywood	2.71	4	6.71
5950	5/8" text 1-11 cedar plywood	4.11	4	8.11
6000	1" x 4" vert T.&G. redwood	4.69	5.40	10.09
6050	1" x 8" vert T.&G. redwood	4.28	4.88	9.16
6100	1" x 5" rabbetted cedar bev siding	3.58	4.38	7.96
6150	1" x 6" cedar drop siding	3.62	4.41	8.03
6200	1" x 12" rough sawn cedar	2.99	4.03	7.02
6250	1" x 12" sawn cedar, 1" x 4" battens	3.98	4.33	8.31
6300	1" x 10" redwood shiplap siding	4.21	4.20	8.41
6350	18" no. 1 red cedar shingles, 5-1/2" exposed	3.57	5.20	8.77
6500	6" exposed	3.44	5.05	8.49
6550	6-1/2" exposed	3.31	4.88	8.19
6600	7" exposed	3.18	4.72	7.90
6650	7-1/2" exposed	3.18	4.59	7.77
6700	8" wide aluminum siding	2.97	3.57	6.54
6750	8" plain vinyl siding	2.38	3.46	5.84
6800	8" insulated vinyl siding	2.46	3.46	5.92
7000	3-5/8" metal studs, 16 Ga. 24" OC insul wall, 5/8" text 1-11 fir plywood	2.52	3.55	6.07
7050	5/8" text 1-11 cedar plywood	3.92	3.55	7.47
7100	1" x 4" vert T.&G. redwood	4.50	5.05	9.55
7150	1" x 8" vert T.&G. redwood	4.09	4.56	8.65
7200	1" x 5" rabbetted cedar bev siding	3.39	4.06	7.45
7250	1" x 6" cedar drop siding	3.42	4.09	7.51
7300	1" x 12" rough sawn cedar	2.80	3.71	6.51
7350	1" x 12" sawn cedar 1" x 4" battens	3.79	4.01	7.80
7400	1" x 10" redwood shiplap siding	4.02	3.88	7.90
7450	18" no. 1 red cedar shingles, 5-1/2" exposed	3.51	5.05	8.56
7500	6" exposed	3.38	4.89	8.27
7550	6-1/2" exposed	3.25	4.73	7.98
7600	7" exposed	3.12	4.56	7.68
7650	7-1/2" exposed	2.86	4.23	7.09
7700	8" wide aluminum siding	2.78	3.25	6.03
7750	8" plain vinyl siding	2.19	3.27	5.46
7800	8" insul. vinyl siding	2.27	3.27	5.54

SHELL

B

303

B2010 Exterior Walls

The table below lists costs for some typical stucco walls including all the components as demonstrated in the component block below. Prices are presented for backup walls using wood studs, metal studs and CMU.

Exterior Stucco Wall

B2010 151	Stucco Wall	COST PER S.F.		
		MAT.	INST.	TOTAL
2100	Cement stucco, 7/8" th., plywood sheathing, stud wall, 2" x 4", 16" O.C.	2.08	6.20	8.28
2200	24" O.C.	1.98	6.05	8.03
2300	2" x 6", 16" O.C.	2.42	6.40	8.82
2400	24" O.C.	2.26	6.15	8.41
2500	No sheathing, metal lath on stud wall, 2" x 4", 16" O.C.	1.44	5.40	6.84
2600	24" O.C.	1.34	5.25	6.59
2700	2" x 6", 16" O.C.	1.78	5.55	7.33
2800	24" O.C.	1.62	5.35	6.97
2900	1/2" gypsum sheathing, 3-5/8" metal studs, 16" O.C.	1.82	5.60	7.42
2950	24" O.C.	1.63	5.30	6.93
3000	Cement stucco, 5/8" th., 2 coats on std. CMU block, 8"x 16", 8" thick	2.02	7.75	9.77
3100	10" Thick	2.69	8	10.69
3200	12" Thick	2.75	9.20	11.95
3300	Std. light Wt. block 8" x 16", 8" Thick	2.43	7.50	9.93
3400	10" Thick	3.06	7.70	10.76
3500	12" Thick	3.53	8.75	12.28
3600	3 coat stucco, self furring metal lath 3.4 Lb/SY, on 8" x 16", 8" thick	2.06	8.20	10.26
3700	10" Thick	2.73	8.45	11.18
3800	12" Thick	2.79	9.65	12.44
3900	Lt. Wt. block, 8" Thick	2.47	7.95	10.42
4000	10" Thick	3.10	8.15	11.25
4100	12" Thick	3.57	9.20	12.77

B2010 151	E.I.F.S.	COST PER S.F.		
		MAT.	INST.	TOTAL
5100	E.I.F.S., plywood sheathing, stud wall, 2" x 4", 16" O.C., 1" EPS	3.38	7.60	10.98
5110	2" EPS	3.65	7.60	11.25
5120	3" EPS	3.80	7.60	11.40
5130	4" EPS	4.06	7.60	11.66
5140	2" x 6", 16" O.C., 1" EPS	3.72	7.80	11.52
5150	2" EPS	3.99	7.80	11.79
5160	3" EPS	4.14	7.80	11.94
5170	4" EPS	4.40	7.80	12.20
5180	Cement board sheathing, 3-5/8" metal studs, 16" O.C., 1" EPS	3.92	9.05	12.97
5190	2" EPS	4.18	9.05	13.23
5200	3" EPS	4.34	9.05	13.39
5210	4" EPS	5.50	9.05	14.55
5220	6" metal studs, 16" O.C., 1" EPS	4.22	9.15	13.37
5230	2" EPS	4.48	9.15	13.63

SHELL B

Important: See the Reference Section for critical supporting data - Location Factors & Historical Cost Indexes

B20 Exterior Enclosure

B2010 Exterior Walls

B2010 151	E.I.F.S.	COST PER S.F.		
		MAT.	INST.	TOTAL
5240	3" EPS	4.64	9.15	13.79
5250	4" EPS	5.80	9.15	14.95
5260	CMU block, 8" x 8" x 16", 1" EPS	3.58	10.50	14.08
5270	2" EPS	3.85	10.50	14.35
5280	3" EPS	4	10.50	14.50
5290	4" EPS	4.26	10.50	14.76
5300	8" x 10" x 16", 1" EPS	4.25	10.70	14.95
5310	2" EPS	4.52	10.70	15.22
5320	3" EPS	4.67	10.70	15.37
5330	4" EPS	4.93	10.70	15.63
5340	8" x 12" x 16", 1" EPS	4.31	11.95	16.26
5350	2" EPS	4.58	11.95	16.53
5360	3" EPS	4.73	11.95	16.68
5370	4" EPS	4.99	11.95	16.94

SHELL

B

B2010 Exterior Walls

Description: The table below lists costs, $/S.F., for channel girts with sag rods and connector angles top and bottom for various column spacings, building heights and wind loads. Additive costs are shown for wind columns.

SHELL B

B2010 154			Metal Siding Support					
	BLDG. HEIGHT (FT.)	WIND LOAD (P.S.F.)	COL. SPACING (FT.)		INTERMEDIATE COLUMNS	COST PER S.F.		
						MAT.	INST.	TOTAL
3000	18	20	20			1.11	2.78	3.89
3100					wind cols.	.57	.20	.77
3200		20	25			1.21	2.80	4.01
3300					wind cols.	.46	.16	.62
3400		20	30			1.34	2.86	4.20
3500					wind cols.	.38	.13	.51
3600		20	35			1.47	2.90	4.37
3700					wind cols.	.33	.12	.45
3800		30	20			1.23	2.82	4.05
3900					wind cols.	.57	.20	.77
4000		30	25			1.34	2.86	4.20
4100					wind cols.	.46	.16	.62
4200		30	30			1.48	2.90	4.38
4300					wind cols.	.52	.19	.71
4600	30	20	20			1.08	1.95	3.03
4700					wind cols.	.81	.29	1.10
4800		20	25			1.21	2	3.21
4900					wind cols.	.65	.23	.88
5000		20	30			1.37	2.05	3.42
5100					wind cols.	.64	.23	.87
5200	30	20	35			1.53	2.11	3.64
5300					wind cols.	.63	.23	.86
5400		30	20			1.23	2	3.23
5500					wind cols.	.96	.35	1.31
5600		30	25			1.36	2.05	3.41
5700					wind cols.	.91	.33	1.24
5800		30	30			1.54	2.11	3.65
5900					wind cols.	.86	.31	1.17

B2010 156	Exterior Wall Specialties	MAT.	INST.	TOTAL
0100	Canopies, wall hung, prefinished aluminum, 8' x 10'	1,625	1,150	2,775

Important: See the Reference Section for critical supporting data - Location Factors & Historical Cost Indexes

The table below lists window systems by material, type and size. Prices between sizes listed can be interpolated with reasonable accuracy. Prices include frame, hardware, and casing as illustrated in the component block below.

B2020 102	Wood, Steel & Aluminum							
	MATERIAL	TYPE	GLAZING	SIZE	DETAIL	COST PER UNIT		
						MAT.	INST.	TOTAL
3000	Wood	double hung	std. glass	2'-8" x 4'-6"		178	163	341
3050				3'-0" x 5'-6"		234	188	422
3100			insul. glass	2'-8" x 4'-6"		191	163	354
3150				3'-0" x 5'-6"		252	188	440
3200		sliding	std. glass	3'-4" x 2'-7"		215	136	351
3250				4'-4" x 3'-3"		245	149	394
3300				5'-4" x 6'-0"		310	180	490
3350			insul. glass	3'-4" x 2'-7"		262	159	421
3400				4'-4" x 3'-3"		299	173	472
3450				5'-4" x 6'-0"		375	205	580
3500		awning	std. glass	2'-10" x 1'-9"		160	82	242
3600				4'-4" x 2'-8"		262	79.50	341.50
3700			insul. glass	2'-10" x 1'-9"		196	94.50	290.50
3800				4'-4" x 2'-8"		320	87	407
3900		casement	std. glass	1'-10" x 3'-2"	1 lite	205	108	313
3950				4'-2" x 4'-2"	2 lite	420	143	563
4000				5'-11" x 5'-2"	3 lite	645	184	829
4050				7'-11" x 6'-3"	4 lite	910	222	1,132
4100				9'-11" x 6'-3"	5 lite	1,225	257	1,482
4150			insul. glass	1'-10" x 3'-2"	1 lite	215	108	323
4200				4'-2" x 4'-2"	2 lite	440	143	583
4250				5'-11" x 5'-2"	3 lite	685	184	869
4300				7'-11" x 6'-3"	4 lite	970	222	1,192
4350				9'-11" x 6'-3"	5 lite	1,275	257	1,532
4400		picture	std. glass	4'-6" x 4'-6"		340	184	524
4450				5'-8" x 4'-6"		380	205	585
4500		picture	insul. glass	4'-6" x 4'-6"		415	214	629
4550				5'-8" x 4'-6"		465	238	703
4600		fixed bay	std. glass	8' x 5'		1,200	335	1,535
4650				9'-9" x 5'-4"		845	475	1,320
4700			insul. glass	8' x 5'		1,825	335	2,160
4750				9'-9" x 5'-4"		915	475	1,390
4800		casement bay	std. glass	8' x 5'		1,200	385	1,585
4850			insul. glass	8' x 5'		1,350	385	1,735
4900		vert. bay	std. glass	8' x 5'		1,350	385	1,735
4950			insul. glass	8' x 5'		1,400	385	1,785

B | SHELL

B2020 102		Wood, Steel & Aluminum						
	MATERIAL	TYPE	GLAZING	SIZE	DETAIL	COST PER UNIT		
						MAT.	INST.	TOTAL
5000	Steel	double hung	1/4" tempered	2'-8" x 4'-6"		620	132	752
5050				3'-4" x 5'-6"		945	201	1,146
5100			insul. glass	2'-8" x 4'-6"		630	151	781
5150				3'-4" x 5'-6"		965	230	1,195
5202		horiz. pivoted	std. glass	2' x 2'		202	44	246
5250				3' x 3'		455	99	554
5300				4' x 4'		810	176	986
5350				6' x 4'		1,225	264	1,489
5400			insul. glass	2' x 2'		207	50.50	257.50
5450				3' x 3'		465	113	578
5500				4' x 4'		825	201	1,026
5550				6' x 4'		1,250	300	1,550
5600		picture window	std. glass	3' x 3'		271	99	370
5650				6' x 4'		720	264	984
5700			insul. glass	3' x 3'		280	113	393
5750				6' x 4'		750	300	1,050
5800		industrial security	std. glass	2'-9" x 4'-1"		590	123	713
5850				4'-1" x 5'-5"		1,175	243	1,418
5900			insul. glass	2'-9" x 4'-1"		600	141	741
5950				4'-1" x 5'-5"		1,175	278	1,453
6000		comm. projected	std. glass	3'-9" x 5'-5"		980	223	1,203
6050				6'-9" x 4'-1"		1,325	305	1,630
6100			insul. glass	3'-9" x 5'-5"		1,000	256	1,256
6150				6'-9" x 4'-1"		1,350	345	1,695
6200		casement	std. glass	4'-2" x 4'-2"	2 lite	755	191	946
6250			insul. glass	4'-2" x 4'-2"		775	218	993
6300			std. glass	5'-11" x 5'-2"	3 lite	1,450	335	1,785
6350			insul. glass	5'-11" x 5'-2"		1,475	385	1,860
6400	Aluminum	projecting	std. glass	3'-1" x 3'-2"		213	99	312
6450				4'-5" x 5'-3"		300	124	424
6500			insul. glass	3'-1" x 3'-2"		256	119	375
6550				4'-5" x 5'-3"		360	149	509
6600		sliding	std. glass	3' x 2'		161	99	260
6650				5' x 3'		204	110	314
6700				8' x 4'		294	165	459
6750				9' x 5'		445	247	692
6800			insul. glass	3' x 2'		178	99	277
6850				5' x 3'		286	110	396
6900				8' x 4'		475	165	640
6950				9' x 5'		710	247	957
7000		single hung	std. glass	2' x 3'		143	99	242
7050				2'-8" x 6'-8"		305	124	429
7100				3'-4" x 5'0"		196	110	306
7150			insul. glass	2' x 3'		173	99	272
7200				2'-8" x 6'-8"		390	124	514
7250				3'-4" x 5'		275	110	385
7300		double hung	std. glass	2' x 3'		208	66	274
7350				2'-8" x 6'-8"		615	196	811
7400				3'-4" x 5'		575	183	758
7450			insul. glass	2' x 3'		214	75.50	289.50
7500				2'-8" x 6'-8"		635	224	859
7550				3'-4" x 5'-0"		595	210	805

Important: See the Reference Section for critical supporting data - Location Factors & Historical Cost Indexes

SHELL B

B2020 Exterior Windows

B2020 102			Wood, Steel & Aluminum					

	MATERIAL	TYPE	GLAZING	SIZE	DETAIL	COST PER UNIT		
						MAT.	INST.	TOTAL
7600	Aluminum	casements	std. glass	3'-1" x 3'-2"		170	107	277
7650				4'-5" x 5'-3"		410	260	670
7700			insul. glass	3'-1" x 3'-2"		180	123	303
7750				4'-5" x 5'-3"		435	297	732
7800		hinged swing	std. glass	3' x 4'		385	132	517
7850				4' x 5'		640	220	860
7900			insul. glass	3' x 4'		400	151	551
7950				4' x 5'		665	252	917
8002		folding type	std. glass	3'-0" x 4'-0"		215	106	321
8050				4'-0" x 5'-0"		300	106	406
8100			insul. glass	3'-0" x 4'-0"		269	106	375
8150				4'-0" x 5'-0"		390	106	496
8200		picture unit	std. glass	2'-0" x 3'-0"		110	66	176
8250				2'-8" x 6'-8"		325	196	521
8300				3'-4" x 5'-0"		305	183	488
8350			insul. glass	2'-0" x 3'-0"		116	75.50	191.50
8400				2'-8" x 6'-8"		345	224	569
8450				3'-4" x 5'-0"		320	210	530
8500		awning type	std. glass	3'-0" x 3'-0"	2 lite	350	70.50	420.50
8550				3'-0" x 4'-0"	3 lite	410	99	509
8600				3'-0" x 5'-4"	4 lite	495	99	594
8650				4'-0" x 5'-4"	4 lite	545	110	655
8700			insul. glass	3'-0" x 3'-0"	2 lite	375	70.50	445.50
8750				3'-0" x 4'-0"	3 lite	475	99	574
8800				3'-0" x 5'-4"	4 lite	585	99	684
8850				4'-0" x 5'-4"	4 lite	650	110	760
8900		jalousie type	std. glass	1'-7" x 3'-2"		131	99	230
8950				2'-3" x 4'-0"		186	99	285
9000				3'-1" x 2'-0"		147	99	246
9050				3'-1" x 5'-3"		266	99	365
9051								
9052								

B2020 118	Tubular Aluminum Framing	COST/S.F. OPNG.		
		MAT.	INST.	TOTAL
1100	Alum flush tube frame,for 1/4"glass,1-3/4"x4",5'x6'opng, no inter horizntls	9.25	8.90	18.15
1150	One intermediate horizontal	12.60	10.40	23
1200	Two intermediate horizontals	15.95	11.95	27.90
1250	5' x 20' opening, three intermediate horizontals	8.30	7.40	15.70
1400	1-3/4" x 4-1/2", 5' x 6' opening, no intermediate horizontals	10.50	8.90	19.40
1450	One intermediate horizontal	14	10.40	24.40
1500	Two intermediate horizontals	17.55	11.95	29.50
1550	5' x 20' opening, three intermediate horizontals	9.35	7.40	16.75
1700	For insulating glass, 2"x4-1/2", 5'x6' opening, no intermediate horizontals	11.65	9.35	21
1750	One intermediate horizontal	15.25	10.95	26.20
1800	Two intermediate horizontals	18.80	12.60	31.40
1850	5' x 20' opening, three intermediate horizontals	10.15	7.75	17.90
2000	Thermal break frame, 2-1/4"x4-1/2", 5'x6'opng, no intermediate horizontals	12.45	9.45	21.90
2050	One intermediate horizontal	16.75	11.35	28.10
2100	Two intermediate horizontals	21	13.25	34.25
2150	5' x 20' opening, three intermediate horizontals	11.35	8.10	19.45

B SHELL

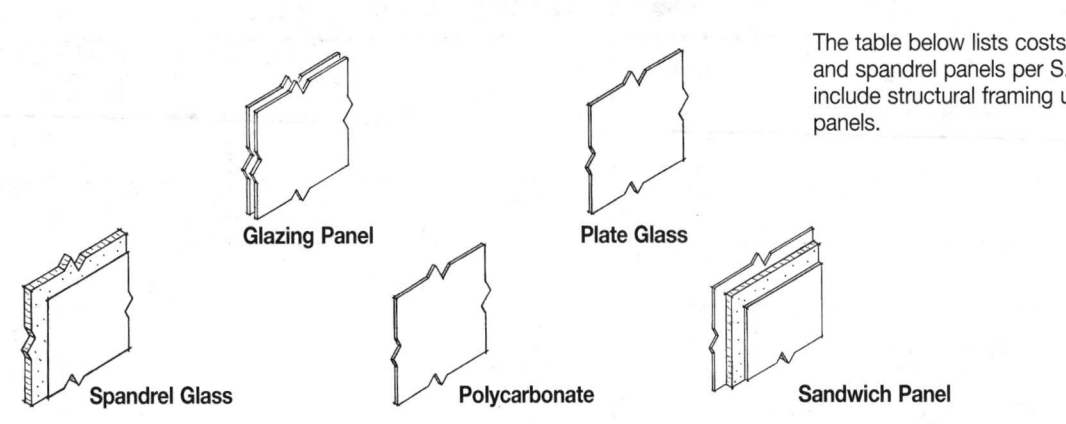

Glazing Panel

Plate Glass

Spandrel Glass

Polycarbonate

Sandwich Panel

The table below lists costs of curtain wall and spandrel panels per S.F. Costs do not include structural framing used to hang the panels.

SHELL B

B2020 120	Curtain Wall Panels	COST PER S.F.		
		MAT.	INST.	TOTAL
1000	Glazing panel, insulating, 1/2" thick, 2 lites 1/8" float, clear	7.15	7.65	14.80
1100	Tinted	10.65	7.65	18.30
1200	5/8" thick units, 2 lites 3/16" float, clear	8.65	8.05	16.70
1300	Tinted	8.90	8.05	16.95
1400	1" thick units, 2 lites, 1/4" float, clear	12.25	9.70	21.95
1500	Tinted	14.80	9.70	24.50
1600	Heat reflective film inside	17.30	8.55	25.85
1700	Light and heat reflective glass, tinted	19.75	8.55	28.30
2000	Plate glass, 1/4" thick, clear	5.20	6.05	11.25
2050	Tempered	6.10	6.05	12.15
2100	Tinted	4.94	6.05	10.99
2200	3/8" thick, clear	8.15	9.70	17.85
2250	Tempered	12.30	9.70	22
2300	Tinted	9.90	9.70	19.60
2400	1/2" thick, clear	16.10	13.20	29.30
2450	Tempered	18.45	13.20	31.65
2500	Tinted	17.30	13.20	30.50
2600	3/4" thick, clear	22	21	43
2650	Tempered	26	21	47
3000	Spandrel glass, panels, 1/4" plate glass insul w/fiberglass, 1" thick	10.55	6.05	16.60
3100	2" thick	12.30	6.05	18.35
3200	Galvanized steel backing, add	3.64		3.64
3300	3/8" plate glass, 1" thick	17.65	6.05	23.70
3400	2" thick	19.40	6.05	25.45
4000	Polycarbonate, masked, clear or colored, 1/8" thick	6.10	4.27	10.37
4100	3/16" thick	7.35	4.40	11.75
4200	1/4" thick	8.15	4.69	12.84
4300	3/8" thick	15	4.84	19.84
5000	Facing panel, textured al, 4' x 8' x 5/16" plywood backing, sgl face	2.40	2.31	4.71
5100	Double face	3.62	2.31	5.93
5200	4' x 10' x 5/16" plywood backing, single face	2.55	2.31	4.86
5300	Double face	3.75	2.31	6.06
5400	4' x 12' x 5/16" plywood backing, single face	3.29	2.31	5.60
5500	Sandwich panel, 22 Ga. galv., both sides 2" insulation, enamel exterior	7.55	4.03	11.58
5600	Polyvinylidene floride exterior finish	7.95	4.03	11.98
5700	26 Ga., galv. both sides, 1" insulation, colored 1 side	3.97	3.81	7.78
5800	Colored 2 sides	5.10	3.81	8.91

B2030 Exterior Doors

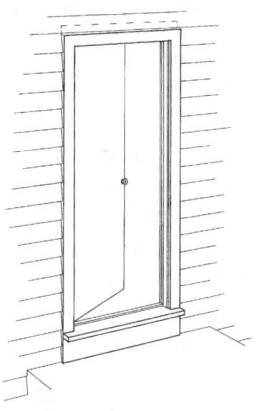

Exterior Door System

The table below lists exterior door systems by material, type and size. Prices between sizes listed can be interpolated with reasonable accuracy. Prices are per opening for a complete door system including frame and required hardware.

Wood doors in this table are designed with wood frames and metal doors with hollow metal frames. Depending upon quality the total material and installation cost of a wood door frame is about the same as a hollow metal frame.

B2030 105				Wood, Steel & Aluminum				
	MATERIAL	TYPE	DOORS	SPECIFICATION	OPENING	COST PER OPNG.		
						MAT.	INST.	TOTAL
2350	Birch	solid core	single door	hinged	2'-6" x 6'-8"	860	203	1,063
2400					2'-6" x 7'-0"	860	205	1,065
2450					2'-8" x 7'-0"	870	205	1,075
2500					3'-0" x 7'-0"	880	208	1,088
2550			double door	hinged	2'-6" x 6'-8"	1,600	370	1,970
2600					2'-6" x 7'-0"	1,600	375	1,975
2650					2'-8" x 7'-0"	1,600	375	1,975
2700					3'-0" x 7'-0"	1,625	380	2,005
2750	Wood	combination	storm & screen	hinged	3'-0" x 6'-8"	282	50	332
2800					3'-0" x 7'-0"	310	55	365
2850		overhead	panels, H.D.	manual oper.	8'-0" x 8'-0"	525	375	900
2900					10'-0" x 10'-0"	785	415	1,200
2950					12'-0" x 12'-0"	1,125	500	1,625
3000					14'-0" x 14'-0"	1,750	575	2,325
3050					20'-0" x 16'-0"	3,600	1,150	4,750
3100				electric oper.	8'-0" x 8'-0"	1,050	560	1,610
3150					10'-0" x 10'-0"	1,325	600	1,925
3200					12'-0" x 12'-0"	1,650	685	2,335
3250					14'-0" x 14'-0"	2,275	760	3,035
3300					20'-0" x 16'-0"	4,325	1,525	5,850
3350	Steel 18 Ga.	hollow metal	1 door w/frame	no label	2'-6" x 7'-0"	845	211	1,056
3400					2'-8" x 7'-0"	845	211	1,056
3450					3'-0" x 7'-0"	845	211	1,056
3500					3'-6" x 7'-0"	920	222	1,142
3550		hollow metal	1 door w/frame	no label	4'-0" x 8'-0"	1,050	223	1,273
3600			2 doors w/frame	no label	5'-0" x 7'-0"	1,650	390	2,040
3650					5'-4" x 7'-0"	1,650	390	2,040
3700					6'-0" x 7'-0"	1,650	390	2,040
3750					7'-0" x 7'-0"	1,800	415	2,215
3800					8'-0" x 8'-0"	2,025	415	2,440
3850			1 door w/frame	"A" label	2'-6" x 7'-0"	1,000	255	1,255
3900					2'-8" x 7'-0"	1,000	258	1,258
3950					3'-0" x 7'-0"	1,000	258	1,258
4000					3'-6" x 7'-0"	1,075	263	1,338
4050					4'-0" x 8'-0"	1,125	276	1,401
4100			2 doors w/frame	"A" label	5'-0" x 7'-0"	1,950	470	2,420

B SHELL

B2030 Exterior Doors

| B2030 105 | Wood, Steel & Aluminum |

	MATERIAL	TYPE	DOORS	SPECIFICATION	OPENING	COST PER OPNG.		
						MAT.	INST.	TOTAL
4150	Steel 18 Ga.				5'-4" x 7'-0"	1,950	480	2,430
4200					6'-0" x 7'-0"	1,950	480	2,430
4250					7'-0" x 7'-0"	2,100	490	2,590
4300					8'-0" x 8'-0"	2,100	490	2,590
4350	Steel 24 Ga.	overhead	sectional	manual oper.	8'-0" x 8'-0"	425	375	800
4400					10'-0" x 10'-0"	585	415	1,000
4450					12'-0" x 12'-0"	790	500	1,290
4500					20'-0" x 14'-0"	1,850	1,075	2,925
4550				electric oper.	8'-0" x 8'-0"	960	560	1,520
4600					10'-0" x 10'-0"	1,125	600	1,725
4650					12'-0" x 12'-0"	1,325	685	2,010
4700					20'-0" x 14'-0"	2,575	1,450	4,025
4750	Steel	overhead	rolling	manual oper.	8'-0" x 8'-0"	790	620	1,410
4800					10'-0" x 10'-0"	1,050	705	1,755
4850					12'-0" x 12'-0"	1,350	825	2,175
4900					14'-0" x 14'-0"	1,800	1,225	3,025
4950					20'-0" x 12'-0"	2,200	1,100	3,300
5000					20'-0" x 16'-0"	2,550	1,650	4,200
5050				electric oper.	8'-0" x 8'-0"	1,550	820	2,370
5100					10'-0" x 10'-0"	1,800	905	2,705
5150					12'-0" x 12'-0"	2,100	1,025	3,125
5200					14'-0" x 14'-0"	2,550	1,425	3,975
5250					20'-0" x 12'-0"	2,950	1,300	4,250
5300					20'-0" x 16'-0"	3,300	1,850	5,150
5350				fire rated	10'-0" x 10'-0"	1,375	900	2,275
5400			rolling grill	manual oper.	10'-0" x 10'-0"	1,725	990	2,715
5450					15'-0" x 8'-0"	2,025	1,225	3,250
5500		vertical lift	1 door w/frame	motor operator	16'-0" x 16'-0"	17,300	4,150	21,450
5550					32'-0" x 24'-0"	34,000	2,775	36,775
5600	St. Stl. & glass	revolving	stock unit	manual oper.	6'-0" x 7'-0"	31,700	6,550	38,250
5650				auto Cntrls.	6'-10" x 7'-0"	44,400	6,975	51,375
5700	Bronze	revolving	stock unit	manual oper.	6'-10" x 7'-0"	36,900	13,200	50,100
5750				auto Cntrls.	6'-10" x 7'-0"	49,600	13,600	63,200
5800	St. Stl. & glass	balanced	standard	economy	3'-0" x 7'-0"	6,850	1,100	7,950
5850				premium	3'-0" x 7'-0"	11,900	1,425	13,325
6000	Aluminum	combination	storm & screen	hinged	3'-0" x 6'-8"	227	53.50	280.50
6050					3'-0" x 7'-0"	250	59	309
6100		overhead	rolling grill	manual oper.	12'-0" x 12'-0"	2,775	1,725	4,500
6150				motor oper.	12'-0" x 12'-0"	3,650	1,925	5,575
6200	Alum. & Fbrgls.	overhead	heavy duty	manual oper.	12'-0" x 12'-0"	1,000	500	1,500
6250				electric oper.	12'-0" x 12'-0"	1,525	685	2,210
6300	Alum. & glass	w/o transom	narrow stile	w/panic Hrdwre.	3'-0" x 7'-0"	1,050	720	1,770
6350				dbl. door, Hrdwre.	6'-0" x 7'-0"	1,900	1,175	3,075
6400			wide stile	hdwre.	3'-0" x 7'-0"	1,375	705	2,080
6450				dbl. door, Hdwre.	6'-0" x 7'-0"	2,725	1,400	4,125
6500			full vision	hdwre.	3'-0" x 7'-0"	1,550	1,125	2,675
6550				dbl. door, Hdwre.	6'-0" x 7'-0"	2,300	1,600	3,900
6600			non-standard	hdwre.	3'-0" x 7'-0"	970	705	1,675
6650				dbl. door, Hdwre.	6'-0" x 7'-0"	1,950	1,400	3,350
6700			bronze fin.	hdwre.	3'-0" x 7'-0"	1,025	705	1,730
6750				dbl. door, Hrdwre.	6'-0" x 7'-0"	2,075	1,400	3,475
6800			black fin.	hdwre.	3'-0" x 7'-0"	1,725	705	2,430

Important: See the Reference Section for critical supporting data - Location Factors & Historical Cost Indexes

B2030 Exterior Doors

B2030 105			Wood, Steel & Aluminum				

	MATERIAL	TYPE	DOORS	SPECIFICATION	OPENING	COST PER OPNG.		
						MAT.	INST.	TOTAL
6850	Alum. & glass			dbl. door, Hdwre.	6'-0" x 7'-0"	3,450	1,400	4,850
6900		w/transom	narrow stile	hdwre.	3'-0" x 10'-0"	1,375	815	2,190
6950				dbl. door, Hdwre.	6'-0" x 10'-0"	1,925	1,425	3,350
7000			wide stile	hdwre.	3'-0" x 10'-0"	1,600	980	2,580
7050				dbl. door, Hdwre.	6'-0" x 10'-0"	2,100	1,700	3,800
7100			full vision	hdwre.	3'-0" x 10'-0"	1,750	1,100	2,850
7150				dbl. door, Hdwre.	6'-0" x 10'-0"	2,225	1,875	4,100
7200			non-standard	hdwre.	3'-0" x 10'-0"	1,025	760	1,785
7250				dbl. door, Hdwre.	6'-0" x 10'-0"	2,050	1,525	3,575
7300			bronze fin.	hdwre.	3'-0" x 10'-0"	1,100	760	1,860
7350				dbl. door, Hdwre.	6'-0" x 10'-0"	2,175	1,525	3,700
7400			black fin.	hdwre.	3'-0" x 10'-0"	1,775	760	2,535
7450				dbl. door, Hdwre.	6'-0" x 10'-0"	3,575	1,525	5,100
7500		revolving	stock design	minimum	6'-10" x 7'-0"	16,900	2,625	19,525
7550				average	6'-0" x 7'-0"	20,900	3,300	24,200
7600				maximum	6'-10" x 7'-0"	27,500	4,400	31,900
7650				min., automatic	6'-10" x 7'-0"	29,600	3,050	32,650
7700				avg., automatic	6'-10" x 7'-0"	33,600	3,725	37,325
7750				max., automatic	6'-10" x 7'-0"	40,200	4,825	45,025
7800		balanced	standard	economy	3'-0" x 7'-0"	5,200	1,100	6,300
7850				premium	3'-0" x 7'-0"	6,475	1,400	7,875
7900		mall front	sliding panels	alum. fin.	16'-0" x 9'-0"	2,550	560	3,110
7950					24'-0" x 9'-0"	3,700	1,050	4,750
8000				bronze fin.	16'-0" x 9'-0"	2,975	650	3,625
8050					24'-0" x 9'-0"	4,300	1,225	5,525
8100			fixed panels	alum. fin.	48'-0" x 9'-0"	6,875	805	7,680
8150				bronze fin.	48'-0" x 9'-0"	8,000	940	8,940
8200		sliding entrance	5' x 7' door	electric oper.	12'-0" x 7'-6"	6,500	1,050	7,550
8250		sliding patio	temp. glass	economy	6'-0" x 7'-0"	850	187	1,037
8300			temp. glass	economy	12'-0" x 7'-0"	2,150	249	2,399
8350				premium	6'-0" x 7'-0"	1,275	281	1,556
8400					12'-0" x 7'-0"	3,225	375	3,600

B SHELL

B3010 Roof Coverings

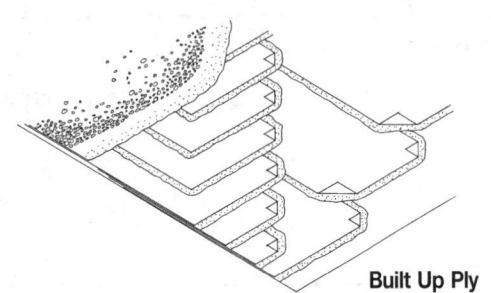

Multiple ply roofing is the most popular covering for minimum pitch roofs.

Built Up Ply

B3010 105	Built-Up	COST PER S.F.		
		MAT.	INST.	TOTAL
1200	Asphalt flood coat w/gravel; not incl. insul, flash., nailers			
1300				
1400	Asphalt base sheets & 3 plies #15 asphalt felt, mopped	.45	1.23	1.68
1500	On nailable deck	.49	1.29	1.78
1600	4 plies #15 asphalt felt, mopped	.63	1.36	1.99
1700	On nailable deck	.57	1.43	2
1800	Coated glass base sheet, 2 plies glass (type IV), mopped	.50	1.23	1.73
1900	For 3 plies	.59	1.36	1.95
2000	On nailable deck	.55	1.43	1.98
2300	4 plies glass fiber felt (type IV), mopped	.71	1.36	2.07
2400	On nailable deck	.64	1.43	2.07
2500	Organic base sheet & 3 plies #15 organic felt, mopped	.47	1.38	1.85
2600	On nailable deck	.49	1.43	1.92
2700	4 plies #15 organic felt, mopped	.60	1.23	1.83
2750				
2800	Asphalt flood coat, smooth surface			
2900	Asphalt base sheet & 3 plies #15 asphalt felt, mopped	.47	1.13	1.60
3000	On nailable deck	.44	1.18	1.62
3100	Coated glass fiber base sheet & 2 plies glass fiber felt, mopped	.44	1.08	1.52
3200	On nailable deck	.42	1.13	1.55
3300	For 3 plies, mopped	.53	1.18	1.71
3400	On nailable deck	.50	1.23	1.73
3700	4 plies glass fiber felt (type IV), mopped	.62	1.18	1.80
3800	On nailable deck	.59	1.23	1.82
3900	Organic base sheet & 3 plies #15 organic felt, mopped	.47	1.13	1.60
4000	On nailable decks	.43	1.18	1.61
4100	4 plies #15 organic felt, mopped	.54	1.23	1.77
4200	Coal tar pitch with gravel surfacing			
4300	4 plies #15 tarred felt, mopped	1.12	1.29	2.41
4400	3 plies glass fiber felt (type IV), mopped	.93	1.43	2.36
4500	Coated glass fiber base sheets 2 plies glass fiber felt, mopped	.93	1.43	2.36
4600	On nailable decks	.82	1.51	2.33
4800	3 plies glass fiber felt (type IV), mopped	1.27	1.29	2.56
4900	On nailable decks	1.16	1.36	2.52

B3010 Roof Coverings

Full Adhered
Single Ply Membrane

Preformed Metal

Flat Seam

Batten Seam

Standing Seam

B3010 120	Single Ply Membrane	COST PER S.F.		
		MAT.	INST.	TOTAL
1000	CSPE (Chlorosulfonated polyethylene), 35 mils, fully adhered	1.36	.71	2.07
2000	EPDM (Ethylene propylene diene monomer), 45 mils, fully adhered	.89	.71	1.60
4000	Modified bit., SBS modified, granule surface cap sheet, mopped, 150 mils	.50	1.36	1.86
4500	APP modified, granule surface cap sheet, torched, 180 mils	.55	.92	1.47
6000	Reinforced PVC, 48 mils, loose laid and ballasted with stone	1.03	.36	1.39
6200	Fully adhered with adhesive	1.28	.71	1.99

B3010 130	Preformed Metal Roofing	COST PER S.F.		
		MAT.	INST.	TOTAL
0200	Corrugated roofing, aluminum, mill finish, .0175" thick, .272 P.S.F.	.68	1.21	1.89
0250	.0215 thick, .334 P.S.F.	.88	1.21	2.09

B3010 135	Formed Metal	COST PER S.F.		
		MAT.	INST.	TOTAL
1000	Batten seam, formed copper roofing, 3"min slope, 16 oz., 1.2 P.S.F.	4.50	4.01	8.51
1100	18 oz., 1.35 P.S.F.	5	4.41	9.41
2000	Zinc copper alloy, 3" min slope, .020" thick, .88 P.S.F.	5.75	3.66	9.41
3000	Flat seam, copper, 1/4" min. slope, 16 oz., 1.2 P.S.F.	4	3.66	7.66
3100	18 oz., 1.35 P.S.F.	4.50	3.81	8.31
5000	Standing seam, copper, 2-1/2" min. slope, 16 oz., 1.25 P.S.F.	4.30	3.41	7.71
5100	18 oz., 1.40 P.S.F.	4.85	3.66	8.51
6000	Zinc copper alloy, 2-1/2" min. slope, .020" thick, .87 P.S.F.	5.65	3.66	9.31
6100	.032" thick, 1.39 P.S.F.	7.70	4.01	11.71

B SHELL

B3010   Roof Coverings

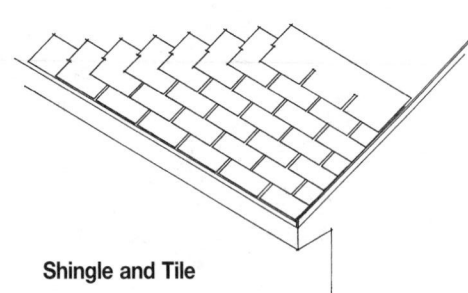

Shingle and Tile

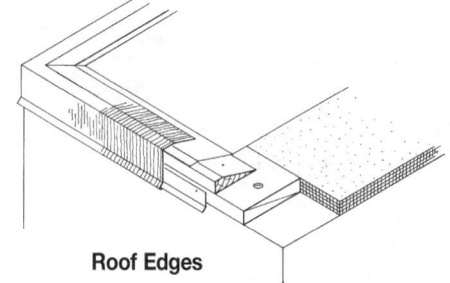

Roof Edges

Shingles and tiles are practical in applications where the roof slope is more than 3-1/2″ per foot of rise. Lines 1100 through 6000 list the various materials and the weight per square foot.

The table below lists the costs for various types of roof perimeter edge treatments.

Roof edge systems include the cost per L.F. for a 2″ x 8″ treated wood nailer fastened at 4′-0″ O.C. and a diagonally cut 4″ x 6″ treated wood cant.

Roof edge and base flashing are assumed to be made from the same material.

B3010 140	Shingle & Tile	COST PER S.F.		
		MAT.	INST.	TOTAL
1095	Asphalt roofing			
1100	Strip shingles,4″ slope,inorganic class A 210-235 lb/Sq.	.33	.76	1.09
1150	Organic, class C, 235-240 lb./sq.	.43	.83	1.26
1200	Premium laminated multi-layered, class A, 260-300 lb./Sq.	.57	1.13	1.70
1545	Metal roofing			
1550	Alum., shingles, colors, 3″min slope, .019″thick, 0.4 PSF	1.87	.81	2.68
1850	Steel, colors, 3″ min slope, 26 gauge, 1.0 PSF	2.18	1.71	3.89
2795	Slate roofing			
2800	4″ min. slope, shingles, 3/16″ thick, 8.0 PSF	6	2.14	8.14
3495	Wood roofing			
3500	4″ min slope, cedar shingles, 16″ x 5″, 5″ exposure 1.6 PSF	1.71	1.61	3.32
4000	Shakes, 18″, 8-1/2″ exposure, 2.8 PSF	1.12	1.99	3.11
5095	Tile roofing			
5100	Aluminum, mission, 3″ min slope, .019″ thick, 0.65 PSF	4.75	1.55	6.30
6000	Clay, Americana, 3″ minimum slope, 8 PSF	5.85	2.27	8.12

SHELL B

B3010 Roof Coverings

B3010 320	Roof Deck Rigid Insulation	COST PER S.F.		
		MAT.	INST.	TOTAL
0100	Fiberboard low density, 1/2" thick, R1.39			
0150	1" thick R2.78	.37	.45	.82
0300	1 1/2" thick R4.17	.55	.45	1
0350	2" thick R5.56	.75	.45	1.20
0370	Fiberboard high density, 1/2" thick R1.3	.22	.36	.58
0380	1" thick R2.5	.40	.45	.85
0390	1 1/2" thick R3.8	.65	.45	1.10
0410	Fiberglass, 3/4" thick R2.78	.51	.36	.87
0450	15/16" thick R3.70	.67	.36	1.03
0550	1-5/16" thick R5.26	1.16	.36	1.52
0650	2 7/16" thick R10	1.41	.45	1.86
1510	Polyisocyanurate 2#/CF density, 1" thick R7.14	.36	.26	.62
1550	1 1/2" thick R10.87	.39	.29	.68
1600	2" thick R14.29	.50	.33	.83
1650	2 1/2" thick R16.67	.55	.34	.89
1700	3" thick R21.74	.78	.36	1.14
1750	3 1/2" thick R25	.81	.36	1.17
1800	Tapered for drainage	.42	.26	.68
1810	Expanded polystyrene, 1#/CF density, 3/4" thick R2.89	.21	.24	.45
1820	2" thick R7.69	.41	.29	.70
1830	Extruded polystyrene, 15 PSI compressive strength, 1" thick R5	.25	.24	.49
1835	2" thick R10	.40	.29	.69
1840	3" thick R15	.85	.36	1.21
2550	40 PSI compressive strength, 1" thick R5	.39	.24	.63
2600	2" thick R10	.76	.29	1.05
2650	3" thick R15	1.11	.36	1.47
2700	4" thick R20	1.49	.36	1.85
2750	Tapered for drainage	.55	.26	.81
2810	60 PSI compressive strength, 1" thick R5	.46	.25	.71
2850	2" thick R10	.83	.30	1.13
2900	Tapered for drainage	.67	.26	.93
3050	Composites with 2" EPS			
3060	1" Fiberboard	.86	.38	1.24
3070	7/16" oriented strand board	1.02	.45	1.47
3080	1/2" plywood	1.10	.45	1.55
3090	1" perlite	.90	.45	1.35
4000	Composites with 1-1/2" polyisocyanurate			
4010	1" fiberboard	.92	.45	1.37
4020	1" perlite	.97	.43	1.40
4030	7/16" oriented strand board	1.11	.45	1.56

B3010 420		Roof Edges				COST PER L.F.		
	EDGE TYPE	DESCRIPTION	SPECIFICATION	FACE HEIGHT		MAT.	INST.	TOTAL
1000	Aluminum	mill finish	.050" thick	4"		8.40	7.25	15.65
1100				6"		9.10	7.50	16.60
1300		duranodic	.050" thick	4"		8.95	7.25	16.20
1400				6"		9.65	7.50	17.15
1600		painted	.050" thick	4"		9.55	7.25	16.80
1700				6"		10.45	7.50	17.95
2000	Copper	plain	16 oz.	4"		6.55	7.25	13.80
2100				6"		7.40	7.50	14.90
2300			20 oz.	4"		7.10	8.25	15.35
2400				6"		7.80	7.95	15.75

SHELL

B

B3010 Roof Coverings

B3010 420 — Roof Edges

	EDGE TYPE	DESCRIPTION	SPECIFICATION	FACE HEIGHT		COST PER L.F.		
						MAT.	INST.	TOTAL
2700	Sheet Metal	galvanized	20 Ga.	4″		7.25	8.55	15.80
2800				6″		8.45	8.55	17
3000			24 Ga.	4″		6.55	7.25	13.80
3100				6″		7.40	7.25	14.65

B3010 430 — Flashing

	MATERIAL	BACKING	SIDES	SPECIFICATION	QUANTITY	COST PER S.F.		
						MAT.	INST.	TOTAL
0040	Aluminum	none		.019″		.85	2.50	3.35
0050				.032″		1.14	2.50	3.64
0300		fabric	2	.004″		1.02	1.10	2.12
0400		mastic		.004″		1.02	1.10	2.12
0700	Copper	none		16 oz.	<500 lbs.	2.53	3.15	5.68
0800				24 oz.	<500 lbs.	5.15	3.45	8.60
2000	Copper lead	fabric	1	2 oz.		1.55	1.10	2.65
3500	PVC black	none		.010″		.15	1.27	1.42
3700				.030″		.32	1.27	1.59
4200	Neoprene			1/16″		1.63	1.27	2.90
4500	Stainless steel	none		.015″	<500 lbs.	3.48	3.15	6.63
4600	Copper clad				>2000 lbs.	3.36	2.34	5.70
5000	Plain			32 ga.		2.38	2.34	4.72

B3010 Roof Coverings

B3010 610 — Gutters

	SECTION	MATERIAL	THICKNESS	SIZE	FINISH	COST PER L.F.		
						MAT.	INST.	TOTAL
0050	Box	aluminum	.027"	5"	enameled	1.20	3.61	4.81
0100					mill	1.33	3.61	4.94
0200			.032"	5"	enameled	1.36	3.61	4.97
0500		copper	16 Oz.	4"	lead coated	7.95	3.61	11.56
0600					mill	3.88	3.61	7.49
1000		steel galv.	28 Ga.	5"	enameled	1.14	3.61	4.75
1200			26 Ga.	5"	mill	1.05	3.61	4.66
1800		vinyl		4"	colors	.94	3.40	4.34
1900				5"	colors	1.10	3.40	4.50
2300		hemlock or fir		4"x5"	treated	8	3.74	11.74
3000	Half round	copper	16 Oz.	4"	lead coated	5.25	3.61	8.86
3100					mill	3.80	3.61	7.41
3600		steel galv.	28 Ga.	5"	enameled	1.14	3.61	4.75
4102		stainless steel		5"	mill	5.50	3.61	9.11
5000		vinyl		4"	white	.75	3.40	4.15

B3010 630 — Downspouts

	MATERIALS	SECTION	SIZE	FINISH	THICKNESS	COST PER V.L.F.		
						MAT.	INST.	TOTAL
0100	Aluminum	rectangular	2"x3"	embossed mill	.020"	.81	2.28	3.09
0150				enameled	.020"	1	2.28	3.28
0250			3"x4"	enameled	.024"	1.75	3.09	4.84
0300		round corrugated	3"	enameled	.020"	.87	2.28	3.15
0350			4"	enameled	.025"	1.66	3.09	4.75
0500	Copper	rectangular corr.	2"x3"	mill	16 Oz.	3.60	2.28	5.88
0600		smooth		mill	16 Oz.	5.25	2.28	7.53
0700		rectangular corr.	3"x4"	mill	16 Oz.	5.35	2.98	8.33
1300	Steel	rectangular corr.	2"x3"	galvanized	28 Ga.	.59	2.28	2.87
1350				epoxy coated	24 Ga.	1.11	2.28	3.39
1400		smooth		galvanized	28 Ga.	.69	2.28	2.97
1450		rectangular corr.	3"x4"	galvanized	28 Ga.	1.65	2.98	4.63
1500				epoxy coated	24 Ga.	1.84	2.98	4.82
1550		smooth		galvanized	28 Ga.	1.35	2.98	4.33
1652		round corrugated	3"	galvanized	28 Ga.	.80	2.28	3.08
1700			4"	galvanized	28 Ga.	1.18	2.98	4.16
1750			5"	galvanized	28 Ga.	1.54	3.33	4.87
2000	Steel pipe	round	4"	black	X.H.	5.85	21.50	27.35
2552	S.S. tubing sch.5	rectangular	3"x4"	mill		23.50	2.98	26.48

B3010 630 — Gravel Stop

	MATERIALS	SECTION	SIZE	FINISH	THICKNESS	COST PER L.F.		
						MAT.	INST.	TOTAL
5100	Aluminum	extruded	4"	mill	.050"	3.54	2.98	6.52
5200			4"	duranodic	.050"	4.06	2.98	7.04
5300			8"	mill	.050"	4.94	3.46	8.40
5400			8"	duranodic	.050"	5.50	3.46	8.96
6000			12"-2 pc.	duranodic	.050"	6.90	4.33	11.23
6100	Stainless	formed	6"	mill	24 Ga.	7.85	3.21	11.06

B3020 Roof Openings

Roof Hatch

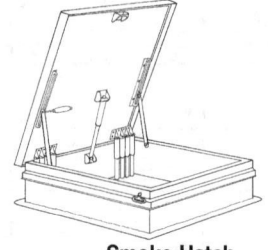

Smoke Hatch

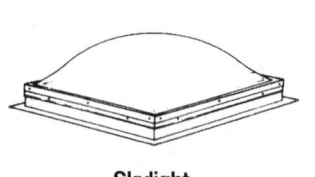

Skylight

B3020 210	Hatches	COST PER OPNG.		
		MAT.	INST.	TOTAL
0200	Roof hatches, with curb, and 1" fiberglass insulation, 2'-6"x3'-0", aluminum	450	145	595
0300	Galvanized steel 165 lbs.	375	145	520
0400	Primed steel 164 lbs.	405	145	550
0500	2'-6"x4'-6" aluminum curb and cover, 150 lbs.	640	161	801
0600	Galvanized steel 220 lbs.	535	161	696
0650	Primed steel 218 lbs.	635	161	796
0800	2'x6"x8'-0" aluminum curb and cover, 260 lbs.	1,275	220	1,495
0900	Galvanized steel, 360 lbs.	1,025	220	1,245
0950	Primed steel 358 lbs.	1,050	220	1,270
1200	For plexiglass panels, add to the above	380		380
2100	Smoke hatches, unlabeled not incl. hand winch operator, 2'-6"x3', galv	455	175	630
2200	Plain steel, 160 lbs.	490	175	665
2400	2'-6"x8'-0", galvanized steel, 360 lbs.	1,125	241	1,366
2500	Plain steel, 350 lbs.	1,150	241	1,391
3000	4'-0"x8'-0", double leaf low profile, aluminum cover, 359 lb.	1,575	181	1,756
3100	Galvanized steel 475 lbs.	1,375	181	1,556
3200	High profile, aluminum cover, galvanized curb, 361 lbs.	1,425	181	1,606

B3020 210	Skylights	COST PER S.F.		
		MAT.	INST.	TOTAL
5100	Skylights, plastic domes, insul curbs, nom. size to 10 S.F., single glaze	14.55	9.05	23.60
5200	Double glazing	21	11.15	32.15
5300	10 S.F. to 20 S.F., single glazing	8.70	3.67	12.37
5400	Double glazing	17	4.60	21.60
5500	20 S.F. to 30 S.F., single glazing	11	3.12	14.12
5600	Double glazing	15.65	3.67	19.32
5700	30 S.F. to 65 S.F., single glazing	15.20	2.38	17.58
5800	Double glazing	11.75	3.12	14.87
6000	Sandwich panels fiberglass, 9-1/16" thick, 2 S.F. to 10 S.F.	16.10	7.25	23.35
6100	10 S.F. to 18 S.F.	14.45	5.45	19.90
6200	2-3/4" thick, 25 S.F. to 40 S.F.	23.50	4.91	28.41
6300	40 S.F. to 70 S.F.	19.05	4.39	23.44
6301				

For information about Means Estimating Seminars, see yellow pages 11 and 12 in back of book

SHELL B

Important: See the Reference Section for critical supporting data - Location Factors & Historical Cost Indexes

C INTERIORS

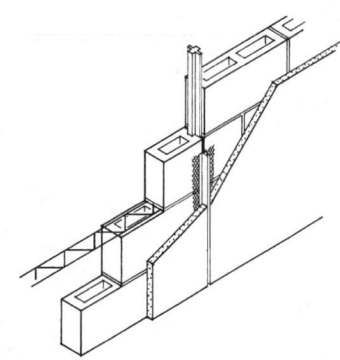

The Concrete Block Partition Systems are defined by weight and type of block, thickness, type of finish and number of sides finished. System components include joint reinforcing on alternate courses and vertical control joints.

C1010 102 — Concrete Block Partitions - Regular Weight

	TYPE	THICKNESS (IN.)	TYPE FINISH	SIDES FINISHED		MAT.	INST.	TOTAL
1000	Hollow	4	none	0		1.11	4.10	5.21
1010			gyp. plaster 2 coat	1		1.45	6	7.45
1020				2		1.80	7.85	9.65
1200			portland - 3 coat	1		1.35	6.25	7.60
1400			5/8" drywall	1		1.46	5.45	6.91
1500		6	none	0		1.59	4.40	5.99
1510			gyp. plaster 2 coat	1		1.93	6.30	8.23
1520				2		2.28	8.15	10.43
1700			portland - 3 coat	1		1.83	6.55	8.38
1900			5/8" drywall	1		1.94	5.75	7.69
1910				2		2.29	7.10	9.39
2000		8	none	0		1.72	4.70	6.42
2010			gyp. plaster 2 coat	1		2.06	6.60	8.66
2020			gyp. plaster 2 coat	2		2.41	8.45	10.86
2200			portland - 3 coat	1		1.96	6.85	8.81
2400			5/8" drywall	1		2.07	6.05	8.12
2410				2		2.42	7.40	9.82
2500		10	none	0		2.39	4.92	7.31
2510			gyp. plaster 2 coat	1		2.73	6.80	9.53
2520				2		3.08	8.65	11.73
2700			portland - 3 coat	1		2.63	7.10	9.73
2900			5/8" drywall	1		2.74	6.25	8.99
2910				2		3.09	7.60	10.69
3000	Solid	2	none	0		.97	4.06	5.03
3010			gyp. plaster	1		1.31	5.95	7.26
3020				2		1.66	7.80	9.46
3200			portland - 3 coat	1		1.21	6.20	7.41
3400			5/8" drywall	1		1.32	5.40	6.72
3410				2		1.67	6.75	8.42
3500		4	none	0		1.49	4.25	5.74
3510			gyp. plaster	1		1.90	6.15	8.05
3520				2		2.18	8	10.18
3700			portland - 3 coat	1		1.73	6.40	8.13
3900			5/8" drywall	1		1.84	5.60	7.44
3910				2		2.19	6.95	9.14

Important: See the Reference Section for critical supporting data - Location Factors & Historical Cost Indexes

C10 Interior Construction

C1010 Partitions

C1010 102 — Concrete Block Partitions - Regular Weight

	TYPE	THICKNESS (IN.)	TYPE FINISH	SIDES FINISHED		COST PER S.F.		
						MAT.	INST.	TOTAL
4002	Solid	6	none	0		1.73	4.57	6.30
4010			gyp. plaster	1		2.07	6.45	8.52
4020				2		2.42	8.30	10.72
4200			portland - 3 coat	1		1.97	6.75	8.72
4400			5/8" drywall	1		2.08	5.90	7.98
4410				2		2.43	7.25	9.68

C1010 102 — Concrete Block Partitions - Lightweight

	TYPE	THICKNESS (IN.)	TYPE FINISH	SIDES FINISHED		COST PER S.F.		
						MAT.	INST.	TOTAL
5000	Hollow	4	none	0		1.29	4.01	5.30
5010			gyp. plaster	1		1.63	5.90	7.53
5020				2		1.98	7.75	9.73
5200			portland - 3 coat	1		1.53	6.15	7.68
5400			5/8" drywall	1		1.64	5.35	6.99
5410				2		1.99	6.70	8.69
5500		6	none	0		1.73	4.30	6.03
5510			gyp. plaster	1		2.07	6.20	8.27
5520			gyp. plaster	2		2.42	8.05	10.47
5700			portland - 3 coat	1		1.97	6.45	8.42
5900			5/8" drywall	1		2.08	5.65	7.73
5910				2		2.43	7	9.43
6000		8	none	0		2.10	4.59	6.69
6010			gyp. plaster	1		2.44	6.50	8.94
6020				2		2.79	8.35	11.14
6200			portland - 3 coat	1		2.34	6.75	9.09
6400	Hollow	8	5/8" drywall	1		2.45	5.95	8.40
6410				2		2.80	7.30	10.10
6500		10	none	0		2.73	4.79	7.52
6510			gyp. plaster	1		3.07	6.70	9.77
6520				2		3.42	8.55	11.97
6700			portland - 3 coat	1		2.97	6.95	9.92
6900			5/8" drywall	1		3.08	6.15	9.23
6910				2		3.43	7.50	10.93
7000	Solid	4	none	0		1.40	4.20	5.60
7010			gyp. plaster	1		1.74	6.10	7.84
7020				2		2.09	7.95	10.04
7200			portland - 3 coat	1		1.64	6.35	7.99
7400			5/8" drywall	1		1.75	5.55	7.30
7410				2		2.10	6.90	9
7500		6	none	0		1.94	4.51	6.45
7510			gyp. plaster	1		2.34	6.45	8.79
7520				2		2.63	8.25	10.88
7700			portland - 3 coat	1		2.18	6.65	8.83
7900			5/8" drywall	1		2.29	5.85	8.14
7910				2		2.64	7.20	9.84
8000		8	none	0		2.61	4.83	7.44
8010			gyp. plaster	1		2.95	6.70	9.65
8020				2		3.30	8.55	11.85
8200			portland - 3 coat	1		2.85	7	9.85
8400			5/8" drywall	1		2.96	6.20	9.16
8410				2		3.31	7.55	10.86

INTERIORS

C

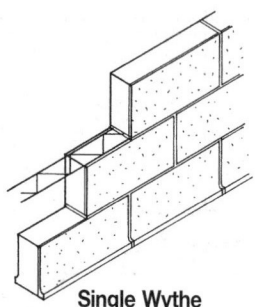

Single Wythe

Structural facing tile
8W series
8" x 16"

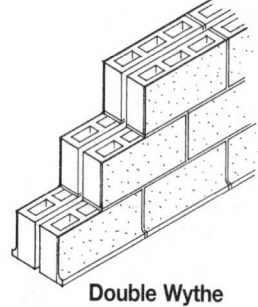

Double Wythe

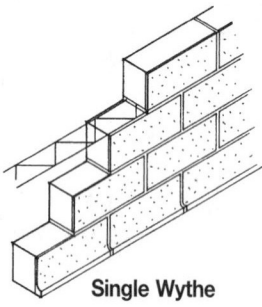

Single Wythe

Structural facing tile
6T series
5-1/3" x 12"

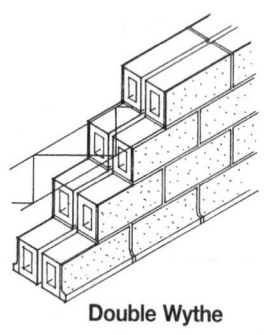

Double Wythe

C1010 120	Tile Partitions	COST PER S.F.		
		MAT.	INST.	TOTAL
1000	8W series 8"x16", 4" thick wall, reinf every 2 courses, glazed 1 side	7.95	4.93	12.88
1100	Glazed 2 sides	10.10	5.25	15.35
1200	Glazed 2 sides, using 2 wythes of 2" thick tile	13.80	9.45	23.25
1300	6" thick wall, horizontal reinf every 2 courses, glazed 1 side	11.20	5.15	16.35
1400	Glazed 2 sides, each face different color, 2" and 4" tile	14.85	9.65	24.50
1500	8" thick wall, glazed 2 sides using 2 wythes of 4" thick tile	15.90	9.85	25.75
1600	10" thick wall, glazed 2 sides using 1 wythe of 4" tile & 1 wythe of 6" tile	19.15	10.10	29.25
1700	Glazed 2 sides cavity wall, using 2 wythes of 4" thick tile	15.90	9.85	25.75
1800	12" thick wall, glazed 2 sides using 2 wythes of 6" thick tile	22.50	10.30	32.80
1900	Glazed 2 sides cavity wall, using 2 wythes of 4" thick tile	15.90	9.85	25.75
2100	6T series 5-1/3"x12" tile, 4" thick, non load bearing glazed one side,	7.45	7.75	15.20
2200	Glazed two sides	12.55	8.70	21.25
2300	Glazed two sides, using two wythes of 2" thick tile	12.80	15.10	27.90
2400	6" thick, glazed one side	11.55	8.10	19.65
2500	Glazed two sides	14.40	9.20	23.60
2600	Glazed two sides using 2" thick tile and 4" thick tile	13.85	15.30	29.15
2700	8" thick, glazed one side	16	9.45	25.45
2800	Glazed two sides using two wythes of 4" thick tile	14.90	15.50	30.40
2900	Glazed two sides using 6" thick tile and 2" thick tile	17.95	15.65	33.60
3000	10" thick cavity wall, glazed two sides using two wythes of 4" tile	14.90	15.50	30.40
3100	12" thick, glazed two sides using 4" thick tile and 8" thick tile	23.50	17.20	40.70
3200	2" thick facing tile, glazed one side, on 6" concrete block	8.35	8.95	17.30
3300	On 8" concrete block	8.50	9.25	17.75
3400	On 10" concrete block	9.15	9.45	18.60

Important: See the Reference Section for critical supporting data - Location Factors & Historical Cost Indexes

C1010 Partitions

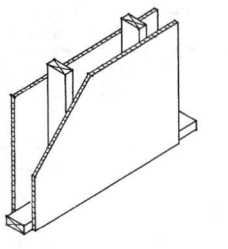

Gypsum board, single layer each side on wood studs.

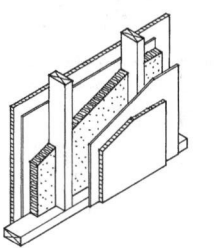

Gypsum board, sound deadening board each side, with 1-1/2″ insulation on wood studs.

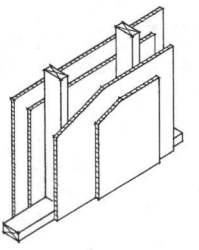

Gypsum board, two layers each side on wood studs.

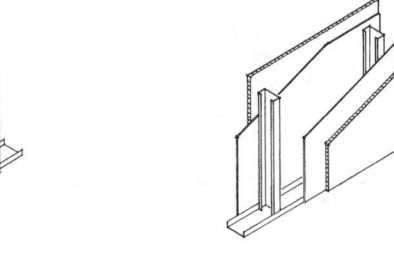

Gypsum board, single layer each side on metal studs.

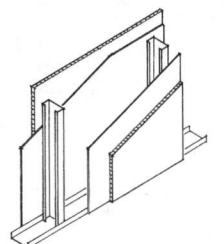

Gypsum board, sound deadening board each side on metal studs.

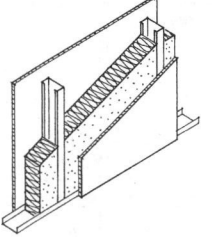

Gypsum board two layers one side, single layer opposite side, with 3-1/2″ insulation on metal studs.

C1010 124			Drywall Partitions/Wood Stud Framing					
	FACE LAYER	BASE LAYER	FRAMING	OPPOSITE FACE	INSULATION	COST PER S.F.		
						MAT.	INST.	TOTAL
1200	5/8″ FR drywall	none	2 x 4, @ 16″ O.C.	same	0	.97	2.23	3.20
1250				5/8″ reg. drywall	0	.97	2.23	3.20
1300				nothing	0	.69	1.49	2.18
1400		1/4″ SD gypsum	2 x 4 @ 16″ O.C.	same	1-1/2″ fiberglass	1.90	3.44	5.34
1450				5/8″ FR drywall	1-1/2″ fiberglass	1.66	3.02	4.68
1500				nothing	1-1/2″ fiberglass	1.38	2.28	3.66
1600		resil. channels	2 x 4 @ 16″, O.C.	same	1-1/2″ fiberglass	1.67	4.36	6.03
1650				5/8″ FR drywall	1-1/2″ fiberglass	1.54	3.48	5.02
1700				nothing	1-1/2″ fiberglass	1.26	2.74	4
1800		5/8″ FR drywall	2 x 4 @ 24″ O.C.	same	0	1.35	2.82	4.17
1850				5/8″ FR drywall	0	1.11	2.45	3.56
1900				nothing	0	.83	1.71	2.54
2200		5/8″ FR drywall	2 rows-2 x 4	same	2″ fiberglass	2.36	4.08	6.44
2250			16″O.C.	5/8″ FR drywall	2″ fiberglass	2.12	3.71	5.83
2300				nothing	2″ fiberglass	1.84	2.97	4.81
2400	5/8″ WR drywall	none	2 x 4, @ 16″ O.C.	same	0	1.09	2.23	3.32
2450				5/8″ FR drywall	0	1.03	2.23	3.26
2500				nothing	0	.75	1.49	2.24
2600		5/8″ FR drywall	2 x 4, @ 24″ O.C.	same	0	1.47	2.82	4.29
2650				5/8″ FR drywall	0	1.17	2.45	3.62
2700				nothing	0	.89	1.71	2.60
2800	5/8 VF drywall	none	2 x 4, @ 16″ O.C.	same	0	1.85	2.41	4.26
2850				5/8″ FR drywall	0	1.41	2.32	3.73
2900				nothing	0	1.13	1.58	2.71

C1010 Partitions

C1010 124		**Drywall Partitions/Wood Stud Framing**						
	FACE LAYER	BASE LAYER	FRAMING	OPPOSITE FACE	INSULATION	COST PER S.F.		
						MAT.	INST.	TOTAL
3000	5/8 VF drywall	5/8" FR drywall	2 x 4 , 24" O.C.	same	0	2.23	3	5.23
3050				5/8" FR drywall	0	1.55	2.54	4.09
3100				nothing	0	1.27	1.80	3.07
3200	1/2" reg drywall	3/8" reg drywall	2 x 4, @ 16" O.C.	same	0	1.37	2.97	4.34
3252	1/2" reg. drywall	3/8" reg. drywall	2x4, @ 16"O.C.	5/8"FR drywall	0	1.17	2.60	3.77
3300				nothing	0	.89	1.86	2.75

C1010 124		**Drywall Partitions/Metal Stud Framing**						
	FACE LAYER	BASE LAYER	FRAMING	OPPOSITE FACE	INSULATION	COST PER S.F.		
						MAT.	INST.	TOTAL
5200	5/8" FR drywall	none	1-5/8" @ 24" O.C.	same	0	.69	1.97	2.66
5250				5/8" reg. drywall	0	.69	1.97	2.66
5300				nothing	0	.41	1.23	1.64
5400			3-5/8" @ 24" O.C.	same	0	.72	1.98	2.70
5450				5/8" reg. drywall	0	.72	1.98	2.70
5500				nothing	0	.44	1.24	1.68
5600		1/4" SD gypsum	1-5/8" @ 24" O.C.	same	0	1.17	2.81	3.98
5650				5/8" FR drywall	0	.93	2.39	3.32
5700				nothing	0	.65	1.65	2.30
5800			2-1/2" @ 24" O.C.	same	0	1.18	2.82	4
5850				5/8" FR drywall	0	.94	2.40	3.34
5900				nothing	0	.66	1.66	2.32
6000		5/8" FR drywall	2-1/2" @ 16" O.C.	same	0	1.34	3.18	4.52
6050				5/8" FR drywall	0	1.10	2.81	3.91
6100				nothing	0	.82	2.07	2.89
6200			3-5/8" @ 24" O.C.	same	0	1.20	2.72	3.92
6250				5/8"FR drywall	3-1/2" fiberglass	1.33	2.58	3.91
6300				nothing	0	.68	1.61	2.29
6400	5/8" WR drywall	none	1-5/8" @ 24" O.C.	same	0	.81	1.97	2.78
6450				5/8" FR drywall	0	.75	1.97	2.72
6500				nothing	0	.47	1.23	1.70
6600			3-5/8" @ 24" O.C.	same	0	.84	1.98	2.82
6650				5/8" FR drywall	0	.78	1.98	2.76
6700				nothing	0	.50	1.24	1.74
6800		5/8" FR drywall	2-1/2" @ 16" O.C.	same	0	1.46	3.18	4.64
6850				5/8" FR drywall	0	1.16	2.81	3.97
6900				nothing	0	.88	2.07	2.95
7000			3-5/8" @ 24" O.C.	same	0	1.32	2.72	4.04
7050				5/8"FR drywall	3-1/2" fiberglass	1.39	2.58	3.97
7100				nothing	0	.74	1.61	2.35
7200	5/8" VF drywall	none	1-5/8" @ 24" O.C.	same	0	1.57	2.15	3.72
7250				5/8" FR drywall	0	1.13	2.06	3.19
7300				nothing	0	.85	1.32	2.17
7400			3-5/8" @ 24" O.C.	same	0	1.60	2.16	3.76
7450				5/8" FR drywall	0	1.16	2.07	3.23
7500				nothing	0	.88	1.33	2.21
7600		5/8" FR drywall	2-1/2" @ 16" O.C.	same	0	2.22	3.36	5.58
7650				5/8" FR drywall	0	1.54	2.90	4.44
7700				nothing	0	1.26	2.16	3.42
7800			3-5/8" @ 24" O.C.	same	0	2.08	2.90	4.98
7850				5/8"FR drywall	3-1/2" fiberglass	1.77	2.67	4.44
7900				nothing	0	1.12	1.70	2.82

Important: See the Reference Section for critical supporting data - Location Factors & Historical Cost Indexes

C1010 128	Drywall Components	COST PER S.F.		
		MAT.	INST.	TOTAL
0060	Metal studs, 24" O.C. including track, load bearing, 20 gage, 2-1/2"	.36	.70	1.06
0080	3-5/8"	.43	.71	1.14
0100	4"	.39	.73	1.12
0120	6"	.58	.74	1.32
0140	Metal studs, 24" O.C. including track, load bearing, 18 gage, 2-1/2"	.36	.70	1.06
0160	3-5/8"	.43	.71	1.14
0180	4"	.39	.73	1.12
0200	6"	.58	.74	1.32
0220	16 gage, 2-1/2"	.42	.80	1.22
0240	3-5/8"	.50	.81	1.31
0260	4"	.53	.83	1.36
0280	6"	.67	.85	1.52
0300	Non load bearing, 25 gage, 1-5/8"	.13	.49	.62
0340	3-5/8"	.16	.50	.66
0360	4"	.17	.50	.67
0380	6"	.26	.52	.78
0400	20 gage, 2-1/2"	.22	.62	.84
0420	3-5/8"	.25	.63	.88
0440	4"	.27	.63	.90
0460	6"	.33	.64	.97
0540	Wood studs including blocking, shoe and double top plate, 2"x4", 12"O.C.	.51	.94	1.45
0560	16" O.C.	.41	.75	1.16
0580	24" O.C.	.31	.60	.91
0600	2"x6", 12" O.C.	.81	1.07	1.88
0620	16" O.C.	.66	.83	1.49
0640	24" O.C.	.50	.65	1.15
0642	Furring one side only, steel channels, 3/4", 12" O.C.	.18	1.45	1.63
0644	16" O.C.	.16	1.29	1.45
0646	24" O.C.	.11	.98	1.09
0647	1-1/2", 12" O.C.	.27	1.63	1.90
0648	16" O.C.	.24	1.42	1.66
0649	24" O.C.	.16	1.12	1.28
0650	Wood strips, 1" x 3", on wood, 12" O.C.	.21	.68	.89
0651	16" O.C.	.16	.51	.67
0652	On masonry, 12" O.C.	.21	.75	.96
0653	16" O.C.	.16	.56	.72
0654	On concrete, 12" O.C.	.21	1.44	1.65
0655	16" O.C.	.16	1.08	1.24
0665	Gypsum board, one face only, exterior sheathing, 1/2"	.31	.66	.97
0680	Interior, fire resistant, 1/2"	.24	.37	.61
0700	5/8"	.24	.37	.61
0720	Sound deadening board 1/4"	.24	.42	.66
0740	Standard drywall 3/8"	.21	.37	.58
0760	1/2"	.23	.37	.60
0780	5/8"	.24	.37	.61
0800	Tongue & groove coreboard 1"	.50	1.56	2.06
0820	Water resistant, 1/2"	.24	.37	.61
0840	5/8"	.30	.37	.67
0860	Add for the following:, foil backing	.11		.11
0880	Fiberglass insulation, 3-1/2"	.37	.23	.60
0900	6"	.45	.28	.73
0920	Rigid insulation 1"	.37	.37	.74
0940	Resilient furring @ 16" O.C.	.16	1.18	1.34
0960	Taping and finishing	.04	.37	.41
0980	Texture spray	.06	.43	.49
1000	Thin coat plaster	.08	.47	.55
1040	2"x4" staggered studs 2"x6" plates & blocking	.59	.81	1.40

INTERIORS

C

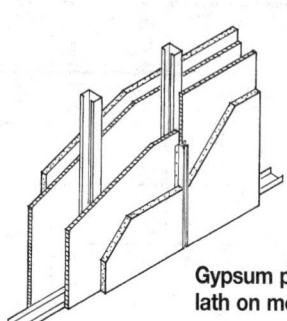

Gypsum plaster and gypsum lath on metal studs.

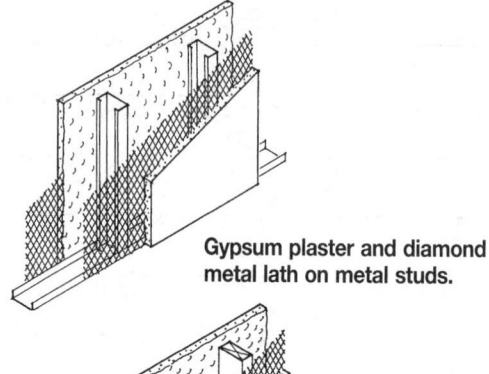

Gypsum plaster and diamond metal lath on metal studs.

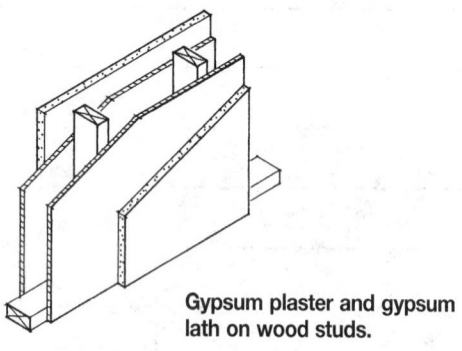

Gypsum plaster and gypsum lath on wood studs.

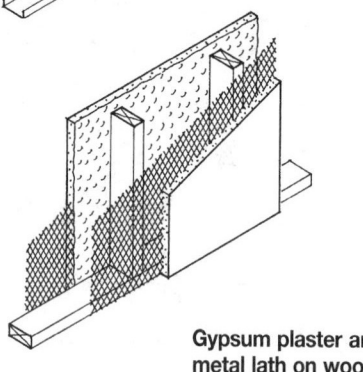

Gypsum plaster and diamond metal lath on wood studs.

INTERIORS C

C1010 140				Plaster Partitions/Metal Stud Framing				
	TYPE	FRAMING	LATH	OPPOSITE FACE		COST PER S.F.		
						MAT.	INST.	TOTAL
1000	2 coat gypsum	2-1/2" @ 16"O.C.	3/8" gypsum	same		2.08	5.45	7.53
1010				nothing		1.17	3.19	4.36
1100		3-1/4" @ 24"O.C.	1/2" gypsum	same		2.05	5.30	7.35
1110				nothing		1.14	2.96	4.10
1500	2 coat vermiculite	2-1/2" @ 16"O.C.	3/8" gypsum	same		1.93	6	7.93
1510				nothing		1.10	3.44	4.54
1600		3-1/4" @ 24"O.C.	1/2" gypsum	same		1.90	5.80	7.70
1610				nothing		1.07	3.21	4.28
2000	3 coat gypsum	2-1/2" @ 16"O.C.	3/8" gypsum	same		1.97	6.20	8.17
2010				nothing		1.12	3.56	4.68
2020			3.4lb. diamond	same		1.64	6.20	7.84
2030				nothing		.96	3.56	4.52
2040			2.75lb. ribbed	same		1.64	6.20	7.84
2050				nothing		.96	3.56	4.52
2100		3-1/4" @ 24"O.C.	1/2" gypsum	same		1.94	6.05	7.99
2110				nothing		1.09	3.33	4.42
2120			3.4lb. ribbed	same		1.86	6.05	7.91
2130				nothing		1.05	3.33	4.38

Important: See the Reference Section for critical supporting data - Location Factors & Historical Cost Indexes

C1010 Partitions

C1010 140				Plaster Partitions/Metal Stud Framing			

	TYPE	FRAMING	LATH	OPPOSITE FACE		COST PER S.F.		
						MAT.	INST.	TOTAL
3500	3 coat gypsum	2-1/2" @ 16"O.C.	3/8" gypsum	same		2.50	8	10.50
3510	W/med. Keenes			nothing		1.39	4.43	5.82
3520			3.4lb. diamond	same		2.17	8	10.17
3530				nothing		1.23	4.43	5.66
3540			2.75lb. ribbed	same		2.17	8	10.17
3550				nothing		1.23	4.43	5.66
3600		3-1/4" @ 24"O.C.	1/2" gypsum	same		2.47	7.80	10.27
3610				nothing		1.36	4.20	5.56
3620			3.4lb. ribbed	same		2.39	7.80	10.19
3630				nothing		1.32	4.20	5.52
4000	3 coat gypsum	2-1/2" @ 16"O.C.	3/8" gypsum	same		2.51	8.80	11.31
4010	W/hard Keenes			nothing		1.39	4.86	6.25
4022	3 coat gypsum	2-1/2" @ 16"O.C.	3.4 lb. diamond	same		2.18	8.80	10.98
4032	W/hard Keenes			nothing		1.23	4.86	6.09
4040			2.75lb. ribbed	same		2.18	8.80	10.98
4050				nothing		1.23	4.86	6.09
4100		3-1/4" @ 24"O.C.	1/2" gypsum	same		2.48	8.60	11.08
4110				nothing		1.36	4.63	5.99
4120			3.4lb. ribbed	same		2.40	8.60	11
4130				nothing		1.32	4.63	5.95

C1010 140				Plaster Partitions/Wood Stud Framing			

	TYPE	FRAMING	LATH	OPPOSITE FACE		COST PER S.F.		
						MAT.	INST.	TOTAL
5000	2 coat gypsum	2"x4" @ 16"O.C.	3/8" gypsum	same		2.28	5.35	7.63
5010				nothing		1.38	3.12	4.50
5100		2"x4" @ 24"O.C.	1/2" gypsum	same		2.18	5.25	7.43
5110				nothing		1.28	2.99	4.27
5500	2 coat vermiculite	2"x4" @ 16"O.C.	3/8" gypsum	same		2.13	5.85	7.98
5510				nothing		1.31	3.37	4.68
5600		2"x4" @ 24"O.C.	1/2" gypsum	same		2.03	5.80	7.83
5610				nothing		1.21	3.24	4.45
6000	3 coat gypsum	2"x4" @ 16"O.C.	3/8" gypsum	same		2.17	6.10	8.27
6010				nothing		1.33	3.49	4.82
6020			3.4lb. diamond	same		1.86	6.15	8.01
6030				nothing		1.18	3.51	4.69
6040			2.75lb. ribbed	same		1.84	6.20	8.04
6050				nothing		1.17	3.53	4.70
6100		2"x4" @ 24"O.C.	1/2" gypsum	same		2.07	6	8.07
6110				nothing		1.23	3.36	4.59
6120			3.4lb. ribbed	same		1.76	6.05	7.81
6130				nothing		1.08	3.40	4.48
7500	3 coat gypsum	2"x4" @ 16"O.C.	3/8" gypsum	same		2.70	7.85	10.55
7510	W/med Keenes			nothing		1.60	4.36	5.96
7520			3.4lb. diamond	same		2.39	7.90	10.29
7530				nothing		1.45	4.38	5.83
7540			2.75lb. ribbed	same		2.37	7.95	10.32
7550				nothing		1.44	4.40	5.84
7600		2"x4" @ 24"O.C.	1/2" gypsum	same		2.60	7.75	10.35
7610				nothing		1.50	4.23	5.73
7620			3.4lb. ribbed	same		2.54	7.90	10.44
7630				nothing		1.47	4.30	5.77

C INTERIORS

C1010 Partitions

C1010 140		Plaster Partitions/Wood Stud Framing				COST PER S.F.		
	TYPE	FRAMING	LATH	OPPOSITE FACE		MAT.	INST.	TOTAL
8000	3 coat gypsum	2"x4" @ 16"O.C.	3/8" gypsum	same		2.71	8.70	11.41
8010	W/hard Keenes			nothing		1.60	4.79	6.39
8020			3.4lb. diamond	same		2.40	8.75	11.15
8030				nothing		1.45	4.81	6.26
8040			2.75lb. ribbed	same		2.38	8.75	11.13
8050				nothing		1.44	4.83	6.27
8100		2"x4" @ 24"O.C.	1/2" gypsum	same		2.61	8.60	11.21
8110				nothing		1.50	4.66	6.16
8120			3.4lb. ribbed	same		2.55	8.70	11.25
8130				nothing		1.47	4.73	6.20

C1010 144	Plaster Partition Components	COST PER S.F.		
		MAT.	INST.	TOTAL
0060	Metal studs, 16" O.C., including track, non load bearing, 25 gage, 1-5/8"	.19	.76	.95
0080	2-1/2"	.19	.76	.95
0100	3-1/4"	.21	.78	.99
0120	3-5/8"	.21	.78	.99
0140	4"	.23	.79	1.02
0160	6"	.34	.79	1.13
0180	Load bearing, 20 gage, 2-1/2"	.49	.97	1.46
0200	3-5/8"	.59	.99	1.58
0220	4"	.62	1.01	1.63
0240	6"	.78	1.03	1.81
0260	16 gage 2-1/2"	.58	1.10	1.68
0280	3-5/8"	.69	1.13	1.82
0300	4"	.73	1.15	1.88
0320	6"	.91	1.17	2.08
0340	Wood studs, including blocking, shoe and double plate, 2"x4" , 12" O.C.	.51	.94	1.45
0360	16" O.C.	.41	.75	1.16
0380	24" O.C.	.31	.60	.91
0400	2"x6" , 12" O.C.	.81	1.07	1.88
0420	16" O.C.	.66	.83	1.49
0440	24" O.C.	.50	.65	1.15
0460	Furring one face only, steel channels, 3/4" , 12" O.C.	.18	1.45	1.63
0480	16" O.C.	.16	1.29	1.45
0500	24" O.C.	.11	.98	1.09
0520	1-1/2" , 12" O.C.	.27	1.63	1.90
0540	16" O.C.	.24	1.42	1.66
0560	24"O.C.	.16	1.12	1.28
0580	Wood strips 1"x3", on wood., 12" O.C.	.21	.68	.89
0600	16"O.C.	.16	.51	.67
0620	On masonry, 12" O.C.	.21	.75	.96
0640	16" O.C.	.16	.56	.72
0660	On concrete, 12" O.C.	.21	1.44	1.65
0680	16" O.C.	.16	1.08	1.24
0700	Gypsum lath. plain or perforated, nailed to studs, 3/8" thick	.43	.45	.88
0720	1/2" thick	.43	.47	.90
0740	Clipped to studs, 3/8" thick	.44	.51	.95
0760	1/2" thick	.44	.54	.98
0780	Metal lath, diamond painted, nailed to wood studs, 2.5 lb.	.18	.45	.63
0800	3.4 lb.	.28	.47	.75
0820	Screwed to steel studs, 2.5 lb.	.18	.47	.65
0840	3.4 lb.	.28	.51	.79
0860	Rib painted, wired to steel, 2.75 lb	.28	.51	.79
0880	3.4 lb	.40	.54	.94

INTERIORS C

Important: See the Reference Section for critical supporting data - Location Factors & Historical Cost Indexes

C1010 Partitions

C1010 144	Plaster Partition Components	COST PER S.F.		
		MAT.	INST.	TOTAL
0900	4.0 lb	.41	.58	.99
0910				
0920	Gypsum plaster, 2 coats	.38	1.79	2.17
0940	3 coats	.54	2.16	2.70
0960	Perlite or vermiculite plaster, 2 coats	.40	2.04	2.44
0980	3 coats	.66	2.56	3.22
1000	Stucco, 3 coats, 1" thick, on wood framing	.45	3.58	4.03
1020	On masonry	.25	2.79	3.04
1100	Metal base galvanized and painted 2-1/2" high	.49	1.42	1.91

INTERIORS

C

Folding Accordion   Folding Leaf   Movable and Borrow Lites

C1010 154	Partitions	COST PER S.F.		
		MAT.	INST.	TOTAL
0360	Folding accordion, vinyl covered, acoustical, 3 lb. S.F., 17 ft max. hgt	23	7.45	30.45
0380	5 lb. per S.F. 27 ft max height	32	7.85	39.85
0400	5.5 lb. per S.F., 17 ft. max height	36	8.30	44.30
0420	Commercial, 1.75 lb per S.F., 8 ft. max height	16.05	3.32	19.37
0440	2.0 Lb per S.F., 17 ft. max height	17.25	4.98	22.23
0460	Industrial, 4.0 lb. per S.F. 27 ft. max height	29.50	9.95	39.45
0480	Vinyl clad wood or steel, electric operation 6 psf	41.50	4.67	46.17
0500	Wood, non acoustic, birch or mahogany	21.50	2.49	23.99
0560	Folding leaf, aluminum framed acoustical 12 ft.high.,5.5 lb per S.F.,min.	31	12.45	43.45
0580	Maximum	37.50	25	62.50
0600	6.5 lb. per S.F., minimum	32.50	12.45	44.95
0620	Maximum	40	25	65
0640	Steel acoustical, 7.5 per S.F., vinyl faced, minimum	46	12.45	58.45
0660	Maximum	56	25	81
0680	Wood acoustic type, vinyl faced to 18' high 6 psf, minimum	43	12.45	55.45
0700	Average	51.50	16.60	68.10
0720	Maximum	66.50	25	91.50
0740	Formica or hardwood faced, minimum	44.50	12.45	56.95
0760	Maximum	47.50	25	72.50
0780	Wood, low acoustical type to 12 ft. high 4.5 psf	32.50	14.95	47.45
0840	Demountable, trackless wall, cork finish, semi acous, 1-5/8"th, min	30.50	2.30	32.80
0860	Maximum	28.50	3.93	32.43
0880	Acoustic, 2" thick, minimum	24.50	2.45	26.95
0900	Maximum	42.50	3.32	45.82
0920	In-plant modular office system, w/prehung steel door			
0940	3" thick honeycomb core panels			
0960	12' x 12', 2 wall	8.85	.46	9.31
0970	4 wall	8.25	.61	8.86
0980	16' x 16', 2 wall	8.30	.32	8.62
0990	4 wall	5.75	.33	6.08
1000	Gypsum, demountable, 3" to 3-3/4" thick x 9' high, vinyl clad	4.72	1.73	6.45
1020	Fabric clad	11.75	1.89	13.64
1040	1.75 system, vinyl clad hardboard, paper honeycomb core panel			
1060	1-3/4" to 2-1/2" thick x 9' high	7.95	1.73	9.68
1080	Unitized gypsum panel system, 2" to 2-1/2" thick x 9' high			
1100	Vinyl clad gypsum	10.20	1.73	11.93
1120	Fabric clad gypsum	16.85	1.89	18.74
1140	Movable steel walls, modular system			
1160	Unitized panels, 48" wide x 9' high			
1180	Baked enamel, pre-finished	11.55	1.38	12.93
1200	Fabric clad	16.65	1.48	18.13

INTERIORS C

C1020   Interior Doors

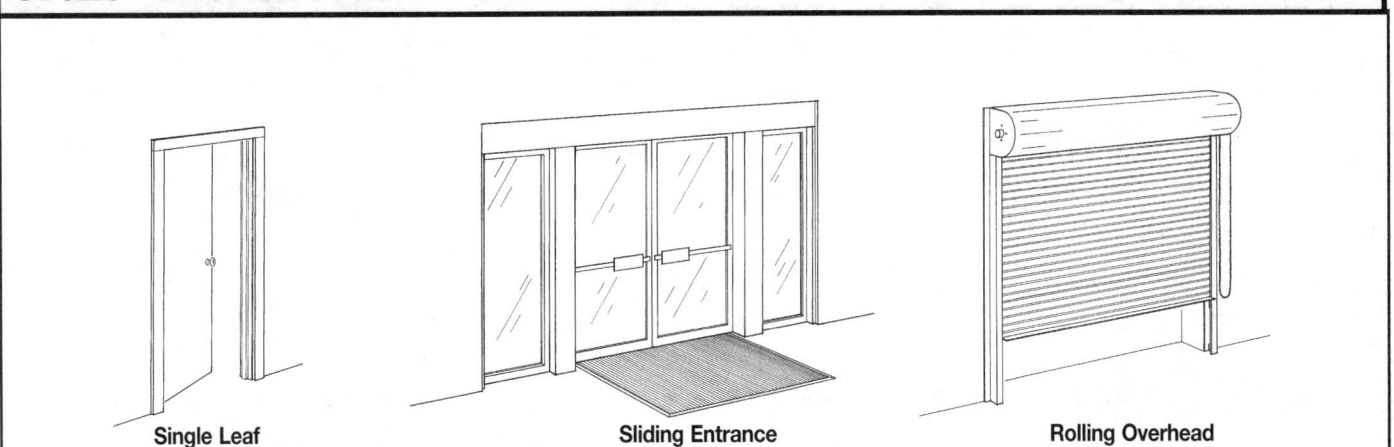

| Single Leaf | Sliding Entrance | Rolling Overhead |

C1020 102	Special Doors	COST PER OPNG.		
		MAT.	INST.	TOTAL
2500	Single leaf, wood, 3'-0"x7'-0"x1 3/8", birch, solid core	291	141	432
2510	Hollow core	253	135	388
2530	Hollow core, lauan	235	135	370
2540	Louvered pine	355	135	490
2550	Paneled pine	365	135	500
2600	Hollow metal, comm. quality, flush, 3'-0"x7'-0"x1-3/8"	390	147	537
2650	3'-0"x10'-0" openings with panel	635	191	826
2700	Metal fire, comm. quality, 3'-0"x7'-0"x1-3/8"	460	153	613
2800	Kalamein fire, comm. quality, 3'-0"x7'-0"x1-3/4"	385	281	666
3200	Double leaf, wood, hollow core, 2 - 3'-0"x7'-0"x1-3/8"	405	258	663
3300	Hollow metal, comm. quality, B label, 2'-3'-0"x7'-0"x1-3/8"	870	325	1,195
3400	6'-0"x10'-0" opening, with panel	1,225	395	1,620
3500	Double swing door system, 12'-0"x7'-0", mill finish	6,225	2,150	8,375
3700	Black finish	6,775	2,225	9,000
3800	Sliding entrance door and system mill finish	7,325	1,850	9,175
3900	Bronze finish	8,025	2,000	10,025
4000	Black finish	8,400	2,075	10,475
4100	Sliding panel mall front, 16'x9' opening, mill finish	2,550	560	3,110
4200	Bronze finish	3,325	730	4,055
4300	Black finish	4,075	895	4,970
4400	24'x9' opening mill finish	3,700	1,050	4,750
4500	Bronze finish	4,800	1,375	6,175
4600	Black finish	5,925	1,675	7,600
4700	48'x9' opening mill finish	6,875	805	7,680
4800	Bronze finish	8,950	1,050	10,000
4900	Black finish	11,000	1,300	12,300
5000	Rolling overhead steel door, manual, 8' x 8' high	790	620	1,410
5100	10' x 10' high	1,050	705	1,755
5200	20' x 10' high	2,125	990	3,115
5300	12' x 12' high	1,350	825	2,175
5400	Motor operated, 8' x 8' high	1,550	820	2,370
5500	10' x 10' high	1,800	905	2,705
5600	20' x 10' high	2,875	1,200	4,075
5700	12' x 12' high	2,100	1,025	3,125
5800	Roll up grille, aluminum, manual, 10' x 10' high, mill finish	1,925	1,200	3,125
5900	Bronze anodized	3,050	1,200	4,250
6000	Motor operated, 10' x 10' high, mill finish	2,800	1,400	4,200
6100	Bronze anodized	3,925	1,400	5,325
6200	Steel, manual, 10' x 10' high	1,725	990	2,715
6300	15' x 8' high	2,025	1,225	3,250
6400	Motor operated, 10' x 10' high	2,600	1,200	3,800
6500	15' x 8' high	2,900	1,425	4,325

INTERIORS

C

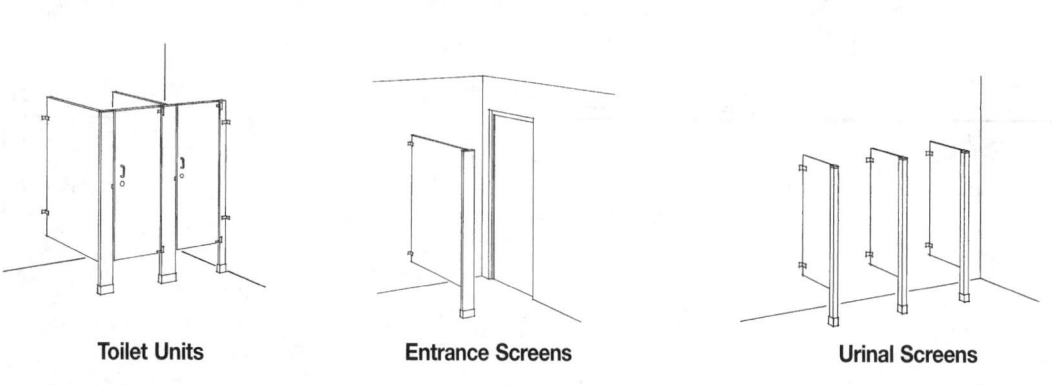

Toilet Units   Entrance Screens   Urinal Screens

C1030 110	Toilet Partitions	COST PER UNIT		
		MAT.	INST.	TOTAL
0380	Toilet partitions, cubicles, ceiling hung, marble	1,500	370	1,870
0400	Painted metal	430	187	617
0420	Plastic laminate	585	187	772
0460	Stainless steel	1,100	187	1,287
0480	Handicap addition	345		345
0520	Floor and ceiling anchored, marble	1,625	296	1,921
0540	Painted metal	405	149	554
0560	Plastic laminate	540	149	689
0600	Stainless steel	1,275	149	1,424
0620	Handicap addition	286		286
0660	Floor mounted marble	960	246	1,206
0680	Painted metal	390	107	497
0700	Plastic laminate	545	107	652
0740	Stainless steel	1,300	107	1,407
0760	Handicap addition	286		286
0780	Juvenile deduction	38.50		38.50
0820	Floor mounted with handrail marble	920	246	1,166
0840	Painted metal	405	125	530
0860	Plastic laminate	730	125	855
0900	Stainless steel	1,250	125	1,375
0920	Handicap addition	286		286
0960	Wall hung, painted metal	510	107	617
1020	Stainless steel	1,225	107	1,332
1040	Handicap addition	286		286
1080	Entrance screens, floor mounted, 54" high, marble	605	82	687
1100	Painted metal	182	50	232
1140	Stainless steel	670	50	720
1300	Urinal screens, floor mounted, 24" wide, laminated plastic	255	93.50	348.50
1320	Marble	530	110	640
1340	Painted metal	181	93.50	274.50
1380	Stainless steel	540	93.50	633.50
1428	Wall mounted wedge type, painted metal	221	74.50	295.50
1460	Stainless steel	455	74.50	529.50
1500	Partitions, shower stall, single wall, painted steel, 2'-8" x 2'-8"	705	173	878
1510	Fiberglass, 2'-8" x 2'-8"	575	192	767
1520	Double wall, enameled steel, 2'-8" x 2'-8"	805	173	978
1530	Stainless steel, 2'-8" x 2'-8"	1,350	173	1,523
1560	Tub enclosure, sliding panels, tempered glass, aluminum frame	310	216	526
1570	Chrome/brass frame-deluxe	890	289	1,179

INTERIORS C

C1030 Fittings

C1030 210	Bath and Toilet Accessories, EACH	COST EACH		
		MAT.	INST.	TOTAL
0100	Specialties, bathroom accessories, st. steel, curtain rod, 5' long, 1" diam	29.50	28.50	58
0120	Dispenser, towel, surface mounted	42	23.50	65.50
0140	Grab bar, 1-1/4" diam., 12" long	23.50	15.55	39.05
0160	Mirror, framed with shelf, 18" x 24"	109	18.70	127.70
0170	72" x 24"	298	62.50	360.50
0180	Toilet tissue dispenser, surface mounted, single roll	12.95	12.45	25.40
0200	Towel bar, 18" long	35	16.25	51.25
0300	Medicine cabinets, sliding mirror doors, 20" x 16" x 4-3/4", unlighted	93.50	53.50	147
0310	24" x 19" x 8-1/2", lighted	147	74.50	221.50

C1030 230	Bath and Toilet Accessories, L.F.	COST PER L.F.		
		MAT.	INST.	TOTAL
0100	Partitions, hospital curtain, ceiling hung, polyester oxford cloth	15.25	3.65	18.90
0110	Designer oxford cloth	29.50	4.62	34.12

C1030 310	Storage Specialties, EACH	COST EACH		
		MAT.	INST.	TOTAL
0200	Lockers, steel, single tier, 5' to 6' high, per opening, min.	100	31	131
0210	Maximum	191	36	227
0600	Shelving, metal industrial, braced, 3' wide, 1' deep	33	8.45	41.45
0610	2' deep	51.50	9	60.50

C1030 420	Fabricated Cabinets & Counters, L.F.	COST PER L.F.		
		MAT.	INST.	TOTAL
0110	Household, base, hardwood, one top drawer & one door below x 12" wide	139	30	169
0120	Four drawer x 24" wide	320	33.50	353.50
0130	Wall, hardwood, 30" high with one door x 12" wide	132	34	166
0140	Two doors x 48" wide	310	40.50	350.50
0150	Counter top-laminated plastic, stock, economy	8.45	12.45	20.90
0160	Custom-square edge, 7/8" thick	11.55	28	39.55
0170	School, counter, wood, 32" high	165	37.50	202.50
0180	Metal, 84" high	255	50	305

C1030 510	Identifying/Visual Aid Specialties, EACH	COST EACH		
		MAT.	INST.	TOTAL
0100	Control boards, magnetic, porcelain finish, framed, 24" x 18"	175	93.50	268.50
0110	96" x 48"	755	149	904
0120	Directory boards, outdoor, black plastic, 36" x 24"	370	375	745
0130	36" x 36"	770	500	1,270
0140	Indoor, economy, open faced, 18" x 24"	139	107	246
0500	Signs, interior electric exit sign, wall mounted, 6"	61	53	114
0510	Street, reflective alum., dbl. face, 4 way, w/bracket	117	25	142
0520	Letters, cast aluminum, 1/2" deep, 4" high	19.15	21	40.15
0530	1" deep, 10" high	34	21	55
0540	Plaques, cast aluminum, 20" x 30"	895	187	1,082
0550	Cast bronze, 36" x 48"	3,525	375	3,900

C1030 520	Identifying/Visual Aid Specialties, S.F.	COST PER S.F.		
		MAT.	INST.	TOTAL
0100	Bulletin board, cork sheets, no frame, 1/4" thick	6.80	2.58	9.38
0120	Aluminum frame, 1/4" thick, 3' x 5'	8.25	3.12	11.37
0200	Chalkboards, wall hung, alum, frame & chalktrough	9.30	1.66	10.96
0210	Wood frame & chalktrough	8.45	1.78	10.23
0220	Sliding board, one board with back panel	33	1.59	34.59
0230	Two boards with back panel	51	1.59	52.59

C INTERIORS

C1030 Fittings

C1030 520	Identifying/Visual Aid Specialties, S.F.	COST PER S.F.		
		MAT.	INST.	TOTAL
0240	Liquid chalk type, alum. frame & chalktrough	11.15	1.66	12.81
0250	Wood frame & chalktrough	19.40	1.66	21.06

C1030 910	Other Fittings, EACH	COST EACH		
		MAT.	INST.	TOTAL
0500	Mail boxes, horizontal, rear loaded, aluminum, 5" x 6" x 15" deep	34	11	45
0510	Front loaded, aluminum, 10" x 12" x 15" deep	118	18.70	136.70
0520	Vertical, front loaded, aluminum, 15" x 5" x 6" deep	26.50	11	37.50
0530	Bronze, duranodic finish	48.50	11	59.50
0540	Letter slot, post office	234	46.50	280.50
0550	Mail counter, window, post office, with grille	565	187	752
0700	Turnstiles, one way, 4' arm, 46" diam., manual	280	149	429
0710	Electric	935	625	1,560
0720	3 arm, 5'-5' diam. & 7' high, manual	2,425	745	3,170
0730	Electric	3,150	1,250	4,400

INTERIORS

C

Important: See the Reference Section for critical supporting data - Location Factors & Historical Cost Indexes

C2010 Stair Construction

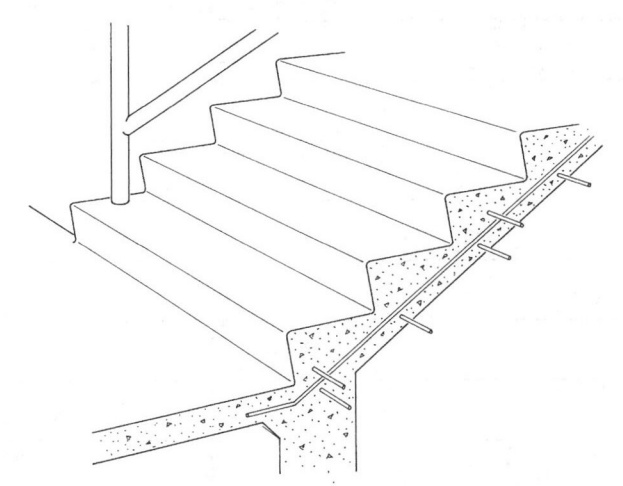

The table below lists the cost per flight for 4'-0" wide stairs. Side walls are not included. Railings are included in the prices.

C2010 110	Stairs	COST PER FLIGHT		
		MAT.	INST.	TOTAL
0470	Stairs, C.I.P. concrete, w/o landing, 12 risers, w/o nosing	740	1,600	2,340
0480	With nosing	1,125	1,825	2,950
0550	W/landing, 12 risers, w/o nosing	825	1,975	2,800
0560	With nosing	1,200	2,200	3,400
0570	16 risers, w/o nosing	1,025	2,475	3,500
0580	With nosing	1,550	2,775	4,325
0590	20 risers, w/o nosing	1,225	3,000	4,225
0600	With nosing	1,875	3,375	5,250
0610	24 risers, w/o nosing	1,425	3,500	4,925
0620	With nosing	2,200	3,925	6,125
0630	Steel, grate type w/nosing & rails, 12 risers, w/o landing	1,800	585	2,385
0640	With landing	2,350	900	3,250
0660	16 risers, with landing	2,950	1,100	4,050
0680	20 risers, with landing	3,550	1,275	4,825
0700	24 risers, with landing	4,150	1,500	5,650
0701				
0710	Cement fill metal pan & picket rail, 12 risers, w/o landing	2,350	585	2,935
0720	With landing	3,100	1,000	4,100
0740	16 risers, with landing	3,875	1,200	5,075
0760	20 risers, with landing	4,650	1,400	6,050
0780	24 risers, with landing	5,450	1,600	7,050
0790	Cast iron tread & pipe rail, 12 risers, w/o landing	2,350	585	2,935
0800	With landing	3,100	1,000	4,100
1120	Wood, prefab box type, oak treads, wood rails 3'-6" wide, 14 risers	955	310	1,265
1150	Prefab basement type, oak treads, wood rails 3'-0" wide, 14 risers	735	250	985

C INTERIORS

C3010 105	Paint & Covering	COST PER S.F.		
		MAT.	INST.	TOTAL
0060	Painting, interior on plaster and drywall, brushwork, primer & 1 coat	.10	.51	.61
0080	Primer & 2 coats	.15	.67	.82
0100	Primer & 3 coats	.20	.82	1.02
0120	Walls & ceilings, roller work, primer & 1 coat	.10	.34	.44
0140	Primer & 2 coats	.15	.43	.58
0160	Woodwork incl. puttying, brushwork, primer & 1 coat	.10	.73	.83
0180	Primer & 2 coats	.15	.97	1.12
0200	Primer & 3 coats	.20	1.32	1.52
0260	Cabinets and casework, enamel, primer & 1 coat	.10	.82	.92
0280	Primer & 2 coats	.15	1.01	1.16
0300	Masonry or concrete, latex, brushwork, primer & 1 coat	.18	.69	.87
0320	Primer & 2 coats	.24	.98	1.22
0340	Addition for block filler	.11	.84	.95
0380	Fireproof paints, intumescent, 1/8" thick 3/4 hour	1.77	.67	2.44
0400	3/16" thick 1 hour	3.94	1.01	4.95
0420	7/16" thick 2 hour	5.10	2.35	7.45
0440	1-1/16" thick 3 hour	8.30	4.70	13
0480	Miscellaneous metal brushwork, exposed metal, primer & 1 coat	.08	.82	.90
0500	Gratings, primer & 1 coat	.19	1.03	1.22
0600	Pipes over 12" diameter	.48	3.29	3.77
0700	Structural steel, brushwork, light framing 300-500 S.F./Ton	.06	1.30	1.36
0720	Heavy framing 50-100 S.F./Ton	.06	.65	.71
0740	Spraywork, light framing 300-500 S.F./Ton	.06	.29	.35
0760	Heavy framing 50-100 S.F./Ton	.06	.32	.38
0800	Varnish, interior wood trim, no sanding sealer & 1 coat	.07	.82	.89
0820	Hardwood floor, no sanding 2 coats	.13	.17	.30
0840	Wall coatings, acrylic glazed coatings, minimum	.28	.63	.91
0860	Maximum	.57	1.08	1.65
0880	Epoxy coatings, minimum	.35	.63	.98
0900	Maximum	1.09	1.93	3.02
0940	Exposed epoxy aggregate, troweled on, 1/16" to 1/4" aggregate, minimum	.54	1.40	1.94
0960	Maximum	1.17	2.53	3.70
0980	1/2" to 5/8" aggregate, minimum	1.08	2.53	3.61
1000	Maximum	1.84	4.11	5.95
1020	1" aggregate, minimum	1.87	3.65	5.52
1040	Maximum	2.86	6	8.86
1060	Sprayed on, minimum	.51	1.11	1.62
1080	Maximum	.92	2.27	3.19
1100	High build epoxy 50 mil, minimum	.59	.84	1.43
1120	Maximum	1.02	3.46	4.48
1140	Laminated epoxy with fiberglass minimum	.65	1.11	1.76
1160	Maximum	1.16	2.27	3.43
1180	Sprayed perlite or vermiculite 1/16" thick, minimum	.23	.11	.34
1200	Maximum	.66	.51	1.17
1260	Wall coatings, vinyl plastic, minimum	.30	.45	.75
1280	Maximum	.73	1.37	2.10
1300	Urethane on smooth surface, 2 coats, minimum	.22	.29	.51
1320	Maximum	.50	.49	.99
1340	3 coats, minimum	.30	.39	.69
1360	Maximum	.66	.70	1.36
1380	Ceramic-like glazed coating, cementitious, minimum	.43	.75	1.18
1400	Maximum	.72	.95	1.67
1420	Resin base, minimum	.30	.51	.81
1440	Maximum	.48	1	1.48
1460	Wall coverings, aluminum foil	.88	1.19	2.07
1480	Copper sheets, .025" thick, phenolic backing	6.10	1.37	7.47
1500	Vinyl backing	4.70	1.37	6.07
1520	Cork tiles, 12"x12", light or dark, 3/16" thick	2.75	1.37	4.12

INTERIORS C

Important: See the Reference Section for critical supporting data - Location Factors & Historical Cost Indexes

C3010 Wall Finishes

C3010 105	Paint & Covering	COST PER S.F.		
		MAT.	INST.	TOTAL
1540	5/16" thick	3.12	1.40	4.52
1560	Basketweave, 1/4" thick	4.80	1.37	6.17
1580	Natural, non-directional, 1/2" thick	6.95	1.37	8.32
1600	12"x36", granular, 3/16" thick	1.03	.85	1.88
1620	1" thick	1.34	.89	2.23
1640	12"x12", polyurethane coated, 3/16" thick	3.23	1.37	4.60
1660	5/16" thick	4.60	1.40	6
1661	Paneling, prefinished plywood, birch	1.23	1.78	3.01
1662	Mahogany, African	2.31	1.87	4.18
1663	Philippine (lauan)	.99	1.49	2.48
1664	Oak or cherry	2.97	1.87	4.84
1665	Rosewood	4.21	2.34	6.55
1666	Teak	2.97	1.87	4.84
1667	Chestnut	4.39	1.99	6.38
1668	Pecan	1.89	1.87	3.76
1669	Walnut	4.80	1.87	6.67
1670	Wood board, knotty pine, finished	1.67	3.12	4.79
1671	Rough sawn cedar	2.05	3.12	5.17
1672	Redwood	4.42	3.12	7.54
1673	Aromatic cedar	3.47	3.35	6.82
1680	Cork wallpaper, paper backed, natural	1.84	.68	2.52
1700	Color	2.28	.68	2.96
1720	Gypsum based, fabric backed, minimum	.72	.41	1.13
1740	Average	1.06	.46	1.52
1760	Maximum	1.18	.51	1.69
1780	Vinyl wall covering, fabric back, light weight	.61	.51	1.12
1800	Medium weight	.75	.68	1.43
1820	Heavy weight	1.19	.75	1.94
1840	Wall paper, double roll, solid pattern, avg. workmanship	.31	.51	.82
1860	Basic pattern, avg. workmanship	.67	.61	1.28
1880	Basic pattern, quality workmanship	1.61	.75	2.36
1900	Grass cloths with lining paper, minimum	.67	.82	1.49
1920	Maximum	2.15	.94	3.09
1940	Ceramic tile, thin set, 4-1/4" x 4-1/4"	2.30	3.26	5.56
1942				

C3010 105	Paint Trim	COST PER L.F.		
		MAT.	INST.	TOTAL
2040	Painting, wood trim, to 6" wide, enamel, primer & 1 coat	.10	.41	.51
2060	Primer & 2 coats	.15	.52	.67
2080	Misc. metal brushwork, ladders	.39	4.11	4.50
2100	Pipes, to 4" dia.	.06	.87	.93
2120	6" to 8" dia.	.13	1.73	1.86
2140	10" to 12" dia.	.38	2.59	2.97
2160	Railings, 2" pipe	.14	2.06	2.20
2180	Handrail, single	.10	.82	.92

C INTERIORS

C3020 410	Tile & Covering	COST PER S.F.		
		MAT.	INST.	TOTAL
0060	Carpet tile, nylon, fusion bonded, 18" x 18" or 24" x 24", 24 oz.	3.44	.48	3.92
0080	35 oz.	4	.48	4.48
0100	42 oz.	4.77	.48	5.25
0140	Carpet, tufted, nylon, roll goods, 12' wide, 26 oz.	3.11	.55	3.66
0160	36 oz.	4.94	.55	5.49
0180	Woven, wool, 36 oz.	7.85	.55	8.40
0200	42 oz.	7.95	.55	8.50
0220	Padding, add to above, minimum	.43	.25	.68
0240	Maximum	.74	.25	.99
0260	Composition flooring, acrylic, 1/4" thick	1.36	3.66	5.02
0280	3/8" thick	1.79	4.23	6.02
0300	Epoxy, minimum	2.41	2.82	5.23
0320	Maximum	2.93	3.88	6.81
0340	Epoxy terrazzo, minimum	5.25	4	9.25
0360	Maximum	8.15	5.35	13.50
0380	Mastic, hot laid, 1-1/2" thick, minimum	3.53	2.76	6.29
0400	Maximum	4.53	3.66	8.19
0420	Neoprene 1/4" thick, minimum	3.48	3.49	6.97
0440	Maximum	4.79	4.42	9.21
0460	Polyacrylate with ground granite 1/4", minimum	2.94	2.59	5.53
0480	Maximum	5.70	3.96	9.66
0500	Polyester with colored quart 2 chips 1/16", minimum	2.84	1.78	4.62
0520	Maximum	4.43	2.82	7.25
0540	Polyurethane with vinyl chips, minimum	6.80	1.78	8.58
0560	Maximum	9.75	2.21	11.96
0600	Concrete topping, granolithic concrete, 1/2" thick	.18	1.65	1.83
0620	1" thick	.36	1.69	2.05
0640	2" thick	.72	1.94	2.66
0660	Heavy duty 3/4" thick, minimum	2.46	2.56	5.02
0680	Maximum	2.46	3.04	5.50
0700	For colors, add to above, minimum	.35	.59	.94
0720	Maximum	.70	.65	1.35
0740	Exposed aggregate finish, minimum	4.22	.54	4.76
0760	Maximum	13.40	.73	14.13
0780	Abrasives, .25 P.S.F. add to above, minimum	.18	.40	.58
0800	Maximum	.21	.40	.61
0820	Dust on coloring, add, minimum	.66	.26	.92
0840	Maximum	2.31	.54	2.85
0860	1/2" integral, minimum	.22	1.65	1.87
0880	Maximum	.22	1.65	1.87
0900	Dustproofing, add, minimum	.09	.18	.27
0920	Maximum	.16	.26	.42
0930	Paint	.24	.98	1.22
0940	Hardeners, metallic add, minimum	.32	.40	.72
0960	Maximum	.73	.59	1.32
0980	Non-metallic, minimum	.19	.40	.59
1000	Maximum	.52	.59	1.11
1020	Integral topping, 3/16" thick	.06	.97	1.03
1040	1/2" thick	.17	1.02	1.19
1060	3/4" thick	.26	1.14	1.40
1080	1" thick	.34	1.30	1.64
1100	Terrazzo, minimum	2.68	6.15	8.83
1120	Maximum	5	13.30	18.30
1280	Resilient, asphalt tile, 1/8" thick on concrete, minimum	1.06	.86	1.92
1300	Maximum	1.17	.86	2.03
1320	On wood, add for felt underlay	.20		.20
1340	Cork tile, minimum	3.80	1.09	4.89
1360	Maximum	9.05	1.09	10.14

Important: See the Reference Section for critical supporting data - Location Factors & Historical Cost Indexes

INTERIORS C

C3020 Floor Finishes

C3020 410	Tile & Covering	COST PER S.F.		
		MAT.	INST.	TOTAL
1380	Polyethylene, in rolls, minimum	2.54	1.25	3.79
1400	Maximum	4.94	1.25	6.19
1420	Polyurethane, thermoset, minimum	3.91	3.44	7.35
1440	Maximum	4.64	6.85	11.49
1460	Rubber, sheet goods, minimum	3.50	2.86	6.36
1480	Maximum	5.75	3.82	9.57
1500	Tile, minimum	3.92	.86	4.78
1520	Maximum	6.65	1.25	7.90
1580	Vinyl, composition tile, minimum	.80	.69	1.49
1600	Maximum	2.05	.69	2.74
1620	Tile, minimum	1.75	.69	2.44
1640	Maximum	4.93	.69	5.62
1660	Sheet goods, minimum	2.04	1.37	3.41
1680	Maximum	3.93	1.72	5.65
1720	Tile, ceramic natural clay	4.29	3.38	7.67
1730	Marble, synthetic 12"x12"x5/8"	7.35	10.30	17.65
1740	Porcelain type, minimum	4.61	3.38	7.99
1760	Maximum	6.35	3.99	10.34
1800	Quarry tile, mud set, minimum	3.20	4.42	7.62
1820	Maximum	5.65	5.65	11.30
1840	Thin set, deduct		.88	.88
1860	Terrazzo precast, minimum	3.83	4.58	8.41
1880	Maximum	8.60	4.58	13.18
1900	Non-slip, minimum	16.35	23	39.35
1920	Maximum	18.65	31.50	50.15
1960	Wood, block, end grain factory type, creosoted, 2" thick	3.09	1.27	4.36
1980	2-1/2" thick	3.20	2.99	6.19
2000	3" thick	3.15	2.99	6.14
2020	Natural finish, 2" thick	3.30	2.99	6.29
2040	Fir, vertical grain, 1"x4", no finish, minimum	2.38	1.47	3.85
2060	Maximum	2.53	1.47	4
2080	Prefinished white oak, prime grade, 2-1/4" wide	6.45	2.20	8.65
2100	3-1/4" wide	8.25	2.02	10.27
2120	Maple strip, sanded and finished, minimum	3.64	3.19	6.83
2140	Maximum	4.30	3.19	7.49
2160	Oak strip, sanded and finished, minimum	3.69	3.19	6.88
2180	Maximum	4.57	3.19	7.76
2200	Parquetry, sanded and finished, minimum	3.68	3.33	7.01
2220	Maximum	6.35	4.73	11.08
2260	Add for sleepers on concrete, treated, 24" O.C., 1"x2"	.65	1.93	2.58
2280	1"x3"	.81	1.50	2.31
2300	2"x4"	.82	.76	1.58
2340	Underlayment, plywood, 3/8" thick	.58	.50	1.08
2350	1/2" thick	.70	.52	1.22
2360	5/8" thick	.77	.53	1.30
2370	3/4" thick	.94	.57	1.51
2380	Particle board, 3/8" thick	.42	.50	.92
2390	1/2" thick	.44	.52	.96
2400	5/8" thick	.59	.53	1.12
2410	3/4" thick	.64	.57	1.21
2420	Hardboard, 4' x 4', .215" thick	.45	.50	.95

INTERIORS

C

C3030 Ceiling Finishes

Plaster Partitions are defined as follows: type of plaster, type and spacing of framing, type of lath and treatment of opposite face.

Included in the system components are expansion joints. Metal studs are assumed to be nonload bearing.

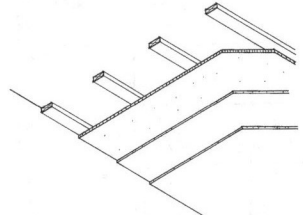

2 Coats of Plaster on Gypsum Lath on Wood Furring

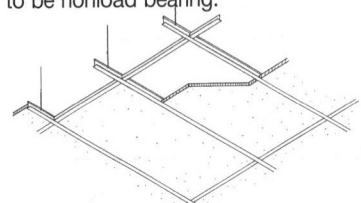

Fiberglass Board on Exposed Suspended Grid System

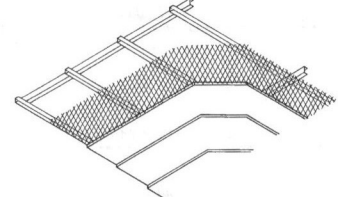

Plaster and Metal Lath on Metal Furring

C3030 105					Plaster Ceilings			
						COST PER S.F.		
	TYPE	**LATH**	**FURRING**	**SUPPORT**		**MAT.**	**INST.**	**TOTAL**
2400	2 coat gypsum	3/8" gypsum	1"x3" wood, 16" O.C.	wood		1.07	3.81	4.88
2500	Painted			masonry		1.07	3.89	4.96
2600				concrete		1.07	4.35	5.42
2700	3 coat gypsum	3.4# metal	1"x3" wood, 16" O.C.	wood		1.08	4.07	5.15
2800	Painted			masonry		1.08	4.15	5.23
2900				concrete		1.08	4.61	5.69
3000	2 coat perlite	3/8" gypsum	1"x3" wood, 16" O.C.	wood		1.09	3.98	5.07
3100	Painted			masonry		1.09	4.06	5.15
3200				concrete		1.09	4.52	5.61
3300	3 coat perlite	3.4# metal	1"x3" wood, 16" O.C.	wood		.94	4.07	5.01
3400	Painted			masonry		.94	4.15	5.09
3500				concrete		.94	4.61	5.55
3600	2 coat gypsum	3/8" gypsum	3/4" CRC, 12" O.C.	1-1/2" CRC, 48"O.C.		1.09	4.64	5.73
3700	Painted		3/4" CRC, 16" O.C.	1-1/2" CRC, 48"O.C.		1.07	4.19	5.26
3800			3/4" CRC, 24" O.C.	1-1/2" CRC, 48"O.C.		1.02	3.82	4.84
3900	2 coat perlite	3/8" gypsum	3/4" CRC, 12" O.C.	1-1/2" CRC, 48"O.C		1.11	4.99	6.10
4000	Painted		3/4" CRC, 16" O.C.	1-1/2" CRC, 48"O.C.		1.09	4.54	5.63
4100			3/4" CRC, 24" O.C.	1-1/2" CRC, 48"O.C.		1.04	4.17	5.21
4200	3 coat gypsum	3.4# metal	3/4" CRC, 12" O.C.	1-1/2" CRC, 36" O.C.		1.47	6	7.47
4300	Painted		3/4" CRC, 16" O.C.	1-1/2" CRC, 36" O.C.		1.45	5.55	7
4400			3/4" CRC, 24" O.C.	1-1/2" CRC, 36" O.C.		1.40	5.20	6.60
4500	3 coat perlite	3.4# metal	3/4" CRC, 12" O.C.	1-1/2" CRC,36" O.C.		1.59	6.60	8.19
4600	Painted		3/4" CRC, 16" O.C.	1-1/2" CRC, 36" O.C.		1.57	6.15	7.72
4700			3/4" CRC, 24" O.C.	1-1/2" CRC, 36" O.C.		1.52	5.80	7.32

C3030 105					Drywall Ceilings			
						COST PER S.F.		
	TYPE	**FINISH**	**FURRING**	**SUPPORT**		**MAT.**	**INST.**	**TOTAL**
4800	1/2" F.R. drywall	painted and textured	1"x3" wood, 16" O.C.	wood		.61	2.34	2.95
4900				masonry		.61	2.42	3.03
5000				concrete		.61	2.88	3.49
5100	5/8" F.R. drywall	painted and textured	1"x3" wood, 16" O.C.	wood		.61	2.34	2.95
5200				masonry		.61	2.42	3.03
5300				concrete		.61	2.88	3.49

INTERIORS C

C3030 Ceiling Finishes

C3030 105 — Drywall Ceilings

	TYPE	FINISH	FURRING	SUPPORT		COST PER S.F.		
						MAT.	INST.	TOTAL
5400	1/2" F.R. drywall	painted and textured	7/8"resil. channels	24" O.C.		.55	2.28	2.83
5500			1"x2" wood	stud clips		.70	2.20	2.90
5602	1/2" F.R. drywall	painted	1-5/8" metal studs	24" O.C.		.58	2.03	2.61
5700	5/8" F.R. drywall	painted and textured	1-5/8"metal studs	24" O.C.		.58	2.03	2.61
5702								

C3030 105 — Acoustical Ceilings

	TYPE	TILE	GRID	SUPPORT		COST PER S.F.		
						MAT.	INST.	TOTAL
5800	5/8" fiberglass board	24" x 48"	tee	suspended		1.29	1.08	2.37
5900		24" x 24"	tee	suspended		1.40	1.18	2.58
6000	3/4" fiberglass board	24" x 48"	tee	suspended		1.96	1.10	3.06
6100		24" x 24"	tee	suspended		2.07	1.20	3.27
6500	5/8" mineral fiber	12" x 12"	1"x3" wood, 12" O.C.	wood		1.08	1.44	2.52
6600				masonry		1.08	1.54	2.62
6700				concrete		1.08	2.15	3.23
6800	3/4" mineral fiber	12" x 12"	1"x3" wood, 12" O.C.	wood		1.56	1.44	3
6900				masonry		1.56	1.44	3
7000				concrete		1.56	1.44	3
7100	3/4"mineral fiber on	12" x 12"	25 ga. channels	runners		1.90	1.95	3.85
7102	5/8" F.R. drywall							
7200	5/8" plastic coated	12" x 12"		adhesive backed		1.23	.37	1.60
7201	Mineral fiber							
7202								
7300	3/4" plastic coated	12" x 12"		adhesive backed		1.71	.37	2.08
7301	Mineral fiber							
7302								
7400	3/4" mineral fiber	12" x 12"	conceal 2" bar &	suspended		1.82	1.89	3.71
7401			channels					
7402								

C3030 135 — Acoustical Ceiling Components

		COST PER S.F.		
		MAT.	INST.	TOTAL
2480	Ceiling boards, eggcrate, acrylic, 1/2" x 1/2" x 1/2" cubes	1.55	.75	2.30
2500	Polystyrene, 3/8" x 3/8" x 1/2" cubes	1.30	.73	2.03
2520	1/2" x 1/2" x 1/2" cubes	1.74	.75	2.49
2540	Fiberglass boards, plain, 5/8" thick	.57	.60	1.17
2560	3/4" thick	1.24	.62	1.86
2580	Grass cloth faced, 3/4" thick	1.85	.75	2.60
2600	1" thick	1.90	.77	2.67
2620	Luminous panels, prismatic, acrylic	1.88	.93	2.81
2640	Polystyrene	.96	.93	1.89
2660	Flat or ribbed, acrylic	3.28	.93	4.21
2680	Polystyrene	2.24	.93	3.17
2700	Drop pan, white, acrylic	4.81	.93	5.74
2720	Polystyrene	4.02	.93	4.95
2740	Mineral fiber boards, 5/8" thick, standard	.67	.55	1.22
2760	Plastic faced	1.05	.93	1.98
2780	2 hour rating	.91	.55	1.46
2800	Perforated aluminum sheets, .024 thick, corrugated painted	1.91	.76	2.67
2820	Plain	3.33	.75	4.08
3080	Mineral fiber, plastic coated, 12" x 12" or 12" x 24", 5/8" thick	.87	.37	1.24
3100	3/4" thick	1.35	.37	1.72

INTERIORS

C

C3030 Ceiling Finishes

C3030 135	Acoustical Ceiling Components	COST PER S.F.		
		MAT.	INST.	TOTAL
3120	Fire rated, 3/4" thick, plain faced	1.17	.37	1.54
3140	Mylar faced	1.20	.37	1.57
3160	Add for ceiling primer	.13		.13
3180	Add for ceiling cement	.36		.36
3240	Suspension system, furring, 1" x 3" wood 12" O.C.	.21	1.07	1.28
3260	T bar suspension system, 2' x 4' grid	.71	.47	1.18
3280	2' x 2' grid	.82	.57	1.39
3300	Concealed Z bar suspension system 12" module	.40	.72	1.12
3320	Add to above for 1-1/2" carrier channels 4' O.C.	.24	.79	1.03
3340	Add to above for carrier channels for recessed lighting	.37	.81	1.18

C3030 140	Plaster Ceiling Components	COST PER S.F.		
		MAT.	INST.	TOTAL
0060	Plaster, gypsum incl. finish, 2 coats			
0080	3 coats	.54	2.39	2.93
0100	Perlite, incl. finish, 2 coats	.40	2.39	2.79
0120	3 coats	.66	2.97	3.63
0140	Thin coat on drywall	.08	.47	.55
0200	Lath, gypsum, 3/8" thick	.43	.63	1.06
0220	1/2" thick	.43	.65	1.08
0240	5/8" thick	.48	.76	1.24
0260	Metal, diamond, 2.5 lb.	.18	.51	.69
0280	3.4 lb.	.28	.54	.82
0300	Flat rib, 2.75 lb.	.28	.51	.79
0320	3.4 lb.	.40	.54	.94
0440	Furring, steel channels, 3/4" galvanized , 12" O.C.	.18	1.63	1.81
0460	16" O.C.	.16	1.18	1.34
0480	24" O.C.	.11	.81	.92
0500	1-1/2" galvanized , 12" O.C.	.27	1.80	2.07
0520	16" O.C.	.24	1.31	1.55
0540	24" O.C.	.16	.88	1.04
0560	Wood strips, 1"x3", on wood, 12" O.C.	.21	1.07	1.28
0580	16" O.C.	.16	.80	.96
0600	24" O.C.	.11	.54	.65
0620	On masonry, 12" O.C.	.21	1.17	1.38
0640	16" O.C.	.16	.88	1.04
0660	24" O.C.	.11	.59	.70
0680	On concrete, 12" O.C.	.21	1.78	1.99
0700	16" O.C.	.16	1.34	1.50
0720	24" O.C.	.11	.89	1
0940	Paint on plaster or drywall, roller work, primer + 1 coat	.10	.34	.44
0960	Primer + 2 coats	.15	.43	.58

For information about Means Estimating Seminars, see yellow pages 11 and 12 in back of book

INTERIORS C

Important: See the Reference Section for critical supporting data - Location Factors & Historical Cost Indexes

D SERVICES

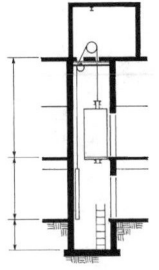

| Hydraulic | Traction Geared | Traction Gearless |

D1010 110	Hydraulic	COST EACH		
		MAT.	INST.	TOTAL
1300	Pass. elev., 1500 lb., 2 Floors, 100 FPM	34,300	9,125	43,425
1400	5 Floors, 100 FPM	50,500	24,300	74,800
1600	2000 lb., 2 Floors, 100 FPM	34,800	9,125	43,925
1700	5 floors, 100 FPM	51,000	24,300	75,300
1900	2500 lb., 2 Floors, 100 FPM	35,500	9,125	44,625
2000	5 floors, 100 FPM	52,000	24,300	76,300
2200	3000 lb., 2 Floors, 100 FPM	37,300	9,125	46,425
2300	5 floors, 100 FPM	53,500	24,300	77,800
2500	3500 lb., 2 Floors, 100 FPM	39,400	9,125	48,525
2600	5 floors, 100 FPM	56,000	24,300	80,300
2800	4000 lb., 2 Floors, 100 FPM	40,200	9,125	49,325
2900	5 floors, 100 FPM	56,500	24,300	80,800
3100	4500 lb., 2 Floors, 100 FPM	41,500	9,125	50,625
3200	5 floors, 100 FPM	58,000	24,300	82,300
4000	Hospital elevators, 3500 lb., 2 Floors, 100 FPM	51,000	9,125	60,125
4100	5 floors, 100 FPM	76,000	24,300	100,300
4300	4000 lb., 2 Floors, 100 FPM	51,000	9,125	60,125
4400	5 floors, 100 FPM	76,000	24,300	100,300
4600	4500 lb., 2 Floors, 100 FPM	57,000	9,125	66,125
4800	5 floors, 100 FPM	82,000	24,300	106,300
4900	5000 lb., 2 Floors, 100 FPM	60,000	9,125	69,125
5000	5 floors, 100 FPM	84,500	24,300	108,800
6700	Freight elevators (Class "B"), 3000 lb., 2 Floors, 50 FPM	53,500	11,600	65,100
6800	5 floors, 100 FPM	82,000	30,500	112,500
7000	4000 lb., 2 Floors, 50 FPM	56,000	11,600	67,600
7100	5 floors, 100 FPM	84,500	30,500	115,000
7500	10,000 lb., 2 Floors, 50 FPM	69,500	11,600	81,100
7600	5 floors, 100 FPM	98,500	30,500	129,000
8100	20,000 lb., 2 Floors, 50 FPM	85,000	11,600	96,600
8200	5 Floors, 100 FPM	113,500	30,500	144,000

D1010 140	Traction Geared Elevators	COST EACH		
		MAT.	INST.	TOTAL
1300	Passenger, 2000 Lb., 5 floors, 200 FPM	78,500	22,900	101,400
1500	15 floors, 350 FPM	122,000	69,500	191,500
1600	2500 Lb., 5 floors, 200 FPM	81,000	22,900	103,900
1800	15 floors, 350 FPM	124,500	69,500	194,000
1900	3000 Lb., 5 floors, 200 FPM	82,500	22,900	105,400
2100	15 floors, 350 FPM	126,000	69,500	195,500
2200	3500 Lb., 5 floors, 200 FPM	84,500	22,900	107,400
2400	15 floors, 350 FPM	128,000	69,500	197,500
2500	4000 Lb., 5 floors, 200 FPM	84,500	22,900	107,400
2700	15 floors, 350 FPM	128,000	69,500	197,500

Important: See the Reference Section for critical supporting data - Location Factors & Historical Cost Indexes

D1010 Elevators and Lifts

D1010 140	Traction Geared Elevators	COST EACH		
		MAT.	INST.	TOTAL
2800	4500 Lb., 5 floors, 200 FPM	86,500	22,900	109,400
3000	15 floors, 350 FPM	130,000	69,500	199,500
3100	5000 Lb., 5 floors, 200 FPM	89,000	22,900	111,900
3300	15 floors, 350 FPM	132,500	69,500	202,000
4000	Hospital, 3500 Lb., 5 floors, 200 FPM	94,500	22,900	117,400
4200	15 floors, 350 FPM	163,000	69,500	232,500
4300	4000 Lb., 5 floors, 200 FPM	94,500	22,900	117,400
4500	15 floors, 350 FPM	163,000	69,500	232,500
4600	4500 Lb., 5 floors, 200 FPM	99,500	22,900	122,400
4800	15 floors, 350 FPM	168,000	69,500	237,500
4900	5000 Lb., 5 floors, 200 FPM	101,500	22,900	124,400
5100	15 floors, 350 FPM	169,500	69,500	239,000
6000	Freight, 4000 Lb., 5 floors, 50 FPM class 'B'	84,500	23,700	108,200
6200	15 floors, 200 FPM class 'B'	156,000	82,500	238,500
6300	8000 Lb., 5 floors, 50 FPM class 'B'	101,500	23,700	125,200
6500	15 floors, 200 FPM class 'B'	173,000	82,500	255,500
7000	10,000 Lb., 5 floors, 50 FPM class 'B'	118,000	23,700	141,700
7200	15 floors, 200 FPM class 'B'	316,500	82,500	399,000
8000	20,000 Lb., 5 floors, 50 FPM class 'B'	130,500	23,700	154,200
8200	15 floors, 200 FPM class 'B'	329,000	82,500	411,500

D1010 150	Traction Gearless Elevators	COST EACH		
		MAT.	INST.	TOTAL
1700	Passenger, 2500 Lb., 10 floors, 200 FPM	156,000	59,500	215,500
1900	30 floors, 600 FPM	272,000	152,500	424,500
2000	3000 Lb., 10 floors, 200 FPM	157,500	59,500	217,000
2200	30 floors, 600 FPM	273,500	152,500	426,000
2300	3500 Lb., 10 floors, 200 FPM	159,000	59,500	218,500
2500	30 floors, 600 FPM	275,000	152,500	427,500
2700	50 floors, 800 FPM	359,000	245,500	604,500
2800	4000 lb., 10 floors, 200 FPM	159,500	59,500	219,000
3000	30 floors, 600 FPM	275,500	152,500	428,000
3200	50 floors, 800 FPM	359,500	245,500	605,000
3300	4500 lb., 10 floors, 200 FPM	161,500	59,500	221,000
3500	30 floors, 600 FPM	277,500	152,500	430,000
3700	50 floors, 800 FPM	361,500	245,500	607,000
3800	5000 lb., 10 floors, 200 FPM	163,500	59,500	223,000
4000	30 floors, 600 FPM	279,500	152,500	432,000
4200	50 floors, 800 FPM	363,500	245,500	609,000
6000	Hospital, 3500 Lb., 10 floors, 200 FPM	181,500	59,500	241,000
6200	30 floors, 600 FPM	346,500	152,500	499,000
6400	4000 Lb., 10 floors, 200 FPM	181,500	59,500	241,000
6600	30 floors, 600 FPM	346,500	152,500	499,000
6800	4500 Lb., 10 floors, 200 FPM	186,500	59,500	246,000
7000	30 floors, 600 FPM	351,500	152,500	504,000
7200	5000 Lb., 10 floors, 200 FPM	188,500	59,500	248,000
7400	30 floors, 600 FPM	353,500	152,500	506,000

D1090 510	Chutes	COST EACH		
		MAT.	INST.	TOTAL
0100	Chutes, linen or refuse incl. sprinklers, galv. steel, 18" diam., per floor	810	247	1,057
0110	Aluminized steel, 36" diam., per floor	1,100	310	1,410
0120	Mail, 8-3/4" x 3-1/2", aluminum & glass, per floor	620	173	793
0130	Bronze or stainless, per floor	960	192	1,152

D | SERVICES

Minimum Plumbing Fixture Requirements

Type of Building or Occupancy (2)	Water Closets (14) (Fixtures per Person)		Urinals (5,10) (Fixtures per Person)		Lavatories (Fixtures per Person)		Bathtubs or Showers (Fixtures per Person)	Drinking Fountains (Fixtures per Person) (3, 13)
	Male	Female	Male	Female	Male	Female		
Assembly Places- Theatres, Auditoriums, Convention Halls, etc.-for permanent employee use	1: 1 - 15 2: 16 - 35 3: 36 - 55	1: 1 - 15 2: 16 - 35 3: 36 - 55	0: 1 - 9 1: 10 - 50		1 per 40	1 per 40		
	Over 55, add 1 fixture for each additional 40 persons		Add one fixture for each additional 50 males					
Assembly Places- Theatres, Auditoriums, Convention Halls, etc. - for public use	1: 1 - 100 2: 101 - 200 3: 201 - 400	3: 1 - 50 4: 51 - 100 8: 101 - 200 11: 201 - 400	1: 1 - 100 2: 101 - 200 3: 201 - 400 4: 401 - 600		1: 1 - 200 2: 201 - 400 3: 401 - 750	1: 1 - 200 2: 201 - 400 3: 401 - 750		1: 1 - 150 2: 151 - 400 3: 401 - 750
	Over 400, add 1 fixture for each additional 500 males and 1 for each additional 125 females		Over 600, add 1 fixture for each additional 300 males		Over 750, add 1 fixture for each additional 500 persons			Over 750, add one fixture for each additional 500 persons
Dormitories (9) School or Labor	1 per 10	1 per 8	1 per 25		1 per 12	1 per 12	1 per 8	1 per 150 (12)
	Add 1 fixture for each additional 25 males (over 10) and 1 for each additional 20 females (over 8)		Over 150, add 1 fixture for each additional 50 males		Over 12 add 1 fixture for each additional 20 males and 1 for each 15 additional females		For females add 1 bathtub per 30. Over 150, add 1 per 20	
Dormitories- for Staff Use	1: 1 - 15 2: 16 - 35 3: 36 - 55	1: 1 - 15 3: 16 - 35 4: 36 - 55	1 per 50		1 per 40	1 per 40	1 per 8	
	Over 55, add 1 fixture for each additional 40 persons							

Dwellings:
  Single Dwelling
  Multiple Dwelling
  or Apartment House
	1 per dwelling
	1 per dwelling or apartment unit
					1 per dwelling
					1 per dwelling or apartment unit
							1 per dwelling
							1 per dwelling or apartment unit	

Hospital Waiting rooms	1 per room				1 per room			1 per 150 (12)
Hospitals- for employee use	1: 1 - 15 2: 16 - 35 3: 36 - 55	1: 1 - 15 3: 16 - 35 4: 36 - 55	0: 1 - 9 1: 10 - 50		1 per 40	1 per 40		
	Over 55, add 1 fixture for each additional 40 persons		Add 1 fixture for each additional 50 males					

Hospitals:
  Individual Room
  Ward Room
	1 per room
	1 per 8 patients
					1 per room
					1 per 10 patients
							1 per room
							1 per 20 patients
								1 per 150 (12)

Industrial (6) Warehouses Workshops, Foundries and similar establishments- for employee use	1: 1 -10 2: 11 - 25 3: 26 - 50 4: 51 - 75 5: 76 - 100	1: 1 -10 2: 11 - 25 3: 26 - 50 4: 51 - 75 5: 76 - 100			Up to 100, per 10 persons Over 100, 1 per 15 persons (7, 8)		1 shower for each 15 persons exposed to excessive heat or to skin contamination with poisonous, infectious or irritating material	1 per 150 (12)
	Over 100, add 1 fixture for each additional 30 persons							
Institutional - Other than Hospitals or Penal Institutions (on each occupied floor)	1 per 25	1 per 20	0: 1 - 9 1: 10 - 50		1 per 10	1 per 10	1 per 8	1 per 150 (12)
			Add 1 fixture for each additional 50 males					
Institutional - Other than Hospitals or Penal Institutions (on each occupied floor)- for employee use	1: 1 - 15 2: 16 - 35 3: 36 - 55	1: 1 - 15 3: 16 - 35 4: 36 - 55	0: 1 - 9 1: 10 - 50		1 per 40	1 per 40	1 per 8	1 per 150 (12)
	Over 55, add 1 fixture for each additional 40 persons		Add 1 fixture for each additional 50 males					
Office or Public Buildings	1: 1 - 100 2: 101 - 200 3: 201 - 400	3: 1 - 50 4: 51 - 100 8: 101 - 200 11: 201 - 400	1: 1 - 100 2: 101 - 200 3: 201 - 400 4: 401 - 600		1: 1 - 200 2: 201 - 400 3: 401 - 750	1: 1 - 200 2: 201 - 400 3: 401 - 750		1 per 150 (12)
	Over 400, add 1 fixture for each additional 500 males and 1 for each additional 150 females		Over 600, add 1 fixture for each additional 300 males		Over 750, add 1 fixture for each additional 500 persons			
Office or Public Buildings - for employee use	1: 1 - 15 2: 16 - 35 3: 36 - 55	1: 1 - 15 3: 16 - 35 4: 36 - 55	0: 1 - 9 1: 10 - 50		1 per 40	1 per 40		
	Over 55, add 1 fixture for each additional 40 persons		Add 1 fixture for each additional 50 males					

Important: See the Reference Section for critical supporting data - Location Factors & Historical Cost Indexes

D2010 Plumbing Fixtures

Minimum Plumbing Fixture Requirements

Type of Building or Occupancy	Water Closets (14) (Fixtures per Person)		Urinals (5, 10) (Fixtures per Person)		Lavatories (Fixtures per Person)		Bathtubs or Showers (Fixtures per Person)	Drinking Fountains (Fixtures per Person) (3, 13)
	Male	Female	Male	Female	Male	Female		
Penal Institutions - for employee use	1: 1 - 15 2: 16 - 35 3: 36 - 55 Over 55, add 1 fixture for each additional 40 persons	1: 1 - 15 3: 16 - 35 4: 36 - 55	0: 1 - 9 1: 10 - 50 Add 1 fixture for each additional 50 males		1 per 40	1 per 40		1 per 150 (12)
Penal Institutions - for prison use								
Cell	1 per cell				1 per cell			1 per cellblock floor
Exercise room	1 per exercise room		1 per exercise room		1 per exercise room			1 per exercise room
Restaurants, Pubs and Lounges (11)	1: 1 - 50 2: 51 - 150 3: 151 - 300 Over 300, add 1 fixture for each additional 200 persons	1: 1 - 50 2: 51 - 150 4: 151 - 300	1: 1 - 150 Over 150, add 1 fixture for each additional 150 males		1: 1 - 150 2: 151 - 200 3: 201 - 400 Over 400, add 1 fixture for each additional 400 persons	1: 1 - 150 2: 151 - 200 3: 201 - 400		
Schools - for staff use All Schools	1: 1 - 15 2: 16 - 35 3: 36 - 55 Over 55, add 1 fixture for each additional 40 persons	1: 1 - 15 3: 16 - 35 4: 36 - 55	1 per 50		1 per 40	1 per 40		
Schools - for student use: Nursery	1: 1 - 20 2: 21 - 50 Over 50, add 1 fixture for each additional 50 persons	1: 1 - 20 2: 21 - 50			1: 1 - 25 2: 26 - 50 Over 50, add 1 fixture for each additional 50 persons	1: 1 - 25 2: 26 - 50		1 per 150 (12)
Elementary	1 per 30	1 per 25	1 per 75		1 per 35	1 per 35		1 per 150 (12)
Secondary	1 per 40	1 per 30	1 per 35		1 per 40	1 per 40		1 per 150 (12)
Others (Colleges, Universities, Adult Centers, etc.	1 per 40	1 per 30	1 per 35		1 per 40	1 per 40		1 per 150 (12)
Worship Places Educational and Activities Unit	1 per 150	1 per 75	1 per 150		1 per 2 water closets			1 per 150 (12)
Worship Places Principal Assembly Place	1 per 150	1 per 75	1 per 150		1 per 2 water closets			1 per 150 (12)

Notes:
1. The figures shown are based upon one (1) fixture being the minimum required for the number of persons indicated or any fraction thereof.
2. Building categories not shown on this table shall be considered separately by the Administrative Authority.
3. Drinking fountains shall not be installed in toilet rooms.
4. Laundry trays. One (1) laundry tray or one (1) automatic washer standpipe for each dwelling unit or one (1) laundry trays or one (1) automatic washer standpipes, or combination thereof, for each twelve (12) apartments. Kitchen sinks, one (1) for each dwelling or apartment unit.
5. For each urinal added in excess of the minimum required, one water closet may be deducted. The number of water closets shall not be reduced to less than two-thirds (2/3) of the minimum requirement.
6. As required by ANSI Z4.1-1968, Sanitation in Places of Employment.
7. Where there is exposure to skin contamination with poisonous, infectious, or irritating materials, provide one (1) lavatory for each five (5) persons.
8. Twenty-four (24) lineal inches of wash sink or eighteen (18) inches of a circular basin, when provided with water outlets for such space shall be considered equivalent to one (1) lavatory.
9. Laundry trays, one (1) for each fifty (50) persons. Service sinks, one (1) for each hundred (100) persons.
10. General.. In applying this schedule of facilities, consideration shall be given to the accessibility of the fixtures. Conformity purely on a numerical basis may not result in an installation suited to the need of the individual establishment. For example, schools should be provided with toilet facilities on each floor having classrooms.
    a. Surrounding materials, wall and floor space to a point two (2) feet in front of urinal lip and four (4) feet above the floor, and at least two (2) feet to each side of the urinal shall be lined with non-absorbent materials.
    b. Trough urinals shall be prohibited.
11. A restaurant is defined as a business which sells food to be consumed on the premises.
    a. The number of occupants for a drive-in restaurant shall be considered as equal to the number of parking stalls.
    b. Employee toilet facilities shall not to be included in the above restaurant requirements. Hand washing facilities shall be available in the kitchen for employees.
12. Where food is consumed indoors, water stations may be substituted for drinking fountains. Offices, or public buildings for use by more than six (6) persons shall have one (1) drinking fountain for the first one hundred fifty (150) persons and one additional fountain for each three hundred (300) persons thereafter.
13. There shall be a minimum of one (1) drinking fountain per occupied floor in schools, theaters, auditoriums, dormitories, offices of public building.
14. The total number of water closets for females shall be at least equal to the total number of water closets and urinals required for males.

SERVICES

D

D2010 Plumbing Fixtures

One Piece Wall Hung Water Closet

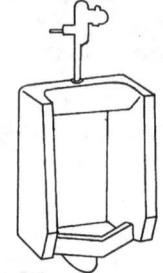

Wall Hung Urinal

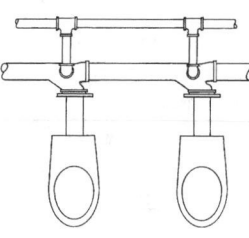

Side By Side Water Closet Group

Floor Mount Water Closet

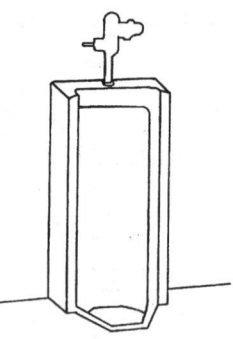

Stall Type Urinal

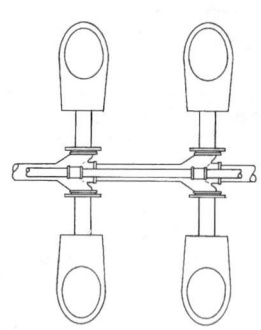

Back to Back Water Closet Group

D2010 110	Water Closet Systems	COST EACH		
		MAT.	INST.	TOTAL
1800	Water closet, vitreous china, elongated			
1840	Tank type, wall hung, one piece	760	450	1,210
1880	Close coupled two piece	890	450	1,340
1920	Floor mount, one piece	710	490	1,200
1960	One piece low profile	605	490	1,095
2000	Two piece close coupled	320	490	810
2040	Bowl only with flush valve			
2080	Wall hung	695	515	1,210
2120	Floor mount	525	500	1,025
2122				

D2010 120	Water Closets, Group	COST EACH		
		MAT.	INST.	TOTAL
1760	Water closets, battery mount, wall hung, side by side, first closet	735	530	1,265
1800	Each additional water closet, add	700	500	1,200
3000	Back to back, first pair of closets	1,325	715	2,040
3100	Each additional pair of closets, back to back	1,300	705	2,005

D2010 210	Urinal Systems	COST EACH		
		MAT.	INST.	TOTAL
2000	Urinal, vitreous china, wall hung	415	515	930
2040	Stall type	655	615	1,270

Important: See the Reference Section for critical supporting data - Location Factors & Historical Cost Indexes

D2010 Plumbing Fixtures

Systems are complete with trim and rough-in (supply, waste and vent) to connect to supply branches and waste mains.

Vanity Top   **Wall Hung**   **Counter Top Single Bowl**   **Counter Top Double Bowl**

D2010 310	Lavatory Systems	COST EACH		
		MAT.	INST.	TOTAL
1560	Lavatory w/trim, vanity top, PE on CI, 20" x 18", Vanity top by others.	269	460	729
1600	19" x 16" oval	183	460	643
1640	18" round	249	460	709
1680	Cultured marble, 19" x 17"	197	460	657
1720	25" x 19"	227	460	687
1760	Stainless, self-rimming, 25" x 22"	268	460	728
1800	17" x 22"	263	460	723
1840	Steel enameled, 20" x 17"	183	475	658
1880	19" round	188	475	663
1920	Vitreous china, 20" x 16"	285	485	770
1960	19" x 16"	277	485	762
2000	22" x 13"	335	485	820
2040	Wall hung, PE on CI, 18" x 15"	505	510	1,015
2080	19" x 17"	490	510	1,000
2120	20" x 18"	460	510	970
2160	Vitreous china, 18" x 15"	405	525	930
2200	19" x 17"	385	525	910
2240	24" x 20"	635	525	1,160
2241				
2243				

D2010 410	Kitchen Sink Systems	COST EACH		
		MAT.	INST.	TOTAL
1720	Kitchen sink w/trim, countertop, PE on CI, 24"x21", single bowl	315	505	820
1760	30" x 21" single bowl	370	505	875
1800	32" x 21" double bowl	420	545	965
1840	42" x 21" double bowl	625	555	1,180
1880	Stainless steel, 19" x 18" single bowl	405	505	910
1920	25" x 22" single bowl	435	505	940
1960	33" x 22" double bowl	595	545	1,140
2000	43" x 22" double bowl	670	555	1,225
2040	44" x 22" triple bowl	935	575	1,510
2080	44" x 24" corner double bowl	645	555	1,200
2120	Steel, enameled, 24" x 21" single bowl	212	505	717
2160	32" x 21" double bowl	261	545	806
2240	Raised deck, PE on CI, 32" x 21", dual level, double bowl	500	690	1,190
2280	42" x 21" dual level, triple bowl	705	755	1,460

SERVICES

D

D2010 Plumbing Fixtures

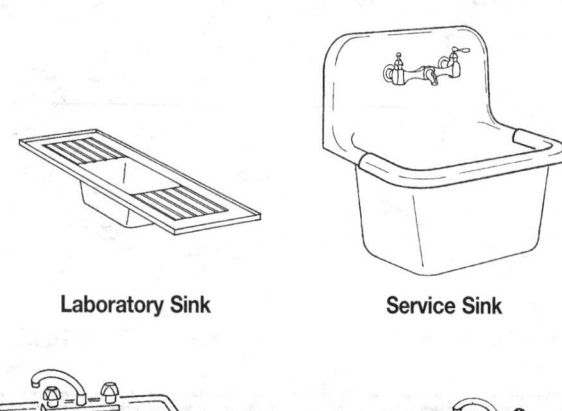

Laboratory Sink

Service Sink

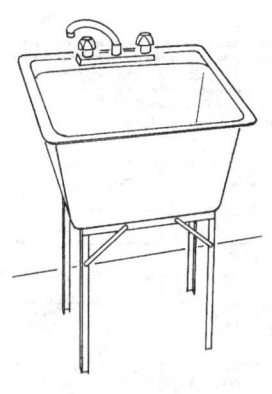

Single Compartment Sink

Double Compartment Sink

Systems are complete with trim and rough-in (supply, waste and vent) to connect to supply branches and waste mains.

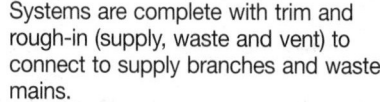

D2010 420	Laundry Sink Systems	COST EACH		
		MAT.	INST.	TOTAL
1740	Laundry sink w/trim, PE on CI, black iron frame			
1760	24" x 20", single compartment	360	495	855
1840	48" x 21" double compartment	585	540	1,125
1920	Molded stone, on wall, 22" x 21" single compartment	227	495	722
1960	45"x 21" double compartment	375	540	915
2040	Plastic, on wall or legs, 18" x 23" single compartment	189	485	674
2080	20" x 24" single compartment	220	485	705
2120	36" x 23" double compartment	250	525	775
2160	40" x 24" double compartment	320	525	845
2200	Stainless steel, countertop, 22" x 17" single compartment	405	495	900
2240	19" x 22" single compartment	485	495	980
2280	33" x 22" double compartment	495	540	1,035

D2010 430	Laboratory Sink Systems	COST EACH		
		MAT.	INST.	TOTAL
1580	Laboratory sink w/trim, polyethylene, single bowl,			
1600	Double drainboard, 54" x 24" O.D.	975	645	1,620
1640	Single drainboard, 47" x 24"O.D.	1,000	645	1,645
1680	70" x 24" O.D.	1,025	645	1,670
1760	Flanged, 14-1/2" x 14-1/2" O.D.	385	580	965
1800	18-1/2" x 18-1/2" O.D.	345	580	925
1840	23-1/2" x 20-1/2" O.D.	395	580	975
1920	Polypropylene, cup sink, oval, 7" x 4" O.D.	179	515	694
1960	10" x 4-1/2" O.D.	192	515	707
1961				

Important: See the Reference Section for critical supporting data - Location Factors & Historical Cost Indexes

D2010 Plumbing Fixtures

D2010 440	Service Sink Systems	COST EACH		
		MAT.	INST.	TOTAL
4260	Service sink w/trim, PE on CI, corner floor, 28" x 28", w/rim guard	755	650	1,405
4300	Wall hung w/rim guard, 22" x 18"	805	755	1,560
4340	24" x 20"	845	755	1,600
4380	Vitreous china, wall hung 22" x 20"	775	755	1,530
4383				

D2010 Plumbing Fixtures

Recessed Bathtub

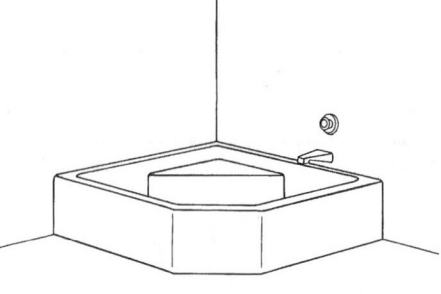

Corner Bathtub

Systems are complete with trim and rough-in (supply, waste and vent) to connect to supply branches and waste mains.

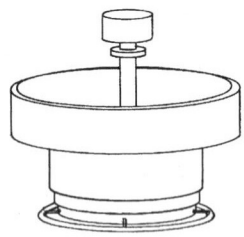

Circular Wash Fountain

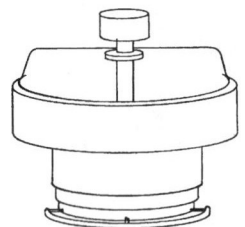

Semi-Circular Wash Fountain

D2010 510	Bathtub Systems	COST EACH		
		MAT.	INST.	TOTAL
1960	Bathtub, recessed, P.E. on Cl., 42" x 37"	965	530	1,495
2000	48" x 42"	1,650	570	2,220
2040	72" x 36"	1,475	635	2,110
2080	Mat bottom, 5' long	560	555	1,115
2120	5'-6" long	1,125	570	1,695
2160	Corner, 48" x 42"	1,575	555	2,130
2200	Formed steel, enameled, 4'-6" long	460	510	970
2240	5' long	460	520	980

D2010 610	Group Wash Fountain Systems	COST EACH		
		MAT.	INST.	TOTAL
1740	Group wash fountain, precast terrazzo			
1760	Circular, 36" diameter	2,800	835	3,635
1800	54" diameter	3,375	915	4,290
1840	Semi-circular, 36" diameter	2,525	835	3,360
1880	54" diameter	3,100	915	4,015
1960	Stainless steel, circular, 36" diameter	3,225	775	4,000
2000	54" diameter	4,225	865	5,090
2040	Semi-circular, 36" diameter	2,800	775	3,575
2080	54" diameter	3,650	865	4,515
2160	Thermoplastic, circular, 36" diameter	2,525	630	3,155
2200	54" diameter	2,950	735	3,685
2240	Semi-circular, 36" diameter	2,375	630	3,005
2280	54" diameter	2,800	735	3,535

Important: See the Reference Section for critical supporting data - Location Factors & Historical Cost Indexes

D2010 Plumbing Fixtures

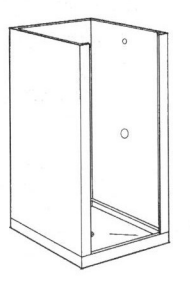

Square Shower Stall

Corner Angle Shower Stall

Systems are complete with trim and rough-in (supply, waste and vent) to connect to supply branches and waste mains.

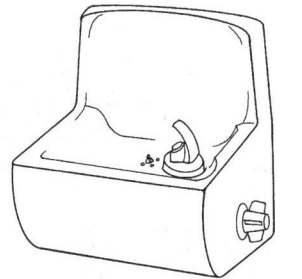

Wall Mounted, Low Back

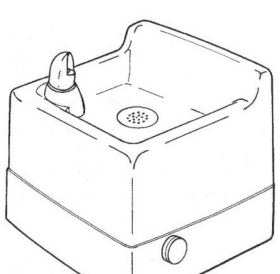

Wall Mounted, No Back

D2010 710	Shower Systems	COST EACH		
		MAT.	INST.	TOTAL
1560	Shower, stall, baked enamel, molded stone receptor, 30" square	465	770	1,235
1600	32" square	450	770	1,220
1640	Terrazzo receptor, 32" square	850	770	1,620
1680	36" square	995	815	1,810
1720	36" corner angle	915	815	1,730
1800	Fiberglass one piece, three walls, 32" square	460	705	1,165
1840	36" square	505	705	1,210
1880	Polypropylene, molded stone receptor, 30" square	440	770	1,210
1920	32" square	485	770	1,255
1960	Built-in head, arm, bypass, stops and handles	105	199	304

D2010 810	Drinking Fountain Systems	COST EACH		
		MAT.	INST.	TOTAL
1740	Drinking fountain, one bubbler, wall mounted			
1760	Non recessed			
1800	Bronze, no back	1,150	305	1,455
1840	Cast iron, enameled, low back	525	305	830
1880	Fiberglass, 12" back	640	305	945
1920	Stainless steel, no back	965	305	1,270
1960	Semi-recessed, poly marble	765	305	1,070
2040	Stainless steel	915	305	1,220
2080	Vitreous china	530	305	835
2120	Full recessed, poly marble	855	305	1,160
2200	Stainless steel	750	305	1,055
2240	Floor mounted, pedestal type, aluminum	535	415	950
2320	Bronze	1,300	415	1,715
2360	Stainless steel	1,300	415	1,715

SERVICES

D

D2010 Plumbing Fixtures

Wall Hung Water Cooler

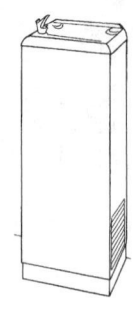

Floor Mounted Water Cooler

Systems are complete with trim and rough-in (supply, waste and vent) to connect to supply branches and waste mains.

D2010 820	Water Cooler Systems	COST EACH		
		MAT.	INST.	TOTAL
1840	Water cooler, electric, wall hung, 8.2 GPH	645	390	1,035
1880	Dual height, 14.3 GPH	950	400	1,350
1920	Wheelchair type, 7.5 G.P.H.	1,575	390	1,965
1960	Semi recessed, 8.1 G.P.H.	915	390	1,305
2000	Full recessed, 8 G.P.H.	1,375	420	1,795
2040	Floor mounted, 14.3 G.P.H.	680	340	1,020
2080	Dual height, 14.3 G.P.H.	1,000	415	1,415
2120	Refrigerated compartment type, 1.5 G.P.H.	1,225	340	1,565

D2010 Plumbing Fixtures

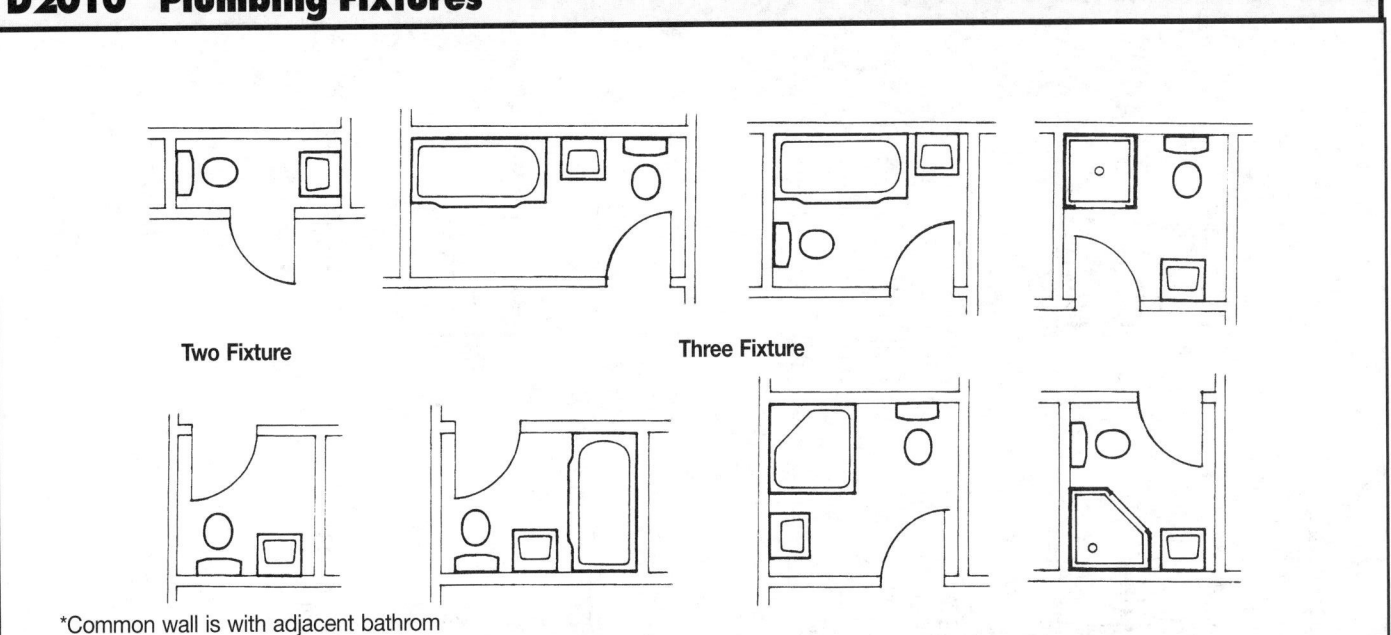

Two Fixture   Three Fixture

*Common wall is with adjacent bathrom

D2010 952	Two Fixture Bathroom, Two Wall Plumbing	COST EACH		
		MAT.	INST.	TOTAL
1180	Bathroom, lavatory & water closet, 2 wall plumbing, stand alone	885	1,275	2,160
1200	Share common plumbing wall*	825	1,100	1,925

D2010 952	Two Fixture Bathroom, One Wall Plumbing	COST EACH		
		MAT.	INST.	TOTAL
2220	Bathroom, lavatory & water closet, one wall plumbing, stand alone	845	1,150	1,995
2240	Share common plumbing wall*	750	970	1,720

D2010 953	Three Fixture Bathroom, One Wall Plumbing	COST EACH		
		MAT.	INST.	TOTAL
1150	Bathroom, three fixture, one wall plumbing			
1160	Lavatory, water closet & bathtub			
1170	Stand alone	1,275	1,475	2,750
1180	Share common plumbing wall *	1,075	1,075	2,150

D2010 953	Three Fixture Bathroom, Two Wall Plumbing	COST EACH		
		MAT.	INST.	TOTAL
2130	Bathroom, three fixture, two wall plumbing			
2140	Lavatory, water closet & bathtub			
2160	Stand alone	1,300	1,500	2,800
2180	Long plumbing wall common *	1,125	1,200	2,325
3610	Lavatory, bathtub & water closet			
3620	Stand alone	1,375	1,700	3,075
3640	Long plumbing wall common *	1,300	1,550	2,850
4660	Water closet, corner bathtub & lavatory			
4680	Stand alone	2,300	1,525	3,825
4700	Long plumbing wall common *	2,175	1,150	3,325
6100	Water closet, stall shower & lavatory			
6120	Stand alone	1,300	1,950	3,250
6140	Long plumbing wall common *	1,250	1,800	3,050
7060	Lavatory, corner stall shower & water closet			
7080	Stand alone	1,675	1,775	3,450
7100	Short plumbing wall common *	1,525	1,300	2,825

D SERVICES

D2010 Plumbing Fixtures

Four Fixture

Five Fixture

*Common wall is with adjacent bathrom

D2010 954	Four Fixture Bathroom, Two Wall Plumbing	COST EACH		
		MAT.	INST.	TOTAL
1140	Bathroom, four fixture, two wall plumbing			
1150	Bathtub, water closet, stall shower & lavatory			
1160	Stand alone	1,525	1,875	3,400
1180	Long plumbing wall common *	1,400	1,500	2,900
2260	Bathtub, lavatory, corner stall shower & water closet			
2280	Stand alone	2,000	1,925	3,925
2320	Long plumbing wall common *	1,850	1,550	3,400
3620	Bathtub, stall shower, lavatory & water closet			
3640	Stand alone	1,750	2,300	4,050
3660	Long plumbing wall (opp. door) common *	1,600	1,925	3,525
4680	Bathroom, four fixture, three wall plumbing			
4700	Bathtub, stall shower, lavatory & water closet			
4720	Stand alone	2,275	2,525	4,800
4761	Long plumbing wall (opposite door) common*	2,225	2,350	4,575

D2010 955	Five Fixture Bathroom, Two Wall Plumbing	COST EACH		
		MAT.	INST.	TOTAL
1320	Bathroom, five fixture, two wall plumbing			
1340	Bathtub, water closet, stall shower & two lavatories			
1360	Stand alone	2,200	2,850	5,050
1400	One short plumbing wall common *	2,075	2,475	4,550
2360	Bathroom, five fixture, three wall plumbing			
2380	Water closet, bathtub, two lavatories & stall shower			
2400	Stand alone	2,675	2,900	5,575
2440	One short plumbing wall common *	2,525	2,525	5,050
4080	Bathroom, five fixture, one wall plumbing			
4100	Bathtub, two lavatories, corner stall shower & water closet			
4120	Stand alone	2,525	2,625	5,150
4160	Share common wall *	2,100	1,800	3,900

Important: See the Reference Section for critical supporting data - Location Factors & Historical Cost Indexes

D2020 Domestic Water Distribution

Installation includes piping and fittings within 10' of heater. Gas heaters require vent piping (not included with these units).

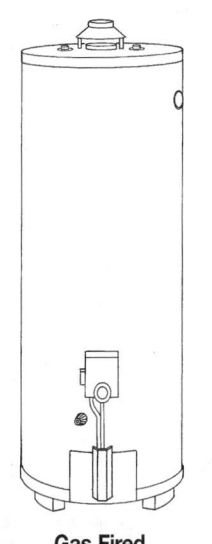

Gas Fired

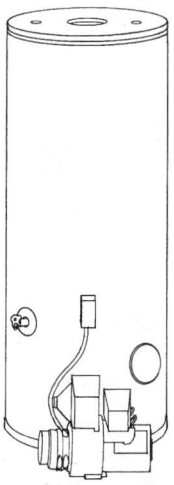

Oil Fired

D2020 220	Gas Fired Water Heaters - Residential Systems	COST EACH		
		MAT.	INST.	TOTAL
2200	Gas fired water heater, residential, 100° F rise			
2260	30 gallon tank, 32 GPH	600	900	1,500
2300	40 gallon tank, 32 GPH	685	1,000	1,685
2340	50 gallon tank, 63 GPH	785	1,000	1,785
2380	75 gallon tank, 63 GPH	1,075	1,125	2,200
2420	100 gallon tank, 63 GPH	1,500	1,200	2,700
2422				

D2020 230	Oil Fired Water Heaters - Residential Systems	COST EACH		
		MAT.	INST.	TOTAL
2200	Oil fired water heater, residential, 100° F rise			
2220	30 gallon tank, 103 GPH	1,075	835	1,910
2260	50 gallon tank, 145 GPH	1,425	935	2,360
2300	70 gallon tank, 164 GPH	1,875	1,050	2,925
2340	85 gallon tank, 181 GPH	4,375	1,075	5,450

SERVICES

D

D2020 Domestic Water Distribution

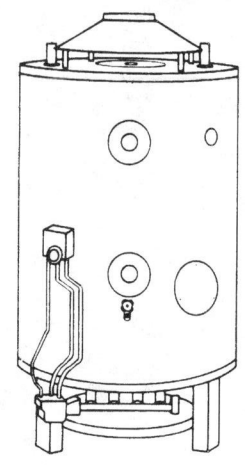

Systems below include piping and fittings within 10' of heater. Electric water heaters do not require venting. Gas fired heaters require vent piping (not included in these prices).

Electric

Gas Fired

D2020 240	Electric Water Heaters - Commercial Systems	COST EACH		
		MAT.	INST.	TOTAL
1800	Electric water heater, commercial, 100° F rise			
1820	50 gallon tank, 9 KW 37 GPH	2,500	755	3,255
1860	80 gal, 12 KW 49 GPH	3,175	935	4,110
1900	36 KW 147 GPH	4,325	1,000	5,325
1940	120 gal, 36 KW 147 GPH	4,650	1,075	5,725
1980	150 gal, 120 KW 490 GPH	14,400	1,175	15,575
2020	200 gal, 120 KW 490 GPH	15,200	1,200	16,400
2060	250 gal, 150 KW 615 GPH	16,900	1,400	18,300
2100	300 gal, 180 KW 738 GPH	18,400	1,475	19,875
2140	350 gal, 30 KW 123 GPH	13,600	1,575	15,175
2180	180 KW 738 GPH	18,900	1,575	20,475
2220	500 gal, 30 KW 123 GPH	17,800	1,875	19,675
2260	240 KW 984 GPH	25,900	1,875	27,775
2300	700 gal, 30 KW 123 GPH	21,400	2,150	23,550
2340	300 KW 1230 GPH	33,000	2,150	35,150
2380	1000 gal, 60 KW 245 GPH	25,900	2,950	28,850
2420	480 KW 1970 GPH	25,100	2,975	28,075
2460	1500 gal, 60 KW 245 GPH	37,800	3,675	41,475
2500	480 KW 1970 GPH	52,000	3,675	55,675

D2020 250	Gas Fired Water Heaters - Commercial Systems	COST EACH		
		MAT.	INST.	TOTAL
1760	Gas fired water heater, commercial, 100° F rise			
1780	75.5 MBH input, 63 GPH	1,850	1,175	3,025
1820	95 MBH input, 86 GPH	2,950	1,175	4,125
1860	100 MBH input, 91 GPH	3,150	1,225	4,375
1900	115 MBH input, 110 GPH	2,700	1,250	3,950
1980	155 MBH input, 150 GPH	3,800	1,400	5,200
2020	175 MBH input, 168 GPH	3,375	1,500	4,875
2060	200 MBH input, 192 GPH	3,475	1,700	5,175
2100	240 MBH input, 230 GPH	3,700	1,850	5,550
2140	300 MBH input, 278 GPH	4,050	2,100	6,150
2180	390 MBH input, 374 GPH	4,775	2,125	6,900
2220	500 MBH input, 480 GPH	6,425	2,300	8,725
2260	600 MBH input, 576 GPH	10,000	2,475	12,475

SERVICES D

Important: See the Reference Section for critical supporting data - Location Factors & Historical Cost Indexes

D2020 Domestic Water Distribution

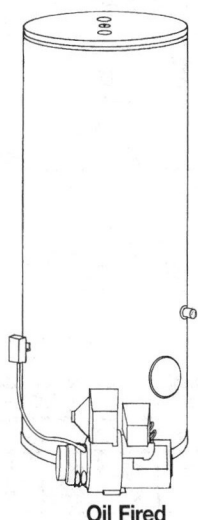

Oil Fired

Oil fired water heater systems include piping and fittings within 10' of heater. Oil fired heaters require vent piping (not included in these systems).

D2020 260	Oil Fired Water Heaters - Commercial Systems	COST EACH		
		MAT.	INST.	TOTAL
1800	Oil fired water heater, commercial, 100° F rise			
1820	103 MBH output, 116 GPH	1,200	1,025	2,225
1900	134 MBH output, 161 GPH	2,025	1,300	3,325
1940	161 MBH output, 192 GPH	2,225	1,500	3,725
1980	187 MBH output, 224 GPH	2,500	1,725	4,225
2060	262 MBH output, 315 GPH	3,050	1,975	5,025
2100	341 MBH output, 409 GPH	3,725	2,100	5,825
2140	420 MBH output, 504 GPH	4,150	2,275	6,425
2180	525 MBH output, 630 GPH	7,050	2,550	9,600
2220	630 MBH output, 756 GPH	5,800	2,550	8,350
2260	735 MBH output, 880 GPH	8,000	2,975	10,975
2300	840 MBH output, 1000 GPH	9,125	2,975	12,100
2340	1050 MBH output, 1260 GPH	10,000	3,075	13,075
2380	1365 MBH output, 1640 GPH	12,000	3,475	15,475
2420	1680 MBH output, 2000 GPH	13,400	4,375	17,775
2460	2310 MBH output, 2780 GPH	17,600	5,400	23,000
2500	2835 MBH output, 3400 GPH	20,000	5,400	25,400
2540	3150 MBH output, 3780 GPH	19,000	7,425	26,425

SERVICES

D

Design Assumptions: Vertical conductor size is based on a maximum rate of rainfall of 4″ per hour. To convert roof area to other rates multiply "Max. S.F. Roof Area" shown by four and divide the result by desired local rate. The answer is the local roof area that may be handled by the indicated pipe diameter.

Basic cost is for roof drain, 10′ of vertical leader and 10′ of horizontal, plus connection to the main.

Pipe Dia.	Max. S.F. Roof Area	Gallons per Min.
2″	544	23
3″	1610	67
4″	3460	144
5″	6280	261
6″	10,200	424
8″	22,000	913

D2040 210	Roof Drain Systems	COST PER SYSTEM		
		MAT.	INST.	TOTAL
1880	Roof drain, DWV PVC, 2″ diam., piping, 10′ high	113	435	548
1920	For each additional foot add	2.91	13.25	16.16
1960	3″ diam., 10′ high	155	510	665
2000	For each additional foot add	4.38	14.75	19.13
2040	4″ diam., 10′ high	190	570	760
2080	For each additional foot add	5.35	16.25	21.60
2120	5″ diam., 10′ high	385	660	1,045
2160	For each additional foot add	7.95	18.15	26.10
2200	6″ diam., 10′ high	595	730	1,325
2240	For each additional foot add	9.30	20	29.30
2280	8″ diam., 10′ high	1,425	1,250	2,675
2320	For each additional foot add	17.95	25.50	43.45
3940	C.I., soil, single hub, service wt., 2″ diam. piping, 10′ high	258	475	733
3980	For each additional foot add	5.50	12.40	17.90
4120	3″ diam., 10′ high	355	515	870
4160	For each additional foot add	7.45	13	20.45
4200	4″ diam., 10′ high	415	560	975
4240	For each additional foot add	9.30	14.20	23.50
4280	5″ diam., 10′ high	470	620	1,090
4320	For each additional foot add	12.50	15.95	28.45
4360	6″ diam., 10′ high	720	665	1,385
4400	For each additional foot add	15.25	16.65	31.90
4440	8″ diam., 10′ high	1,525	1,350	2,875
4480	For each additional foot add	23	28	51
6040	Steel galv. sch 40 threaded, 2″ diam. piping, 10′ high	275	460	735
6080	For each additional foot add	4.52	12.20	16.72
6120	3″ diam., 10′ high	500	660	1,160
6160	For each additional foot add	7.35	18.15	25.50
6200	4″ diam., 10′ high	720	850	1,570
6240	For each additional foot add	10.70	21.50	32.20
6280	5″ diam., 10′ high	925	1,075	2,000
6320	For each additional foot add	22	30	52
6360	6″ diam, 10′ high	1,600	1,325	2,925
6400	For each additional foot add	27.50	39	66.50
6440	8″ diam., 10′ high	3,150	1,800	4,950
6480	For each additional foot add	42.50	45	87.50

SERVICES D

D2090 Other Plumbing Systems

D2090 910	Piping - Installed - Unit Costs	COST PER L.F.		
		MAT.	INST.	TOTAL
0840	Cast iron, soil, B & S, service weight, 2" diameter	5.50	12.40	17.90
0860	3" diameter	7.45	13	20.45
0880	4" diameter	9.30	14.20	23.50
0900	5" diameter	12.50	15.95	28.45
0920	6" diameter	15.25	16.65	31.90
0940	8" diameter	23	28	51
0960	10" diameter	38.50	30.50	69
0980	12" diameter	54	34.50	88.50
1040	No hub, 1-1/2" diameter	6.25	11	17.25
1060	2" diameter	6.40	11.65	18.05
1080	3" diameter	8.35	12.20	20.55
1100	4" diameter	10.40	13.45	23.85
1120	5" diameter	14.50	14.65	29.15
1140	6" diameter	18.05	15.35	33.40
1160	8" diameter	28	24	52
1180	10" diameter	46.50	27	73.50
1220	Copper tubing, hard temper, solder, type K, 1/2" diameter	1.87	5.55	7.42
1260	3/4" diameter	2.90	5.85	8.75
1280	1" diameter	3.59	6.55	10.14
1300	1-1/4" diameter	4.37	7.75	12.12
1320	1-1/2" diameter	5.40	8.65	14.05
1340	2" diameter	8.25	10.85	19.10
1360	2-1/2" diameter	12.05	13	25.05
1380	3" diameter	16.55	14.45	31
1400	4" diameter	26	20.50	46.50
1480	5" diameter	68.50	24.50	93
1500	6" diameter	99	32	131
1520	8" diameter	174	35.50	209.50
1560	Type L, 1/2" diameter	1.69	5.35	7.04
1600	3/4" diameter	2.27	5.70	7.97
1620	1" diameter	2.97	6.40	9.37
1640	1-1/4" diameter	3.82	7.50	11.32
1660	1-1/2" diameter	4.68	8.35	13.03
1680	2" diameter	7	10.30	17.30
1700	2-1/2" diameter	10.30	12.60	22.90
1720	3" diameter	13.90	13.95	27.85
1740	4" diameter	22	20	42
1760	5" diameter	62	23	85
1780	6" diameter	78	30.50	108.50
1800	8" diameter	134	33.50	167.50
1840	Type M, 1/2" diameter	1.39	5.15	6.54
1880	3/4" diameter	1.84	5.55	7.39
1900	1" diameter	2.45	6.20	8.65
1920	1-1/4" diameter	3.21	7.25	10.46
1940	1-1/2" diameter	4.21	8.05	12.26
1960	2" diameter	6.50	9.85	16.35
1980	2-1/2" diameter	9.15	12.20	21.35
2000	3" diameter	11.75	13.45	25.20
2020	4" diameter	19.80	19.50	39.30
2060	6" diameter	76.50	29	105.50
2080	8" diameter	126	32	158
2120	Type DWV, 1-1/4" diameter	3.17	7.25	10.42
2160	1-1/2" diameter	3.82	8.05	11.87
2180	2" diameter	4.91	9.85	14.76
2200	3" diameter	8.30	13.45	21.75
2220	4" diameter	14.55	19.50	34.05
2240	5" diameter	43	21.50	64.50
2260	6" diameter	60.50	29	89.50

SERVICES

D

D2090 Other Plumbing Systems

D2090 910	Piping - Installed - Unit Costs	COST PER L.F.		
		MAT.	INST.	TOTAL
2280	8" diameter	136	32	168
2800	Plastic,PVC, DWV, schedule 40, 1-1/4" diameter	2.60	10.30	12.90
2820	1-1/2" diameter	2.73	12.05	14.78
2830	2" diameter	2.91	13.25	16.16
2840	3" diameter	4.38	14.75	19.13
2850	4" diameter	5.35	16.25	21.60
2890	6" diameter	9.30	20	29.30
3010	Pressure pipe 200 PSI, 1/2" diameter	2.33	8.05	10.38
3030	3/4" diameter	2.38	8.50	10.88
3040	1" diameter	2.54	9.45	11.99
3050	1-1/4" diameter	2.76	10.30	13.06
3060	1-1/2" diameter	2.87	12.05	14.92
3070	2" diameter	3.12	13.25	16.37
3080	2-1/2" diameter	4.42	13.95	18.37
3090	3" diameter	4.90	14.75	19.65
3100	4" diameter	6.15	16.25	22.40
3110	6" diameter	10.65	20	30.65
3120	8" diameter	19.20	25.50	44.70
4000	Steel, schedule 40, threaded, black, 1/2" diameter	1.45	6.90	8.35
4020	3/4" diameter	1.62	7.10	8.72
4030	1" diameter	2.07	8.20	10.27
4040	1-1/4" diameter	2.46	8.75	11.21
4050	1-1/2" diameter	2.76	9.75	12.51
4060	2" diameter	3.50	12.20	15.70
4070	2-1/2" diameter	5.55	15.60	21.15
4080	3" diameter	6.95	18.15	25.10
4090	4" diameter	10.25	21.50	31.75
4100	5" diameter	19.35	30	49.35
4110	6" diameter	24	39	63
4120	8" diameter	34	45	79
4130	10" diameter	51.50	53	104.50
4140	12" diameter	65	67.50	132.50
4200	Galvanized, 1/2" diameter	1.66	6.90	8.56
4220	3/4" diameter	2.12	7.10	9.22
4230	1" diameter	2.52	8.20	10.72
4240	1-1/4" diameter	3.12	8.75	11.87
4250	1-1/2" diameter	3.37	9.75	13.12
4260	2" diameter	4.52	12.20	16.72
4270	2-1/2" diameter	5.90	15.60	21.50
4280	3" diameter	7.35	18.15	25.50
4290	4" diameter	10.70	21.50	32.20
4300	5" diameter	22	30	52
4310	6" diameter	27.50	39	66.50
4320	8" diameter	42.50	45	87.50
4330	10" diameter	64.50	53	117.50
4340	12" diameter	81	67.50	148.50
5010	Flanged, black, 1" diameter	6.95	12.40	19.35
5020	1-1/4" diameter	7.70	13.60	21.30
5030	1-1/2" diameter	7.95	14.95	22.90
5040	2" diameter	9	19.30	28.30
5050	2-1/2" diameter	10.80	24	34.80
5060	3" diameter	12.20	27.50	39.70
5070	4" diameter	16.35	33.50	49.85
5080	5" diameter	25	41	66
5090	6" diameter	27.50	52	79.50
5100	8" diameter	41.50	68.50	110
5110	10" diameter	75.50	81.50	157
5120	12" diameter	98	93	191

Important: See the Reference Section for critical supporting data - Location Factors & Historical Cost Indexes

D2090 Other Plumbing Systems

D2090 910	Piping - Installed - Unit Costs	COST PER L.F.		
		MAT.	INST.	TOTAL
5720	Grooved joints, black, 3/4" diameter	2.17	6.10	8.27
5730	1" diameter	2.31	6.90	9.21
5740	1-1/4" diameter	2.87	7.50	10.37
5750	1-1/2" diameter	3.20	8.50	11.70
5760	2" diameter	3.75	10.85	14.60
5770	2-1/2" diameter	5.20	13.70	18.90
5900	3" diameter	6.35	15.60	21.95
5910	4" diameter	9.35	17.35	26.70
5920	5" diameter	14	21	35
5930	6" diameter	16.65	29	45.65
5940	8" diameter	25.50	33	58.50
5950	10" diameter	40.50	39	79.50
5960	12" diameter	48	45	93

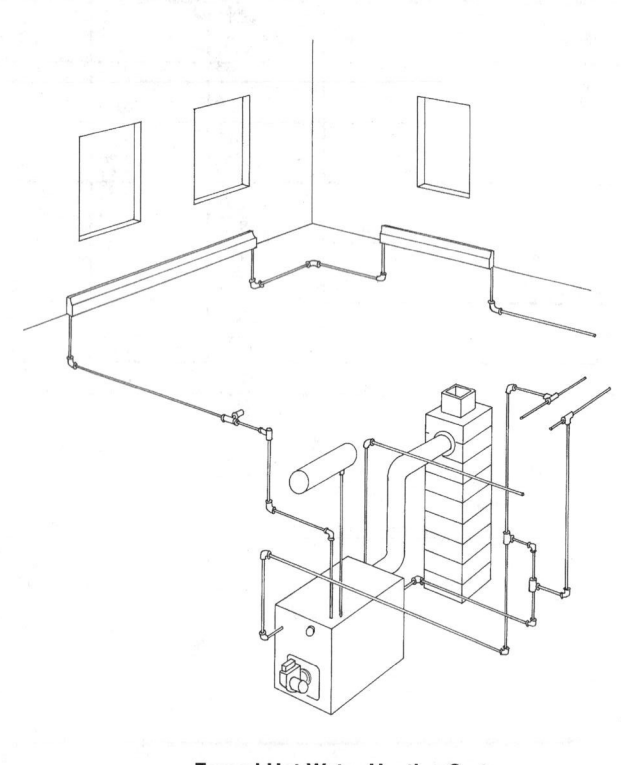

Forced Hot Water Heating System

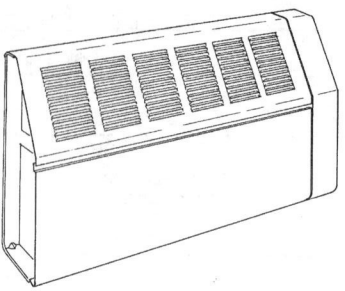

Fin Tube Radiation

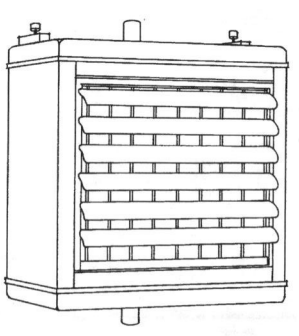

Terminal Unit Heater

D3010 510	Apartment Building Heating - Fin Tube Radiation	COST PER S.F.		
		MAT.	INST.	TOTAL
1740	Heating systems, fin tube radiation, forced hot water			
1760	1,000 S.F. area, 10,000 C.F. volume	4.38	4.23	8.61
1800	10,000 S.F. area, 100,000 C.F. volume	1.45	2.53	3.98
1840	20,000 S.F. area, 200,000 C.F. volume	1.61	2.85	4.46
1880	30,000 S.F. area, 300,000 C.F. volume	1.50	2.76	4.26

D3010 520	Commercial Building Heating - Fin Tube Radiation	COST PER S.F.		
		MAT.	INST.	TOTAL
1940	Heating systems, fin tube radiation, forced hot water			
1960	1,000 S.F. bldg, one floor	9.05	9.53	18.58
2000	10,000 S.F., 100,000 C.F., total two floors	2.47	3.63	6.10
2040	100,000 S.F., 1,000,000 C.F., total three floors	1.13	1.65	2.78
2080	1,000,000 S.F., 10,000,000 C.F., total five floors	.54	.87	1.41

D3010 530	Commercial Bldg. Heating - Terminal Unit Heaters	COST PER S.F.		
		MAT.	INST.	TOTAL
1860	Heating systems, terminal unit heaters, forced hot water			
1880	1,000 S.F. bldg., one floor	8.88	8.73	17.61
1920	10,000 S.F. bldg, 100,000 C.F. total two floors	2.21	3.01	5.22
1960	100,000 S.F. bldg., 1,000,000 C.F. total three floors	1.17	1.52	2.69
2000	1,000,000 S.F. bldg., 10,000,000 C.F. total five floors	.72	.97	1.69

D3020 Heat Generating Systems

Small Electric Boiler Systems Considerations:

1. Terminal units are fin tube baseboard radiation rated at 720 BTU/hr with 200° water temperature or 820 BTU/hr steam.
2. Primary use being for residential or smaller supplementary areas, the floor levels are based on 7-1/2" ceiling heights.
3. All distribution piping is copper for boilers through 205 MBH. All piping for larger systems is steel pipe.

Large Electric Boiler System Considerations:

1. Terminal units are all unit heaters of the same size. Quantities are varied to accommodate total requirements.
2. All air is circulated through the heaters a minimum of three times per hour.
3. As the capacities are adequate for commercial use, gross output rating by floor levels are based on 10' ceiling height.
4. All distribution piping is black steel pipe.

D3020 102	Small Heating Systems, Hydronic, Electric Boilers	COST PER S.F.		
		MAT.	INST.	TOTAL
1100	Small heating systems, hydronic, electric boilers			
1120	Steam, 1 floor, 1480 S.F., 61 M.B.H.	8.65	5.39	14.04
1160	3,000 S.F., 123 M.B.H.	5.10	4.72	9.82
1200	5,000 S.F., 205 M.B.H.	3.86	4.35	8.21
1240	2 floors, 12,400 S.F., 512 M.B.H.	2.75	4.33	7.08
1280	3 floors, 24,800 S.F., 1023 M.B.H.	2.40	4.27	6.67
1320	34,750 S.F., 1,433 M.B.H.	2.12	4.15	6.27
1360	Hot water, 1 floor, 1,000 S.F., 41 M.B.H.	7.85	3.02	10.87
1400	2,500 S.F., 103 M.B.H.	4.92	5.40	10.32
1440	2 floors, 4,850 S.F., 205 M.B.H.	4.52	6.50	11.02
1480	3 floors, 9,700 S.F., 410 M.B.H.	4.39	6.70	11.09

D3020 104	Large Heating Systems, Hydronic, Electric Boilers	COST PER S.F.		
		MAT.	INST.	TOTAL
1230	Large heating systems, hydronic, electric boilers			
1240	9,280 S.F., 150 K.W., 510 M.B.H., 1 floor	3.03	2.06	5.09
1280	14,900 S.F., 240 K.W., 820 M.B.H., 2 floors	3.15	3.39	6.54
1320	18,600 S.F., 296 K.W., 1,010 M.B.H., 3 floors	3.05	3.68	6.73
1360	26,100 S.F., 420 K.W., 1,432 M.B.H., 4 floors	3.18	3.62	6.80
1400	39,100 S.F., 666 K.W., 2,273 M.B.H., 4 floors	2.81	3.01	5.82
1440	57,700 S.F., 900 K.W., 3,071 M.B.H., 5 floors	2.63	2.97	5.60
1480	111,700 S.F., 1,800 K.W., 6,148 M.B.H., 6 floors	2.31	2.56	4.87
1520	149,000 S.F., 2,400 K.W., 8,191 M.B.H., 8 floors	2.26	2.56	4.82
1560	223,300 S.F., 3,600 K.W., 12,283 M.B.H., 14 floors	2.29	2.89	5.18

SERVICES

D

D3020 Heat Generating Systems

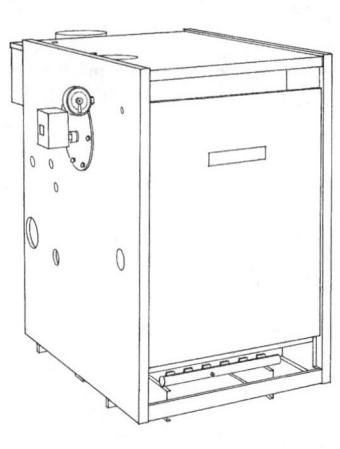

Boiler Selection: The maximum allowable working pressures are limited by ASME "Code for Heating Boilers" to 15 PSI for steam and 160 PSI for hot water heating boilers, with a maximum temperature limitation of 250° F. Hot water boilers are generally rated for a working pressure of 30 PSI. High pressure boilers are governed by the ASME "Code for Power Boilers" which is used almost universally for boilers operating over 15 PSIG. High pressure boilers used for a combination of heating/process loads are usually designed for 150 PSIG.

Boiler ratings are usually indicated as either Gross or Net Output. The Gross Load is equal to the Net Load plus a piping and pickup allowance. When this allowance cannot be determined, divide the gross output rating by 1.25 for a value equal to or greater than the next heat loss requirement of the building.

Table below lists installed cost per boiler and includes insulating jacket, standard controls, burner and safety controls. Costs do not include piping or boiler base pad. Outputs are Gross.

D3020 106	Boilers, Hot Water & Steam	COST EACH		
		MAT.	**INST.**	**TOTAL**
0600	Boiler, electric, steel, hot water, 12 K.W., 41 M.B.H.	3,725	930	4,655
0620	30 K.W., 103 M.B.H.	4,100	1,000	5,100
0640	60 K.W., 205 M.B.H.	5,200	1,100	6,300
0660	120 K.W., 410 M.B.H.	6,625	1,350	7,975
0680	210 K.W., 716 M.B.H.	11,000	2,025	13,025
0700	510 K.W., 1,739 M.B.H.	20,800	3,750	24,550
0720	720 K.W., 2,452 M.B.H.	26,200	4,225	30,425
0740	1,200 K.W., 4,095 M.B.H.	37,000	4,850	41,850
0760	2,100 K.W., 7,167 M.B.H.	58,000	6,100	64,100
0780	3,600 K.W., 12,283 M.B.H.	79,500	10,300	89,800
0820	Steam, 6 K.W., 20.5 M.B.H.	8,700	1,000	9,700
0840	24 K.W., 81.8 M.B.H.	9,250	1,100	10,350
0860	60 K.W., 205 M.B.H.	10,700	1,200	11,900
0880	150 K.W., 512 M.B.H.	13,900	1,850	15,750
0900	510 K.W., 1,740 M.B.H.	22,600	4,575	27,175
0920	1,080 K.W., 3,685 M.B.H.	34,700	6,575	41,275
0940	2,340 K.W., 7,984 M.B.H.	65,500	10,300	75,800
0980	Gas, cast iron, hot water, 80 M.B.H.	1,125	1,150	2,275
1000	100 M.B.H.	1,300	1,225	2,525
1020	163 M.B.H.	1,775	1,675	3,450
1040	280 M.B.H.	2,500	1,850	4,350
1060	544 M.B.H.	4,375	3,275	7,650
1080	1,088 M.B.H.	7,375	4,175	11,550
1100	2,000 M.B.H.	12,700	6,500	19,200
1120	2,856 M.B.H.	17,200	8,325	25,525
1140	4,720 M.B.H.	50,500	11,500	62,000
1160	6,970 M.B.H.	65,000	18,700	83,700
1180	For steam systems under 2,856 M.B.H., add 8%			
1240	Steel, hot water, 72 M.B.H.	1,925	610	2,535
1260	101 M.B.H.	2,200	680	2,880
1280	132 M.B.H.	2,500	720	3,220
1300	150 M.B.H.	2,900	815	3,715
1320	240 M.B.H.	4,450	940	5,390
1340	400 M.B.H.	6,350	1,525	7,875
1360	640 M.B.H.	8,625	2,050	10,675
1380	800 M.B.H.	10,100	2,450	12,550
1400	960 M.B.H.	12,500	2,725	15,225
1420	1,440 M.B.H.	17,400	3,500	20,900

Important: See the Reference Section for critical supporting data - Location Factors & Historical Cost Indexes

D3020 Heat Generating Systems

D3020 106	Boilers, Hot Water & Steam	COST EACH		
		MAT.	INST.	TOTAL
1440	2,400 M.B.H.	28,100	6,125	34,225
1460	3,000 M.B.H.	34,700	8,150	42,850
1520	Oil, cast iron, hot water, 109 M.B.H.	1,600	1,400	3,000
1540	173 M.B.H.	2,000	1,675	3,675
1560	236 M.B.H.	3,125	1,950	5,075
1580	1,084 M.B.H.	7,500	4,425	11,925
1600	1,600 M.B.H.	9,925	6,350	16,275
1620	2,480 M.B.H.	13,800	8,125	21,925
1640	3,550 M.B.H.	17,800	9,750	27,550
1660	Steam systems same price as hot water			
1700	Steel, hot water, 103 M.B.H.	1,400	765	2,165
1720	137 M.B.H.	2,700	895	3,595
1740	225 M.B.H.	3,125	1,000	4,125
1760	315 M.B.H.	4,000	1,275	5,275
1780	420 M.B.H.	4,825	1,750	6,575
1800	630 M.B.H.	6,300	2,350	8,650
1820	735 M.B.H.	6,300	2,550	8,850
1840	1,050 M.B.H.	10,200	3,350	13,550
1860	1,365 M.B.H.	11,900	3,800	15,700
1880	1,680 M.B.H.	13,400	4,250	17,650
1900	2,310 M.B.H.	14,400	5,850	20,250
1920	2,835 M.B.H.	18,600	7,200	25,800
1940	3,150 M.B.H.	19,500	9,400	28,900

D3020 Heat Generating Systems

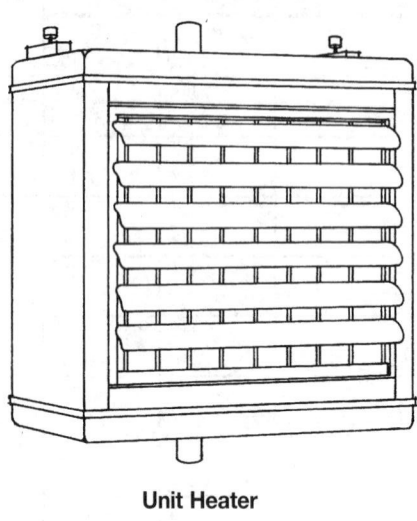

Unit Heater

Fossil Fuel Boiler System Considerations:

1. Terminal units are horizontal unit heaters. Quantities are varied to accommodate total heat loss per building.
2. Unit heater selection was determined by their capacity to circulate the building volume a minimum of three times per hour in addition to the BTU output.
3. Systems shown are forced hot water. Steam boilers cost slightly more than hot water boilers. However, this is compensated for by the smaller size or fewer terminal units required with steam.
4. Floor levels are based on 10' story heights.
5. MBH requirements are gross boiler output.

D3020 108	Heating Systems, Unit Heaters	COST PER S.F.		
		MAT.	INST.	TOTAL
1260	Heating systems, hydronic, fossil fuel, terminal unit heaters,			
1280	Cast iron boiler, gas, 80 M.B.H., 1,070 S.F. bldg.	5.75	6.50	12.25
1320	163 M.B.H., 2,140 S.F. bldg.	3.88	4.40	8.28
1360	544 M.B.H., 7,250 S.F. bldg.	2.71	3.17	5.88
1400	1,088 M.B.H., 14,500 S.F. bldg.	2.38	3	5.38
1440	3,264 M.B.H., 43,500 S.F. bldg.	1.95	2.28	4.23
1480	5,032 M.B.H., 67,100 S.F. bldg.	2.39	2.31	4.70
1520	Oil, 109 M.B.H., 1,420 S.F. bldg.	5.90	5.75	11.65
1560	235 M.B.H., 3,150 S.F. bldg.	4.12	4.20	8.32
1600	940 M.B.H., 12,500 S.F. bldg.	2.87	2.75	5.62
1640	1,600 M.B.H., 21,300 S.F. bldg.	2.79	2.62	5.41
1680	2,480 M.B.H., 33,100 S.F. bldg.	2.77	2.37	5.14
1720	3,350 M.B.H., 44,500 S.F. bldg.	2.38	2.42	4.80
1760	Coal, 148 M.B.H., 1,975 S.F. bldg.	4.90	3.76	8.66
1800	300 M.B.H., 4,000 S.F. bldg.	3.79	2.95	6.74
1840	2,360 M.B.H., 31,500 S.F. bldg.	2.66	2.43	5.09
1880	Steel boiler, gas, 72 M.B.H., 1,020 S.F. bldg.	5.70	4.78	10.48
1920	240 M.B.H., 3,200 S.F. bldg.	4.02	3.84	7.86
1960	480 M.B.H., 6,400 S.F. bldg.	3.26	2.85	6.11
2000	800 M.B.H., 10,700 S.F. bldg.	2.80	2.54	5.34
2040	1,960 M.B.H., 26,100 S.F. bldg.	2.48	2.27	4.75
2080	3,000 M.B.H., 40,000 S.F. bldg.	2.45	2.32	4.77
2120	Oil, 97 M.B.H., 1,300 S.F. bldg.	5.45	5.45	10.90
2160	315 M.B.H., 4,550 S.F. bldg.	3.39	2.78	6.17
2200	525 M.B.H., 7,000 S.F. bldg.	3.60	2.84	6.44
2240	1,050 M.B.H., 14,000 S.F. bldg.	3.06	2.86	5.92
2280	2,310 M.B.H. 30,800 S.F. bldg.	2.68	2.47	5.15
2320	3,150 M.B.H., 42,000 S.F. bldg.	2.48	2.53	5.01

SERVICES D

Important: See the Reference Section for critical supporting data - Location Factors & Historical Cost Indexes

D3020 Heat Generating Systems

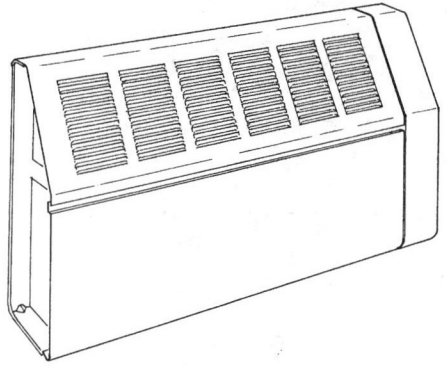

Fin Tube Radiator

Fossil Fuel Boiler System Considerations:

1. Terminal units are commercial steel fin tube radiation. Quantities are varied to accommodate total heat loss per building.
2. Systems shown are forced hot water. Steam boilers cost slightly more than hot water boilers. However, this is compensated for by the smaller size or fewer terminal units required with steam.
3. Floor levels are based on 10' story heights.
4. MBH requirements are gross boiler output.

D3020 110	Heating System, Fin Tube Radiation	COST PER S.F.		
		MAT.	INST.	TOTAL
3230	Heating systems, hydronic, fossil fuel, fin tube radiation			
3240	Cast iron boiler, gas, 80 MBH, 1,070 S.F. bldg.	7.10	9.72	16.82
3280	169 M.B.H., 2,140 S.F. bldg.	4.48	6.15	10.63
3320	544 M.B.H., 7,250 S.F. bldg.	3.69	5.35	9.04
3360	1,088 M.B.H., 14,500 S.F. bldg.	3.44	5.25	8.69
3400	3,264 M.B.H., 43,500 S.F. bldg.	3.09	4.59	7.68
3440	5,032 M.B.H., 67,100 S.F. bldg.	3.55	4.65	8.20
3480	Oil, 109 M.B.H., 1,420 S.F. bldg.	8.35	10.60	18.95
3520	235 M.B.H., 3,150 S.F. bldg.	5.10	6.30	11.40
3560	940 M.B.H., 12,500 S.F. bldg.	3.95	4.99	8.94
3600	1,600 M.B.H., 21,300 S.F. bldg.	3.94	4.96	8.90
3640	2,480 M.B.H., 33,100 S.F. bldg.	3.94	4.71	8.65
3680	3,350 M.B.H., 44,500 S.F. bldg.	3.53	4.74	8.27
3720	Coal, 148 M.B.H., 1,975 S.F. bldg.	5.85	5.85	11.70
3760	300 M.B.H., 4,000 S.F. bldg.	4.74	5.05	9.79
3800	2,360 M.B.H., 31,500 S.F. bldg.	3.78	4.72	8.50
3840	Steel boiler, gas, 72 M.B.H., 1,020 S.F. bldg.	8.10	9.20	17.30
3880	240 M.B.H., 3,200 S.F. bldg.	5.05	6.10	11.15
3920	480 M.B.H., 6,400 S.F. bldg.	4.27	5.05	9.32
3960	800 M.B.H., 10,700 S.F. bldg.	4.19	5.35	9.54
4000	1,960 M.B.H., 26,100 S.F. bldg.	3.60	4.56	8.16
4040	3,000 M.B.H., 40,000 S.F. bldg.	3.54	4.59	8.13
4080	Oil, 97 M.B.H., 1,300 S.F. bldg.	6.30	8.70	15
4120	315 M.B.H., 4,550 S.F. bldg.	4.31	4.75	9.06
4160	525 M.B.H., 7,000 S.F. bldg.	4.70	5.10	9.80
4200	1,050 M.B.H., 14,000 S.F. bldg.	4.16	5.10	9.26
4240	2,310 M.B.H., 30,800 S.F. bldg.	3.79	4.74	8.53
4280	3,150 M.B.H., 42,000 S.F. bldg.	3.59	4.81	8.40

SERVICES

D

D3030 Cooling Generating Systems

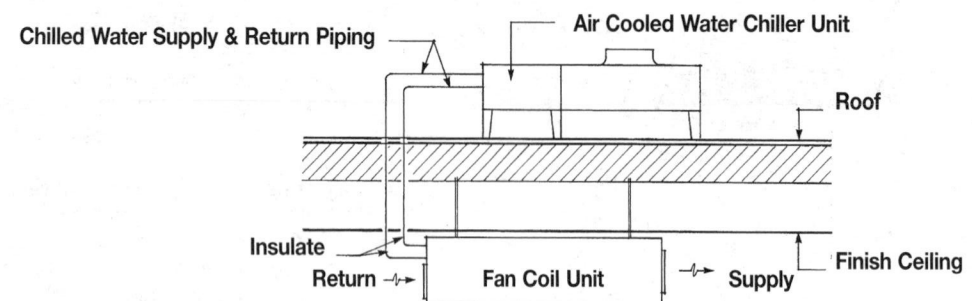

Chilled Water Supply & Return Piping — Air Cooled Water Chiller Unit

Roof

Insulate

Return → Fan Coil Unit → Supply

Finish Ceiling

*Cooling requirements would lead to a choice of multiple chillers.

D3030 110	Chilled Water, Air Cooled Condenser Systems	COST PER S.F.		
		MAT.	INST.	TOTAL
1180	Packaged chiller, air cooled, with fan coil unit			
1200	Apartment corridors, 3,000 S.F., 5.50 ton	5.13	5.30	10.43
1240	6,000 S.F., 11.00 ton	3.84	4.24	8.08
1280	10,000 S.F., 18.33 ton	3.43	3.23	6.66
1320	20,000 S.F., 36.66 ton	2.86	2.34	5.20
1360	40,000 S.F., 73.33 ton	3.05	2.37	5.42
1440	Banks and libraries, 3,000 S.F., 12.50 ton	7.45	6.25	13.70
1480	6,000 S.F., 25.00 ton	7.05	5.10	12.15
1520	10,000 S.F., 41.66 ton	5.90	3.56	9.46
1560	20,000 S.F., 83.33 ton	5.60	3.24	8.84
1600	40,000 S.F., 167 ton*			
1680	Bars and taverns, 3,000 S.F., 33.25 ton	14.95	7.40	22.35
1720	6,000 S.F., 66.50 ton	12.40	5.75	18.15
1760	10,000 S.F., 110.83 ton	11.90	1.84	13.74
1800	20,000 S.F., 220 ton*			
1840	40,000 S.F., 440 ton*			
1920	Bowling alleys, 3,000 S.F., 17.00 ton	9.60	7.05	16.65
1960	6,000 S.F., 34.00 ton	8.05	4.93	12.98
2000	10,000 S.F., 56.66 ton	6.55	3.59	10.14
2040	20,000 S.F., 113.33 ton	7.05	3.28	10.33
2080	40,000 S.F., 227 ton*			
2160	Department stores, 3,000 S.F., 8.75 ton	7.45	5.90	13.35
2200	6,000 S.F., 17.50 ton	5.25	4.57	9.82
2240	10,000 S.F., 29.17 ton	4.50	3.31	7.81
2280	20,000 S.F., 58.33 ton	3.61	2.45	6.06
2320	40,000 S.F., 116.66 ton	4.30	2.45	6.75
2400	Drug stores, 3,000 S.F., 20.00 ton	11.05	7.25	18.30
2440	6,000 S.F., 40.00 ton	9.50	5.40	14.90
2480	10,000 S.F., 66.66 ton	8.70	4.64	13.34
2520	20,000 S.F., 133.33 ton	8.55	3.44	11.99
2560	40,000 S.F., 267 ton*			
2640	Factories, 2,000 S.F., 10.00 ton	6.40	6	12.40
2680	6,000 S.F., 20.00 ton	5.95	4.83	10.78
2720	10,000 S.F., 33.33 ton	4.98	3.39	8.37
2760	20,000 S.F., 66.66 ton	4.79	3.08	7.87
2800	40,000 S.F., 133.33 ton	4.76	2.52	7.28
2880	Food supermarkets, 3,000 S.F., 8.50 ton	7.30	5.90	13.20
2920	6,000 S.F., 17.00 ton	5.15	4.54	9.69
2960	10,000 S.F., 28.33 ton	4.31	3.21	7.52
3000	20,000 S.F., 56.66 ton	3.48	2.39	5.87
3040	40,000 S.F., 113.33 ton	4.19	2.45	6.64
3120	Medical centers, 3,000 S.F., 7.00 ton	6.40	5.70	12.10

Important: See the Reference Section for critical supporting data - Location Factors & Historical Cost Indexes

SERVICES D

D3030 Cooling Generating Systems

D3030 110	Chilled Water, Air Cooled Condenser Systems	COST PER S.F.		
		MAT.	INST.	TOTAL
3160	6,000 S.F., 14.00 ton	4.48	4.40	8.88
3200	10,000 S.F., 23.33 ton	4.13	3.37	7.50
3240	20,000 S.F., 46.66 ton	3.50	2.47	5.97
3280	40,000 S.F., 93.33 ton	3.71	2.41	6.12
3360	Offices, 3,000 S.F., 9.50 ton	6.15	5.85	12
3400	6,000 S.F., 19.00 ton	5.75	4.77	10.52
3440	10,000 S.F., 31.66 ton	4.79	3.36	8.15
3480	20,000 S.F., 63.33 ton	4.74	3.13	7.87
3520	40,000 S.F., 126.66 ton	4.64	2.54	7.18
3600	Restaurants, 3,000 S.F., 15.00 ton	8.55	6.50	15.05
3640	6,000 S.F., 30.00 ton	7.45	4.92	12.37
3680	10,000 S.F., 50.00 ton	6.10	3.63	9.73
3720	20,000 S.F., 100.00 ton	7	3.42	10.42
3760	40,000 S.F., 200 ton*			
3840	Schools and colleges, 3,000 S.F., 11.50 ton	7.05	6.15	13.20
3880	6,000 S.F., 23.00 ton	6.60	4.98	11.58
3920	10,000 S.F., 38.33 ton	5.55	3.49	9.04
3960	20,000 S.F., 76.66 ton	5.40	3.26	8.66
4000	40,000 S.F., 153 ton*			

SERVICES

D

Reciprocating Package Chiller — Condenser Water — Cooling Tower — Cooling Tower Water Makeup — Chilled Water Supply & Return Piping — Roof Structure — Finish Ceiling — Insulate — Return — Supply — Fan Coil Unit

*Cooling requirements would lead to a choice of multiple chillers

D3030 115	Chilled Water, Cooling Tower Systems	COST PER S.F.		
		MAT.	INST.	TOTAL
1300	Packaged chiller, water cooled, with fan coil unit			
1320	Apartment corridors, 4,000 S.F., 7.33 ton	5.83	5.18	11.01
1360	6,000 S.F., 11.00 ton	4.39	4.47	8.86
1400	10,000 S.F., 18.33 ton	3.99	3.31	7.30
1440	20,000 S.F., 26.66 ton	3.14	2.47	5.61
1480	40,000 S.F., 73.33 ton	3.18	2.55	5.73
1520	60,000 S.F., 110.00 ton	3.09	2.61	5.70
1600	Banks and libraries, 4,000 S.F., 16.66 ton	9.30	5.70	15
1640	6,000 S.F., 25.00 ton	7.60	5	12.60
1680	10,000 S.F., 41.66 ton	6.30	3.75	10.05
1720	20,000 S.F., 83.33 ton	5.90	3.60	9.50
1760	40,000 S.F., 166.66 ton	5.60	4.44	10.04
1800	60,000 S.F., 250.00 ton	5.50	4.69	10.19
1880	Bars and taverns, 4,000 S.F., 44.33 ton	15.55	7.20	22.75
1920	6,000 S.F., 66.50 ton	15.35	7.45	22.80
1960	10,000 S.F., 110.83 ton	12.85	5.95	18.80
2000	20,000 S.F., 221.66 ton	11.85	6.35	18.20
2040	40,000 S.F., 440 ton*			
2080	60,000 S.F., 660 ton*			
2160	Bowling alleys, 4,000 S.F., 22.66 ton	10.80	6.25	17.05
2200	6,000 S.F., 34.00 ton	9	5.50	14.50
2240	10,000 S.F., 56.66 ton	7.75	4.05	11.80
2280	20,000 S.F., 113.33 ton	7	3.80	10.80
2320	40,000 S.F., 226.66 ton	6.60	4.57	11.17
2360	60,000 S.F., 340 ton			
2440	Department stores, 4,000 S.F., 11.66 ton	6.30	5.65	11.95
2480	6,000 S.F., 17.50 ton	6.70	4.63	11.33
2520	10,000 S.F., 29.17 ton	4.81	3.47	8.28
2560	20,000 S.F., 58.33 ton	3.95	2.58	6.53
2600	40,000 S.F., 116.66 ton	4.01	2.68	6.69
2640	60,000 S.F., 175.00 ton	4.42	4.38	8.80
2720	Drug stores, 4,000 S.F., 26.66 ton	11.55	6.45	18
2760	6,000 S.F., 40.00 ton	10	5.50	15.50
2800	10,000 S.F., 66.66 ton	9.65	5.05	14.70
2840	20,000 S.F., 133.33 ton	8	3.95	11.95
2880	40,000 S.F., 266.67 ton	7.70	5	12.70
2920	60,000 S.F., 400 ton*			
3000	Factories, 4,000 S.F., 13.33 ton	7.80	5.45	13.25
3040	6,000 S.F., 20.00 ton	6.70	4.68	11.38
3080	10,000 S.F., 33.33 ton	5.35	3.59	8.94
3120	20,000 S.F., 66.66 ton	5.20	3.24	8.44
3160	40,000 S.F., 133.33 ton	4.47	2.78	7.25
3200	60,000 S.F., 200.00 ton	4.80	4.51	9.31
3280	Food supermarkets, 4,000 S.F., 11.33 ton	6.20	5.60	11.80

Important: See the Reference Section for critical supporting data - Location Factors & Historical Cost Indexes

D3030 Cooling Generating Systems

D3030 115	Chilled Water, Cooling Tower Systems	COST PER S.F.		
		MAT.	INST.	TOTAL
3320	6,000 S.F., 17.00 ton	5.85	4.54	10.39
3360	10,000 S.F., 28.33 ton	4.71	3.44	8.15
3400	20,000 S.F., 56.66 ton	4.03	2.59	6.62
3440	40,000 S.F., 113.33 ton	3.96	2.69	6.65
3480	60,000 S.F., 170.00 ton	4.36	4.38	8.74
3560	Medical centers, 4.000 S.F., 9.33 ton	5.25	5.15	10.40
3600	6,000 S.F., 14.00 ton	5.65	4.44	10.09
3640	10,000 S.F., 23.33 ton	4.44	3.33	7.77
3680	20,000 S.F., 46.66 ton	3.52	2.52	6.04
3720	40,000 S.F., 93.33 ton	3.61	2.63	6.24
3760	60,000 S.F., 140.00 ton	3.99	4.33	8.32
3840	Offices, 4,000 S.F., 12.66 ton	7.50	5.35	12.85
3880	6,000 S.F., 19.00 ton	6.55	4.77	11.32
3920	10,000 S.F., 31.66 ton	5.25	3.61	8.86
3960	20,000 S.F., 63.33 ton	5.05	3.24	8.29
4000	40,000 S.F., 126.66 ton	4.65	4.24	8.89
4040	60,000 S.F., 190.00 ton	4.62	4.45	9.07
4120	Restaurants, 4,000 S.F., 20.00 ton	9.60	5.80	15.40
4160	6,000 S.F., 30.00 ton	8.05	5.15	13.20
4200	10,000 S.F., 50.00 ton	7.05	3.88	10.93
4240	20,000 S.F., 100.00 ton	6.70	3.78	10.48
4280	40,000 S.F., 200.00 ton	5.85	4.30	10.15
4320	60,000 S.F., 300.00 ton	6.20	4.85	11.05
4400	Schools and colleges, 4,000 S.F., 15.33 ton	8.70	5.60	14.30
4440	6,000 S.F., 23.00 ton	7.10	4.92	12.02
4480	10,000 S.F., 38.33 ton	5.90	3.66	9.56
4520	20,000 S.F., 76.66 ton	5.60	3.56	9.16
4560	40,000 S.F., 153.33 ton	5.25	4.33	9.58
4600	60,000 S.F., 230.00 ton	5.05	4.52	9.57

SERVICES

D

D3030 Cooling Generating Systems

Rooftop, Single Zone System

*Above normal capacity

D3030 202	Rooftop Single Zone Unit Systems	COST PER S.F.		
		MAT.	INST.	TOTAL
1260	Rooftop, single zone, air conditioner			
1280	Apartment corridors, 500 S.F., .92 ton	4.60	2.55	7.15
1320	1,000 S.F., 1.83 ton	4.55	2.51	7.06
1360	1500 S.F., 2.75 ton	3.05	2.08	5.13
1400	3,000 S.F., 5.50 ton	2.81	2	4.81
1440	5,000 S.F., 9.17 ton	2.97	1.82	4.79
1480	10,000 S.F., 18.33 ton	2.59	1.71	4.30
1560	Banks or libraries, 500 S.F., 2.08 ton	10.35	5.70	16.05
1600	1,000 S.F., 4.17 ton	6.95	4.73	11.68
1640	1,500 S.F., 6.25 ton	6.35	4.55	10.90
1680	3,000 S.F., 12.50 ton	6.75	4.14	10.89
1720	5,000 S.F., 20.80 ton	5.85	3.89	9.74
1760	10,000 S.F., 41.67 ton	5.85	3.87	9.72
1840	Bars and taverns, 500 S.F. 5.54 ton	15.55	7.60	23.15
1880	1,000 S.F., 11.08 ton	16.55	6.50	23.05
1920	1,500 S.F., 16.62 ton	12.45	6.10	18.55
1960	3,000 S.F., 33.25 ton	14.65	5.80	20.45
2000	5,000 S.F., 55.42 ton	13.70	5.80	19.50
2040	10,000 S.F., 110.83 ton*			
2080	Bowling alleys, 500 S.F., 2.83 ton	9.40	6.40	15.80
2120	1,000 S.F., 5.67 ton	8.65	6.20	14.85
2160	1,500 S.F., 8.50 ton	9.15	5.60	14.75
2200	3,000 S.F., 17.00 ton	7.10	5.40	12.50
2240	5,000 S.F., 28.33 ton	8.20	5.25	13.45
2280	10,000 S.F., 56.67 ton	7.70	5.25	12.95
2360	Department stores, 500 S.F., 1.46 ton	7.25	4.01	11.26
2400	1,000 S.F., 2.92 ton	4.85	3.32	8.17
2480	3,000 S.F., 8.75 ton	4.73	2.89	7.62
2520	5,000 S.F., 14.58 ton	3.65	2.78	6.43
2560	10,000 S.F., 29.17 ton	4.22	2.71	6.93
2640	Drug stores, 500 S.F., 3.33 ton	11.05	7.55	18.60
2680	1,000 S.F., 6.67 ton	10.20	7.25	17.45
2720	1,500 S.F., 10.00 ton	10.80	6.60	17.40
2760	3,000 S.F., 20.00 ton	9.40	6.25	15.65
2800	5,000 S.F., 33.33 ton	9.65	6.20	15.85
2840	10,000 S.F., 66.67 ton	9.05	6.20	15.25
2920	Factories, 500 S.F., 1.67 ton	8.25	4.58	12.83
3000	1,500 S.F., 5.00 ton	5.10	3.63	8.73
3040	3,000 S.F., 10.00 ton	5.40	3.31	8.71
3080	5,000 S.F., 16.67 ton	4.16	3.18	7.34
3120	10,000 S.F., 33.33 ton	4.82	3.09	7.91
3200	Food supermarkets, 500 S.F., 1.42 ton	7.05	3.89	10.94

Important: See the Reference Section for critical supporting data - Location Factors & Historical Cost Indexes

D3030 Cooling Generating Systems

D3030 202	Rooftop Single Zone Unit Systems	COST PER S.F.		
		MAT.	INST.	TOTAL
3240	1,000 S.F., 2.83 ton	4.66	3.20	7.86
3280	1,500 S.F., 4.25 ton	4.34	3.09	7.43
3320	3,000 S.F., 8.50 ton	4.59	2.82	7.41
3360	5,000 S.F., 14.17 ton	3.54	2.71	6.25
3400	10,000 S.F., 28.33 ton	4.10	2.63	6.73
3480	Medical centers, 500 S.F., 1.17 ton	5.80	3.21	9.01
3520	1,000 S.F., 2.33 ton	5.80	3.20	9
3560	1,500 S.F., 3.50 ton	3.88	2.65	6.53
3640	5,000 S.F., 11.67 ton	3.77	2.32	6.09
3680	10,000 S.F., 23.33 ton	3.29	2.18	5.47
3760	Offices, 500 S.F., 1.58 ton	7.90	4.34	12.24
3800	1,000 S.F., 3.17 ton	5.25	3.60	8.85
3840	1,500 S.F., 4.75 ton	4.84	3.46	8.30
3880	3,000 S.F., 9.50 ton	5.15	3.15	8.30
3920	5,000 S.F., 15.83 ton	3.96	3.01	6.97
3960	10,000 S.F., 31.67 ton	4.58	2.94	7.52
4000	Restaurants, 500 S.F., 2.50 ton	12.45	6.85	19.30
4040	1,000 S.F., 5.00 ton	7.65	5.45	13.10
4080	1,500 S.F., 7.50 ton	8.10	4.96	13.06
4120	3,000 S.F., 15.00 ton	6.25	4.77	11.02
4160	5,000 S.F., 25.00 ton	7.05	4.68	11.73
4200	10,000 S.F., 50.00 ton	6.80	4.64	11.44
4240	Schools and colleges, 500 S.F., 1.92 ton	9.55	5.25	14.80
4280	1,000 S.F., 3.83 ton	6.35	4.35	10.70
4360	3,000 S.F., 11.50 ton	6.20	3.81	10.01
4400	5,000 S.F., 19.17 ton	5.40	3.59	8.99

D3030 Cooling Generating Systems

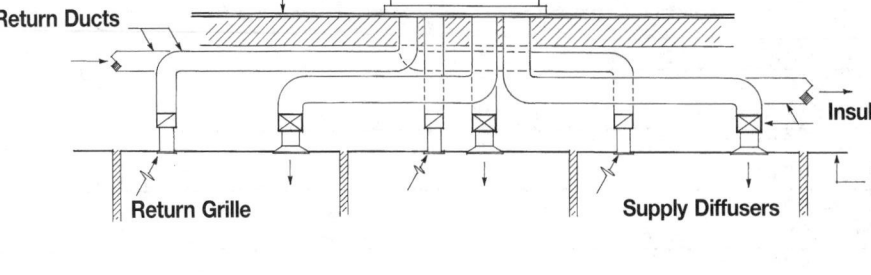

Roof

Rooftop Unit

Return Ducts

Insulated Supply Ducts

Finish Ceiling

Return Grille

Supply Diffusers

*Note A: Small single zone unit recommended.

*Note B: A combination of multizone units recommended.

D3030 206	Rooftop Multizone Unit Systems	COST PER S.F.		
		MAT.	INST.	TOTAL
1240	Rooftop, multizone, air conditioner			
1260	Apartment corridors, 1,500 S.F., 2.75 ton. See Note A.			
1280	3,000 S.F., 5.50 ton	8.60	3.24	11.84
1320	10,000 S.F., 18.30 ton	6.35	3.13	9.48
1360	15,000 S.F., 27.50 ton	5.50	3.11	8.61
1400	20,000 S.F., 36.70 ton	5.55	3.11	8.66
1440	25,000 S.F., 45.80 ton	5.55	3.13	8.68
1520	Banks or libraries, 1,500 S.F., 6.25 ton	19.55	7.35	26.90
1560	3,000 S.F., 12.50 ton	16.50	7.20	23.70
1600	10,000 S.F., 41.67 ton	12.65	7.15	19.80
1640	15,000 S.F., 62.50 ton	8.45	7.05	15.50
1680	20,000 S.F., 83.33 ton	8.45	7.10	15.55
1720	25,000 S.F., 104.00 ton	7.95	7.05	15
1800	Bars and taverns, 1,500 S.F., 16.62 ton	42	10.60	52.60
1840	3,000 S.F., 33.24 ton	32	10.25	42.25
1880	10,000 S.F., 110.83 ton	19.40	10.20	29.60
1920	15,000 S.F., 165 ton, See Note B			
1960	20,000 S.F., 220 ton, See Note B			
2000	25,000 S.F., 275 ton, See Note B			
2080	Bowling alleys, 1,500 S.F., 8.50 ton	26.50	10	36.50
2120	3,000 S.F., 17.00 ton	22.50	9.80	32.30
2160	10,000 S.F., 56.70 ton	17.20	9.70	26.90
2200	15,000 S.F., 85.00 ton	11.50	9.60	21.10
2240	20,000 S.F., 113.00 ton	10.80	9.55	20.35
2280	25,000 S.F., 140.00 ton see Note B			
2360	Department stores, 1,500 S.F., 4.37 ton, See Note A.			
2400	3,000 S.F., 8.75 ton	13.70	5.15	18.85
2440	10,000 S.F., 29.17 ton	8.85	4.95	13.80
2520	20,000 S.F., 58.33 ton	5.90	4.95	10.85
2560	25,000 S.F., 72.92 ton	5.90	4.95	10.85
2640	Drug stores, 1,500 S.F., 10.00 ton	31.50	11.80	43.30
2680	3,000 S.F., 20.00 ton	23	11.35	34.35
2720	10,000 S.F., 66.66 ton	13.55	11.30	24.85
2760	15,000 S.F., 100.00 ton	12.75	11.30	24.05
2800	20,000 S.F., 135 ton, See Note B			
2840	25,000 S.F., 165 ton, See Note B			
2920	Factories, 1,500 S.F., 5 ton, See Note A			
3000	10,000 S.F., 33.33 ton	10.10	5.65	15.75
3040	15,000 S.F., 50.00 ton	10.10	5.70	15.80
3080	20,000 S.F., 66.66 ton	6.75	5.65	12.40

Important: See the Reference Section for critical supporting data - Location Factors & Historical Cost Indexes

D3030 Cooling Generating Systems

D3030 206	Rooftop Multizone Unit Systems	COST PER S.F.		
		MAT.	INST.	TOTAL
3120	25,000 S.F., 83.33 ton	6.75	5.65	12.40
3200	Food supermarkets, 1,500 S.F., 4.25 ton, See Note A			
3240	3,000 S.F., 8.50 ton	13.30	5	18.30
3280	10,000 S.F., 28.33 ton	8.45	4.81	13.26
3320	15,000 S.F., 42.50 ton	8.60	4.84	13.44
3360	20,000 S.F., 56.67 ton	5.75	4.81	10.56
3400	25,000 S.F., 70.83 ton	5.75	4.81	10.56
3480	Medical centers, 1,500 S.F., 3.5 ton, See Note A			
3520	3,000 S.F., 7.00 ton	10.95	4.13	15.08
3560	10,000 S.F., 23.33 ton	7.40	3.97	11.37
3600	15,000 S.F., 35.00 ton	7.05	3.97	11.02
3640	20,000 S.F., 46.66 ton	7.10	3.99	11.09
3680	25,000 S.F., 58.33 ton	4.74	3.96	8.70
3760	Offices, 1,500 S.F., 4.75 ton, See Note A			
3800	3,000 S.F., 9.50 ton	14.85	5.60	20.45
3840	10,000 S.F., 31.66 ton	9.60	5.40	15
3920	20,000 S.F., 63.33 ton	6.40	5.40	11.80
3960	25,000 S.F., 79.16 ton	6.40	5.40	11.80
4000	Restaurants, 1,500 S.F., 7.50 ton	23.50	8.85	32.35
4040	3,000 S.F., 15.00 ton	19.80	8.65	28.45
4080	10,000 S.F., 50.00 ton	15.15	8.55	23.70
4120	15,000 S.F., 75.00 ton	10.15	8.50	18.65
4160	20,000 S.F., 100.00 ton	9.60	8.45	18.05
4200	25,000 S.F., 125 ton, See Note B			
4240	Schools and colleges, 1,500 S.F., 5.75 ton	18	6.80	24.80
4320	10,000 S.F., 38.33 ton	11.60	6.55	18.15
4360	15,000 S.F., 57.50 ton	7.80	6.50	14.30
4400	20,000 S.F., 76.66 ton	7.75	6.50	14.25
4440	25,000 S.F., 95.83 ton	7.35	6.50	13.85

SERVICES

D

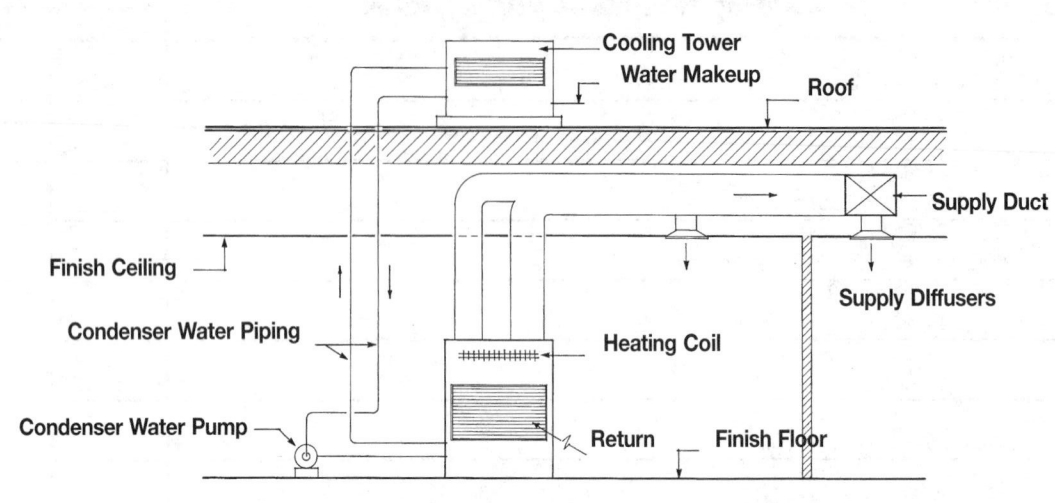

Self-Contained Water Cooled System

D3030 210	Self-contained, Water Cooled Unit Systems	COST PER S.F.		
		MAT.	INST.	TOTAL
1280	Self-contained, water cooled unit	2.84	2.05	4.89
1300	Apartment corridors, 500 S.F., .92 ton	2.85	2.05	4.90
1320	1,000 S.F., 1.83 ton	2.84	2.03	4.87
1360	3,000 S.F., 5.50 ton	2.26	1.71	3.97
1400	5,000 S.F., 9.17 ton	2.33	1.59	3.92
1440	10,000 S.F., 18.33 ton	1.96	1.43	3.39
1520	Banks or libraries, 500 S.F., 2.08 ton	6.25	2.06	8.31
1560	1,000 S.F., 4.17 ton	5.15	3.89	9.04
1600	3,000 S.F., 12.50 ton	5.30	3.62	8.92
1640	5,000 S.F., 20.80 ton	4.47	3.24	7.71
1680	10,000 S.F., 41.66 ton	3.88	3.25	7.13
1760	Bars and taverns, 500 S.F., 5.54 ton	12.95	3.46	16.41
1800	1,000 S.F., 11.08 ton	13.70	6.15	19.85
1840	3,000 S.F., 33.25 ton	10.35	4.90	15.25
1880	5,000 S.F., 55.42 ton	9.75	5.10	14.85
1920	10,000 S.F., 110.00 ton	9.55	5.05	14.60
2000	Bowling alleys, 500 S.F., 2.83 ton	8.50	2.80	11.30
2040	1,000 S.F., 5.66 ton	7	5.30	12.30
2080	3,000 S.F., 17.00 ton	6.10	4.42	10.52
2120	5,000 S.F., 28.33 ton	5.50	4.26	9.76
2160	10,000 S.F., 56.66 ton	5.15	4.38	9.53
2200	Department stores, 500 S.F., 1.46 ton	4.38	1.44	5.82
2240	1,000 S.F., 2.92 ton	3.60	2.73	6.33
2280	3,000 S.F., 8.75 ton	3.71	2.53	6.24
2320	5,000 S.F., 14.58 ton	3.13	2.27	5.40
2360	10,000 S.F., 29.17 ton	2.83	2.19	5.02
2440	Drug stores, 500 S.F., 3.33 ton	10	3.29	13.29
2480	1,000 S.F., 6.66 ton	8.20	6.25	14.45
2520	3,000 S.F., 20.00 ton	7.15	5.20	12.35
2560	5,000 S.F., 33.33 ton	7.50	5.15	12.65
2600	10,000 S.F., 66.66 ton	6.05	5.15	11.20
2680	Factories, 500 S.F., 1.66 ton	4.98	1.65	6.63
2720	1,000 S.F. 3.37 ton	4.15	3.16	7.31
2760	3,000 S.F., 10.00 ton	4.24	2.91	7.15
2800	5,000 S.F., 16.66 ton	3.57	2.61	6.18
2840	10,000 S.F., 33.33 ton	3.22	2.52	5.74
2920	Food supermarkets, 500 S.F., 1.42 ton	4.25	1.40	5.65
2960	1,000 S.F., 2.83 ton	4.40	3.16	7.56

Important: See the Reference Section for critical supporting data - Location Factors & Historical Cost Indexes

D30 HVAC

D3030 Cooling Generating Systems

D3030 210	Self-contained, Water Cooled Unit Systems	COST PER S.F.		
		MAT.	INST.	TOTAL
3000	3,000 S.F., 8.50 ton	3.49	2.65	6.14
3040	5,000 S.F., 14.17 ton	3.60	2.47	6.07
3080	10,000 S.F., 28.33 ton	2.74	2.14	4.88
3160	Medical centers, 500 S.F., 1.17 ton	3.49	1.15	4.64
3200	1,000 S.F., 2.33 ton	3.62	2.60	6.22
3240	3,000 S.F., 7.00 ton	2.88	2.18	5.06
3280	5,000 S.F., 11.66 ton	2.97	2.02	4.99
3320	10,000 S.F., 23.33 ton	2.51	1.82	4.33
3400	Offices, 500 S.F., 1.58 ton	4.74	1.56	6.30
3440	1,000 S.F., 3.17 ton	4.91	3.52	8.43
3480	3,000 S.F., 9.50 ton	4.02	2.74	6.76
3520	5,000 S.F., 15.83 ton	3.39	2.46	5.85
3560	10,000 S.F., 31.67 ton	3.06	2.38	5.44
3640	Restaurants, 500 S.F., 2.50 ton	7.50	2.47	9.97
3680	1,000 S.F., 5.00 ton	6.15	4.68	10.83
3720	3,000 S.F., 15.00 ton	6.35	4.35	10.70
3760	5,000 S.F., 25.00 ton	5.35	3.90	9.25
3800	10,000 S.F., 50.00 ton	3.27	3.59	6.86
3880	Schools and colleges, 500 S.F., 1.92 ton	5.75	1.89	7.64
3920	1,000 S.F., 3.83 ton	4.72	3.58	8.30
3960	3,000 S.F., 11.50 ton	4.87	3.33	8.20
4000	5,000 S.F., 19.17 ton	4.11	2.99	7.10
4040	10,000 S.F., 38.33 ton	3.57	2.99	6.56

D3030 Cooling Generating Systems

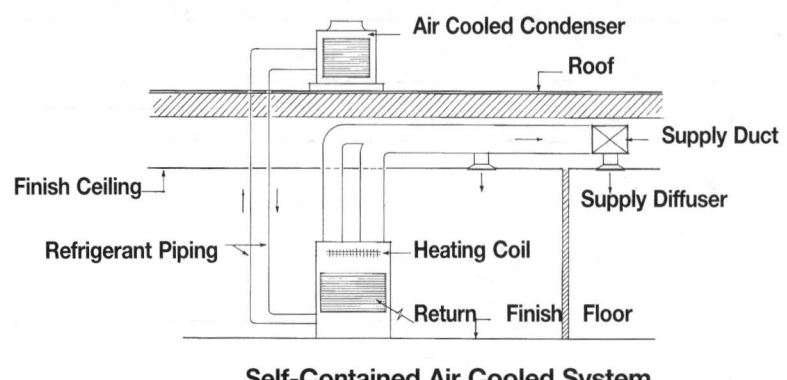

Self-Contained Air Cooled System

D3030 214	Self-contained, Air Cooled Unit Systems	COST PER S.F.		
		MAT.	INST.	TOTAL
1300	Self-contained, air cooled unit			
1320	Apartment corridors, 500 S.F., .92 ton	4.45	2.60	7.05
1360	1,000 S.F., 1.83 ton	4.41	2.58	6.99
1400	3,000 S.F., 5.50 ton	3.60	2.41	6.01
1440	5,000 S.F., 9.17 ton	3.08	2.30	5.38
1480	10,000 S.F., 18.33 ton	2.55	2.12	4.67
1560	Banks or libraries, 500 S.F., 2.08 ton	9.75	3.28	13.03
1600	1,000 S.F., 4.17 ton	8.15	5.50	13.65
1640	3,000 S.F., 12.50 ton	7	5.25	12.25
1680	5,000 S.F., 20.80 ton	5.80	4.80	10.60
1720	10,000 S.F., 41.66 ton	5.30	4.71	10.01
1800	Bars and taverns, 500 S.F., 5.54 ton	17.80	7.15	24.95
1840	1,000 S.F., 11.08 ton	18.30	10.45	28.75
1880	3,000 S.F., 33.25 ton	13.95	9.10	23.05
1920	5,000 S.F., 55.42 ton	13.40	9.10	22.50
1960	10,000 S.F., 110.00 ton	13.55	9.05	22.60
2040	Bowling alleys, 500 S.F., 2.83 ton	13.35	4.49	17.84
2080	1,000 S.F., 5.66 ton	11.10	7.50	18.60
2120	3,000 S.F., 17.00 ton	7.90	6.55	14.45
2160	5,000 S.F., 28.33 ton	7.25	6.40	13.65
2200	10,000 S.F., 56.66 ton	7	6.40	13.40
2240	Department stores, 500 S.F., 1.46 ton	6.85	2.31	9.16
2280	1,000 S.F., 2.92 ton	5.75	3.85	9.60
2320	3,000 S.F., 8.75 ton	4.90	3.66	8.56
2360	5,000 S.F., 14.58 ton	4.90	3.66	8.56
2400	10,000 S.F., 29.17 ton	3.75	3.29	7.04
2480	Drug stores, 500 S.F., 3.33 ton	15.65	5.30	20.95
2520	1,000 S.F., 6.66 ton	13.05	8.80	21.85
2560	3,000 S.F., 20.00 ton	9.30	7.70	17
2600	5,000 S.F., 33.33 ton	8.55	7.50	16.05
2640	10,000 S.F., 66.66 ton	8.30	7.55	15.85
2720	Factories, 500 S.F., 1.66 ton	7.95	2.66	10.61
2760	1,000 S.F., 3.33 ton	6.60	4.41	11.01
2800	3,000 S.F., 10.00 ton	5.60	4.19	9.79
2840	5,000 S.F., 16.66 ton	4.61	3.85	8.46
2880	10,000 S.F., 33.33 ton	4.27	3.76	8.03
2960	Food supermarkets, 500 S.F., 1.42 ton	6.70	2.25	8.95
3000	1,000 S.F., 2.83 ton	6.80	4	10.80
3040	3,000 S.F., 8.50 ton	5.55	3.74	9.29
3080	5,000 S.F., 14.17 ton	4.76	3.56	8.32
3120	10,000 S.F., 28.33 ton	3.64	3.21	6.85
3200	Medical centers, 500 S.F., 1.17 ton	5.55	1.86	7.41

Important: See the Reference Section for critical supporting data - Location Factors & Historical Cost Indexes

D30 HVAC

D3030 Cooling Generating Systems

D3030 214	Self-contained, Air Cooled Unit Systems	COST PER S.F.		
		MAT.	INST.	TOTAL
3240	1,000 S.F., 2.33 ton	5.65	3.31	8.96
3280	3,000 S.F., 7.00 ton	4.59	3.08	7.67
3320	5,000 S.F., 16.66 ton	3.94	2.93	6.87
3360	10,000 S.F., 23.33 ton	3.26	2.70	5.96
3440	Offices, 500 S.F., 1.58 ton	7.45	2.51	9.96
3480	1,000 S.F., 3.16 ton	7.65	4.48	12.13
3520	3,000 S.F., 9.50 ton	5.35	3.98	9.33
3560	5,000 S.F., 15.83 ton	4.42	3.65	8.07
3600	10,000 S.F., 31.66 ton	4.07	3.57	7.64
3680	Restaurants, 500 S.F., 2.50 ton	11.70	3.96	15.66
3720	1,000 S.F., 5.00 ton	9.80	6.60	16.40
3760	3,000 S.F., 15.00 ton	8.40	6.30	14.70
3800	5,000 S.F., 25.00 ton	6.40	5.65	12.05
3840	10,000 S.F., 50.00 ton	6.40	5.65	12.05
3920	Schools and colleges, 500 S.F., 1.92 ton	9	3.03	12.03
3960	1,000 S.F., 3.83 ton	7.60	5.05	12.65
4000	3,000 S.F., 11.50 ton	6.45	4.81	11.26
4040	5,000 S.F., 19.17 ton	5.35	4.42	9.77
4080	10,000 S.F., 38.33 ton	4.88	4.33	9.21

D3030 Cooling Generating Systems

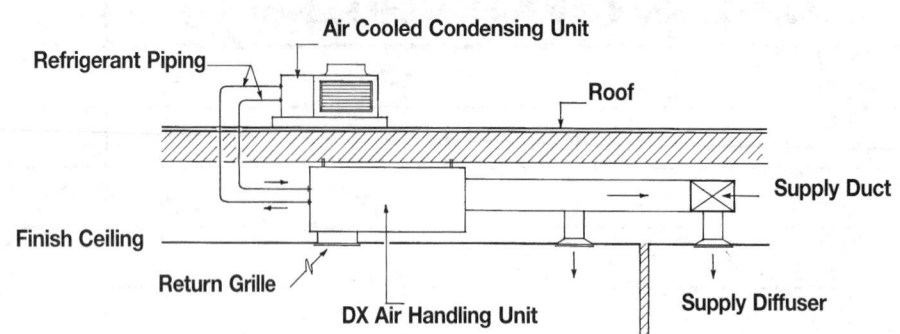

*Cooling requirements would lead to more than one system.

D3030 218	Split Systems With Air Cooled Condensing Units	COST PER S.F.		
		MAT.	INST.	TOTAL
1260	Split system, air cooled condensing unit			
1280	Apartment corridors, 1,000 S.F., 1.83 ton	1.68	1.63	3.31
1320	2,000 S.F., 3.66 ton	1.35	1.64	2.99
1360	5,000 S.F., 9.17 ton	1.80	2.06	3.86
1400	10,000 S.F., 18.33 ton	2.18	2.24	4.42
1440	20,000 S.F., 36.66 ton	1.93	2.28	4.21
1520	Banks and libraries, 1,000 S.F., 4.17 ton	3.10	3.73	6.83
1560	2,000 S.F., 8.33 ton	4.12	4.69	8.81
1600	5,000 S.F., 20.80 ton	4.96	5.10	10.06
1640	10,000 S.F., 41.66 ton	4.40	5.20	9.60
1680	20,000 S.F., 83.32 ton	4.11	5.40	9.51
1760	Bars and taverns, 1,000 S.F., 11.08 ton	10.25	7.35	17.60
1800	2,000 S.F., 22.16 ton	15.30	8.80	24.10
1840	5,000 S.F., 55.42 ton	10.70	8.25	18.95
1880	10,000 S.F., 110.84 ton	9.75	8.55	18.30
1920	20,000 S.F., 220 ton*			
2000	Bowling alleys, 1,000 S.F., 5.66 ton	4.18	7	11.18
2040	2,000 S.F., 11.33 ton	5.60	6.35	11.95
2080	5,000 S.F., 28.33 ton	6.80	6.95	13.75
2120	10,000 S.F., 56.66 ton	6	7.05	13.05
2160	20,000 S.F., 113.32 ton	5.75	7.60	13.35
2320	Department stores, 1,000 S.F., 2.92 ton	2.26	2.56	4.82
2360	2,000 S.F., 5.83 ton	2.16	3.59	5.75
2400	5,000 S.F., 14.58 ton	2.89	3.28	6.17
2440	10,000 S.F., 29.17 ton	3.49	3.57	7.06
2480	20,000 S.F., 58.33 ton	3.08	3.64	6.72
2560	Drug stores, 1,000 S.F., 6.66 ton	4.93	8.20	13.13
2600	2,000 S.F., 13.32 ton	6.60	7.50	14.10
2640	5,000 S.F., 33.33 ton	7.05	8.30	15.35
2680	10,000 S.F., 66.66 ton	6.65	8.65	15.30
2720	20,000 S.F., 133.32 ton*			
2800	Factories, 1,000 S.F., 3.33 ton	2.57	2.92	5.49
2840	2,000 S.F., 6.66 ton	2.46	4.10	6.56
2880	5,000 S.F., 16.66 ton	3.98	4.09	8.07
2920	10,000 S.F., 33.33 ton	3.51	4.15	7.66
2960	20,000 S.F., 66.66 ton	3.34	4.34	7.68
3040	Food supermarkets, 1,000 S.F., 2.83 ton	2.18	2.49	4.67
3080	2,000 S.F., 5.66 ton	2.10	3.49	5.59
3120	5,000 S.F., 14.66 ton	2.81	3.18	5.99
3160	10,000 S.F., 28.33 ton	3.39	3.47	6.86
3200	20,000 S.F., 56.66 ton	2.99	3.53	6.52
3280	Medical centers, 1,000 S.F., 2.33 ton	1.84	2.01	3.85

Important: See the Reference Section for critical supporting data - Location Factors & Historical Cost Indexes

D3030 Cooling Generating Systems

D3030 218	Split Systems With Air Cooled Condensing Units	COST PER S.F.		
		MAT.	INST.	TOTAL
3320	2,000 S.F., 4.66 ton	1.72	2.88	4.60
3360	5,000 S.F., 11.66 ton	2.31	2.63	4.94
3400	10,000 S.F., 23.33 ton	2.79	2.86	5.65
3440	20,000 S.F., 46.66 ton	2.46	2.91	5.37
3520	Offices, 1,000 S.F., 3.17 ton	2.44	2.78	5.22
3560	2,000 S.F., 6.33 ton	2.34	3.90	6.24
3600	5,000 S.F., 15.83 ton	3.14	3.56	6.70
3640	10,000 S.F., 31.66 ton	3.78	3.88	7.66
3680	20,000 S.F., 63.32 ton	3.16	4.12	7.28
3760	Restaurants, 1,000 S.F., 5.00 ton	3.71	6.15	9.86
3800	2,000 S.F., 10.00 ton	4.95	5.65	10.60
3840	5,000 S.F., 25.00 ton	6	6.15	12.15
3880	10,000 S.F., 50.00 ton	5.30	6.25	11.55
3920	20,000 S.F., 100.00 ton	5.10	6.75	11.85
4000	Schools and colleges, 1,000 S.F., 3.83 ton	2.84	3.43	6.27
4040	2,000 S.F., 7.66 ton	3.80	4.30	8.10
4080	5,000 S.F., 19.17 ton	4.58	4.69	9.27
4120	10,000 S.F., 38.33 ton	4.05	4.78	8.83

SERVICES

D

D30 HVAC

D3090 Other HVAC Systems/Equip

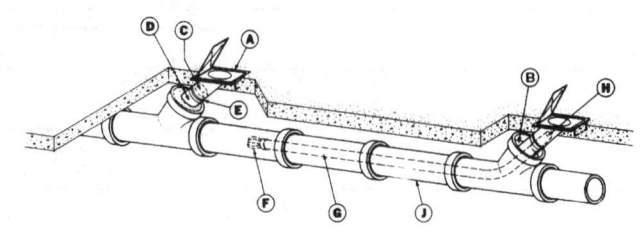

Vitrified Clay Garage Exhaust System

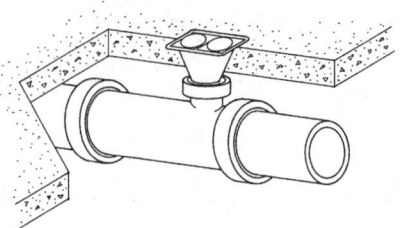

Dual Exhaust System

D3090 320	Garage Exhaust Systems	COST PER BAY		
		MAT.	INST.	TOTAL
1040	Garage, single exhaust, 3" outlet, cars & light trucks, one bay	1,675	865	2,540
1060	Additional bays up to seven bays	310	126	436
1500	4" outlet, trucks, one bay	1,700	865	2,565
1520	Additional bays up to six bays	315	126	441
1600	5" outlet, diesel trucks, one bay	1,925	865	2,790
1650	Additional single bays up to six	555	146	701
1700	Two adjoining bays	1,925	865	2,790
2000	Dual exhaust, 3" outlets, pair of adjoining bays	1,975	965	2,940
2100	Additional pairs of adjoining bays	575	146	721

SERVICES D

Important: See the Reference Section for critical supporting data - Location Factors & Historical Cost Indexes

Reference-System Classification

System Classification
Rules for installation of sprinkler systems vary depending on the classification of occupancy falling into one of three categories as follows:

Light Hazard Occupancy
The protection area allotted per sprinkler should not exceed 225 S.F., with the maximum distance between lines and sprinklers on lines being 15'. The sprinklers do not need to be staggered. Branch lines should not exceed eight sprinklers on either side of a cross main. Each large area requiring more than 100 sprinklers and without a sub-dividing partition should be supplied by feed mains or risers sized for ordinary hazard occupancy.
Maximum system area = 52,000 S.F.

Included in this group are:
Churches	Nursing Homes
Clubs	Offices
Educational	Residential
Hospitals	Restaurants
Institutional	Theaters and Auditoriums
Libraries	(except stages and prosceniums)
(except large stack rooms)	Unused Attics
Museums	

Ordinary Hazard Occupancy
The protection area allotted per sprinkler shall not exceed 130 S.F. of noncombustible ceiling and 130 S.F. of combustible ceiling. The maximum allowable distance between sprinkler lines and sprinklers on line is 15'. Sprinklers shall be staggered if the distance between heads exceeds 12'. Branch lines should not exceed eight sprinklers on either side of a cross main.
Maximum system area = 52,000 S.F.

Included in this group are:
Group 1	Group 2
Automotive Parking and Showrooms	Cereal Mills
Bakeries	Chemical Plants—Ordinary
Beverage manufacturing	Confectionery Products
Canneries	Distilleries
Dairy Products Manufacturing/Processing	Dry Cleaners
Electronic Plans	Feed Mills
Glass and Glass Products Manufacturing	Horse Stables
Laundries	Leather Goods Manufacturing
Restaurant Service Areas	Libraries—Large Stack Room Areas
	Machine Shops
	Metal Working
	Mercantile
	Paper and Pulp Mills
	Paper Process Plants
	Piers and Wharves
	Post Offices
	Printing and Publishing
	Repair Garages
	Stages
	Textile Manufacturing
	Tire Manufacturing
	Tobacco Products Manufacturing
	Wood Machining
	Wood Product Assembly

Extra Hazard Occupancy
The protection area allotted per sprinkler shall not exceed 100 S.F. of noncombustible ceiling and 100 S.F. of combustible ceiling. The maximum allowable distance between lines and between sprinklers on lines is 12'. Sprinklers on alternate lines shall be staggered if the distance between sprinklers on lines exceeds 8'. Branch lines should not exceed six sprinklers on either side of a cross main.
Maximum system area:
   Design by pipe schedule = 25,000 S.F.
   Design by hydraulic calculation = 40,000 S.F.

Included in this group are:
Group 1	Group 2
Aircraft hangars	Asphalt Saturating
Combustible Hydraulic Fluid Use Area	Flammable Liquids Spraying
Die Casting	Flow Coating
Metal Extruding	Manufactured/Modular Home Building Assemblies (where finished enclosure is present and has combustible interiors)
Plywood/Particle Board Manufacturing	
Printing (inkls with flash points < 100 degrees F	
Rubber Reclaiming, Compounding, Drying, Milling, Vulcanizing	Open Oil Quenching
Saw Mills	Plastics Processing
Textile Picking, Opening, Blending, Garnetting, Carding, Combing of Cotton, Synthetics, Wood Shoddy, or Burlap	Solvent Cleaning
Upholstering with Plastic Foams	Varnish and Paint Dipping

SERVICES

D

387

Reference-Sprinkler Systems (Automatic)

Sprinkler systems may be classified by type as follows:

1. **Wet Pipe System.** A system employing automatic sprinklers attached to a piping system containing water and connected to a water supply so that water discharges immediately from sprinklers opened by a fire.

2. **Dry Pipe System.** A system employing automatic sprinklers attached to a piping system containing air under pressure, the release of which as from the opening of sprinklers permits the water pressure to open a valve known as a "dry pipe valve". The water then flows into the piping system and out the opened sprinklers.

3. **Pre-Action System.** A system employing automatic sprinklers attached to a piping system containing air that may or may not be under pressure, with a supplemental heat responsive system of generally more sensitive characteristics than the automatic sprinklers themselves, installed in the same areas as the sprinklers; actuation of the heat responsive system, as from a fire, opens a valve which permits water to flow into the sprinkler piping system and to be discharged from any sprinklers which may be open.

4. **Deluge System.** A system employing open sprinklers attached to a piping system connected to a water supply through a valve which is opened by the operation of a heat responsive system installed in the same areas as the sprinklers. When this valve opens, water flows into the piping system and discharges from all sprinklers attached thereto.

5. **Combined Dry Pipe and Pre-Action Sprinkler System.** A system employing automatic sprinklers attached to a piping system containing air under pressure with a supplemental heat responsive system of generally more sensitive characteristics than the automatic sprinklers themselves, installed in the same areas as the sprinklers; operation of the heat responsive system, as from a fire, actuates tripping devices which open dry pipe valves simultaneously and without loss of air pressure in the system. Operation of the heat responsive system also opens approved air exhaust valves at the end of the feed main which facilitates the filling of the system with water which usually precedes the opening of sprinklers. The heat responsive system also serves as an automatic fire alarm system.

6. **Limited Water Supply System.** A system employing automatic sprinklers and conforming to these standards but supplied by a pressure tank of limited capacity.

7. **Chemical Systems.** Systems using halon, carbon dioxide, dry chemical or high expansion foam as selected for special requirements. Agent may extinguish flames by chemically inhibiting flame propagation, suffocate flames by excluding oxygen, interrupting chemical action of oxygen uniting with fuel or sealing and cooling the combustion center.

8. **Firecycle System.** Firecycle is a fixed fire protection sprinkler system utilizing water as its extinguishing agent. It is a time delayed, recycling, preaction type which automatically shuts the water off when heat is reduced below the detector operating temperature and turns the water back on when that temperature is exceeded. The system senses a fire condition through a closed circuit electrical detector system which controls water flow to the fire automatically. Batteries supply up to 90 hour emergency power supply for system operation. The piping system is dry (until water is required) and is monitored with pressurized air. Should any leak in the system piping occur, an alarm will sound, but water will not enter the system until heat is sensed by a firecycle detector.

Area coverage sprinkler systems may be laid out and fed from the supply in any one of several patterns as shown below. It is desirable, if possible, to utilize a central feed and achieve a shorter flow path from the riser to the furthest sprinkler. This permits use of the smallest sizes of pipe possible with resulting savings.

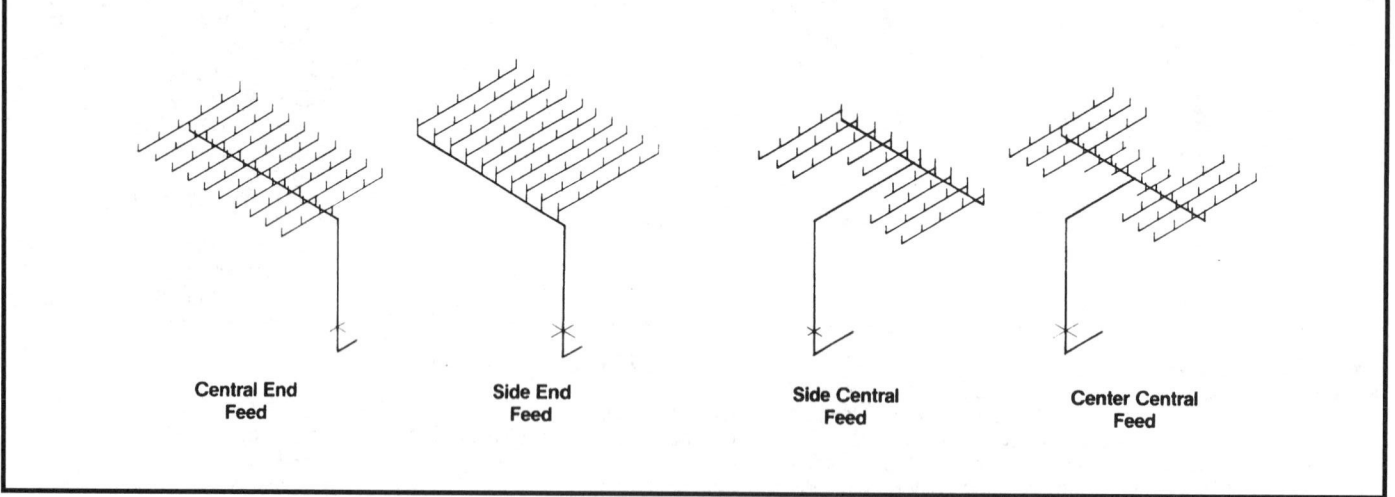

Central End Feed   Side End Feed   Side Central Feed   Center Central Feed

Important: See the Reference Section for critical supporting data - Location Factors & Historical Cost Indexes

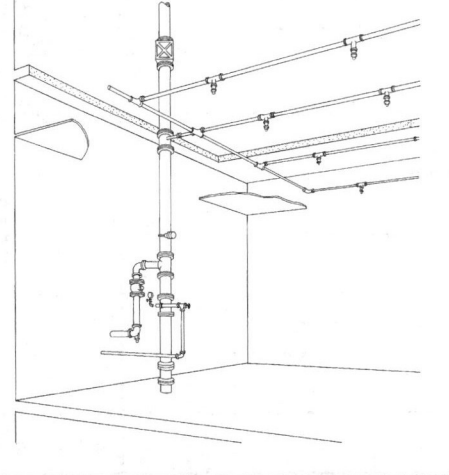

Wet Pipe System. A system employing automatic sprinklers attached to a piping system containing water and connected to a water supply so that water discharges immediately from sprinklers opened by heat from a fire.

All areas are assumed to be open.

D4010 305	Wet Pipe Sprinkler Systems	COST PER S.F.		
		MAT.	INST.	TOTAL
0520	Wet pipe sprinkler systems, steel, black, sch. 40 pipe			
0530	Light hazard, one floor, 500 S.F.	1.16	1.96	3.12
0560	1000 S.F.	1.76	2.05	3.81
0580	2000 S.F.	1.83	2.09	3.92
0600	5000 S.F.	.87	1.47	2.34
0620	10,000 S.F.	.57	1.22	1.79
0640	50,000 S.F.	.36	1.14	1.50
0660	Each additional floor, 500 S.F.	.57	1.67	2.24
0680	1000 S.F.	.55	1.57	2.12
0700	2000 S.F.	.49	1.42	1.91
0720	5000 S.F.	.34	1.21	1.55
0740	10,000 S.F.	.33	1.12	1.45
0760	50,000 S.F.	.26	.90	1.16
1000	Ordinary hazard, one floor, 500 S.F.	1.26	2.10	3.36
1020	1000 S.F.	1.72	2.04	3.76
1040	2000 S.F.	1.90	2.18	4.08
1060	5000 S.F.	.99	1.55	2.54
1080	10,000 S.F.	.69	1.61	2.30
1100	50,000 S.F.	.63	1.61	2.24
1140	Each additional floor, 500 S.F.	.68	1.88	2.56
1160	1000 S.F.	.51	1.56	2.07
1180	2000 S.F.	.57	1.57	2.14
1200	5000 S.F.	.57	1.47	2.04
1220	10,000 S.F.	.45	1.52	1.97
1240	50,000 S.F.	.48	1.43	1.91
1500	Extra hazard, one floor, 500 S.F.	3.29	3.28	6.57
1520	1000 S.F.	2.13	2.87	5
1540	2000 S.F.	2.01	2.92	4.93
1560	5000 S.F.	1.23	2.56	3.79
1580	10,000 S.F.	1.18	2.40	3.58
1600	50,000 S.F.	1.24	2.34	3.58
1660	Each additional floor, 500 S.F.	.87	2.35	3.22
1680	1000 S.F.	.84	2.25	3.09
1700	2000 S.F.	.75	2.25	3
1720	5000 S.F.	.61	1.99	2.60
1740	10,000 S.F.	.75	1.82	2.57
1760	50,000 S.F.	.74	1.75	2.49
2020	Grooved steel, black sch. 40 pipe, light hazard, one floor, 2000 S.F.	1.86	1.76	3.62
2060	10,000 S.F.	.71	1.09	1.80
2100	Each additional floor, 2000 S.F.	.53	1.15	1.68

SERVICES

D

D4010 305	Wet Pipe Sprinkler Systems	COST PER S.F.		
		MAT.	INST.	TOTAL
2150	10,000 S.F.	.34	.94	1.28
2200	Ordinary hazard, one floor, 2000 S.F.	1.89	1.87	3.76
2250	10,000 S.F.	.68	1.34	2.02
2300	Each additional floor, 2000 S.F.	.56	1.26	1.82
2350	10,000 S.F.	.44	1.25	1.69
2400	Extra hazard, one floor, 2000 S.F.	2.04	2.40	4.44
2450	10,000 S.F.	.95	1.76	2.71
2500	Each additional floor, 2000 S.F.	.80	1.85	2.65
2550	10,000 S.F.	.65	1.57	2.22
3050	Grooved steel black sch. 10 pipe, light hazard, one floor, 2000 S.F.	1.83	1.75	3.58
3100	10,000 S.F.	.56	1.03	1.59
3150	Each additional floor, 2000 S.F.	.50	1.13	1.63
3200	10,000 S.F.	.32	.93	1.25
3250	Ordinary hazard, one floor, 2000 S.F.	1.86	1.86	3.72
3300	10,000 S.F.	.66	1.30	1.96
3350	Each additional floor, 2000 S.F.	.53	1.25	1.78
3400	10,000 S.F.	.42	1.21	1.63
3450	Extra hazard, one floor, 2000 S.F.	2.01	2.38	4.39
3500	10,000 S.F.	.91	1.72	2.63
3550	Each additional floor, 2000 S.F.	.77	1.83	2.60
3600	10,000 S.F.	.63	1.54	2.17
4050	Copper tubing, type M, light hazard, one floor, 2000 S.F.	1.88	1.73	3.61
4100	10,000 S.F.	.67	1.05	1.72
4150	Each additional floor, 2000 S.F.	.56	1.14	1.70
4200	10,000 S.F.	.43	.95	1.38
4250	Ordinary hazard, one floor, 2000 S.F.	1.95	1.95	3.90
4300	10,000 S.F.	.80	1.24	2.04
4350	Each additional floor, 2000 S.F.	.64	1.27	1.91
4400	10,000 S.F.	.54	1.12	1.66
4450	Extra hazard, one floor, 2000 S.F.	2.14	2.42	4.56
4500	10,000 S.F.	1.48	1.90	3.38
4550	Each additional floor, 2000 S.F.	.90	1.87	2.77
4600	10,000 S.F.	.96	1.70	2.66
5050	Copper tubing, type M, T-drill system, light hazard, one floor			
5060	2000 S.F.	1.87	1.59	3.46
5100	10,000 S.F.	.63	.88	1.51
5150	Each additional floor, 2000 S.F.	.55	1	1.55
5200	10,000 S.F.	.39	.78	1.17
5250	Ordinary hazard, one floor, 2000 S.F.	1.87	1.64	3.51
5300	10,000 S.F.	.75	1.10	1.85
5350	Each additional floor, 2000 S.F.	.54	1.03	1.57
5400	10,000 S.F.	.51	1.01	1.52
5450	Extra hazard, one floor, 2000 S.F.	1.97	1.99	3.96
5500	10,000 S.F.	1.21	1.38	2.59
5550	Each additional floor, 2000 S.F.	.78	1.47	2.25
5600	10,000 S.F.	.69	1.18	1.87

SERVICES D

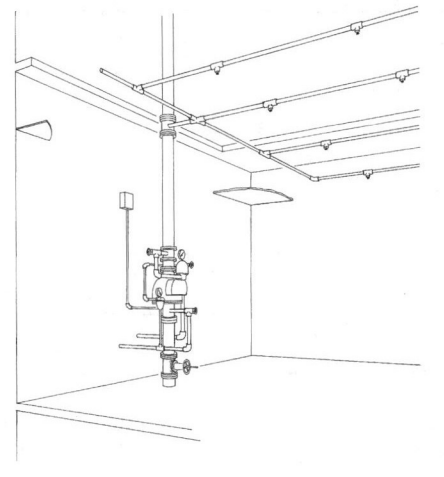

Dry Pipe System: A system employing automatic sprinklers attached to a piping system containing air under pressure, the release of which as from the opening of sprinklers permits the water pressure to open a valve known as a "dry pipe valve". The water then flows into the piping system and out the opened sprinklers.

All areas are assumed to be open.

D4010 400	Dry Pipe Sprinkler Systems	COST PER S.F.		
		MAT.	INST.	TOTAL
0520	Dry pipe sprinkler systems, steel, black, sch. 40 pipe			
0530	Light hazard, one floor, 500 S.F.	4.51	3.87	8.38
0560	1000 S.F.	2.45	2.27	4.72
0580	2000 S.F.	2.18	2.30	4.48
0600	5000 S.F.	1.08	1.57	2.65
0620	10,000 S.F.	.74	1.28	2.02
0640	50,000 S.F.	.49	1.16	1.65
0660	Each additional floor, 500 S.F.	.81	1.86	2.67
0680	1000 S.F.	.68	1.54	2.22
0700	2000 S.F.	.63	1.44	2.07
0720	5000 S.F.	.49	1.23	1.72
0740	10,000 S.F.	.46	1.13	1.59
0760	50,000 S.F.	.40	1.03	1.43
1000	Ordinary hazard, one floor, 500 S.F.	4.57	3.92	8.49
1020	1000 S.F.	2.47	2.31	4.78
1040	2000 S.F.	2.25	2.42	4.67
1060	5000 S.F.	1.24	1.66	2.90
1080	10,000 S.F.	.92	1.68	2.60
1100	50,000 S.F.	.84	1.66	2.50
1140	Each additional floor, 500 S.F.	.87	1.91	2.78
1160	1000 S.F.	.75	1.74	2.49
1180	2000 S.F.	.76	1.60	2.36
1200	5000 S.F.	.68	1.35	2.03
1220	10,000 S.F.	.58	1.34	1.92
1240	50,000 S.F.	.61	1.23	1.84
1500	Extra hazard, one floor, 500 S.F.	6.15	4.89	11.04
1520	1000 S.F.	3.54	3.55	7.09
1540	2000 S.F.	2.44	3.08	5.52
1560	5000 S.F.	1.40	2.33	3.73
1580	10,000 S.F.	1.42	2.20	3.62
1600	50,000 S.F.	1.49	2.15	3.64
1660	Each additional floor, 500 S.F.	1.15	2.38	3.53
1680	1000 S.F.	1.12	2.28	3.40
1700	2000 S.F.	1.03	2.28	3.31
1720	5000 S.F.	.86	2	2.86
1740	10,000 S.F.	1.02	1.82	2.84
1760	50,000 S.F.	1.03	1.77	2.80
2020	Grooved steel, black, sch. 40 pipe, light hazard, one floor, 2000 S.F.	2.16	1.99	4.15
2060	10,000 S.F.	.75	1.10	1.85
2100	Each additional floor, 2000 S.F.	.67	1.17	1.84

D4010 Sprinklers

D4010 400	Dry Pipe Sprinkler Systems	COST PER S.F.		
		MAT.	INST.	TOTAL
2150	10,000 S.F.	.47	.95	1.42
2200	Ordinary hazard, one floor, 2000 S.F.	2.24	2.11	4.35
2250	10,000 S.F.	.91	1.41	2.32
2300	Each additional floor, 2000 S.F.	.75	1.29	2.04
2350	10,000 S.F.	.63	1.27	1.90
2400	Extra hazard, one floor, 2000 S.F.	2.48	2.64	5.12
2450	10,000 S.F.	1.27	1.85	3.12
2500	Each additional floor, 2000 S.F.	1.08	1.88	2.96
2550	10,000 S.F.	.93	1.61	2.54
3050	Grooved steel black sch. 10 pipe, light hazard, one floor, 2000 S.F.	2.13	1.98	4.11
3100	10,000 S.F.	.73	1.09	1.82
3150	Each additional floor, 2000 S.F.	.64	1.15	1.79
3200	10,000 S.F.	.45	.94	1.39
3250	Ordinary hazard, one floor, 2000 S.F.	2.21	2.10	4.31
3300	10,000 S.F.	.89	1.37	2.26
3350	Each additional floor, 2000 S.F.	.72	1.28	2
3400	10,000 S.F.	.61	1.23	1.84
3450	Extra hazard, one floor, 2000 S.F.	2.45	2.62	5.07
3500	10,000 S.F.	1.23	1.81	3.04
3550	Each additional floor, 2000 S.F.	1.05	1.86	2.91
3600	10,000 S.F.	.91	1.58	2.49
4050	Copper tubing, type M, light hazard, one floor, 2000 S.F.	2.18	1.96	4.14
4100	10,000 S.F.	.84	1.11	1.95
4150	Each additional floor, 2000 S.F.	.70	1.16	1.86
4200	10,000 S.F.	.56	.96	1.52
4250	Ordinary hazard, one floor, 2000 S.F.	2.30	2.19	4.49
4300	10,000 S.F.	1.03	1.31	2.34
4350	Each additional floor, 2000 S.F.	.90	1.35	2.25
4400	10,000 S.F.	.73	1.14	1.87
4450	Extra hazard, one floor, 2000 S.F.	2.58	2.66	5.24
4500	10,000 S.F.	1.81	1.99	3.80
4550	Each additional floor, 2000 S.F.	1.18	1.90	3.08
4600	10,000 S.F.	1.24	1.74	2.98
5050	Copper tubing, type M, T-drill system, light hazard, one floor			
5060	2000 S.F.	2.17	1.82	3.99
5100	10,000 S.F.	.80	.94	1.74
5150	Each additional floor, 2000 S.F.	.69	1.02	1.71
5200	10,000 S.F.	.52	.79	1.31
5250	Ordinary hazard, one floor, 2000 S.F.	2.22	1.88	4.10
5300	10,000 S.F.	.98	1.17	2.15
5350	Each additional floor, 2000 S.F.	.73	1.06	1.79
5400	10,000 S.F.	.67	.98	1.65
5450	Extra hazard, one floor, 2000 S.F.	2.41	2.23	4.64
5500	10,000 S.F.	1.54	1.47	3.01
5550	Each additional floor, 2000 S.F.	1.01	1.47	2.48
5600	10,000 S.F.	.97	1.22	2.19

Important: See the Reference Section for critical supporting data - Location Factors & Historical Cost Indexes

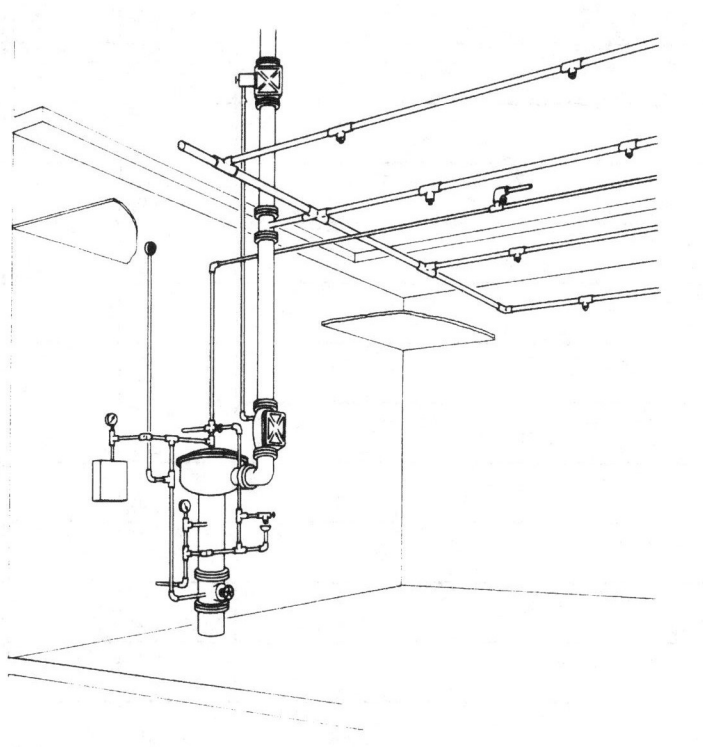

Preaction System: A system employing automatic sprinklers attached to a piping system containing air that may or may not be under pressure, with a supplemental heat responsive system of generally more sensitive characteristics than the automatic sprinklers themselves, installed in the same areas as the sprinklers. Actuation of the heat responsive system, as from a fire, opens a valve which permits water to flow into the sprinkler piping system and to be discharged from those sprinklers which were opened by heat from the fire.

All areas are assumed to be opened.

D4010 415	Preaction Sprinkler Systems	COST PER S.F.		
		MAT.	INST.	TOTAL
0520	Preaction sprinkler systems, steel, black, sch. 40 pipe			
0530	Light hazard, one floor, 500 S.F.	4.47	3.08	7.55
0560	1000 S.F.	2.54	2.32	4.86
0580	2000 S.F.	2.20	2.30	4.50
0600	5000 S.F.	1.11	1.56	2.67
0620	10,000 S.F.	.76	1.28	2.04
0640	50,000 S.F.	.51	1.16	1.67
0660	Each additional floor, 500 S.F.	1	1.66	2.66
0680	1000 S.F.	.76	1.54	2.30
0700	2000 S.F.	.71	1.44	2.15
0720	5000 S.F.	.52	1.22	1.74
0740	10,000 S.F.	.48	1.13	1.61
0760	50,000 S.F.	.45	1.06	1.51
1000	Ordinary hazard, one floor, 500 S.F.	1.72	2.24	3.96
1020	1000 S.F.	2.50	2.31	4.81
1040	2000 S.F.	2.39	2.41	4.80
1060	5000 S.F.	1.23	1.64	2.87
1080	10,000 S.F.	.88	1.67	2.55
1100	50,000 S.F.	.79	1.64	2.43
1140	Each additional floor, 500 S.F.	1.12	1.91	3.03
1160	1000 S.F.	.73	1.58	2.31
1180	2000 S.F.	.68	1.58	2.26
1200	5000 S.F.	.73	1.45	2.18
1220	10,000 S.F.	.60	1.53	2.13
1240	50,000 S.F.	.62	1.44	2.06
1500	Extra hazard, one floor, 500 S.F.	6.10	4.30	10.40
1520	1000 S.F.	3.36	3.25	6.61
1540	2000 S.F.	2.38	3.07	5.45

SERVICES

D

D4010 415	Preaction Sprinkler Systems	COST PER S.F.		
		MAT.	INST.	TOTAL
1560	5000 S.F.	1.41	2.51	3.92
1580	10,000 S.F.	1.36	2.43	3.79
1600	50,000 S.F.	1.40	2.37	3.77
1660	Each additional floor, 500 S.F.	1.31	2.38	3.69
1680	1000 S.F.	1.06	2.27	3.33
1700	2000 S.F.	.97	2.27	3.24
1720	5000 S.F.	.79	2	2.79
1740	10,000 S.F.	.90	1.83	2.73
1760	50,000 S.F.	.87	1.74	2.61
2020	Grooved steel, black, sch. 40 pipe, light hazard, one floor, 2000 S.F.	2.24	1.99	4.23
2060	10,000 S.F.	.77	1.10	1.87
2100	Each additional floor of 2000 S.F.	.75	1.17	1.92
2150	10,000 S.F.	.49	.95	1.44
2200	Ordinary hazard, one floor, 2000 S.F.	2.27	2.10	4.37
2250	10,000 S.F.	.87	1.40	2.27
2300	Each additional floor, 2000 S.F.	.78	1.28	2.06
2350	10,000 S.F.	.59	1.26	1.85
2400	Extra hazard, one floor, 2000 S.F.	2.42	2.63	5.05
2450	10,000 S.F.	1.15	1.82	2.97
2500	Each additional floor, 2000 S.F.	1.02	1.87	2.89
2550	10,000 S.F.	.80	1.58	2.38
3050	Grooved steel, black, sch. 10 pipe light hazard, one floor, 2000 S.F.	2.21	1.98	4.19
3100	10,000 S.F.	.75	1.09	1.84
3150	Each additional floor, 2000 S.F.	.72	1.15	1.87
3200	10,000 S.F.	.47	.94	1.41
3250	Ordinary hazard, one floor, 2000 S.F.	2.21	1.96	4.17
3300	10,000 S.F.	.74	1.35	2.09
3350	Each additional floor, 2000 S.F.	.75	1.27	2.02
3400	10,000 S.F.	.57	1.22	1.79
3450	Extra hazard, one floor, 2000 S.F.	2.39	2.61	5
3500	10,000 S.F.	1.10	1.78	2.88
3550	Each additional floor, 2000 S.F.	.99	1.85	2.84
3600	10,000 S.F.	.78	1.55	2.33
4050	Copper tubing, type M, light hazard, one floor, 2000 S.F.	2.26	1.96	4.22
4100	10,000 S.F.	.86	1.11	1.97
4150	Each additional floor, 2000 S.F.	.78	1.16	1.94
4200	10,000 S.F.	.47	.95	1.42
4250	Ordinary hazard, one floor, 2000 S.F.	2.33	2.18	4.51
4300	10,000 S.F.	.99	1.30	2.29
4350	Each additional floor, 2000 S.F.	.78	1.19	1.97
4400	10,000 S.F.	.63	1.04	1.67
4450	Extra hazard, one floor, 2000 S.F.	2.52	2.65	5.17
4500	10,000 S.F.	1.67	1.96	3.63
4550	Each additional floor, 2000 S.F.	1.12	1.89	3.01
4600	10,000 S.F.	1.11	1.71	2.82
5050	Copper tubing, type M, T-drill system, light hazard, one floor			
5060	2000 S.F.	2.25	1.82	4.07
5100	10,000 S.F.	.82	.94	1.76
5150	Each additional floor, 2000 S.F.	.77	1.02	1.79
5200	10,000 S.F.	.54	.79	1.33
5250	Ordinary hazard, one floor, 2000 S.F.	2.25	1.87	4.12
5300	10,000 S.F.	.94	1.16	2.10
5350	Each additional floor, 2000 S.F.	.77	1.06	1.83
5400	10,000 S.F.	.66	1.02	1.68
5450	Extra hazard, one floor, 2000 S.F.	2.35	2.22	4.57
5500	10,000 S.F.	1.40	1.44	2.84
5550	Each additional floor, 2000 S.F.	.95	1.46	2.41
5600	10,000 S.F.	.84	1.19	2.03

Important: See the Reference Section for critical supporting data - Location Factors & Historical Cost Indexes

D4010 Sprinklers

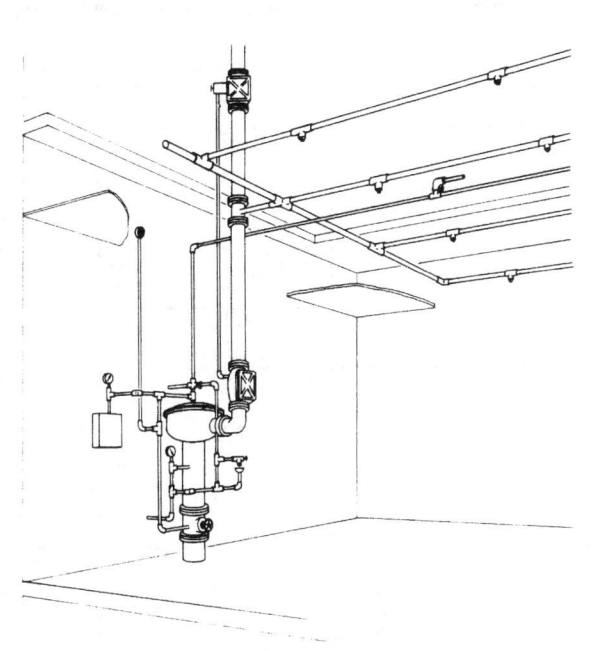

Deluge System: A system employing open sprinklers attached to a piping system connected to a water supply through a valve which is opened by the operation of a heat responsive system installed in the same areas as the sprinklers. When this valve opens, water flows into the piping system and discharges from all sprinklers attached thereto.

D4010 425	Deluge Sprinkler Systems	COST PER S.F.		
		MAT.	INST.	TOTAL
0520	Deluge sprinkler systems, steel, black, sch. 40 pipe			
0530	Light hazard, one floor, 500 S.F.	5.70	3.12	8.82
0560	1000 S.F.	3.14	2.23	5.37
0580	2000 S.F.	2.63	2.31	4.94
0600	5000 S.F.	1.28	1.56	2.84
0620	10,000 S.F.	.84	1.28	2.12
0640	50,000 S.F.	.52	1.16	1.68
0660	Each additional floor, 500 S.F.	1	1.66	2.66
0680	1000 S.F.	.76	1.54	2.30
0700	2000 S.F.	.71	1.44	2.15
0720	5000 S.F.	.52	1.22	1.74
0740	10,000 S.F.	.48	1.13	1.61
0760	50,000 S.F.	.45	1.06	1.51
1000	Ordinary hazard, one floor, 500 S.F.	6.10	3.57	9.67
1020	1000 S.F.	3.12	2.33	5.45
1040	2000 S.F.	2.80	2.42	5.22
1060	5000 S.F.	1.40	1.64	3.04
1080	10,000 S.F.	.96	1.67	2.63
1100	50,000 S.F.	.84	1.68	2.52
1140	Each additional floor, 500 S.F.	1.12	1.91	3.03
1160	1000 S.F.	.73	1.58	2.31
1180	2000 S.F.	.68	1.58	2.26
1200	5000 S.F.	.67	1.33	2
1220	10,000 S.F.	.59	1.36	1.95
1240	50,000 S.F.	.59	1.33	1.92
1500	Extra hazard, one floor, 500 S.F.	7.30	4.34	11.64
1520	1000 S.F.	4.18	3.37	7.55
1540	2000 S.F.	2.79	3.08	5.87
1560	5000 S.F.	1.49	2.31	3.80
1580	10,000 S.F.	1.39	2.23	3.62
1600	50,000 S.F.	1.43	2.18	3.61
1660	Each additional floor, 500 S.F.	1.31	2.38	3.69

SERVICES

D

D4010 425	Deluge Sprinkler Systems	COST PER S.F.		
		MAT.	INST.	TOTAL
1680	1000 S.F.	1.06	2.27	3.33
1700	2000 S.F.	.97	2.27	3.24
1720	5000 S.F.	.79	2	2.79
1740	10,000 S.F.	.92	1.90	2.82
1760	50,000 S.F.	.92	1.86	2.78
2000	Grooved steel, black, sch. 40 pipe, light hazard, one floor			
2020	2000 S.F.	2.66	2	4.66
2060	10,000 S.F.	.86	1.11	1.97
2100	Each additional floor, 2,000 S.F.	.75	1.17	1.92
2150	10,000 S.F.	.49	.95	1.44
2200	Ordinary hazard, one floor, 2000 S.F.	2.27	2.10	4.37
2250	10,000 S.F.	.95	1.40	2.35
2300	Each additional floor, 2000 S.F.	.78	1.28	2.06
2350	10,000 S.F.	.59	1.26	1.85
2400	Extra hazard, one floor, 2000 S.F.	2.83	2.64	5.47
2450	10,000 S.F.	1.24	1.83	3.07
2500	Each additional floor, 2000 S.F.	1.02	1.87	2.89
2550	10,000 S.F.	.80	1.58	2.38
3000	Grooved steel, black, sch. 10 pipe, light hazard, one floor			
3050	2000 S.F.	2.34	1.90	4.24
3100	10,000 S.F.	.83	1.09	1.92
3150	Each additional floor, 2000 S.F.	.72	1.15	1.87
3200	10,000 S.F.	.47	.94	1.41
3250	Ordinary hazard, one floor, 2000 S.F.	2.65	2.10	4.75
3300	10,000 S.F.	.82	1.35	2.17
3350	Each additional floor, 2000 S.F.	.75	1.27	2.02
3400	10,000 S.F.	.57	1.22	1.79
3450	Extra hazard, one floor, 2000 S.F.	2.80	2.62	5.42
3500	10,000 S.F.	1.18	1.78	2.96
3550	Each additional floor, 2000 S.F.	.99	1.85	2.84
3600	10,000 S.F.	.78	1.55	2.33
4000	Copper tubing, type M, light hazard, one floor			
4050	2000 S.F.	2.67	1.97	4.64
4100	10,000 S.F.	.94	1.11	2.05
4150	Each additional floor, 2000 S.F.	.78	1.16	1.94
4200	10,000 S.F.	.47	.95	1.42
4250	Ordinary hazard, one floor, 2000 S.F.	2.74	2.19	4.93
4300	10,000 S.F.	1.07	1.30	2.37
4350	Each additional floor, 2000 S.F.	.78	1.19	1.97
4400	10,000 S.F.	.63	1.04	1.67
4450	Extra hazard, one floor, 2000 S.F.	2.93	2.66	5.59
4500	10,000 S.F.	1.77	1.97	3.74
4550	Each additional floor, 2000 S.F.	1.12	1.89	3.01
4600	10,000 S.F.	1.11	1.71	2.82
5000	Copper tubing, type M, T-drill system, light hazard, one floor			
5050	2000 S.F.	2.66	1.83	4.49
5100	10,000 S.F.	.90	.94	1.84
5150	Each additional floor, 2000 S.F.	.80	1.03	1.83
5200	10,000 S.F.	.54	.79	1.33
5250	Ordinary hazard, one floor, 2000 S.F.	2.66	1.88	4.54
5300	10,000 S.F.	1.02	1.16	2.18
5350	Each additional floor, 2000 S.F.	.76	1.05	1.81
5400	10,000 S.F.	.66	1.02	1.68
5450	Extra hazard, one floor, 2000 S.F.	2.76	2.23	4.99
5500	10,000 S.F.	1.48	1.44	2.92
5550	Each additional floor, 2000 S.F.	.95	1.46	2.41
5600	10,000 S.F.	.84	1.19	2.03

Important: See the Reference Section for critical supporting data - Location Factors & Historical Cost Indexes

D4010 Sprinklers

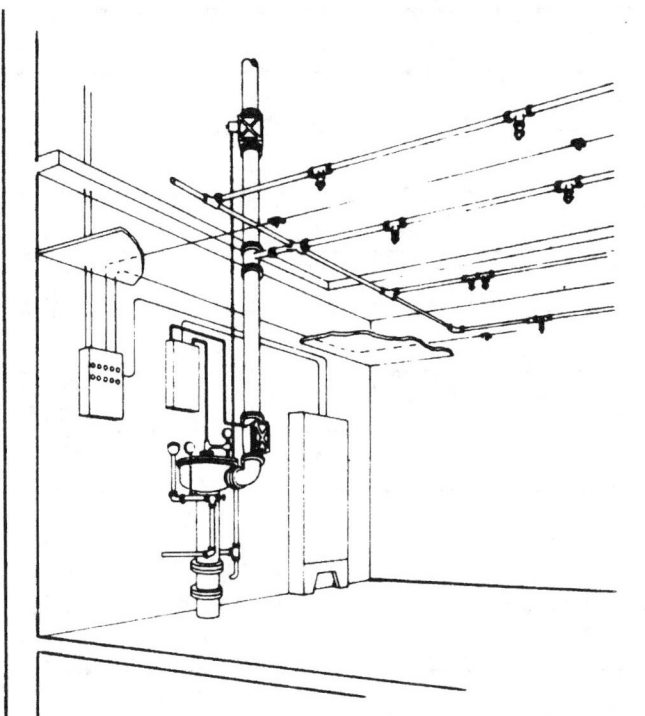

Firecycle is a fixed fire protection sprinkler system utilizing water as its extinguishing agent. It is a time delayed, recycling, preaction type which automatically shuts the water off when heat is reduced below the detector operating temperature and turns the water back on when that temperature is exceeded.

The system senses a fire condition through a closed circuit electrical detector system which controls water flow to the fire automatically. Batteries supply up to 90 hours emergency power supply for system operation. The piping system is dry (until water is required) and is monitored with pressurized air. Shouldany leak in the system piping occur, an alarm will sound, but water will not enter the system until heat is sensed by a Firecycle detector.

D4010 435	Firecycle Sprinkler Systems	COST PER S.F.		
		MAT.	INST.	TOTAL
0520	Firecycle sprinkler systems, steel black sch. 40 pipe			
0530	Light hazard, one floor, 500 S.F.	18.95	6.15	25.10
0560	1000 S.F.	9.75	3.84	13.59
0580	2000 S.F.	5.75	2.83	8.58
0600	5000 S.F.	2.54	1.78	4.32
0620	10,000 S.F.	1.50	1.40	2.90
0640	50,000 S.F.	.68	1.18	1.86
0660	Each additional floor of 500 S.F.	1.03	1.67	2.70
0680	1000 S.F.	.78	1.54	2.32
0700	2000 S.F.	.61	1.43	2.04
0720	5000 S.F.	.53	1.23	1.76
0740	10,000 S.F.	.52	1.14	1.66
0760	50,000 S.F.	.47	1.06	1.53
1000	Ordinary hazard, one floor, 500 S.F.	19.10	6.40	25.50
1020	1000 S.F.	9.70	3.83	13.53
1040	2000 S.F.	5.85	2.94	8.79
1060	5000 S.F.	2.66	1.86	4.52
1080	10,000 S.F.	1.62	1.79	3.41
1100	50,000 S.F.	1.08	1.87	2.95
1140	Each additional floor, 500 S.F.	1.15	1.92	3.07
1160	1000 S.F.	.75	1.58	2.33
1180	2000 S.F.	.78	1.45	2.23
1200	5000 S.F.	.68	1.34	2.02
1220	10,000 S.F.	.58	1.34	1.92
1240	50,000 S.F.	.60	1.27	1.87
1500	Extra hazard, one floor, 500 S.F.	20.50	7.35	27.85
1520	1000 S.F.	10.50	4.76	15.26
1540	2000 S.F.	5.95	3.60	9.55
1560	5000 S.F.	2.75	2.53	5.28
1580	10,000 S.F.	2.08	2.54	4.62

SERVICES

D

D4010 Sprinklers

D4010 435	Firecycle Sprinkler Systems	COST PER S.F.		
		MAT.	INST.	TOTAL
1600	50,000 S.F.	1.71	2.80	4.51
1660	Each additional floor, 500 S.F.	1.34	2.39	3.73
1680	1000 S.F.	1.08	2.27	3.35
1700	2000 S.F.	.99	2.27	3.26
1720	5000 S.F.	.80	2.01	2.81
1740	10,000 S.F.	.94	1.84	2.78
1760	50,000 S.F.	.93	1.80	2.73
2020	Grooved steel, black, sch. 40 pipe, light hazard, one floor			
2030	2000 S.F.	5.80	2.52	8.32
2060	10,000 S.F.	1.63	1.70	3.33
2100	Each additional floor, 2000 S.F.	.77	1.17	1.94
2150	10,000 S.F.	.53	.96	1.49
2200	Ordinary hazard, one floor, 2000 S.F.	5.80	2.63	8.43
2250	10,000 S.F.	1.73	1.60	3.33
2300	Each additional floor, 2000 S.F.	.80	1.28	2.08
2350	10,000 S.F.	.63	1.27	1.90
2400	Extra hazard, one floor, 2000 S.F.	5.95	3.16	9.11
2450	10,000 S.F.	1.87	1.93	3.80
2500	Each additional floor, 2000 S.F.	1.04	1.87	2.91
2550	10,000 S.F.	.84	1.59	2.43
3050	Grooved steel, black, sch. 10 pipe light hazard, one floor,			
3060	2000 S.F.	5.75	2.51	8.26
3100	10,000 S.F.	1.49	1.21	2.70
3150	Each additional floor, 2000 S.F.	.74	1.15	1.89
3200	10,000 S.F.	.51	.95	1.46
3250	Ordinary hazard, one floor, 2000 S.F.	5.80	2.62	8.42
3300	10,000 S.F.	1.59	1.48	3.07
3350	Each additional floor, 2000 S.F.	.77	1.27	2.04
3400	10,000 S.F.	.61	1.23	1.84
3450	Extra hazard, one floor, 2000 S.F.	5.95	3.14	9.09
3500	10,000 S.F.	1.83	1.89	3.72
3550	Each additional floor, 2000 S.F.	1.01	1.85	2.86
3600	10,000 S.F.	.82	1.56	2.38
4060	Copper tubing, type M, light hazard, one floor, 2000 S.F.	5.80	2.49	8.29
4100	10,000 S.F.	1.60	1.23	2.83
4150	Each additional floor, 2000 S.F.	.80	1.16	1.96
4200	10,000 S.F.	.62	.97	1.59
4250	Ordinary hazard, one floor, 2000 S.F.	5.90	2.71	8.61
4300	10,000 S.F.	1.73	1.42	3.15
4350	Each additional floor, 2000 S.F.	.80	1.19	1.99
4400	10,000 S.F.	.66	1.04	1.70
4450	Extra hazard, one floor, 2000 S.F.	6.05	3.18	9.23
4500	10,000 S.F.	2.47	2.10	4.57
4550	Each additional floor, 2000 S.F.	1.14	1.89	3.03
4600	10,000 S.F.	1.15	1.72	2.87
5060	Copper tubing, type M, T-drill system, light hazard, one floor 2000 S.F.	5.80	2.35	8.15
5100	10,000 S.F.	1.56	1.06	2.62
5150	Each additional floor, 2000 S.F.	.88	1.08	1.96
5200	10,000 S.F.	.58	.80	1.38
5250	Ordinary hazard, one floor, 2000 S.F.	5.80	2.40	8.20
5300	10,000 S.F.	1.68	1.28	2.96
5350	Each additional floor, 2000 S.F.	.78	1.05	1.83
5400	10,000 S.F.	.70	1.03	1.73
5450	Extra hazard, one floor, 2000 S.F.	5.90	2.75	8.65
5500	10,000 S.F.	2.13	1.55	3.68
5550	Each additional floor, 2000 S.F.	.97	1.46	2.43
5600	10,000 S.F.	.88	1.20	2.08

Important: See the Reference Section for critical supporting data - Location Factors & Historical Cost Indexes

SERVICES D

Roof

Roof connections with hose gate valves (for combustible roof)

Hose connections on each floor (size based on class of service)

Check valve

Siamese inlet connections (for fire department use)

D4020 310	Wet Standpipe Risers, Class I	COST PER FLOOR		
		MAT.	INST.	TOTAL
0550	Wet standpipe risers, Class I, steel black sch. 40, 10' height			
0560	4" diameter pipe, one floor	2,025	2,175	4,200
0580	Additional floors	485	670	1,155
0600	6" diameter pipe, one floor	3,550	3,850	7,400
0620	Additional floors	875	1,075	1,950
0640	8" diameter pipe, one floor	5,300	4,625	9,925
0660	Additional floors	1,175	1,325	2,500

D4020 310	Wet Standpipe Risers, Class II	COST PER FLOOR		
		MAT.	INST.	TOTAL
1030	Wet standpipe risers, Class II, steel black sch. 40, 10' height			
1040	2" diameter pipe, one floor	865	780	1,645
1060	Additional floors	282	305	587
1080	2-1/2" diameter pipe, one floor	1,125	1,150	2,275
1100	Additional floors	315	350	665

D4020 310	Wet Standpipe Risers, Class III	COST PER FLOOR		
		MAT.	INST.	TOTAL
1530	Wet standpipe risers, Class III, steel black sch. 40, 10' height			
1540	4" diameter pipe, one floor	2,100	2,175	4,275
1560	Additional floors	400	560	960
1580	6" diameter pipe, one floor	3,625	3,850	7,475
1600	Additional floors	905	1,075	1,980
1620	8" diameter pipe, one floor	5,350	4,625	9,975
1640	Additional floors	1,225	1,325	2,550

SERVICES

D

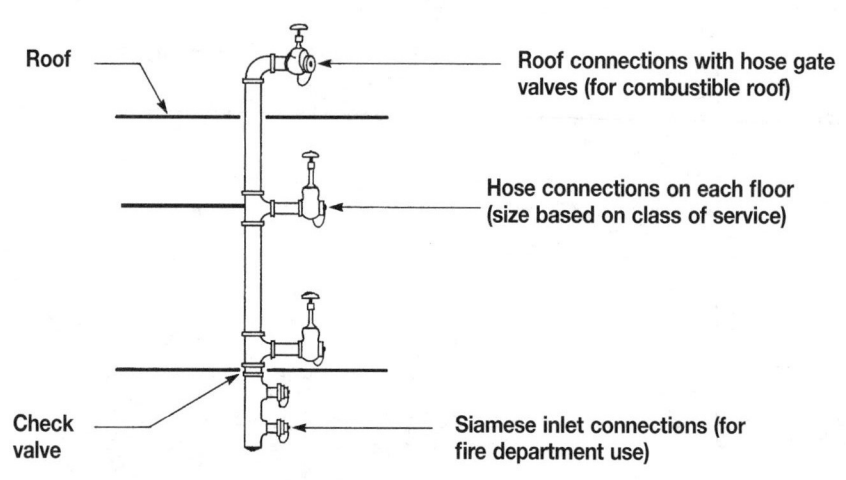

Roof

Roof connections with hose gate valves (for combustible roof)

Hose connections on each floor (size based on class of service)

Check valve

Siamese inlet connections (for fire department use)

D4020 330	Dry Standpipe Risers, Class I	COST PER FLOOR		
		MAT.	INST.	TOTAL
0530	Dry standpipe riser, Class I, steel black sch. 40, 10' height			
0540	4" diameter pipe, one floor	1,275	1,750	3,025
0560	Additional floors	440	635	1,075
0580	6" diameter pipe, one floor	2,600	3,050	5,650
0600	Additional floors	830	1,025	1,855
0620	8" diameter pipe, one floor	3,825	3,725	7,550
0640	Additional floors	1,150	1,275	2,425

D4020 330	Dry Standpipe Risers, Class II	COST PER FLOOR		
		MAT.	INST.	TOTAL
1030	Dry standpipe risers, Class II, steel black sch. 40, 10' height			
1040	2" diameter pipe, one floor	710	815	1,525
1060	Additional floor	237	266	503
1080	2-1/2" diameter pipe, one floor	845	955	1,800
1100	Additional floors	269	315	584

D4020 330	Dry Standpipe Risers, Class III	COST PER FLOOR		
		MAT.	INST.	TOTAL
1530	Dry standpipe risers, Class III, steel black sch. 40, 10' height			
1540	4" diameter pipe, one floor	1,300	1,725	3,025
1560	Additional floors	365	570	935
1580	6" diameter pipe, one floor	2,625	3,050	5,675
1600	Additional floors	860	1,025	1,885
1620	8" diameter pipe, one floor	3,850	3,725	7,575
1640	Additional floor	1,175	1,275	2,450

Important: See the Reference Section for critical supporting data - Location Factors & Historical Cost Indexes

D4020 410	Standpipe Equipment	COST EACH		
		MAT.	INST.	TOTAL
0100	Adapters, reducing, 1 piece, FxM, hexagon, cast brass, 2-1/2" x 1-1/2"	31.50		31.50
0200	Pin lug, 1-1/2" x 1"	9.90		9.90
0250	3" x 2-1/2"	39		39
0300	For polished chrome, add 75% mat.			
0400	Cabinets, D.S. glass in door, recessed, steel box, not equipped			
0500	Single extinguisher, steel door & frame	62	98.50	160.50
0550	Stainless steel door & frame	128	98.50	226.50
0600	Valve, 2-1/2" angle, steel door & frame	69	65.50	134.50
0650	Aluminum door & frame	91.50	65.50	157
0700	Stainless steel door & frame	120	65.50	185.50
0750	Hose rack assy, 2-1/2" x 1-1/2" valve & 100' hose, steel door & frame	140	131	271
0800	Aluminum door & frame	207	131	338
0850	Stainless steel door & frame	269	131	400
0900	Hose rack assy,& extinguisher,2-1/2"x1-1/2" valve & hose,steel door & frame	153	158	311
0950	Aluminum	264	158	422
1000	Stainless steel	290	158	448
1550	Compressor, air, dry pipe system, automatic, 200 gal., 1/3 H.P.	770	335	1,105
1600	520 gal., 1 H.P.	805	335	1,140
1650	Alarm, electric pressure switch (circuit closer)	144	16.85	160.85
2500	Couplings, hose, rocker lug, cast brass, 1-1/2"	29		29
2550	2-1/2"	38.50		38.50
3000	Escutcheon plate, for angle valves, polished brass, 1-1/2"	11.55		11.55
3050	2-1/2"	27.50		27.50
3500	Fire pump, electric, w/controller, fittings, relief valve			
3550	4" pump, 30 H.P., 500 G.P.M.	14,900	2,450	17,350
3600	5" pump, 40 H.P., 1000 G.P.M.	21,800	2,775	24,575
3650	5" pump, 100 H.P., 1000 G.P.M.	24,400	3,100	27,500
3700	For jockey pump system, add	2,675	395	3,070
5000	Hose, per linear foot, synthetic jacket, lined,			
5100	300 lb. test, 1-1/2" diameter	1.62	.30	1.92
5150	2-1/2" diameter	2.70	.36	3.06
5200	500 lb. test, 1-1/2" diameter	1.67	.30	1.97
5250	2-1/2" diameter	2.90	.36	3.26
5500	Nozzle, plain stream, polished brass, 1-1/2" x 10"	24		24
5550	2-1/2" x 15" x 13/16" or 1-1/2"	48.50		48.50
5600	Heavy duty combination adjustable fog and straight stream w/handle 1-1/2"	273		273
5650	2-1/2" direct connection	390		390
6000	Rack, for 1-1/2" diameter hose 100 ft. long, steel	35.50	39.50	75
6050	Brass	55	39.50	94.50
6500	Reel, steel, for 50 ft. long 1-1/2" diameter hose	75.50	56.50	132
6550	For 75 ft. long 2-1/2" diameter hose	122	56.50	178.50
7050	Siamese, w/plugs & chains, polished brass, sidewalk, 4" x 2-1/2" x 2-1/2"	340	315	655
7100	6" x 2-1/2" x 2-1/2"	565	395	960
7200	Wall type, flush, 4" x 2-1/2" x 2-1/2"	297	158	455
7250	6" x 2-1/2" x 2-1/2"	405	171	576
7300	Projecting, 4" x 2-1/2" x 2-1/2"	271	158	429
7350	6" x 2-1/2" x 2-1/2"	445	171	616
7400	For chrome plate, add 15% mat.			
8000	Valves, angle, wheel handle, 300 Lb., rough brass, 1-1/2"	29.50	36.50	66
8050	2-1/2"	51.50	62.50	114
8100	Combination pressure restricting, 1-1/2"	47	36.50	83.50
8150	2-1/2"	100	62.50	162.50
8200	Pressure restricting, adjustable, satin brass, 1-1/2"	76	36.50	112.50
8250	2-1/2"	107	62.50	169.50
8300	Hydrolator, vent and drain, rough brass, 1-1/2"	45	36.50	81.50
8350	2-1/2"	45	36.50	81.50
8400	Cabinet assy, incls. adapter, rack, hose, and nozzle	445	237	682

SERVICES

D

D4090 Other Fire Protection Systems

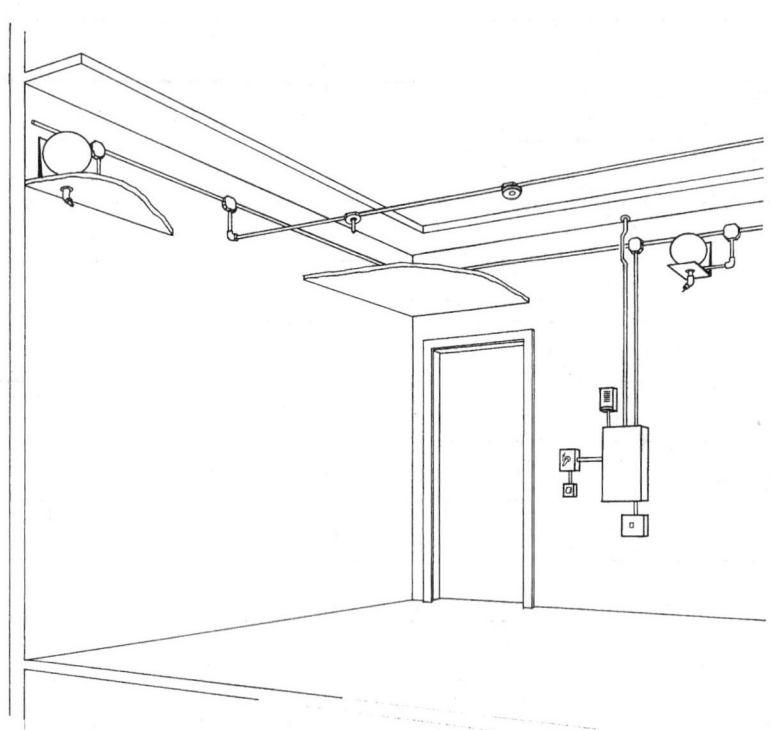

General: Automatic fire protection (suppression) systems other than water sprinklers may be desired for special environments, high risk areas, isolated locations or unusual hazards. Some typical applications would include:

1. Paint dip tanks
2. Securities vaults
3. Electronic data processing
4. Tape and data storage
5. Transformer rooms
6. Spray booths
7. Petroleum storage
8. High rack storage

Piping and wiring costs are highly variable and are not included.

SERVICES D

D4090 410	FM200 Fire Suppression	COST EACH		
		MAT.	INST.	TOTAL
0020	Detectors with brackets			
0040	Fixed temperature heat detector	31	53	84
0060	Rate of temperature rise detector	37	53	90
0080	Ion detector (smoke) detector	82.50	68.50	151
0200	Extinguisher agent			
0240	200 lb FM200, container	6,375	195	6,570
0280	75 lb carbon dioxide cylinder	1,050	130	1,180
0320	Dispersion nozzle			
0340	FM200 1-1/2" dispersion nozzle	55	31	86
0380	Carbon dioxide 3" x 5" dispersion nozzle	55	24	79
0420	Control station			
0440	Single zone control station with batteries	1,400	425	1,825
0470	Multizone (4) control station with batteries	2,700	845	3,545
0500	Electric mechanical release	138	214	352
0550	Manual pull station	49.50	72.50	122
0640	Battery standby power 10" x 10" x 17"	760	106	866
0740	Bell signalling device	54.50	53	107.50

D4090 410	FM200 Systems	COST PER C.F.		
		MAT.	INST.	TOTAL
0820	Average FM200 system, minimum			1.38
0840	Maximum			2.75

Important: See the Reference Section for critical supporting data - Location Factors & Historical Cost Indexes

D5010 Electrical Service/Distribution

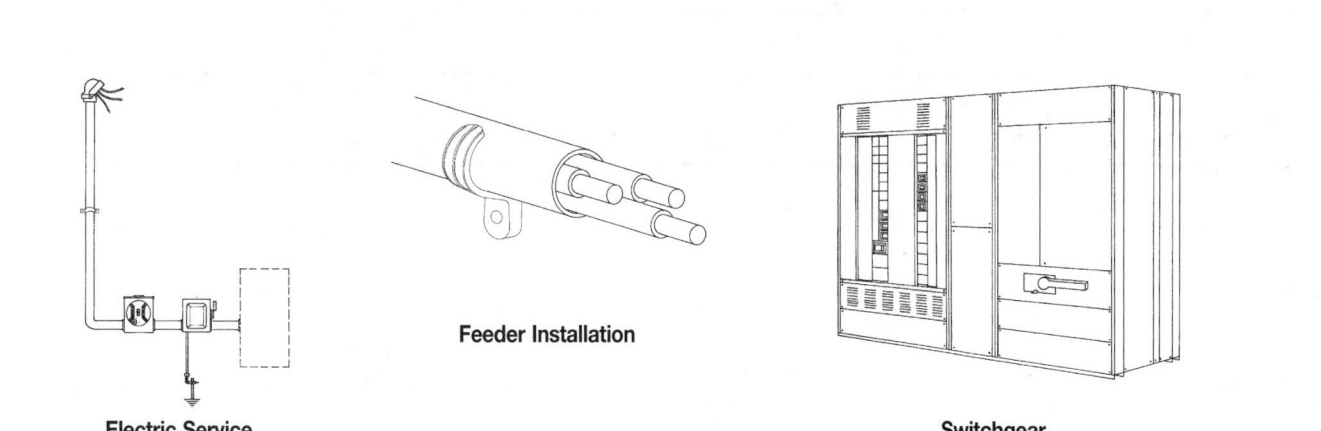

Electric Service

Feeder Installation

Switchgear

D5010 120	Electric Service, 3 Phase - 4 Wire	COST EACH		
		MAT.	INST.	TOTAL
0200	Service installation, includes breakers, metering, 20' conduit & wire			
0220	3 phase, 4 wire, 120/208 volts, 60 A	630	655	1,285
0240	100 A	780	785	1,565
0280	200 A	1,075	1,200	2,275
0320	400 A	2,525	2,225	4,750
0360	600 A	4,750	3,000	7,750
0400	800 A	6,125	3,625	9,750
0440	1000 A	7,800	4,150	11,950
0480	1200 A	9,450	4,250	13,700
0520	1600 A	18,300	6,125	24,425
0560	2000 A	20,300	6,950	27,250
0570	Add 25% for 277/480 volt			

D5010 230	Feeder Installation	COST PER L.F.		
		MAT.	INST.	TOTAL
0200	Feeder installation 600 V, including RGS conduit and XHHW wire, 60 A	3.37	7.90	11.27
0240	100 A	6.10	10.45	16.55
0280	200 A	12.50	16.15	28.65
0320	400 A	25	32.50	57.50
0360	600 A	52	52.50	104.50
0400	800 A	66	63	129
0440	1000 A	83.50	80.50	164
0480	1200 A	97.50	82.50	180
0520	1600 A	132	126	258
0560	2000 A	167	161	328
1200	Branch installation 600 V, including EMT conduit and THW wire, 15 A	.68	4.03	4.71
1240	20 A	.68	4.03	4.71
1280	30 A	1.03	4.96	5.99
1320	50 A	1.69	5.70	7.39
1360	65 A	2.01	6.05	8.06
1400	85 A	3.06	7.15	10.21
1440	100 A	3.78	7.60	11.38
1480	130 A	4.87	8.55	13.42
1520	150 A	5.85	9.80	15.65
1560	200 A	7.50	11	18.50

D5010 240	Switchgear	COST EACH		
		MAT.	INST.	TOTAL
0200	Switchgear inst., incl. swbd., panels & circ bkr, 400 A, 120/208volt	3,475	2,675	6,150
0240	600 A	8,825	3,575	12,400

SERVICES

D

D5010 Electrical Service/Distribution

D5010 240	Switchgear	COST EACH		
		MAT.	INST.	TOTAL
0280	800 A	11,100	5,125	16,225
0320	1200 A	14,100	7,875	21,975
0360	1600 A	19,100	11,000	30,100
0400	2000 A	24,200	14,100	38,300
0410	Add 20% for 277/480 volt			

D5020 Lighting and Branch Wiring

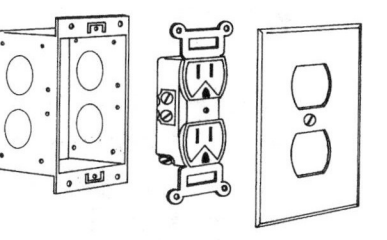

Duplex Wall Receptacle

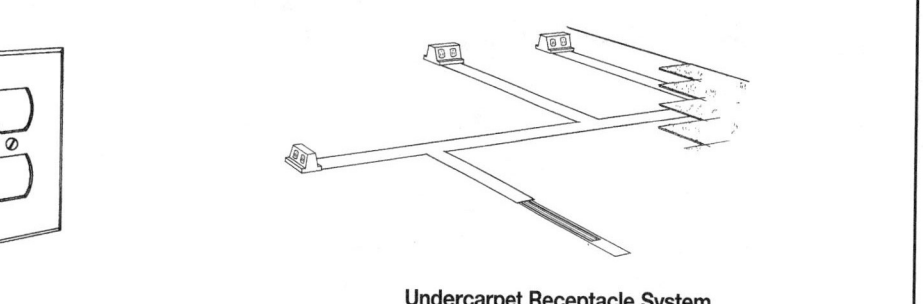

Undercarpet Receptacle System

D5020 110	Receptacle (by Wattage)	COST PER S.F.		
		MAT.	INST.	TOTAL
0190	Receptacles include plate, box, conduit, wire & transformer when required			
0200	2.5 per 1000 S.F., .3 watts per S.F.	.27	.95	1.22
0240	With transformer	.31	1	1.31
0280	4 per 1000 S.F., .5 watts per S.F.	.31	1.09	1.40
0320	With transformer	.37	1.16	1.53
0360	5 per 1000 S.F., .6 watts per S.F.	.36	1.30	1.66
0400	With transformer	.44	1.39	1.83
0440	8 per 1000 S.F., .9 watts per S.F.	.38	1.43	1.81
0480	With transformer	.49	1.56	2.05
0520	10 per 1000 S.F., 1.2 watts per S.F.	.39	1.57	1.96
0560	With transformer	.57	1.78	2.35
0600	16.5 per 1000 S.F., 2.0 watts per S.F.	.47	1.95	2.42
0640	With transformer	.77	2.32	3.09
0680	20 per 1000 S.F., 2.4 watts per S.F.	.50	2.13	2.63
0720	With transformer	.85	2.55	3.40

D5020 115	Receptacles	COST PER S.F.		
		MAT.	INST.	TOTAL
0200	Receptacle systems, underfloor duct, 5' on center, low density	3.99	2.09	6.08
0240	High density	4.29	2.68	6.97
0280	7' on center, low density	3.16	1.79	4.95
0320	High density	3.46	2.38	5.84
0400	Poke thru fittings, low density	.89	.96	1.85
0440	High density	1.78	1.89	3.67
0520	Telepoles, using Romex, low density	.66	.63	1.29
0560	High density	1.32	1.25	2.57
0600	Using EMT, low density	.69	.83	1.52
0640	High density	1.39	1.65	3.04
0720	Conduit system with floor boxes, low density	.72	.71	1.43
0760	High density	1.44	1.43	2.87
0840	Undercarpet power system, 3 conductor with 5 conductor feeder, low density	1.30	.24	1.54
0880	High density	2.53	.52	3.05

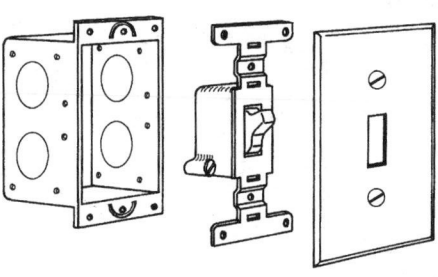

Description: Table D5020 130 includes the cost for switch, plate, box, conduit in slab or EMT exposed and copper wire. Add 20% for exposed conduit.

No power required for switches.

Federal energy guidelines recommend the maximum lighting area controlled per switch shall not exceed 1000 S.F. and that areas over 500 S.F. shall be so controlled that total illumination can be reduced by at least 50%.

D5020 130	Wall Switch by Sq. Ft.	COST PER S.F.		
		MAT.	INST.	TOTAL
0200	Wall switches, 1.0 per 1000 S.F.	.04	.15	.19
0240	1.2 per 1000 S.F.	.04	.18	.22
0280	2.0 per 1000 S.F.	.05	.24	.29
0320	2.5 per 1000 S.F.	.07	.30	.37
0360	5.0 per 1000 S.F.	.18	.66	.84
0400	10.0 per 1000 S.F.	.35	1.33	1.68

D5020 135	Miscellaneous Power	COST PER S.F.		
		MAT.	INST.	TOTAL
0200	Miscellaneous power, to .5 watts	.02	.07	.09
0240	.8 watts	.03	.11	.14
0280	1 watt	.04	.13	.17
0320	1.2 watts	.05	.17	.22
0360	1.5 watts	.06	.20	.26
0400	1.8 watts	.07	.22	.29
0440	2 watts	.09	.27	.36
0480	2.5 watts	.10	.33	.43
0520	3 watts	.12	.39	.51

D5020 140	Central A. C. Power (by Wattage)	COST PER S.F.		
		MAT.	INST.	TOTAL
0200	Central air conditioning power, 1 watt	.05	.17	.22
0220	2 watts	.06	.19	.25
0240	3 watts	.07	.21	.28
0280	4 watts	.11	.29	.40
0320	6 watts	.19	.39	.58
0360	8 watts	.22	.42	.64
0400	10 watts	.29	.48	.77

D50 Electrical

D5020　Lighting and Branch Wiring

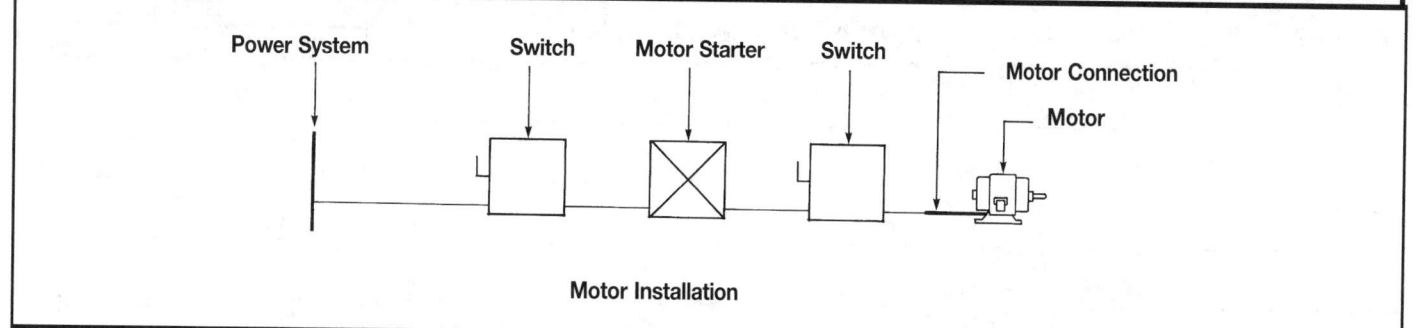

Power System　　Switch　　Motor Starter　　Switch　　Motor Connection　　Motor

Motor Installation

D5020 145	Motor Installation	COST EACH		
		MAT.	INST.	TOTAL
0200	Motor installation, single phase, 115V, to and including 1/3 HP motor size	455	645	1,100
0240	To and incl. 1 HP motor size	475	645	1,120
0280	To and incl. 2 HP motor size	510	685	1,195
0320	To and incl. 3 HP motor size	570	695	1,265
0360	230V, to and including 1 HP motor size	455	650	1,105
0400	To and incl. 2 HP motor size	480	650	1,130
0440	To and incl. 3 HP motor size	540	700	1,240
0520	Three phase, 200V, to and including 1-1/2 HP motor size	535	715	1,250
0560	To and incl. 3 HP motor size	575	780	1,355
0600	To and incl. 5 HP motor size	615	870	1,485
0640	To and incl. 7-1/2 HP motor size	630	885	1,515
0680	To and incl. 10 HP motor size	1,025	1,100	2,125
0720	To and incl. 15 HP motor size	1,350	1,225	2,575
0760	To and incl. 20 HP motor size	1,675	1,425	3,100
0800	To and incl. 25 HP motor size	1,700	1,425	3,125
0840	To and incl. 30 HP motor size	2,650	1,675	4,325
0880	To and incl. 40 HP motor size	3,225	2,000	5,225
0920	To and incl. 50 HP motor size	5,675	2,325	8,000
0960	To and incl. 60 HP motor size	5,825	2,450	8,275
1000	To and incl. 75 HP motor size	7,500	2,800	10,300
1040	To and incl. 100 HP motor size	15,600	3,300	18,900
1080	To and incl. 125 HP motor size	15,800	3,625	19,425
1120	To and incl. 150 HP motor size	19,000	4,250	23,250
1160	To and incl. 200 HP motor size	22,800	5,175	27,975
1240	230V, to and including 1-1/2 HP motor size	510	710	1,220
1280	To and incl. 3 HP motor size	545	770	1,315
1320	To and incl. 5 HP motor size	585	860	1,445
1360	To and incl. 7-1/2 HP motor size	585	860	1,445
1400	To and incl. 10 HP motor size	920	1,050	1,970
1440	To and incl. 15 HP motor size	1,050	1,150	2,200
1480	To and incl. 20 HP motor size	1,550	1,375	2,925
1520	To and incl. 25 HP motor size	1,675	1,425	3,100
1560	To and incl. 30 HP motor size	1,725	1,425	3,150
1600	To and incl. 40 HP motor size	3,175	1,950	5,125
1640	To and incl. 50 HP motor size	3,275	2,050	5,325
1680	To and incl. 60 HP motor size	5,675	2,325	8,000
1720	To and incl. 75 HP motor size	6,850	2,625	9,475
1760	To and incl. 100 HP motor size	7,800	2,950	10,750
1800	To and incl. 125 HP motor size	15,800	3,400	19,200
1840	To and incl. 150 HP motor size	16,900	3,875	20,775
1880	To and incl. 200 HP motor size	18,300	4,300	22,600
1960	460V, to and including 2 HP motor size	610	715	1,325
2000	To and incl. 5 HP motor size	650	780	1,430
2040	To and incl. 10 HP motor size	680	860	1,540
2080	To and incl. 15 HP motor size	910	985	1,895
2120	To and incl. 20 HP motor size	940	1,050	1,990

D　SERVICES

D5020 Lighting and Branch Wiring

D5020 145	Motor Installation	COST EACH		
		MAT.	INST.	TOTAL
2160	To and incl. 25 HP motor size	1,025	1,100	2,125
2200	To and incl. 30 HP motor size	1,325	1,200	2,525
2240	To and incl. 40 HP motor size	1,675	1,275	2,950
2280	To and incl. 50 HP motor size	1,850	1,425	3,275
2320	To and incl. 60 HP motor size	2,825	1,675	4,500
2360	To and incl. 75 HP motor size	3,200	1,850	5,050
2400	To and incl. 100 HP motor size	3,450	2,050	5,500
2440	To and incl. 125 HP motor size	5,850	2,325	8,175
2480	To and incl. 150 HP motor size	7,200	2,600	9,800
2520	To and incl. 200 HP motor size	8,275	2,950	11,225
2600	575V, to and including 2 HP motor size	610	715	1,325
2640	To and incl. 5 HP motor size	650	780	1,430
2680	To and incl. 10 HP motor size	680	860	1,540
2720	To and incl. 20 HP motor size	910	985	1,895
2760	To and incl. 25 HP motor size	940	1,050	1,990
2800	To and incl. 30 HP motor size	1,325	1,200	2,525
2840	To and incl. 50 HP motor size	1,425	1,250	2,675
2880	To and incl. 60 HP motor size	2,775	1,650	4,425
2920	To and incl. 75 HP motor size	2,825	1,675	4,500
2960	To and incl. 100 HP motor size	3,200	1,850	5,050
3000	To and incl. 125 HP motor size	5,775	2,300	8,075
3040	To and incl. 150 HP motor size	5,850	2,325	8,175
3080	To and incl. 200 HP motor size	7,300	2,650	9,950

D5020 155	Motor Feeder	COST PER L.F.		
		MAT.	INST.	TOTAL
0200	Motor feeder systems, single phase, feed up to 115V 1HP or 230V 2 HP	1.53	5	6.53
0240	115V 2HP, 230V 3HP	1.59	5.10	6.69
0280	115V 3HP	1.75	5.30	7.05
0360	Three phase, feed to 200V 3HP, 230V 5HP, 460V 10HP, 575V 10HP	1.59	5.40	6.99
0440	200V 5HP, 230V 7.5HP, 460V 15HP, 575V 20HP	1.68	5.50	7.18
0520	200V 10HP, 230V 10HP, 460V 30HP, 575V 30HP	1.92	5.80	7.72
0600	200V 15HP, 230V 15HP, 460V 40HP, 575V 50HP	2.52	6.65	9.17
0680	200V 20HP, 230V 25HP, 460V 50HP, 575V 60HP	3.62	8.45	12.07
0760	200V 25HP, 230V 30HP, 460V 60HP, 575V 75HP	4.34	8.60	12.94
0840	200V 30HP	4.44	8.85	13.29
0920	230V 40HP, 460V 75HP, 575V 100HP	5.60	9.70	15.30
1000	200V 40HP	6.30	10.35	16.65
1080	230V 50HP, 460V 100HP, 575V 125HP	7.25	11.45	18.70
1160	200V 50HP, 230V 60HP, 460V 125HP, 575V 150HP	8.50	12.10	20.60
1240	200V 60HP, 460V 150HP	10.25	14.20	24.45
1320	230V 75HP, 575V 200HP	11.95	14.80	26.75
1400	200V 75HP	13.10	15.15	28.25
1480	230V 100HP, 460V 200HP	18.50	17.65	36.15
1560	200V 100HP	25	22	47
1640	230V 125HP	28	24	52
1720	200V 125HP, 230V 150HP	29	27.50	56.50
1800	200V 150HP	33	30	63
1880	200V 200HP	49.50	44	93.50
1960	230V 200HP	43	33	76

D5020 Lighting and Branch Wiring

Fluorescent Fixture   Incandescent Fixture

D5020 210	Fluorescent Fixtures (by Wattage)	COST PER S.F.		
		MAT.	INST.	TOTAL
0190	Fluorescent fixtures recess mounted in ceiling			
0200	1 watt per S.F., 20 FC, 5 fixtures per 1000 S.F.	.57	1.25	1.82
0240	2 watts per S.F., 40 FC, 10 fixtures per 1000 S.F.	1.13	2.43	3.56
0280	3 watts per S.F., 60 FC, 15 fixtures per 1000 S.F	1.70	3.68	5.38
0320	4 watts per S.F., 80 FC, 20 fixtures per 1000 S.F.	2.26	4.89	7.15
0400	5 watts per S.F., 100 FC, 25 fixtures per 1000 S.F.	2.83	6.10	8.93

D5020 216	Incandescent Fixture (by Wattage)	COST PER S.F.		
		MAT.	INST.	TOTAL
0190	Incandescent fixture recess mounted, type A			
0200	1 watt per S.F., 8 FC, 6 fixtures per 1000 S.F.	.66	1	1.66
0240	2 watt per S.F., 16 FC, 12 fixtures per 1000 S.F.	1.30	1.99	3.29
0280	3 watt per S.F., 24 FC, 18 fixtures, per 1000 S.F.	1.95	2.95	4.90
0320	4 watt per S.F., 32 FC, 24 fixtures per 1000 S.F.	2.60	3.95	6.55
0400	5 watt per S.F., 40 FC, 30 fixtures per 1000 S.F.	3.27	4.95	8.22

SERVICES

D

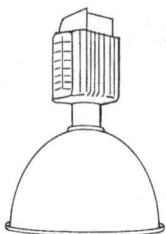

High Bay Fixture

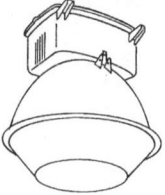

Low Bay Fixture

D5020 226	H.I.D. Fixture, High Bay (by Wattage)	COST PER S.F.		
		MAT.	INST.	TOTAL
0190	High intensity discharge fixture, 16' above work plane			
0200	1 watt/S.F., type D, 23 FC, 1 fixture/1000 S.F.	.70	1	1.70
0240	Type E, 42 FC, 1 fixture/1000 S.F.	.83	1.05	1.88
0280	Type G, 52 FC, 1 fixture/1000 S.F.	.83	1.05	1.88
0320	Type C, 54 FC, 2 fixture/1000 S.F.	.99	1.17	2.16
0400	2 watt/S.F., type D, 45 FC, 2 fixture/1000 S.F.	1.39	1.99	3.38
0440	Type E, 84 FC, 2 fixture/1000 S.F.	1.66	2.12	3.78
0480	Type G, 105 FC, 2 fixture/1000 S.F.	1.66	2.12	3.78
0520	Type C, 108 FC, 4 fixture/1000 S.F.	1.95	2.34	4.29
0600	3 watt/S.F., type D, 68 FC, 3 fixture/1000 S.F.	2.08	2.97	5.05
0640	Type E, 126 FC, 3 fixture/1000 S.F.	2.49	3.18	5.67
0680	Type G, 157 FC, 3 fixture/1000 S.F.	2.49	3.18	5.67
0720	Type C, 162 FC, 6 fixture/1000 S.F.	2.94	3.49	6.43
0800	4 watt/ S.F., type D, 91 FC, 4 fixture/1000 S.F.	3.02	4.68	7.70
0840	Type E, 168 FC, 4 fixture/1000 S.F.	3.32	4.26	7.58
0880	Type G, 210 FC, 4 fixture/1000 S.F.	3.32	4.26	7.58
0920	Type C, 243 FC, 9 fixture/1000 S.F.	4.28	4.88	9.16
1000	5 watt/S.F., type D, 113 FC, 5 fixture/1000 S.F.	3.47	4.97	8.44
1040	Type E, 210 FC, 5 fixture/1000 S.F.	4.15	5.30	9.45
1080	Type G, 262 FC, 5 fixture/1000 S.F.	4.15	5.30	9.45
1120	Type C, 297 FC, 11 fixture/1000 S.F.	5.25	6.05	11.30

D5020 234	H.I.D. Fixture, High Bay (by Wattage)	COST PER S.F.		
		MAT.	INST.	TOTAL
0190	High intensity discharge fixture, 30' above work plane			
0200	1 watt/S.F., type D, 23 FC, 1 fixture/1000 S.F.	.77	1.29	2.06
0240	Type E, 37 FC, 1 fixture/1000 S.F.	.90	1.33	2.23
0280	Type G, 45 FC., 1 fixture/1000 S.F.	.90	1.33	2.23
0320	Type F, 50 FC, 1 fixture/1000 S.F.	.78	1.04	1.82
0400	2 watt/S.F., type D, 40 FC, 2 fixtures/1000 S.F.	1.55	2.51	4.06
0440	Type E, 74 FC, 2 fixtures/1000 S.F.	1.81	2.64	4.45
0480	Type G, 92 FC, 2 fixtures/1000 S.F.	1.81	2.64	4.45
0520	Type F, 100 FC, 2 fixtures/1000 S.F.	1.56	2.10	3.66
0600	3 watt/S.F., type D, 60 FC, 3 fixtures/1000 S.F.	2.34	3.80	6.14
0640	Type E, 110 FC, 3 fixtures/1000 S.F.	2.72	4.02	6.74
0680	Type G, 138FC, 3 fixtures/1000 S.F.	2.72	4.02	6.74
0720	Type F, 150 FC, 3 fixtures/1000 S.F.	2.34	3.15	5.49
0800	4 watt/ S.F., type D, 80 FC, 4 fixtures/1000 S.F.	3.10	5.05	8.15
0840	Type E, 148 FC, 4 fixtures/1000 S.F.	3.63	5.35	8.98
0880	Type G, 185 FC, 4 fixtures/1000 S.F.	3.63	5.35	8.98
0920	Type F, 200 FC, 4 fixtures/1000 S.F.	3.10	4.22	7.32
1000	5 watt/ S.F., type D, 100 FC 5 fixtures/1000 S.F.	3.88	6.30	10.18
1040	Type E, 185 FC, 5 fixtures/1000 S.F.	4.52	6.65	11.17
1080	Type G, 230 FC, 5 fixtures/1000 S.F.	4.52	6.65	11.17

Important: See the Reference Section for critical supporting data - Location Factors & Historical Cost Indexes

D50 Electrical

D5020 Lighting and Branch Wiring

D5020 234	H.I.D. Fixture, High Bay (by Wattage)	COST PER S.F.		
		MAT.	INST.	TOTAL
1120	Type F, 250 FC, 5 fixtures/1000 S.F.	3.88	5.25	9.13

D5020 238	H.I.D. Fixture, Low Bay (by Wattage)	COST PER S.F.		
		MAT.	INST.	TOTAL
0190	High intensity discharge fixture, 8'-10' above work plane			
0200	1 watt/S.F., type H, 19 FC, 4 fixtures/1000 S.F.	1.69	2.13	3.82
0240	Type J, 30 FC, 4 fixtures/1000 S.F.	2.01	2.15	4.16
0280	Type K, 29 FC, 5 fixtures/1000 S.F.	1.94	1.95	3.89
0360	2 watt/S.F. type H, 33 FC, 7 fixtures/1000 S.F.	3.07	4.11	7.18
0400	Type J, 52 FC, 7 fixtures/1000 S.F.	3.58	3.96	7.54
0440	Type K, 63 FC, 11 fixtures/1000 S.F.	4.20	4.07	8.27
0520	3 watt/S.F., type H, 51 FC, 11 fixtures/1000 S.F.	4.77	6.30	11.07
0560	Type J, 81 FC, 11 fixtures/1000 S.F.	5.55	5.95	11.50
0600	Type K, 92 FC, 16 fixtures/1000 S.F.	6.15	6	12.15
0680	4 watt/S.F., type H, 65 FC, 14 fixtures/1000 S.F.	6.15	8.25	14.40
0720	Type J, 103 FC, 14 fixtures/1000 S.F.	7.15	7.85	15
0760	Type K, 127 FC, 22 fixtures/1000 S.F.	8.40	8.15	16.55
0840	5 watt/S.F., type H, 84 FC, 18 fixtures/1000 S.F.	7.85	10.40	18.25
0880	Type J, 133 FC, 18 fixtures/1000 S.F.	9.15	9.95	19.10
0920	Type K, 155 FC, 27 fixtures/1000 S.F.	10.35	10.05	20.40

D5020 242	H.I.D. Fixture, Low Bay (by Wattage)	COST PER S.F.		
		MAT.	INST.	TOTAL
0190	High intensity discharge fixture, mounted 16' above work plane			
0200	1 watt/S.F., type H, 19 FC, 4 fixtures/1000 S.F.	1.78	2.43	4.21
0240	Type J, 28 FC, 4 fixt./1000 S.F.	2.10	2.45	4.55
0280	Type K, 27 FC, 5 fixt./1000 S.F.	2.18	2.75	4.93
0360	2 watts/S.F., type H, 30 FC, 7 fixt/1000 S.F.	3.24	4.70	7.94
0400	Type J, 48 FC, 7 fixt/1000 S.F.	3.81	4.73	8.54
0440	Type K, 58 FC, 11 fixt/1000 S.F.	4.67	5.60	10.27
0520	3 watts/S.F., type H, 47 FC, 11 fixt/1000 S.F.	5	7.15	12.15
0560	Type J, 75 FC, 11 fixt/1000 S.F.	5.90	7.20	13.10
0600	Type K, 85 FC, 16 fixt/1000 S.F.	6.85	8.35	15.20
0680	4 watts/S.F., type H, 60 FC, 14 fixt/1000 S.F.	6.50	9.40	15.90
0720	Type J, 95 FC, 14 fixt/1000 S.F.	7.65	9.45	17.10
0760	Type K, 117 FC, 22 fixt/1000 S.F.	9.30	11.20	20.50
0840	5 watts/S.F., type H, 77 FC, 18 fixt/1000 S.F.	8.35	12.20	20.55
0880	Type J, 122 FC, 18 fixt/1000 S.F.	9.75	11.95	21.70
0920	Type K, 143 FC, 27 fixt/1000 S.F.	11.50	13.90	25.40

SERVICES

D

D5020 Lighting and Branch Wiring

Light Pole

D5020 246	Light Pole (Installed)	COST EACH		
		MAT.	INST.	TOTAL
0200	Light pole, aluminum, 20' high, 1 arm bracket	815	765	1,580
0240	2 arm brackets	905	765	1,670
0280	3 arm brackets	995	790	1,785
0320	4 arm brackets	1,075	790	1,865
0360	30' high, 1 arm bracket	1,475	965	2,440
0400	2 arm brackets	1,575	965	2,540
0440	3 arm brackets	1,650	995	2,645
0480	4 arm brackets	1,750	995	2,745
0680	40' high, 1 arm bracket	1,800	1,300	3,100
0720	2 arm brackets	1,875	1,300	3,175
0760	3 arm brackets	1,975	1,325	3,300
0800	4 arm brackets	2,050	1,325	3,375
0840	Steel, 20' high, 1 arm bracket	1,000	820	1,820
0880	2 arm brackets	1,075	820	1,895
0920	3 arm brackets	1,100	845	1,945
0960	4 arm brackets	1,175	845	2,020
1000	30' high, 1 arm bracket	1,150	1,025	2,175
1040	2 arm brackets	1,225	1,025	2,250
1080	3 arm brackets	1,250	1,050	2,300
1120	4 arm brackets	1,350	1,050	2,400
1320	40' high, 1 arm bracket	1,500	1,375	2,875
1360	2 arm brackets	1,575	1,375	2,950
1400	3 arm brackets	1,600	1,425	3,025
1440	4 arm brackets	1,700	1,425	3,125

SERVICES D

Important: See the Reference Section for critical supporting data - Location Factors & Historical Cost Indexes

D5030 Communications and Security

D5030 810	Communication & Alarm Systems	COST EACH		
		MAT.	INST.	TOTAL
0200	Communication & alarm systems, includes outlets, boxes, conduit & wire			
0210	Sound system, 6 outlets	4,400	5,350	9,750
0220	12 outlets	6,225	8,575	14,800
0240	30 outlets	10,800	16,200	27,000
0280	100 outlets	32,600	54,000	86,600
0320	Fire detection systems, 12 detectors	2,250	4,450	6,700
0360	25 detectors	3,950	7,500	11,450
0400	50 detectors	7,575	14,800	22,375
0440	100 detectors	13,800	26,500	40,300
0480	Intercom systems, 6 stations	2,500	3,650	6,150
0560	25 stations	8,650	14,000	22,650
0640	100 stations	33,100	51,500	84,600
0680	Master clock systems, 6 rooms	3,825	5,975	9,800
0720	12 rooms	5,675	10,200	15,875
0760	20 rooms	7,600	14,400	22,000
0800	30 rooms	12,200	26,500	38,700
0840	50 rooms	19,600	44,700	64,300
0920	Master TV antenna systems, 6 outlets	2,175	3,775	5,950
0960	12 outlets	4,050	7,025	11,075
1000	30 outlets	8,100	16,300	24,400
1040	100 outlets	26,900	53,000	79,900

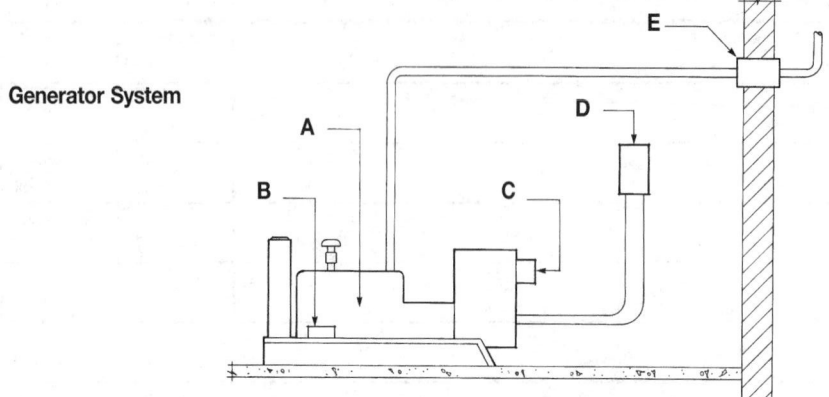

Generator System

A: Engine
B: Battery
C: Charger
D: Transfer Switch
E: Muffler

D5090 210	Generators (by kW)	COST PER kW		
		MAT.	INST.	TOTAL
0190	Generator sets, include battery, charger, muffler & transfer switch			
0200	Gas/gasoline operated, 3 phase, 4 wire, 277/480 volt, 7.5 kW	880	198	1,078
0240	11.5 kW	815	149	964
0280	20 kW	550	98	648
0320	35 kW	375	63.50	438.50
0360	80 kW	283	38.50	321.50
0400	100 kW	248	37.50	285.50
0440	125 kW	405	35	440
0480	185 kW	360	26.50	386.50
0560	Diesel engine with fuel tank, 30 kW	585	74	659
0600	50 kW	435	58.50	493.50
0640	75 kW	375	46.50	421.50
0680	100 kW	315	39.50	354.50
0760	150 kW	252	31.50	283.50
0840	200 kW	213	25.50	238.50
0880	250 kW	200	21	221
0960	350 kW	176	17.45	193.45
1040	500 kW	190	13.60	203.60

For information about Means Estimating Seminars, see yellow pages 11 and 12 in back of book

SERVICES D

E EQUIPMENT & FURNISHINGS

E10 Equipment

E1010 Commercial Equipment

E1010 110 — Security/Vault, EACH

		COST EACH		
		MAT.	INST.	TOTAL
0100	Bank equipment, drive up window, drawer & mike, no glazing, economy	4,025	990	5,015
0110	Deluxe	7,925	1,975	9,900
0120	Night depository, economy	5,875	990	6,865
0130	Deluxe	8,475	1,975	10,450
0140	Pneumatic tube systems, 2 station, standard	21,200	3,200	24,400
0150	Teller, automated, 24 hour, single unit	39,200	3,200	42,400
0160	Teller window, bullet proof glazing, 44" x 60"	2,850	605	3,455
0170	Pass through, painted steel, 72" x 40"	3,050	1,225	4,275
0300	Safe, office type, 1 hr. rating, 34" x 20" x 20"	1,650		1,650
0310	4 hr. rating, 62" x 33" x 20"	6,975		6,975
0320	Data storage, 4 hr. rating, 63" x 44" x 16"	9,375		9,375
0330	Jewelers, 63" x 44" x 16"	16,000		16,000
0340	Money, "B" label, 9" x 14" x 14"	495		495
0350	Tool and torch resistive, 24" x 24" x 20"	7,050	146	7,196
0500	Security gates-scissors type, painted steel, single, 6' high, 5-1/2' wide	126	247	373
0510	Double gate, 7-1/2' high, 14' wide	445	495	940

E1010 510 — Mercantile Equipment, EACH

		COST EACH		
		MAT.	INST.	TOTAL
0015	Barber equipment, chair, hydraulic, economy	450	15.55	465.55
0020	Deluxe	2,975	23.50	2,998.50
0100	Checkout counter, single belt	2,225	58.50	2,283.50
0110	Double belt, power take-away	3,825	65	3,890
0200	Display cases, freestanding, glass and aluminum, 3'-6" x 3' x 1'-0" deep	475	93.50	568.50
0220	Wall mounted, glass and aluminum, 3' x 4' x 1'-4" deep	1,150	149	1,299
0320	Frozen food, chest type, 12 ft. long	5,800	238	6,050

E1010 610 — Laundry/Dry Cleaning, EACH

		COST EACH		
		MAT.	INST.	TOTAL
0100	Laundry equipment, dryers, gas fired, residential, 16 lb. capacity	580	145	725
0110	Commercial, 30 lb. capacity, single	2,650	145	2,795
0120	Dry cleaners, electric, 20 lb. capacity	31,200	4,275	35,475
0130	30 lb. capacity	42,700	5,700	48,400
0140	Ironers, commercial, 120" with canopy, 8 roll	152,000	12,200	164,200
0150	Institutional, 110", single roll	27,000	2,125	29,125
0160	Washers, residential, 4 cycle	665	145	810
0170	Commercial, coin operated, deluxe	2,800	145	2,945

E1020 110	Ecclesiastical Equipment, EACH	COST EACH		
		MAT.	INST.	TOTAL
0090	Church equipment, altar, wood, custom, plain	1,900	267	2,167
0100	Granite, custom, deluxe	24,000	3,700	27,700
0110	Baptistry, fiberglass, economy	2,750	965	3,715
0120	Bells & carillons, keyboard operation	11,100	6,350	17,450
0130	Confessional, wood, single, economy	2,400	625	3,025
0140	Double, deluxe	15,000	1,875	16,875
0150	Steeples, translucent fiberglas, 30" square, 15' high	3,600	1,275	4,875

E1020 130	Ecclesiastical Equipment, L.F.	COST PER L.F.		
		MAT.	INST.	TOTAL
0100	Arch. equip., church equip. pews, bench type, hardwood, economy	66	18.70	84.70
0110	Deluxe	132	25	157

E1020 210	Library Equipment, EACH	COST EACH		
		MAT.	INST.	TOTAL
0110	Library equipment, carrels, metal, economy	223	74.50	297.50
0120	Hardwood, deluxe	885	93.50	978.50

E1020 230	Library Equipment, L.F.	COST PER L.F.		
		MAT.	INST.	TOTAL
0100	Library equipment, book shelf, metal, single face, 90" high x 10" shelf	172	31	203
0110	Double face, 90" high x 10" shelf	223	67.50	290.50
0120	Charging desk, built-in, with counter, plastic laminate	475	53.50	528.50

E1020 310	Theater and Stage Equipment, EACH	COST EACH		
		MAT.	INST.	TOTAL
0200	Movie equipment, changeover, economy	405		405
0210	Film transport, incl. platters and autowind, economy	4,400		4,400
0220	Lamphouses, incl. rectifiers, xenon, 1000W	5,550	212	5,762
0230	4000W	7,850	282	8,132
0240	Projector mechanisms, 35 mm, economy	9,700		9,700
0250	Deluxe	13,300		13,300
0260	Sound systems, incl. amplifier, single, economy	2,900	470	3,370
0270	Dual, Dolby/super sound	14,400	1,050	15,450
0280	Projection screens, wall hung, manual operation, 50 S.F., economy	255	74.50	329.50
0290	Electric operation, 100 S.F., deluxe	1,725	375	2,100

E1020 320	Theater and Stage Equipment, S.F.	COST PER S.F.		
		MAT.	INST.	TOTAL
0090	Movie equipment, projection screens, rigid in wall, acrylic, 1/4" thick	36.50	3.73	40.23
0100	1/2" thick	42.50	5.60	48.10
0110	Stage equipment, curtains, velour, medium weight	6.75	1.25	8
0120	Silica based yarn, fireproof	13.45	14.95	28.40
0130	Stages, portable with steps, folding legs, 8" high	16.70		16.70
0140	Telescoping platforms, aluminum, deluxe	30	19.40	49.40

E1020 330	Theater and Stage Equipment, L.F.	COST PER L.F.		
		MAT.	INST.	TOTAL
0100	Stage equipment, curtain track, heavy duty	42	41.50	83.50
0110	Lights, border, quartz, colored	155	21	176

E1020 610	Detention Equipment, EACH	COST EACH		
		MAT.	INST.	TOTAL
0110	Detention equipment, cell front rolling door, 7/8" bars, 5' x 7' high	4,725	1,050	5,775
0120	Cells, prefab., including front, 5' x 7' x 7' deep	8,150	1,400	9,550

E EQUIPMENT & FURNISHINGS

E1020 Institutional Equipment

E1020 610	Detention Equipment, EACH	COST EACH		
		MAT.	INST.	TOTAL
0130	Doors and frames, 3' x 7', single plate	3,550	525	4,075
0140	Double plate	4,400	525	4,925
0150	Toilet apparatus, incl wash basin	2,575	645	3,220
0160	Visitor cubicle, vision panel, no intercom	2,600	1,050	3,650

E1020 710	Laboratory Equipment, EACH	COST EACH		
		MAT.	INST.	TOTAL
0110	Laboratory equipment, glassware washer, distilled water, economy	6,575	475	7,050
0120	Deluxe	8,450	855	9,305
0140	Radio isotope	12,100		12,100

E1020 720	Laboratory Equipment, S.F.	COST PER S.F.		
		MAT.	INST.	TOTAL
0100	Arch. equip., lab equip., counter tops, acid proof, economy	23.50	9.10	32.60
0110	Stainless steel	76.50	9.10	85.60

E1020 730	Laboratory Equipment, L.F.	COST PER L.F.		
		MAT.	INST.	TOTAL
0110	Laboratory equipment, cabinets, wall, open	102	37.50	139.50
0120	Base, drawer units	320	41.50	361.50
0130	Fume hoods, not incl. HVAC, economy	555	138	693
0140	Deluxe incl. fixtures	1,300	310	1,610

E1020 810	Medical Equipment, EACH	COST EACH		
		MAT.	INST.	TOTAL
0100	Dental equipment, central suction system, economy	3,125	360	3,485
0110	Compressor-air, deluxe	23,800	775	24,575
0120	Chair, hydraulic, economy	3,925	775	4,700
0130	Deluxe	8,325	1,550	9,875
0140	Drill console with accessories, economy	3,525	242	3,767
0150	Deluxe	10,100	242	10,342
0160	X-ray unit, portable	4,725	96.50	4,821.50
0170	Panoramic unit	21,400	645	22,045
0300	Medical equipment, autopsy table, standard	7,775	435	8,210
0310	Deluxe	11,600	725	12,325
0320	Incubators, economy	3,150		3,150
0330	Deluxe	17,700		17,700
0700	Station, scrub-surgical, single, economy	1,250	145	1,395
0710	Dietary, medium, with ice	14,300		14,300
0720	Sterilizers, general purpose, single door, 20" x 20" x 28"	13,600		13,600
0730	Floor loading, double door, 28" x 67" x 52"	181,500		181,500
0740	Surgery tables, standard	12,300	705	13,005
0750	Deluxe	28,000	990	28,990
0770	Tables, standard, with base cabinets, economy	1,150	249	1,399
0780	Deluxe	3,425	375	3,800
0790	X-ray, mobile, economy	11,900		11,900
0800	Stationary, deluxe	194,000		194,000

E1030 110	Vehicular Service Equipment, EACH	COST EACH		
		MAT.	INST.	TOTAL
0110	Automotive equipment, compressors, electric, 1-1/2 H.P., std. controls	355	705	1,060
0120	5 H.P., dual controls	1,800	1,050	2,850
0130	Hoists, single post, 4 ton capacity, swivel arms	3,850	2,650	6,500
0140	Dual post, 12 ton capacity, adjustable frame	6,600	7,050	13,650
0150	Lube equipment, 3 reel type, with pumps	6,600	2,125	8,725
0160	Product dispenser, 6 nozzles, w/vapor recovery, not incl. piping, installed	16,500		16,500
0800	Scales, dial type, built in floor, 5 ton capacity, 8' x 6' platform	6,325	2,250	8,575
0810	10 ton capacity, 9' x 7' platform	9,625	3,200	12,825
0820	Truck (including weigh bridge), 20 ton capacity, 24' x 10'	9,700	3,725	13,425

E1030 210	Parking Control Equipment, EACH	COST EACH		
		MAT.	INST.	TOTAL
0110	Parking equipment, automatic gates, 8 ft. arm, one way	3,325	770	4,095
0120	Traffic detectors, single treadle	2,225	355	2,580
0130	Booth for attendant, economy	5,225		5,225
0140	Deluxe	23,300		23,300
0150	Ticket printer/dispenser, rate computing	7,400	605	8,005
0160	Key station on pedestal	675	206	881

E1030 310	Loading Dock Equipment, EACH	COST EACH		
		MAT.	INST.	TOTAL
0110	Dock bumpers, rubber blocks, 4-1/2" thick, 10" high, 14" long	46.50	14.35	60.85
0120	6" thick, 20" high, 11" long	126	28.50	154.50
0130	Dock boards, H.D., 5' x 5', aluminum, 5000 lb. capacity	1,225		1,225
0140	16,000 lb. capacity	1,450		1,450
0150	Dock levelers, hydraulic, 7' x 8', 10 ton capacity	8,025	1,025	9,050
0160	Dock lifters, platform, 6' x 6', portable, 3000 lb. capacity	7,850		7,850
0170	Dock shelters, truck, scissor arms, economy	1,250	375	1,625
0180	Deluxe	1,825	745	2,570

E EQUIPMENT & FURNISHINGS

E

419

E1090 210	Solid Waste Handling Equipment, EACH	COST EACH		
		MAT.	INST.	TOTAL
0110	Waste handling, compactors, single bag, 250 lbs./hr., hand fed	6,825	440	7,265
0120	Heavy duty industrial, 5 C.Y. capacity	15,900	2,125	18,025
0130	Incinerator, electric, 100 lbs./hr., economy	11,900	2,125	14,025
0140	Gas, 2000 lbs./hr., deluxe	187,000	16,500	203,500
0150	Shredder, no baling, 35 tons/hr.	230,500		230,500
0160	Incl. baling, 50 tons/day	461,000		461,000

E1090 350	Food Service Equipment, EACH	COST EACH		
		MAT.	INST.	TOTAL
0110	Kitchen equipment, bake oven, single deck	4,350	97.50	4,447.50
0120	Broiler, without oven	4,175	97.50	4,272.50
0130	Commercial dish washer, semiautomatic, 50 racks/hr.	6,225	600	6,825
0140	Automatic, 275 racks/hr.	25,700	2,575	28,275
0150	Cooler, beverage, reach-in, 6 ft. long	3,250	130	3,380
0160	Food warmer, counter, 1.65 kw	745		745
0170	Fryers, with submerger, single	3,350	112	3,462
0180	Double	4,525	156	4,681
0190	Kettles, steam jacketed, 20 gallons	6,475	179	6,654
0200	Range, restaurant type, burners, 2 ovens and 24" griddle	5,175	130	5,305
0210	Range hood, incl. carbon dioxide system, economy	3,350	260	3,610
0220	Deluxe	24,300	780	25,080

E1090 360	Food Service Equipment, S.F.	COST PER S.F.		
		MAT.	INST.	TOTAL
0110	Refrigerators, prefab, walk-in, 7'-6" high, 6' x 6'	107	13.65	120.65
0120	12' x 20'	67	6.80	73.80

E1090 410	Residential Equipment, EACH	COST EACH		
		MAT.	INST.	TOTAL
0110	Arch. equip., appliances, range, cook top, 4 burner, economy	210	70.50	280.50
0120	Built in, single oven 30" wide, economy	350	70.50	420.50
0130	Standing, single oven-21" wide, economy	315	58.50	373.50
0140	Double oven-30" wide, deluxe	1,600	58.50	1,658.50
0150	Compactor, residential, economy	390	74.50	464.50
0160	Deluxe	425	125	550
0170	Dish washer, built-in, 2 cycles, economy	279	214	493
0180	4 or more cycles, deluxe	575	430	1,005
0190	Garbage disposer, sink type, economy	42.50	85.50	128
0200	Deluxe	201	85.50	286.50
0210	Refrigerator, no frost, 10 to 12 C.F., economy	495	58.50	553.50
0220	21 to 29 C.F., deluxe	3,050	195	3,245

E1090 710	School Equipment, EACH	COST EACH		
		MAT.	INST.	TOTAL
0110	School equipment, basketball backstops, wall mounted, wood, fixed	1,025	660	1,685
0120	Suspended type, electrically operated	3,625	1,275	4,900
0130	Bleachers-telescoping, manual operation, 15 tier, economy (per seat)	56	23.50	79.50
0140	Power operation, 30 tier, deluxe (per seat)	213	41.50	254.50
0150	Weight lifting gym, universal, economy	2,500	585	3,085
0160	Deluxe	13,900	1,175	15,075
0170	Scoreboards, basketball, 1 side, economy	2,175	590	2,765
0180	4 sides, deluxe	24,000	8,125	32,125
0800	Vocational shop equipment, benches, metal	258	149	407
0810	Wood	440	149	589
0820	Dust collector, not incl. ductwork, 6' diam.	2,850	395	3,245
0830	Planer, 13" x 6"	1,325	187	1,512

Important: See the Reference Section for critical supporting data - Location Factors & Historical Cost Indexes

E1090 720	School Equipment, S.F.	COST PER S.F.		
		MAT.	INST.	TOTAL
0110	School equipment, gym mats, naugahyde cover, 2" thick	3.45		3.45
0120	Wrestling, 1" thick, heavy duty	4.62		4.62

E1090 810	Athletic, Recreational, and Therapeutic Equipment, EACH	COST EACH		
		MAT.	INST.	TOTAL
0110	Sauna, prefabricated, incl. heater and controls, 7' high, 6' x 4'	4,250	570	4,820
0120	10' x 12'	8,225	1,250	9,475
0130	Heaters, wall mounted, to 200 C.F.	660		660
0140	Floor standing, to 1000 C.F., 12500 W	1,800	141	1,941
0610	Shooting range incl. bullet traps, controls, separators, ceilings, economy	12,600	4,950	17,550
0620	Deluxe	22,500	8,350	30,850
0650	Sport court, squash, regulation, in existing building, economy			15,500
0660	Deluxe			29,000
0670	Racketball, regulation, in existing building, economy	2,050	7,075	9,125
0680	Deluxe	27,500	14,100	41,600
0700	Swimming pool equipment, diving stand, stainless steel, 1 meter	4,150	277	4,427
0710	3 meter	5,100	1,875	6,975
0720	Diving boards, 16 ft. long, aluminum	1,600	277	1,877
0730	Fiberglass	1,200	277	1,477
0740	Filter system, sand, incl. pump, 6000 gal./hr.	1,150	480	1,630
0750	Lights, underwater, 12 volt with transformer, 300W	128	1,050	1,178
0760	Slides, fiberglass with aluminum handrails & ladder, 6' high, straight	2,350	465	2,815
0780	12' high, straight with platform	2,100	625	2,725

E1090 820	Athletic, Recreational, and Therapeutic Equipment, S.F.	COST PER S.F.		
		MAT.	INST.	TOTAL
0110	Swimming pools, residential, vinyl liner, metal sides	9.75	4.98	14.73
0120	Concrete sides	11.65	9.30	20.95
0130	Gunite shell, plaster finish, 350 S.F.	19.35	19.20	38.55
0140	800 S.F.	15.55	11.15	26.70
0150	Motel, gunite shell, plaster finish	24	24.50	48.50
0160	Municipal, gunite shell, tile finish, formed gutters	100	41.50	141.50

E EQUIPMENT & FURNISHINGS

E

E2010 Fixed Furnishings

E2010 310	Window Treatment, EACH	COST EACH		
		MAT.	INST.	TOTAL
0110	Furnishings, blinds, exterior, aluminum, louvered, 1'-4" wide x 3'-0" long	43.50	37.50	81
0120	1'-4" wide x 6'-8" long	87	41.50	128.50
0130	Hemlock, solid raised, 1-'4" wide x 3'-0" long	57	37.50	94.50
0140	1'-4" wide x 6'-9" long	107	41.50	148.50
0150	Polystyrene, louvered, 1'-3" wide x 3'-3" long	44	37.50	81.50
0160	1'-3" wide x 6'-8" long	95.50	41.50	137
0200	Interior, wood folding panels, louvered, 7" x 20" (per pair)	44.50	22	66.50
0210	18" x 40" (per pair)	138	22	160

E2010 320	Window Treatment, S.F.	COST PER S.F.		
		MAT.	INST.	TOTAL
0110	Furnishings, blinds-interior, venetian-aluminum, stock, 2" slats, economy	1.71	.63	2.34
0120	Custom, 1" slats, deluxe	7.40	.85	8.25
0130	Vertical, PVC or cloth, T&B track, economy	7.40	.81	8.21
0140	Deluxe	9.85	.93	10.78
0150	Draperies, unlined, economy	2.29		2.29
0160	Lightproof, deluxe	6.85		6.85
0510	Shades, mylar, wood roller, single layer, non-reflective	5.90	.55	6.45
0520	Metal roller, triple layer, heat reflective	8.60	.55	9.15
0530	Vinyl, light weight, 4 ga.	.42	.55	.97
0540	Heavyweight, 6 ga.	1.32	.55	1.87
0550	Vinyl coated cotton, lightproof decorator shades	1.65	.55	2.20
0560	Woven aluminum, 3/8" thick, light and fireproof	4.33	1.07	5.40

E2010 420	Fixed Floor Grilles and Mats, S.F.	COST PER S.F.		
		MAT.	INST.	TOTAL
0110	Floor mats, recessed, inlaid black rubber, 3/8" thick, solid	22	1.88	23.88
0120	Colors, 1/2" thick, perforated	36.50	1.88	38.38
0130	Link-including nosings, steel-galvanized, 3/8" thick	7.15	1.88	9.03
0140	Vinyl, in colors	19.65	1.88	21.53

E2010 510	Fixed Multiple Seating, EACH	COST EACH		
		MAT.	INST.	TOTAL
0110	Seating, painted steel, upholstered, economy	116	21.50	137.50
0120	Deluxe	370	26.50	396.50
0400	Seating, lecture hall, pedestal type, economy	122	34	156
0410	Deluxe	435	51.50	486.50
0500	Auditorium chair, veneer construction	137	34	171
0510	Fully upholstered, spring seat	174	34	208

Important: See the Reference Section for critical supporting data - Location Factors & Historical Cost Indexes

E2020 Moveable Furnishings

E2020 210	Furnishings/Each	COST EACH		
		MAT.	INST.	TOTAL
0200	Hospital furniture, beds, manual, economy	910		910
0210	Deluxe	1,575		1,575
0220	All electric, economy	1,525		1,525
0230	Deluxe	3,775		3,775
0240	Patient wall systems, no utilities, economy, per room	910		910
0250	Deluxe, per room	1,675		1,675
0300	Hotel furnishings, standard room set, economy, per room	1,750		1,750
0310	Deluxe, per room	8,675		8,675
0500	Office furniture, standard employee set, economy, per person	395		395
0510	Deluxe, per person	2,325		2,325
0550	Posts, portable, pedestrian traffic control, economy	108		108
0560	Deluxe	315		315
0700	Restaurant furniture, booth, molded plastic, stub wall and 2 seats, economy	365	187	552
0710	Deluxe	1,575	249	1,824
0720	Upholstered seats, foursome, single-economy	1,225	74.50	1,299.50
0730	Foursome, double-deluxe	2,525	125	2,650

E2020 220	Furniture and Accessories, L.F.	COST PER L.F.		
		MAT.	INST.	TOTAL
0210	Dormitory furniture, desk top (built-in),laminated plastc, 24"deep, economy	26	14.95	40.95
0220	30" deep, deluxe	144	18.70	162.70
0230	Dressing unit, built-in, economy	193	62.50	255.50
0240	Deluxe	575	93.50	668.50
0310	Furnishings, cabinets, hospital, base, laminated plastic	208	74.50	282.50
0320	Stainless steel	260	74.50	334.50
0330	Counter top, laminated plastic, no backsplash	36	18.70	54.70
0340	Stainless steel	99.50	18.70	118.20
0350	Nurses station, door type, laminated plastic	241	74.50	315.50
0360	Stainless steel	278	74.50	352.50
0710	Restaurant furniture, bars, built-in, back bar	154	74.50	228.50
0720	Front bar	212	74.50	286.50
0910	Wardrobes & coatracks, standing, steel, single pedestal, 30" x 18" x 63"	61		61
0920	Double face rack, 39" x 26" x 70"	63		63
0930	Wall mounted rack, steel frame & shelves, 12" x 15" x 26"	56.50	5.30	61.80
0940	12" x 15" x 50"	44.50	2.78	47.28

For information about Means Estimating Seminars, see yellow pages 11 and 12 in back of book

EQUIPMENT & FURNISHINGS

E

F SPECIAL CONSTRUCTION & DEMOLITION

F1010 120	Air-Supported Structures, S.F.	COST PER S.F.		
		MAT.	INST.	TOTAL
0110	Air supported struc., polyester vinyl fabric, 24oz., warehouse, 5000 S.F.	17	.23	17.23
0120	50,000 S.F.	7.75	.19	7.94
0130	Tennis, 7,200 S.F.	15.85	.19	16.04
0140	24,000 S.F.	10.75	.19	10.94
0150	Woven polyethylene, 6 oz., shelter, 3,000 S.F.	6.70	.39	7.09
0160	24,000 S.F.	5.65	.19	5.84
0170	Teflon coated fiberglass, stadium cover, economy	41	.10	41.10
0180	Deluxe	47.50	.14	47.64
0190	Air supported storage tank covers, reinf. vinyl fabric, 12 oz., 400 S.F.	15.30	.33	15.63
0200	18,000 S.F.	4.71	.30	5.01

F1010 310	Pre-Engineered Structures, EACH	COST EACH		
		MAT.	INST.	TOTAL
0600	Radio towers, guyed, 40 lb. section, 50' high, 70 MPH basic wind speed	1,750	990	2,740
0610	90 lb. section, 400' high, wind load 70 MPH basic wind speed	22,800	10,800	33,600
0620	Self supporting, 60' high, 70 MPH basic wind speed	3,675	1,900	5,575
0630	190' high, wind load 90 MPH basic wind speed	21,800	7,575	29,375
0700	Shelters, aluminum frame, acrylic glazing, 8' high, 3' x 9'	2,800	865	3,665
0710	9' x 12'	4,675	1,350	6,025

F1010 320	Other Special Structures, S.F.	COST PER S.F.		
		MAT.	INST.	TOTAL
0110	Swimming pool enclosure, transluscent, freestanding, economy	12.45	3.74	16.19
0120	Deluxe	179	10.65	189.65
0510	Tension structures, steel frame, polyester vinyl fabric, 12,000 S.F.	10.90	1.71	12.61
0520	20,800 S.F.	11.30	1.54	12.84

F10 Special Construction

F1020 Integrated Construction

F1020 110	Integrated Construction, EACH	COST EACH		
		MAT.	INST.	TOTAL
0110	Integrated ceilings, radiant electric, 2' x 4' panel, manila finish	91	16.95	107.95
0120	ABS plastic finish	106	32.50	138.50

F1020 120	Integrated Construction, S.F.	COST PER S.F.		
		MAT.	INST.	TOTAL
0110	Integrated ceilings, Luminaire, suspended, 5' x 5' modules, 50% lighted	4.07	8.90	12.97
0120	100% lighted	6.05	16.05	22.10
0130	Dimensionaire, 2' x 4' module tile system, no air bar	2.46	1.49	3.95
0140	With air bar, deluxe	3.62	1.49	5.11
0220	Pedestal access floors, stl pnls, no stringers, vinyl covering, >6000 S.F.	13.65	1.66	15.31
0230	With stringers, high pressure laminate covering, under 6000 S.F.	14.80	1.87	16.67
0240	Carpet covering, under 6000 S.F.	16.65	1.87	18.52
0250	Aluminum panels, no stringers, no covering	27	1.49	28.49

F1020 230	Other Integrated Construction, S.F.	COST PER S.F.		
		MAT.	INST.	TOTAL
0110	Anechoic chambers, 7' high, 100 cps cutoff, 25 S.F.			2,000
0120	200 cps cutoff, 100 S.F.			1,025
0130	Audiometric rooms, under 500 S.F.	44	15.25	59.25
0140	Over 500 S.F.	42	12.45	54.45

F1020 250	Special Purpose Room, EACH	COST EACH		
		MAT.	INST.	TOTAL
0110	Portable booth, acoustical, 27 db 1000 hz., 15 S.F. floor	3,325		3,325
0120	55 S.F. flr.	6,825		6,825

F1020 260	Special Purpose Room, S.F.	COST PER S.F.		
		MAT.	INST.	TOTAL
0300	Darkrooms, shell, not including door, 240 S.F., 8' high	38	6.25	44.25
0310	64 S.F., 12' high	88	11.70	99.70
0510	Music practice room, modular, perforated steel, under 500 S.F.	27	10.65	37.65
0520	Over 500 S.F.	23.50	9.35	32.85

F1020 330	Special Construction, L.F.	COST PER L.F.		
		MAT.	INST.	TOTAL
0110	Spec. const., air curtains, shipping & receiving, 8'high x 5'wide, economy	305	86.50	391.50
0120	20' high x 8' wide, heated, deluxe	296	216	512
0130	Customer entrance, 10' high x 5' wide, economy	299	86.50	385.50
0140	12' high x 4' wide, heated, deluxe	445	108	553

F SPECIAL CONSTRUCTION & DEMOLITION

F1030 120	Sound, Vibration, and Seismic Construction, S.F.	COST PER S.F.		
		MAT.	INST.	TOTAL
0020	Special construction, acoustical, enclosure, 4" thick, 8 psf panels	27	15.55	42.55
0030	Reverb chamber, 4" thick, parallel walls	38.50	18.70	57.20
0110	Sound absorbing panels, 2'-6" x 8', painted metal	10.25	5.20	15.45
0120	Vinyl faced	8	4.67	12.67
0130	Flexible transparent curtain, clear	6.15	6.05	12.20
0140	With absorbing foam, 75% coverage	8.60	6.05	14.65
0150	Strip entrance, 2/3 overlap	4.26	9.60	13.86
0160	Full overlap	5.25	11.30	16.55
0200	Audio masking system, plenum mounted, over 10,000 S.F.	.59	.19	.78
0210	Ceiling mounted, under 5,000 S.F.	1.20	.35	1.55

F1030 210	Radiation Protection, EACH	COST EACH		
		MAT.	INST.	TOTAL
0110	Shielding, lead x-ray protection, radiography room, 1/16" lead, economy	5,050	2,725	7,775
0120	Deluxe	6,600	4,550	11,150
0210	Deep therapy x-ray, 1/4" lead, economy	17,100	8,550	25,650
0220	Deluxe	23,100	11,400	34,500

F1030 220	Radiation Protection, S.F.	COST PER S.F.		
		MAT.	INST.	TOTAL
0110	Shielding, lead, gypsum board, 5/8" thick, 1/16" lead	4.94	4.54	9.48
0120	1/8" lead	10.20	5.20	15.40
0130	Lath, 1/16" thick	4.75	5.05	9.80
0140	1/8" thick	9.95	5.70	15.65
0150	Radio frequency, copper, prefab screen type, economy	26.50	4.15	30.65
0160	Deluxe	34	5.15	39.15

F1030 910	Other Special Construction Systems, EACH	COST EACH		
		MAT.	INST.	TOTAL
0110	Disappearing stairways, folding, pine, 8'-6" ceiling	98	93.50	191.50
0200	Fire escape, galvanized steel, 8'-0" to 10'-6" ceiling	1,250	745	1,995
0220	Automatic electric, wood, 8' to 9' ceiling	6,125	745	6,870
0300	Fireplace prefabricated, freestanding or wall hung, economy	1,075	287	1,362
0310	Deluxe	3,125	415	3,540
0320	Woodburning stoves, cast iron, economy	815	575	1,390
0330	Deluxe	2,425	935	3,360

Important: See the Reference Section for critical supporting data - Location Factors & Historical Cost Indexes

F1040 Special Facilities

F1040 210	Ice Rinks, EACH	COAT EACH		
		MAT.	INST.	TOTAL
0100	Ice skating rink, 85' x 200', 55° system, 5 mos., 100 ton			451,000
0110	90° system, 12 mos., 135 ton			500,500
0120	Dash boards, acrylic screens, polyethylene coated plywood	126,500	25,300	151,800
0130	Fiberglass and aluminum construction	176,000	25,300	201,300

F1040 510	Liquid & Gas Storage Tanks, EACH	COST EACH		
		MAT.	INST.	TOTAL
0100	Tanks, steel, ground level, 100,000 gal.			99,500
0110	10,000,000 gal.			1,962,000
0120	Elevated water, 50,000 gal.		138,000	188,500
0130	1,000,000 gal.		644,500	881,000
0150	Cypress wood, ground level, 3,000 gal.	6,350	7,425	13,775
0160	Redwood, ground level, 45,000 gal.	49,100	20,200	69,300

F1040 810	Special Structures, EACH	COST EACH		
		MAT.	INST.	TOTAL
0110	Kiosks, round, 5' diam., 1/4" fiberglass wall, 8' high	6,050		6,050
0120	Rectangular, 5' x 9', 1" insulated dbl. wall fiberglass, 7'-6" high	10,500		10,500
0200	Silos, concrete stave, industrial, 12' diam., 35' high	17,800	15,500	33,300
0210	25' diam., 75' high	58,500	34,000	92,500
0220	Steel prefab, 30,000 gal., painted, economy	12,600	4,175	16,775
0230	Epoxy-lined, deluxe	26,200	8,375	34,575

F1040 820	Special Structures, S.F.	COST PER SF		
		MAT.	INST.	TOTAL
0110	Comfort stations, prefab, mobile on steel frame, economy	40		40
0120	Permanent on concrete slab, deluxe	227	35.50	262.50
0210	Domes, bulk storage, wood framing, wood decking, 50' diam.	28.50	1.27	29.77
0220	116' diam.	17	1.48	18.48
0230	Steel framing, metal decking, 150' diam.	30.50	8.70	39.20
0240	400' diam.	25	6.65	31.65
0250	Geodesic, wood framing, wood panels, 30' diam.	19	1.36	20.36
0260	60' diam.	14.75	.90	15.65
0270	Aluminum framing, acrylic panels, 40' diam.			82.50
0280	Aluminum panels, 400' diam.			33
0310	Garden house, prefab, wood, shell only, 48 S.F.	21	3.74	24.74
0320	200 S.F.	38	15.55	53.55
0410	Greenhouse, shell-stock, residential, lean-to, 8'-6" long x 3'-10" wide	36	22	58
0420	Freestanding, 8'-6" long x 13'-6" wide	28	6.90	34.90
0430	Commercial-truss frame, under 2000 S.F., deluxe			40.50
0440	Over 5,000 S.F., economy			24.50
0450	Institutional-rigid frame, under 500 S.F., deluxe			105
0460	Over 2,000 S.F., economy			41.50
0510	Hangar, prefab, galv. roof and walls, bottom rolling doors, economy	9.35	3.45	12.80
0520	Electric bifolding doors, deluxe	10.20	4.49	14.69

SPECIAL CONSTRUCTION & DEMOLITION

F

429

F1040 910	Special Construction, EACH	COST EACH		
		MAT.	INST.	TOTAL
0110	Special construction, bowling alley incl. pinsetter, scorer etc., economy	38,600	7,475	46,075
0120	Deluxe	48,300	8,300	56,600
0130	For automatic scorer, economy, add	6,600		6,600
0140	Deluxe, add	7,475		7,475
0200	Chimney, metal, high temp. steel jacket, factory lining, 24" diameter	9,550	3,675	13,225
0210	60" diameter	69,500	15,900	85,400
0220	Poured concrete, brick lining, 10' diam., 200' high			1,210,000
0230	20' diam., 500' high	5,350,000		5,350,000
0240	Radial brick, 3'-6" I.D., 75' high			199,000
0250	7' I.D., 175' high			595,000
0300	Control tower, modular, 12' x 10', incl. instrumentation, economy			463,500
0310	Deluxe			618,000
0400	Garage costs, residential, prefab, wood, single car economy	3,500	745	4,245
0410	Two car deluxe	10,500	1,500	12,000
0500	Hangars, prefab, galv. steel, bottom rolling doors, economy (per plane)	9,250	3,825	13,075
0510	Electrical bi-folding doors, deluxe (per plane)	12,900	5,250	18,150

For information about Means Estimating Seminars, see yellow pages 11 and 12 in back of book

SPECIAL CONSTRUCTION & DEMOLITION F

Important: See the Reference Section for critical supporting data - Location Factors & Historical Cost Indexes

G BUILDING SITEWORK

G1030 805	Trenching	COST PER L.F.		
		MAT.	INST.	TOTAL
1310	Trenching, backhoe, 0 to 1 slope, 2' wide, 2' deep, 3/8 C.Y. bucket		2.49	2.49
1330	4' deep, 3/8 C.Y. bucket		4.32	4.32
1360	10' deep, 1 C.Y. bucket		8.90	8.90
1400	4' wide, 2' deep, 3/8 C.Y. bucket		5.09	5.09
1420	4' deep, 1/2 C.Y. bucket		7.26	7.26
1450	10' deep, 1 C.Y. bucket		18	18
1480	18' deep, 2-1/2 C.Y. bucket		27.95	27.95
1520	6' wide, 6' deep, 5/8 C.Y. bucket		17.20	17.20
1540	10' deep, 1 C.Y. bucket		24.25	24.25
1570	20' deep, 3-1/2 C.Y. bucket		39	39
1640	8' wide, 12' deep, 1-1/4 C.Y. bucket		28.85	28.85
1680	24' deep, 3-1/2 C.Y. bucket		86.50	86.50
1730	10' wide, 20' deep, 3-1/2 C.Y. bucket		67.50	67.50
1740	24' deep, 3-1/2 C.Y. bucket		109	109
3500	1 to 1 slope, 2' wide, 2' deep, 3/8 C.Y. bucket		3.79	3.79
3540	4' deep, 3/8 C.Y. bucket		8.91	8.91
3600	10' deep, 1 C.Y. bucket		38.85	38.85
3800	4' wide, 2' deep, 3/8 C.Y. bucket		6.45	6.45
3840	4' deep, 1/2 C.Y. bucket		10.51	10.51
3900	10' deep, 1 C.Y. bucket		41.30	41.30
4030	6' wide, 6' deep, 5/8 C.Y. bucket		23.35	23.35
4050	10' deep, 1 C.Y. bucket		48.45	48.45
4080	20' deep, 3-1/2 C.Y. bucket		127.50	127.50
4500	8' wide, 12' deep, 1-1/4 C.Y. bucket		55	55
4650	24' deep, 3-1/2 C.Y. bucket		263	263
4800	10' wide, 20' deep, 3-1/2 C.Y. bucket		152.50	152.50
4850	24' deep, 3-1/2 C.Y. bucket		282	282

G1030 815	Pipe Bedding	COST PER L.F.		
		MAT.	INST.	TOTAL
1440	Pipe bedding, side slope 0 to 1, 1' wide, pipe size 6" diameter	.29	.47	.76
1460	2' wide, pipe size 8" diameter	.63	1.03	1.66
1500	Pipe size 12" diameter	.67	1.08	1.75
1600	4' wide, pipe size 20" diameter	1.66	2.69	4.35
1660	Pipe size 30" diameter	1.75	2.85	4.60
1680	6' wide, pipe size 32" diameter	3.07	4.98	8.05
1740	8' wide, pipe size 60" diameter	5.10	8.30	13.40
1780	12' wide, pipe size 84" diameter	10.05	16.30	26.35

BUILDING SITEWORK G

Important: See the Reference Section for critical supporting data - Location Factors & Historical Cost Indexes

G2010 Roadways

G2010 210	Roadway Pavement	COST PER L.F.		
		MAT.	INST.	TOTAL
1500	Roadways, bituminous conc. paving, 2-1/2" thick, 20' wide	38	41	79
1580	30' wide	46.50	47	93.50
1800	3" thick, 20' wide	39.50	40	79.50
1880	30' wide	49	47.50	96.50
2100	4" thick, 20' wide	43	40	83
2180	30' wide	54.50	47	101.50
2400	5" thick, 20' wide	50	41	91
2480	30' wide	65	48.50	113.50
3000	6" thick, 20' wide	71	42	113
3080	30' wide	96	49.50	145.50
3300	8" thick 20' wide	87	42.50	129.50
3380	30' wide	121	50	171
3600	12" thick 20' wide	107	45	152
3700	32' wide	159	55.50	214.50

BUILDING SITEWORK

G

G2020 Parking Lots

G2020 210	Parking Lots	COST PER CAR		
		MAT.	INST.	TOTAL
1500	Parking lot, 90° angle parking, 3" bituminous paving, 6" gravel base	310	222	532
1540	10" gravel base	370	259	629
1560	4" bituminous paving, 6" gravel base	365	221	586
1600	10" gravel base	425	257	682
1620	6" bituminous paving, 6" gravel base	470	229	699
1660	10" gravel base	530	266	796
1800	60° angle parking, 3" bituminous paving, 6" gravel base	355	243	598
1840	10" gravel base	425	286	711
1860	4" bituminous paving, 6" gravel base	420	241	661
1900	10" gravel base	490	284	774
1920	6" bituminous paving, 6" gravel base	545	252	797
1960	10" gravel base	610	295	905
2200	45° angle parking, 3" bituminous paving, 6" gravel base	405	263	668
2240	10" gravel base	480	310	790
2260	4" bituminous paving, 6" gravel base	475	262	737
2300	10" gravel base	555	310	865
2320	6" bituminous paving, 6" gravel base	615	273	888
2360	10" gravel base	695	325	1,020

G2040 Site Development

G2040 810	Flagpoles	COST EACH		
		MAT.	INST.	TOTAL
0110	Flagpoles, on grade, aluminum, tapered, 20' high	665	435	1,100
0120	70' high	7,250	1,100	8,350
0130	Fiberglass, tapered, 23' high	1,000	435	1,435
0140	59' high	6,450	970	7,420
0150	Concrete, internal halyard, 20' high	1,275	350	1,625
0160	100' high	12,800	870	13,670

G2040 950	Other Site Development, EACH	COST EACH		
		MAT.	INST.	TOTAL
0110	Grandstands, permanent, closed deck, steel, economy (per seat)			33
0120	Deluxe (per seat)			88
0130	Composite design, economy (per seat)			82.50
0140	Deluxe (per seat)			154

BUILDING SITEWORK

G

G3030 210	Manholes & Catch Basins	COST PER EACH		
		MAT.	INST.	TOTAL
1920	Manhole/catch basin, brick, 4' I.D. riser, 4' deep	710	1,150	1,860
1980	10' deep	1,550	2,725	4,275
3200	Block, 4' I.D. riser, 4' deep	650	930	1,580
3260	10' deep	1,275	2,250	3,525
4620	Concrete, cast-in-place, 4' I.D. riser, 4' deep	785	1,625	2,410
4680	10' deep	1,750	3,925	5,675
5820	Concrete, precast, 4' I.D. riser, 4' deep	855	865	1,720
5880	10' deep	1,650	2,025	3,675
6200	6' I.D. riser, 4' deep	1,325	1,200	2,525
6260	10' deep	2,675	2,875	5,550

For information about Means Estimating Seminars, see yellow pages 11 and 12 in back of book

Reference Section

All the reference information is in one section making it easy to find what you need to know . . . and easy to use the book on a daily basis. This section is visually identified by a vertical gray bar on the edge of pages.

In this Reference Section, we've included General Conditions; Historical Cost Indexes for cost comparisons over time; Location Factors; a Glossary; and an explanation of all Abbreviations used in the book.

Table of Contents

437

General Conditions, Overhead & Profit

The total building costs in the Commercial/Industrial/Institutional section include a 10% allowance for general conditions and a 15% allowance for the general contractor's overhead and profit and contingencies.

General contractor overhead includes indirect costs such as permits, Workers' Compensation, insurances, supervision and bonding fees. Overhead will vary with the size of project, the contractor's operating procedures and location. Profits will vary with economic activity and local conditions.

Contingencies provide for unforeseen construction difficulties which include material shortages and weather. In all situations, the appraiser should give consideration to possible adjustment of the 25% factor used in developing the Commercial/Industrial/Institutional models.

Architectural Fees

Tabulated below are typical percentage fees by project size, for good professional architectural service. Fees may vary from those listed depending upon degree of design difficulty and economic conditions in any particular area.

Rates can be interpolated horizontally and vertically. Various portions of the same project requiring different rates should be adjusted proportionately. For alterations, add 50% to the fee for the first $500,000 of project cost and add 25% to the fee for project cost over $500,000.

Architectural fees tabulated below include Engineering fees.

Insurance Exclusions

Many insurance companies exclude from coverage such items as architect's fees, excavation, foundations below grade, underground piping and site preparation. Since exclusions vary among insurance companies, it is recommended that for greatest accuracy each exclusion be priced separately using the unit-in-place section.

As a rule of thumb, exclusions can be calculated at 9% of total building cost plus the appropriate allowance for architect's fees.

Building Types	Total Project Size in Thousands of Dollars						
	100	250	500	1,000	5,000	10,000	50,000
Factories, garages, warehouses, repetitive housing	9.0%	8.0%	7.0%	6.2%	5.3%	4.9%	4.5%
Apartments, banks, schools, libraries, offices, municipal buildings	12.2	12.3	9.2	8.0	7.0	6.6	6.2
Churches, hospitals, homes, laboratories, museums, research	15.0	13.6	12.7	11.9	9.5	8.8	8.0
Memorials, monumental work, decorative furnishings	—	16.0	14.5	13.1	10.0	9.0	8.3

Location Factors

Costs shown in *Means cost data publications* are based on National Averages for materials and installation. To adjust these costs to a specific location, simply multiply the base cost by the factor for that city. The data is arranged alphabetically by state and postal zip code numbers. For a city not listed, use the factor for a nearby city with similar economic characteristics.

STATE/ZIP	CITY	Residential	Commercial
ALABAMA			
350-352	Birmingham	.85	.86
354	Tuscaloosa	.80	.78
355	Jasper	.76	.77
356	Decatur	.79	.80
357-358	Huntsville	.81	.82
359	Gadsden	.80	.81
360-361	Montgomery	.82	.80
362	Anniston	.73	.74
363	Dothan	.79	.77
364	Evergreen	.79	.77
365-366	Mobile	.81	.82
367	Selma	.79	.77
368	Phenix City	.82	.80
369	Butler	.79	.77
ALASKA			
995-996	Anchorage	1.25	1.24
997	Fairbanks	1.25	1.24
998	Juneau	1.24	1.23
999	Ketchikan	1.30	1.29
ARIZONA			
850,853	Phoenix	.92	.89
852	Mesa/Tempe	.87	.84
855	Globe	.88	.85
856-857	Tucson	.90	.87
859	Show Low	.89	.85
860	Flagstaff	.92	.88
863	Prescott	.90	.86
864	Kingman	.89	.85
865	Chambers	.88	.84
ARKANSAS			
716	Pine Bluff	.80	.80
717	Camden	.70	.70
718	Texarkana	.74	.74
719	Hot Springs	.69	.69
720-722	Little Rock	.81	.81
723	West Memphis	.79	.79
724	Jonesboro	.79	.79
725	Batesville	.75	.75
726	Harrison	.76	.76
727	Fayetteville	.69	.66
728	Russellville	.77	.74
729	Fort Smith	.83	.80
CALIFORNIA			
900-902	Los Angeles	1.08	1.08
903-905	Inglewood	1.06	1.06
906-908	Long Beach	1.07	1.07
910-912	Pasadena	1.07	1.07
913-916	Van Nuys	1.09	1.09
917-918	Alhambra	1.08	1.08
919-921	San Diego	1.10	1.06
922	Palm Springs	1.09	1.05
923-924	San Bernardino	1.08	1.04
925	Riverside	1.11	1.07
926-927	Santa Ana	1.09	1.06
928	Anaheim	1.10	1.08
930	Oxnard	1.13	1.08
931	Santa Barbara	1.11	1.08
932-933	Bakersfield	1.11	1.06
934	San Luis Obispo	1.14	1.08
935	Mojave	1.09	1.05
936-938	Fresno	1.12	1.08
939	Salinas	1.12	1.12
940-941	San Francisco	1.21	1.24
942,956-958	Sacramento	1.11	1.10
943	Palo Alto	1.15	1.18
944	San Mateo	1.16	1.19
945	Vallejo	1.11	1.14
946	Oakland	1.16	1.19
947	Berkeley	1.16	1.18
948	Richmond	1.14	1.17
949	San Rafael	1.26	1.20
950	Santa Cruz	1.16	1.14
951	San Jose	1.22	1.20
952	Stockton	1.13	1.09
953	Modesto	1.13	1.09

STATE/ZIP	CITY	Residential	Commercial
CALIFORNIA (CONT'D)			
954	Santa Rosa	1.14	1.17
955	Eureka	1.10	1.09
959	Marysville	1.10	1.09
960	Redding	1.11	1.10
961	Susanville	1.11	1.10
COLORADO			
800-802	Denver	.99	.95
803	Boulder	.88	.84
804	Golden	.97	.93
805	Fort Collins	.98	.92
806	Greeley	.90	.84
807	Fort Morgan	.97	.91
808-809	Colorado Springs	.94	.92
810	Pueblo	.94	.92
811	Alamosa	.89	.87
812	Salida	.89	.87
813	Durango	.88	.86
814	Montrose	.86	.84
815	Grand Junction	.90	.85
816	Glenwood Springs	.95	.91
CONNECTICUT			
060	New Britain	1.04	1.05
061	Hartford	1.04	1.05
062	Willimantic	1.03	1.04
063	New London	1.05	1.04
064	Meriden	1.03	1.04
065	New Haven	1.04	1.05
066	Bridgeport	1.02	1.05
067	Waterbury	1.05	1.05
068	Norwalk	1.01	1.05
069	Stamford	1.04	1.08
D.C.			
200-205	Washington	.93	.95
DELAWARE			
197	Newark	1.00	1.01
198	Wilmington	1.00	1.01
199	Dover	1.00	1.01
FLORIDA			
320,322	Jacksonville	.83	.82
321	Daytona Beach	.87	.86
323	Tallahassee	.75	.77
324	Panama City	.70	.72
325	Pensacola	.84	.82
326,344	Gainesville	.84	.81
327-328,347	Orlando	.86	.84
329	Melbourne	.90	.89
330-332,340	Miami	.83	.85
333	Fort Lauderdale	.83	.85
334,349	West Palm Beach	.86	.83
335-336,346	Tampa	.80	.82
337	St. Petersburg	.81	.83
338	Lakeland	.79	.81
339,341	Fort Myers	.79	.79
342	Sarasota	.78	.80
GEORGIA			
300-303,399	Atlanta	.85	.90
304	Statesboro	.72	.74
305	Gainesville	.76	.80
306	Athens	.77	.82
307	Dalton	.68	.67
308-309	Augusta	.76	.78
310-312	Macon	.81	.81
313-314	Savannah	.80	.81
315	Waycross	.74	.74
316	Valdosta	.76	.76
317	Albany	.77	.79
318-319	Columbus	.78	.78
HAWAII			
967	Hilo	1.27	1.23
968	Honolulu	1.27	1.23

Location Factors

STATE/ZIP	CITY	Residential	Commercial
STATES & POSS.			
969	Guam	1.37	1.33
IDAHO			
832	Pocatello	.94	.93
833	Twin Falls	.79	.78
834	Idaho Falls	.83	.82
835	Lewiston	1.09	1.01
836-837	Boise	.94	.93
838	Coeur d'Alene	.95	.88
ILLINOIS			
600-603	North Suburban	1.11	1.09
604	Joliet	1.11	1.10
605	South Suburban	1.10	1.09
606	Chicago	1.13	1.12
609	Kankakee	1.00	1.00
610-611	Rockford	1.05	1.04
612	Rock Island	1.06	.97
613	La Salle	1.06	.98
614	Galesburg	1.09	1.01
615-616	Peoria	1.09	1.02
617	Bloomington	1.05	1.00
618-619	Champaign	1.04	1.01
620-622	East St. Louis	1.00	1.00
623	Quincy	.99	.97
624	Effingham	1.02	.99
625	Decatur	1.01	.98
626-627	Springfield	1.01	.98
628	Centralia	.99	.99
629	Carbondale	.97	.97
INDIANA			
460	Anderson	.95	.93
461-462	Indianapolis	.98	.96
463-464	Gary	1.04	1.02
465-466	South Bend	.94	.92
467-468	Fort Wayne	.92	.93
469	Kokomo	.93	.92
470	Lawrenceburg	.93	.90
471	New Albany	.93	.89
472	Columbus	.96	.93
473	Muncie	.94	.93
474	Bloomington	.96	.93
475	Washington	.93	.93
476-477	Evansville	.95	.95
478	Terre Haute	.96	.95
479	Lafayette	.92	.92
IOWA			
500-503,509	Des Moines	.97	.93
504	Mason City	.86	.81
505	Fort Dodge	.84	.78
506-507	Waterloo	.88	.82
508	Creston	.89	.84
510-511	Sioux City	.95	.88
512	Sibley	.80	.78
513	Spencer	.80	.78
514	Carroll	.84	.80
515	Council Bluffs	.96	.89
516	Shenandoah	.82	.77
520	Dubuque	.99	.88
521	Decorah	.88	.79
522-524	Cedar Rapids	1.01	.92
525	Ottumwa	.94	.86
526	Burlington	.92	.86
527-528	Davenport	.98	.96
KANSAS			
660-662	Kansas City	.96	.94
664-666	Topeka	.86	.85
667	Fort Scott	.85	.83
668	Emporia	.81	.81
669	Belleville	.87	.81
670-672	Wichita	.89	.86
673	Independence	.82	.79
674	Salina	.85	.81
675	Hutchinson	.79	.76
676	Hays	.84	.80
677	Colby	.85	.81
678	Dodge City	.84	.81
679	Liberal	.78	.75
KENTUCKY			
400-402	Louisville	.95	.92
403-405	Lexington	.87	.84

STATE/ZIP	CITY	Residential	Commercial
KENTUCKY (CONT'D)			
406	Frankfort	.92	.86
407-409	Corbin	.78	.73
410	Covington	.98	.95
411-412	Ashland	.96	.97
413-414	Campton	.77	.73
415-416	Pikeville	.82	.83
417-418	Hazard	.76	.73
420	Paducah	.97	.92
421-422	Bowling Green	.96	.91
423	Owensboro	.91	.89
424	Henderson	.94	.92
425-426	Somerset	.75	.72
427	Elizabethtown	.94	.90
LOUISIANA			
700-701	New Orleans	.86	.85
703	Thibodaux	.85	.85
704	Hammond	.84	.83
705	Lafayette	.84	.81
706	Lake Charles	.83	.83
707-708	Baton Rouge	.82	.81
710-711	Shreveport	.81	.81
712	Monroe	.79	.79
713-714	Alexandria	.78	.78
MAINE			
039	Kittery	.86	.88
040-041	Portland	.91	.93
042	Lewiston	.92	.93
043	Augusta	.88	.88
044	Bangor	.93	.93
045	Bath	.89	.89
046	Machias	.87	.87
047	Houlton	.89	.89
048	Rockland	.86	.86
049	Waterville	.86	.85
MARYLAND			
206	Waldorf	.87	.87
207-208	College Park	.90	.90
209	Silver Spring	.89	.89
210-212	Baltimore	.91	.91
214	Annapolis	.89	.90
215	Cumberland	.87	.88
216	Easton	.73	.73
217	Hagerstown	.90	.88
218	Salisbury	.76	.77
219	Elkton	.82	.83
MASSACHUSETTS			
010-011	Springfield	1.04	1.02
012	Pittsfield	.99	.99
013	Greenfield	1.02	1.00
014	Fitchburg	1.08	1.04
015-016	Worcester	1.10	1.06
017	Framingham	1.06	1.06
018	Lowell	1.08	1.08
019	Lawrence	1.09	1.09
020-022, 024	Boston	1.14	1.15
023	Brockton	1.06	1.08
025	Buzzards Bay	1.02	1.04
026	Hyannis	1.04	1.05
027	New Bedford	1.06	1.07
MICHIGAN			
480,483	Royal Oak	1.03	1.02
481	Ann Arbor	1.04	1.03
482	Detroit	1.07	1.06
484-485	Flint	.99	1.00
486	Saginaw	.97	.98
487	Bay City	.96	.97
488-489	Lansing	1.01	.98
490	Battle Creek	1.01	.95
491	Kalamazoo	1.00	.94
492	Jackson	.99	.96
493,495	Grand Rapids	.88	.85
494	Muskegon	.95	.92
496	Traverse City	.87	.84
497	Gaylord	.87	.88
498-499	Iron Mountain	.98	.95
MINNESOTA			
550-551	Saint Paul	1.09	1.07
553-555	Minneapolis	1.11	1.08

STATE/ZIP	CITY	Residential	Commercial
556-558	Duluth	1.04	1.05
559	Rochester	1.03	.99
560	Mankato	1.00	.99
561	Windom	.90	.89
562	Willmar	.93	.92
563	St. Cloud	1.11	1.03
564	Brainerd	1.06	.99
565	Detroit Lakes	.88	.95
566	Bemidji	.91	.98
567	Thief River Falls	.87	.94
MISSISSIPPI			
386	Clarksdale	.70	.66
387	Greenville	.80	.77
388	Tupelo	.71	.71
389	Greenwood	.72	.68
390-392	Jackson	.80	.76
393	Meridian	.76	.75
394	Laurel	.72	.69
395	Biloxi	.84	.80
396	McComb	.72	.70
397	Columbus	.70	.71
MISSOURI			
630-631	St. Louis	1.00	1.03
633	Bowling Green	.92	.94
634	Hannibal	.99	.93
635	Kirksville	.86	.90
636	Flat River	.94	.97
637	Cape Girardeau	.93	.96
638	Sikeston	.90	.92
639	Poplar Bluff	.90	.92
640-641	Kansas City	1.04	1.01
644-645	St. Joseph	.90	.94
646	Chillicothe	.82	.86
647	Harrisonville	.98	.96
648	Joplin	.84	.86
650-651	Jefferson City	.98	.92
652	Columbia	.99	.93
653	Sedalia	.99	.92
654-655	Rolla	.95	.89
656-658	Springfield	.86	.88
MONTANA			
590-591	Billings	.93	.91
592	Wolf Point	.92	.90
593	Miles City	.91	.89
594	Great Falls	.92	.91
595	Havre	.90	.89
596	Helena	.91	.90
597	Butte	.90	.89
598	Missoula	.89	.88
599	Kalispell	.88	.87
NEBRASKA			
680-681	Omaha	.92	.91
683-685	Lincoln	.88	.83
686	Columbus	.74	.73
687	Norfolk	.84	.83
688	Grand Island	.88	.83
689	Hastings	.82	.78
690	Mccook	.79	.75
691	North Platte	.86	.82
692	Valentine	.77	.74
693	Alliance	.75	.71
NEVADA			
889-891	Las Vegas	1.05	1.04
893	Ely	.93	.94
894-895	Reno	.95	1.00
897	Carson City	.96	.99
898	Elko	.90	.93
NEW HAMPSHIRE			
030	Nashua	.94	.95
031	Manchester	.94	.95
032-033	Concord	.93	.94
034	Keene	.78	.79
035	Littleton	.82	.83
036	Charleston	.77	.77
037	Claremont	.76	.77
038	Portsmouth	.93	.92

STATE/ZIP	CITY	Residential	Commercial
NEW JERSEY			
070-071	Newark	1.14	1.12
072	Elizabeth	1.09	1.08
073	Jersey City	1.11	1.10
074-075	Paterson	1.12	1.12
076	Hackensack	1.10	1.10
077	Long Branch	1.10	1.08
078	Dover	1.11	1.09
079	Summit	1.08	1.06
080,083	Vineland	1.11	1.08
081	Camden	1.11	1.08
082,084	Atlantic City	1.11	1.08
085-086	Trenton	1.12	1.10
087	Point Pleasant	1.10	1.08
088-089	New Brunswick	1.12	1.10
NEW MEXICO			
870-872	Albuquerque	.88	.90
873	Gallup	.88	.91
874	Farmington	.88	.91
875	Santa Fe	.88	.90
877	Las Vegas	.88	.90
878	Socorro	.87	.89
879	Truth/Consequences	.87	.87
880	Las Cruces	.84	.84
881	Clovis	.89	.89
882	Roswell	.90	.90
883	Carrizozo	.91	.91
884	Tucumcari	.90	.90
NEW YORK			
100-102	New York	1.35	1.35
103	Staten Island	1.31	1.31
104	Bronx	1.30	1.30
105	Mount Vernon	1.20	1.20
106	White Plains	1.19	1.19
107	Yonkers	1.22	1.22
108	New Rochelle	1.20	1.20
109	Suffern	1.14	1.14
110	Queens	1.30	1.30
111	Long Island City	1.31	1.31
112	Brooklyn	1.31	1.31
113	Flushing	1.32	1.32
114	Jamaica	1.30	1.30
115,117,118	Hicksville	1.26	1.26
116	Far Rockaway	1.32	1.32
119	Riverhead	1.27	1.27
120-122	Albany	.97	.97
123	Schenectady	.98	.98
124	Kingston	1.11	1.09
125-126	Poughkeepsie	1.13	1.11
127	Monticello	1.09	1.07
128	Glens Falls	.95	.93
129	Plattsburgh	.95	.93
130-132	Syracuse	.99	.97
133-135	Utica	.91	.94
136	Watertown	.92	.95
137-139	Binghamton	.94	.94
140-142	Buffalo	1.05	1.02
143	Niagara Falls	1.07	1.03
144-146	Rochester	.99	1.00
147	Jamestown	.98	.94
148-149	Elmira	.95	.93
NORTH CAROLINA			
270,272-274	Greensboro	.75	.76
271	Winston-Salem	.74	.75
275-276	Raleigh	.76	.76
277	Durham	.75	.76
278	Rocky Mount	.68	.68
279	Elizabeth City	.70	.70
280	Gastonia	.74	.75
281-282	Charlotte	.74	.75
283	Fayetteville	.75	.75
284	Wilmington	.73	.75
285	Kinston	.67	.67
286	Hickory	.66	.67
287-288	Asheville	.73	.75
289	Murphy	.66	.67
NORTH DAKOTA			
580-581	Fargo	.79	.84
582	Grand Forks	.77	.82
583	Devils Lake	.76	.81
584	Jamestown	.76	.81
585	Bismarck	.80	.84

STATE/ZIP	CITY	Residential	Commercial
NORTH DAKOTA (CONT'D)			
586	Dickinson	.81	.85
587	Minot	.82	.87
588	Williston	.76	.80
OHIO			
430-432	Columbus	.98	.96
433	Marion	.92	.93
434-436	Toledo	1.02	1.01
437-438	Zanesville	.93	.92
439	Steubenville	.97	.97
440	Lorain	1.05	.98
441	Cleveland	1.09	1.03
442-443	Akron	1.02	1.01
444-445	Youngstown	1.01	.98
446-447	Canton	.97	.96
448-449	Mansfield	.96	.94
450	Hamilton	1.00	.94
451-452	Cincinnati	1.00	.94
453-454	Dayton	.94	.93
455	Springfield	.95	.93
456	Chillicothe	1.02	.96
457	Athens	.92	.91
458	Lima	.96	.95
OKLAHOMA			
730-731	Oklahoma City	.82	.84
734	Ardmore	.83	.82
735	Lawton	.84	.83
736	Clinton	.80	.82
737	Enid	.83	.82
738	Woodward	.82	.81
739	Guymon	.69	.68
740-741	Tulsa	.84	.81
743	Miami	.86	.83
744	Muskogee	.75	.73
745	Mcalester	.76	.77
746	Ponca City	.82	.81
747	Durant	.79	.81
748	Shawnee	.79	.81
749	Poteau	.85	.81
OREGON			
970-972	Portland	1.08	1.06
973	Salem	1.06	1.05
974	Eugene	1.05	1.04
975	Medford	1.05	1.04
976	Klamath Falls	1.05	1.04
977	Bend	1.06	1.05
978	Pendleton	1.03	1.01
979	Vale	.98	.96
PENNSYLVANIA			
150-152	Pittsburgh	1.04	1.02
153	Washington	1.02	1.00
154	Uniontown	1.01	.99
155	Bedford	1.03	.96
156	Greensburg	1.02	1.00
157	Indiana	1.05	.98
158	Dubois	1.04	.97
159	Johnstown	1.04	.97
160	Butler	1.01	.98
161	New Castle	1.01	.98
162	Kittanning	1.02	.99
163	Oil City	.91	.95
164-165	Erie	.98	.97
166	Altoona	1.04	.96
167	Bradford	.99	.97
168	State College	.96	.96
169	Wellsboro	.93	.94
170-171	Harrisburg	.98	.97
172	Chambersburg	.96	.95
173-174	York	.97	.95
175-176	Lancaster	.95	.93
177	Williamsport	.91	.90
178	Sunbury	.95	.94
179	Pottsville	.95	.94
180	Lehigh Valley	1.04	1.03
181	Allentown	1.01	1.00
182	Hazleton	.97	.96
183	Stroudsburg	1.01	1.00
184-185	Scranton	.95	.98
186-187	Wilkes-Barre	.93	.96
188	Montrose	.93	.96
189	Doylestown	.94	1.06

STATE/ZIP	CITY	Residential	Commercial
PENNSYLVANIA (CONT'D)			
190-191	Philadelphia	1.13	1.11
193	Westchester	1.08	1.06
194	Norristown	1.09	1.07
195-196	Reading	.97	.98
PUERTO RICO			
009	San Juan	.86	.86
RHODE ISLAND			
028	Newport	1.02	1.04
029	Providence	1.02	1.04
SOUTH CAROLINA			
290-292	Columbia	.72	.75
293	Spartanburg	.71	.73
294	Charleston	.73	.75
295	Florence	.71	.73
296	Greenville	.70	.73
297	Rock Hill	.64	.67
298	Aiken	.80	.83
299	Beaufort	.68	.70
SOUTH DAKOTA			
570-571	Sioux Falls	.88	.81
572	Watertown	.84	.78
573	Mitchell	.83	.77
574	Aberdeen	.84	.78
575	Pierre	.84	.79
576	Mobridge	.84	.77
577	Rapid City	.85	.79
TENNESSEE			
370-372	Nashville	.86	.86
373-374	Chattanooga	.82	.81
375,380-381	Memphis	.84	.84
376	Johnson City	.80	.79
377-379	Knoxville	.80	.80
382	Mckenzie	.69	.69
383	Jackson	.68	.75
384	Columbia	.76	.76
385	Cookeville	.68	.68
TEXAS			
750	Mckinney	.88	.81
751	Waxahackie	.82	.82
752-753	Dallas	.89	.85
754	Greenville	.78	.72
755	Texarkana	.87	.77
756	Longview	.84	.74
757	Tyler	.91	.80
758	Palestine	.72	.72
759	Lufkin	.76	.76
760-761	Fort Worth	.83	.82
762	Denton	.87	.78
763	Wichita Falls	.80	.80
764	Eastland	.73	.72
765	Temple	.77	.76
766-767	Waco	.81	.80
768	Brownwood	.72	.71
769	San Angelo	.79	.76
770-772	Houston	.87	.88
773	Huntsville	.73	.73
774	Wharton	.75	.76
775	Galveston	.86	.87
776-777	Beaumont	.82	.83
778	Bryan	.81	.82
779	Victoria	.78	.78
780	Laredo	.76	.77
781-782	San Antonio	.82	.83
783-784	Corpus Christi	.80	.79
785	Mc Allen	.78	.76
786-787	Austin	.78	.81
788	Del Rio	.68	.68
789	Giddings	.72	.71
790-791	Amarillo	.81	.81
792	Childress	.75	.78
793-794	Lubbock	.78	.80
795-796	Abilene	.79	.79
797	Midland	.78	.79
798-799,885	El Paso	.79	.78
UTAH			
840-841	Salt Lake City	.90	.89
842,844	Ogden	.90	.88

Location Factors

STATE/ZIP	CITY	Residential	Commercial
UTAH(CONT'D)			
843	Logan	.91	.89
845	Price	.81	.80
846-847	Provo	.90	.89
VERMONT			
050	White River Jct.	.73	.72
051	Bellows Falls	.73	.73
052	Bennington	.72	.71
053	Brattleboro	.74	.73
054	Burlington	.85	.86
056	Montpelier	.84	.85
057	Rutland	.87	.86
058	St. Johnsbury	.74	.75
059	Guildhall	.73	.74
VIRGINIA			
220-221	Fairfax	.89	.90
222	Arlington	.89	.90
223	Alexandria	.89	.90
224-225	Fredericksburg	.83	.84
226	Winchester	.78	.79
227	Culpeper	.78	.79
228	Harrisonburg	.75	.75
229	Charlottesville	.83	.82
230-232	Richmond	.86	.84
233-235	Norfolk	.82	.82
236	Newport News	.82	.81
237	Portsmouth	.81	.81
238	Petersburg	.86	.84
239	Farmville	.74	.72
240-241	Roanoke	.76	.76
242	Bristol	.79	.74
243	Pulaski	.73	.72
244	Staunton	.76	.74
245	Lynchburg	.80	.76
246	Grundy	.72	.72
WASHINGTON			
980-981,987	Seattle	1.00	1.05
982	Everett	.97	1.03
983-984	Tacoma	1.05	1.03
985	Olympia	1.05	1.03
986	Vancouver	1.11	1.04
988	Wenatchee	.94	.97
989	Yakima	1.02	1.00
990-992	Spokane	.99	.98
993	Richland	1.00	.99
994	Clarkston	.99	.98
WEST VIRGINIA			
247-248	Bluefield	.89	.89
249	Lewisburg	.90	.90
250-253	Charleston	.93	.93
254	Martinsburg	.75	.76
255-257	Huntington	.93	.95
258-259	Beckley	.90	.90
260	Wheeling	.93	.95
261	Parkersburg	.92	.94
262	Buckhannon	.97	.94
263-264	Clarksburg	.97	.94
265	Morgantown	.97	.94
266	Gassaway	.93	.93
267	Romney	.90	.90
268	Petersburg	.95	.92
WISCONSIN			
530,532	Milwaukee	1.02	1.01
531	Kenosha	1.02	1.01
534	Racine	1.06	1.00
535	Beloit	1.00	.98
537	Madison	1.00	.98
538	Lancaster	.91	.89
539	Portage	.98	.96
540	New Richmond	1.03	.95
541-543	Green Bay	1.00	.97
544	Wausau	.98	.94
545	Rhinelander	.98	.94
546	La Crosse	.98	.95
547	Eau Claire	1.04	.96
548	Superior	1.03	.97
549	Oshkosh	.97	.94

STATE/ZIP	CITY	Residential	Commercial
WYOMING			
820	Cheyenne	.86	.81
821	Yellowstone Nat. Pk.	.80	.77
822	Wheatland	.83	.78
823	Rawlins	.81	.77
824	Worland	.78	.76
825	Riverton	.81	.78
826	Casper	.86	.82
827	Newcastle	.79	.75
828	Sheridan	.83	.80
829-831	Rock Springs	.83	.78
CANADIAN FACTORS (reflect Canadian currency)			
ALBERTA			
	Calgary	.99	.96
	Edmonton	.99	.96
	Fort McMurray	.98	.95
	Lethbridge	.98	.95
	Lloydminster	.98	.95
	Medicine Hat	.98	.95
	Red Deer	.98	.95
BRITISH COLUMBIA			
	Kamloops	1.02	1.03
	Prince George	1.04	1.05
	Vancouver	1.05	1.06
	Victoria	1.04	1.05
MANITOBA			
	Brandon	.97	.96
	Portage la Prairie	.97	.96
	Winnipeg	.97	.96
NEW BRUNSWICK			
	Bathurst	.93	.91
	Dalhousie	.93	.91
	Fredericton	.95	.93
	Moncton	.92	.90
	Newcastle	.93	.91
	Saint John	.96	.94
NEWFOUNDLAND			
	Corner Brook	.95	.94
	St. John's	.94	.93
NORTHWEST TERRITORIES			
	Yellowknife	.91	.90
NOVA SCOTIA			
	Dartmouth	.96	.95
	Halifax	.96	.95
	New Glasgow	.96	.95
	Sydney	.94	.93
	Yarmouth	.96	.95
ONTARIO			
	Barrie	1.08	1.07
	Brantford	1.10	1.08
	Cornwall	1.08	1.06
	Hamilton	1.12	1.08
	Kingston	1.08	1.07
	Kitchener	1.05	1.03
	London	1.08	1.06
	North Bay	1.07	1.05
	Oshawa	1.09	1.07
	Ottawa	1.09	1.07
	Owen Sound	1.08	1.07
	Peterborough	1.07	1.06
	Sarnia	1.10	1.08
	St. Catharines	1.04	1.02
	Sudbury	1.04	1.02
	Thunder Bay	1.05	1.03
	Toronto	1.12	1.11
	Windsor	1.06	1.04
PRINCE EDWARD ISLAND			
	Charlottetown	.92	.90
	Summerside	.92	.90

Location Factors

STATE/ZIP	CITY	Residential	Commercial
QUEBEC			
	Cap-de-la-Madeleine	1.03	1.02
	Charlesbourg	1.03	1.02
	Chicoutimi	1.02	1.01
	Gatineau	1.01	1.00
	Laval	1.02	1.01
	Montreal	1.08	1.01
	Quebec	1.10	1.02
	Sherbrooke	1.02	1.01
	Trois Rivieres	1.03	1.02
SASKATCHEWAN			
	Moose Jaw	.91	.91
	Prince Albert	.91	.91
	Regina	.92	.92
	Saskatoon	.91	.91
YUKON			
	Whitehorse	.91	.90

Historical Cost Indexes

The following tables are the revised Historical Cost Indexes based on a 30-city national average with a base of 100 on January 1, 1993.

The indexes may be used to:

1. Estimate and compare construction costs for different years in the same city.

2. Estimate and compare construction costs in different cities for the same year.

3. Estimate and compare construction costs in different cities for different years.

4. Compare construction trends in any city with the national average.

EXAMPLES

1. Estimate and compare construction costs for different years in the same city.

   A. To estimate the construction cost of a building in Lexington, KY in 1970, knowing that it cost $900,000 in 2002.

   Index Lexington, KY in 1970 = 26.9
   Index Lexington, KY in 2002 = 106.1

   $$\frac{\text{Index } 1970}{\text{Index } 2002} \times \text{Cost } 2002 = \text{Cost } 1970$$

   $$\frac{26.9}{106.1} \times \$900,000 = \$228,000$$

   Construction Cost in Lexington, KY in 1970 = $228,000

   B. To estimate the current construction cost of a building in Boston, MA that was built in 1980 for $900,000.

   Index Boston, MA in 1980 = 64.0
   Index Boston, MA in 2002 = 144.6

   $$\frac{\text{Index } 2002}{\text{Index } 1980} \times \text{Cost } 1980 = \text{Cost } 2002$$

   $$\frac{144.6}{64.0} \times \$900,000 = \$2,033,500$$

   Construction Cost in Boston in 2002 = $2,033,500

2. Estimate and compare construction costs in different cities for the same year.

   To compare the construction cost of a building in Topeka, KS in 2002 with the known cost of $800,000 in Baltimore, MD in 2002

   Index Topeka, KS in 2002   = 107.2
   Index Baltimore, MD in 2002 = 115.2

   $$\frac{\text{Index Topeka}}{\text{Index Baltimore}} \times \text{Cost Baltimore} = \text{Cost Topeka}$$

   $$\frac{107.2}{115.2} \times \$800,000 = \$744,500$$

   Construction Cost in Topeka in 2002 = $744,500

3. Estimate and compare construction costs in different cities for different years.

   To compare the construction cost of a building in Detroit, MI in 2002 with the known construction cost of $5,000,000 for the same building in San Francisco, CA in 1980.

   Index Detroit, MI in 2002   = 134.0
   Index San Francisco, CA in 1980 = 75.2

   $$\frac{\text{Index Detroit } 2002}{\text{Index San Francisco } 1980} \times \text{Cost San Francisco } 1980 = \text{Cost Detroit } 2002$$

   $$\frac{134.0}{75.2} \times \$5,000,000 = \$8,909,500$$

   Construction Cost in Detroit in 2002   = $8,909,500

4. Compare construction trends in any city with the national average.

   To compare the construction cost in Las Vegas, NV from 1975 to 2002 with the increase in the National Average during the same time period.

   Index Las Vegas, NV for 1975   = 42.8   For 2002 =   131.1
   Index 30 City Average for 1975 = 43.7   For 2002 =   126.1

   A. National Average escalation   = $\dfrac{\text{Index} - \text{30 City } 2002}{\text{Index} - \text{30 City } 1975}$
      From 1975 to 2002

   $$= \frac{126.1}{43.7}$$

   National Average escalation
   From 1975 to 2002   = 2.89 or increased by 189%

   B. Escalation for Las Vegas, NV   = $\dfrac{\text{Index Las Vegas, NV } 2002}{\text{Index Las Vegas, NV } 1975}$
      From 1975 to 2002

   $$= \frac{131.1}{42.8}$$

   Las Vegas escalation 1975 — 2002 = 3.06 or increased by 206%

   Conclusion: Construction costs in Las Vegas are higher than National average costs and increased at a greater rate from 1975 to 2002 than the National Average.

Year	National 30 City Average	Alabama					Alaska	Arizona		Arkansas		California				
		Birming-ham	Hunts-ville	Mobile	Mont-gomery	Tusca-loosa	Anchor-age	Phoenix	Tuscon	Fort Smith	Little Rock	Anaheim	Bakers-field	Fresno	Los Angeles	Oxnard
Jan 2002	126.1E	109.2E	103.2E	103.4E	101.1E	98.7E	156.4E	112.5E	109.5E	100.4E	101.7E	136.2E	133.5E	135.9E	136.2E	136.0E
2001	122.2	106.0	100.5	100.6	98.3	95.9	152.6	109.0	106.4	97.3	98.7	132.5	129.3	132.8	132.4	132.4
2000	118.9	104.1	98.9	99.1	94.2	94.6	148.3	106.9	104.9	94.4	95.7	129.4	125.2	129.4	129.9	129.7
1999	116.6	101.2	97.4	97.7	92.5	92.8	145.9	105.5	103.3	93.1	94.4	127.9	123.6	126.9	128.7	128.2
1998	113.6	96.2	94.0	94.8	90.3	89.8	143.8	102.1	101.0	90.6	91.4	125.2	120.3	123.9	124.6	124.7
1997	111.5	94.6	92.4	93.3	88.8	88.3	142.0	101.8	100.7	89.3	90.1	124.0	119.1	122.3	124.6	123.5
1996	108.9	90.9	91.4	92.3	87.5	86.5	140.4	98.3	97.4	86.5	86.7	121.7	117.2	120.2	122.4	121.5
1995	105.6	87.8	88.0	88.8	84.5	83.5	138.0	96.1	95.5	85.0	85.6	120.1	115.7	117.5	120.9	120.0
1994	103.0	85.9	86.1	86.8	82.6	81.7	134.0	93.7	93.1	82.4	83.9	118.1	113.6	114.5	119.0	117.5
1993	100.0	82.8	82.7	86.1	82.1	78.7	132.0	90.9	90.9	80.9	82.4	115.0	111.3	112.9	115.6	115.4
1992	97.9	81.7	81.5	84.9	80.9	77.6	128.6	88.8	89.4	79.7	81.2	113.5	108.1	110.3	113.7	113.7
1991	95.7	80.5	78.9	83.9	79.8	76.6	127.4	88.1	88.5	78.3	80.0	111.0	105.8	108.2	110.9	111.4
1990	93.2	79.4	77.6	82.7	78.4	75.2	125.8	86.4	87.0	77.1	77.9	107.7	102.7	103.3	107.5	107.4
1989	91.0	77.5	76.1	81.1	76.8	73.7	123.4	85.3	85.6	75.6	76.4	105.6	101.0	101.7	105.3	105.4
1988	88.5	75.7	74.7	79.7	75.4	72.3	121.4	84.4	84.3	74.2	75.0	102.7	97.6	99.2	102.6	102.3
1987	85.7	74.0	73.8	77.9	74.5	71.7	119.3	79.7	80.6	72.2	72.8	100.9	96.1	98.0	100.0	101.1
1986	83.7	72.8	71.7	76.7	72.3	69.4	117.3	78.9	78.8	70.7	72.3	99.2	94.2	93.8	97.8	100.4
1985	81.8	71.1	70.5	72.7	70.9	67.9	116.0	78.1	77.7	69.6	71.0	95.4	92.2	92.6	94.6	96.6
1984	80.6	69.7	69.0	75.2	69.4	66.3	113.8	79.4	79.2	69.1	70.3	93.4	89.8	90.6	92.1	94.0
1983	78.2	67.4	68.8	72.9	69.4	65.5	104.5	77.5	77.9	66.8	68.7	90.9	88.8	88.3	89.7	91.8
1982	72.1	63.5	63.1	67.3	64.4	61.8	96.1	72.3	71.7	62.2	64.6	83.1	83.0	82.5	80.9	83.4
1981	66.1	60.0	58.1	62.2	61.3	58.1	91.5	69.0	67.5	57.7	60.2	75.7	76.5	75.2	74.8	77.1
1980	60.7	55.2	54.1	56.8	56.8	54.0	91.4	63.7	62.4	53.1	55.8	68.7	69.4	68.7	67.4	69.9
1979	54.9	49.9	48.9	52.5	50.8	48.7	82.5	55.9	56.5	47.8	49.5	62.9	61.9	62.5	61.7	62.9
1975	43.7	40.0	40.9	41.8	40.1	37.8	57.3	44.5	45.2	39.0	38.7	47.6	46.6	47.7	48.3	47.0
1970	27.8	24.1	25.1	25.8	25.4	24.2	43.0	27.2	28.5	24.3	22.3	31.0	30.8	31.1	29.0	31.0
1965	21.5	19.6	19.3	19.6	19.5	18.7	34.9	21.8	22.0	18.7	18.5	23.9	23.7	24.0	22.7	23.9
1960	19.5	17.9	17.5	17.8	17.7	16.9	31.7	19.9	20.0	17.0	16.8	21.7	21.5	21.8	20.6	21.7
1955	16.3	14.8	14.7	14.9	14.9	14.2	26.6	16.7	16.7	14.3	14.5	18.2	18.1	18.3	17.3	18.2
1950	13.5	12.2	12.1	12.3	12.3	11.7	21.9	13.8	13.8	11.8	11.6	15.1	15.0	15.1	14.3	15.1
1945	8.6	7.8	7.7	7.8	7.8	7.5	14.0	8.8	8.8	7.5	7.4	9.6	9.5	9.6	9.1	9.6
1940	6.6	6.0	6.0	6.1	6.0	5.8	10.8	6.8	6.8	5.8	5.7	7.4	7.4	7.4	7.0	7.4

Year	National 30 City Average	California							Colorado			Connecticut				
		River-side	Sacra-mento	San Diego	San Francisco	Santa Barbara	Stockton	Vallejo	Colorado Springs	Denver	Pueblo	Bridge-Port	Bristol	Hartford	New Britain	New Haven
Jan 2002	126.1E	135.1E	138.8E	133.1E	156.5E	135.7E	137.0E	143.4E	116.3E	120.6E	116.0E	132.5E	132.3E	132.4E	132.2E	132.5E
2001	122.2	131.4	135.5	129.7	151.8	131.5	134.1	140.2	113.3	117.2	113.1	128.6	128.5	128.5	128.3	128.6
2000	118.9	128.0	131.5	127.1	146.9	128.7	130.7	137.1	109.8	111.8	108.8	122.7	122.6	122.9	122.4	122.7
1999	116.6	126.5	129.4	124.9	145.1	127.2	127.9	135.3	107.0	109.1	107.1	121.1	121.3	121.2	121.1	121.3
1998	113.6	123.7	125.9	121.3	141.9	123.7	124.7	132.5	103.3	106.5	103.7	119.1	119.4	120.0	119.7	120.0
1997	111.5	122.6	124.7	120.3	139.2	122.4	123.3	130.6	101.1	104.4	102.0	119.2	119.5	119.9	119.7	120.0
1996	108.9	120.6	122.4	118.4	136.8	120.5	121.4	128.1	98.5	101.4	99.7	117.4	117.6	117.9	117.8	118.1
1995	105.6	119.2	119.5	115.4	133.8	119.0	119.0	122.5	96.1	98.9	96.8	116.0	116.5	116.9	116.3	116.5
1994	103.0	116.6	114.8	113.4	131.4	116.4	115.5	119.7	94.2	95.9	94.6	114.3	115.0	115.3	114.7	114.8
1993	100.0	114.3	112.5	111.3	129.6	114.3	115.2	117.5	92.1	93.8	92.6	108.7	107.8	108.4	106.3	106.1
1992	97.9	111.8	110.8	109.5	127.9	112.2	112.7	115.2	90.6	91.5	90.9	106.7	106.0	106.6	104.5	104.4
1991	95.7	109.4	108.5	107.7	125.8	110.0	111.0	113.4	88.9	90.4	89.4	97.6	97.3	98.0	97.3	97.9
1990	93.2	107.0	104.9	105.6	121.8	106.4	105.4	111.4	86.9	88.8	88.8	96.3	95.9	96.6	95.9	96.5
1989	91.0	105.2	103.1	103.6	119.2	104.6	103.7	109.8	85.6	87.0	87.4	94.5	94.2	94.9	94.2	94.6
1988	88.5	102.6	100.5	101.0	115.2	101.4	101.2	107.2	83.8	87.0	85.6	92.8	92.6	93.3	92.6	93.0
1987	85.7	100.3	98.7	99.0	111.0	99.3	99.6	103.2	81.6	84.8	83.8	92.4	93.0	93.3	93.0	92.6
1986	83.7	98.6	94.4	96.9	108.0	97.0	96.1	99.1	81.9	83.1	82.7	88.5	88.7	89.4	88.6	88.8
1985	81.8	94.8	92.2	94.2	106.2	93.2	93.7	97.0	79.6	81.0	80.4	86.1	86.2	87.2	86.1	86.3
1984	80.6	91.8	93.8	92.0	102.8	90.6	91.9	95.6	79.8	84.1	80.5	83.5	83.4	84.4	83.0	83.2
1983	78.2	89.4	92.0	90.3	101.0	88.8	91.0	93.0	77.3	80.5	77.4	79.6	79.9	81.1	79.5	79.3
1982	72.1	82.1	84.9	83.8	91.9	81.8	84.9	85.7	71.2	73.4	75.0	71.9	71.2	73.2	71.4	72.0
1981	66.1	75.5	77.7	74.5	82.8	75.9	77.4	78.5	65.5	66.3	65.4	67.0	66.3	67.3	66.3	67.0
1980	60.7	68.7	71.3	68.1	75.2	71.1	71.2	71.9	60.7	60.9	59.5	61.5	60.7	61.9	60.7	61.4
1979	54.9	62.2	64.0	61.6	67.5	65.0	64.2	65.2	54.9	56.0	54.0	56.0	55.3	56.4	55.4	55.8
1975	43.7	47.3	49.1	47.7	49.8	46.1	47.6	46.5	43.1	42.7	42.5	45.2	45.5	46.0	45.2	46.0
1970	27.8	31.0	32.1	30.7	31.6	31.3	31.8	31.8	27.6	26.1	27.3	29.2	28.2	29.6	28.2	29.3
1965	21.5	23.9	24.8	24.0	23.7	24.1	24.5	24.5	21.3	20.9	21.0	22.4	21.7	22.6	21.7	23.2
1960	19.5	21.7	22.5	21.7	21.5	21.9	22.3	22.2	19.3	19.0	19.1	20.0	19.8	20.0	19.8	20.0
1955	16.3	18.2	18.9	18.2	18.0	18.4	18.7	18.6	16.2	15.9	16.0	16.8	16.6	16.8	16.6	16.8
1950	13.5	15.0	15.6	15.0	14.9	15.2	15.4	15.4	13.4	13.2	13.2	13.9	13.7	13.8	13.7	13.9
1945	8.6	9.6	10.0	9.6	9.5	9.7	9.9	9.8	8.5	8.4	8.4	8.8	8.7	8.8	8.7	8.9
1940	6.6	7.4	7.7	7.4	7.3	7.5	7.6	7.6	6.6	6.5	6.5	6.8	6.8	6.8	6.8	6.8

Historical Cost Indexes

Year	National 30 City Average	Connecticut			Delaware	D.C.	Florida						Georgia			
		Norwalk	Stamford	Water-bury	Wilming-ton	Washing-ton	Fort Lau-derdale	Jackson-ville	Miami	Orlando	Talla-hassee	Tampa	Albany	Atlanta	Colum-bus	Macon
Jan 2002	126.1E	132.0E	135.4E	132.7E	127.9E	118.9E	106.8E	103.3E	107.0E	106.3E	97.3E	102.8E	99.0E	112.8E	98.2E	101.8E
2001	122.2	128.2	131.5	128.8	124.8	115.9	104.3	100.8	104.6	103.8	94.8	100.3	96.4	109.1	95.5	99.0
2000	118.9	121.5	126.4	123.2	117.4	113.8	102.4	99.0	101.8	101.2	93.6	98.9	94.9	106.1	94.1	97.0
1999	116.6	120.0	122.0	121.2	116.6	111.5	101.4	98.0	100.8	100.1	92.5	97.9	93.5	102.9	92.8	95.8
1998	113.6	118.8	120.8	120.1	112.3	109.6	99.4	96.2	99.0	98.4	90.9	96.3	91.4	100.8	90.4	93.3
1997	111.5	118.9	120.9	120.2	110.7	106.4	98.6	95.4	98.2	97.2	89.9	95.5	89.9	98.5	88.7	91.6
1996	108.9	117.0	119.0	118.4	108.6	105.4	95.9	92.6	95.9	95.1	88.0	93.6	86.6	94.3	83.9	88.3
1995	105.6	115.5	117.9	117.3	106.1	102.3	94.0	90.9	93.7	93.5	86.5	92.2	85.0	92.0	82.6	86.8
1994	103.0	113.9	116.4	115.8	105.0	99.6	92.2	88.9	91.8	91.5	84.5	90.2	82.3	89.6	80.6	83.7
1993	100.0	108.8	110.6	104.8	101.5	96.3	87.4	86.1	87.1	88.5	82.1	87.7	79.5	85.7	77.8	80.9
1992	97.9	107.2	109.0	103.1	100.3	94.7	85.7	84.0	85.3	87.1	80.8	86.2	78.2	84.3	76.5	79.6
1991	95.7	100.6	103.2	96.5	94.5	92.9	85.1	82.8	85.2	85.5	79.7	86.3	76.3	82.6	75.4	78.4
1990	93.2	96.3	98.9	95.1	92.5	90.4	83.9	81.1	84.0	82.9	78.4	85.0	75.0	80.4	74.0	76.9
1989	91.0	94.4	97.0	93.4	89.5	87.5	82.4	79.7	82.5	81.6	76.9	83.5	73.4	78.6	72.4	75.3
1988	88.5	92.3	94.9	91.8	87.7	85.0	80.9	78.0	81.0	80.1	75.4	81.9	71.8	76.9	70.8	73.6
1987	85.7	92.0	92.7	92.5	85.1	82.1	78.8	76.0	77.9	76.6	74.4	79.4	72.2	73.6	70.3	72.3
1986	83.7	88.2	89.7	87.8	83.8	80.8	78.9	75.0	79.9	76.2	72.3	79.1	68.5	72.0	67.5	69.8
1985	81.8	85.3	86.8	85.6	81.1	78.8	76.7	73.4	78.3	73.9	70.6	77.3	66.9	70.3	66.1	68.1
1984	80.6	82.6	83.7	82.5	79.7	79.1	73.8	72.6	75.6	73.0	69.6	76.5	65.9	68.6	65.2	67.5
1983	78.2	78.5	79.8	78.7	76.3	76.0	71.5	70.4	72.9	71.1	67.6	73.6	65.6	68.6	64.7	65.3
1982	72.1	71.7	72.0	71.8	69.1	69.4	65.4	65.2	65.7	66.0	62.8	67.7	60.1	61.9	59.3	60.0
1981	66.1	66.2	66.2	67.3	63.4	64.9	60.3	61.0	60.7	62.0	58.6	62.1	56.2	58.3	55.7	56.1
1980	60.7	60.7	60.9	62.3	58.5	59.6	55.3	55.8	56.5	56.7	53.5	57.2	51.7	54.0	51.1	51.3
1979	54.9	55.2	55.6	57.0	53.6	53.9	50.0	51.7	50.8	51.8	48.9	51.8	46.5	48.3	46.1	46.4
1975	43.7	44.7	45.0	46.3	42.9	43.7	42.1	40.3	43.2	41.5	38.1	41.3	37.5	38.4	36.2	36.5
1970	27.8	28.1	28.2	28.9	27.0	26.3	25.7	22.8	27.0	26.2	24.5	24.2	23.8	25.2	22.8	23.4
1965	21.5	21.6	21.7	22.2	20.9	21.8	19.8	17.4	19.3	20.2	18.9	18.6	18.3	19.8	17.6	18.0
1960	19.5	19.7	19.7	20.2	18.9	19.4	18.0	15.8	17.6	18.3	17.2	16.9	16.7	17.1	16.0	16.4
1955	16.3	16.5	16.5	17.0	15.9	16.3	15.1	13.2	14.7	15.4	14.4	14.1	14.0	14.4	13.4	13.7
1950	13.5	13.6	13.7	14.0	13.1	13.4	12.5	11.0	12.2	12.7	11.9	11.7	11.5	11.9	11.0	11.3
1945	8.6	8.7	8.7	8.9	8.4	8.6	7.9	7.0	7.8	8.1	7.6	7.5	7.4	7.6	7.0	7.2
1940	6.6	6.7	6.7	6.9	6.4	6.6	6.1	5.4	6.0	6.2	5.8	5.7	5.7	5.8	5.5	5.6

Year	National 30 City Average	Georgia Savan-nah	Hawaii Hono-lulu	Idaho		Illinois						Indiana				
				Boise	Poca-tello	Chicago	Decatur	Joliet	Peoria	Rock-ford	Spring-field	Ander-son	Evans-ville	Fort Wayne	Gary	Indian-apolis
Jan 2002	126.1E	101.7E	155.1E	117.4E	116.3E	140.9E	123.3E	137.7E	128.8E	132.0E	123.0E	117.5E	119.3E	116.5E	128.1E	120.6E
2001	122.2	99.0	150.0	114.3	113.4	135.8	120.1	133.7	124.3	127.8	119.8	113.4	115.6	112.1	123.4	116.4
2000	118.9	97.5	144.8	112.9	112.1	131.2	115.1	124.6	119.0	122.2	116.2	109.8	111.5	108.4	117.8	113.2
1999	116.6	96.0	143.0	110.2	109.6	129.6	113.0	122.6	116.4	120.7	113.8	107.1	109.3	106.8	112.8	110.6
1998	113.6	93.7	140.4	107.4	107.0	125.2	110.1	119.8	113.8	115.5	111.1	105.0	107.2	104.5	111.1	108.0
1997	111.5	92.0	139.8	104.6	104.7	121.3	107.8	117.5	111.5	113.1	108.9	101.9	104.4	101.8	110.3	105.2
1996	108.9	88.6	134.5	102.2	102.1	118.8	106.6	116.2	109.3	111.5	106.5	100.0	102.1	99.9	107.5	102.7
1995	105.6	87.4	130.3	99.5	98.2	114.2	98.5	110.5	102.3	103.6	98.1	96.4	97.2	95.0	100.7	100.1
1994	103.0	85.3	124.0	94.8	95.0	111.3	97.3	108.9	100.9	102.2	97.0	93.6	95.8	93.6	99.1	97.1
1993	100.0	82.0	122.0	92.2	92.1	107.6	95.6	106.8	98.9	99.6	95.2	91.2	94.3	91.5	96.7	93.9
1992	97.9	80.8	120.0	91.0	91.0	104.3	94.4	104.2	97.3	98.2	94.0	89.5	92.9	89.9	95.0	91.5
1991	95.7	79.5	106.1	89.5	89.4	100.9	92.3	100.0	95.9	95.8	91.5	87.8	91.4	88.3	93.3	89.1
1990	93.2	77.9	104.7	88.2	88.1	98.4	90.9	98.4	93.7	94.0	90.1	84.6	89.3	83.4	88.4	87.1
1989	91.0	76.0	102.8	86.6	86.5	93.7	89.4	92.8	91.6	92.1	88.6	82.3	87.8	81.7	86.5	85.1
1988	88.5	74.5	101.1	83.9	83.7	90.6	87.5	89.8	88.5	87.9	86.8	80.8	85.5	80.0	84.6	83.1
1987	85.7	72.0	99.1	81.4	81.5	86.6	84.4	86.7	86.3	85.2	85.0	78.8	82.4	78.1	81.7	80.4
1986	83.7	70.8	97.5	80.6	80.3	84.4	83.5	85.3	85.1	84.5	83.4	77.0	80.9	76.5	79.6	78.6
1985	81.8	68.9	94.7	78.0	78.0	82.4	81.9	83.4	83.7	83.0	81.5	75.2	79.8	75.0	77.8	77.1
1984	80.6	67.9	90.7	76.6	76.8	80.2	80.4	82.2	83.6	80.6	80.1	73.5	77.4	73.5	77.3	75.9
1983	78.2	66.3	87.2	76.0	75.8	79.0	78.8	80.6	81.5	78.9	78.8	70.9	75.0	71.2	75.2	74.0
1982	72.1	61.1	79.3	71.0	70.1	75.0	72.7	74.4	75.3	72.6	72.7	66.4	69.8	67.1	70.5	68.2
1981	66.1	57.1	73.4	65.4	64.8	68.6	67.2	68.8	70.3	67.3	67.0	61.5	64.3	61.5	65.3	62.4
1980	60.7	52.2	68.9	60.3	59.5	62.8	62.3	63.4	64.5	61.6	61.1	56.5	59.0	56.7	59.8	57.9
1979	54.9	47.2	63.0	54.4	53.7	56.5	56.0	57.0	58.4	54.8	55.6	50.8	53.3	51.2	54.0	51.9
1975	43.7	36.9	44.6	40.8	40.5	45.7	43.1	44.5	44.7	42.7	42.4	39.5	41.7	39.9	41.9	40.6
1970	27.8	21.0	30.4	26.7	26.6	29.1	28.0	28.6	29.0	27.5	27.6	25.4	26.4	25.5	27.1	26.2
1965	21.5	16.4	21.8	20.6	20.5	22.7	21.5	22.1	22.4	21.2	21.3	19.5	20.4	19.7	20.8	20.7
1960	19.5	14.9	19.8	18.7	18.6	20.2	19.6	20.0	20.3	19.2	19.3	17.7	18.7	17.9	18.9	18.4
1955	16.3	12.5	16.6	15.7	15.6	16.9	16.4	16.8	17.0	16.1	16.2	14.9	15.7	15.0	15.9	15.5
1950	13.5	10.3	13.7	13.0	12.9	14.0	13.6	13.9	14.0	13.3	13.4	12.3	12.9	12.4	13.1	12.8
1945	8.6	6.6	8.8	8.3	8.2	8.9	8.6	8.9	9.0	8.5	8.6	7.8	8.3	7.9	8.4	8.1
1940	6.6	5.1	6.8	6.4	6.3	6.9	6.7	6.8	6.9	6.5	6.6	6.0	6.4	6.1	6.5	6.3

Year	National 30 City Average	Indiana Muncie	Indiana South Bend	Indiana Terre Haute	Iowa Cedar Rapids	Iowa Daven-port	Iowa Des Moines	Iowa Sioux City	Iowa Water-loo	Kansas Topeka	Kansas Wichita	Kentucky Lexing-ton	Kentucky Louis-ville	Louisiana Baton Rouge	Louisiana Lake Charles	Louisiana New Orleans
Jan 2002	126.1E	116.8E	115.4E	119.9E	115.8E	120.9E	116.5E	111.3E	103.7	107.2E	108.2E	106.1E	115.5E	102.3E	104.2E	107.5E
2001	122.2	112.8	111.6	115.5	112.5	117.5	113.2	107.7	100.6	104.3	105.2	103.3	112.7	99.4	101.3	104.6
2000	118.9	109.1	106.9	110.8	108.8	112.6	108.9	99.0	98.1	101.0	101.1	101.4	109.3	97.8	99.7	102.1
1999	116.6	105.7	103.7	107.9	104.1	109.2	107.6	96.7	96.4	98.7	99.3	99.7	106.6	96.1	97.6	99.5
1998	113.6	103.6	102.3	106.0	102.5	106.8	103.8	95.1	94.8	97.4	97.4	97.4	101.5	94.7	96.1	97.7
1997	111.5	101.3	101.2	104.6	101.2	105.6	102.5	93.8	93.5	96.3	96.2	96.1	99.9	93.3	96.8	96.2
1996	108.9	99.4	99.1	102.1	99.1	102.0	99.1	90.9	90.7	93.6	93.8	94.4	98.3	91.5	94.9	94.3
1995	105.6	95.6	94.6	97.2	95.4	96.6	94.4	88.2	88.9	91.1	91.0	91.6	94.6	89.6	92.7	91.6
1994	103.0	93.2	93.0	95.8	93.0	92.3	92.4	85.9	86.8	89.4	88.1	89.6	92.2	87.7	89.9	89.5
1993	100.0	91.0	90.7	94.4	91.1	90.5	90.7	84.1	85.2	87.4	86.2	87.3	89.4	86.4	88.5	87.8
1992	97.9	89.4	89.3	93.1	89.8	89.3	89.4	82.9	83.8	86.2	85.0	85.8	88.0	85.2	87.3	86.6
1991	95.7	87.7	87.5	91.7	88.5	88.0	88.4	81.8	81.6	84.4	83.9	84.1	84.4	83.2	85.3	85.8
1990	93.2	83.9	85.1	89.1	87.1	86.5	86.9	80.4	80.3	83.2	82.6	82.8	82.6	82.0	84.1	84.5
1989	91.0	81.9	83.5	86.7	85.3	84.9	85.3	79.0	78.8	81.6	81.2	81.3	80.1	80.6	82.7	83.4
1988	88.5	80.4	82.1	84.9	83.4	83.1	83.6	76.8	77.2	80.1	79.5	79.4	78.2	79.0	81.1	82.0
1987	85.7	78.5	79.6	83.0	82.4	81.1	81.2	75.4	75.6	78.9	78.4	77.3	76.3	77.1	78.0	81.3
1986	83.7	76.9	77.5	81.1	79.5	80.0	79.8	73.8	73.7	76.9	76.2	76.5	75.3	76.1	78.1	79.2
1985	81.8	75.2	76.0	79.2	75.9	77.6	77.1	72.1	72.3	75.0	74.7	75.5	74.5	74.9	78.5	78.2
1984	80.6	73.6	75.1	78.0	80.3	77.8	76.6	74.4	73.7	75.1	74.6	75.6	73.7	77.3	78.3	77.5
1983	78.2	71.2	74.5	76.1	79.6	77.0	75.5	74.6	73.0	73.2	72.4	74.9	74.1	75.1	76.6	74.7
1982	72.1	66.0	68.3	69.9	73.5	71.7	71.1	70.1	67.7	68.4	66.1	69.4	69.2	69.3	70.5	69.2
1981	66.1	61.0	63.1	65.1	68.6	66.7	67.1	65.5	63.5	63.0	62.2	64.7	65.1	64.0	64.8	62.7
1980	60.7	56.1	58.3	59.7	62.7	59.6	61.8	59.3	57.7	58.9	58.0	59.3	59.8	59.1	60.0	57.2
1979	54.9	50.7	52.6	54.0	56.6	54.1	55.6	53.6	52.0	53.5	52.9	53.2	54.5	53.5	54.5	52.6
1975	43.7	39.2	40.3	41.9	43.1	41.0	43.1	41.0	40.0	42.0	43.1	42.7	42.5	40.4	40.6	41.5
1970	27.8	25.3	26.2	27.0	28.2	26.9	27.6	26.6	25.9	27.0	25.5	26.9	25.9	25.0	26.7	27.2
1965	21.5	19.5	20.2	20.9	21.7	20.8	21.7	20.5	20.0	20.8	19.6	20.7	20.3	19.4	20.6	20.4
1960	19.5	17.8	18.3	18.9	19.7	18.8	19.5	18.6	18.2	18.9	17.8	18.8	18.4	17.6	18.7	18.5
1955	16.3	14.9	15.4	15.9	16.6	15.8	16.3	15.6	15.2	15.8	15.0	15.8	15.4	14.8	15.7	15.6
1950	13.5	12.3	12.7	13.1	13.7	13.1	13.5	12.9	12.6	13.1	12.4	13.0	12.8	12.2	13.0	12.8
1945	8.6	7.8	8.1	8.4	8.7	8.3	8.6	8.2	8.0	8.3	7.9	8.3	8.1	7.8	8.3	8.2
1940	6.6	6.0	6.3	6.5	6.7	6.4	6.6	6.3	6.2	6.4	6.1	6.4	6.3	6.0	6.4	6.3

Year	National 30 City Average	Louisiana Shreve-port	Maine Lewis-ton	Maine Portland	Maryland Balti-more	Boston	Brockton	Fall River	Law-rence	Lowell	New Bedford	Pitts-field	Spring-field	Wor-cester	Michigan Ann Arbor	Michigan Dear-born
Jan 2002	126.1E	101.5E	117.4E	117.1E	115.2E	144.6E	135.6E	134.7E	136.7E	136.3E	134.5E	124.5E	128.1E	134.0E	130.1E	134.2E
2001	122.2	98.3	114.6	114.3	111.7	140.9	132.1	131.4	133.1	132.9	131.3	120.3	124.5	130.1	126.8	129.8
2000	118.9	96.2	105.7	105.4	107.7	138.9	129.4	128.4	129.7	130.3	128.3	116.7	121.0	127.3	124.3	125.8
1999	116.6	94.8	105.0	104.7	106.4	136.2	126.7	126.3	127.3	127.4	126.2	115.0	118.8	123.8	117.5	122.7
1998	113.6	92.0	102.8	102.5	104.1	132.8	125.0	124.7	125.0	125.3	124.6	114.2	117.1	122.6	116.0	119.7
1997	111.5	90.1	101.7	101.4	102.2	132.1	124.4	124.7	125.1	125.6	124.7	114.6	117.4	122.9	113.5	117.9
1996	108.9	88.5	99.5	99.2	99.6	128.7	121.6	122.2	121.7	122.1	122.2	112.7	115.0	120.2	112.9	116.2
1995	105.6	86.7	96.8	96.5	96.1	128.6	119.6	117.7	120.5	119.6	117.2	110.7	112.7	114.8	106.1	110.5
1994	103.0	85.0	95.2	94.9	94.4	124.9	114.2	113.7	116.8	116.9	111.6	109.2	111.1	112.9	105.0	109.0
1993	100.0	83.6	93.0	93.0	93.1	121.1	111.7	111.5	114.9	114.3	109.4	107.1	108.9	110.3	102.8	105.8
1992	97.9	82.3	91.7	91.7	90.9	118.0	110.0	109.8	110.9	110.2	107.7	105.7	107.2	108.6	101.5	103.2
1991	95.7	81.6	89.8	89.9	89.1	115.2	105.9	104.7	107.3	105.5	105.1	100.6	103.2	105.7	95.1	98.3
1990	93.2	80.3	88.5	88.6	85.6	110.9	103.7	102.9	105.2	102.8	102.6	98.7	101.4	103.2	93.2	96.5
1989	91.0	78.8	86.7	86.7	83.5	107.2	100.4	99.7	102.9	99.7	99.8	94.7	96.4	99.1	89.6	94.6
1988	88.5	77.0	84.0	84.1	81.2	102.5	96.7	96.4	97.7	96.6	96.4	91.9	92.7	94.4	85.8	91.4
1987	85.7	74.5	81.3	81.4	78.6	97.3	94.5	94.8	94.9	94.8	95.0	89.4	90.5	92.2	87.4	88.7
1986	83.7	74.1	79.0	79.0	75.7	95.1	91.2	90.8	92.5	90.8	90.7	87.1	87.8	89.8	81.7	85.3
1985	81.8	73.6	76.7	77.0	72.7	92.8	88.8	88.7	89.4	88.2	88.6	85.0	85.6	86.7	80.1	82.7
1984	80.6	73.3	75.0	75.1	72.4	88.1	85.5	85.8	86.8	84.1	84.8	82.8	83.5	83.6	78.9	81.1
1983	78.2	72.2	73.1	72.9	70.6	84.0	82.5	82.2	82.7	80.9	81.3	79.5	80.6	81.3	77.7	79.4
1982	72.1	67.0	67.4	67.4	64.7	76.9	75.7	75.0	74.5	73.6	74.2	72.3	73.7	73.1	73.7	75.8
1981	66.1	62.5	62.2	63.1	59.0	67.8	68.7	69.2	68.4	67.1	68.7	66.3	67.0	66.7	68.6	70.1
1980	60.7	58.7	57.3	58.5	53.6	64.0	63.7	64.1	63.4	62.7	63.1	61.8	62.0	62.3	62.9	64.0
1979	54.9	53.1	52.3	52.6	48.2	57.9	58.0	57.5	57.3	57.0	57.4	56.2	55.9	56.0	57.0	57.8
1975	43.7	40.5	41.9	42.1	39.8	46.6	45.8	45.7	46.2	45.7	46.1	45.5	45.8	46.0	44.1	44.9
1970	27.8	26.4	26.5	25.8	25.1	29.2	29.1	29.2	29.1	28.9	29.0	28.6	28.5	28.7	28.5	28.9
1965	21.5	20.3	20.4	19.4	20.2	23.0	22.5	22.5	22.5	22.2	22.4	22.0	22.4	22.1	22.0	22.3
1960	19.5	18.5	18.6	17.6	17.5	20.5	20.4	20.4	20.4	20.2	20.3	20.0	20.1	20.1	20.0	20.2
1955	16.3	15.5	15.6	14.7	14.7	17.2	17.1	17.1	17.1	16.9	17.1	16.8	16.9	16.8	16.7	16.9
1950	13.5	12.8	12.9	12.2	12.1	14.2	14.1	14.1	14.1	14.0	14.1	13.8	14.0	13.9	13.8	14.0
1945	8.6	8.1	8.2	7.8	7.7	9.1	9.0	9.0	9.0	8.9	9.0	8.9	8.9	8.9	8.8	8.9
1940	6.6	6.3	6.3	6.0	6.0	7.0	6.9	7.0	7.0	6.9	6.9	6.8	6.9	6.8	6.8	6.9

Historical Cost Indexes

Year	National 30 City Average	Michigan						Minnesota			Mississippi		Missouri			
		Detroit	Flint	Grand Rapids	Kala-mazoo	Lansing	Sagi-naw	Duluth	Minne-apolis	Roches-ter	Biloxi	Jackson	Kansas City	St. Joseph	St. Louis	Spring-field
Jan 2002	126.1E	134.0E	125.5E	107.2E	118.4E	123.3E	122.9E	131.9E	136.5E	125.4E	101.0E	96.3E	126.7E	118.2E	129.6E	110.2E
2001	122.2	129.4	122.4	104.3	115.2	120.1	119.7	131.4	136.1	124.9	98.7	93.9	121.8	114.7	125.5	106.7
2000	118.9	125.3	119.8	102.7	111.6	117.6	115.9	124.1	131.1	120.1	97.2	92.9	118.2	111.9	122.4	104.3
1999	116.6	122.6	113.7	100.9	106.1	111.5	110.1	120.3	126.5	117.1	95.7	91.8	114.9	106.1	119.8	101.1
1998	113.6	119.5	112.3	99.7	104.9	110.1	108.7	117.7	124.6	115.5	92.1	89.5	108.2	104.2	115.9	98.9
1997	111.5	117.6	110.8	98.9	104.2	108.7	107.5	115.9	121.9	114.1	90.9	88.3	106.4	102.4	113.2	97.3
1996	108.9	116.0	109.7	93.9	103.5	107.8	106.9	115.0	120.4	113.5	87.1	85.4	103.2	99.9	110.1	94.6
1995	105.6	110.1	104.1	91.1	96.4	98.3	102.3	100.3	111.9	102.5	84.2	83.5	99.7	96.1	106.3	89.5
1994	103.0	108.5	102.9	89.7	95.0	97.1	101.1	99.1	109.3	101.1	82.4	81.7	97.3	92.8	103.2	88.2
1993	100.0	105.4	100.7	87.7	92.7	95.3	98.8	99.8	106.7	99.9	80.0	79.4	94.5	90.9	99.9	86.2
1992	97.9	102.9	99.4	86.4	91.4	94.0	97.4	98.5	105.3	98.6	78.8	78.2	92.9	89.5	98.6	84.9
1991	95.7	97.6	93.6	85.3	90.0	92.1	90.3	96.0	102.9	97.3	77.6	76.9	90.9	88.1	96.4	83.9
1990	93.2	96.0	91.7	84.0	87.4	90.2	88.9	94.3	100.5	95.6	76.4	75.6	89.5	86.8	94.0	82.6
1989	91.0	94.0	90.2	82.4	85.8	88.6	87.4	92.5	97.4	93.7	75.0	74.1	87.3	85.1	91.8	80.9
1988	88.5	90.8	87.8	80.7	84.1	86.3	85.6	90.5	94.6	91.9	73.5	72.9	84.9	82.9	89.0	79.2
1987	85.7	87.8	85.4	79.7	82.6	85.2	84.5	88.5	92.0	90.6	72.4	71.3	81.7	82.7	85.5	78.2
1986	83.7	84.3	82.8	77.5	80.9	82.7	81.5	87.2	89.7	88.4	70.6	69.9	79.9	79.4	83.5	75.6
1985	81.8	81.6	80.7	75.8	79.4	80.0	80.5	85.2	87.9	86.5	69.4	68.5	78.2	77.3	80.8	73.2
1984	80.6	79.4	79.5	74.3	78.7	79.0	79.9	84.9	86.9	86.0	68.6	67.5	77.7	76.7	77.7	72.7
1983	78.2	77.0	77.4	74.1	76.2	77.1	78.0	81.6	81.4	82.6	68.0	67.3	76.2	76.8	76.2	71.6
1982	72.1	74.8	73.5	69.9	72.3	73.2	73.7	74.6	75.2	75.6	63.1	62.6	70.2	70.3	69.2	66.5
1981	66.1	68.6	68.2	64.6	67.1	67.5	68.5	68.8	69.6	69.8	58.7	59.1	63.5	65.7	64.5	61.4
1980	60.7	62.5	63.1	59.8	61.7	60.3	63.3	63.4	64.1	64.8	54.3	54.3	59.1	60.4	59.7	56.7
1979	54.9	56.2	56.4	53.7	55.5	55.8	56.8	59.2	58.5	58.7	49.3	49.3	54.4	54.6	54.9	51.3
1975	43.7	45.8	44.9	41.8	43.7	44.1	44.8	45.2	46.0	45.3	37.9	37.7	42.7	45.4	44.8	41.1
1970	27.8	29.7	28.5	26.7	28.1	27.8	28.7	28.9	29.5	28.9	24.4	21.3	25.3	28.3	28.4	26.0
1965	21.5	22.1	21.9	20.6	21.7	21.5	22.2	22.3	23.4	22.3	18.8	16.4	20.6	21.8	21.8	20.0
1960	19.5	20.1	20.0	18.7	19.7	19.5	20.1	20.3	20.5	20.2	17.1	14.9	19.0	19.8	19.5	18.2
1955	16.3	16.8	16.7	15.7	16.5	16.3	16.9	17.0	17.2	17.0	14.3	12.5	16.0	16.6	16.3	15.2
1950	13.5	13.9	13.8	13.0	13.6	13.5	13.9	14.0	14.2	14.0	11.8	10.3	13.2	13.7	13.5	12.6
1945	8.6	8.9	8.8	8.3	8.7	8.6	8.9	8.9	9.0	8.9	7.6	6.6	8.4	8.8	8.6	8.0
1940	6.6	6.8	6.8	6.4	6.7	6.6	6.9	6.9	7.0	6.9	5.8	5.1	6.5	6.7	6.7	6.2

Year	National 30 City Average	Montana		Nebraska		Nevada		New Hamphire		New Jersey					NM	NY
		Billings	Great Falls	Lincoln	Omaha	Las Vegas	Reno	Man-chester	Nashua	Camden	Jersey City	Newark	Pater-son	Trenton	Albu-querque	Albany
Jan 2002	126.1E	114.7E	115.3E	104.0E	114.1E	131.1E	125.2E	120.0E	119.7E	135.7E	138.7E	140.6E	140.3E	138.1E	113.1E	122.5E
2001	122.2	117.5	117.7	101.3	111.6	127.8	122.5	116.2	116.2	133.4	136.7	136.9	136.8	136.0	111.4	119.2
2000	118.9	113.7	113.9	98.8	107.0	125.8	118.2	111.9	111.9	128.4	130.5	132.6	132.4	130.6	109.0	116.5
1999	116.6	112.1	112.7	96.6	104.8	121.9	114.4	109.6	109.6	125.3	128.8	131.6	129.5	129.6	106.7	114.6
1998	113.6	109.7	109.1	94.9	101.2	118.1	111.4	110.1	110.0	124.3	127.7	128.9	128.5	127.9	103.8	113.0
1997	111.5	107.4	107.6	93.3	99.5	114.6	109.8	108.6	108.6	121.9	125.1	126.4	126.4	125.2	100.8	110.0
1996	108.9	108.0	107.5	91.4	97.4	111.8	108.9	106.5	106.4	119.1	122.5	122.5	123.8	121.8	98.6	108.3
1995	105.6	104.7	104.9	85.6	93.4	108.5	105.0	100.9	100.8	107.0	112.2	111.9	112.1	111.2	96.3	103.6
1994	103.0	100.2	100.7	84.3	91.0	105.3	102.2	97.8	97.7	105.4	110.6	110.1	110.6	108.2	93.4	102.4
1993	100.0	97.9	97.8	82.3	88.7	102.8	99.9	95.4	95.4	103.7	109.0	108.2	109.0	106.4	90.1	99.8
1992	97.9	95.5	96.4	81.1	87.4	99.4	98.4	90.4	90.4	102.0	107.5	107.0	107.8	103.9	87.5	98.5
1991	95.7	94.2	95.1	80.1	86.4	97.5	95.8	87.8	87.8	95.0	98.5	95.6	100.0	96.6	86.2	96.1
1990	93.2	92.9	93.8	78.8	85.0	96.3	94.5	86.3	86.3	93.0	93.5	93.8	97.2	94.5	84.9	93.2
1989	91.0	91.3	92.2	77.3	83.6	94.8	92.3	84.8	84.8	90.4	91.4	91.8	95.5	91.9	83.3	88.4
1988	88.5	89.5	90.4	75.8	82.0	93.0	90.5	83.2	83.2	87.7	89.5	89.7	92.5	89.5	81.7	86.8
1987	85.7	85.4	87.3	75.3	79.8	91.5	88.3	82.4	82.5	86.1	88.3	88.7	90.3	87.9	78.1	84.5
1986	83.7	86.2	87.3	72.6	78.8	90.1	87.2	79.7	79.7	84.6	86.1	86.7	87.8	86.1	78.9	81.7
1985	81.8	83.9	84.3	71.5	77.5	87.6	85.0	78.1	78.1	81.3	83.3	83.5	84.5	82.8	76.7	79.5
1984	80.6	83.2	82.9	70.9	77.6	85.9	83.3	76.4	75.5	78.2	80.3	80.1	81.3	78.5	75.9	77.3
1983	78.2	80.3	80.0	72.1	77.3	83.0	81.9	73.0	72.2	74.6	76.5	76.7	76.5	74.2	73.1	74.2
1982	72.1	74.3	74.1	67.8	72.5	76.9	75.8	67.0	66.7	67.9	69.8	70.4	70.1	69.1	67.1	69.2
1981	66.1	69.5	70.3	64.1	68.2	70.2	68.5	61.5	61.1	62.5	65.0	65.1	64.8	63.4	63.3	64.4
1980	60.7	63.9	64.6	58.5	63.5	64.6	63.0	56.6	56.0	58.6	60.6	60.1	60.0	58.9	59.0	59.5
1979	54.9	57.5	58.9	52.9	56.2	58.4	56.7	51.2	50.8	53.5	55.6	55.3	54.7	54.0	54.3	54.2
1975	43.7	43.1	43.8	40.9	43.2	42.8	41.9	41.3	40.8	42.3	43.4	44.2	43.9	43.8	40.3	43.9
1970	27.8	28.5	28.9	26.4	26.8	29.4	28.0	26.2	25.6	27.2	27.8	29.0	27.8	27.4	26.4	28.3
1965	21.5	22.0	22.3	20.3	20.6	22.4	21.6	20.6	19.7	20.9	21.4	23.8	21.4	21.3	20.6	22.3
1960	19.5	20.0	20.3	18.5	18.7	20.2	19.6	18.0	17.9	19.0	19.4	19.4	19.4	19.2	18.5	19.3
1955	16.3	16.7	17.0	15.5	15.7	16.9	16.4	15.1	15.0	16.0	16.3	16.3	16.3	16.1	15.6	16.2
1950	13.5	13.9	14.0	12.8	13.0	14.0	13.6	12.5	12.4	13.2	13.5	13.5	13.5	13.3	12.9	13.4
1945	8.6	8.8	9.0	8.1	8.3	8.9	8.7	8.0	7.9	8.4	8.6	8.6	8.6	8.5	8.2	8.5
1940	6.6	6.8	6.9	6.3	6.4	6.9	6.7	6.2	6.1	6.5	6.6	6.6	6.6	6.6	6.3	6.6

Historical Cost Indexes

Year	National 30 City Average	Binghamton	Buffalo	New York	Rochester	Schenectady	Syracuse	Utica	Yonkers	Charlotte	Durham	Greensboro	Raleigh	Winston-Salem	Fargo (N.Dakota)	Akron (Ohio)
Jan 2002	126.1E	118.8E	127.8E	169.4E	125.9E	123.3E	121.6E	118.0E	152.9E	94.2E	95.6E	95.4E	95.9E	95.0E	105.6E	126.6E
2001	122.2	116.0	125.2	164.4	123.1	120.1	118.6	115.3	151.4	91.5	92.9	92.9	93.3	92.3	103.1	123.5
2000	118.9	112.4	122.3	159.2	120.0	117.5	115.1	112.4	144.8	90.3	91.6	91.6	91.9	91.0	97.9	117.8
1999	116.6	108.6	120.2	155.9	116.8	114.7	113.7	108.5	140.6	89.3	90.2	90.3	90.5	90.1	96.8	116.0
1998	113.6	109.0	119.1	154.4	117.2	114.2	113.6	108.6	141.4	88.2	89.1	89.2	89.4	89.0	95.2	113.1
1997	111.5	107.2	115.7	150.3	115.2	111.2	110.6	107.0	138.8	86.8	87.7	87.8	87.9	87.6	93.5	110.6
1996	108.9	105.5	114.0	148.0	113.3	109.5	108.2	105.2	137.2	85.0	85.9	86.0	86.1	85.8	91.7	107.9
1995	105.6	99.3	110.1	140.7	106.6	104.6	104.0	97.7	129.4	81.8	82.6	82.6	82.7	82.6	88.4	103.4
1994	103.0	98.1	107.2	137.1	105.2	103.5	102.3	96.5	128.1	80.3	81.1	81.0	81.2	81.1	87.0	102.2
1993	100.0	95.7	102.2	133.3	102.0	100.8	99.6	94.1	126.3	78.2	78.9	78.9	78.9	78.8	85.8	100.3
1992	97.9	93.7	100.1	128.2	99.3	99.3	98.1	90.5	123.9	77.1	77.7	77.8	77.8	77.6	83.1	98.0
1991	95.7	89.5	96.8	124.4	96.0	96.7	95.5	88.7	121.5	76.0	76.6	76.8	76.7	76.6	82.6	96.8
1990	93.2	87.0	94.1	118.1	94.6	93.9	91.0	85.6	111.4	74.8	75.4	75.5	75.5	75.3	81.3	94.6
1989	91.0	85.4	91.8	114.8	92.6	89.6	88.4	84.1	102.1	73.3	74.0	74.0	74.0	73.8	79.7	92.9
1988	88.5	83.6	89.4	106.5	87.9	87.7	86.5	82.4	99.9	71.5	72.2	72.2	72.2	72.0	78.2	91.1
1987	85.7	82.4	85.8	102.6	85.8	85.3	85.1	82.1	98.9	70.6	71.2	71.2	71.1	71.0	77.4	90.1
1986	83.7	80.2	84.4	99.8	83.7	82.4	83.3	79.2	96.6	68.4	69.1	69.1	69.1	68.9	75.4	87.7
1985	81.8	77.5	83.2	94.9	81.6	80.1	81.0	77.0	92.8	66.9	67.6	67.7	67.6	67.5	73.4	86.8
1984	80.6	75.4	81.3	91.1	80.6	78.5	79.0	76.1	89.5	66.3	66.6	66.8	66.2	66.0	72.2	82.8
1983	78.2	73.2	78.2	86.5	77.1	75.3	76.5	74.5	85.4	64.6	65.1	65.7	64.9	64.6	70.5	79.4
1982	72.1	67.5	70.3	78.3	71.3	69.8	70.2	68.4	77.8	58.7	60.0	60.5	59.5	59.4	66.1	73.2
1981	66.1	62.6	65.2	71.6	65.6	64.7	64.6	63.5	71.0	55.3	56.6	56.7	55.9	56.3	61.7	67.7
1980	60.7	58.0	60.6	66.0	60.5	60.3	61.6	58.5	65.9	51.1	52.2	52.5	51.7	50.8	57.4	62.3
1979	54.9	52.8	56.0	60.2	55.0	54.9	56.2	53.2	60.0	45.9	46.9	47.1	46.4	45.9	52.2	56.3
1975	43.7	42.5	45.1	49.5	44.8	43.7	44.8	42.7	47.1	36.1	37.0	37.0	37.3	36.1	39.3	44.7
1970	27.8	27.0	28.9	32.8	29.2	27.8	28.5	26.9	30.0	20.9	23.7	23.6	23.4	23.0	25.8	28.3
1965	21.5	20.8	22.2	25.5	22.8	21.4	21.9	20.7	23.1	16.0	18.2	18.2	18.0	17.7	19.9	21.8
1960	19.5	18.9	19.9	21.5	19.7	19.5	19.9	18.8	21.0	14.4	16.6	16.6	16.4	16.1	18.1	19.9
1955	16.3	15.8	16.7	18.1	16.5	16.3	16.7	15.8	17.6	12.1	13.9	13.9	13.7	13.5	15.2	16.6
1950	13.5	13.1	13.8	14.9	13.6	13.5	13.8	13.0	14.5	10.0	11.5	11.5	11.4	11.2	12.5	13.7
1945	8.6	8.3	8.8	9.5	8.7	8.6	8.8	8.3	9.3	6.4	7.3	7.3	7.2	7.1	8.0	8.8
1940	6.6	6.4	6.8	7.4	6.7	6.7	6.8	6.4	7.2	4.9	5.6	5.6	5.6	5.5	6.2	6.8

New York columns: Binghamton–Yonkers. North Carolina columns: Charlotte–Winston-Salem. Fargo = N.Dakota. Akron = Ohio.

Year	National 30 City Average	Canton	Cincinnati	Cleveland	Columbus	Dayton	Lorain	Springfield	Toledo	Youngstown	Lawton	Oklahoma City	Tulsa	Eugene	Portland	Allentown
Jan 2002	126.1E	120.5E	118.8E	129.2E	120.2E	116.8E	123.6E	117.5E	127.5E	123.1E	104.9E	105.1E	102.3E	131.2E	133.1E	126.2E
2001	122.2	117.7	115.9	125.9	117.1	114.0	120.4	114.7	123.8	119.9	102.0	102.1	99.7	129.8	131.2	123.5
2000	118.9	112.7	110.1	121.3	112.5	109.0	115.0	109.2	115.7	114.1	98.7	98.9	97.5	126.3	127.4	119.9
1999	116.6	110.6	107.9	118.7	109.5	107.1	112.0	106.4	113.6	111.9	97.5	97.4	96.2	120.9	124.3	117.8
1998	113.6	108.5	105.0	114.8	106.8	104.5	109.4	104.1	111.2	109.0	94.3	94.5	94.5	120.0	122.2	115.7
1997	111.5	106.4	102.9	112.9	104.7	102.6	107.4	102.2	108.5	107.2	92.9	93.2	93.0	118.1	119.6	113.9
1996	108.9	103.2	100.2	110.1	101.1	98.8	103.8	98.4	105.1	104.1	90.9	91.3	91.2	114.4	116.1	112.0
1995	105.6	98.8	97.1	106.4	99.1	94.6	97.1	92.0	100.6	100.1	85.3	88.0	89.0	112.2	114.3	108.3
1994	103.0	97.7	95.0	104.8	95.4	93.0	96.0	90.1	99.3	98.9	84.0	86.5	87.6	107.6	109.2	106.4
1993	100.0	95.9	92.3	101.9	93.9	90.6	93.9	87.7	98.3	97.2	81.3	83.9	84.8	107.2	108.8	103.6
1992	97.9	94.5	90.6	98.7	92.6	89.2	92.2	86.4	97.1	96.0	80.1	82.7	83.4	101.1	102.6	101.6
1991	95.7	93.4	88.6	97.2	90.6	87.7	91.1	84.9	94.2	92.7	78.3	80.7	81.4	99.5	101.1	98.9
1990	93.2	92.1	86.7	95.1	88.1	85.9	89.4	83.4	92.9	91.8	77.1	79.9	80.0	98.0	99.4	95.0
1989	91.0	90.3	84.6	93.6	85.8	82.6	87.9	81.0	91.4	88.4	75.8	78.5	78.4	95.2	97.0	92.4
1988	88.5	89.1	82.9	92.1	83.9	81.0	85.2	79.5	89.6	86.9	74.3	77.0	76.8	93.5	95.3	89.8
1987	85.7	87.7	80.9	89.6	81.5	79.5	86.9	78.0	86.1	85.4	72.6	74.4	74.7	91.4	91.9	86.9
1986	83.7	84.9	78.8	87.3	79.7	78.0	81.6	76.1	84.6	83.8	71.6	74.1	73.9	90.0	91.7	84.7
1985	81.8	83.7	78.2	86.2	77.7	76.3	80.7	74.3	84.7	82.7	71.2	73.9	73.8	88.7	90.1	81.9
1984	80.6	81.3	77.4	82.5	76.1	75.9	79.5	73.9	83.5	80.3	70.6	72.9	72.8	89.5	89.3	78.4
1983	78.2	76.9	74.9	77.9	74.5	71.6	77.2	71.9	80.4	77.3	69.1	71.5	72.2	89.0	89.2	75.0
1982	72.1	70.8	70.1	71.6	68.6	65.2	70.8	66.5	74.6	71.7	63.5	65.2	66.2	82.3	83.6	68.2
1981	66.1	65.8	65.3	66.0	63.9	60.6	65.7	61.3	69.4	67.2	57.0	60.3	61.5	74.2	74.5	62.7
1980	60.7	60.6	59.8	61.0	58.6	56.6	60.8	56.2	64.0	61.5	52.2	55.5	57.2	68.9	68.4	58.7
1979	54.9	55.2	54.3	54.8	52.7	51.2	55.2	51.0	57.9	56.0	47.7	49.8	51.4	61.9	60.6	54.1
1975	43.7	44.0	43.7	44.6	42.2	40.5	42.7	40.0	44.9	44.5	38.6	38.6	39.2	44.6	44.9	42.3
1970	27.8	27.8	28.2	29.6	26.7	27.0	27.5	25.4	29.1	29.6	24.1	23.4	24.1	30.4	30.0	27.3
1965	21.5	21.5	20.7	21.1	20.2	20.0	21.1	19.6	21.3	21.0	18.5	17.8	20.6	23.4	22.6	21.0
1960	19.5	19.5	19.3	19.6	18.7	18.1	19.2	17.8	19.4	19.1	16.9	16.1	18.1	21.2	21.1	19.1
1955	16.3	16.3	16.2	16.5	15.7	15.2	16.1	14.9	16.2	16.0	14.1	13.5	15.1	17.8	17.7	16.0
1950	13.5	13.5	13.3	13.6	13.0	12.5	13.3	12.3	13.4	13.2	11.7	11.2	12.5	14.7	14.6	13.2
1945	8.6	8.6	8.5	8.7	8.3	8.0	8.5	7.8	8.6	8.4	7.4	7.1	8.0	9.4	9.3	8.4
1940	6.6	6.7	6.5	6.7	6.4	6.2	6.5	6.1	6.6	6.5	5.7	5.5	6.1	7.2	7.2	6.5

Ohio columns: Canton–Youngstown. Oklahoma columns: Lawton, Oklahoma City, Tulsa. Oregon columns: Eugene, Portland. Allentown = PA.

Historical Cost Indexes

Year	National 30 City Average	Pennsylvania Erie	Harris-burg	Phila-delphia	Pitts-burgh	Reading	Scranton	RI Provi-dence	South Carolina Charles-ton	Colum-bia	South Dakota Rapid City	Sioux Falls	Chatta-nooga	Tennessee Knox-ville	Memphis	Nash-ville
Jan 2002	126.1E	121.7E	121.4E	139.9E	128.0E	123.1E	123.7E	130.6E	94.7E	93.8E	99.6E	102.5E	101.8E	100.2E	106.2E	108.3E
2001	122.2	118.9	118.4	136.9	124.1	120.4	121.1	127.8	92.1	91.3	96.8	99.7	98.2	96.5	102.6	104.3
2000	118.9	114.6	114.0	132.1	120.9	115.9	117.7	122.8	89.9	89.1	94.2	98.0	96.9	95.0	100.8	100.8
1999	116.6	113.8	112.8	129.8	119.6	114.8	116.3	121.6	89.0	88.1	92.4	95.8	95.9	93.4	99.7	98.6
1998	113.6	109.8	110.5	126.6	117.1	112.5	113.7	120.3	88.0	87.2	90.7	93.6	94.9	92.4	97.2	96.8
1997	111.5	108.5	108.8	123.3	113.9	110.8	112.3	118.9	86.5	85.6	89.3	92.1	93.6	90.8	96.1	94.9
1996	108.9	106.8	105.7	120.3	110.8	108.7	109.9	117.2	84.9	84.1	87.2	90.0	91.6	88.8	94.0	91.9
1995	105.6	99.8	100.6	117.1	106.3	103.6	103.8	111.1	82.7	82.2	84.0	84.7	89.2	86.0	91.2	87.6
1994	103.0	98.2	99.4	115.2	103.7	102.2	102.6	109.6	81.2	80.6	82.7	83.1	87.4	84.1	89.0	84.8
1993	100.0	94.9	96.6	107.4	99.0	98.6	99.8	108.2	78.4	77.9	81.2	81.9	85.1	82.3	86.8	81.9
1992	97.9	92.1	94.8	105.3	96.5	96.7	97.8	106.9	77.1	76.8	80.0	80.7	83.6	77.7	85.4	80.7
1991	95.7	90.5	92.5	101.7	93.2	94.1	94.8	96.1	75.0	75.8	78.7	79.7	81.9	76.6	83.0	79.1
1990	93.2	88.5	89.5	98.5	91.2	90.8	91.3	94.1	73.7	74.5	77.2	78.4	79.9	75.1	81.3	77.1
1989	91.0	86.6	86.5	94.2	89.4	87.9	88.8	92.4	72.1	72.8	75.7	77.0	78.5	73.7	81.2	74.8
1988	88.5	84.9	84.0	89.8	87.7	85.3	86.4	89.1	70.5	71.3	74.2	75.5	77.0	72.1	79.7	73.1
1987	85.7	82.6	81.8	86.8	85.6	82.9	83.3	86.3	70.2	70.2	73.6	74.9	74.4	70.2	77.5	71.0
1986	83.7	81.5	79.6	84.9	83.6	79.7	81.3	85.2	67.4	68.1	71.0	72.3	74.2	69.2	75.3	68.1
1985	81.8	79.6	77.2	82.2	81.5	77.7	80.0	83.0	65.9	66.9	69.7	71.2	72.5	67.7	74.3	66.7
1984	80.6	78.3	75.4	79.0	78.9	75.9	78.8	80.3	64.3	65.5	68.7	70.4	71.4	66.4	73.3	66.3
1983	78.2	76.3	72.2	74.4	75.5	74.0	75.0	76.3	65.3	65.2	67.7	69.7	69.3	64.1	72.8	65.1
1982	72.1	71.0	66.3	69.0	70.9	67.7	68.4	70.0	59.9	60.0	64.4	67.6	64.5	59.8	66.5	62.2
1981	66.1	64.6	61.7	63.4	66.3	62.5	62.3	64.3	56.3	55.9	59.7	63.6	59.9	55.9	60.3	57.3
1980	60.7	59.4	56.9	58.7	61.3	57.7	58.4	59.2	50.6	51.1	54.8	57.1	55.0	51.6	55.9	53.1
1979	54.9	54.0	52.6	54.2	55.6	53.8	53.5	53.3	45.7	44.4	49.7	51.5	49.8	46.4	51.0	47.0
1975	43.7	43.4	42.2	44.5	44.5	43.6	41.9	42.7	35.0	36.4	37.7	39.6	39.1	37.3	40.7	37.0
1970	27.8	27.5	25.8	27.5	28.7	27.1	27.0	27.3	22.8	22.9	24.8	25.8	24.9	22.0	23.0	22.8
1965	21.5	21.1	20.1	21.7	22.4	20.9	20.8	21.9	17.6	17.6	19.1	19.9	19.2	17.1	18.3	17.4
1960	19.5	19.1	18.6	19.4	19.7	19.0	18.9	19.1	16.0	16.0	17.3	18.1	17.4	15.5	16.6	15.8
1955	16.3	16.1	15.6	16.3	16.5	15.9	15.9	16.0	13.4	13.4	14.5	15.1	14.6	13.0	13.9	13.3
1950	13.5	13.3	12.9	13.5	13.6	13.2	13.1	13.2	11.1	11.1	12.0	12.5	12.1	10.7	11.5	10.9
1945	8.6	8.5	8.2	8.6	8.7	8.4	8.3	8.4	7.1	7.1	7.7	8.0	7.7	6.8	7.3	7.0
1940	6.6	6.5	6.3	6.6	6.7	6.4	6.5	6.5	5.4	5.5	5.9	6.2	6.0	5.3	5.6	5.4

Year	National 30 City Average	Texas Abilene	Ama-rillo	Austin	Beau-mont	Corpus Christi	Dallas	El Paso	Fort Worth	Houston	Lubbock	Odessa	San Antonio	Waco	Wichita Falls	Utah Ogden
Jan 2002	126.1E	99.2E	101.6E	102.4E	104.9E	99.2E	107.2E	98.3E	103.7E	111.2E	100.6E	96.0E	104.0E	100.4E	100.6E	110.7E
2001	122.2	93.4	98.4	99.8	102.3	96.6	103.8	95.5	100.9	107.8	97.6	93.4	100.5	97.8	97.3	107.7
2000	118.9	93.4	98.1	99.1	101.6	96.9	102.7	92.4	99.9	106.0	97.6	93.4	99.4	97.3	96.9	104.6
1999	116.6	91.8	94.5	96.0	99.7	94.0	101.0	90.7	97.6	104.6	96.0	92.1	98.0	94.8	95.5	103.3
1998	113.6	89.8	92.7	94.2	97.9	91.8	97.9	88.4	94.5	101.3	93.3	90.1	94.8	92.7	92.9	98.5
1997	111.5	88.4	91.3	92.8	96.8	90.3	96.1	87.0	93.3	100.1	91.9	88.8	93.4	91.4	91.5	96.1
1996	108.9	86.8	89.6	90.9	95.3	88.5	94.1	86.7	91.5	97.9	90.2	87.1	92.3	89.7	89.9	94.1
1995	105.6	85.2	87.4	89.3	93.7	87.4	91.4	85.2	89.5	95.9	88.4	85.6	88.9	86.4	86.8	92.2
1994	103.0	83.3	85.5	87.0	91.8	84.6	89.6	82.2	87.4	93.4	87.0	83.9	87.0	84.9	85.4	89.4
1993	100.0	81.4	83.4	84.9	90.0	82.8	87.8	80.2	85.5	91.1	84.8	81.9	85.0	83.0	83.5	87.1
1992	97.9	80.3	82.3	83.8	88.9	81.6	86.2	79.0	84.1	89.8	83.6	80.7	83.9	81.8	82.4	85.1
1991	95.7	79.2	81.1	82.8	87.8	80.6	85.9	78.0	83.3	87.9	82.7	79.8	83.3	80.6	81.5	84.1
1990	93.2	78.0	80.1	81.3	86.5	79.3	84.5	76.7	82.1	85.4	81.5	78.6	80.7	79.6	80.3	83.4
1989	91.0	76.6	78.7	79.7	85.1	77.8	82.5	75.2	80.6	84.0	80.0	77.0	78.7	78.1	78.7	81.9
1988	88.5	75.0	77.2	78.3	83.9	76.3	81.1	73.6	79.8	82.5	78.4	75.5	77.1	76.5	77.4	80.4
1987	85.7	75.9	76.4	76.0	81.3	75.6	79.9	72.2	77.7	81.3	77.3	74.6	75.7	75.2	77.4	79.1
1986	83.7	72.5	74.1	75.3	80.5	73.5	78.9	70.7	76.7	80.3	75.4	72.4	74.3	72.8	74.5	77.3
1985	81.8	71.1	72.5	74.5	79.3	72.3	77.6	69.4	75.1	79.6	74.0	71.2	73.9	71.7	73.3	75.2
1984	80.6	70.5	71.6	72.8	78.7	71.6	79.8	68.3	77.1	80.7	72.7	69.8	73.1	71.3	72.4	74.9
1983	78.2	68.3	70.6	71.0	75.0	70.2	77.8	66.3	74.9	79.8	71.0	69.0	71.8	69.4	70.1	76.1
1982	72.1	62.7	65.5	64.9	65.7	64.7	70.7	62.6	68.2	72.1	65.6	62.5	66.0	64.3	64.8	69.7
1981	66.1	57.9	60.1	59.3	62.4	59.6	63.1	58.6	61.0	65.1	61.0	58.3	60.5	59.2	60.0	65.3
1980	60.7	53.4	55.2	54.5	57.6	54.5	57.9	53.1	57.0	59.4	55.6	57.2	55.0	54.9	55.4	62.2
1979	54.9	48.6	49.8	49.0	52.3	48.7	51.5	48.5	51.4	53.3	50.3	51.1	49.2	49.7	49.4	53.3
1975	43.7	37.6	39.0	39.0	39.6	38.1	40.7	38.0	40.4	41.2	38.9	37.9	39.0	38.6	38.0	40.0
1970	27.8	24.5	24.9	24.9	25.7	24.5	25.5	23.7	25.9	25.4	25.1	24.6	23.3	24.8	24.5	26.8
1965	21.5	18.9	19.2	19.2	19.9	18.9	19.9	19.0	19.9	20.0	19.4	19.0	18.5	19.2	18.9	20.6
1960	19.5	17.1	17.4	17.4	18.1	17.1	18.2	17.0	18.1	18.2	17.6	17.3	16.8	17.4	17.2	18.8
1955	16.3	14.4	14.6	14.6	15.1	14.4	15.3	14.3	15.2	15.2	14.8	14.5	14.1	14.6	14.4	15.7
1950	13.5	11.9	12.1	12.1	12.5	11.9	12.6	11.8	12.5	12.6	12.2	12.0	11.6	12.1	11.9	13.0
1945	8.6	7.6	7.7	7.7	8.0	7.6	8.0	7.5	8.0	8.0	7.8	7.6	7.4	7.7	7.6	8.3
1940	6.6	5.9	5.9	5.9	6.1	5.8	6.2	5.8	6.2	6.2	6.0	5.9	5.7	5.9	5.8	6.4

HISTORICAL COST INDEXES

Historical Cost Indexes

Year	National 30 City Average	Utah Salt Lake City	Vermont Burlington	Vermont Rutland	Virginia Alexandria	Virginia Newport News	Virginia Norfolk	Virginia Richmond	Virginia Roanoke	Washington Seattle	Washington Spokane	Washington Tacoma	West Virginia Charleston	West Virginia Huntington	Wisconsin Green Bay	Wisconsin Kenosha
Jan 2002	126.1E	112.2E	108.7E	108.1E	113.5E	102.4E	102.9E	106.2E	95.1E	132.6E	123.2E	129.6E	116.7E	119.6E	121.6E	126.6E
2001	122.2	109.1	105.7	105.2	110.8	99.8	100.3	102.9	92.1	127.9	120.3	125.7	114.6	117.5	119.1	123.6
2000	118.9	106.5	98.9	98.3	108.1	96.5	97.6	100.2	90.7	124.6	118.3	122.9	111.5	114.4	114.6	119.1
1999	116.6	104.5	98.2	97.7	106.1	95.6	96.5	98.8	89.8	123.3	116.7	121.6	110.6	113.4	112.1	115.8
1998	113.6	99.5	97.8	97.3	104.1	93.7	93.9	97.0	88.3	119.4	114.3	118.3	106.7	109.0	109.5	112.9
1997	111.5	97.2	96.6	96.3	101.2	91.6	91.7	92.9	86.9	118.1	111.7	117.2	105.3	107.7	105.6	109.1
1996	108.9	94.9	95.1	94.8	99.7	90.2	90.4	91.6	85.5	115.2	109.2	114.3	103.1	104.8	103.8	106.4
1995	105.6	93.1	91.1	90.8	96.3	86.0	86.4	87.8	82.8	113.7	107.4	112.8	95.8	97.2	97.6	97.9
1994	103.0	90.2	89.5	89.3	93.9	84.6	84.8	86.3	81.4	109.9	104.0	108.3	94.3	95.3	96.3	96.2
1993	100.0	87.9	87.6	87.6	91.6	82.9	83.0	84.3	79.5	107.3	103.9	106.7	92.6	93.5	94.0	94.3
1992	97.9	86.0	86.1	86.1	90.1	81.0	81.6	82.0	78.3	105.1	101.4	103.7	91.4	92.3	92.0	92.1
1991	95.7	84.9	84.2	84.2	88.2	77.6	77.9	79.8	77.3	102.2	100.0	102.2	89.7	88.6	88.6	89.8
1990	93.2	84.3	83.0	82.9	86.1	76.3	76.7	77.6	76.1	100.1	98.5	100.5	86.1	86.8	86.7	87.8
1989	91.0	82.8	81.4	81.3	83.4	74.9	75.2	76.0	74.2	96.1	96.8	98.4	84.5	84.8	84.3	85.6
1988	88.5	81.3	79.7	79.7	81.0	73.4	73.7	74.1	72.3	94.2	95.0	96.6	82.8	83.1	82.0	83.0
1987	85.7	79.8	79.0	79.0	77.9	71.0	71.7	72.7	70.2	91.9	92.4	92.9	81.1	81.3	80.2	81.7
1986	83.7	78.1	76.4	76.4	76.8	70.3	70.5	71.0	67.8	90.5	91.9	92.8	79.9	80.2	77.8	79.3
1985	81.8	75.9	74.8	74.9	75.1	68.7	68.8	69.5	67.2	88.3	89.0	91.2	77.7	77.7	76.7	77.4
1984	80.6	75.2	73.7	73.7	75.0	67.9	68.3	68.8	66.2	89.2	88.1	90.0	75.4	75.8	75.0	75.1
1983	78.2	74.5	70.9	71.5	72.9	65.9	66.5	68.9	66.0	87.6	86.9	88.8	73.8	74.4	73.4	73.2
1982	72.1	68.1	64.9	65.4	67.0	60.6	60.9	63.7	61.0	80.7	80.6	81.7	67.6	68.4	67.5	69.1
1981	66.1	61.9	59.3	59.7	61.6	56.0	56.8	58.2	55.6	75.7	72.9	73.8	62.6	63.3	63.6	63.5
1980	60.7	57.0	55.3	58.3	57.3	52.5	52.4	54.3	51.3	67.9	66.3	66.7	57.7	58.3	58.6	58.3
1979	54.9	52.5	49.8	51.2	51.7	47.6	47.6	49.1	46.8	61.1	60.1	60.0	51.8	52.5	52.7	53.2
1975	43.7	40.1	41.8	43.9	41.7	37.2	36.9	37.1	37.1	44.9	44.4	44.5	41.0	40.0	40.9	40.5
1970	27.8	26.1	25.4	26.8	26.2	23.9	21.5	22.0	23.7	28.8	29.3	29.6	26.1	25.8	26.4	26.5
1965	21.5	20.0	19.8	20.6	20.2	18.4	17.1	17.2	18.3	22.4	22.5	22.8	20.1	19.9	20.3	20.4
1960	19.5	18.4	18.0	18.8	18.4	16.7	15.4	15.6	16.6	20.4	20.8	20.8	18.3	18.1	18.4	18.6
1955	16.3	15.4	15.1	15.7	15.4	14.0	12.9	13.1	13.9	17.1	17.4	17.4	15.4	15.2	15.5	15.6
1950	13.5	12.7	12.4	13.0	12.7	11.6	10.7	10.8	11.5	14.1	14.4	14.4	12.7	12.5	12.8	12.9
1945	8.6	8.1	7.9	8.3	8.1	7.4	6.8	6.9	7.3	9.0	9.2	9.2	8.1	8.0	8.1	8.2
1940	6.6	6.3	6.1	6.4	6.2	5.7	5.3	5.3	5.7	7.0	7.1	7.1	6.2	6.2	6.3	6.3

Year	National 30 City Average	Wisconsin Madison	Wisconsin Milwaukee	Wisconsin Racine	Wyoming Cheyenne	Canada Calgary	Canada Edmonton	Canada Hamilton	Canada London	Canada Montreal	Canada Ottawa	Canada Quebec	Canada Toronto	Canada Vancouver	Canada Winnipeg
Jan 2002	126.1E	123.2E	127.3E	126.2E	101.6E	121.2E	121.2E	136.1E	133.6E	127.4E	134.7E	128.6E	139.5E	133.8E	120.5E
2001	122.2	120.6	123.9	123.3	99.0	117.5	117.4	131.5	129.1	124.4	130.2	125.5	134.7	130.2	117.2
2000	118.9	116.6	120.5	118.7	98.1	115.9	115.8	130.0	127.5	122.8	128.8	124.1	133.1	128.4	115.6
1999	116.6	115.9	117.4	115.5	96.9	115.3	115.2	128.1	125.6	120.8	126.8	121.9	131.2	127.1	115.2
1998	113.6	110.8	113.4	112.7	95.4	112.5	112.4	126.3	123.9	119.0	124.7	119.6	128.5	123.8	113.7
1997	111.5	106.1	110.1	109.3	93.2	110.7	110.6	124.4	121.9	114.6	122.8	115.4	125.7	121.9	111.4
1996	108.9	104.4	107.1	106.5	91.1	109.1	109.0	122.6	120.2	112.9	121.0	113.6	123.9	119.0	109.7
1995	105.6	96.5	103.9	97.8	87.6	107.4	107.4	119.9	117.5	110.8	118.2	111.5	121.6	116.2	107.6
1994	103.0	94.5	100.6	96.1	85.4	106.7	106.6	116.5	114.2	109.5	115.0	110.2	117.8	115.1	105.5
1993	100.0	91.3	96.7	93.8	82.9	104.7	104.5	113.7	111.9	106.9	112.0	107.0	114.9	109.2	102.8
1992	97.9	89.2	93.9	91.7	81.7	103.4	103.2	112.4	110.7	104.2	110.8	103.6	113.7	108.0	101.6
1991	95.7	86.2	91.6	89.3	80.3	102.1	102.0	108.2	106.7	101.8	106.9	100.4	109.0	106.8	98.6
1990	93.2	84.3	88.9	87.3	79.1	98.0	97.1	103.8	101.4	99.0	102.7	96.8	104.6	103.2	95.1
1989	91.0	81.8	86.4	85.0	77.8	95.6	94.7	98.2	96.5	94.9	98.5	92.5	98.5	97.6	92.9
1988	88.5	79.9	84.1	82.4	76.3	93.9	93.0	95.4	93.2	90.6	94.0	88.3	94.8	95.9	90.0
1987	85.7	78.1	81.1	81.1	76.7	90.2	89.3	89.7	90.1	87.2	89.6	85.0	90.4	94.2	87.1
1986	83.7	76.1	78.9	78.6	73.4	91.2	90.1	88.9	87.9	84.8	87.7	82.4	89.3	93.2	85.3
1985	81.8	74.3	77.4	77.0	72.3	90.2	89.1	85.8	84.9	82.4	83.9	79.9	86.5	89.8	83.4
1984	80.6	72.5	76.3	74.7	73.8	89.6	88.1	85.1	84.1	81.6	82.9	80.0	85.2	89.2	82.5
1983	78.2	71.9	74.0	73.0	73.8	84.6	82.8	80.1	79.1	76.4	77.6	75.8	79.9	83.3	77.9
1982	72.1	65.9	70.5	68.6	68.3	76.2	74.1	73.7	71.8	70.2	72.3	69.6	71.8	76.4	70.7
1981	66.1	61.4	64.9	63.3	62.7	70.6	68.1	68.8	67.1	64.6	65.8	64.3	67.5	70.3	65.4
1980	60.7	56.8	58.8	58.1	56.9	64.9	63.3	63.7	61.9	59.2	60.9	59.3	60.9	65.0	61.7
1979	54.9	51.4	53.0	52.8	51.2	58.9	58.5	58.1	56.6	54.4	55.5	53.9	56.0	59.2	56.1
1975	43.7	40.7	43.3	40.7	40.6	42.2	41.6	42.9	41.6	39.7	41.5	39.0	42.2	42.4	39.2
1970	27.8	26.5	29.4	26.5	26.0	28.9	28.6	28.5	27.8	25.6	27.6	26.0	25.6	26.0	23.1
1965	21.5	20.6	21.8	20.4	20.0	22.3	22.0	22.0	21.4	18.7	21.2	20.1	19.4	20.5	17.5
1960	19.5	18.1	19.0	18.6	18.2	20.2	20.0	20.0	19.5	17.0	19.3	18.2	17.6	18.6	15.8
1955	16.3	15.2	15.9	15.6	15.2	17.0	16.8	16.7	16.3	14.3	16.2	15.3	14.8	15.5	13.3
1950	13.5	12.5	13.2	12.9	12.6	14.0	13.9	13.8	13.5	11.8	13.4	12.6	12.2	12.8	10.9
1945	8.6	8.0	8.4	8.2	8.0	9.0	8.8	8.8	8.6	7.5	8.5	8.0	7.8	8.2	7.0
1940	6.6	6.2	6.5	6.3	6.2	6.9	6.8	6.8	6.6	5.8	6.6	6.2	6.0	6.3	5.4

Accent lighting–Fixtures or directional beams of light arranged to bring attention to an object or area.

Acoustical material–A material fabricated for the sole purpose of absorbing sound.

Addition–An expansion to an existing structure generally in the form of a room, floor(s) or wing(s). An increase in the floor area or volume of a structure.

Aggregate–Materials such as sand, gravel, stone, vermiculite, perlite and fly ash (slag) which are essential components in the production of concrete, mortar or plaster.

Air conditioning system–An air treatment to control the temperature, humidity and cleanliness of air and to provide for its distribution throughout the structure.

Air curtain–A stream of air directed downward to prevent the loss of hot or cool air from the structure and inhibit the entrance of dust and insects. Generally installed at loading platforms.

Anodized aluminum–Aluminum treated by an electrolytic process to produce an oxide film that is corrosion resistant.

Arcade–A covered passageway between buildings often with shops and offices on one or both sides.

Ashlar–A square-cut building stone or a wall constructed of cut stone.

Asphalt paper–A sheet paper material either coated or saturated with asphalt to increase its strength and resistance to water.

Asphalt shingles–Shingles manufactured from saturated roofing felts, coated with asphalt and topped with mineral granules to prevent weathering.

Assessment ratio–The ratio between the market value and assessed valuation of a property.

Back-up–That portion of a masonry wall behind the exterior facing-usually load bearing.

Balcony–A platform projecting from a building either supported from below or cantilevered. Usually protected by railing.

Bay–Structural component consisting of beams and columns occurring consistently throughout the structure.

Beam–A horizontal structural framing member that transfers loads to vertical members (columns) or bearing walls.

Bid–An offer to perform work described in a contract at a specified price.

Board foot–A unit or measure equal in volume to a board one foot long, one foot wide and one inch thick.

Booster pump–A supplemental pump installed to increase or maintain adequate pressure in the system.

Bowstring roof–A roof supported by trusses fabricated in the shape of a bow and tied together by a straight member.

Brick veneer–A facing of brick laid against, but not bonded to, a wall.

Bridging–A method of bracing joists for stiffness, stability and load distribution.

Broom finish–A method of finishing concrete by lightly dragging a broom over freshly placed concrete.

Built-up roofing–Installed on flat or extremely low-pitched roofs, a roof covering composed of plies or laminations of saturated roofing felts alternated with layers of coal tar pitch or asphalt and surfaced with a layer of gravel or slag in a thick coat of asphalt and finished with a capping sheet.

Caisson–A watertight chamber to expedite work on foundations or structure below water level.

Cantilever–A structural member supported only at one end.

Carport–An automobile shelter having one or more sides open to the weather.

Casement window–A vertical opening window having one side fixed.

Caulking–A resilient compound generally having a silicone or rubber base used to prevent infiltration of water or outside air.

Cavity wall–An exterior wall usually of masonry having an inner and outer wall separated by a continuous air space for thermal insulation.

Cellular concrete–A lightweight concrete consisting of cement mixed with gas-producing materials which, in turn, forms bubbles resulting in good insulating qualities.

Chattel mortgage–A secured interest in a property as collateral for payment of a note.

Clapboard–A wood siding used as exterior covering in frame construction. It is applied horizontally with grain running lengthwise. The thickest section of the board is on the bottom.

Clean room–An environmentally controlled room usually found in medical facilities or precision manufacturing spaces where it is essential to eliminate dust, lint or pathogens.

Close studding–A method of construction whereby studs are spaced close together and the intervening spaces are plastered.

Cluster housing–A closely grouped series of houses resulting in a high density land use.

Cofferdam–A watertight enclosure used for foundation construction in waterfront areas. Water is pumped out of the cofferdam allowing free access to the work area.

Collar joint–Joint between the collar beam and roof rafters.

Column–A vertical structural member supporting horizontal members (beams) along the direction of its longitudinal axis.

Combination door (windows)–Door (windows) having interchangeable screens and glass for seasonal use.

Common area–Spaces either inside or outside the building designated for use by the occupant of the building but not the general public.

Common wall–A wall used jointly by two dwelling units.

Compound wall–A wall constructed of more than one material.

Concrete–A composite material consisting of sand, coarse aggregate (gravel, stone or slag), cement and water that when mixed and allowed to harden forms a hard stone-like material.

Concrete floor hardener–A mixture of chemicals applied to the surface of concrete to produce a dense, wear-resistant bearing surface.

Conduit–A tube or pipe used to protect electric wiring.

Coping–The top cover or capping on a wall.

Craneway–Steel or concrete column and beam supports and rails on which a crane travels.

Curtain wall–A non-bearing exterior wall not supported by beams or girders of the steel frame.

Dampproofing–Coating of a surface to prevent the passage of water or moisture.

Dead load–Total weight of all structural components plus permanently attached fixtures and equipment.

Decibel–Unit of acoustical measurement.

Decorative block–A concrete masonry unit having a special treatment on its face.

Deed restriction–A restriction on the use of a property as set forth in the deed.

Depreciation–A loss in property value caused by physical aging, functional or economic obsolescence.

Direct heating–Heating of spaces by means of exposed heated surfaces (stove, fire, radiators, etc.).

Distributed load–A load that acts evenly over a structural member.

Dock bumper–A resilient material attached to a loading dock to absorb the impact of trucks backing in.

Dome–A curved roof shape spanning an area.

Dormer–A structure projecting from a sloping roof.

Double-hung window–A window having two vertical sliding sashes-one covering the upper section and one covering the lower.

Downspout–A vertical pipe for carrying rain water from the roof to the ground.

Drain tile–A hollow tile used to drain water-soaked soil.

Dressed size–In lumber about 3/8″ - 1/2″ less than nominal size after planing and sawing.

Drop panel–The depressed surface on the bottom side of a flat concrete slab which surrounds a column.

Dry pipe sprinkler system–A sprinkler system that is activated with water only in case of fire. Used in areas susceptible to freezing or to avoid the hazards of leaking pipes.

Drywall–An interior wall constructed of gypsum board, plywood or wood paneling. No water is required for application.

Duct–Pipe for transmitting warm or cold air. A pipe containing electrical cable or wires.

Dwelling–A structure designed as living quarters for one or more families.

Easement–A right of way or free access over land owned by another.

Eave–The lower edge of a sloping roof that overhangs the side wall of the structure.

Economic rent–That rent on a property sufficient to pay all operating costs exclusive of services and utilities.

Economic life–The term during which a structure is expected to be profitable. Generally shorter than the physical life of the structure.

Economic obsolescence–Loss in value due to unfavorable economic influences occurring from outside the structure itself.

Effective age–The age a structure appears to be based on observed physical condition determined by degree of maintenance and repair.

Elevator–A mechanism for vertical transport of personnel or freight equipped with car or platform.

Ell–A secondary wing or addition to a structure at right angles to the main structure.

Eminent domain–The right of the state to take private property for public use.

Envelope–The shape of a building indicating volume.

Equity–Value of an owner's interest calculated by subtracting outstanding mortgages and expenses from the value of the property.

Escheat–The assumption of ownership by the state, of property whose owner cannot be determined.

Facade–The exterior face of a building sometimes decorated with elaborate detail.

Face brick–Brick manufactured to present an attractive appearance.

Facing block–A concrete masonry unit having a decorative exterior finish.

Feasibility study–A detailed evaluation of a proposed project to determine its financial potential.

Felt paper–Paper sheathing used on walls as an insulator and to prevent infiltration of moisture.

Fenestration–Pertaining to the density and arrangement of windows in a structure.

Fiberboard–A building material composed of wood fiber compressed with a binding agent, produced in sheet form.

Fiberglass–Fine spun filaments of glass processed into various densities to produce thermal and acoustical insulation.

Field house–A long structure used for indoor athletic events.

Finish floor–The top or wearing surface of the floor system.

Fireproofing–The use of fire-resistant materials for the protection of structural members to ensure structural integrity in the event of fire.

Fire stop–A material or member used to seal an opening to prevent the spread of fire.

Flashing–A thin, impervious material such as copper or sheet metal used to prevent air or water penetration.

Float finish–A concrete finish accomplished by using a flat tool with a handle on the back.

Floor drain–An opening installed in a floor for removing excess water into a plumbing system.

Floor load–The live load the floor system has been designed for and which may be applied safely.

Flue–A heat-resistant enclosed passage in a chimney to remove gaseous products of combustion from a fireplace or boiler.

Footing–The lowest portion of the foundation wall that transmits the load directly to the soil.

Foundation–Below grade wall system that supports the structure.

Foyer–A portion of the structure that serves as a transitional space between the interior and exterior.

Frontage–That portion of a lot that is placed facing a street, public way or body of water.

Frost action–The effects of freezing and thawing on materials and the resultant structural damage.

Functional obsolescence–An inadequacy caused by outmoded design, dated construction materials or over or undersized areas, all of which cause excessive operating costs.

Furring strips–Wood or metal channels used for attaching gypsum or metal lath to masonry walls as a finish or for leveling purposes.

Gambrel roof (mansard)–A roof system having two pitches on each side.

Garden apartment–Two or three story apartment building with common outside areas.

General contractor–The prime contractor responsible for work on the construction site.

Girder–A principal horizontal supporting member.

Girt–A horizontal framing member for adding rigid support to columns which also acts as a support for sheathing or siding.

Grade beam–That portion of the foundation system that directly supports the exterior walls of the building.

Gross area–The total enclosed floor area of a building.

Ground area–The area computed by the exterior dimensions of the structure.

Ground floor–The floor of a building in closest proximity to the ground.

Grout–A mortar containing a high water content used to fill joints and cavities in masonry work.

Gutter–A wood or sheet metal conduit set along the building eaves used to channel rainwater to leaders or downspouts.

Gunite–A concrete mix placed pneumatically.

Gypsum–Hydrated calcium sulphate used as both a retarder in Portland cement and as a major ingredient in plaster of Paris. Used in sheets as a substitute for plaster.

Hall–A passageway providing access to various parts of a building–a large room for entertainment or assembly.

Hangar–A structure for the storage or repair of aircraft.

Head room–The distance between the top of the finished floor to the bottom of the finished ceiling.

Hearth–The floor of a fireplace and the adjacent area of fireproof material.

Heat pump–A mechanical device for providing either heating or air conditioning.

Hip roof–A roof whose four sides meet at a common point with no gabled ends.

I-beam–A structural member having a cross-section resembling the letter "I".

Insulation–Material used to reduce the effects of heat, cold or sound.

Jack rafter–An unusually short rafter generally found in hip roofs.

Jalousie–Adjustable glass louvers which pivot simultaneously in a common frame.

Jamb–The vertical member on either side of a door frame or window frame.

Joist–Parallel beams of timber, concrete or steel used to support floor and ceiling.

Junction box–A box that protects splices or joints in electrical wiring.

Kalamein door–Solid core wood doors clad with galvanized sheet metal.

Kip–A unit of weight equal to 1,000 pounds.

Lally column–A concrete-filled steel pipe used as a vertical support.

Laminated beam–A beam built up by gluing together several pieces of timber.

Landing–A platform between flights of stairs.

Lath–Strips of wood or metal used as a base for plaster.

Lead-lined door, sheetrock–Doors or sheetrock internally lined with sheet lead to provide protection from X-ray radiation.

Lean-to–A small shed or building addition with a single pitched roof attached to the exterior wall of the main building.

Lintel–A horizontal framing member used to carry a load over a wall opening.

Live load–The moving or movable load on a structure composed of furnishings, equipment or personnel weight.

Load bearing partition–A partition that supports a load in addition to its own weight.

Loading dock leveler–An adjustable platform for handling off-on loading of trucks.

Loft building–A commercial/industrial type building containing large, open unpartitioned floor areas.

Louver window–A window composed of a series of sloping, overlapping blades or slats that may be adjusted to admit varying degrees of air or light.

Main beam–The principal load bearing beam used to transmit loads directly to columns.

Mansard roof–A roof having a double pitch on all four sides, the lower level having the steeper pitch.

Mansion–An extremely large and imposing residence.

Masonry–The utilization of brick, stone, concrete or block for walls and other building components.

Mastic–A sealant or adhesive compound generally used as either a binding agent or floor finish.

Membrane fireproofing–A coating of metal lath and plaster to provide resistance to fire and heat.

Mesh reinforcement–An arrangement of wire tied or welded at their intersection to provide strength and resistance to cracking.

Metal lath–A diamond-shaped metallic base for plaster.

Mezzanine–A low story situated between two main floors.

Mill construction–A heavy timber construction that achieves fire resistance by using large wood structural members, noncombustible bearing and non-bearing walls and omitting concealed spaces under floors and roof.

Mixed occupancy–Two or more classes of occupancy in a single structure.

Molded brick–A specially shaped brick used for decorative purposes.

Monolithic concrete–Concrete poured in a continuous process so there are no joints.

Movable partition–A non-load bearing demountable partition that can be relocated and can be either ceiling height or partial height.

Moving walk–A continually moving horizontal passenger carrying device.

Multi-zone system–An air conditioning system that is capable of handling several individual zones simultaneously.

Net floor area–The usable, occupied area of a structure excluding stairwells, elevator shafts and wall thicknesses.

Non-combustible construction–Construction in which the walls, partitions and the structural members are of non-combustible materials.

Nurses call system–An electrically operated system for use by patients or personnel for summoning a nurse.

Occupancy rate–The number of persons per room, per dwelling unit, etc.

One-way joist connection–A framing system for floors and roofs in a concrete building consisting of a series of parallel joists supported by girders between columns.

Open web joist–A lightweight prefabricated metal chord truss.

Parapet–A low guarding wall at the point of a drop. That portion of the exterior wall that extends above the roof line.

Parging–A thin coat of mortar applied to a masonry wall used primarily for waterproofing purposes.

Parquet–Inlaid wood flooring set in a simple design.

Peaked roof–A roof of two or more slopes that rises to a peak.

Penthouse–A structure occupying usually half the roof area and used to house elevator, HVAC equipment, or other mechanical or electrical systems.

Pier–A column designed to support a concentrated load.

Pilaster–A column usually formed of the same material and integral with, but projecting from, a wall.

Pile–A concrete, wood or steel column usually less than 2′ in diameter which is driven or otherwise introduced into the soil to carry a vertical load or provide vertical support.

Pitched roof–A roof having one or more surfaces with a pitch greater than 10°.

Plenum–The space between the suspended ceiling and the floor above.

Plywood–Structural wood made of three or more layers of veneer and bonded together with glue.

Porch–A structure attached to a building to provide shelter for an entranceway or to serve as a semi-enclosed space.

Post and beam framing–A structural framing system where beams rest on posts rather than bearing walls.

Post-tensioned concrete–Concrete that has the reinforcing tendons tensioned after the concrete has set.

Pre-cast concrete–Concrete structural components fabricated at a location other than in-place.

Pre-stressed concrete–Concrete that has the reinforcing tendons tensioned prior to the concrete setting.

Purlins–A horizontal structural member supporting the roof deck and resting on the trusses, girders, beams or rafters.

Rafters–Structural members supporting the roof deck and covering.

Resilient flooring–A manufactured interior floor covering material in either sheet or tile form, that is resilient.

Ridge–The horizontal line at the junction of two sloping roof edges.

Rigid frame–A structural framing system in which all columns and beams are rigidly connected; there are no hinged joints.

Roof drain–A drain designed to accept rainwater and discharge it into a leader or downspout.

Rough floor–A layer of boards or plywood nailed to the floor joists that serves as a base for the finished floor.

Sanitary sewer–A sewer line designed to carry only liquid or waterborne waste from the structure to a central treatment plant.

Sawtooth roof–Roof shape found primarily in industrial roofs that creates the appearance of the teeth of a saw.

Semi-rigid frame–A structural system wherein the columns and beams are so attached that there is some flexibility at the joints.

Septic tank–A covered tank in which waste matter is decomposed by natural bacterial action.

Service elevator–A combination passenger and freight elevator.

Sheathing–The first covering of exterior studs by boards, plywood or particle board.

Glossary

Shoring–Temporary bracing for structural components during construction.

Span–The horizontal distance between supports.

Specification–A detailed list of materials and requirements for construction of a building.

Storm sewer–A system of pipes used to carry rainwater or surface waters.

Stucco–A cement plaster used to cover exterior wall surfaces usually applied over a wood or metal lath base.

Stud–A vertical wooden structural component.

Subfloor–A system of boards, plywood or particle board laid over the floor joists to form a base.

Superstructure–That portion of the structure above the foundation or ground level.

Terrazzo–A durable floor finish made of small chips of colored stone or marble, embedded in concrete, then polished to a high sheen.

Tile–A thin piece of fired clay, stone or concrete used for floor, roof or wall finishes.

Unit cost–The cost per unit of measurement.

Vault–A room especially designed for storage.

Veneer–A thin surface layer covering a base of common material.

Vent–An opening serving as an outlet for air.

Wainscot–The lower portion of an interior wall whose surface differs from that of the upper wall.

Wall bearing construction–A structural system where the weight of the floors and roof are carried directly by the masonry walls rather than the structural framing system.

Waterproofing–Any of a number of materials applied to various surfaces to prevent the infiltration of water.

Weatherstrip–A thin strip of metal, wood or felt used to cover the joint between the door or window sash and the jamb, casing or sill to keep out air, dust, rain, etc.

Wing–A building section or addition projecting from the main structure.

X-ray protection–Lead encased in sheetrock or plaster to prevent the escape of radiation.

A	Area Square Feet; Ampere
ABS	Acrylonitrile Butadiene Stryrene; Asbestos Bonded Steel
A.C.	Alternating Current; Air-Conditioning; Asbestos Cement; Plywood Grade A & C
A.C.I.	American Concrete Institute
AD	Plywood, Grade A & D
Addit.	Additional
Adj.	Adjustable
af	Audio-frequency
A.G.A.	American Gas Association
Agg.	Aggregate
A.H.	Ampere Hours
A hr	Ampere-hour
A.H.U.	Air Handling Unit
A.I.A.	American Institute of Architects
AIC	Ampere Interrupting Capacity
Allow.	Allowance
alt.	Altitude
Alum.	Aluminum
a.m.	Ante Meridiem
Amp.	Ampere
Anod.	Anodized
Approx.	Approximate
Apt.	Apartment
Asb.	Asbestos
A.S.B.C.	American Standard Building Code
Asbe.	Asbestos Worker
A.S.H.R.A.E.	American Society of Heating, Refrig. & AC Engineers
A.S.M.E.	American Society of Mechanical Engineers
A.S.T.M.	American Society for Testing and Materials
Attchmt.	Attachment
Avg.	Average
A.W.G.	American Wire Gauge
AWWA	American Water Works Assoc.
Bbl.	Barrel
B. & B.	Grade B and Better; Balled & Burlapped
B. & S.	Bell and Spigot
B. & W.	Black and White
b.c.c.	Body-centered Cubic
B.C.Y.	Bank Cubic Yards
BE	Bevel End
B.F.	Board Feet
Bg. cem.	Bag of Cement
BHP	Boiler Horsepower; Brake Horsepower
B.I.	Black Iron
Bit.; Bitum.	Bituminous
Bk.	Backed
Bkrs.	Breakers
Bldg.	Building
Blk.	Block
Bm.	Beam
Boil.	Boilermaker
B.P.M.	Blows per Minute
BR	Bedroom
Brg.	Bearing
Brhe.	Bricklayer Helper
Bric.	Bricklayer
Brk.	Brick
Brng.	Bearing
Brs.	Brass
Brz.	Bronze
Bsn.	Basin
Btr.	Better
BTU	British Thermal Unit
BTUH	BTU per Hour
B.U.R.	Built-up Roofing
BX	Interlocked Armored Cable
c	Conductivity, Copper Sweat
C	Hundred; Centigrade
C/C	Center to Center, Cedar on Cedar
Cab.	Cabinet
Cair.	Air Tool Laborer
Calc	Calculated
Cap.	Capacity
Carp.	Carpenter
C.B.	Circuit Breaker
C.C.A.	Chromate Copper Arsenate
C.C.F.	Hundred Cubic Feet
cd	Candela
cd/sf	Candela per Square Foot
CD	Grade of Plywood Face & Back
CDX	Plywood, Grade C & D, exterior glue
Cefi.	Cement Finisher
Cem.	Cement
CF	Hundred Feet
C.F.	Cubic Feet
CFM	Cubic Feet per Minute
c.g.	Center of Gravity
CHW	Chilled Water; Commercial Hot Water
C.I.	Cast Iron
C.I.P.	Cast in Place
Circ.	Circuit
C.L.	Carload Lot
Clab.	Common Laborer
C.L.F.	Hundred Linear Feet
CLF	Current Limiting Fuse
CLP	Cross Linked Polyethylene
cm	Centimeter
CMP	Corr. Metal Pipe
C.M.U.	Concrete Masonry Unit
CN	Change Notice
Col.	Column
CO$_2$	Carbon Dioxide
Comb.	Combination
Compr.	Compressor
Conc.	Concrete
Cont.	Continuous; Continued
Corr.	Corrugated
Cos	Cosine
Cot	Cotangent
Cov.	Cover
C/P	Cedar on Paneling
CPA	Control Point Adjustment
Cplg.	Coupling
C.P.M.	Critical Path Method
CPVC	Chlorinated Polyvinyl Chloride
C.Pr.	Hundred Pair
CRC	Cold Rolled Channel
Creos.	Creosote
Crpt.	Carpet & Linoleum Layer
CRT	Cathode-ray Tube
CS	Carbon Steel, Constant Shear Bar Joist
Csc	Cosecant
C.S.F.	Hundred Square Feet
CSI	Construction Specifications Institute
C.T.	Current Transformer
CTS	Copper Tube Size
Cu	Copper, Cubic
Cu. Ft.	Cubic Foot
cw	Continuous Wave
C.W.	Cool White; Cold Water
Cwt.	100 Pounds
C.W.X.	Cool White Deluxe
C.Y.	Cubic Yard (27 cubic feet)
C.Y./Hr.	Cubic Yard per Hour
Cyl.	Cylinder
d	Penny (nail size)
D	Deep; Depth; Discharge
Dis.;Disch.	Discharge
Db.	Decibel
Dbl.	Double
DC	Direct Current
DDC	Direct Digital Control
Demob.	Demobilization
d.f.u.	Drainage Fixture Units
D.H.	Double Hung
DHW	Domestic Hot Water
Diag.	Diagonal
Diam.	Diameter
Distrib.	Distribution
Dk.	Deck
D.L.	Dead Load; Diesel
DLH	Deep Long Span Bar Joist
Do.	Ditto
Dp.	Depth
D.P.S.T.	Double Pole, Single Throw
Dr.	Driver
Drink.	Drinking
D.S.	Double Strength
D.S.A.	Double Strength A Grade
D.S.B.	Double Strength B Grade
Dty.	Duty
DWV	Drain Waste Vent
DX	Deluxe White, Direct Expansion
dyn	Dyne
e	Eccentricity
E	Equipment Only; East
Ea.	Each
E.B.	Encased Burial
Econ.	Economy
EDP	Electronic Data Processing
EIFS	Exterior Insulation Finish System
E.D.R.	Equiv. Direct Radiation
Eq.	Equation
Elec.	Electrician; Electrical
Elev.	Elevator; Elevating
EMT	Electrical Metallic Conduit; Thin Wall Conduit
Eng.	Engine, Engineered
EPDM	Ethylene Propylene Diene Monomer
EPS	Expanded Polystyrene
Eqhv.	Equip. Oper., Heavy
Eqlt.	Equip. Oper., Light
Eqmd.	Equip. Oper., Medium
Eqmm.	Equip. Oper., Master Mechanic
Eqol.	Equip. Oper., Oilers
Equip.	Equipment
ERW	Electric Resistance Welded
E.S.	Energy Saver
Est.	Estimated
esu	Electrostatic Units
E.W.	Each Way
EWT	Entering Water Temperature
Excav.	Excavation
Exp.	Expansion, Exposure
Ext.	Exterior
Extru.	Extrusion
f.	Fiber stress
F	Fahrenheit; Female; Fill
Fab.	Fabricated
FBGS	Fiberglass
F.C.	Footcandles
f.c.c.	Face-centered Cubic
f'c.	Compressive Stress in Concrete; Extreme Compressive Stress
F.E.	Front End
FEP	Fluorinated Ethylene Propylene (Teflon)
F.G.	Flat Grain
F.H.A.	Federal Housing Administration
Fig.	Figure
Fin.	Finished
Fixt.	Fixture
Fl. Oz.	Fluid Ounces
Flr.	Floor
F.M.	Frequency Modulation; Factory Mutual
Fmg.	Framing
Fndtn.	Foundation
Fori.	Foreman, Inside
Foro.	Foreman, Outside

457

Abbreviation	Definition
Fount.	Fountain
FPM	Feet per Minute
FPT	Female Pipe Thread
Fr.	Frame
F.R.	Fire Rating
FRK	Foil Reinforced Kraft
FRP	Fiberglass Reinforced Plastic
FS	Forged Steel
FSC	Cast Body; Cast Switch Box
Ft.	Foot; Feet
Ftng.	Fitting
Ftg.	Footing
Ft. Lb.	Foot Pound
Furn.	Furniture
FVNR	Full Voltage Non-Reversing
FXM	Female by Male
Fy.	Minimum Yield Stress of Steel
g	Gram
G	Gauss
Ga.	Gauge
Gal.	Gallon
Gal./Min.	Gallon per Minute
Galv.	Galvanized
Gen.	General
G.F.I.	Ground Fault Interrupter
Glaz.	Glazier
GPD	Gallons per Day
GPH	Gallons per Hour
GPM	Gallons per Minute
GR	Grade
Gran.	Granular
Grnd.	Ground
H	High; High Strength Bar Joist; Henry
H.C.	High Capacity
H.D.	Heavy Duty; High Density
H.D.O.	High Density Overlaid
Hdr.	Header
Hdwe.	Hardware
Help.	Helpers Average
HEPA	High Efficiency Particulate Air Filter
Hg	Mercury
HIC	High Interrupting Capacity
HM	Hollow Metal
H.O.	High Output
Horiz.	Horizontal
H.P.	Horsepower; High Pressure
H.P.F.	High Power Factor
Hr.	Hour
Hrs./Day	Hours per Day
HSC	High Short Circuit
Ht.	Height
Htg.	Heating
Htrs.	Heaters
HVAC	Heating, Ventilation & Air-Conditioning
Hvy.	Heavy
HW	Hot Water
Hyd.;Hydr.	Hydraulic
Hz.	Hertz (cycles)
I.	Moment of Inertia
I.C.	Interrupting Capacity
ID	Inside Diameter
I.D.	Inside Dimension; Identification
I.F.	Inside Frosted
I.M.C.	Intermediate Metal Conduit
In.	Inch
Incan.	Incandescent
Incl.	Included; Including
Int.	Interior
Inst.	Installation
Insul.	Insulation/Insulated
I.P.	Iron Pipe
I.P.S.	Iron Pipe Size
I.P.T.	Iron Pipe Threaded
I.W.	Indirect Waste
J	Joule
J.I.C.	Joint Industrial Council
K	Thousand; Thousand Pounds; Heavy Wall Copper Tubing, Kelvin
K.A.H.	Thousand Amp. Hours
KCMIL	Thousand Circular Mils
KD	Knock Down
K.D.A.T.	Kiln Dried After Treatment
kg	Kilogram
kG	Kilogauss
kgf	Kilogram Force
kHz	Kilohertz
Kip.	1000 Pounds
KJ	Kiljoule
K.L.	Effective Length Factor
K.L.F.	Kips per Linear Foot
Km	Kilometer
K.S.F.	Kips per Square Foot
K.S.I.	Kips per Square Inch
kV	Kilovolt
kVA	Kilovolt Ampere
K.V.A.R.	Kilovar (Reactance)
KW	Kilowatt
KWh	Kilowatt-hour
L	Labor Only; Length; Long; Medium Wall Copper Tubing
Lab.	Labor
lat	Latitude
Lath.	Lather
Lav.	Lavatory
lb.; #	Pound
L.B.	Load Bearing; L Conduit Body
L. & E.	Labor & Equipment
lb./hr.	Pounds per Hour
lb./L.F.	Pounds per Linear Foot
lbf/sq.in.	Pound-force per Square Inch
L.C.L.	Less than Carload Lot
Ld.	Load
LE	Lead Equivalent
LED	Light Emitting Diode
L.F.	Linear Foot
Lg.	Long; Length; Large
L & H	Light and Heat
LH	Long Span Bar Joist
L.H.	Labor Hours
L.L.	Live Load
L.L.D.	Lamp Lumen Depreciation
L-O-L	Lateralolet
lm	Lumen
lm/sf	Lumen per Square Foot
lm/W	Lumen per Watt
L.O.A.	Length Over All
log	Logarithm
L.P.	Liquefied Petroleum; Low Pressure
L.P.F.	Low Power Factor
LR	Long Radius
L.S.	Lump Sum
Lt.	Light
Lt. Ga.	Light Gauge
L.T.L.	Less than Truckload Lot
Lt. Wt.	Lightweight
L.V.	Low Voltage
M	Thousand; Material; Male; Light Wall Copper Tubing
M²CA	Meters Squared Contact Area
m/hr; M.H.	Man-hour
mA	Milliampere
Mach.	Machine
Mag. Str.	Magnetic Starter
Maint.	Maintenance
Marb.	Marble Setter
Mat; Mat'l.	Material
Max.	Maximum
MBF	Thousand Board Feet
MBH	Thousand BTU's per hr.
MC	Metal Clad Cable
M.C.F.	Thousand Cubic Feet
M.C.F.M.	Thousand Cubic Feet per Minute
M.C.M.	Thousand Circular Mils
M.C.P.	Motor Circuit Protector
MD	Medium Duty
M.D.O.	Medium Density Overlaid
Med.	Medium
MF	Thousand Feet
M.F.B.M.	Thousand Feet Board Measure
Mfg.	Manufacturing
Mfrs.	Manufacturers
mg	Milligram
MGD	Million Gallons per Day
MGPH	Thousand Gallons per Hour
MH, M.H.	Manhole; Metal Halide; Man-Hour
MHz	Megahertz
Mi.	Mile
MI	Malleable Iron; Mineral Insulated
mm	Millimeter
Mill.	Millwright
Min., min.	Minimum, minute
Misc.	Miscellaneous
ml	Milliliter, Mainline
M.L.F.	Thousand Linear Feet
Mo.	Month
Mobil.	Mobilization
Mog.	Mogul Base
MPH	Miles per Hour
MPT	Male Pipe Thread
MRT	Mile Round Trip
ms	Millisecond
M.S.F.	Thousand Square Feet
Mstz.	Mosaic & Terrazzo Worker
M.S.Y.	Thousand Square Yards
Mtd.	Mounted
Mthe.	Mosaic & Terrazzo Helper
Mtng.	Mounting
Mult.	Multi; Multiply
M.V.A.	Million Volt Amperes
M.V.A.R.	Million Volt Amperes Reactance
MV	Megavolt
MW	Megawatt
MXM	Male by Male
MYD	Thousand Yards
N	Natural; North
nA	Nanoampere
NA	Not Available; Not Applicable
N.B.C.	National Building Code
NC	Normally Closed
N.E.M.A.	National Electrical Manufacturers Assoc.
NEHB	Bolted Circuit Breaker to 600V.
N.L.B.	Non-Load-Bearing
NM	Non-Metallic Cable
nm	Nanometer
No.	Number
NO	Normally Open
N.O.C.	Not Otherwise Classified
Nose.	Nosing
N.P.T.	National Pipe Thread
NQOD	Combination Plug-on/Bolt on Circuit Breaker to 240V.
N.R.C.	Noise Reduction Coefficient
N.R.S.	Non Rising Stem
ns	Nanosecond
nW	Nanowatt
OB	Opposing Blade
OC	On Center
OD	Outside Diameter
O.D.	Outside Dimension
ODS	Overhead Distribution System
O.G.	Ogee
O.H.	Overhead
O & P	Overhead and Profit
Oper.	Operator
Opng.	Opening
Orna.	Ornamental
OSB	Oriented Strand Board
O. S. & Y.	Outside Screw and Yoke
Ovhd.	Overhead
OWG	Oil, Water or Gas

Abbreviations

Oz.	Ounce	SCFM	Standard Cubic Feet per Minute	THW.	Insulated Strand Wire
P.	Pole;Applied Load; Projection	Scaf.	Scaffold	THWN;	Nylon Jacketed Wire
p.	Page	Sch.; Sched.	Schedule	T.L.	Truckload
Pape.	Paperhanger	S.C.R.	Modular Brick	T.M.	Track Mounted
P.A.P.R.	Powered Air Purifying Respirator	S.D.	Sound Deadening	Tot.	Total
PAR	Parabolic Reflector	S.D.R.	Standard Dimension Ratio	T-O-L	Threadolet
Pc., Pcs.	Piece, Pieces	S.E.	Surfaced Edge	T.S.	Trigger Start
P.C.	Portland Cement;Power Connector	Sel.	Select	Tr.	Trade
P.C.F.	Pounds per Cubic Foot	S.E.R.;	Service Entrance Cable	Transf.	Transformer
P.C.M.	Phase Contract Microscopy	S.E.U.	Service Entrance Cable	Trhv.	Truck Driver, Heavy
P.E.	Professional Engineer;	S.F.	Square Foot	Trlr	Trailer
	Porcelain Enamel;	S.F.C.A.	Square Foot Contact Area	Trlt.	Truck Driver, Light
	Polyethylene; Plain End	S.F.G.	Square Foot of Ground	TV	Television
Perf.	Perforated	S.F. Hor.	Square Foot Horizontal	T.W.	Thermoplastic Water Resistant
Ph.	Phase	S.F.R.	Square Feet of Radiation		Wire
P.I.	Pressure Injected	S.F. Shlf.	Square Foot of Shelf	UCI	Uniform Construction Index
Pile.	Pile Driver	S4S	Surface 4 Sides	UF	Underground Feeder
Pkg.	Package	Shee.	Sheet Metal Worker	UGND	Underground Feeder
Pl.	Plate	Sin.	Sine	U.H.F.	Ultra High Frequency
Plah.	Plasterer Helper	Skwk.	Skilled Worker	U.L.	Underwriters Laboratory
Plas.	Plasterer	SL	Saran Lined	Unfin.	Unfinished
Pluh.	Plumbers Helper	S.L.	Slimline	URD	Underground Residential
Plum.	Plumber	Sldr.	Solder		Distribution
Ply.	Plywood	SLH	Super Long Span Bar Joist	US	United States
p.m.	Post Meridiem	S.N.	Solid Neutral	USP	United States Primed
Pntd.	Painted	S-O-L	Socketolet	UTP	Unshielded Twisted Pair
Pord.	Painter, Ordinary	sp	Standpipe	V	Volt
pp	Pages	S.P.	Static Pressure; Single Pole; Self-	V.A.	Volt Amperes
PP; PPL	Polypropylene		Propelled	V.C.T.	Vinyl Composition Tile
P.P.M.	Parts per Million	Spri.	Sprinkler Installer	VAV	Variable Air Volume
Pr.	Pair	spwg	Static Pressure Water Gauge	VC	Veneer Core
P.E.S.B.	Pre-engineered Steel Building	Sq.	Square; 100 Square Feet	Vent.	Ventilation
Prefab.	Prefabricated	S.P.D.T.	Single Pole, Double Throw	Vert.	Vertical
Prefin.	Prefinished	SPF	Spruce Pine Fir	V.F.	Vinyl Faced
Prop.	Propelled	S.P.S.T.	Single Pole, Single Throw	V.G.	Vertical Grain
PSF; psf	Pounds per Square Foot	SPT	Standard Pipe Thread	V.H.F.	Very High Frequency
PSI; psi	Pounds per Square Inch	Sq. Hd.	Square Head	VHO	Very High Output
PSIG	Pounds per Square Inch Gauge	Sq. In.	Square Inch	Vib.	Vibrating
PSP	Plastic Sewer Pipe	S.S.	Single Strength; Stainless Steel	V.L.F.	Vertical Linear Foot
Pspr.	Painter, Spray	S.S.B.	Single Strength B Grade	Vol.	Volume
Psst.	Painter, Structural Steel	sst	Stainless Steel	VRP	Vinyl Reinforced Polyester
P.T.	Potential Transformer	Sswk.	Structural Steel Worker	W	Wire;Watt;Wide;West
P. & T.	Pressure & Temperature	Sswl.	Structural Steel Welder	w/	With
Ptd.	Painted	St.;Stl.	Steel	W.C.	Water Column;Water Closet
Ptns.	Partitions	S.T.C.	Sound Transmission Coefficient	W.F.	Wide Flange
Pu	Ultimate Load	Std.	Standard	W.G.	Water Gauge
PVC	Polyvinyl Chloride	STK	Select Tight Knot	Wldg.	Welding
Pvmt.	Pavement	STP	Standard Temperature & Pressure	W. Mile	Wire Mile
Pwr.	Power	Stpi.	Steamfitter, Pipefitter	W-O-L	Weldolet
Q	Quantity Heat Flow	Str.	Strength; Starter; Straight	W.R.	Water Resistant
Quan.;Qty.	Quantity	Strd.	Stranded	Wrck.	Wrecker
Q.C.	Quick Coupling	Struct.	Structural	W.S.P.	Water, Steam, Petroleum
r	Radius of Gyration	Sty.	Story	WT.,Wt.	Weight
R	Resistance	Subj.	Subject	WWF	Welded Wire Fabric
R.C.P.	Reinforced Concrete Pipe	Subs.	Subcontractors	XFER	Transfer
Rect.	Rectangle	Surf.	Surface	XFMR	Transformer
Reg.	Regular	Sw.	Switch	XHD	Extra Heavy Duty
Reinf.	Reinforced	Swbd.	Switchboard	XHHW; XLPE	Cross-Linked Polyethylene Wire
Req'd.	Required	S.Y.	Square Yard		Insulation
Res.	Resistant	Syn.	Synthetic	XLP	Cross-linked Polyethylene
Resi.	Residential	S.Y.P.	Southern Yellow Pine	Y	Wye
Rgh.	Rough	Sys.	System	yd	Yard
RGS	Rigid Galvanized Steel	t.	Thickness	yr	Year
R.H.W.	Rubber, Heat & Water Resistant;	T	Temperature;Ton	Δ	Delta
	Residential Hot Water	Tan	Tangent	%	Percent
rms	Root Mean Square	T.C.	Terra Cotta	~	Approximately
Rnd.	Round	T & C	Threaded and Coupled	Ø	Phase
Rodm.	Rodman	T.D.	Temperature Difference	@	At
Rofc.	Roofer, Composition	T.E.M.	Transmission Electron Microscopy	#	Pound; Number
Rofp.	Roofer, Precast	TFE	Tetrafluoroethylene (Teflon)	<	Less Than
Rohe.	Roofer Helpers (Composition)	T. & G.	Tongue & Groove;	>	Greater Than
Rots.	Roofer,Tile & Slate		Tar & Gravel		
R.O.W.	Right of Way	Th.;Thk.	Thick		
RPM	Revolutions per Minute	Thn.	Thin		
R.S.	Rapid Start	Thrded	Threaded		
Rsr	Riser	Tilf.	Tile Layer, Floor		
RT	Round Trip	Tilh.	Tile Layer, Helper		
S.	Suction; Single Entrance; South	THHN	Nylon Jacketed Wire		

RESIDENTIAL COST ESTIMATE

OWNER'S NAME: _____   APPRAISER: _____

RESIDENCE ADDRESS: _____   PROJECT: _____

CITY, STATE, ZIP CODE: _____   DATE: _____

CLASS OF CONSTRUCTION
- ☐ ECONOMY
- ☐ AVERAGE
- ☐ CUSTOM
- ☐ LUXURY

RESIDENCE TYPE
- ☐ 1 STORY
- ☐ 1-1/2 STORY
- ☐ 2 STORY
- ☐ 2-1/2 STORY
- ☐ 3 STORY
- ☐ BI-LEVEL
- ☐ TRI-LEVEL

CONFIGURATION
- ☐ DETACHED
- ☐ TOWN/ROW HOUSE
- ☐ SEMI-DETACHED

OCCUPANCY
- ☐ ONE STORY
- ☐ TWO FAMILY
- ☐ THREE FAMILY
- ☐ OTHER _____

EXTERIOR WALL SYSTEM
- ☐ WOOD SIDING—WOOD FRAME
- ☐ BRICK VENEER—WOOD FRAME
- ☐ STUCCO ON WOOD FRAME
- ☐ PAINTED CONCRETE BLOCK
- ☐ SOLID MASONRY (AVERAGE & CUSTOM)
- ☐ STONE VENEER—WOOD FRAME
- ☐ SOLID BRICK (LUXURY)
- ☐ SOLID STONE (LUXURY)

*LIVING AREA (Main Building)		*LIVING AREA (Wing or Ell) ( )		*LIVING AREA (Wing or Ell) ( )	
First Level	_____ S.F.	First Level	_____ S.F.	First Level	_____ S.F.
Second Level	_____ S.F.	Second Level	_____ S.F.	Second Level	_____ S.F.
Third Level	_____ S.F.	Third Level	_____ S.F.	Third Level	_____ S.F.
Total	_____ S.F.	Total	_____ S.F.	Total	_____ S.F.

*Basement Area is not part of living area.

MAIN BUILDING	COSTS PER S.F. LIVING AREA
Cost per Square Foot of Living Area, from Page _____	$
Basement Addition: _____ % Finished, _____ % Unfinished	+
Roof Cover Adjustment: _____ Type, Page _____ (Add or Deduct)	( )
Central Air Conditioning: ☐ Separate Ducts ☐ Heating Ducts, Page _____	+
Heating System Adjustment: _____ Type, Page _____ (Add or Deduct)	( )
Main Building: Adjusted Cost per S.F. of Living Area	$

MAIN BUILDING TOTAL COST | $ _____ /S.F. | x | _____ S.F. | x | _____ | = | $ _____ |

Cost per S.F. Living Area   Living Area   Town/Row House Multiplier (Use 1 for Detached)   TOTAL COST

WING OR ELL ( ) _____ STORY	COSTS PER S.F. LIVING AREA
Cost per Square Foot of Living Area, from Page _____	$
Basement Addition: _____ % Finished, _____ % Unfinished	+
Roof Cover Adjustment: _____ Type, Page _____ (Add or Deduct)	( )
Central Air Conditioning: ☐ Separate Ducts ☐ Heating Ducts, Page _____	+
Heating System Adjustment: _____ Type, Page _____ (Add or Deduct)	( )
Wing or Ell ( ): Adjusted Cost per S.F. of Living Area	$

WING OR ELL (   ) TOTAL COST | $ _____ /S.F. | x | _____ S.F. | = | $ _____ |

Cost per S.F. Living Area   Living Area   TOTAL COST

WING OR ELL ( ) _____ STORY	COSTS PER S.F. LIVING AREA
Cost per Square Foot of Living Area, from Page _____	$
Basement Addition: _____ % Finished, _____ % Unfinished	+
Roof Cover Adjustment: _____ Type, Page _____ (Add or Deduct)	( )
Central Air Conditioning: ☐ Separate Ducts ☐ Heating Ducts, Page _____	+
Heating System Adjustment: _____ Type, Page _____ (Add or Deduct)	( )
Wing or Ell ( ): Adjusted Cost per S.F. of Living Area	$

WING OR ELL (   ) TOTAL COST | $ _____ /S.F. | x | _____ S.F. | = | $ _____ |

Cost per S.F. Living Area   Living Area   TOTAL COST

TOTAL THIS PAGE   [_____]

FORMS

RESIDENTIAL
COST ESTIMATE

		QUANTITY	UNIT COST	$
Total Page 1				
Additional Bathrooms: _____ Full, _____ Half				
Finished Attic: _____ Ft. x _____ Ft.		S.F.		+
Breezeway: ☐ Open ☐ Enclosed _____ Ft. x _____ Ft.		S.F.		+
Covered Porch: ☐ Open ☐ Enclosed _____ Ft. x _____ Ft.		S.F.		+
Fireplace: ☐ Interior Chimney ☐ Exterior Chimney ☐ No. of Flues ☐ Additional Fireplaces				+
Appliances:				+
Kitchen Cabinets Adjustment: (+/−)				
☐ Garage ☐ Carport: _____ Car(s) Description _____ (+/−)				
Miscellaneous:				+

ADJUSTED TOTAL BUILDING COST $ _____

REPLACEMENT COST

ADJUSTED TOTAL BUILDING COST	$ _____
Site Improvements	
(A) Paving & Sidewalks	$ _____
(B) Landscaping	$ _____
(C) Fences	$ _____
(D) Swimming Pool	$ _____
(E) Miscellaneous	$ _____
TOTAL	$ _____
Location Factor	x _____
Location Replacement Cost	$ _____
Depreciation	−$ _____
LOCAL DEPRECIATED COST	$ _____

INSURANCE COST

ADJUSTED TOTAL BUILDING COST	$ _____
Insurance Exclusions	
(A) Footings, Site Work, Underground Piping	−$ _____
(B) Architects' Fees	−$ _____
Total Building Cost Less Exclusion	$ _____
Location Factor	x _____
LOCAL INSURABLE REPLACEMENT COST	$ _____

SKETCH AND ADDITIONAL CALCULATIONS

CII APPRAISAL

5. DATE: _____

1. SUBJECT PROPERTY: _____
6. APPRAISER: _____

2. BUILDING: _____

3. ADDRESS: _____

4. BUILDING USE: _____
7. YEAR BUILT: _____

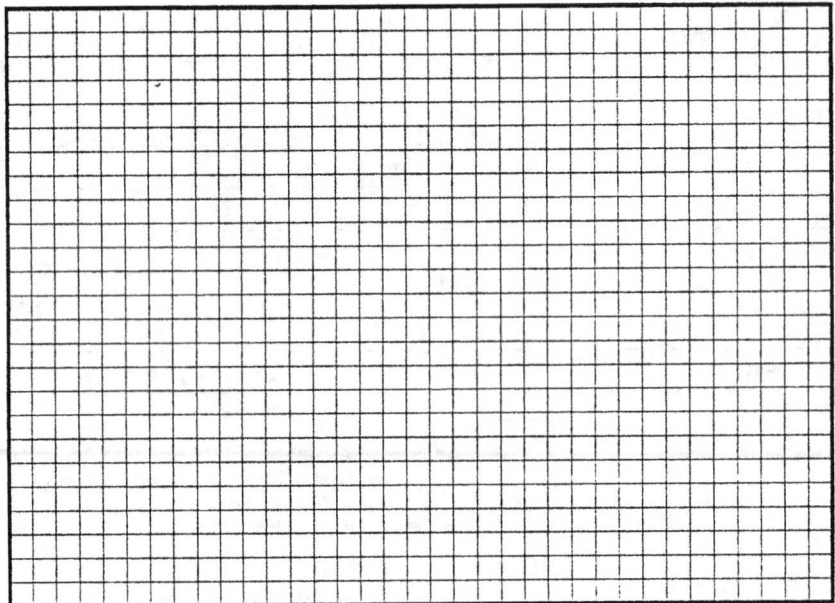

8. EXTERIOR WALL CONSTRUCTION: _____

9. FRAME: _____

10. GROUND FLOOR AREA: _____ S.F. 11. GROSS FLOOR AREA (EXCL. BASEMENT): _____ S.F.

12. NUMBER OF STORIES: _____ 13. STORY HEIGHT: _____

14. PERIMETER: _____ L.F. 15. BASEMENT AREA: _____ S.F.

16. GENERAL COMMENTS: _____

_____

_____

_____

_____

_____

_____

_____

	NO.	DESCRIPTION		UNIT	UNIT COST	NEW SF COST	MODEL SF COST	+/- CHANGE
A SUBSTRUCTURE								
A	1010	Standard Foundations	Bay size:	S.F. Gnd.				
A	1030	Slab on Grade Material:	Thickness:	S.F. Slab				
A	2010	Basement Excavation Depth:	Area:	S.F. Gnd.				
A	2020	Basement Walls		L.F. Walls				
B SHELL								
	B10 Superstructure							
B	1010	Floor Construction Elevated floors:		S.F. Floor				
				S.F. Floor				
B	1020	Roof Construction		S.F. Roof				
	B20 Exterior Enclosure							
B	2010	Exterior walls Material:	Thickness: % of wall	S.F. Walls				
		Material:	Thickness: % of wall	S.F. Walls				
B	2020	Exterior Windows Type:	% of wall	S.F. Wind.				
		Type:	% of wall Each	S.F. Wind.				
B	2030	Exterior Doors Type:	Number:	Each				
		Type:	Number:	Each				
	B30 Roofing							
B	3010	Roof Coverings Material:		S.F. Roof				
		Material:		S.F. Roof				
B	3020	Roof Openings		S.F. Opng.				
C INTERIORS								
C	1010	Partitions: Material:	Density:	S.F. Part.				
		Material:	Density:					
C	1020	Interior Doors Type:	Number:	Each				
C	1030	Fittings		Each				
C	2010	Stair Construction		Flight				
C	3010	Wall Finishes Material:	% of Wall	S.F. Walls				
		Material:		S.F. Walls				
C	3020	Floor Finishes Material:		S.F. Floor				
		Material:		S.F. Floor				
		Material:		S.F. Floor				
C	3030	Ceiling Finishes Material:		S.F. Ceil.				

	NO.	SYSTEM/COMPONENT	DESCRIPTION	UNIT	UNIT COST	NEW S.F. COST	MODEL S.F. COST	+/- CHANGE
D	**SERVICES**							
	D10 Conveying							
D	1010	Elevators & Lifts	Type: Capacity: Stops:	Each				
D	1020	Escalators & Moving Walks	Type:	Each				
	D20 Plumbing							
D	2010	Plumbing		Each				
D	2020	Domestic Water Distribution		S.F. Floor				
D	2040	Rain Water Drainage		S.F. Roof				
	D30 HVAC							
D	3010	Energy Supply		Each				
D	3020	Heat Generating Systems	Type:	S.F. Floor				
D	3030	Cooling Generating Systems	Type:	S.F. Floor				
D	3090	Other HVAC Sys. & Equipment		Each				
	D40 Fire Protection							
D	4010	Sprinklers		S.F. Floor				
D	4020	Standpipes		S.F. Floor				
	D50 Electrical							
D	5010	Electrical Service/Distribution		S.F. Floor				
D	5020	Lighting & Branch Wiring		S.F. Floor				
D	5030	Communications & Security		S.F. Floor				
D	5090	Other Electrical Systems		S.F. Floor				
E	**EQUIPMENT & FURNISHINGS**							
E	1010	Commercial Equipment		Each				
E	1020	Institutional Equipment		Each				
E	1030	Vehicular Equipment		Each				
E	1090	Other Equipment		Each				
F	**SPECIAL CONSTRUCTION**							
F	1020	Integrated Construction		S.F.				
F	1040	Special Facilities		S.F.				
G	**BUILDING SITEWORK**							

ITEM		Total Change
17	Model 020 total ____ $ ____	
18	Adjusted S.F. cost $ ____ item 17 +/- changes	
19	Building area - from item 11 ____ S.F. x adjusted S.F. cost	$..
20	Basement area - from item 15 ____ S.F. x S.F. cost $	$..
21	Base building sub-total - item 20 + item 19	$....
22	Miscellaneous addition (quality, etc.)	$......
23	Sub-total - item 22 + 21	$.......
24	General conditions -25 % of item 23	$.....
25	Sub-total - item 24 + item 23	$.....
26	Architects fees ____ % of item 25	$
27	Sub-total - item 26 + item 27	$......
28	Location modifier	x....
29	Local replacement cost - item 28 x item 27	$.....
30	Depreciation ____ % of item 29	$
31	Depreciated local replacement cost - item 29 less item 30	$....
32	Exclusions	$........
33	Net depreciated replacement cost - item 31 less item 32	$....

Base Cost per square foot floor area: *(from square foot table)*

Specify source: Page: _____   Model # _____   Area _____ S.F.
Exterior wall _____   Frame _____

Adjustments for exterior wall variation:

Size Adjustment:
(Interpolate)

Height Adjustment: _____   +   _____   =   _____

Adjusted Base Cost per square foot: [   ]

Building Cost $ _____   x   _____   =   _____
      Adjusted Base Cost per square foot   *Floor Area*

Basement Cost $ _____   x   _____   =   _____
      Basement Cost   *Basement Area*

Lump Sum Additions

TOTAL BUILDING COST *(Sum of above costs)*   _____

Modifications: *(complexity, workmanship, size)*   +/- _____ %   _____

Location Modifier: City _____   Date _____   x   _____

Local cost of replacement   _____

Less depreciation: _____

Local cost of replacement less depreciation $ _____

R.S. Means Company, Inc., a CMD company, is the leading provider of construction cost data in North America and supplies comprehensive construction cost guides, related technical publications and education services.

CMD, a leading worldwide provider of total construction information solutions, is comprised of three synergistic product groups designed specifically to help construction professionals advance their businesses with timely, accurate and actionable project, product and cost data. CMD is a division of Cahners Business Information, a member of the Reed Elsevier plc group of companies.

The *Project, Product, and Cost & Estimating* divisions offer a variety of innovative products designed for the full spectrum of design, construction and manufacturing professionals. Together with Cahners, CMD created *Buildingteam.com,* a valuable Internet portal of the construction community. Through it's *International* companies, CMD's reputation for quality construction market data is growing worldwide.

Project Data
CMD provides complete, accurate and relevant project information through all stages of construction. Customers are supplied industry data through leads, project reports, contact lists, plans and specifications surveys, market penetration analyses and sales evaluation reports. Any of these products can pinpoint a county, look at a state, or cover the country. Data is delivered via paper, e-mail, CD-ROM or the Internet.

Building Product Information
The First Source suite of products is the only integrated building product information system offered to the commercial construction industry for comparing and specifying building products. These print and online resources include *First Source for Products,* SPEC-DATA™, MANU-SPEC™, CADBlocks, First Source Exchange (www.firstsourceexchange.com), and Manufacturer Catalogs. Written by industry professionals and organized using CSI's MasterFormat™, construction professionals use this information to make better design decisions.

Cost Information
R.S. Means, the undisputed market leader and authority on construction costs, publishes current cost and estimating information in annual cost books and on the CostWorks CD-ROM. R.S. Means furnishes the construction industry with a rich library of complementary reference books and a series of professional seminars that are designed to sharpen professional skills and maximize the effective use of cost estimating and management tools. R.S. Means also provides construction cost consulting for Owners, Manufacturers Designers and Contractors.

Buildingteam.com
Combining CMD's project, product and cost data with news and information from Cahners' *Building Design & Construction* and *Consulting-Specifying Engineer,* this industry-focused site offers easy and unlimited access to vital information for all construction professionals.

International
BIMSA/Mexico provides construction project news, product information, cost-data, seminars and consulting services to construction professionals in Mexico. Its subsidiary, PRISMA, provides job costing software.

Byggfakta Scandinavia AB, founded in 1936, is the parent company for the leaders of customized construction market data for Denmark, Estonia, Finland, Norway and Sweden. Each company fully covers the local construction market and provides information across several platforms including subscription, ad-hoc basis, electronically and on paper.

CMD Canada serves the Canadian construction market with reliable and comprehensive project and product information services that cover all facets of construction. Core services include: Buildcore, product selection and specification tools available in print and on the Internet; CMD Building Reports, a national construction project lead service; CanaData, statistical and forecasting information; *Daily Commercial News,* a construction newspaper reporting on news and projects in Ontario; and *Journal of Commerce,* reporting news in British Columbia and Alberta.

Cordell Building Information Services, with its complete range of project and cost and estimating services, is Australia's specialist in the construction information industry. Cordell provides in-depth and historical information on all aspects of construction projects and estimation, including several customized reports, construction and sales leads, and detailed cost information among others.

For more information, please visit our website at www.cmdg.com.

CMD Corporate Office
30 Technology Parkway South
Norcross, GA 30092-2912
(800) 793-0304
(700) 417-4002 (fax)
info@cmdg.com
www.cmdg.com

RSMeans
CMD

Means Project Cost Report

By filling out and returning the Project Description, you can receive a discount of $20.00 off any one of the Means products advertised in the following pages. The cost information required includes all items marked (✔) except those where no costs occurred. The sum of all major items should equal the Total Project Cost.

$20.00 Discount per product for each report you submit.

DISCOUNT PRODUCTS AVAILABLE—FOR U.S. CUSTOMERS ONLY—STRICTLY CONFIDENTIAL

Project Description (No remodeling projects, please.)

✔ Type Building _____

✔ Location _____

Capacity _____

✔ Frame _____

✔ Exterior _____

✔ Basement: full ☐ partial ☐ none ☐ crawl ☐

✔ Height in Stories _____

✔ Total Floor Area _____

Ground Floor Area _____

✔ Volume in C.F. _____

% Air Conditioned _____ Tons _____

Comments _____

Owner _____

Architect _____

General Contractor _____

✔ Bid Date _____

Typical Bay Size _____

✔ Labor Force: _____ % Union _____ % Non-Union

✔ Project Description (Circle one number in each line)
1. Economy 2. Average 3. Custom 4. Luxury
1. Square 2. Rectangular 3. Irregular 4. Very Irregular

	✔ Total Project Cost		$			
A	✔ General Conditions		$			
B	✔ Site Work		$			
BS	Site Clearing & Improvement					
BE	Excavation	(		C.Y.)		
BF	Caissons & Piling	(		L.F.)		
BU	Site Utilities					
BP	Roads & Walks Exterior Paving	(		S.Y.)		
C	✔ Concrete		$			
C	Cast in Place	(		C.Y.)		
CP	Precast	(		S.F.)		
D	✔ Masonry		$			
DB	Brick	(		M)		
DC	Block	(		M)		
DT	Tile	(		S.F.)		
DS	Stone	(		S.F.)		
E	✔ Metals		$			
ES	Structural Steel	(		Tons)		
EM	Misc. & Ornamental Metals					
F	✔ Wood & Plastics		$			
FR	Rough Carpentry	(		MBF)		
FF	Finish Carpentry					
FM	Architectural Millwork					
G	✔ Thermal & Moisture Protection		$			
GW	Waterproofing-Dampproofing	(		S.F.)		
GN	Insulation	(		S.F.)		
GR	Roofing & Flashing	(		S.F.)		
GM	Metal Siding/Curtain Wall	(		S.F.)		
H	✔ Doors and Windows		$			
HD	Doors	(		Ea.)		
HW	Windows	(		S.F.)		
HH	Finish Hardware					
HG	Glass & Glazing	(		S.F.)		
HS	Storefronts	(		S.F.)		

J	✔ Finishes				$	
JL	Lath & Plaster		(		S.Y.)	
JD	Drywall		(		S.F.)	
JM	Tile & Marble		(		S.F.)	
JT	Terrazzo		(		S.F.)	
JA	Acoustical Treatment		(		S.F.)	
JC	Carpet		(		S.Y.)	
JF	Hard Surface Flooring		(		S.F.)	
JP	Painting & Wall Covering		(		S.F.)	
K	✔ Specialties				$	
KB	Bathroom Partitions & Access.		(		S.F.)	
KF	Other Partitions		(		S.F.)	
KL	Lockers		(		Ea.)	
L	✔ Equipment				$	
LK	Kitchen					
LS	School					
LO	Other					
M	✔ Furnishings				$	
MW	Window Treatment					
MS	Seating		(		Ea.)	
N	✔ Special Construction				$	
NA	Acoustical		(		S.F.)	
NB	Prefab. Bldgs.		(		S.F.)	
NO	Other					
P	✔ Conveying Systems				$	
PE	Elevators		(		Ea.)	
PS	Escalators		(		Ea.)	
PM	Material Handling					
Q	✔ Mechanical				$	
QP	Plumbing	(No. of fixtures			)	
QS	Fire Protection (Sprinklers)					
QF	Fire Protection (Hose Standpipes)					
QB	Heating, Ventilating & A.C.					
QH	Heating & Ventilating	(BTU Output			)	
QA	Air Conditioning		(		Tons)	
R	✔ Electrical				$	
RL	Lighting		(		S.F.)	
RP	Power Service					
RD	Power Distribution					
RA	Alarms					
RG	Special Systems					
S	✔ Mech./Elec. Combined				$	

Product Name _____

Product Number _____

Your Name _____

Title _____

Company _____

☐ Company

☐ Home   Street Address _____

City, State, Zip _____

☐ Please send _____ forms.

Please specify the Means product you wish to receive. Complete the address information as requested and return this form with your check (product cost less $20.00) to address below.

R.S. Means Company, Inc.,
Square Foot Costs Department
P.O. Box 800
Kingston, MA 02364-9988

R.S. Means Company, Inc. . . . a tradition of excellence in Construction Cost Information and Services since 1942.

For more information
visit Means Web Site
at www.rsmeans.com

Book Selection Guide

The following table provides definitive information on the content of each cost data publication. The number of lines of data provided in each unit price or assemblies division, as well as the number of reference tables and crews is listed for each book. The presence of other elements such as an historical cost index, city cost indexes, square foot models or cross-referenced index is also indicated. You can use the table to help select the Means' book that has the quantity and type of information you most need in your work.

Unit Cost Divisions	Building Construction Costs	Mechanical	Electrical	Repair & Remodel.	Square Foot	Site Work Landsc.	Assemblies	Interior	Concrete Masonry	Open Shop	Heavy Construc.	Light Commercial	Facil. Construc.	Plumbing	Western Construction Costs	Residential
1	1103	590	668	866		1091		492	991	1100	1133	625	1617	663	1095	420
2	3141	1462	508	1881		9148		851	1655	3091	5620	984	5014	1825	3114	1065
3	1511	91	71	767		1315		198	1932	1505	1489	227	1379	43	1510	266
4	905	27	0	695		753		624	1164	882	651	428	1111	0	877	345
5	1928	178	176	978		790		989	708	1895	1070	810	1877	95	1911	780
6	1356	96	83	1290		491		1299	337	1347	610	1456	1353	59	1734	1529
7	1365	180	91	1329		490		546	451	1364	354	1036	1357	188	1364	803
8	1902	60	0	1907		337		1821	723	1883	9	1187	2035	0	1903	1134
9	1690	50	0	1513		188		1788	324	1638	114	1434	1887	50	1685	1293
10	992	58	31	565		218		823	196	996	0	467	997	250	992	250
11	1157	434	212	622		151		916	34	1051	95	237	1220	370	1050	116
12	352	0	0	48		228		1537	30	344	0	68	1538	0	343	72
13	1343	1181	460	567		448		1029	106	1317	307	561	2053	1080	1298	236
14	364	43	0	261		36		330	0	364	40	18	362	17	362	9
15	2514	14716	763	2254		1828		1466	80	2516	2030	1563	12320	10884	2546	1100
16	1521	540	10211	1202		879		1290	62	1538	916	1312	9896	481	1463	705
17	429	354	429	0		0		0	0	429	0	0	429	356	429	0
Totals	23573	20060	13703	16745		18391		15999	8793	23260	14438	12413	46445	16361	23676	10123

Assembly Divisions	Building Construction Costs	Mechanical	Electrical	Repair & Remodel.	Square Foot	Site Work Landsc.	Assemblies	Interior	Concrete Masonry	Open Shop	Heavy Construc.	Light Commercial	Facil. Construc.	Plumbing	Western Construction Costs	Asm Div	Residential
A		28	0	333	169	787	863	0	777		765	168	95	2		1	841
B		0	0	1331	2716	0	6707	495	2515		0	2218	265	3		2	584
C		0	0	1150	954	0	1490	1809	182		0	850	343	3		3	1544
D		2269	1546	1513	2172	0	4422	1402	0		0	1608	1511	2106		4	2049
E		0	0	129	295	0	348	153	0		0	297	153	1		5	1081
F		0	0	0	148	0	150	0	0		0	149	52	0		6	1077
G		622	163	631	104	2436	715	0	692		554	104	125	680		7	709
																8	1686
																9	238
																10	0
																11	0
																12	0
Totals		2919	1709	5087	6558	3223	14695	3859	4166		1319	5394	2544	2795			9809

Reference Section	Building Construction Costs	Mechanical	Electrical	Repair & Remodel.	Square Foot	Site Work Landsc.	Assemblies	Interior	Concrete Masonry	Open Shop	Heavy Construc.	Light Commercial	Facil. Construc.	Plumbing	Western Construction Costs	Residential
Tables	129	45	85	69	4	80	219	46	71	130	61	57	104	51	131	42
Models					102							43				32
Crews	410	410	410	391		410		410	410	393	410	393	391	410	410	393
City Cost Indexes	yes	yes	yes	yes	yes	yes	yes	yes	yes	yes	yes	yes	yes	yes	yes	yes
Historical Cost Indexes	yes	yes	yes	yes	yes	yes	yes	yes	yes	yes	yes	yes	yes	yes	yes	no
Index	yes	yes	yes	yes	no	yes	yes	yes	yes	yes	yes	yes	yes	yes	yes	yes

Annual Cost Guides

For more information
visit Means Web Site
at www.rsmeans.com

Means Building Construction Cost Data 2002

Available in Both Softbound and Looseleaf Editions

The "Bible" of the industry comes in the standard softcover edition or the looseleaf edition.

Many customers enjoy the convenience and flexibility of the looseleaf binder, which increases the usefulness of *Means Building Construction Cost Data 2002* by making it easy to add and remove pages. You can insert your own cost information pages, so everything is in one place. Copying pages for faxing is easier also. Whichever edition you prefer, softbound or the convenient looseleaf edition, you'll be eligible to receive a FREE e-mail subscription to *The Change Notice* quarterly newsletter.

$99.95 per copy, Softbound
Catalog No. 60012

$124.95 per copy, Looseleaf
Catalog No. 61012

Means Building Construction Cost Data 2002

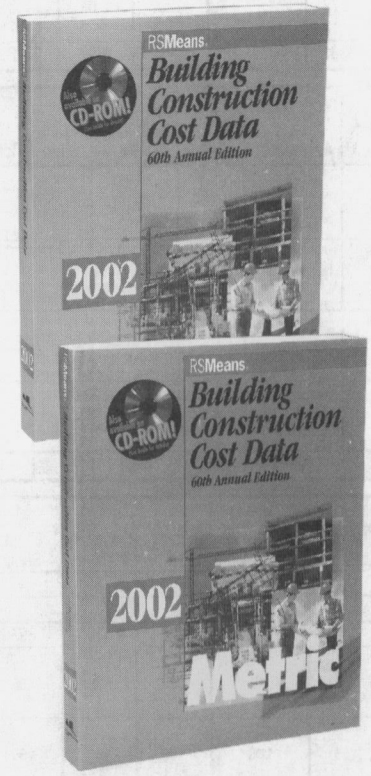

Offers you unchallenged unit price reliability in an easy-to-use arrangement. Whether used for complete, finished estimates or for periodic checks, it supplies more cost facts better and faster than any comparable source. Over 23,000 unit prices for 2002. The City Cost Indexes cover over 930 areas, for indexing to any project location in North America. Order and get *The Change Notice* e-mail quarterly newsletter sent to you FREE. You'll have year-long access to the Means Estimating **Hotline** FREE with your subscription. Expert assistance when using Means data is just a phone call away.

$99.95 per copy
Over 650 pages, illustrated, available Oct. 2001
Catalog No. 60012

Means Building Construction Cost Data 2002

Metric Version

The Federal Government has stated that all federal construction projects must now use metric documentation. The *Metric Version* of *Means Building Construction Cost Data 2002* is presented in metric measurements covering all construction areas. Don't miss out on these billion dollar opportunities. Make the switch to metric today.

$99.95 per copy
Over 650 pages, illus., available Nov. 2001
Catalog No. 63012

For more information
visit Means Web Site
at www.rsmeans.com

Annual Cost Guides

Means Mechanical Cost Data 2002

• HVAC • Controls

Total unit and systems price guidance for mechanical construction...materials, parts, fittings, and complete labor cost information. Includes prices for piping, heating, air conditioning, ventilation, and all related construction.

Plus new 2002 unit costs for:
• Over 2500 installed HVAC/controls assemblies
• "On Site" Location Factors for over 930 cities and towns in the U.S. and Canada
• Crews, labor and equipment

$99.95 per copy
Over 600 pages, illustrated, available Oct. 2001
Catalog No. 60022

Means Plumbing Cost Data 2002

Comprehensive unit prices and assemblies for plumbing, irrigation systems, commercial and residential fire protection, point-of-use water heaters, and the latest approved materials. This publication and its companion, *Means Mechanical Cost Data*, provide full-range cost estimating coverage for all the mechanical trades.

$99.95 per copy
Over 500 pages, illustrated, available Nov. 2001
Catalog No. 60212

Means Electrical Cost Data 2002

Pricing information for every part of electrical cost planning: More than 15,000 unit and systems costs with design tables; clear specifications and drawings; engineering guides and illustrated estimating procedures; complete labor-hour and materials costs for better scheduling and procurement; the latest electrical products and construction methods.
• A Variety of Special Electrical Systems including Cathodic Protection
• Costs for maintenance, demolition, HVAC/mechanical, specialties, equipment, and more

$99.95 per copy
Over 450 pages, illustrated, available Oct. 2001
Catalog No. 60032

Means Electrical Change Order Cost Data 2002

You are provided with electrical unit prices exclusively for pricing change orders based on the recent, direct experience of contractors and suppliers. Analyze and check your own change order estimates against the experience others have had doing the same work. It also covers productivity analysis and change order cost justifications. With useful information for calculating the effects of change orders and dealing with their administration.

$99.95 per copy
Over 450 pages, available Oct. 2001
Catalog No. 60232

Means Facilities Maintenance & Repair Cost Data 2002

Published in a looseleaf format, *Means Facilities Maintenance & Repair Cost Data* gives you a complete system to manage and plan your facility repair and maintenance costs and budget efficiently. Guidelines for auditing a facility and developing an annual maintenance plan. Budgeting is included, along with reference tables on cost and management and information on frequency and productivity of maintenance operations.

The only nationally recognized source of maintenance and repair costs. Developed in cooperation with the Army Corps of Engineers.

$219.95 per copy
Over 600 pages, illustrated, available Dec. 2001
Catalog No. 60302

Means Square Foot Costs 2002

It's Accurate and Easy To Use!

• **Updated 2002 price information,** based on nationwide figures from suppliers, estimators, labor experts and contractors.

• "How-to-Use" Sections, with **clear examples** of commercial, residential, industrial, and institutional structures.

• Realistic graphics, offering true-to-life illustrations of building projects.

• Extensive information on using square foot cost data, including **sample estimates** and **alternate pricing methods.**

$109.95 per copy
Over 450 pages, illustrated, available Oct. 2001
Catalog No. 60052

Annual Cost Guides

For more information
visit Means Web Site
at www.rsmeans.com

Means Repair & Remodeling Cost Data 2002

Commercial/Residential

You can use this valuable tool to estimate commercial and residential renovation and remodeling.

Includes: New costs for hundreds of unique methods, materials and conditions that only come up in repair and remodeling. PLUS:

- Unit costs for over 16,000 construction components
- Installed costs for over 90 assemblies
- Costs for 300+ construction crews
- Over 930 "On Site" localization factors for the U.S. and Canada.

$87.95 per copy
Over 600 pages, illustrated, available Nov. 2001
Catalog No. 60042

Means Facilities Construction Cost Data 2002

For the maintenance and construction of commercial, industrial, municipal, and institutional properties. Costs are shown for new and remodeling construction and are broken down into materials, labor, equipment, overhead, and profit. Special emphasis is given to sections on mechanical, electrical, furnishings, site work, building maintenance, finish work, and demolition. More than 45,000 unit costs, plus assemblies and reference sections are included.

$241.95 per copy
Over 1150 pages, illustrated, available Nov. 2001
Catalog No. 60202

Means Residential Cost Data 2002

Contains square foot costs for 30 basic home models with the look of today, plus hundreds of custom additions and modifications you can quote right off the page. With costs for the 100 residential systems you're most likely to use in the year ahead. Complete with blank estimating forms, sample estimates and step-by-step instructions.

Means Light Commercial Cost Data 2002

Specifically addresses the light commercial market, which is an increasingly specialized niche in the industry. Aids you, the owner/designer/contractor, in preparing all types of estimates, from budgets to detailed bids. Includes new advances in methods and materials. Assemblies section allows you to evaluate alternatives in the early stages of design/planning.

Over 13,000 unit costs for 2002 ensure you have the prices you need...when you need them.

$87.95 per copy
Over 550 pages, illustrated, available Dec. 2001
Catalog No. 60172

$87.95 per copy
Over 600 pages, illustrated, available Dec. 2001
Catalog No. 60182

Means Assemblies Cost Data 2002

Means Assemblies Cost Data 2002 takes the guesswork out of preliminary or conceptual estimates. Now you don't have to try to calculate the assembled cost by working up individual components costs. We've done all the work for you.

Presents detailed illustrations, descriptions, specifications and costs for every conceivable building assembly—240 types in all—arranged in the easy-to-use UNIFORMAT II system. Each illustrated "assembled" cost includes a complete grouping of materials and associated installation costs including the installing contractor's overhead and profit.

$164.95 per copy
Over 550 pages, illustrated, available Oct. 2001
Catalog No. 60062

Means Site Work & Landscape Cost Data 2002

Means Site Work & Landscape Cost Data 2002 is organized to assist you in all your estimating needs. Hundreds of fact-filled pages help you make accurate cost estimates efficiently.

Updated for 2002!

- Demolition features—including ceilings, doors, electrical, flooring, HVAC, millwork, plumbing, roofing, walls and windows
- State-of-the-art segmental retaining walls
- Flywheel trenching costs and details
- Updated Wells section
- Landscape materials, flowers, shrubs and trees

$99.95 per copy
Over 600 pages, illustrated, available Oct. 2001
Catalog No. 60282

For more information
visit Means Web Site
at www.rsmeans.com

Annual Cost Guides

Means Open Shop Building Construction Cost Data 2002

The latest costs for accurate budgeting and estimating of new commercial and residential construction... renovation work... change orders... cost engineering. *Means Open Shop BCCD* will assist you to...

• Develop benchmark prices for change orders
• Plug gaps in preliminary estimates, budgets
• Estimate complex projects
• Substantiate invoices on contracts
• Price ADA-related renovations

$99.95 per copy
Over 650 pages, illustrated, available Nov. 2001
Catalog No. 60152

Means Heavy Construction Cost Data 2002

A comprehensive guide to heavy construction costs. Includes costs for highly specialized projects such as tunnels, dams, highways, airports, and waterways. Information on different labor rates, equipment, and material costs is included. Has unit price costs, systems costs, and numerous reference tables for costs and design. Valuable not only to contractors and civil engineers, but also to government agencies and city/town engineers.

$99.95 per copy
Over 450 pages, illustrated, available Nov. 2001
Catalog No. 60162

Means Building Construction Cost Data 2002
Western Edition

This regional edition provides more precise cost information for western North America. Labor rates are based on union rates from 13 western states and western Canada. Included are western practices and materials not found in our national edition: tilt-up concrete walls, glu-lam structural systems, specialized timber construction, seismic restraints, landscape and irrigation systems.

$99.95 per copy
Over 600 pages, illustrated, available Dec. 2001
Catalog No. 60222

Means Heavy Construction Cost Data 2002
Metric Version

Make sure you have the Means industry standard metric costs for the federal, state, municipal and private marketplace. With thousands of up-to-date metric unit prices in tables by CSI standard divisions. Supplies you with assemblies costs using the metric standard for reliable cost projections in the design stage of your project. Helps you determine sizes, material amounts, and has tips for handling metric estimates.

$99.95 per copy
Over 450 pages, illustrated, available Nov. 2001
Catalog No. 63162

Means Construction Cost Indexes 2002

Who knows what 2002 holds? What materials and labor costs will change unexpectedly? By how much?

• Breakdowns for 305 major cities.
• National averages for 30 key cities.
• Expanded five major city indexes.
• Historical construction cost indexes.

$218.00 per year
$54.50 individual quarters
Catalog No. 60142 A,B,C,D

Means Interior Cost Data 2002

Provides you with prices and guidance needed to make accurate interior work estimates. Contains costs on materials, equipment, hardware, custom installations, furnishings, labor costs . . . every cost factor for new and remodel commercial and industrial interior construction, including updated information on office furnishings, plus more than 50 reference tables. For contractors, facility managers, owners.

$99.95 per copy
Over 550 pages, illustrated, available Oct. 2001
Catalog No. 60092

Means Concrete & Masonry Cost Data 2002

Provides you with cost facts for virtually all concrete/masonry estimating needs, from complicated formwork to various sizes and face finishes of brick and block, all in great detail. The comprehensive unit cost section contains more than 8,500 selected entries. Also contains an assemblies cost section, and a detailed reference section which supplements the cost data.

$92.95 per copy
Over 450 pages, illustrated, available Nov. 2001
Catalog No. 60112

Means Labor Rates for the Construction Industry 2002

Complete information for estimating labor costs, making comparisons and negotiating wage rates by trade for over 300 cities (United States and Canada). With 46 construction trades listed by local union number in each city, and historical wage rates included for comparison. No similar book is available through the trade.

Each city chart lists the county and is alphabetically arranged with handy visual flip tabs for quick reference.

$219.95 per copy
Over 300 pages, available Dec. 2001
Catalog No. 60122

Reference Books

For more information
visit Means Web Site
at www.rsmeans.com

Preventive Maintenance Guidelines for School Facilities

NEW!

By John C. Maciha

This new publication is a complete PM program for K-12 schools that ensures sustained security, safety, property integrity, user satisfaction, and reasonable ongoing expenditures.

Includes schedules for weekly, monthly, semiannual, and annual maintenance. Comes as a 3-ring binder with hard copy and electronic forms available at Means website, and a laminated wall chart.

$149.95 per copy
Over 225 pages, Hardcover
Catalog No. 67326

Means Spanish/English Construction Dictionary

By R.S. Means, The International Conference of Building Officials (ICBO), and Rolf Jensen & Associates (RJA)

Designed to facilitate communication among Spanish- and English-speaking construction personnel—improving performance and job-site safety. Features the most common words and phrases used in the construction industry, with easy-to-follow pronunciations. Includes extensive building systems and tools illustrations.

$22.95 per copy
250 pages, illustrated, Softcover
Catalog No. 67327

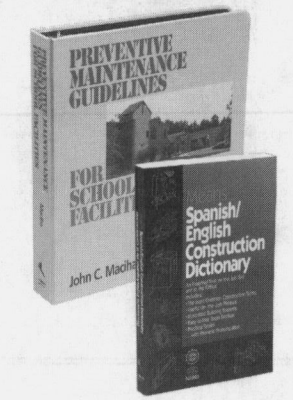

Project Scheduling & Management for Construction

New 2nd Edition

By David R. Pierce, Jr.

A comprehensive yet easy-to-follow guide to construction project scheduling and control—from vital project management principles through the latest scheduling, tracking, and controlling techniques. The author is a leading authority on scheduling with years of field and teaching experience at leading academic institutions. Spend a few hours with this book and come away with a solid understanding of this essential management topic.

$64.95 per copy
Over 250 pages; illustrated, Hardcover Catalog No. 67247A

Residential & Light Commercial Construction Standards

A Unique Collection of Industry Standards That Define Quality in Construction
For Contractors & Subcontractors, Owners, Developers, Architects & Engineers, Attorneys & Insurance Personnel

Compiled from the nation's major building codes, professional associations, and product manufacturer institutes, this unique resource enables you to:
• Set a standard for subcontractors and employees
• Protect yourself against defect claims
• Resolve disputes
• Answer client questions with authority

$59.95 per copy
Over 500 pages, illustrated, Softcover Catalog No. 67322

The Building Professional's Guide to Contract Documents

New 3rd Edition

By Waller S. Poage, AIA, CSI, CVS

This comprehensive treatment of Contract Documents is an important reference for owners, design professionals, contractors, and students.

• Structure your Documents for Maximum Efficiency
• Effectively communicate construction requirements to all concerned
• Understand the Roles and Responsibilities of Construction Professionals
• Improve Methods of Project Delivery

$64.95 per copy, 400 pages
Diagrams and construction forms, Hardcover
Catalog No. 67261A

Cyberplaces: The Internet Guide for AECs & Facility Managers

2nd Edition

By Paul Doherty, AIA, CSI

Completely updated with a section on the New Economy and e-commerce—such as AEC arbitrage, aggregate purchasing, e-supply chains, and more. Plus the best Knowledge Management and Internet tools, including PDAs and the newest Web software for scheduling and project management.

The **Web site** keeps you current on the best Internet tools for building design, construction, and management.

$59.95 per copy
Over 700 pages, illustrated, Softcover
Catalog No. 67317A

For more information
visit Means Web Site
at www.rsmeans.com

Reference Books

Value Engineering: Practical Applications

. . . For Design, Construction, Maintenance & Operations
By Alphonse Dell'Isola, PE

A tool for immediate application—for engineers, architects, facility managers, owners, and contractors. Includes: Making the Case for VE—The Management Briefing, Integrating VE into Planning and Budgeting, Conducting Life Cycle Costing, Integrating VE into the Design Process, Using VE Methodology in Design Review and Consultant Selection, Case Studies, VE Workbook, and a Life Cycle Costing program on disk.

$79.95 per copy
Over 450 pages, illustrated, Hardcover
Catalog No. 67319

Facilities Operations & Engineering Reference

By the Association for Facilities Engineering and R.S. Means

An all-in-one technical referance for planning and managing facility projects and solving day-to-day operations problems. Selected as the official Certified Plant Engineer reference, this handbook covers financial analysis, maintenance, HVAC and energy efficiency, and more.

$99.95 per copy
Over 700 pages, illustrated, Hardcover
Catalog No. 67318

Concrete Repair and Maintenance Illustrated

By Peter H. Emmons

Illustrated guide to analyzing, repairing, and maintaining concrete structures for optimal performance and cost–effectiveness. From parking garages, to roads and bridges, to structural concrete, this book describes the causes, effects, and remedies for concrete failure.

$69.95 per copy
Catalog No. 67146

HVAC: Design Criteria, Options, Selection

Expanded 2nd Edition

By William H. Rowe III, AIA, PE

Includes Indoor Air Quality, CFC Removal, Energy Efficient Systems, and Special Systems by Building Type. Helps you solve a wide range of HVAC system design and selection problems effectively and economically. Gives you clear explanations of the latest ASHRAE standards.

$84.95 per copy
Over 600 pages, illustrated, Hardcover
Catalog No. 67306

Total Productive Facilities Management

By Richard W. Sievert, Jr.

This book provides a comprehensive program to:
- Achieve business goals by optimizing facility resources
- Implement best practices through benchmarking, evaluation & project management
- Increase your value to the organization

$39.98 per copy
Over 270 pages, illustrated, Hardcover
Catalog No. 67321

Facilities Maintenance Management

By Gregory H. Magee, PE

Now you can get successful management methods and techniques for all aspects of facilities maintenance. This comprehensive reference explains and demonstrates successful management techniques for all aspects of maintenance, repair, and improvements for buildings, machinery, equipment, and grounds. Plus, guidance for outsourcing and managing internal staffs.

$86.95 per copy
Over 280 pages with illustrations, Hardcover
Catalog No. 67249

Cost Planning & Estimating for Facilities Maintenance

In this unique book, a team of facilities management authorities shares their expertise at:
- Evaluating and budgeting maintenance operations
- Maintaining & repairing key building components
- Applying *Means Facilities Maintenance & Repair Cost Data* to your estimating

Covers special maintenance requirements of the 10 major building types.

$82.95 per copy
Over 475 pages, Hardcover
Catalog No. 67314

Maintenance Management Audit

This annual audit program is essential for every organization in need of a proper assessment of its maintenance operation. Downloadable electronic forms from Means Web Site allow managers to identify and correct problems and enhance productivity.

Now $32.48 per copy, limited quantity
125 pages, spiral bound, illustrated, Hardcover
Catalog No. 67299

Reference Books

For more information
visit Means Web Site
at www.rsmeans.com

Builder's Essentials: Plan Reading & Material Takeoff

By Wayne J. DelPico

For Residential and Light Commercial Construction

A valuable tool for understanding plans and specs, and accurately calculating material quantities.
Step-by-step instructions and takeoff procedures based on a full set of working drawings.

$35.95 per copy
Over 420 pages, Softcover
Catalog No. 67307

Planning & Managing Interior Projects

New 2nd Edition
By Carol E. Farren, CFM

Addresses changes in technology and business, guiding you through commercial design and construction from initial client meetings to post-project administration. Includes: evaluating space requirements, alternative work models, telecommunications and data management, and environmental issues.

$69.95 per copy
Over 400 pages, illustrated, Hardcover
Catalog No. 67245A

Builder's Essentials: Best Business Practices for Builders & Remodelers:

An Easy-to-Use Checklist System
By Thomas N. Frisby

A comprehensive guide covering all aspects of running a construction business, with more than 40 user-friendly checklists. This book provides expert guidance on: increasing your revenue and keeping more of your profit; planning for long-term growth; keeping good employees and managing subcontractors.

$29.95 per copy
Over 220 pages, Softcover
Catalog No. 67329

Builder's Essentials: Framing & Rough Carpentry 2nd Edition

By Scot Simpson

A complete, training manual for apprentice and experienced carpenters. Develop and improve your skills with "framer-friendly," easy-to-follow instructions, and step-by-step illustrations. Learn proven techniques for framing walls, floors, roofs, stairs, doors, and windows. Updated guidance on standards, building codes, safety requirements, and more.

$24.95 per copy
Over 150 pages, Softcover
Catalog No. 67298A

How to Estimate with Means Data & CostWorks

By R.S. Means and Saleh Mubarak New 2nd Edition!

Learn estimating techniques using Means cost data. Includes an instructional version of Means CostWorks CD–ROM with Sample Building Plans.
The step-by-step guide takes you through all the major construction items. Over 300 sample estimating problems are included.

$59.95 per copy
Over 190 pages, Softcover
Catalog No. 67324

Means Heavy Construction Handbook

Informed guidance for planning, estimating and performing today's heavy construction projects. Provides expert advice on every aspect of heavy construction work including hazardous waste remediation and estimating. To assist planning, estimating, performing, or overseeing work.

$74.95 per copy
Over 430 pages, illustrated, Hardcover
Catalog No. 67148

Superintending for Contractors:

How to Bring Jobs in On-Time, On-Budget
By Paul J. Cook

This book examines the complex role of the superintendent/field project manager, and provides guidelines for the efficient organization of this job. Includes administration of contracts, change orders, purchase orders, and more.

$35.95 per copy
Over 220 pages, illustrated, Softcover
Catalog No. 67233

HVAC Systems Evaluation

By Harold R. Colen, PE

You get direct comparisons of how each type of system works, with the relative costs of installation, operation, maintenance and applications by type of building. With requirements for hooking up electrical power to HVAC components. Contains experienced advice for repairing operational problems in existing HVAC systems, ductwork, fans, cooling coils, and much more!

$84.95 per copy
Over 500 pages, illustrated, Hardcover
Catalog No. 67281

Means Landscape Estimating Methods

3rd Edition
By Sylvia H. Fee

This revised edition offers expert guidance for preparing accurate estimates for new landscape construction and grounds maintenance. Includes a complete project estimate featuring the latest equipment and methods, and **two totally new chapters—Life Cycle Costing, and Landscape Maintenance Estimating.**

$62.95 per copy
Over 300 pages, illustrated, Hardcover
Catalog No. 67295A

Means Environmental Remediation Estimating Methods

By Richard R. Rast

Guidelines for estimating 50 standard remediation technologies. Use it to prepare preliminary budgets, develop estimates, compare costs and solutions, estimate liability, review quotes, negotiate settlements.

A valuable support tool for *Means Environmental Remediation Unit Price* and *Assemblies* books.

$99.95 per copy
Over 600 pages, illustrated, Hardcover
Catalog No. 64777

For more information
visit Means Web Site
at www.rsmeans.com

Reference Books

Means Estimating Handbook

This comprehensive reference covers a full spectrum of technical data for estimating, with information on sizing, productivity, equipment requirements, codes, design standards, and engineering factors.
Means Estimating Handbook will help you: evaluate architectural plans and specifications, prepare accurate quantity takeoffs, prepare estimates from conceptual to detailed, and evaluate change orders.

$99.95 per copy
Over 900 pages, Hardcover
Catalog No. 67276

Means Repair and Remodeling Estimating
3rd Edition
By Edward B. Wetherill & R.S. Means

Focuses on the unique problems of estimating renovations of existing structures. It helps you determine the true costs of remodeling through careful evaluation of architectural details and a site visit.
New section on disaster restoration costs.

$69.95 per copy
Over 450 pages, illustrated, Hardcover
Catalog No. 67265A

Facilities Planning & Relocation
New, lower price and user-friendly format.
By David D. Owen

A complete system for planning space needs and managing relocations. Includes step-by-step manual, over 50 forms, and extensive reference section on materials and furnishings.

$89.95 per copy
Over 450 pages, Softcover
Catalog No. 67301

Means Square Foot & Assemblies Estimating Methods
New 3rd Edition!

Develop realistic Square Foot and Assemblies Costs for budgeting and construction funding. The new edition features updated guidance on square foot and assemblies estimating using UNIFORMAT II. An essential reference for anyone who performs conceptual estimates.

$69.95 per copy
Over 300 pages, illustrated, Hardcover
Catalog No. 67145B

Means Electrical Estimating Methods
2nd Edition

Expanded version includes sample estimates and cost information in keeping with the latest version of the CSI MasterFormat. Contains new coverage of Fiber Optic and Uninterruptible Power Supply electrical systems, broken down by components and explained in detail. A practical companion to *Means Electrical Cost Data*.

$64.95 per copy
Over 325 pages, Hardcover
Catalog No. 67230A

Means Mechanical Estimating Methods
2nd Edition

This guide assists you in making a review of plans, specs, and bid packages with suggestions for takeoff procedures, listings, substitutions and pre-bid scheduling. Includes suggestions for budgeting labor and equipment usage. Compares materials and construction methods to allow you to select the best option.

$64.95 per copy
Over 350 pages, illustrated, Hardcover
Catalog No. 67294

Successful Estimating Methods:
From Concept to Bid
By John D. Bledsoe, PhD, PE

A highly practical, all-in-one guide to the tips and practices of today's successful estimator. Presents techniques for all types of estimates, and advanced topics such as life cycle cost analysis, value engineering, and automated estimating. Estimate spreadsheets available at Means Web site.

$64.95 per copy
Over 300 pages, illustrated, Hardcover
Catalog No. 67287

The ADA in Practice
By Deborah S. Kearney, PhD

Helps you meet and budget for the requirements of the Americans with Disabilities Act. Shows how to do the job right by understanding what the law requires, allows, and enforces. Includes a "buyers guide" for 70 ADA-compliant products.

*Winner of IFMA's "Distinguished Author of the Year" award.

$59.98 per copy
Over 600 pages, illustrated, Softcover
Catalog No. 67147A

Means Productivity Standards for Construction
Expanded Edition *(Formerly Man-Hour Standards)*

Here is the working encyclopedia of labor productivity information for construction professionals, with labor requirements for thousands of construction functions in CSI MasterFormat.
Completely updated, with over 3,000 new work items.

$69.98 per copy
Over 800 pages, Hardcover
Catalog No. 67236A

Means ADA Compliance Pricing Guide

Gives you detailed cost estimates for budgeting modification projects, including estimates for each of 260 alternates. You get the 75 most commonly needed modifications for ADA compliance, each an assembly estimate with detailed cost breakdown.

$59.98 per copy
Over 350 pages, illustrated, Softcover
Catalog No. 67310

Reference Books

For more information
visit Means Web Site
at www.rsmeans.com

Means Unit Price Estimating Methods
2nd Edition
$59.95 per copy
Catalog No. 67303

Means Plumbing Estimating Methods
New 2nd Edition
**By Joseph J. Galeno &
Sheldon T. Greene**
$59.95 per copy
Catalog No. 67283A

Business Management for Contractors
How to Make Profits in Today's Market
By Paul J. Cook
Now $17.98 per copy, limited quantity
Catalog No. 67250

Risk Management for Building Professionals
By Thomas E. Papageorge, RA
Now $15.98 per copy, limited quantity
Catalog No. 67254

Contractor's Business Handbook
By Michael S. Milliner
Now $21.48 per copy, limited quantity
Catalog No. 67255

Basics for Builders: How to Survive and Prosper in Construction
By Thomas N. Frisby
$34.95 per copy
Catalog No. 67273

Means Scheduling Manual
3rd Edition
$32.48 per copy
Catalog No. 67291

Successful Interior Projects Through Effective Contract Documents
By Joel Downey & Patricia K. Gilbert
Now $24.98 per copy, limited quantity
Catalog No. 67313

Estimating for Contractors:
How to Make Estimates that Win Jobs
By Paul J. Cook
$35.95 per copy
Catalog No. 67160

Means Forms for Contractors
For general and specialty contractors
$49.98 per copy
Catalog No. 67288

Understanding Building Automation Systems
**By Reinhold A. Carlson, PE &
Robert Di Giandomenico**
$29.98 per copy, limited quantity
Catalog No. 67284

Illustrated Construction Dictionary, Condensed
$59.95 per copy
Catalog No. 67282

Managing Construction Purchasing
By John G. McConville, CCC, CPE
Now $19.98 per copy, limited quantity
Catalog No. 67302

Hazardous Material & Hazardous Waste
**By Francis J. Hopcroft, PE,
David L. Vitale, M. Ed., &
Donald L. Anglehart, Esq.**
Now $44.98 per copy, limited quantity
Catalog No. 67258

Fundamentals of the Construction Process
By Kweku K. Bentil, AIC
Now $34.98 per copy, limited quantity
Catalog No. 67260

Construction Delays
By Theodore J. Trauner, Jr., PE, PP
$29.48 per copy
Catalog No. 67278

Interior Home Improvement Costs
New 7th Edition
$19.95 per copy
Catalog No. 67308C

Exterior Home Improvement Costs
New 7th Edition
$19.95 per copy
Catalog No. 67309C

Means Facilities Maintenance Standards
By Roger W. Liska, PE, AIC
Now $69.95 per copy
Catalog No. 67246

How to Estimate with Metric Units
Now $9.98 per copy, limited quantity
Catalog No. 67304

Quantity Takeoff for Contractors
How to Get Accurate Material Counts
By Paul J. Cook
Now $17.98 per copy, limited quantity
Catalog No. 67262

Consulting Services Group

Solutions For Your Construction and Facilities Cost Management Problems

We are leaders in the cost engineering field, supporting the unique construction and facilities management costing challenges of clients from the Federal Government, Fortune 1000 Corporations, Building Product Manufacturers, and some of the World's Largest Design and Construction Firms.

Developers of the Dept. of Defense Tri-Services Estimating Database

Recipient of SEARS 1997 Chairman's Award for Innovation

Research...

- **Custom Database Development**—Means expertise in construction cost engineering and database management can be put to work creating customized cost databases and applications.
- **Data Licensing & Integration**—To enhance applications dealing with construction, any segment of Means vast database can be licensed for use, and harnessed in a format compatible with a previously developed proprietary system.
- **Cost Modeling**—Pre-built custom cost models provide organizations that expend countless hours estimating repetitive work with a systematic time-saving estimating solution.
- **Database Auditing & Maintenance**—For clients with in-house data, Means can help organize it, and by linking it with Means database, fill in any gaps that exist and provide necessary updates to maintain current and relevant proprietary cost data.

Estimating...

- **Estimating Service**—Means expertise is available to perform construction cost estimates, as well as to develop baseline schedules and establish management plans for projects of all sizes and types. Conceptual, budget and detailed estimates are available.
- **Benchmarking**—Means can run baseline estimates on existing project estimates. Gauging estimating accuracy and identifying inefficiencies improves the success ratio, precision and productivity of estimates.
- **Project Feasibility Studies**—The Consulting Services Group can assist in the review and clarification of the most sound and practical construction approach in terms of time, cost and use.
- **Litigation Support**—Means is available to provide opinions of value and to supply expert interpretations or testimony in the resolution of construction cost claims, litigation and mediation.

Training...

- **Core Curriculum**—Means educational programs, delivered on-site, are designed to sharpen professional skills and to maximize effective use of cost estimating and management tools. On-site training cuts down on travel expenses and time away from the office.
- **Custom Curriculum**—Means can custom-tailor courses to meet the specific needs and requirements of clients. The goal is to simultaneously boost skills and broaden cost estimating and management knowledge while focusing on applications that bring immediate benefits to unique operations, challenges, or markets.
- **Staff Assessments and Development Programs**—In addition to custom curricula, Means can work with a client's Human Resources Department or with individual operating units to create programs consistent with long-term employee development objectives.

For more information and a copy of our capabilities brochure, please call 1-800-448-8182 and ask for the Consulting Services Group, or reach us at www.rsmeans.com

MeansData™

CONSTRUCTION COSTS FOR SOFTWARE APPLICATIONS

Your construction estimating software is only as good as your cost data.

Software Integration

A proven construction cost database is a mandatory part of any estimating package. We have linked MeansData™ directly into the industry's leading software applications. The following list of software providers can offer you MeansData™ as an added feature for their estimating systems. Visit them on-line at *www.rsmeans.com/demo/* for more information and free demos. Or call their numbers listed below.

3D International
713-871-7000 venegas@3di.com

4Clicks-Solutions, LLC
719-574-7721
mbrown@4clicks-solutions.com

ACT
Applied Computer Technologies
Facility Management Software
919-859-1335 info@srs.net

AEPCO, Inc.
301-670-4642 blueworks@aepco.com

**American Contractor/
Maxwell Systems**
800-333-8435 info@amercon.com

ArenaSoft Estimating
888-370-8806 info@arenasoft.com

Ares Corporation
650-401-7100
sales@arescorporation.com

**AssetWork
CSI-Maximus**
Facility Management Software
800-659-9001 info@assetworks.com

Benchmark, Inc.
800-393-9193 sales@benchmark-inc.com

BSD
Building Systems Design, Inc.
888-273-7638 bsd@bsdsoftlink.com

CProjects, Inc.
203-262-6248 sales@cprojects.com

cManagement
800-945-7093 sales@cmanagement.com

CDCI
Construction Data Controls, Inc.
800-285-3929 sales@cdci.com

CMS
Computerized Micro Solutions
800-255-7407 cms@proest.com

Conac Group
800-663-2338 sales@conac.com

Eagle Point Software
800-678-6565 sales@eaglepoint.com

Estimating Systems, Inc.
800-967-8572 pulsar@capecod.net

G2 Estimator
A Div. of Valli Info. Syst., Inc.
800-657-6312 info@g2estimator.com

G/C EMUNI, Inc.
514-953-5148 rpa@gcei.ca

Geac Commercial Systems, Inc.
800-554-9865 info@geac.com

Hard Dollar
800-637-7496 sales@harddollar.com

IQ Beneco
801-565-1122 mdover@beneco.com

Luqs International
888-682-5573 info@luqs.com

MC²
Management Computer Controls
800-225-5622 vkeys@mc2-ice.com

Prism Computer Corporation
Facility Management Software
800-774-7622 famis@prismcc.com

Quest Solutions, Inc.
800-452-2342 info@questsolutions.com

Sanders Software, Inc.
800-280-9760 hsander@vallnet.com

Sinisoft, Inc.
877-669-4949 info.usa@sinistre.com

Timberline Software Corp.
800-628-6583
product.info@timberline.com

TMA Systems, Inc.
Facility Management Software
800-862-1130 sales@tmasys.com

US Cost, Inc.
800-372-4003
sales@uscost.com

Vertigraph, Inc.
800-989-4243
info-request@vertigraph.com

Wendlware
714-895-7222 sales@corecon.com

Winestimator, Inc.
800-950-2374 sales@winest.com

DemoSource™

One-stop shopping for the latest cost estimating software for just $19.95. This evaluation tool includes product literature and demo diskettes for ten or more estimating systems, all of which link to MeansData™. **Call 1-800-334-3509 to order.**

**FOR MORE INFORMATION ON ELECTRONIC PRODUCTS CALL
1-800-448-8182 OR FAX 1-800-632-6732.**

MeansData™ is a registered trademark of R.S. Means Co., Inc., *CMD*.

For more information
visit Means Web Site
at www.rsmeans.com

Seminars

Developing and Managing Facility Assessment Programs

This two-day program instructs attendees on organizing, staffing, conducting and reporting the results of a facilities audit as the basis for sound capital planning and budgeting.

Some Of What You'll Learn:
- Conducting regular condition inspections to identify deferred maintenance problems, cost estimating, formulating accurate budgets, and planning staff workloads
- The management process required for planning, conducting and documenting the physical condition and functional adequacy of buildings or other facilities
- Gathering reliable data to identify and gauge deferred maintenance requirements
- Essential steps for conducting facilities inspection programs

Who Should Attend: Plant Managers, Facility Executives, Real Estate Planners and Developers, Estimators... and others who plan for capital renewal and replacement and/or whose financial decisions and budgets are based on the conditions of facilities.

Repair and Remodeling Estimating

This two-day seminar emphasizes all the underlying considerations unique to repair/remodeling estimating and presents the correct methods for generating accurate, reliable R&R project costs using the unit price and assemblies methods.

Some Of What You'll Learn:
- Estimating considerations like labor hours, building code compliance, working within existing structures, purchasing materials in smaller quantities, and unforeseen deficiencies
- Identifies problems and provides solutions to estimating building alterations
- Rules for factoring in minimum labor costs, accurate productivity estimates, and allowances for project contingencies
- R&R estimating examples are calculated using unit price and assemblies data

Who Should Attend: Facilities Managers, Plant Engineers, Architects, Contractors, Estimators, Builders... and others who are concerned with the proper preparation and/or evaluation of repair and remodeling estimates.

Mechanical and Electrical Estimating

This two-day course teaches attendees how to prepare more accurate and complete mechanical/electrical estimates, avoiding the pitfalls of omission and double-counting, while understanding the composition and rationale within the Means mechanical/electrical database.

Some Of What You'll Learn:
- The unique way mechanical and electrical systems are interrelated
- M&E estimates, conceptual, planning, budgeting and bidding stages
- Order of magnitude, square foot, assemblies and unit price estimating
- Comparative cost analysis of equipment and design alternatives

Who Should Attend: Architects, Engineers, Facilities Managers, Mechanical and Electrical Contractors... and others needing a highly reliable method for developing, understanding and evaluating mechanical and electrical contracts.

Unit Price Estimating

This interactive two-day seminar teaches attendees how to interpret project information and process it into final, detailed estimates with the greatest accuracy level. The single most important credential an estimator can take to the job is the ability to visualize construction in the mind's eye, and thereby estimate accurately.

Some Of What You'll Learn:
- Interpreting the design in terms of cost
- The most detailed, time-tested methodology for accurate "pricing"
- Key cost drivers: material, labor, equipment, staging, and subcontracts
- Understanding direct and indirect costs for accurate job cost accounting and change order management

Who Should Attend: Corporate and Government Estimators and Purchasers, Architects, Engineers... and others needing to produce accurate project estimates.

Square Foot and Assemblies Cost Estimating

This two-day course teaches attendees how to quickly deliver accurate square foot estimates using limited budget and design information.

Some Of What You'll Learn:
- Using square foot costing to get the estimate done faster
- Taking advantage of a "systems" or "assemblies" format
- The Means "building assemblies/square foot cost approach"
- How to create a very reliable preliminary systems estimate using bare-bones design information

Who Should Attend: Facilities Managers, Facilities Engineers, Estimators, Planners, Developers, Construction Finance Professionals... and others needing to make quick, accurate construction cost estimates at commercial, government, educational and medical facilities.

Scheduling and Project Management

This two-day course teaches attendees the most current and proven scheduling and management techniques needed to bring projects in on time and on budget.

Some Of What You'll Learn:
- Crucial phases of planning and scheduling
- How to establish project priorities, develop realistic schedules and management techniques
- Critical Path and Precedence Methods
- Special emphasis on cost control

Who Should Attend: Construction Project Managers, Supervisors, Engineers, Estimators, Contractors... and others who want to improve their project planning, scheduling and management skills.

Facilities Maintenance and Repair Estimating

This two-day course teaches attendees how to plan, budget, and estimate the cost of ongoing and preventive maintenance and repair for existing buildings and grounds.

Some Of What You'll Learn:
- The most financially favorable maintenance, repair and replacement scheduling and estimating
- Auditing and value engineering facilities
- Preventive planning and facilities upgrading
- Determining both in-house and contract-out service costs
- An annual, asset-protecting M&R plan

Who Should Attend: Facility Managers, Maintenance Supervisors, Buildings and Grounds Superintendents, Plant Managers, Planners, Estimators... and others involved in facilities planning and budgeting.

Managing Facilities Construction and Maintenance

This two-day program teaches attendees the management techniques needed to effectively plan, organize, control and get the most out of limited facilities resources.

Some Of What You'll Learn:
- Helping an organization allocate physical resources
- Evaluating the impact of today's facility decisions on tomorrow's budgets
- Developing budgets, reducing expenditures and assessing productivity
- Common problems and challenges encountered during new construction, renovation or maintenance projects

Who Should Attend: Facility Managers, Maintenance Managers, Plant Managers, Facility Project Managers, Facilities Foremen and Supervisors... and others involved in or aspiring to a role in facilities management.

Advanced Project Management

This two-day seminar will teach you how to effectively manage and control the entire design-build process and allow you to take home tangible skills that will be immediately applicable on existing projects.

Some Of What You'll Learn:
- Value engineering, bonding, fast-tracking and bid package creation
- How estimates and schedules can be integrated to provide advanced project management tools
- Cost engineering, quality control, productivity measurement and improvement
- Front-loading a project and predicting its cash flow

Who Should Attend: Owners, Project Managers, Architectural and Engineering Managers, Construction Managers, Contractors... and anyone else who is responsible for the timely design and completion of construction projects.

11

Seminars

For more information
visit Means Web Site
at www.rsmeans.com

2002
Means
Seminar
Schedule

Location	Dates
Las Vegas, NV	March 11-14
Washington, DC	April 22-25
Denver, CO	May 20-23
San Francisco, CA	June 10-13
Cape Cod, MA	September 9-12
Washington, DC	September 23-26
San Diego, CA	October 21-24
Orlando, FL	November 4-7
Atlantic City, NJ	November 18-21
Las Vegas, NV	December 2-5

Note: Call for exact dates and details.

Registration Information

Register Early... Save up to $150! Register 30 days before the start date of a seminar and save $150 off your total fee. *Note: This discount can be applied only once per order.*

How to Register Register by phone today! Means toll-free number for making reservations is: **1-800-448-8182.**

Individual Seminar Registration Fee $875 To register by mail, complete the registration form and return with your full fee to: Seminar Division, R.S. Means Company, Inc., 63 Smiths Lane, Kingston, MA 02364.

Federal Government Pricing All Federal Government employees save 25% off regular seminar price. Other promotional discounts cannot be combined with Federal Government discount.

Team Discount Program Two to four seminar registrations: $760 per person–Five or more seminar registrations: $710 per person–Ten or more seminar registrations: Call for pricing.

Consecutive Seminar Offer One individual signing up for two separate courses at the same location during the designated time period pays only $1,400. You get the second course for only $525 (**a 40% discount**). Payment must be received at least ten days prior to seminar dates to confirm attendance.

Refunds Cancellations will be accepted up to ten days prior to the seminar start. There are no refunds for cancellations postmarked later than ten working days prior to the first day of the seminar. A $150 processing fee will be charged for all cancellations. Written notice or telegram is required for all cancellations. Substitutions can be made at anytime before the session starts. **No-shows are subject to the full seminar fee.**

AACE Approved Courses The R.S. Means Construction Estimating and Management Seminars described and offered to you here have each been approved for 14 hours (1.4 recertification credits) of credit by the AACE International Certification Board toward meeting the continuing education requirements for re-certification as a Certified Cost Engineer/Certified Cost Consultant.

AIA Continuing Education R.S. Means is registered with the AIA Continuing Education System (AIA/CES) and is committed to developing quality learning activities in accordance with the CES criteria. R.S. Means seminars meet the AIA/CES criteria for Quality Level 2. AIA members will receive (28) learning units (LUs) for each two day R.S. Means Course.

Daily Course Schedule The first day of each seminar session begins at 8:30 A.M. and ends at 4:30 P.M. The second day is 8:00 A.M.–4:00 P.M. Participants are urged to bring a hand-held calculator since many actual problems will be worked out in each session.

Continental Breakfast Your registration includes the cost of a continental breakfast, a morning coffee break, and an afternoon break. These informal segments will allow you to discuss topics of mutual interest with other members of the seminar. (You are free to make your own lunch and dinner arrangements.)

Hotel/Transportation Arrangements R.S. Means has arranged to hold a block of rooms at each hotel hosting a seminar. To take advantage of special group rates when making your reservation be sure to mention that you are attending the Means Seminar. You are of course free to stay at the lodging place of your choice. (**Hotel reservations and transportation arrangements should be made directly by seminar attendees.**)

Important Class sizes are limited, so please register as soon as possible.

Registration Form   Call 1-800-448-8182 x5115 to register or FAX 1-800-632-6732. Visit our Web site www.rsmeans.com

Please register the following people for the Means Construction Seminars as shown here. Full payment or deposit is enclosed, and we understand that we must make our own hotel reservations if overnight stays are necessary.

☐ Full payment of $ _____ enclosed.

☐ Bill me

Name of Registrant(s)
(To appear on certificate of completion)

_____

_____

P.O. #: _____
GOVERNMENT AGENCIES MUST SUPPLY PURCHASE ORDER NUMBER

Firm Name _____

Address _____

City/State/Zip _____

Telephone No. Fax No. _____

E-Mail Address _____

Charge our registration(s) to: ☐ MasterCard ☐ VISA ☐ American Express ☐ Discover

Account No. _____ Exp. Date _____

Cardholder's Signature _____

Seminar Name   City   Dates

_____

_____

Please mail check to: R.S. Means Company, Inc., 63 Smiths Lane, P.O.Box 800, Kingston, MA 02364 USA